INTEGRATED NEUROSCIENCE
A Clinical Problem Solving Approach

INTEGRATED NEUROSCIENCE
A Clinical Problem Solving Approach

by

Elliott M. Marcus, M.D.
Professor Emeritus of Neurology, University of Massachusetts School of Medicine; Lecturer in Neurology, Tufts University School of Medicine; Chairman Emeritus, Dept. of Neurology, St. Vincent Hospital and Fallon Clinic

and

Stanley Jacobson, Ph.D.
Professor of Anatomy & Cellular Biology
Tufts University Health Science Campus, Boston, MA

With Contributions by **Brian Curtis, Ph.D.**
University of Illinois School of Medicine at Peoria

Illustrations by **Mary Gauthier Delaplane**
Boston University School of Medicine

KLUWER ACADEMIC PUBLISHERS
Boston / Dordrecht / London

Distributors for North, Central and South America:
Kluwer Academic Publishers
101 Philip Drive
Assinippi Park
Norwell, Massachusetts 02061 USA
Telephone (781) 871-6600
Fax (781) 681-9045
E-Mail < kluwer@wkap.com>

Distributors for all other countries:
Kluwer Academic Publishers Group
Post Office Box 322
3300 AH Dordrecht, THE NETHERLANDS
Telephone 31 786 576 000
Fax 31 786 576 254
E-Mail < services@wkap.nl>

 Electronic Services < http://www.wkap.nl>

Library of Congress Cataloging-in-Publication Data

Marcus, Elliott M., 1932-
   Integrated neuroscience : a clinical problem solving approach / by Elliott M. Marcus
and Stanley Jacobson ; with contributions by Brian Curtis ; illustrations by Mary
Gauthier Delaplane.
   p. ; cm.
Includes bibliographical references and index.
ISBN 1-40207-164-7 (alk. paper)
  1. Neurosciences. 2. Nervous system—Diseases. I. Jacobson, Stanley, 1937 – II. Title.
[DNLM: 1. Nervous System Diseases—diagnosis. 2. Nervous System
Diseases—therapy. 3. Nervous System—anatomy & histology. 4. Nervous System
Physiology. WL 140 M322i 2002]
RC341 .M29 2002
612.8—dc21   2002073000

Copyright © 2003 by Kluwer Academic Publishers.

All rights reserved. No part of this work may be reproduced, stored in a retrieval system, or transmitted in any form or by any means, electronic, mechanical, photocopying, microfilming, recording, or otherwise, without the written permission from the Publisher, with the exception of any material supplied specifically for the purpose of being entered and executed on a computer system, for exclusive use by the purchaser of the work

Permission for books published in Europe: permissions@wkap.nl
Permissions for books published in the United States of America: permissions@wkap.com

Printed on acid-free paper.
Printed in the United States of America

The Publisher offers discounts on this book for course use and bulk purchases. For further information, send email to <joanne.tracy@wkap.com> .

TABLE OF CONTENTS

Preface
Dedication
SECTION I: INTRODUCTION TO BASIC NEUROBIOLOGY

1. Overview of the nervous system		1-21
The neuron	1-1	
The central nervous system	1-3	
Spinal Cord	1-3	
Brain Stem	1-4	
Diencephalon	1-6	
Cerebrum	1-8	
CNS pathways	1-14	
Glands associated with the brain	1-17	
Blood supply	1-17	
Meninges	1-18	
Ventricular system	1-19	
2. Overview of localization of function and neurological diagnosis		1-28
Where is the disease located? What is the nature of the pathology?	2-1	
The neurological history and examination	2-8	
Diagnostic studies in Neurology	2-11	
3. Neurocytology	1-31	
Neurons,	3-1	
Myelin	3-8	
Axoplasmic flow	3-10	
Synapse	3-14	
Supporting Cells of the Central Nervous System	3-20	
Response of Nervous system to Injury	3-24	
Degeneration	3-24	
Regeneration	3-26	
4. Neuroembryology		1-23
Introduction	4-1	
Histogenesis	4-2	
Neuronal Necrosis and apoptosis	4-4	
Differentiation of CNS areas		
Ventricular System	4-5	
Spinal Cord	4-5	
Brain	4-7	
Cranial Nerves	4-10	
Cerebral cortex	4-13	
Abnormal development	4-16	
5. Basic Physiology		1-28
I. Cell physiology	5-1	
Cell Membrane	5-1	
Channels	5-2	
Potassium Channels	5-5	

II. Nerve Physiology 5-11
Action Potential 5-10
Sodium Channels 5-11
Saltatory Conduction 5-15
Calcium channels 5-17

SECTION II: REGIONAL APPROACH TO NEUROANATOMY AND LOCALIZATION

6. Skeletal Muscle and Nerve-Muscle Junction 1-25
Gross Structure and function 6-1
Molecular Architecture of Contraction 6-5
Nerve Muscle Junction 6-9
Disease of Muscle 6-12
Muscular Dystrophies 6-12
Myotonic Dystrophy
  Congenital Myopathies 6-16
  Acquired Disorders of Muscle 6-18
Disease of the Neuromuscular Junction 6-19

7. Spinal Cord: Structure and Physiology 1-27
  Anatomy 7-1
  Segmental functions 7 4
Reflexes 7-6
Membrane basis of integration 7-9
Stretch receptors 7-12
Lamination in spinal cord gray matter 7-17
Nociception and pain 7-19
Tracts 7-22
  descending 7-23
  Ascending 7-23

8. Disease of Peripheral Nerve and Nerve Root 1-25
Introduction 8-1
  Mononeuropathies 8-4
  Polyneuropathies 8-15
  Radiculopathies 8-19

9. Clinical Considerations of the Spinal Cord 1-29
  Spinal cord compression and transection syndromes 9-1
  Extrinsic disorders 9-3
  Intrinsic diseases:
    Segmental 9-10
    System 9-17
    Multifocal 9-17

10. Case History Problem Solving Part I: Spinal Cord, Nerve Root 1-9
  Lesion Diagrams 10-1
  Case Histories 10-2

11. Functional Anatomy of the Brain Stem 1-33
  Introduction 11-1
  Functional Localization in Coronal Sections of the brain stem 11-2

Functional Centers in the Brain Stem	11-20	
Localization of Disease Processed in the Brain Stem	11-25	
Brain Stem and Eye Movements	11-30	
12. Cranial Nerves		1-25
Components	12-1	
Function of each cranial nerve	12- 23	
Effects of Extrinsic Lesions on Cranial Nerves	12-23	
Voluntary Control of the Cranial Nerves	12-23	
13. Brain Stem: Clinical Considerations		1-21
Differentiation of extrinsic and intrinsic lesions		
Extrinsic disorders	13-2	
Intrinsic disorders	13-8	
Vascular	13-8	
Neoplastic	13-19	
Multiple Sclerosis	13-19	
14. Case History Problem Solving Part II: Brain Stem and Cranial Nerves		1-9
Lesion Diagrams	14-1	
Case Histories	14-2	
15. Diencephlaon		1-21
introduction	15-1	
Nuclei of the thalamus	15-2	
Functional organization of Thalamus	15-3	
White Matter of the diencephalon	15-7	
Relationship between the thalamus and the cerebral cortex	15-8	
Subthalamus	15-9	
Major Sensory Pathways of the CNS	15-12	

SECTION III: MAJOR SYSTEMS

16. Hypothalamus, Neuroendocrine and Autonomic systems		1-16
Introduction	16-1	
Hypothalamus	16-2	
Neuroendocrine systems ,	16-5	
Hormnes produced in by hypothalamus	16-7	
Hormones produced by Adenohypophysis	16-7	
Hypothalamus and the Autonomic system	16- 10	
Hypothalamus and Emotions	16-11	
Autonomic Nervous system	16-12	
17. Cerebral cortex, Cytoarchitecture, Functional localization		1-28
Anatomy	17-1	
Functional localization	17-8	
Subcortical white matter	17-16	
Neurophysiology and Seizure Correlations	17-24	
18. Motor System I: Reflex Activity and Cortical Motor Function		1-26
Reflexes and central pattern generators	18-1	
Motor functions of cerebral cortex	18-8	

Primary Motor Cortex	18-10	
Premotor Cortex	18-16	
Pyramidal Tract	18-17	
Cortical control of eye movements	18-19	
Prefrontal cortex	18-23	
Gait disorders in the elderly	18-25	
19. Motor System III: Basal Ganglia and Movement Disorders		1-24
Structure, connections, transmitters	19-1	
Parkinson's disease	19-7	
Choreiform disorders	19-15	
Other dyskinesias	19-21	
20. Motor System III: Cerebellum and Movement	1-22	
Anatomy	20-1	
Syndromes of the lobes	20-6	
Vascular Syndromes	20-14	
Spinocerebellar Degenerations	20-16	
Overview of Tremor	20-21	
21. Somatosensory Function and the Parietal Lobe		1-10
PostCentral gyrus and Syndromes	21-1	
Parietal lobules: Dominant Hemisphere syndromes	21-7	
Non-Dominant Hemisphere Syndromes	21-8	
22. Limbic System		1-27
Anatomy	22-1	
Emotions	22-6	
Hippocampus	22-8	
Temporal lobe and Seizures	22-14	
Prefrontal Cortex and Emotions	22-20	
23. Visual system		1-20
The eye	23-1	
The retina	23-3	
Pathways, Lesions and Syndromes	23-7	
Occipital Cortex and Syndromes	23-12	
24. Speech, Language, Cerebral Dominance and the Aphasias		1-18
Language dominance and development	24-1	
Aphasia:		
nnonfluent	24-6	
nonfluent	24-9	
Visual Agnosia	24-15	
Aparaxia	24-17	
Nondominant functions	24-17	
25. Case History Problem Solving III: Cortical Localization		1-9
Case Histories	25-1	
26. Disease of the Cerebral Hemispheres: I: Vascular Syndromes		1-31
Ischemic occlusive	26-1	
Embolic disease		

Carotid	26-4	
Middle Cerebral Artery	26-11	
Anterior Cerebral Artery	26-16	
Posterior Cerebral Artery	26-18	
Emboism	26-19	
Intracerebral hemorrhage	26-22	
Subarachnoid hemorrhage	26-24	

27. Disease of the Cerebral Hemispheres: II. Non Vascular 1-34
    Trauma 27-1
    Neoplasms 27-7
    Infections 27-18
    (Non bacterial on CD ROM)
    System disorders .... (CD ROM)
    Disorders of myelin .... (CD ROM)
28. Case History Problem Solving IV: Cerebral Hemispheres 1-11
Lesion diagrams
 Case Histories

SECTION IV: COMPLEX FUNCTIONS

29. Alterations in Consciousness: Seizures, Coma, and Sleep 1-28
    Basic definitions 29-1
    Seizures and Epilepsy 29-3
      Focal Epilepsy 29-3
      General Epilepsy 29-6
    Sleep 29-23
    Coma 29-28
30. Learning, Memory, Amnesia, Dementias 1-20
Definitions 30-1
Stages of Human Memory 30-2
Disorders of Recent Memory 30-5
   Wernicke'Korsakoff 30-5
   Diencephalic Mechanisms 30-6
Hippocampal Lesion 30-7
Progressive Dementia 30-11
31. Case History Problem Solving: V: Part I General Case Histories 1-11
    Case Histories 31-1
31 A. Case History Problem Solving V: Part II General Case Histories 1-13
Imaging Correlations Lesion Diagrams 31A-1
Case Histories 31A-7

TABLE OF CONTENTS – CD

ATLAS PDF Format
    Brain Stem Myelin Stain Still Born Levels –medulla, pons

    Gross Brain Labeled Levels

        Coronal
        Horizontal
        MRI from similar levels
    Myelin Stained & Labeled Sections
        Coronal
        Horizontal
        Sagittal
    Descriptive Atlas of Brain
        Spinal Cord Levels; Cervical, Thoracic, Lumbar, Sacral
        Brain Stem: Medulla Levels – Junction with spinal cord, Narrow, and Wide,
            Pons Levels - Facial Colliculus and Trigeminal Nerve Rootlet
            Midbrain Levels - Inferior Colliculus, Superior Colliculus
            Thalamus: Levels - Anterior tubercle, Mid-thalamic, Posterior thalamus
    Blood Supply Overview
    Localization in Cerebrum
    Skull

CASES IN LONG FORMAT - PDF Format

Part I	Spinal Cord PNS, Embryology	1-35
Part II	Brain Stem	1-36
Part III	Cerebral Hemispheres: Motor system, Cerebellum Basal Ganglia	72-110
Part IV	Cerebrum: Vision, Disease Memory	109-157

CRANIAL NERVE Illustrations in PDF Format
    Illustrations of all the Cranial Nerves from chapter 12 in Color

SPECIAL SENSES
    Vestibular & Cochlear Nerves
    Visual System - Chapter 23

PATHWAYS IN THE CNS PDF Format

SUPPLEMENTAL
    Psychiatric Overview
    Survey of Pathology

Preface

INTEGRATED NEUROSCIENCES

This textbook takes as a premise that, in order to make intelligent diagnosis and provide a rational treatment in disorders of the nervous system, it is necessary to develop the capacity to answer the basic questions of clinical neurology: (1) Where is the disease process located? (2) What is the nature of the disease process?

The purpose of this textbook is to enable the medical student to acquire the basic information of the neurosciences and neurology and most importantly the ability to apply that information to the solution of clinical problems. The authors also suggest that hospital trips be a part of any Clinical Neurosciences Course so that the student can put into actual practice what he has learned in the classroom.

We believe that this textbook will be of value to the student throughout the four years of the medical school curriculum.

Medical, psychiatry and neurology residents may also find this text of value as an introduction or review.

It is more true in neurology than in any other system of medicine that a firm knowledge of basic science material, that is, the anatomy, physiology and pathology of the nervous system, enables the student and physician to readily arrive at the diagnosis of where the disease process is located and the nature of the most likely pathology. Subsequently that knowledge may be applied to problem solving in clinical situations.

The two authors have a long experience in teaching neuroscience courses at the first or second year medical student level in which clinical information and clinical problem solving are integral to the course. In addition the first author has developed a case history problem solving seminar in which all medical students at the University of Massachusetts participate during their clinical neurology clerkship rotation. This provides the students an opportunity to refresh their problem solving skills and to review and update that basic science material essential for clinical neurology. The second author has had extensive experience in utilizing sections of this text in neuroscience courses for advanced undergraduate college students and ancillary health profession students.

At these several levels, we have observed that this approach reinforces the subject matter learned by markedly increasing the interest of the students in both basic and clinical science material.

This text is an updated version of an earlier integrated textbook originally developed by the authors along with Dr Brian Curtis and published by W. B. Saunders in 1972 as *"An Introduction to the Neurosciences"*.

The present text provides an updated approach to lesion localization in neurology, utilizing the techniques of computerized axial tomography (CT scanning), magnetic resonance imaging (MRI) and magnetic resonance angiography (MRA) that were not available in 1972. In addition, the other modern clinical techniques of evoked potentials, positron emission tomography (PET), single photon emission computerized tomography (SPECT) and functional MRI neuroimaging are discussed and illustrated. Multiple illustrations demonstrating the value of these techniques in clinical neurology and neuroanatomical localization have been provided. The clinical case illustrations have been utilized both in the body of the text and in special problem solving chapters.

An anatomical atlas including MRI images is provided on the accompanying CD. Neuropathology illustrations in color will also be found on the CD ROM

There are specific review and problem solving chapters with a strong emphasis on clinical case history problem solving. These clinical problem solving exercises are found as specific chapters with part I (chapter 10) covering diseases of spinal cord, nerve root, peripheral nerve and muscle, part II (chapter 14) covering disorders of the brain stem and cranial nerves, part III (chapter 25) covering cortical localization and part IV (chapter 28) covering diseases of the cerebral hemispheres with an emphasis on cerebral vascular disease. Finally part V (chapter31) provides case history problem solving that encompasses all regions of the brain. This final part V, also provides an appendix in which various clinical neuro images are provided for identification of location of lesion and of the type pathology. A series of clinical cases is then provided in which the advanced student is requested to select the appropriate study that was utilized for the particular cases. Review questions for many of the chapters will be found on the CD ROM.

The case history problem solving exercises are designed to be utilized in weekly case history problem solving discussion groups, with an instructor, usually a neurologist, neurosurgeon, neurology or neurosurgical resident. Some medical school courses may provide a physical diagnosis session devoted to the neurological history and examination.

In Chapter 2 we have included an outline of a complete and abbreviated neurological history and an overview of the diagnostic studies to be utilized at each level of the neural axis.

The emphasis through out the text however is on clinical diagnosis. What is the diagnosis before the laboratory and imaging studies were selected. Thus for each of the illustrative case histories, a provisional clinical diagnostic impression is provided before the results of the ancillary studies are presented. This should instill in the student the concept that the history and neurological examination must first be completed, subjected to analysis and a clinical diagnosis or differential impression established before the ancillary studies are selected. Even in this era of modern imaging this is the most efficient and effective approach. It has been said that 75% of neurological diagnosis is dependent on the history. The student should be aware that this stated clinical diagnostic impression does not always correspond to the final diagnosis. The case histories, in all instances, present actual patients. It is our impression that such cases are more instructive and more interesting to students than manufactured, stereotype case histories. Because these are cases based on clinical reality, at times there are minor deviations from the classic picture, or multiple disorders are present.

Many of the case history examples have been abbreviated in the text. More complete versions of these case histories with a commentary providing an analysis of the case will be found on the CD ROM. Bibliographies for the various chapters are provided on the CD-ROM

In general for the problem solving exercises, the cases have been arranged in an increasing order of difficulty. In a number of instances within the text or within the case history problem solving exercises, we have chosen to retain case histories from the earlier version of this text. In some instances, the history and/or findings presented were of a "classical" nature with an opportunity to study the full natural evolution of disease. In most cases, the location of disease process was clearly confirmed by surgery and/or by autopsy. This was more likely to be the case before the development of modern neuro imaging It will also be evident that some cases were retained because those patients continued to have neurological follow-up allowing an overview of the long-term course of particular diseases.

A number of the topics are sometimes covered in other courses and this material has therefore been placed in PDF files on the CD- ROM which accompanies this book. The section of chapter 27 which covers in detail, diffuse disorders of the cerebral hemispheres such as aseptic meningitis, encephalitis, AIDS ,nutritional and toxic disorders is found in a separate file on the CD ROM. Summary information, tables and illustrations for these topics will be found in the textbook. Many of these topics are covered in internal medicine courses.

Supplementary material will also be found for chapter 13. Additional discussion of psychiatric disorders and of complex partial seizures will be found in a PDF file for chapter 22 on the limbic system. Additional discussion of the pathophysiology of the epilepsies will be found in a PDF file for chapter 29 on disorders of consciousness.

It is planned that a web page will be established by the publisher. This will provide a means for sending additional material to the reader. In addition , information regarding the solution of the clinical case history problem solving exercises will be provided.

Most of the case histories utilized in the chapters and in the problem solving exercises, have been drawn from the files of Dr. Marcus. For a number of the cases, our associates at the New England Medical Center, St Vincent Hospital, Fallon Clinic and the University of Massachusetts School of Medicine either requested our opinion or brought the case to our attention, and provided information from their case files. These individual neurologists and neurosurgeons are identified in the specific case histories. We are also indebted to the many referring physicians of those institutions. Some of the cases were presented to Dr. Marcus during morning report by medical house officers at St Vincent Hospital.

In particular, our thanks are due to our associates in Worcester: Drs. Bernard Stone, Alex Danylevich, Robin Davidson, Harold Wilkinson and Gerry McGillicuddy. Drs. Sandra Horowitz, Tom Mullins, Steve Donhowe, Martha Fehr, Lawrence Recht, Paula Raven and Carl Rosenberg, provided additional or followup clinical information from their files for some of the case histories. Our associates at the New England Medical Center: Drs. John Sullivan, Sam Brendler, Peter Carney, John Hills, Huntington Porter, Bertram Selverstone, Thomas Twitchell, C W Watson and Robert Yuan and Thomas Sabin likewise provided access to some of the clinical material for the earlier version of this text.

Dr Milton Weiner at St Vincent Hospital was particularly helpful in providing many of the modern neuroradiological images. Dr. Sam Wolpert and Dr Bertram Selverstone provided this material for the earlier version of the text. The normal MRI's were provided by Dr Val Runge from the University of Kentucky Imaging Center. Dr. Anja Bergman (left handed) had the patience to be our normal case and the images from her brain form the normal MRI's in the basic science chapters and atlas.

Dr. Tom Smith and his associates in pathology provided much of the recent neuropathological material, particularly for the chapters on muscle, peripheral nervous system and dementia. Drs John Hills and Jose Segarra provided access to neuropathological material for the earlier version of the text. Critical review of particular chapters was provided by Dr. Sandra Horowitz, and Dr. David Chad.

Dr Brian Curtis contributed material for inclusion in chapters 5 (cell and nerve physiology), 6 (muscle physiology) 7 (spinal cord physiology) and 23 (the physiology of the visual system). In addition in the section on special senses found on the CD ROM, Dr Curtis has contributed material included in the auditory system.

Many of the new anatomical drawings were provided by Mary Gauthier Delaplane now a medical student at Boston University School of Medicine. Dr Marc Bard provided drawings for the earlier version of this text while a student at Tufts University School of Medicine. Ms Mary Gauthier Delaplane and Mr Seymour Levy provided graphic services and assisted in the layout of text and illustrations. Ms Jane Griesbach and her associates at St. Vincent Hospital provided photographic prints of many of the neuro images which we had selected. Ms Helen Johnson provided typographic assistance in the formative stages of this project.

We have continued to utilize or have modified some of the illustrations which were borrowed with permission from other published sources for the earlier version of this text. We have attempted to contact these original sources for continued permissions. We will acknowledge subsequently any sources which have been inadvertently over looked.

In many of the clinical chapters, various medications are discussed . Before utilizing these medications, the reader should check dosage and indications with other sources and modify as necessary for the individual patient.

It is with great pleasure we extend our thanks to our publishers and particularly our editor Ms Joanne Tracy " she kept our noses to the grindstone". Any faults or errors are those of the authors and we would therefore appreciate any suggestions or comments from our colleagues.

Elliott M. Marcus
Stanley Jacobson

DEDICATION

To our wives and families who demonstrated infinite patience and support.
To our teachers and students.

CHAPTER 1
An Overview of the Central Nervous System

Introduction

Human beings come into this world naked yet equipped with a nervous system that, with experience, is ready to function in almost any environment. One word summarizes the function of the nervous system: response. The central nervous system (brain and spinal cord) monitors and controls the entire body by its peripheral divisions, which are distributed to all the muscles, organs, and tissues. The brain has an advantageous site in the head and above the neck, which can move in about a 140-degree arc. Close to the brain are all of the specialized sense organs, which permit us to see, smell, taste, and hear our world.

The central nervous system is protected by fluid-filled membranes, the meninges, and surrounded by the bony skull and vertebrae. The blood supply to the brain originates from the first major arterial branches from the heart insuring that over 20% of the entire supply of oxygenated blood flows directly into the brain.

The Neuron

The basic conducting element in the nervous system is the nerve cell, or neuron *(Fig. 1-1)*. A neuron has a cell body, dendrite, and axon. The cell body contains many of the organelles vital to maintain the cells structure and function, including the nucleus and nucleolus, and is considered the trophic center of the nerve cell. The dendrites extend from the cell body and increase the receptive surface of the neuron. The axon leaves the cell body and connects to other cells. Axons are covered by a lipoproteinaceous membrane called myelin that insulates the axons from the fluids in the central nervous system. The site of contact between the axon of one nerve cell and the dendrites and cell body of another neuron is the synapse (see Chapter 3).

In the central nervous system the nerve cells are supported by glia and blood vessels; in the peripheral nervous system they are supported by satellite cells, fibroblasts, Schwann cells, and blood vessels. There are three basic categories of neurons:

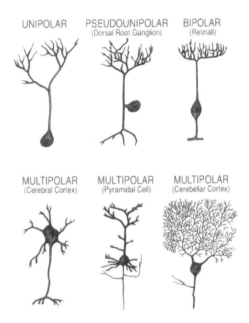

Figure 1-1. Types of Neurons and glia

1) Receptors, the ganglia of the spinal dorsal roots and of the cranial nerves,

2) Effectors, the ventral horn cells, motor cranial nerve nuclei, and motor division of the autonomic nervous system

3) Interneurons, the vast majority of the neurons in the brain.

The cells in the nervous system are classified based on their shapes: unipolar, bipolar, and multipolar (Table 1-1).

The areas in the central nervous system that contain high numbers of neuronal cell bodies are called *gray matter*, while the regions that contain primarily myelinated axons are called *white matter*. In the cerebral cortex and cerebellar cortex the gray matter is on the surface and the white matter inside. In the basal nuclei of the cerebrum, diencephalon, and brain stem, the gray and white matters are intermingled. In the spinal cord gray matter is encircled by white matter *(Fig. 1-2)*.

Neurons are organized into lamina, nuclei,

TABLE 1-1. TYPES OF NEURONS IN THE NERVOUS SYSTEM

Cell Type	% of Nerve Cells	Location in Body
Unipolar	0.05	- Dorsal root ganglia of spinal cord - Cranial nerve ganglia of brain stem - Mesencephalic nucleus of CN V in midbrain
Bipolar	0.05	- Retina, inner ear, and taste buds
Multipolar		
-peripheral	0.1	- Autonomic ganglia
-central	99.8	- Brain and spinal cord

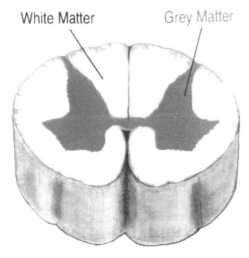

Figure 1-2. A cross section of the spinal cord showing the gray and white matter.

and ganglia. Neurons on the cortex of the cerebrum, cerebellum, and superior colliculus consist of *lamina*, highly organized layers and columns of multipolar neurons with axons that enter and exit the cortex primarily vertically from the inner surfaces. *Nuclei* are found in the brain and spinal cord and consist of groupings of neuronal cell bodies with a common function, e.g., cranial nerve nuclei. Axons enter and leave in all directions. *Ganglia* are an aggregation of neurons found in the peripheral nervous system. They include unipolar sensory ganglia of the spinal cord and brain stem and multipolar autonomic ganglia in autonomic chains or associated with visceral organs.

The Senses

Aristotle distinguished five senses: hearing, sight, smell, taste, and touch. Modern neuroscience, however, includes the *five special senses* (balance, vision, hearing, taste, and smell) and the *four general senses* (pain, temperature, touch, and pressure). Humans have evolved a series of specialized receptors for each of these different sensory functions (Table 1-2).

The special sensory apparatuses are found in the head: the eye and its protective coverings and muscles, the membranous labyrinth in the temporal bone for hearing and balance, the nose with olfactory receptors, and the tongue with taste buds.

The *receptors for general sensation* (mechanicoreceptors, nociceptors, and thermoreceptors) are located primarily in the bodies largest organ, the skin. Certain areas, e.g., the lips, fingers, feet, and genitalia have a proliferation of the tactile mechanicoreceptors. Everywhere except on the soles and palms we have hair, which is an important tactile receptor but is continually being depleted by our concern for grooming. The pain receptors, or free nerve endings in the skin, are located throughout the body, but probably more receptors are in the skin over the face, lips, hands, and feet then over the rest of the body. As you review the receptors in Table 1-2, sense on your own body how the soles are especially good for feeling pressure and placing the body safely in light or darkness and the fingers and face are sensitive to touch and temperature.

Remember that we have only discussed the skin receptors so far, which respond to external stimuli. However, there are also similar receptors within the respiratory, cardiovascular, endocrine, gastrointestinal, and urogenital systems that monitor our internal environment.

Muscles

Muscles form the bulk of the body and consist of three different functional and histological entities: *skeletal, smooth,* and *cardiac*. Skeletal muscles are found in the head, neck, arms, legs, and trunk and permit us to undertake voluntary movements. Smooth, or unstriated, muscles are found in the viscera, blood vessels, and hair folli-

TABLE 1-2. SENSORY RECEPTORS

Class of Receptor	Function	Location	See Chapter
Chemoreceptor	Taste Smell	Taste buds on tongue Olfactory mucosa in nose	11 22
Mechanicoreceptor	Balance Sound Tactile discrimination and pressure	Inner ear (semicircular ducts and vestibule) Inner ear (cochlea) Skin, muscles, Tendons, joints	Vestibular System on CD 5
Nociceptor	Pain	Free nerve endings in skin and organs	5
Photoreceptor	Vision and color	Retina	22
Thermoreceptor	Temperature	Skin, tissues, and organs	5

cles. Cardiac muscles form the heart. Each muscle group has a specialized nerve ending that permits the impulse carried down the motor nerve to contract the muscle through release of a specific chemical.

The general and special sensory receptors in the skin provide the *afferent* nerves that carry sensory information to the spinal cord and brain. Often the brain analyzes the sensory input before the muscles, which are controlled by the efferent nerves carrying information from the brain or spinal cord, make a response. These integrative functions of the central nervous system form the bulk of this book.

The Nervous System

The nervous system consists of the peripheral and central nervous system. The central nervous system (brain and spinal cord) is surrounded by fluid-filled membranes (meninges) and housed in either the skull or vertebrae. In contrast, the peripheral nervous system does not have a bony covering. Peripheral nerves, muscles, and ganglia have axons that connect all tissues and organs in the body.

The sensory information reaches the central nervous system through the sensory divisions of the peripheral nerves; movement of the three muscle groups in response to sensory information travels from the central nervous system via motor peripheral nerves.

PERIPHERAL NERVOUS SYSTEM

Peripheral nerves are found everywhere in the body: skin, muscles, organs, and glands. The neuronal processes originate either from nerve cells whose cell bodies reside in the dorsal root, cranial nerve, and visceral ganglia associated with the spinal cord and brain stem or from the ventral horn cells in the spinal cord and the nuclei of cranial nerves with motor functions in the brain stem.

The peripheral nervous system is divided into a somatic and a visceral division. The *somatic division* innervates the skin and skeletal muscles in the body. The visceral, or *autonomic, division* innervates the cardiac muscles of the heart and the smooth muscles and receptors in the blood vessels and gastrointestinal, respiratory, urogenital, and endocrine organs.

The peripheral nervous system is usually taught as part of Gross Anatomy, so the student may want to review an anatomy text. Chapter 8 provides a general introduction to the peripheral nervous system. The autonomic nervous system is more easily understood if divided into a craniosacral (parasympathetic) and thoracolumbar (sympathetic) system (see Chapter 16).

CENTRAL NERVOUS SYSTEM

Spinal Cord

The spinal cord gray matter lies in the vertebral canal from the upper border of the atlas (first cervical vertebrae) to the lower border of the first lumbar vertebrae in the adult (or third lumbar vertebrae in the neonate). The spinal cord has 32 segments divided into five regions based on the area innervated by those segments (Table 1-3).

Each segment has ganglia on the dorsal, or afferent (sensory), rootlets and a series of ventral, or efferent (motor), rootlets *(Fig. 1-3)*. The

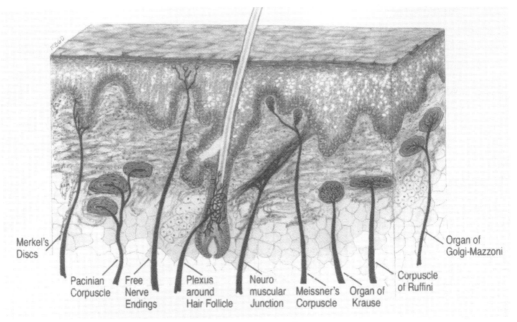

Figure 1-3. Sensory receptors and effectors in the skin.

largest segments of the spinal cord are seen where the segments of the spinal cord innervate the upper limb (C5 to T1) and lower limb (T12 to S2). The thoracic levels are narrower. In the center of the spinal cord is the nearly atrophic ventricular system of the spinal cord, the spinal canal.

Although the spinal cord may look something like a continuation of the peripheral nervous system, there are more nerve cells and different supporting cells in the spinal cord than in the peripheral nervous system. In the spinal cord the gray matter, which is covered by the white mater, is divided into a dorsal sensory horn, ventral motor horn, intermediate zone, and commissural region. The largest neuronal cell bodies are found in the ventral horn *(ventral horn cells)*, whose axons form the efferent division of the peripheral nervous system and innervate the skeletal muscles *(Fig 1-4)*.

The white matter of the spinal cord is divided into three columns: anterior, posterior, and lateral. The pathways interconnecting the spinal cord and brain are found in these columns.

Brain Stem

The brain stem *(Figs. 1-5)* consists of three regions from inferior to superior: *medulla, pons,* and *midbrain*. The brain stem is often the hardest region of the central nervous system for the student to learn because of the presence of the many cranial nerves and associated nuclei. You may initially feel overwhelmed by its intricacy but be patient. Approach neuroanatomy as you would a foreign language: first master the vocabulary and grammar before becoming fluent in conversation.

TABLE 1-3. SPINAL CORD FUNCTIONAL COMPONENTS

Spinal Cord Regions	Functions of Segments
8 cervical	Nerves to the head, neck, and upper extremities.
12 thoracic	Nerves to the thorax, abdomen, and autonomics to these regions.
5 Lumbar	Nerves to the skin; upper buttocks; anterior, medial, and lateral aspect of the thigh and leg; and medial aspect of the foot. L1 and L2 form sympathetic to pelvic plexus.
5 Sacral	Nerves to the skin and buttocks, posterior surface of the thigh and leg, lateral aspect of the foot and to the genitalia via the pudendal plexus.
2 coccygeal	Nerves to the skin and muscles of coccygeal region.

AN OVERVIEW OF THE CENTRAL NERVOUS SYSTEM

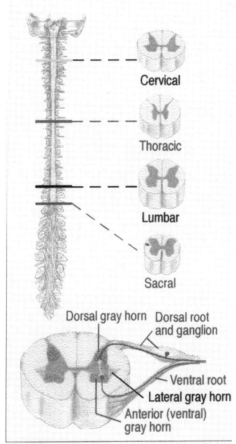

Figure 1-4. Cross sections of the spinal cord with one showing the afferent and efferent pathways.

The distinct gray and white matter in the spinal cord undergoes a slow transition from the upper cervical region to the medulla oblongata until they are completely intermingled. In the center of the brain stem, the ventricular system enlarges. The region of the brain stem anterior to the ventricular system is called the *tegmentum* (ventricle floor), and the posterior region is called the *tectum* (ventricle roof). The zone that lies on the most anterior surface of the medullary tegmentum is called the *basilar zone*.

The relatively new technology of magnetic resonance imaging *(Fig. 1-21)* provides high resolution of the nerve tissue and readily differentiates gray from white matter. It can demonstrate many significant cortical and nuclear areas in the normal brain and spinal cord and is ideally suited for diagnosing most neurologic disorders.

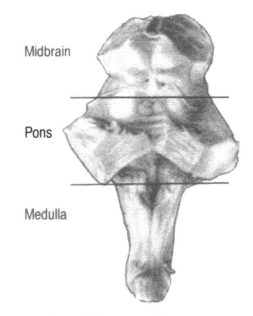

Figure 1-5. The brain stem (posterior view).

In order to understand the functions of the brain stem the student should first appreciate the extensive distribution of the 12 cranial nerves in the head, trunk, and abdomen. Cranial nerves III to XII are located in the brain stem. The key to identifying function and dysfunction in the brain stem is to know what cranial nerve nucleus lies at each level. Table 1-4 lists the location of the cranial nerve nuclei in the appropriate brain stem level. Chapter 11 discusses the nuclei and tracts of the brain stem in greater detail and includes many examples of the consequences of disease in the brain stem.

TABLE 1-4. LOCATION OF CRANIAL NERVES (CN) IN THE BRAIN STEM TEGMENTUM

Medulla	CN VIII to XII and descending nucleus of CN V
Pons	CN V to VII (motor) and chief sensory nucleus of CN V
Midbrain	CN III and IV and mesencephalic nucleus of CN V

Since the tegmentum of the brain stem is an especially important region for the student to master, we have divided this region bilaterally into five zones like an apple with the core, the center, being the reticular formation and the pulp being the surrounding four zones (Table 1-5).

TABLE 1-5. CONTENTS OF THE ZONES IN THE BRAIN STEM TEGMENTUM

Ventricular	Motor and sensory cranial nuclei on the floor of the fourth ventricle or aqueduct.
Medial	Fiber tracts.
Lateral	Ascending and descending fiber tracts and trigeminal nuclei.
Central	Reticular formation with associated nuclei and tracts.
Basilar	Descending pathways, from the cerebrum to brain stem (corticobulbar), cerebellum (corticopontine), and spinal cord (corticospinal).

TABLE 1-6. PHYLOGENETIC REGIONS OF CEREBELLAR CORTEX

Region	Cerebellar Lobes
Archicerebellum (first and smallest part)	Floccular, nodular, and uvula. Also called vestibulo-cerebellum because it receives direct input from CN VIII.
Paleocerebellum (old part)	Vermis and adjacent hemisphere. Also called spinocerebellum as it receives spino- and cuneocerebellar fibers.
Neocerebellum (new and largest part)	Includes cerebellar hemispheres. Also called corticocerebellum because it receives the bulk of its information from the cerebral cortex via the pontine gray.

Cerebellum (Figs. 1-6)

The cerebellum, like the cerebrum, consists of gray matter on the surface and white matter inside. It attaches to the tegmentum of the medulla and pons. The cerebellum is divided into two lateral hemispheres and the midline vermis. From the phylogenetic (evolutionary development) standpoint there are three regions: archicerebellum, paleocerebellum, and neocerebellum. The cerebellum functions at an unconscious level to permit voluntary motor functions to occur smoothly and accurately. Alcohol consumption and many drugs affect this region and temporarily impair, for example, gait and coordination. Table 1-6 lists the three phylogenetic regions of the cerebellum useful in understanding the functional divisions. Chapter 20 will discuss the cerebellum and movement from an integrated structural and functional approach.

In the white matter of the cerebellum are found *four deep cerebellar nuclei:* fastigium, dentatus, emboliformus, and globosus. The cerebellar cortex projects onto these nuclei, and then these nuclei project to other areas of brain stem or diencephalon.

Three major *cerebellar peduncles* (Table 1-7) contain the fiber projections to and from the cerebellum. They can be best visualized when the cerebellum is separated from the medulla and pons by a horizontal section.

Diencephalon

This region, identifiable in the gross brain after a sagittal section separates the cerebral

TABLE 1-7. CEREBELLAR PEDUNCLES

Peduncle	Connections
Inferior peduncle	Interconnects vermis and flocculonuodular with nuclei in medulla and spinal cord including CN V and VIII.
Middle peduncle	Receives input from all cerebral cortical areas that project onto the pontine gray and then decussates and enters cerebellar hemispheres.
Superior peduncle	Originates from dentate nucleus; projects to the midbrain and thalamus.

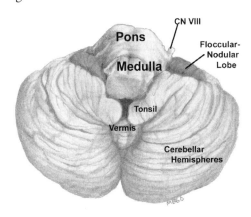

Figure 1-6. The Cerebellum.

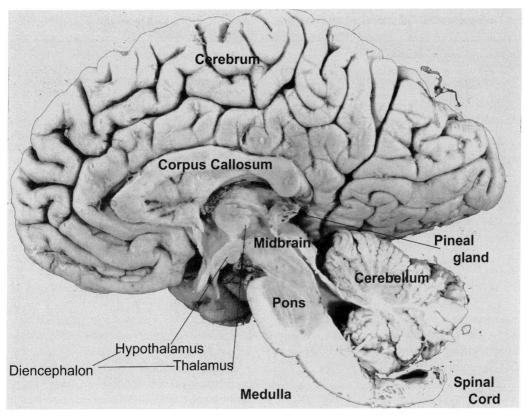

Figure 1-7. The diencephalon (insert cross section is at level of dashed line)

hemispheres, lies in the center of the cerebral hemispheres. The third ventricle separates the right and left diencephalic masses. Their lateral margin is the posterior limb of the internal capsule, and their superior surface is the body of the lateral ventricle and the corpus callosum.

The diencephalon *(Figure 1-7)* stands as the great way station between the brain stem and cerebral cortex. All the ascending pathways terminate in the diencephalon before they are projected onto their respective region of the cerebral cortex. The functional organization in the diencephalon is the basis for much of the function seen in the cerebrum and also contains the multimodal associations that make the cerebrum of the human such a potent analyzer (Table 1-8).

The *dorsal thalamus*, sometimes called the thalamus, is the largest subdivision of the diencephalon, and is divided into three major divisions: the anterior, medial, and lateral nuclear masses, each of which has many subdivisions (nuclei). The functions of this region are to integrate sensory and motor information and to begin to interpret this data according to the perceptions of the emotional areas in the brain. A detailed discussion of the thalamus is found in Chapter 15.

The *epithalamus* is a small zone on the mediodorsal portion of the diencephalon consisting of habenular nuclei, with its associative pathway, the stria medullaris, and the pineal gland; its functions are similar to the hypothalamus.

The *hypothalamus* is the smallest subdivision of the diencephalon and is found inferiorly in the third ventricle. However, by its controlling the pituitary gland and functioning as the "head ganglion" in the autonomic nervous system, it may well be the most important portion and emotional center of the diencephalon. The hypothalamus and the autonomic nervous system are discussed in Chapter 16.

The *metathalamus* in the posterior diencephalon includes the lateral and medial geniculate nuclei, which are the thalamic relay nuclei for vision (from cranial nerve II) and audition (from

TABLE 1-8. DIVISIONS OF THE DIENCEPHALON

Division	Subnuclei	Functions
Dorsal Thalamus	Anterior, medial, and lateral	Termination of most ascending sensory and motor pathways and projects onto sensory, motor and associational areas in cerebral cortex
Epithalamus	Habenula, stria medullaris, and pineal	Similar to hypothalamus
Hypothalamus	Anterior, medial, posterior, and periventricular	Highest subcortical center for emotions as controls autonomics, pineal and pituitary glands
Metathalamus	Lateral geniculate and medial geniculate	Subcortical centers for sight and sound
Subthalamus	Subthalamus and zona incerta	Subcortical unconscious center for movement

cranial nerve VIII), respectively.

The *subthalamus*, found below the thalamus, consists of the subthalamic nucleus and the zona incerta with its fiber pathways. It is an important subcortical motor coordination center between the corpus striatum and the diencephalon.

Cerebrum

The cerebrum, forming the bulk of the brain and thus of the central nervous system, consists of a left and a right hemisphere containing the cortical gray matter, white matter, and basal nuclei (Table 1-9).

The cerebral cortex consists of a corrugated surface, the gyri, separated by narrow spaces or grooves, the sulci *(Figs. 1-8 through 1-17)*. The 10 to 12 billion cortical neurons are found in the gray cortical mantle.

Students handling a preserved brain are often initially struck by the number of the gyri and sulci in the cerebral hemispheres and surprised by its weight (about 1500 gm). If you have a series of brains to observe, note that there is a great variation in size. Remember that the size of the brain is related to the size of the skeleton, muscles, and viscera (but not adipose tissue), so a person of small stature has a proportionally smaller brain. Since women tend to be smaller than men, a woman's brain is usually smaller, although this in no way reflects intelligence or abilities.

One of the first clues about the different functions in the two cerebral hemispheres was observed in stroke victims. Since the left cerebral hemisphere is dominant for language in right-handed people (93% of the population), strokes in that hemisphere usually affect speech. Besides language, dominance includes functions as diverse as initiation of movement and artistic abilities.

Note that the left hemisphere controls the right side of the body. This is because the motor pathway from the left cerebral cortex crosses over in the transition between the spinal cord *and medulla to the right side of the nervous system* (the fibers from the left cerebral hemisphere cross into the right side of the spinal cord while fibers from the right cerebral hemisphere cross into the left spinal cord). Also the sensory fibers from the muscles and skin of the right side of the body cross over to the left side of the nervous system so the same sides of the body are represented in the cerebrum. (Dominant hemispheric function is discussed in Chapter. 21)

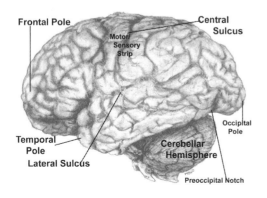

Figure 1-8. Major surface topography of the cerebrum.

TABLE 1-9. COMPONENTS OF THE CEREBRUM

Region	Components
Cerebral cortex consists of gray matter on the surface of each hemisphere.	Divided into lobes based on the relationship to overlying cranial bones: frontal, parietal, occipital, temporal, insular, and cingular gyri.
White matter, found internal to gray matter, forms connections within a hemisphere (associational) to the other hemisphere (commissural) or subcortically (projectional).	Associational bundles (see Table 1-11). Commissural--corpus callosum and anterior commissure Subcortical--afferents (corticopetal) and efferents (corticofugal).
Basal nuclei are found deep inside cerebral hemispheres.	Corpus striatum (caudate, putamen, globus pallidus), amygdala, claustrum)

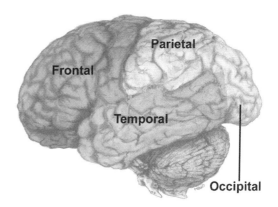

Figure 1-9. Lobes of the brain.

CEREBRAL HEMISPHERES

General Features

The cerebral cortex is divided on its lateral (convexity), medial, and inferior surface into four lobes, named for the overlying cranial bones: the frontal, parietal, occipital, and temporal lobes *(Fig. 1-9)*. Two other important regions in the cerebrum are the insula deep within the lateral sulcus and the cingulate gyrus on the medial surface. Three of the lobes also have poles: the frontal, temporal, and occipital pole.

The frontal lobe is separated from the parietal lobe by the central sulcus and from the temporal lobe by the lateral sulcus. The other lobes can be identified (Fig. 1-9) by first drawing a line from the parieto-occipital sulcus on the medial surface of the hemisphere to the preoccipital notch on the inferior lateral surface of the hemisphere. A line is then extended posteriorly from the lateral sulcus until it intersects the first line. The parietal lobe lies above the lateral sulcus and behind the central sulcus, anterior to the occipital lobe. The temporal lobe lies below the lateral sulcus anterior to the occipital lobe. The occipital lobe is most posteriorly placed.

Functional Localization

The cerebral cortex includes motor, sensory, auditory, and visual regions. In addition, broad areas are involved with multimodal integration, which combines sensory and motor with an emotional content to determine how to respond in any situation. These emotional or limbic areas occupy much of the temporal and frontal lobes. (The *limbic* region denotes the cortical and subcortical brain areas with tracts primarily involved with emotions, including the cingulate gyrus, hippocampal formation, parahippocampal gyrus, and nuclei in the diencephalon; see Chapter 22). Human beings also use language extensively and much of the frontal-parietal-occipital-temporal regions that abut the lateral sulcus in the dominant (left) hemisphere undertake these functions. Similar areas of the right hemisphere are devoted to visual-spatial integration

You might wonder just how we have such detailed information on the organization and functions of the human nervous system. Much of it comes from research on primates or from defects observed in patients after strokes, trauma, tumors, or neurologic diseases like Alzheimer's. l.

Frontal Lobe

Surface Anatomy. On the lateral surface of the cerebral hemisphere, *Figure 1-10* identifies the vertically oriented gyrus; the precentral or motor strip Then note the three anterior-posterior oriented gyri, the superior, middle, and inferior frontal gyri, which are separated from the precentral gyrus by the precentral sulcus and from each other by the superior and inferior frontal sulci. The inferior frontal gyrus is further subdivided into orbital, triangular, and opercular portions. The inferior or orbital surface of the frontal lobe is divided into medial and lateral orbital gyri

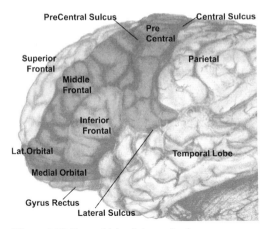

Figure 1-10. Frontal lobe of the cerebral cortex.

and the medially placed gyri rectus. The superior frontal gyrus and precentral gyrus also are found on the medial surface of the hemisphere *(Fig. 1-13)*.

Functional Localization. Identify in Figure 1-10 the motor cortex, precentral gyrus, premotor cortex, and the posterior portions of the superior, middle, and inferior frontal gyri. The precentral gyrus, or motor strip, is best considered as a unit with the postcentral gyrus forming the sensorimotor strip.

Since humans have superb control of their hands and face, these regions in the cerebral cortex are very highly developed. The somatotopic organization places the musculature in the head on the lower third of the motor strip (just above the lateral sulcus); the region controlling the hand, fingers, and distal arm constitutes the middle third of the motor strip, and the musculature of the proximal arm, abdomen, and thorax is controlled by the upper third of the motor strip. The musculature of the lower extremity below the knee is controlled by the portion of the motor strip on the medial surface of the hemisphere.

The corticospinal and corticonuclear pathways, which control the motor cranial and spinal nerves, arise primarily from this region. The inferior frontal gyrus pars opercularis and triangularis, in the hemisphere dominant for language, is Broca's motor speech area (lesions here produce nonfluent motor aphasia). The remainder of the frontal lobe (prefrontal region) is reserved for the most complex functions, e.g., integrating sensory-motor information with emotional responses and forming judgments.

Parietal Lobe

Surface Anatomy. The parietal lobe *Figure 1-11* lies posterior to the frontal lobe and is separated from it by the central sulcus. In Figure 1-11 identify the superiorly and inferiorly oriented gyrus of the postcentral gyrus or sensory strip. Remember that functionally the pre- and postcentral gyrus should be considered a unit, the sensorimotor strip. The remainder of the parietal lobe is divided into a superior and inferior parietal lobule by the intraparietal sulcus. Identify in Figure 1-11 the postcentral sulcus that separates the postcentral gyrus from the inferior and superior parietal lobules. The inferior parietal lobule is further divided into a supramarginal and angular gyrus. Can you see these subdivisions that are significant in language functions? The portion of the parietal lobe above the lateral sulcus is the parietal operculum. The postcentral gyrus and superior parietal lobule extend onto the medial surface of the hemisphere.

Functional Localization. The postcentral gyrus of the parietal lobe is the sensory cortex and postcentral gyri are a functional entity: the sensory-motor strip the postcentral gyrus is functionally organized like the precentral gyrus, except it is primarily sensory. The superior parietal lobule is sensory associational. In the hemisphere dominant for language the supramarginal gyrus is the sensory language area concerned with reading and related functions (lesions here produce aphasia). Deep in the upper bank of the lateral fissure is the taste region.

Temporal Lobe

Surface Anatomy. The temporal lobe *(Figure 1-12)* is found on the lateral and inferior surface of the brain; it lies below the lateral sulcus anterior to the occipital lobe. Deep in the lateral sulcus are seen the transverse gyri of Heschl. The fibers from cranial nerve VIII end here. The temporal lobe is divided into six anteroposteriorly placed gyri: the superior, middle, and inferior temporal gyri on the lateral surface and the occipitotemporal, parahippocampal and hippocampal formation on the inferior surface. The hippocampal formation is the most medial and inferior cortical structure in the temporal lobe

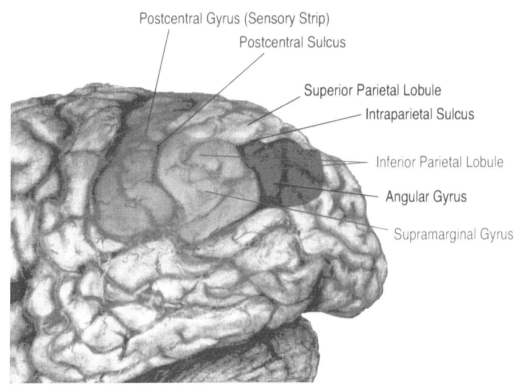

Figure 1-11. Parietal lobe of the Cerebral Cortex.

and is found adjacent to the inferior horn of the lateral ventricle.

Functional Localization. The temporal lobe is divided into lateral and inferior regions with distinct functions. On the lateral surface in the lateral sulcus is the auditory cortex, the transverse temporal gyri of Heschl and adjacent to it are other language areas related to understanding speech. The planum temporale, a center for language is revealed when the frontal and parietal operculum are opened in the dominant hemisphere. This region contains the auditory association cortex. The inferior surface of the temporal lobe is divided into: (1) associative cortex for emotions (especially temporal pole and uncus), (2) new memory (hippocampus), and (3) visual integration (posterior temporal).

Occipital Lobe

Surface Anatomy. The occipital lobe is the most posteriorly placed lobe *(Figure 1-12)*. On its lateral surface the lateral occipital gyri are found. If the cerebral hemispheres are separated, the medial and inferior surface of the occipital lobe can be seen. On the medial surface the occipital lobe is separated from the parietal lobe by the parieto-occipital sulcus *(Fig. 1-13)*, and the prominent calcarine sulcus extends from the parieto-occipital sulcus to the occipital pole. The collateral sulcus is identified parallel to and below the calcarine sulcus. The visual cortex forms the upper and lower bank of the calcarine sulcus.

Functional Localization. The prime function of this region is vision. The calcarine, or primary visual, cortex consists of the cuneus gyrus above the calcarine sulcus and the lingual gyrus below the calcarine sulcus. Information from the optic nerve reaches the visual cortex after an initial synapse in the lateral geniculate of the diencephalon. The remainder of the occipital lobe relates to visual association, including identification of visual images (the mental equivalent of a Rolodex), which is vital in language, whether spoken, written, or read.Cingulate Gyrus

Surface Anatomy. After the cerebral hemispheres are separated by sectioning the major white commissure of the cerebrum the corpus callosum, the medial surface of the hemisphere is seen and the cingulate lobe and the cingulate sul-

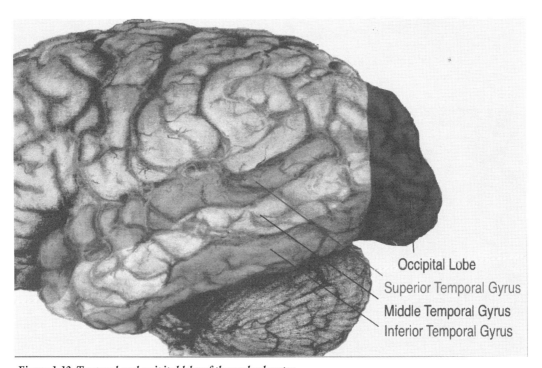

Figure 1-12, Temporal and occipital lobes of the cerebral cortex.

cus are noted above the corpus callosum. If you look carefully you will also note the cingulate lobe is continuous around the back of the corpus callosum with the parahippocampal gyrus of the temporal lobe.

Functional Localization. This associative area in the limbic brain is especially important due to its relationship to the hypothalamus, thalamus,

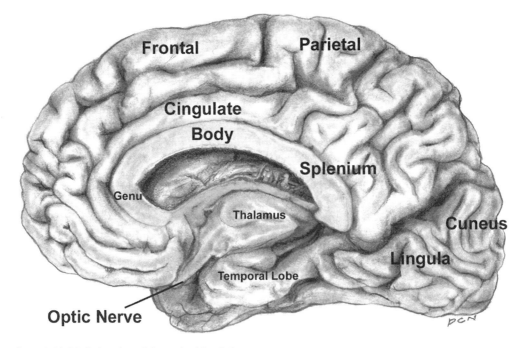

Figure 1-13. Medial surface of the cerebral hemispheres.

and frontal association cortex.

Insular Gyri

This region is found when the lateral sulcus is opened *(Figure 1-14)* and is seen to contain this deeply placed island of cerebral cortex. The functions of this region are probably related to the emotional brain.

Cerebral Functions

Besides all the functions already mentioned, the cerebral hemispheres also have the memory stores, which are the foundation for most conscious and unconscious thought, cultural activity, sexual behavior and all of the other positive or negative traits which make the human being so distinctive. Each cortical region has a memory store that permits the function of that region (Table 1-10).

Basal Nuclei

In a coronal section through a midpoint of the cerebral hemispheres we can identify laterally the cerebral cortex, cortical white matter, and basal nuclei *(Figure 1-15)*. Three important nuclear groups deep within the hemispheres are the corpus striatum, claustrum, and amygdala. This section focuses on the corpus striatum: the caudate nucleus, putamen, and globus pallidus. They are separated from the cerebral cortex by white matter but are functionally linked to the cerebrum. These nuclei are motor associative in function. They carry on at a subconscious level until a dysfunction, such as in Parkinson's disease or Huntington's chorea occurs. The caudate nucleus and putamen are more laterally placed surrounding the white matter, which connects the cortex with the subcortical areas, the internal capsule. The caudate nucleus lies medial and dorsal to the thalamus. The functions of these nuclei will be discussed in further detail in Chapter 20).

Cortical White Matter

The axons entering or leaving the cerebral

TABLE 1-10. OVERVIEW OF CEREBRAL FUNCTIONS

Part	Function
Frontal lobe	Volitional movement of the muscles of the body including limbs, face and voice. Judgmental center.
Parietal lobe	Tactile discrimination from the body, body image, taste, and speech.
Occipital lobe	Vision (sight and interpretation).
Temporal lobe	Emotions and new memories, hearing, language, olfaction, and visual recognition.
Cingulate gyrus	Emotions.

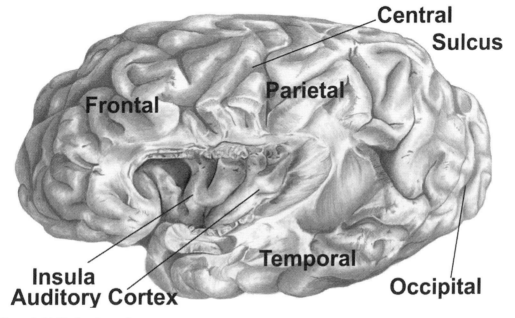

Figure 1-14. The insular gyri.

hemispheres form three distinctive groups of fibers: associational, commissural, and subcortical (Table 1-11).

TABLE 1-11. WHITE MATTER OF THE CEREBRAL HEMISPHERES

Fiber Category	Functional Significance
Short U (e.g., arcuate) or long associational (e.g., uncinate and cingulum) fibers	Associational fibers form the bulk of the white matter in each hemisphere and interconnect diverse areas in a hemisphere providing the multimodal associations essential for cortical functions.
Commissural (e.g., corpus callosum and anterior commissure) fibers	Commissural fibers interconnect the left and right cerebral hemispheres and permit the learning and memory in one hemisphere to be shared with the other.
Subcortical (e.g., corticofugal and corticopetal) fibers	Corticofugal fibers provide connections from cortex to diencephalon, brain stem, cerebellum and spinal cord; corticopetal fibers provide afferents to cerebrum from diencephalon.

Associational Fibers. These two types of fibers (short U and long associational) provide the integrative circuitry for movement, language, memory, and emotions.

Commissural Fibers. The bulk of the frontal, parietal, occipital and temporal lobes are interconnected by the *corpus callosum,* best seen in the medial surface of the hemisphere. The corpus callosum consists of a rostrum, genu, body, and splenium *(Fig. 1-16).* The most rostral part of the temporal lobe and the olfactory cortex of the uncus are interconnected by the *anterior commissure.* The hippocampus, an important area for memory, is also connected by a commissure associated with the fornix.

Subcortical Fibers. This category of fibers includes fiber bundles reaching the cortex from subcortical areas, corticopetal (toward the cortex), and axons leaving the cortex and connecting to subcortical nuclei, corticofugal (away from the cortex).

Corticopetal fibers are the subcortical afferents to each cerebral hemisphere and are primarily from the thalamus. The fibers that enter the cerebral cortex arise in functionally diverse subcortical regions (see Chapter 15, Fig 8A&B).

Corticofugal fibers from the cerebral hemisphere project onto subcortical structures including the basal nuclei, diencephalon, brain stem, cerebellum, and spinal cord. The major subcortical fiber tracts leaving the cerebrum are the corticospinal, corticonuclear, corticopontine, corticomesencephalic, and the fornix.

The *internal capsule* contains the major grouping of corticofugal and corticopetal fibers of the cerebral cortex and consists of an anterior limb, genu, and posterior limb (Figs. 1-15 and 15-6). The anterior limb provides fibers to and from the frontal lobe. The genu provides fibers to and from the lower part of the frontal and parietal lobes (corticonuclear). The posterior limb is the largest portion of the internal capsule and includes the auditory radiations (auditory fibers to the auditory cortex), visual radiations as well as projections from the sensorimotor cortex to the spinal cord, and brain stem (corticospinal, corticonuclear, and corticopontine fibers).

Central Nervous System Pathways

Up to this point, the diverse functional gray units in the brain and spinal cord have been identified. In order for the nervous system to work, an extensive circuitry has been established to interconnect areas throughout the central nervous system. We have just described the intrinsic white matter circuitry within the cerebrum: the associational, commissural, and subcortical fibers. The subcortical fibers consist of a multitude of axonal pathways that provide afferents to the cerebrum and efferents from the cerebrum.

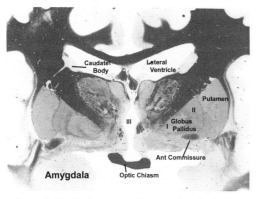

Figure 1-15. The basal nuclei (coronal section).

Many names are used for pathways within the central nervous system, including fasciculus (bundle), lemniscus (ribbon), peduncle (stalk), and tract (trail). Many of the pathways also have a name that indicates its function (e.g., the optic tract connects the eyes and visual regions, the corticospinal (Fig 15-13) tract runs from the cortex to the spinal cord, and the corticonuclear (Fig 15-14) tract connects the cerebral cortex to the cranial nerve nuclei). Unfortunately, other pathways have more obscure names that give no clue to function (e.g., the medial longitudinal fasciculus, lateral lemniscus, and cerebral peduncle). The major tracts and their role in the central nervous system are discussed in Chapter 11. For each pathway the student needs to learn the following:
1. Cells of origin
2. Location of the tract in the brain
3. Site of termination of pathway
4. Function.

We are now going to use as an example of circuitry within the central nervous system a functionally significant major circuit in the brain that provides volitional motor control to the hand muscles.

Motor Control of the Hand

In order to perform a skilled motor task one is dependent on precise sensory input, so we will first describe the sensory portion of this circuit.

Sensory Information from the Hand to the Sensorimotor Cortex. In the sensory system three different neuronal groupings are traversed before the sensory information reaches the cerebral cortex from the periphery: the primary, secondary, and tertiary neurons. The sensory information of importance for precise movement originates from stretch receptors in the muscles, tendons, and joints of the hand and is encoded as electrical impulses that detail the contraction status of the muscles, tendons, and joints. The cells of origin, the primary cell bodies, are in the sensory ganglia attached to the lowest cervical segments of the spinal cord. The information is carried in by the dendrite of the primary sensory cell and its axon enters the cen-

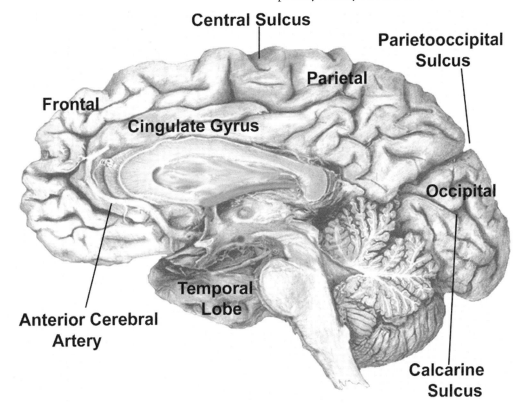

Figure 1-16. Medial surface of a cerebral hemisphere.

tral nervous system, ascends uncrossed and ipsilaterally in the posterior column of the cord until the medullospinal junction. In the medullospinal junction *second-order nerve cells* appear in the posterior column, the cuneate nucleus. The axons from the primary neurons synapse on this nucleus. The axons of the second-order axons cross the midline and ascend contralaterally in one of the major ascending white matter pathways, the medial lemniscus, into the thalamus of the diencephalon where they synapse on the *third-order neurons in the ventral posterior lateral nucleus,* which then send fibers ipsilaterally onto the hand region of the postcentral gyrus.

Motor Control of the Hand from the Motor-Sensory Cortex. In the motor system two distinct neuronal groups are necessary for a movement: one in the motor cortex of the cerebrum (the upper motor neuron) and the other in the spinal cord or brain stem (the lower motor neuron).

Motor control to the hand is carried centrifugally by the corticospinal pathway. The *corticospinal pathway* to the hand originates in the motor cortex of the precentral gyrus and to a lesser degree in the postcentral gyrus. The fibers before exiting the gray matter are covered with myelin and descend on the same side through the internal capsule, cerebral peduncle, pons, and medullary pyramid. The corticospinal fibers cross (decussate) to the other side of the brain in the medullospinal junction and enter the lateral column of the spinal cord. Fibers to the musculature of the hand descend to lower cervical levels and synapse in either the intermediate region of the gray matter of the spinal cord or directly on the ventral horn cells. The axons from the ventral horn cell leave the central nervous system and synapse on the motor endplates of the muscles, which produce the contraction. The actual initiation of the movement is from the prefrontal and premotor region of the frontal lobe (see Chapter 20). Volitional motor control to the muscles controlled by the cranial nerves descend in the corticonuclear and corticomesencephalic pathways; these fibers originate in the lower third of the motor-sensory strip (see Chapter 11).

Interruption of the corticospinal pathway produces upper motor neuron paralysis with increased reflex tone while injury to the spinal cord ventral horn cells produces lower motor neuron paralysis and atrophy of the muscle and decreased reflex tone (Table 1-12).

In summary, it takes three orders of sensory neurons to provide the information to the cerebral cortex and two levels of motor neurons to produce a hand movement in response to the sensory information. Note that the sensory information ascends and crosses over. While the motor information descends and crosses back so that the same side is represented in the cerebral cortex. Of course, the hand movement also needs a conscious decision to be made, which is what the cerebral cortex working as a unit does.

The following Case History illustrates the effects of a lesion in the motor strip on movement.

TABLE 1-12 EFFECTS OF LESIONS IN THE VOLUNTARY UPPER AND LOWER MOTOR PATHWAYS

Location of Lesion	Changes in Reflexes and Muscle Tone
1. Upper Motor Neuron Lesion (In cortex or corticospinal pathway)	Reflexes Increase +3, +4, muscles spastic, Sign of Babinski (extensor plantar response)
2. Lower Motor Neuron Lesion (ventral horn cells, ventral roots, motor cranial nerve nuclei and roots)	Reflexes decrease to absent 0, +1, muscles flaccid and atrophied.

Chapter One, Case History One.

This 45 year old right handed, married, white female, mother of four children and an artist in sculpture and calligraphy was referred for evaluation of progressive weakness of the right lower extremity of two years duration. This weakness was primarily in the foot so that she would stub the toes and stumble. Because she had experienced some minor non-specific back pain for a number of years, the question of a ruptured lumbar disc had been raised as a possible etiology. The patient labeled her back pain as "sciatica" but denied any radiation of the back pain into the leg. She had no sensory symptoms, and no bladder symptoms. Six months before the consultation, the patient had transient mild weakness of the right leg that lasted a week. The patient attributed her symptoms to "menopause" and depression, and her depression had improved with replacement estrogen.

Neurological examination revealed the patient's mental status was normal and the cranial nerves were all intact. There was a minimal drift down of the outstretched right arm, and there was significant weakness in the right lower ex-tremity with ankle dorsiflexion and plantar flexion was less than 30% of normal. Inversion at the ankle was only 10-20% of normal. Toe extension was 50% of normal. and there was minor weakness present in the flexors and extensors at the knee. In walking, the patient had a foot drop gait and had to over lift the foot to clear the floor. Examination of the shoes showed greater wear at the toes of the right shoe with evidence of scuffing of the toe.

The patient's deep tendon stretch reflexes were increased at the patellar and achilles tendons on the right. In addition reflexes in the right arm at the biceps, triceps and radial periosteal were also slightly increased. The plantar response on the right was extensor (sign of Babinski), and the left was equivocal. All sensory modalities were intact. Scoliosis was present with local ten-derness over the lumbar, thoracic and cervical vertebrae, but no tenderness was present over the sciatic and femoral nerves.

Comments

The diagnosis in this case was not certain. It was clear that the symptoms and signs were not related to compression of the lumbar nerve roots by a ruptured disc. Such a compression (lumbar radiculopathy) would have produced a lower motor neuron lesion, a depression of deep tendon stretch reflexes, lumbar radicular pain in the distribution of the sciatic nerve and tenderness over the sciatic nerve, and no increase in reflexes in the upper or lower extremities, and no sign of Babinski.

The fact that reflexes were more active in both upper and lower extremities on the right, with the sign of Babinski on the right, suggested a pro-cess involving the corticospinal tract. The fact that the weakness was greatest in the foot suggested a meningioma involving the parasagittal motor cortex where the foot area is represented. With these clinical considerations in mind

an MRI of the patient's head was obtained and the MRI con-firmed the clinical impression *(Figure 1-17)*. Meningioma in the right cerebral hemisphere,). This tumor was successfully removed by the neurosurgeon with an essentially complete restoring of function.

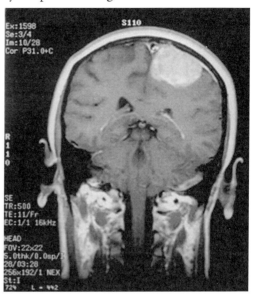

Figure 1-17. Meningioma in the left cerebral hemisphere.

Glands Associated with the Brain

The brain has two glands, the pituitary and the pineal, both of which are attached to the diencephalon. The pituitary (Atlas Figure 4) or hypophysis cerebri is attached to the base of the hypothalamus. Its functions are related to the hypothalamus. The other gland, the pineal (Gross Figure 16), or epiphysis cerebri, is found in the epithalamus. By its location its functions in relationship to light levels and diurnal cycles would not be apparent.

Blood Supply to the Brain

The central nervous system is dependent on a continuous supply of enriched, oxygenated arterial blood. This has been accomplished by prioritizing blood supply to the brain as regards the initial branches of the major artery that leaves the heart, the aorta. Its first major branch is the common carotid, which supplies the neck, head, and brain. Its continuation, the subclavian artery, also supplies the same three areas *(figure 1-18)*.

Blood flow to the brain is divided into the

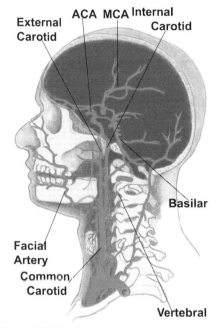

Figure 1-18. The blood supply to the brain.

of the cerebral hemispheres. The *posterior circulation* (subclavian, vertebral, basilar, posterior communicating, and posterior cerebral arteries) supplies the upper cervical spinal cord, cerebellum, brain stem, most of the diencephalon, and inferior and posterior surfaces of the temporal and occipital lobes of the cerebral hemispheres.

At the base of the brain is a set of arteries called the circle of Willis *(Fig. 1-19)* that interconnects the main portions of the anterior and posterior circulation by the formation of anterior and posterior communicating vessels.

Meninges

The brain is enclosed by three membranes, the dura mater, arachnoid, and pia mater *(Fig. 1-20)* (see Chapter 19). These protective fluid-filled membranes are formed by connective tissue with embedded nerves, especially in the dura. The dura mater is the most external membrane, followed by the arachnoid and finally the pia mater, which adheres to the central nervous system.

Dura Mater

The externally located dura mater (figure 1-20) consists of a tough fibrous connective tissue. In the cranial vault the dura forms the periosteum on many of the adjacent bones and also forms an inner, or meningeal, portion that compartmentalizes the brain:

anterior and posterior circulation. The *anterior circulation* (internal carotid, middle cerebral, anterior communicating, and anterior cerebral arteries) supplies the basal ganglia, anterior diencephalon, and the lateral surface of the cerebral hemispheres, anterior two-thirds of the medial surfaces of the cerebral hemisphere including the corpus callosum and the orbital fro0ntal surfaces

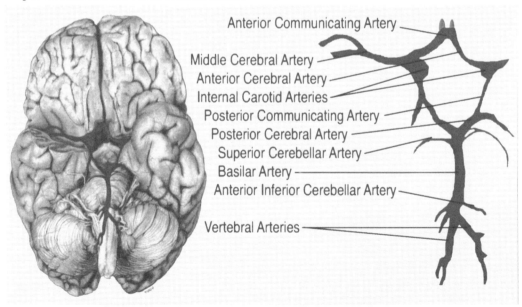

Figure 1-19. Arterial circle of Willis.

AN OVERVIEW OF THE CENTRAL NERVOUS SYSTEM

- *Falx cerebri* (between the cerebral hemispheres)
- *Tentorium cerebelli* (between the cerebellum and cerebrum)
- *Falx cerebelli* (between the cerebellar hemispheres).

The spinal dura forms a sac surrounding the spinal cord. The vertebrae have their own periosteum. The dural sac attaches at the margin of the foramen magnum to the occipital bone and to the inner surface of the second and third cervical vertebrae. It is also continuous with the perineurium on the spinal nerves and covers the filum terminale and becomes continuous with the periosteum on the coccyx as the coccygeal ligaments.

Arachnoid

This thin membrane (figure 1-20) is located between the dura and pia. The space between the pia and arachnoid, the subarachnoid, is filled with the cerebrospinal fluid (CSF). The arachnoid bridges the cerebral sulci and extends from the posterior surface of the medulla to the cerebellum (cisterna magna) and below the neural ending of the spinal cord (lumbar cistern). All of these spaces contain CSF. The arachnoid is usually separated from the dura by the subdural space. The arachnoid membranes form arachnoid granulations (Pacchionian bodies) that permit passage of the CSF from the higher pressure in the ventricular system into the lower pressure of the venous sinuses.

Pia Mater

The pia adheres to the entire central nervous system (figure 1-20) and is continuous with the perineurium of cranial and spinal nerves. It attaches to the blood vessels entering and leaving the central nervous system and fuses with the dura. The cranial pia actually invests the cerebrum and cerebellum and extends into the sulci and fissures. It also forms the non-neural roof of the third ventricle, lateral ventricle, and fourth ventricle. The pia forms the denticulate ligaments that anchor the spinal cord to the dura between the exiting spinal nerve rootlets.

Ventricular System *(Fig. 1-21)*

As already mentioned, the central nervous system is surrounded by cerebrospinal fluid (CSF). Externally, the meninges, especially the subarachnoid space, contain CSF. Internally, the ventricular system and cisterns contain CSF, and a specialized structure, the choroid plexus, excretes CSF. CSF also is found in the extracellular space in the brain.

The ventricular system consists of the lateral ventricles, third ventricle, cerebral aqueduct, fourth ventricle, and spinal canal.

The *lateral ventricles* are found within the cerebral hemispheres and consist of the body and the anterior, posterior, and inferior horns. A horizontal section of the brain shows the relationship between the ventricles and the central nervous system.

The *third ventricle* lies in the midline between the left and right diencephalon and is continuous superiorly with the frontal horn of the lateral ventricles and inferiorly with the cerebral aqueduct. Several recesses associated with the third ventricle are important radiologic landmarks: the optic recess, hypophyseal recess, and pineal recess.

The *cerebral aqueduct* is the narrow ventricular space in the center of the midbrain connecting the third and fourth ventricles.

The *fourth ventricle* forms that portion of the ventricle seen in pontine and medullary levels. The fourth ventricle is continuous laterally (foramina of Luschka) and medially (foramen Magendie) with the subarachnoid space. The

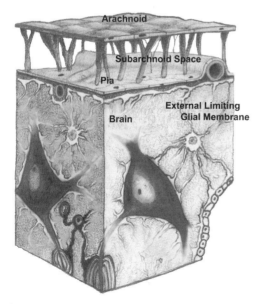

Figure 1-20. The meninges.

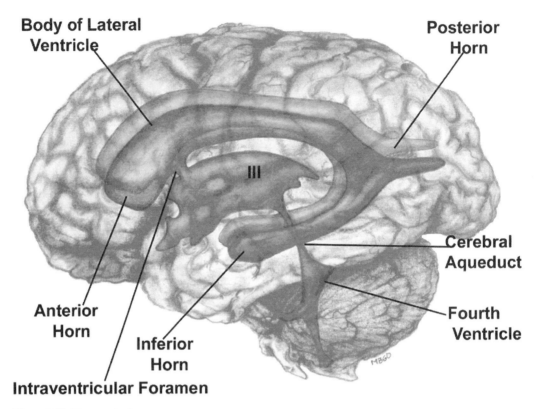

Figure 1-21. The ventricular system.

fourth ventricle is continuous inferiorly with the narrow central canal of the spinal cord.

TABLE 1-12. COMPONENTS OF PLASMA AND CEREBROSPINAL FLUID

	Plasma (mg/dl)	CSF (mg/dl)
Na+	330	310
HCO3-	1200	1310
CA++	10-11	5.3
K+	17	12
HPO4-	3	1.8
SO4-	1.9	0.6
Glucose	100	70
Protein	8000	45

The narrow *spinal canal* is seen in the center of the gray commissure of the spinal cord.

Cerebrospinal Fluid. The ventricular system is lined by ependymal cells (see Fig. 3-27). Mostly the choroid plexus in the lateral ventricles, third ventricle, and fourth ventricle, which are supplied by the choroidal arteries, forms CSF. It is a clear, colorless, and basic fluid that resembles protein-free plasma but with many significant differences (Table 1-12). The total volume of CSF is usually between 100 and 150 ml. The rate of formation is 20 ml/hr, or about 500 ml/day. Excess fluid is readily absorbed into the venous sinuses through the arachnoid granulations because the pressure in the CSF is higher.

Cerebrospinal Fluid Circulation. CSF flows from the lateral ventricles through the foramen of Monro into the third ventricle and then throughout the cerebral aqueduct into the fourth ventricle. In the fourth ventricle it is continuous with the subarachnoid space at the foramens of Luschka (lateral) and Magendie (medial). In the subarachnoid space the fluid passes into the venous sinuses through the arachnoid granulations. An excess in CSF produces *hydrocephalus*, which is frequently a consequence of meningitis or hemorrhage (communicating hydrocephalus) or a blockage in the ventricular system (noncommunicating hydrocephalus) caused by hemorrhage, tumors, or trauma.

CHAPTER 2
An Overview of Localization of Function, and Neurological Diagnosis

In considering the patient with neurological disease, accurate diagnosis is required for a determination of appropriate therapy and for the establishment of prognosis. Diagnosis in neurology seeks to answer two essential questions.

1. Where is the disease (lesion) located?
a) Central nervous system versus
b) Peripheral nervous system versus
c) Neuromuscular junction or muscle.
2. What is the nature of the pathology?

PART I: LOCATION AND PATHOLOGY

A. LOCATION OF DISEASE PROCESS: SYMPTOMS AND SIGNS

Patients with neurological disease come to medical attention because of certain symptoms or complaints elicited in the history and certain signs elicited on examination. These symptoms and signs fall into several categories:

1. Disturbance of mental status (cognitive function) and language function related to disease of cerebral cortex.
2. Disturbance of cranial nerve function due to direct involvement of cranial nerves or of brain stem.
3. Disturbance of motor strength reflecting -

   a) Disease of the lower motor neuron and the motor unit at the level of the anterior horn cell, or the nerve root or the peripheral nerve or the neuromuscular junction or muscle producing atrophy of the muscle, flaccid weakness and a variable loss of reflex activity or

   b) Severe upper neuron disease due to involvement of motor cerebral cortex, or due to the involvement of corticospinal tracts at the level of internal capsule cerebral peduncles, basilar pons, medullary pyramids or lateral columns of spinal cord producing spastic paralysis, increased deep tendon reflexes and a release of the sign of Babinski.

4. Disturbance of motor control and coordination reflecting -

   a) Disease of motor cortex and corticospinal tracts: loss of speed and accuracy.

   b) Disease of the premotor and supplementary motor cortex / motor association cortex resulting in defects in patterns of movement, defects in the visual and tactile control of movement. Apraxia occurs and the release of automatisms may occur.

   c) Disease of cerebellum - resulting in defects in coordination of limb and axial movements and of "balance"

   d) Disease of basal ganglia resulting in
   1) Slowness of movement (akinesia and bradykinesia)
   And/or 2) resting tremor
   And/or 3) other dyskinesias
   And/or 4) loss of righting reflexes and "balance"
   And/or 5) alterations in tone

5. Disturbance of sensation:

   a) Primary modalities (pain, touch, vibration) due to disease of peripheral nerve, nerve root, spinal cord or brain stem or thalamus

   b) Discriminative modalities (position, sense, graphesthesia, stereognosis, tactile localization, awareness of double simultaneous stimulation) due to disease of posterior column system or cerebral cortex.

6. Pain including headache - the localization of pain may reflect disease of peripheral or cranial nerve, nerve root, spinal cord, brain stem, thalamus, or cortex.

Meninges and blood vessels about the head (intra and extracranial) or increased intracranial pressure, or the sinuses or orbits, or dental system, may provide the origin of headache.

B. PRELIMINARY DIFFERENTIATIONS OF LESION LOCATION:

1. Disease of muscle

a) Patients manifest lower motor neuron disease affecting limbs, trunks and in some cases, cranial nerves. In early cases, however, deep tendon reflexes are preserved

b) Involvement is more often proximal than distal.

c) No sensory symptoms or signs are present.

d) No long tract motor or sensory involvement is present.

e) Cognitive function - mental status is normal but may be affected in certain hereditary types.

f) The major disorders are of two types
   1) Hereditary (muscular dystrophy or congenital myopathies)
   2) Acquired (polymyositis or dermatomyositis).

2. Disease of neuromuscular junction

a) Patients manifest intermittent dysfunction of lower motor neuron.

b) Involvement of cranial nerve, motor function is prominent, (extraocular muscles and bulbar motor neurons).

c) If limbs are involved - proximal weakness is often greater than distal weakness.

d) Deep tendon stretch reflexes are preserved.

e) No sensory symptoms are present.

f) No long tract findings are present.

g) Cognitive function is not involved.

h) The major disorder is myasthenia gravis, an autoimmune disorder in which antibodies are produced which block and damage the acetylcholine receptor.

3. Disease of peripheral nerve

a) Motor and/or sensory and/or autonomic function are involved.

b) The pattern of involvement indicates disease of:
   1) A specific peripheral nerve (mononeuropathy), e.g., median, ulnar, radial, femoral, sciatic, peroneal.
   Or 2) a plexus, such as brachial, lumbar sacral (mononeuropathy)
   Or 3) multiple peripheral nerves (mononeuropathy multiplex)
   Or 4) generalized type- polyneuropathy. This is the most common variety.

c) When motor function is involved the pattern is that of a lower motor neuron type lesion with atrophy, flaccid weakness and loss of reflex function.

d) When sensory involvement is present; some or all modalities may be involved. In the generalized type, the sensory pattern may be a glove and stocking type

e) Cranial nerves may be involved but without direct brain stem involvement.

f) Long tract finding are not present

g) Cognitive function is intact.

h) If compression (entrapment) is present, localized pain may be a prominent feature.

4. Nerve root or spinal nerve disease

a) Motor and/or sensory function may be involved.

b) Pain is often a prominent feature.

c) The symptoms and signs (pain, sensory, motor and reflex) follow a segmental (dermatomal) distribution. Refer to Tables 2-1, 2-2, 2-3, 2-4.

d) When motor function is involved; the pattern is that of a lower motor neuron type lesion as above.

e) Some or all sensory modalities are involved, in a dermatomal pattern. *Fig. 2-1*.

f) Cranial nerves are not involved.

g) There are no long tract findings - unless, spinal cord is also involved.

h) Cognitive function is not involved.

Dermatomal-Radicular Patterns

Fig 2-2 compares dermatomal and peripheral nerve patterns.

5. Spinal Cord:

a) Lesions are either
   1) Transverse (segmental)
   2) System involving a system of neurons or fibers over many segments
   3) Multifocal producing spotty lesions over multiple levels of the CNS

Transverse lesions

a) Transverse lesions are usually extrinsic with the implication of compressive mass and possible surgical therapy.

b) Four exceptions are intrinsic lesions, which involve multiple adjacent segments: syringomyelia intrinsic spinal cord tumors (astrocytomas and ependymomas), infarcts due to occlusion of the inferior spinal artery and transverse myelitis.

c) Generally both motor and sensory function are involved.

d) When the motor long fiber systems are involved, the signs of an upper motor neuron lesion are produced: spastic weakness or paralysis, increased deep tendon reflexes and the sign of Babinski.

e) Involvement of the long fiber sensory systems may produce involvement of all modalities below the lesion in a transverse lesion or selective involvement of specific sensory modalities, e.g., position and vibratory sensation when the posterior columns are involved and pain and temperature when the lateral spinothalamic tract is involved. Syringomyelia involves the decussating pain and temperature fibers in the anterior white commissure of the cervical area producing a selective dissociated pain and temperature sensory loss over the shoulders and arms "cape like".

f) Transverse lesions producing damage to the anterior horn cell or anterior root will produce signs of a lower motor neuron lesion at the segmental level of involvement. Fasciculations, twitching of muscle, will be noted in the Segment involved. Fasciculations are indicative of disease of anterior horn cells.

g) Transverse lesions producing damage to the posterior root will produce a local segmental defect at the level of involvement.

h) Cranial nerves are not involved.

i) Mental status, cognitive function, is not involved.

Concept of Level of Lesion at the Spinal Cord Level

In transverse lesions it is then evident that several determinants of level are considered.

1. Local anterior root or anterior horn level indicated by segmental atrophy and segmental flaccid weakness.

2. Local segmental sensory level indicated by segmental (radicular) loss of all modalities of sensation.

3. Local segmental losses or depression of deep tendon reflexes due to loss of the afferent or efferent component of the monosynaptic stretch reflex.

4. Segmental long motor tract level with presence of spastic weakness, increased deep tendon reflexes and release of the sign of Babinski below this level.

5. Segmental long sensory tract levels indicated by posterior column (uncrossed) and lateral spinothalamic (crossed) deficits below this level.

TABLE 2-1: SEGMENTAL SENSORY PATTERNS (FIG. 2-1 AND 2-2. SEE ALSO 8-18 FOR THE UPPER EXTREMITY DETAILS)

The following radicular patterns should be compared to peripheral nerve patterns	
C2	posterior scalp versus trigeminal
C6, C7, C8	the hand
a. C6	thumb and index finger
b. C7	middle finger
c. C8	ring and little (5th) finger
T1	axilla
T5, T6	xiphoid process
T10	umbilicus
T12	above inguinal ligament
L4	medial calf, patella and lateral thigh
L5	lateral calf and medial foot
S1	posterior calf and lateral foot
S2	posterior thigh
S3, 4, 5-	perianal areas.

TABLE 2-2: MOTOR - RADICULAR MUSCLE INNERVATIONS

1. Shoulder muscles,	C4, C5, C6
2. Biceps	C5, C6
3. Triceps	C6, **C7**, C8
4. Intrinsic hand muscles	C7, C8, T1
5. Hip flexors, iliopsoas	L2, **L3**, L4
6. Quadriceps	L2, **L3**, L4
7. Gastrocnemius	LS, L5, **S1**, S2
8. Dorsiflexors of foot (peroneal)	L4, L5, S1

* Bold indicates major innervations

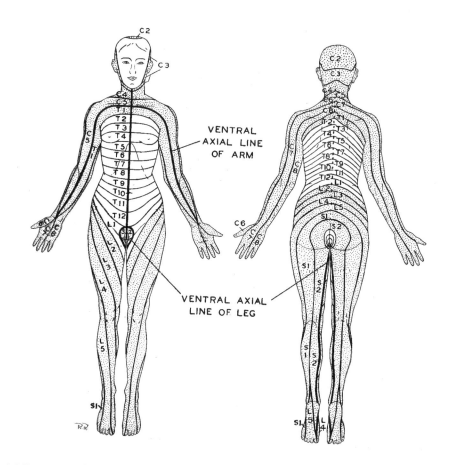

Figure 2-1 Dermatome charts of the human body determined by the pattern of hypalgesia following rupture of an intervertebral disk. From Keegan, J.J., and Garrett, F.D.: Anat. Rec., 102:411, and 1948 (Wiley)

TABLE 2-3: DEEP TENDON STRETCH REFLEXES

1. Jaw	cranial nerve V - pons
2. Biceps	C5, C6
3. Triceps	C6, **C7**, C8
4. Brachioradialis (Radial periosteal)	C5, C6, C7
5. Fingers -	C7, C8, T1
6. Patellar-	L2, **L3, L4**
7. Achilles	L4, L5, **S1**, S2

* Bold indicates major innervations

TABLE 2-4: SUPERFICIAL REFLEXES

1. Upper abdomen -	T7, T8, T9, T10
2. Lower abdomen -	T10, T11, T12
3. Plantar	S1, S2

6. Segmental autonomic level deficits in autonomic function below this level.

6. Brain Stem

a) Cranial nerves findings are present with the specific lower motor neuron or sensory deficits dependent on the level of involvement.

1) CN 12, 11, 10, 9 due to medullary lesions

2) CN 8, 7 and 6 due to pontomedullary involvement

3) Trigeminal - 5 midpontine

4) CN 3 and 4- midbrain

b) Selective involvement of pain in the trigeminal distribution may be present due to involvement of the descending spinal tract and nucleus of the 5th nerve.

c) Long tract motor findings may be present which are unilateral or more often bilateral.

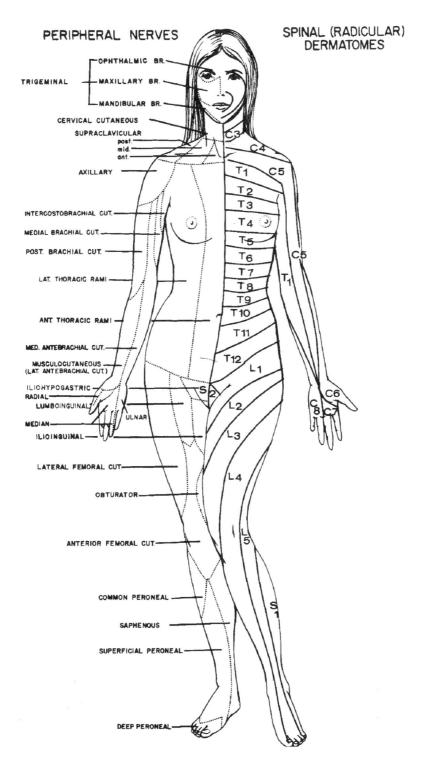

Figure 2-2 Comparison of radicular (dermatome or segmental) and peripheral nerve innervation. A.) anterior

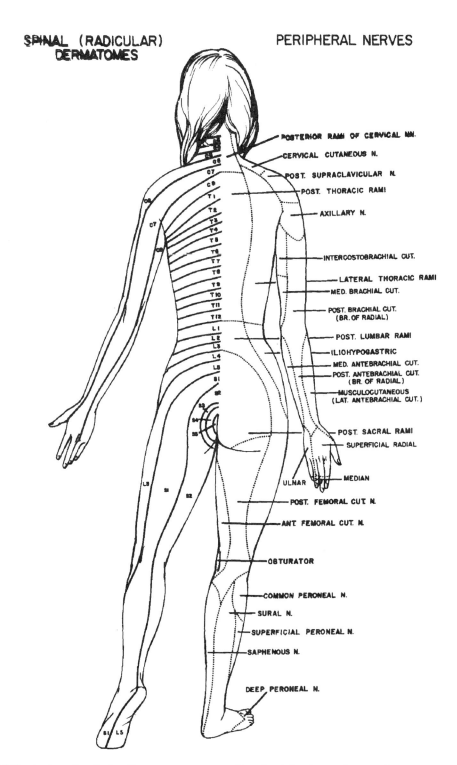

Figure 2-2 Comparison of radicular (dermatome or segmental) and peripheral nerve innervation. B.) posterior view

d) Long tract sensory findings may be present which may be selective, that is dissociated e.g., pain and temperature only when unilateral or bilateral and nonselective. Medial lemniscus or lateral spinothalamic may be involved.

e) Cerebellar pathways may be involved producing alteration in balance, stance, gait and appendicular coordination.

f) In general, mental status, cognitive function is well preserved although level of consciousness may be depressed.

g) Particular combinations of cranial nerve finding ± long tract findings are diagnostic.

1) The jugular foramen syndrome (cranial nerve 9, 10, 11).

2) The cerebellar pontine angle syndrome: (cranial nerves 7 (facial) 8 (auditory and vestibular) and cerebellum plus or minus CN 5, 9, 10, 11.

3) The lateral medullary syndrome (lateral medullary tegmentum) "Wallenberg's syndrome", or syndrome of the posterior inferior cerebellar artery.

4) Ipsilateral peripheral facial weakness and contralateral hemiplegia.

5) Weber's syndrome: Ipsilateral third nerve and contralateral hemiplegia.

h) Vestibular + cochlear symptoms are indicative of cranial nerve 8 disease at the level of the labyrinth or of the cranial nerve 8. If other symptoms are present, brain stem may be involved.

7. Cerebral Cortex

a) Focal (partial) seizures are always indicative of disease involving the cerebral cortex. The specific clinical phenomena of the seizure will depend on the location of the lesion. Rapid secondary generalization of the seizure discharge may obscure the origin.

b) Lateralized deficits in terms of upper motor neuron findings and cortical sensory deficits (somatosensory or visual) may occur based on the location of the lesion.

c) Aphasia is always indicative of involvement of the dominant (usually left) hemisphere, in particular involvement of the speech areas or of the fiber systems inter-connecting the speech areas.

d) Cranial nerve involvement of a lower motor neuron type does not occur. Corticobulbar involvement may produce supranuclear weakness of the lower half of the face. Bilateral corticobulbar damage may produce a pseudobulbar state involving cranial nerves 5; 7, 9, 10, 11 and 12. This is manifested by a hyperactive jaw jerk, spastic speech and emotional lability.

e) Changes in personality and behavior are indicative of disease involving the cerebral cortex particularly the frontal and temporal lobes.

f) Changes in immediate (working) memory often reflect neocortical pathology particularly involving the prefrontal areas. Changes in the ability to consolidate new learning may reflect limbic pathology (medial temporal and medial thalamus).

8. Basal Ganglia

a) There is no direct involvement of the lower motor neuron.

b) Major upper motor neuron descending motor pathways such as the pyramidal tract are not affected.

c) A modulating circuit is dysfunctional. This dysfunction is manifested by alterations in motor function characterized by the following findings:

1) A lack of movement, akinesia or slowness of movement, bradykinesia

2) Excessive movement: tremor or dyskinesia

3) Increased resistance to passive motor: rigidity

4) Alteration in righting reflexes affecting gait and balance

d) No sensory symptoms are present

e) No direct involvement of cranial nerve function is present but the motor effects described above may alter cranial nerve function

f) Cognitive function is usually not directly involved although, many diseases that affect the basal ganglia also affect cerebral cortex or frontal basal ganglia connections and thus are associated with changes in mental status.

9. The Cerebellum

a) Lower motor neuron function is not affected.

b) Upper motor neuron function in terms of long tracts is not affected.

c) A modulating circuit is dysfunctional. The clinical effects of this dysfunction depend on the area of cerebellum involved.

    1) Lateral cerebellar hemisphere relates to the ipsilateral arm and leg. Deficits produce dysmetria of the arms and legs: intention tremor, finger-to-nose and heel-to-shin deficits and an incoordination of movements.

    2) Midline cerebellum relates to the axis of the body. Deficits produce ataxia of trunk.

    3) Floccular nodular (archicerebellum) relates to the vestibular system. Deficits produce a loss of balance in sitting and standing

d) Sensory function of a conscious nature is intact.

e) Cranial nerve function is generally intact, although speech and eye movement are altered by cerebellar dysfunction. At times vestibular function may be altered

f) Certain aspects of cognitive function and motor learning are altered.

C. THE NATURE OF THE PATHOLOGY:

1. The Concept Of Extrinsic Versus Intrinsic: Extrinsic diseases are usually focal mass lesions compressing the spinal cord or brain. Intrinsic disorders arise within the substance of the nervous system. Disorders of muscle, and neuromuscular junction are usually not due to compressive disorders.

2. The common pathological processes related to the site of lesion.

a. At the level of muscle, the common processes are (1) degenerative disorders on a genetic basis: the dystrophies primarily occurring in children (2) genetic congenital myopathies, (3) the metabolic myopathies, often with a genetic basis, (4) the acquired inflammatory myopathies related to auto immune disorders.

b. At the level of the neuromuscular junction, the most common disorder is myasthenia gravis, an autoimmune disorder in which an antibody to the acetylcholine receptor blocks and damages the postsynaptic receptor. Other disorders, involve a paraneoplastic related antibody to calcium channels, which alters the release of acetylcholine from the pre synaptic sites. Botulinum toxin produced by an anaerobic bacterium also alters transmission at the neuromuscular junction.

c. At the level of peripheral nerve, the most common causes of a mononeuropathy are direct trauma and the entrapment syndromes in which compression of a nerve occurs. Vascular disease may also produce a mononeuropathy. As regards polyneuropathies, the most common causes are diabetes mellitus, and B vitamin nutritional deficiencies, followed by toxic and hereditary disorders. In parts of the world the infectious disorder leprosy is the most common cause of a multiple mononeuropathy or of a poly neuropathy.

d. At the level of the nerve root, the most common disorder is compression due to ruptured disks or to the osteophytes of the degenerative disk disorder cervical spondylosis. Tumors (Schwannomas) may also arise from nerve roots and peripheral nerve and in the process compress the nerves.

e. At the level of the spinal cord extrinsic disorders are the more common disorders. The mass may be a ruptured disc, a tumor in the epidural or intradural space, collapsed vertebrae due to metastatic disease, trauma or infection, or an infectious process such as an epidural empyema. Intrinsic processes may be focal, system disorders or multifocal disorders. Examples of focal intrinsic processes are infarcts due to anterior spinal artery occlusion, intrinsic tumors such as gliomas, or a focal enlarging cyst as in syringomyelia or transverse myelitis. Examples of system disorders are amyotrophic lateral sclerosis due to the degeneration of the motor neuron in the anterior horn or posterior lateral column disease of vitamin B12 deficiency. An example of a multifocal disorder is multiple sclerosis, which is the most frequent of the intrinsic disorders.

f. At the level of the brain stem, most disorders are intrinsic. The most frequent disorder is vascular disease. The major type is ischemic occlusive involving the vertebral-basilar arteries producing infarcts. Other intrinsic disorders include tumors, hemorrhage, and demyelinating disease. However the brain stem may be affected by extrinsic processes arising in cerebellum (tumors and hemorrhages) or by tumors arising from the Schwann cells in cranial nerves (primarily the vestibular nerve).

g. At the level of the cerebellum most disorders are intrinsic. Infarcts secondary to occlusive disease are the most common. Other disorders are due to hemorrhage, neoplasms, degenerative disease and multiple sclerosis. In infants and children, tumors are the most common disorder.

h. At the level of the diencephalon, most disorders are intrinsic with the exception of tumors arising from the pituitary and secondarily compressing the hypothalamus. The most common intrinsic disorder is vascular disease (infarcts and hemorrhage). Other disorders are gliomas, and nutritional (Wernicke's encephalopathy due to thiamine deficiency). In children neoplasms are the most common disorder.

i. At the level of the basal ganglia, all disorders to be considered are intrinsic with degenerations being the most common (Parkinson's and Huntington's disease). However, the basal ganglia are also the most frequent site of intracerebral hemorrhage related to the penetrating branches of the middle cerebral artery. These lenticulostriate branches are also subject to occlusion producing infarct involving the internal capsule and basal ganglia.

j. At the level of cerebral cortex and subcortical white matter, the common pathological processes are intrinsic and relate to the age of the patient. In infant's hydrocephalus or congenital malformations or migration disorders may affect cortical function. In children, adolescents and young adults, trauma is a major consideration. In the adult, intrinsic tumors of the glial series are a major consideration. In the older patient, ischemic-occlusive vascular disease is a major neuropathologic process. In the elderly patient, the degenerative disorder, Alzheimer's disease afflicts a significant proportion of the population. In all disorders involving the cerebral cortex, recurrent seizures (epilepsy) are major considerations at all ages. Extrinsic disorders include the previously mentioned head trauma and the extrinsic tumor, the meningioma.

Part II — Neurological History and Examination will be found on CD-ROM.

Part III — Diagnostic Studies in Neurology will be found on CD-ROM.

PART 2: THE NEUROLOGICAL HISTORY AND EXAMINATION

The neurological examination provides a means for the systematic analysis of symptoms and signs.

I. AN OUTLINE OF THE COMPLETE NEUROLOGICAL HISTORY & EXAMINATION

[Note - that for each patient the total history and examination may be tailored so that for some cases abbreviated evaluations may be carried out in a particular area.]

HISTORY: The importance of the detailed history cannot be overemphasized, 75% of neurological diagnosis is dependent on the history.

1. Demographic Data: Age, sex, marital status, handedness, occupation, level of education.

2. Chief Complaint.

3. Present Illness.

4. **Review of Symptoms** Relevant to the Neurological System:

a. Mental Status: Orientation, memory, personality changes, mood changes, delusions, hallucinations, (see below regarding witnesses)

b. Language function.

c. Loss of consciousness: Syncope, Seizures: Details as to the onset, course, duration, confusion or other neurological symptom before, during and after. If relevant may need to obtain information from witnesses and/or relatives, friends, etc.
  d. Alterations in alertness and sleep patterns.
  e. Cranial Nerves:
    I - Anosmia
    II - Blurring or loss of or distortion of vision
    III, IV, VI - Double vision
    V - Numbness or pain in the face
    VII - Weakness of the face, alterations in taste
    VIII - Decrease, loss of or distortion of hearing, Tinnitus, vertigo, dizziness + nausea & vomiting
    IX, X: Dysarthria: (change in voice, hoarseness); dysphagia (Difficulty in swallowing).
    XII - Alterations in tongue movements
  f. Motor system: Weakness, incoordination, clumsiness, unsteadiness - in sitting, standing, walking
  g. Alterations in sensation: paresthesias, dysesthesias, tingling, and numbness
  h. Strokes
  i. Headaches
  j. Trauma to head, neck, back or extremities.
  k. Bladder and bowel function: incontinence, frequency, and spasticity.
5. General Medical History
  a. System Review
  b. Hospital admissions
  c. Surgical history
6. Family History
7. Social History

EXAMINATION:
General Physical Examination to include vital signs
Neurological Examination
1. Mental Status
  a. Level of consciousness: alertness
  b. Orientation: time, place and person
  c. Information: presidents, capitals, historical or local or sports or political information.
  d. Memory:
    1. Immediate recall (working memory): digit span and repetition of words.
    2. New learning - at times, referred to as short term or test of labile phase of remote memory: delayed recall 5 out of 5 in 5 minutes or 3 out of 3 in three minutes.
    3. Remote recall: date of birth, marriage, names and ages of children, siblings, parents, etc.
  e. Insight as regards illness, etc.
  f. Abstract reasoning -
    1. Similarities: concrete versus abstract
    2. Proverb interpretation: concrete or abstract
  g. Calculations: simple additions, subtractions, problems and serial 7s subtractions.
  h. Language and Related Functions:
    1. Fluency
    2. Naming of objects
    3. Repetitions
    4. Ability to follow spoken and written commands
    5. Reading
    6. Writing
    7. Arithmetic
    8. Drawing: constructions
    9. Apraxia testing
    10. Left right orientation
    11. Recognition of objects, pictures
  i. Mood, affect, how appropriate, level of anxiety
  j. Observed hallucinations and perceptual distortions
2. Observed seizure activity describe in detail
3. Cranial Nerves:
  a. I: Sense of smell
  b. II: Fundi, visual fields, blind spot, acuity
  c. III, IV, VI:
    1. Pupillary responses: light, accommodation direct and consensual
    2. Extraocular movements
    3. Nystagmus: spontaneous, induced by eye movement, caloric stimulation, Hallpike maneuvers
  d. V:
    1. Sensation: Touch and pain
    2. Jaw movement

3. Jaw jerk
  e. VII:
   1. Facial movements: upper and lower face
   2. Labial sounds - dysarthria
   3. Taste
  f. VIII: Auditory tested. Vestibular if indicated as in coma, test calorics.
   1. Hearing: Whisper perception; watch tick
   2. Weber, Rinne
  g. IX, X:
   1. Movement of palate
   2. Sensation of pharynx
   3. Gag reflex
   4. Guttural sounds dysarthria
  h. XI: Sternocleidomastoids, trapezii: strength against resistance
  i. XII: Tongue
     1. protrusion, lateral movements, fasciculations and fibrillations
     2. Lingual sounds - dysarthria
4. Motor System:
  a. Atrophy and fasciculations
  b. Motor power: Grade 0-5: Note pattern of weakness: hemiparesis, paraparesis, distal, proximal
  c. Tone: Flaccid, spastic, and rigid
  d. Posture: sitting, standing with eyes open and closed (Romberg test)
  e. Gait:
   1. Standard
   2. Heel to toe, tandem gait
   3. Accessory movements
  f. Coordination:
   1. Finger to finger to nose
   2. Heel to shin
   3. Rapid alternating hand movement
   4. Foot tapping
  g. Spontaneous movements:
   1. Fasciculations
   2. Tremor - at rest; maintained posture, or on movement
   3. Chorea, athetosis, myoclonus
   4. Dystonia
5. Reflexes:
  a. Deep tendon stretch: Grade 0-4 at biceps, triceps, brachioradialis, patella and Achilles.
  b. Superficial
   1. Abdominal
   2. Plantar: Sign of Babinski and associated reflexes
  c. Frontal release signs: Grasp, suck, palmomental
6. Sensation:
  a. Primary: pain, touch vibration
  b. Cortical modalities: position, double simultaneous stimulation, tactile localization, stereognosis, graphesthesia.
7. Meningeal Irritation Signs:
  Kernig's and Brudzinski
8. Vascular:
  a. Carotid and temporal artery pulses and auscultation
  b. Subclavian pulses and auscultation
  c. Radial pulses and lower extremities (peroneal, and posterior tibials)
  d. Bruits over head, orbits, vessels
9. Cervical Spine:
  a. Range of motion: flexion, extension, rotation, and lateral displacements.
  b. Local tenderness
  c. Supraclavicular tenderness
10. Thoracic Spine: Local tenderness
11. Lumbar Spine:
  a. Local tenderness
  b. Mobility for flexion and lateral motion
  c. Sciatic and femoral tenderness
  d. Straight-leg-raising and reverse straight-leg-raising
12. Peripheral Nerves: Palpation + tap
  a. Occipital at occipital notch
  b. Ulnar at olecranon groove
  c. Median at carpal tunnel
  d. Sciatic at sciatic notch
  e. Femoral at femoral canal
  f. Peroneal at fibular head
  g. Post tibial behind medial malleolus
  Note - tenderness, enlargement, Tinel's sign (tingling on palpation)
13. Examination of head, face, tongue, and mouth for bruises, lacerations, hematomas of the scalp.
14. Examination of the limbs and body for bruises and malformations: Cafe au lait spots, vascular nevi, etc.

Clinical Impression:
   Anatomical location of lesion
   Differential Diagnosis as regards pathology
   Laboratory Data already available: if relevant
   Conclusion and plan of diagnostic and therapeutic management

II: ABBREVIATED NEUROLOGICAL EXAMINATION

A. Mental Status and Language Function:
   1. Alertness
   2. Orientation
   3. Delayed recall (5/5 objects in 5 minutes)
   4. Naming of 5 objects
   5. Repetitions: "No ifs, ands or buts"

B. Cranial Nerves II - XII: Fundi, pupils, EOM's, facial movement and sensation, tongue move and gag, shoulder shrug and and SCM on head rotation

C. Motor System:
   1. Resistance against force at shoulder abductors, biceps, triceps, wrist extensors, hand grip and finger abductors; hip flexors, quadriceps, hamstrings, ankle and toe dorsiflexors
   2. Walk a routine gait; walk a tandem gait; walk on toes and heels
   3. Stand on narrow base, eyes open, eyes closed
   4. Other cerebellar - finger to nose, alternating hand motions
   5. Any atrophy or fasciculations
   6. Any tremor or other extra movements noted

D. Reflexes:
   1. Deep tendon reflexes at biceps, triceps, brachioradialis, patella and Achilles
   2. Plantar responses

E. Sensation:
   1. Pain and touch sensation - extremities, shoulders, body
   2. Vibration and position at toes

III: Mini-Mental State Examination (MMSE)
(After Folstein, M.S., Folstein, S.E., and McHugh, P.R. "Mini-Mental State". A Practical Method for Grading the Cognitive State of Patients for the Clinician. J.Psychiatr.Res, 12:189-198, 1975)

                                                        points
1. What is today's date? Month___ Date___ Year___
Day of week?___ Season___   ___ (5)

2. Where are we? City___ County___ State___ Hospital___ Floor___   ___ (5)

3. Repeat after me: ball - flag - tree
Record the number recited initially ___ (3)
Repeat them up to six times
for registration

4. Subtract 7 from 100: 93 - 86 - 79 - 72 - 65
Spell WORLD forward and
backward: D - L - R - O - W
(Write the greater of these
two scores to the right)   ___ (5)

5. Repeat the three words:
ball - flag - tree   ___ (3)

6. Read and obey ("close your eyes") ___ (1)

7. Name these items (pen and watch) ___ (2)

8. Repeat after me
("no ifs and or buts")   ___ (1)

9. Take the piece of paper in your
right hand, fold it in half, and
put it on the floor.   ___ (3)

10. Write a sentence:
                                  ___ (1)

11. Copy this design:

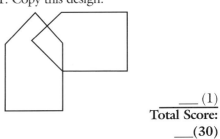

                                  ___ (1)
                          Total Score:
                                  ___ (30)

PART 3: DIAGNOSTIC STUDIES IN NEUROLOGY

Specific laboratory studies are appropriate in providing information about disease affecting particular levels of the nervous system or about particular types of pathology.

A. Muscle and Nerve (fig 2-3):

1. EMG: Small needles are inserted into muscle to record the electrical activity of motor units. Under normal circumstances, at rest, no activity is recorded. With voluntary contraction, a significant number of units are recorded. These units have a range of amplitude and duration. In disease of muscle, voluntary contraction results in motor units of small amplitude and altered duration. In contrast when the anterior horn cell or peripheral nerve has been damaged resulting in denervation, spontaneous activity is present at rest. This consists of small amplitude fibrillations (the contraction of single muscle fibers) and **fasciculations** corresponding to the contraction of all the fibers in a motor unit innervated by a single anterior horn cell.

2. Nerve Conduction Velocity: Stimulation of nerves at two points along the nerve will allow the calculation of speed of conduction over specific segments. A specific site of block may be determined with normal conduction above and delayed conduction below that site, e.g. median nerve at carpal tunnel or ulnar nerve at olecranon groove of the elbow. Alternatively a general modification in conduction may be found in generalized peripheral neuropathies. Those peripheral neuropathies which involve primarily myelin produce a decrease in speed of conduction. Neuropathies that are predominantly axonal do not alter speed of conduction but may alter amplitude of the action potential. Repetitive nerve stimulation recording from appropriate muscle with measurement of the amplitude of muscle action potential may be utilized to study disorder of the neuromuscular junction. A progressive decrement occurs with myasthenia gravis and a progressive increment with Eaton-Lambert Syndrome.

B. Spinal Cord and Nerve Root:
1. Simple radiological studies: (fig 2-4)
Cervical spine, thoracic spine and lumbar sacral spine X-rays provide information about collapse of vertebrae, narrowing of spaces, narrowing of the neural foramina, metastatic involvement of vertebrae, fractures of vertebral elements and osteophyte (spur) formation. Neurofibromas may widen the neural foramina. Intrinsic tumors and syringomyelia may produce an increase in the diameter of the vertebral canal.

2. Computerized axial tomography scanning (CT scan) is now the most frequently employed technique in neurology but is less frequently utilized for spinal cord.

A computer is utilized to determine the differential attenuation of X-rays by the various tissues such as gray vs. white matter vs. blood vs CSF vs bone based on the differential content of water. An X-ray beam is passed through the tissues from multiple sites along a specific plane of section. The computer generates a series of slices usually in the horizontal plane.

Contrast enhancement is the technique of administering radio-opaque dyes intravenously of the types employed in intravenous pyelograms. These dyes do not usually cross the blood brain barrier. When this barrier is damaged as in tumors, brain abscesses, arteriovenous malformation, increased density will occur around or within the lesion. The barrier is also damaged in meningitis and around infarcts.

For the spinal cord; the most frequent use is in the lumbar area to image the nerve roots of the cauda equina.

3. Magnetic Resonance Imaging (MRI scanning) (fig. 2-5, 2-6) also utilizes computer-generated images. *MRI is now the study of choice for imaging spinal cord and nerve root.* Instead of utilizing X-rays, a strong magnetic field and radio frequency waves are employed. Placement of the patient's body in a magnetic field directionally orients the protons of that body. Passage of a brief radiofrequency current alters this directional orientation. When the radiofrequency current ceases the protons realign in the magnetic field. This realignment results in a signal. The signal as in CT scans, depends on the tissue density that is the differential water content of the specific tissue. Bone has little water content; gray matter and white matter have differential water content.

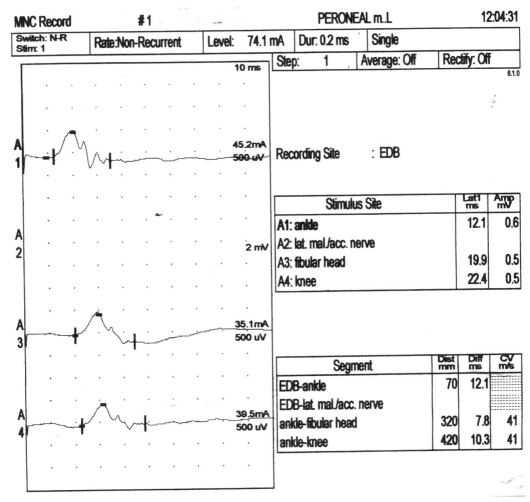

Figure 2-3 Motor nerve conduction velocity (normal). Various points along the course of the peroneal nerve were stimulated, with recordings from extensor digitorum brevis. The differences in latencies are calculated and then divided by the distance to yield a conduction velocity of 41 meters per second. (Courtesy Neurodiagnostic Laboratory University of Massachusetts Hospital)

Edematous and acutely necrotic tissue, as in infarcts and malignant tumors, usually has relatively high water content. In contrast to CT scan, images are obtained in the horizontal coronal and sagittal planes.

As in CT scans, contrast enhancement may be utilized. Gadolinium DTPA that normally does not cross the blood brain barrier is the agent employed.

The MRI procedure may be modified to vary the appearance of cerebrospinal fluid and of tissue water content. Altering the relaxation time (the interval after the application of the radiofrequency wave) will achieve this effect:

Images obtained with a short relaxation time are labeled as T1 and emphasize the normal gray-white anatomical features with the CSF appearing as black. Images produced after longer relaxation times, labeled as T2, produce increased white appearance of CSF and other water content. Demyelinating lesions in multiple sclerosis are often prominent in T2 (MRI).

4. Other Radiographic Techniques Utilized in the Pre CT and MRI Era

Myelography: In this technique a radiocontrast dye is introduced into the subarachnoid space via a lumbar puncture. The spinal cord and nerve roots are then visualized by tilt-

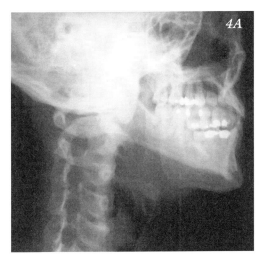

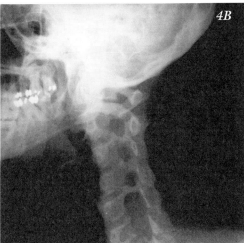

Figure 2-4. Cervical spine x rays. Marked Enlargement of the neural foramen has occurred on the left at C5-6 extending into C6-7 consistent with a neurofibroma/Schwannoma at this level. This 26-year-old female college maintenance worker developed numbness in the left arm extending from the elbow to all of the fingers and pain in the neck and left supraclavicular area. She had absent deep tendon reflexes at left biceps and triceps. A) Right neural foramina on oblique view. B) Left neural foramina on oblique view. (Courtesy of Radiology Department Bay State Medical Center). See figure 2-5.

ing the body. Previously, oil soluble dyes (Pantopaque) were employed and had to be removed at the end of the procedure. Now water-soluble dyes are employed and these are absorbed. At times, CT scan is combined with myelography *(fig. 2-7)*. At times, the dye was allowed to enter the 4th ventricle or the cisterns around the brainstem.

5. Electrophysiological Techniques for Studying Spinal Cord Function:

a. *H. reflex*: submaximal stimulation of a mixed sensory-motor nerve at a voltage intensity which is insufficient to produce a direct orthodromic motor response (the M wave) will produce a muscle contraction after a long latency. This long latency response - the H. wave, involves the activation of the afferent fibers involved in the monosynaptic stretch reflex with activation of the anterior horn cell, anterior root motor fibers to the muscle.

b. *The F response*: Supramaximal stimulation of a motor sensory nerve produces an even longer latency response in the muscles. This depends on antidromic activation of motor neurons, which then induces, the longer latency discharge, activating the muscle fibers.

c. *Evoked Potentials*: There are computer-averaged signals associated with peripheral and central conduction following stimulation of specific sensory systems. The specific technique relevant to spinal cord is the somatosensory evoked potential. *(fig.2-8)*. Stimulation of median nerve in the upper extremity or of the posterior tibial or peroneal nerves in the lower extremities produces a series of waves. These waves are related to specific points in the conduction pathway. In the case of median nerve stimulation the waves relate to brachial plexus, cervical spinal cord and thalamocortical system. These studies are useful in detecting whether abnormalities are present in the posterior column/medial lemniscal system in multiple sclerosis or in spinal cord compression.

C. Brain Stem, Posterior Fossa and Skull Base

1. Radiological studies and other imaging studies:

a) *Skull X-rays* have in large part been replaced by CT scan and MRI scan but are still useful in providing information about skull fractures, enlargement of the pituitary, invagination of the odontoid and intracranial calcifications *(Fig.2- 9)*.

b) *CT scan* discussed previously may provide

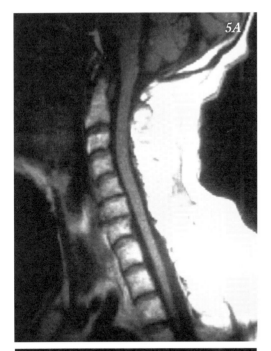

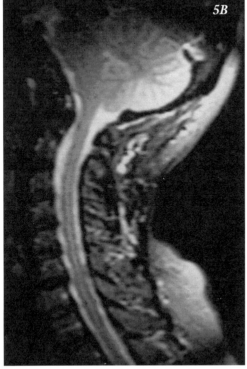

Figure 2-5. Magnetic Resonance Imaging (MRI) of the cervical and upper thoracic spine and spinal cord. A) T1 weighted B) T2 weighted from another patient: sagittal section close to midline. In both cases although degenerative disc disease is demonstrated, the spinal cord remains normal.

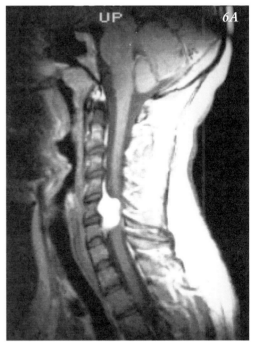

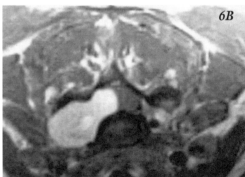

Figure 2-6 Cervical spine. MRI. Same case as figure 2-4. The cause of the enlargement of the neural foramina is now apparent. A large dumbbell tumor is present in the intra spinal bony canal and extends through the foramen to the extraspinal space. A) Sagittal view B) axial/transverse view. (Courtesy Radiology Department Bay State Medical Center)

information regarding tumors, infarcts and hemorrhages affecting the cerebellum and brain stem *(Fig 2-10)*.

c) *MRI* in general has become the standard technique for imaging the brain stem and cerebellum *(Fig 2-11)*.

2. Physiological Techniques

a) *Brain stem auditory evoked potentials* (fig. 2-12): A sequence of waves occurs which have been associated with specific points in the

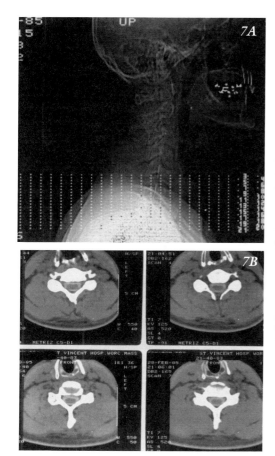

Figure 2-7 CT Computerized Axial Tomography (CT) after metrizamide myelography: normal spinal cord. A) Reference film for level of cross sections in this patient. B) Scans 3, 4, 6, 7- sections through C5-C6 Levels.

auditory conduction system:

Wave I; is associated with electrical activity generated in the cochlea at the origin of the auditory nerve.

Wave II; is associated with the entry of impulses from the auditory nerve into the cochlear nucleus at the medullary pontine junction.

Wave III; relates to signals generated at the level of the superior olivary nucleus in the lower pons.

Wave IV; relates to the nerve impulses generated in the lateral lemniscus.

Wave V; relates nerve impulses generated at the level of the inferior colliculus in the lower midbrain.

Delays are noted after wave I or between

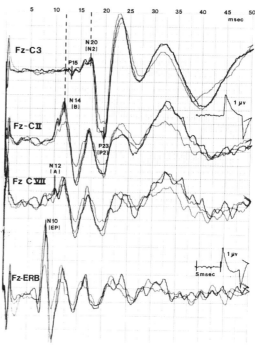

Figure 2-8: Short latency somatosensory evoked potentials (SER) median nerve stimulation at wrist. Recordings from supraclavicular area (FZ- ERB), cervical spine-VII (FZ-Cervical VII), cervical spine-II (FZ-Cervical II) and somatosensory cortical projection area and scalp (F2-C3). The specific wave forms-originate as follows: N10-Brachial plexus at Erb's point, N12-lower cervical spine-root entry area, N14-dorsal columns and dorsal column nuclei-lower medulla. N20-thalamocortical fibers and cortex. P23-Somatosensory cortex. (From Marcus, E.M. and Stone, B. In Evoked Potentials II, Ed. R.N. Nodar and C. Barber: Butterworth Publishers, Boston, 1984).

waves I and III in patients with acoustic neuromas (vestibular Schwannomas).

Somatosensory Evoked Potentials also provide possible information about conduction delays in the medial lemniscus.

b) *Special test of auditory and vestibular function:* audiograms, caloric testing an electronystagmograms.

3) Radiological Techniques No Longer Employed

a). *Pneumoencephalography* (PEG) *(Fig. 2-13):* Air was injected into the subarachnoid space via a lumber puncture. With the patient in the sitting position, the air would rise into the cisterns and ventricular system allowing

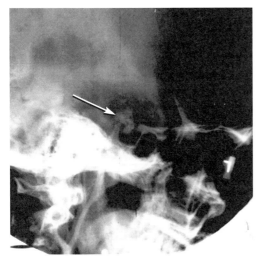

Figure 2-9: Skull X-ray Suprasellar Tumor: Craniopharyngioma. Enlarged sella turcica, and suprasellar calcification with changes in the anterior and posterior bony components (clinoids). This 23-year-old female had a six-year history of intermittent headaches and amenorrhea and recent diplopia. She had bilateral papilledema and an elevated serum prolactin level. Refer to figure 2-13 (A) for a normal comparison (see also figure 27-14 for a CT scan of this case).

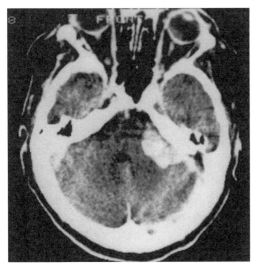

Figure 2-10 Computerized tomographic (CT) scan of the posterior fossa with contrast enhancement (C+). Cerebellar pontine angle tumor in a 65 year old female with a 17 year history of deafness in the left ear and a minor reduction of hearing in the right ear who had had additional findings of mild left peripheral facial weakness, ataxia of gait and dysmetria of left hand. The broad base along the left petrous bone suggested a meningioma. Courtesy of Dr. Tom Mullins.

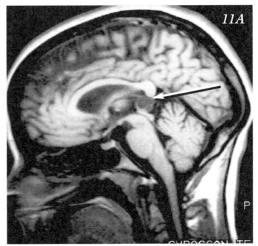

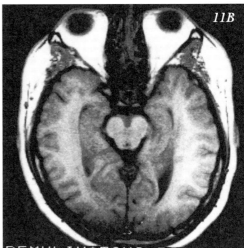

Figure 2-11. MRI: A) T1 sagittal section in a 29-year-old woman with multiple symptoms including headaches and depression but with no neurological findings. MRI/MRA was obtained because sister had an aneurysm and subarachnoid hemorrhage. This study reveals an apparently benign lesion of the pineal. B) MRI B) T1 horizontal /axial section in a young man with multiple sclerosis. No definite demyelinating lesions are demonstrated but the relationship of the mesial temporal areas to the mid brain is well demonstrated.

visualization of structures such as the fourth ventricle and aqueduct of Sylvius, as well as the lateral and third ventricle.

b. *Ventriculography:* In patients with increased intracranial pressure, or with tumors mass lesions in temporal lobe or cerebellum, the PEG was dangerous with the possible complication of herniation. Instead, a needle was introduced into the frontal horn and air or

LOCALIZATION OF FUNCTION, AND NEUROLOGICAL DIAGNOSIS

D. Cerebral Hemispheres

1. Imaging Techniques:

a) *CT scans* are usually employed in cases of acute trauma, intracerebral hemorrhage, subarachnoid hemorrhage, acute infarcts and acute brain abscess. In the case presented at the end of this chapter, an acute hemorrhage was demonstrated involving the leg and proximal arm areas. *(fig 2-29)*. (Refer to fig 1- for location of these areas on lateral surface of the cerebral hemisphere.

b) *MRI* - Has become the preferred technique for imaging patients with brain tumors, malformations, seizure disorders, inflammatory and demyelinating disorders *(fig. 2-14)*. In patients with ischemia and infarctions special diffusion weighted and perfusion studies are of value and are discussed in chapters 26.

Functional MRI - may allow correlation of metabolic activities and normal or disordered function *(fig 2-15)*.

c) As discussed above, pneumoencephalograms and ventriculograms are no longer employed.

d) *Radioisotope Techniques* have a limited value in selected cases.

1. *Radioactive Brain Scans* are no longer

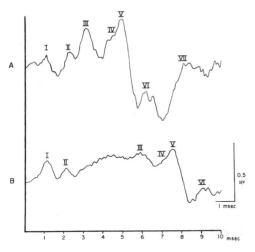

Figure 2-12 Brain Stem auditory evoked potentials (BAER). This 49 year old female had a progressive decrease in hearing in the left ear, decreased sensation on the left side of face for 2 years and additional findings of a left peripheral facial weakness and decrease pain sensation over the face. Imaging studies were normal but a delay was present in the II-III interval on the left suggesting a lesion between the cochlear nucleus and the superior olive. of unknown etiology. A) Normal right ear stimulation, B) abnormal left ear stimulation.

radio-opaque dye was introduced outlining the ventricular system and cisterns.

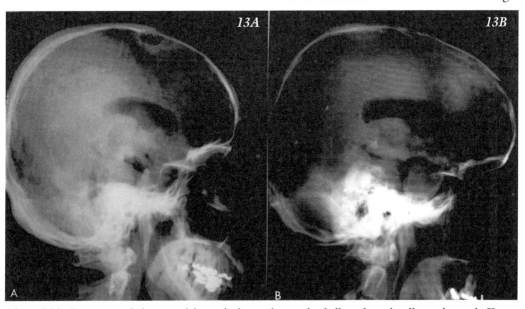

Figure 2-13. Pneumoencephalogram: A large pituitary adenoma has ballooned out the sella turcica on the X-ray. With the injection of air, via a lumbar puncture the extrasellar extension may be seen. A. Normal; B. Abnormal. (Courtesy of Dr. Samuel Wolpert, New England Center Hospitals)

performed. Prior to the development of CT and MRI this technique was of value in detecting lesions in which damage to the blood brain barrier or increased metabolic activity was present. Metastatic tumors, meningiomas, glioblastomas, abscess and acute infarcts were demonstrated but anatomical detail was poor. Several examples will be provided in the text

2. *Radioisotope flow studies* are still occasionally utilized to visualize the passage of a radioisotope injected into the lumbar subarachnoid space into the ventricles, out into the cisterns and through the subarachnoid space over the convexity to be absorbed into the venous sinuses. Normally, the isotope is no longer present in the ventricles at 48 and 72 hours. In the presence of communicating hydrocephalus (primarily normal pressure hydrocephalus), the isotope is still present in the cerebral ventricles at 48 and 72 hours (chapter 18 for an example).

3. *Single photon emission computed tomograph (SPECT)* scanning remains, as a modified form of the radionucleotide brain scan. Unlike the positron emission tomography scan a cyclotron is not necessary. Therefore cost is less, and availability is greater but resolution is poor *(fig 2-16).*

4. *Positron emission tomography (PET) scan* employs positron emitting radio nucleotides combined with computer imaging of the emission to assess metabolic changes in specific areas of the brain. Resolution does yet reach the level of MRI or current high level CT scanning but now approaches that of early CT scans. Cost is high and availability limited since a cyclotron is required to produce the short life radioisotopes required. The major clinical use is in relation to focal (partial) epilepsy (Fig. 2-17). During seizure activity metabolic activity at the focus is increased. Between seizures the metabolic activity at the focus is decreased. The investigational use of PET scanning has provided valuable information about localization of function during the increased metabolic activity of normal cognitive activities, such as reading, motor activities, etc. In addition, information has been provided about the localization of altered metabolic activity in patients with schizophrenia, depression and various disorders of the basal ganglia. Functional MRI may provide a more effective technique for such metabolic correlations. Since positron emission tomography (PET scan) has not been previously considered in detail we will briefly review the technique at this point. This method combines CT with the use of positron emitting radioisotopes, which have been bound to compounds, which have significant biological function as metabolites or transmit-

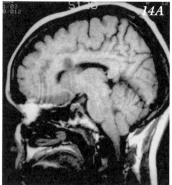 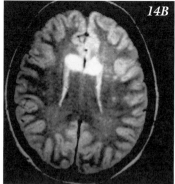 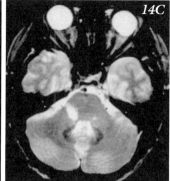

Figure 2-14. MRI. Multiple sclerosis. This 47-year-old female farm owner and manager had a 2-month episode of numbness ("novocaine type sensation") over the entire right trigeminal distribution. Five years previously she had intermittent unsteadiness Her examination demonstrated only a selective decrease in touch sensation over the entire right trigeminal distribution. The MRI studies confirmed a right mid pontine tegmental lesion but also indicated multiple areas of demyelination in the white matter of the cerebral hemispheres particularly involving the corpus callosum. A) T1 sagittal, (B) T2 horizontal views, (C) T2 axial of brain stem.

LOCALIZATION OF FUNCTION, AND NEUROLOGICAL DIAGNOSIS 2-21

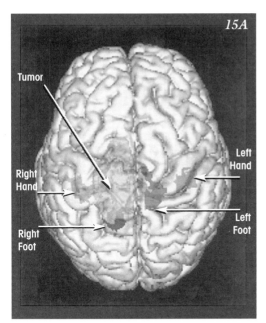

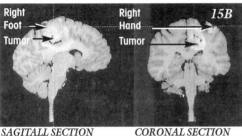

Figure 2-15. Functional MRI. This 33-year-old woman had a 2-year history of focal seizures involving the right foot and arm. A large area of the left premotor and motor cortex is involved by a grade 2 astrocytoma A) The relation of the areas activated by right hand or foot movements to the intrinsic brain tumor are indicated. B) Comparison of normal hemisphere to abnormal hemisphere. Note the displacement of the arm and leg areas by the tumor. Courtesy of Drs. B.R.Buchbinder, H. Jiang, G.R. Cosgrove A Cole, D. Hoch and R. Hill at the Massachusetts General Hospital Epilepsy Center.

ters. In contrast to standard CT scanning (where the source of radiation is the X-ray tube), in PET the positron-emitting isotope taken up by the tissue is the source of radiation. 2-Desoxyglucose is often employed since it is taken up by the neurons, and is phosphorylated as is glucose but is not further metabolized. The isotope of fluorine (18F) is bonded to the desoxy glucose. The uptake of glucose or of desoxy glucose into neurons is proportional to

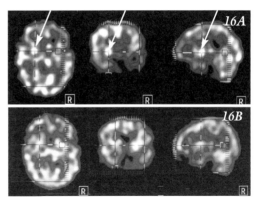

Figure 2-16. SPECT: Areas of increased perfusion during seizure activity are indicated in this scan. A) ictal uptake in areas of right temporal parietal and frontal lobes B) interictal. This 3-year-old child initially had infantile spasms with an EEG pattern consistent with the disorder hypsarrhythmia. Subsequently the EEG abnormalities were right anterior quadrant or parietal. Courtesy of Dr. Paul Marshall Pediatric Neurology University of Massachusetts.

the activity of the neurons. Thus, areas of cerebral cortex undergoing active seizure discharge will show increased activity and increased uptake. In normal individuals with activity of the visual system, e.g., opening the eyes to scan a scene, activity and uptake will increase in the visual projection area of the occipital lobe. With auditory stimulation, on the other hand, activity and uptake will increase in the auditory projection area of Heschl's transverse gyrus of the temporal lobe. Areas of damage will show decreased uptake.

2. Physiological Techniques: NOTE THAT SEVERAL ILLUSTRATIONS OF EEG RECORDS WILL BE FOUND IN AN EEG ATLAS SECTION OF THE CD ROM

a. *Electroencephalography*. This technique provides information about the electrical activity of the cerebral cortex. This activity represents primarily the summated activity of postsynaptic potentials generated in the cerebral cortex. The electroencephalogram in the normal awake adult resting with eyes closed is characterized by the *alpha rhythm (Fig. 2-18)*. This rhythm is composed of a sequence of sinusoidal waves of 8-13 Hz (cps), which is maximal over the parietal-occipital recording

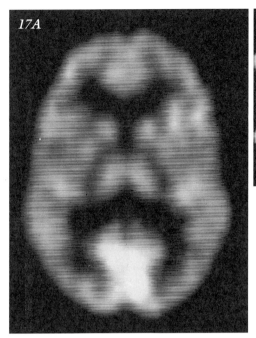
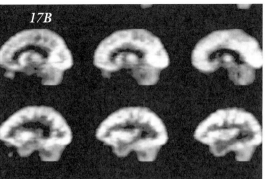

Figure 2-17. PET scans. The most prominent feature is the increased activity in visual cortex most likely reflecting visual activity during the study. MRI demonstrated mesial temporal sclerosis on the right and this study did demonstrate a possible decrease in activity right temporal area. This 22-year-old female had complex partial seizures since puberty. She had prolonged status epilepticus as an infant related to fever. A) Horizontal, B) sagittal. Courtesy of Dr. Cathy Phillips Neurology University of Massachusetts

area. Activity faster than alpha rhythm is referred to as *beta activity* and may be present over frontal areas. Increase amounts of beta activity occur as an effect of various drugs such as barbiturates and benzodiazepines. Alterations occur in this normal background activity related to the following factors: (1) eye opening producing reduction of amplitude *(Fig2-18C)*: (2) *age*: slower activity with infancy and childhood (*delta* 0.5- 3 Hz, *theta* 4-7 Hz), (Fig.2-19-CD ATLAS). (3) *Sleep, (Fig.2-20- CD ATLAS)*, (4) *drugs and anesthesia*: faster and then slower activity (Fig.2-21-CD ATLAS)

Abnormalities may be (1) focal or (2) generalized.

1. **Focal abnormalities** may be categorized as:

A. *Focal spikes* implying focal excessive neuronal discharge involving the cerebral cortex and associated with partial (focal) epilepsy. (See chapter 29).

b. *Focal slow wave* activity, which implies focal cortical damage, as in infarcts, brain tumors and brain abscess. *(Fig. 2-22).*

c. *Focal suppression* of activity implies non-active electrical tissue under the electrodes - this may be seen with a fluid collection (subdural or intra cerebral hemorrhage) or with total destruction of tissue. *(Fig 2-23.)*

2.**Generalized abnormalities:**

a. *Generalized discharges of spike or polyspike – slow wave complexes* are associated with various types of generalized epilepsy. These will be discussed in chapter 29.

1) *Generalized bursts of 3/second spike wave* complexes are associated with absence seizures previously labeled petit mal epilepsy.

2) *Generalized bursts of polyspike and slow wave complexes* are associated with myoclonic seizures.

3) *Generalized polyspike discharges* may be found in the tonic phase of the generalized tonic clonic seizures.

b. *Generalized slow waves* are associated with diffuse disorders: infectious, ischemic, toxic or metabolic encephalopathies *(Fig. 2-24, 25)*.

c. *Generalized periods of suppression* imply a more serious type of diffuse dysfunction, as in anoxia or a deep stage of anesthesia. Figure 17-16 provides an example of a burst suppression pattern.

d. *Total suppression of activity* may be found when neocortical death has occurred - as in

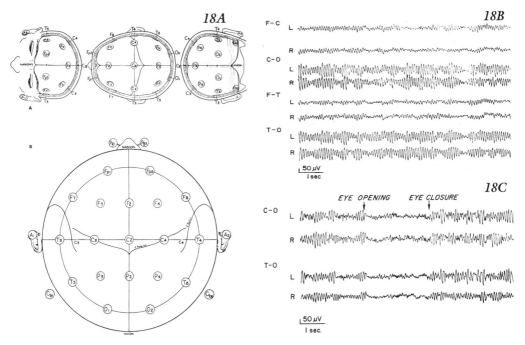

Figure 2-18. A) The International Federation Ten-Twenty Electrode Placement System. Frontal superior and posterior views; and a single plane projection of the head demonstrating the standard positions and the rolandic and sylvian fissures. The term ten-twenty is based on the placement of electrodes at particular percentages of the distance between nasion and inion. From Jasper, H.H.: Electroenceph. Clin. NeuroPhysiol. 10: 374, 1958 (Elsevier).
B) The normal adult electroencephalogram. The patient is awake but in a resting state, recumbent with eyes closed. F = frontal, T = temporal, C = central and 0 = occipital. (Bipolar recordings) These abbreviations will be utilized in subsequent illustrations. C) Effects of Eye Opening and Closure. Opening is associated with a blocking or suppression of the alpha rhythm of 10 cps and with the appearance of a low voltage fast-activity of 20 cps (beta rhythm). With eye closure, there is a return of the alpha rhythm.

severe anoxic encephalopathy or the vegetative state. This total suppression of activity accompanied by an absence of brain stem reflex activity and an absence of spontaneous respiration occurs in brain death. Similar findings of total suppression of activity both electrical and reflex may also occur under condition of deep anesthesia.

b. Specialized Techniques Employing Electroencephalography:

1. *Sphenoidal electrodes* - these are small needle electrodes placed in the sphenoidal sinus area to record from the medial aspects of the temporal lobe.

2. *Video EEG monitoring:* The recording of EEG and behavior onto single videotape allows correlation of electrical activity and of clinical seizure activity. This is usually a prerequisite study prior to epilepsy surgery and is also uti-

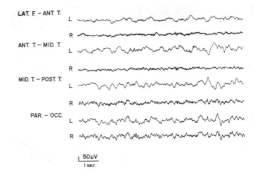

Figure 2-22. Focal 2-3 Hz slow wave activity in the electroencephalogram: left temporal area indicating focal damage in this area. Brain abscess, left temporal lobe in a 14-year-old female secondary to an acute S aureus mastoiditis extending into the petrous ridge. Electroencephalogram. Bipolar recording. (LAT. F. - lateral frontal; ANT- T- = anterior temporal; MID. T. = mid- temporal; POST. T- = posterior temporal; PAR. = parietal; OCC. = occipital). (Listed in text and CD ROM as Figure 27-22).

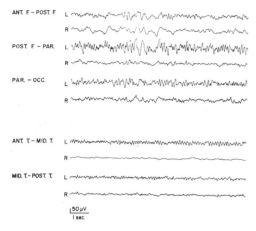

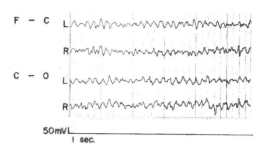

Figure 2-23. Focal suppression of EEG activity right temporal area and focal slow wave activity right frontal. Total right middle cerebral artery occlusion in a 61-year-old female with hypertension. Case 26 1. ANT.F. = Anterior frontal parasagittal; POST.F. = Posterior frontal (parasagittal); PAR. = parietal; OCC. = occipital; ANT.T. = anterior temporal; MID.T. = mid temporal; POST.T. = Posterior temporal. (Listed in text and CD ROM as Figure 26-14).

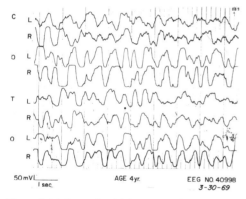

Figure 2-24 Generalized delta 1-2 Hz slow wave activity that persisted despite attempts at arousal this 4-year-old male had acute viral encephalitis.

lized in the analysis of pseudoseizures or other unresolved "spells".

c. *Depth electrode recording:* from medial temporal and other structures may be performed prior to epilepsy surgery often in combination with video monitoring.

d. *Subdural surface grids of electrodes:* may be placed on the cortical surface for better correlation of seizure discharges arising in the frontal or other neocortical areas.

Figure 2-25. Generalized theta 5 Hz slow wave activity. This 67-year-old female had a metabolic encephalopathy due to impaired hepatic function secondary to cirrhosis and at this point was semicomatose in a stuporous state. When the patient was more deeply comatose, the awake activity was even slower at 4 Hz. When the patient was alert during intervening periods of recovery, the dominant activity was in alpha range.

e. *Electrocorticography:* recording directly from the pial surface may be utilized during epilepsy surgery.

f. *Polysomnography* (PSG): this technique is utilized in the evaluation of sleep disorders such as narcolepsy and sleep apnea. EEG activity from the parietal occipital or vertex areas of the scalp is correlated with

(1) Cardiac activity (EKG) rate and rhythm

(2) Respiratory activity rate and rhythm

(3) Oxygen (O2) saturation

(4) Extraocular movements

(5) EMG activity chins and or limb

The use of the PSG and the multiple sleep latency study will be discussed in chapter 29

c. Evoked Potentials:

(1) *Visual (VER or pattern reversal visual evoked potential: PVER):* The time of conduction over the entire visual pathway - to cerebral cortex is measured. Pattern reversal generates a prominent very stable wave at approximately 100ms, the P100 wave *(Fig.2-26).*

(2) *Somatosensory evoked potentials:* As discussed above, these studies may provide information regarding delays in conduction in the thalamocortical system.

3. *Neuropsychological tests.* A variety of tests have been developed. These include the Wechsler Adult Intelligent Score (WAIS) for-

merly termed the Wechsler Bellevue Test of Adult Intelligence. This has a series of separate subtests covering multiple areas of verbal and performance functions. A total, verbal and performance intelligence quotients are derived. A series of tests have been developed to study aphasia and frontal lobe function and are discussed in those chapters. The Wisconsin Card Sorting Test is utilized to study cognitive perseveration. The Wechsler Memory Score provides a quantitative measure of memory function. The Minnesota Multiphasic Personality index, provides information regarding personality, affect, depression etc. Projection tests have also been developed to study personality function the Rorshark and the Thematic Apperception Test. The answers to the pictures provided unless very bizarre may be difficult to score and the results are open to several interpretations.

3. Techniques for the study of the cerebral circulation

a. Magnetic Resonance Angiography (MRA) - Normally in MRI scans, rapidly moving blood is not clearly imaged. However with special software programs, a non-invasive visualization of flow through vessels can be achieved. *(Fig 2-27, 2-28)*. At present, resolution in the range of 2-3 mm can be achieved, allowing visualization of significant aneurysms. This procedure allows imaging of the carotid and other arteries prior to carotid endarterectomy*.

b. *In contrast, cerebral angiography (or arteriography)* is invasive. A catheter must be placed in the femoral artery and advanced into the aorta and then into the carotid or vertebral arteries. Radiopaque dye is then injected to directly image the cerebral vessels. This is the best technique when detailed study of the cerebral vascular is required, e.g., prior to aneurysm surgery. **With the increase resolution of MRA and of CT scan angiography, direct arteriography may be replaced by these non-invasive techniques.**

Under special circumstances, spinal angiography employing selective catheterization of

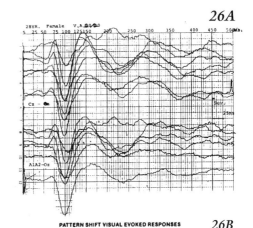

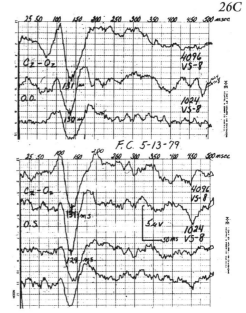

Figure 2-26. Visual evoked potentials. A) Normal. Each tracing represents the summation of 256 trials at different check sizes. Note the stability of the P100 response. B) Normal and abnormal values are obtained by statistical analysis with abnormal defined as >mean + 3 standard deviations. C) A patient with multiple sclerosis who had experienced several episodes of optic neuritis involving first one eye and then the other eye. Note the bilateral prolongation of the P100 responses.

radicular arteries may be performed to visualize spinal cord arteriovenous malformations.

c. *Duplex scans of the extracranial carotid and vertebral arteries:* this technique combines Doppler and ultrasound techniques to image blood flow in the major extra-cranial arteries.

d. *Transcranial Doppler* - may provide gross information regarding flow in the major intracranial vessels.

CEREBROSPINAL FLUID (CSF) EXAMINATION

CSF fluid is usually obtained by a lumbar puncture. The lower end of the spinal cord, the conus medullaris, does not extend below the L2 vertebra. Therefore, introduction of a needle into the subarachnoid space between the L2-L3, or L3-L4, or L4-L5 spinous processes will not damage the spinal cord. CSF pressure when the patient is relaxed, but positioned on one side, will be usually less than 150 mm of CSF. Values in the relaxed state greater than 200 mm are considered abnormal. Respiration, abdominal pressure, flexion of head on chest or thighs and knees onto abdomen will all increase the pressure.

Normally no significant red blood cells (rbc's) should be present. When the puncture is traumatic the first tube collected will contain red cells but these should significantly decrease by the time that the fourth tube is collected.

Normally, less than 7 white blood cells (wbc's) should be present and all should be mononuclears. Any polymorphonuclears are abnormal and should raise the question of infection or inflammatory reaction. Spinal fluid glucose should be no less than 50% of the

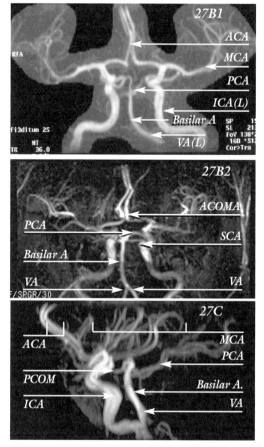

Figure 2-27. Magnetic resonance angiography (MRA).
A) Aortic arch and major arteries in the neck.
B1 & 2) Intracranial circulation coronal submental view demonstrated in two patients 27-B2 labels only the differences from 27-B1 C) Intracranial circulation (lateral view).
ACA= Anterior Cerebral A.
ACOM= Anterior Communications A.
CCA= Commono carotid.
ECA= External carotid A.
ICA= Internal carotid A.
MCA= Middle cerebral artery.
PCA= Posterior cerebral artery.
PCOM= Posterior communications
SCA= Superior cerebral artery.
VA= Vertebral artey.

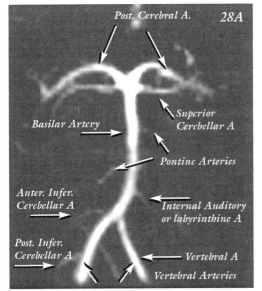

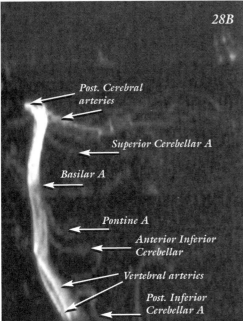

Figure 2-28. Magnetic resonance angiography. Basilar vertebral (posterior) circulation. A) AP view. B) Lateral view. The vertebral, basilar, posterior cerebral, and all of the circumferential cerebellar branches are evident.

blood glucose obtained at the time of the puncture or within two hours prior to the lumbar puncture. In acute bacterial meningitis - the CSF glucose is low due to the interference with the transport system and/or the increased metabolic activities.

Total CSF protein is usually less than 45 mg%. It is increased in a nonspecific manner in many processes affecting the nervous system.

a. Acute inflammation (meningitis and encephalitis).

b. Acute necrosis: infarcts, abscess and tumors

c. Blocks in the CSF - peripheral nerve barrier as in Guillain Barré syndrome. In the latter case no cells are present. This is referred to as albumin- cytologic dissociation and may also be present in patients with diabetes mellitus or myxedema.

d. Blockage of the CSF pathway at a spinal cord level. In patients with blocks in the lower thoracic or lumbar area, the protein level may be very high, with several grams present .The thick yellow fluid may clot in the test tube (Froin's syndrome)

The gamma globulin (IgG) percentage of the total protein may be increased under the following circumstances

1. Production of IgG in the serum is increased.

2. IgG is selectively produced within CNS by plasma cells - as in multiple sclerosis or neurosyphilis. Oligoclonal bands will also be present within the gamma globulin band.

ILLUSTRATIVE CASE HISTORY.

This patient should be compared to the patient of chapter 1. Both began with symptoms of weakness in the leg. However, the time course for evolution of symptoms differed significantly resulting in different diagnoses.

Case 2-1 This 90 year old right handed white male with a past history of hypertension awoke on the morning of admission with weakness of the left side, which was most marked in the leg. The initial admission examination indicated no motor function of the left leg, severe weakness of the left arm and minimal weakness of the left side of the face.

Past history indicated a previous minor "stroke" involving the left side of the body, from which he had made a full functional recovery with only minimal left sided weakness. During year prior to admission, occasional periods of confusion had been present and progressive problems in memory had developed.

Neurological Examination: *Mental Status:* The patient was disoriented for time and place. He was however cooperative and able to follow all commands. He was fluent with no disturbance of language function. His remote memory was excellent. He could repeat the name of the examiner and of his primary physician but could recall neither name after five minutes. *Cranial Nerves:* A minor left central facial weakness was present. *Motor System:* He had no movement of the left leg. There was little function of the left shoulder and elbow. However handgrip was strong and independent finger movements were present the patient was recumbent in bed with external rotation of the left leg into a hemiplegic posture. *Reflexes:* Deep tendon stretch reflexes were absent in the lower extremities and the left upper extremity. The plantar responses were extensor bilaterally (bilateral sign of Babinski. *Sensory System:* No definite abnormalities within limits of testing.

Clinical diagnosis: 1. Cerebrovascular accident involving the superior parasagittal precentral gyrus either due to an anterior cerebral artery occlusion or a cerebral hemorrhage secondary to amyloid angiopathy. 2. Alzheimer's disease.

Laboratory data: A CT scan of the head demonstrated an acute hemorrhage in the superior parasagittal Rolandic area consistent with amyloid angiopathy. *(Fig. 2-29).*

Comment: This patient presents many of the neurological problems, which occur in the elderly. The patient had a one-year history of progressive memory problems, which primarily involved the formation of new memories. At age 90 such memory problems occur in more than 50% of the population. Usually, this represents the development of those degenerative changes in the neurons of the cerebral cortex seen in the process defined as Alzheimer's senile dementia.

The patient had elevated blood pressure for many years and had already experienced one "stroke" affecting the left side 10-15 years previously

In the present episode, the sudden development of symptoms would suggest an additional vascular event. Compare this case to Case 1-1 in which the patient had a gradual development of weakness in the leg secondary to a meningioma.

As regards the localization, the marked involvement of leg and proximal arm with relative sparing of hand and face might suggest a process involving the upper half of the motor cortex (refer to fig 18-12).

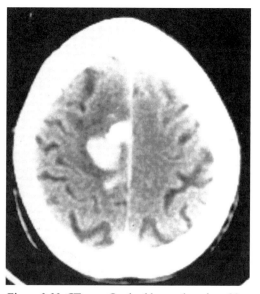

Figure 2-29 . CT scan. Cerebral hemorrhage in a 90-year-old man with left sided weakness primarily involving the right leg and shoulder. See case 2-1 at end of this chapter.

CHAPTER 3
Neurocytology

This chapter focuses on the two major cell types that form the nervous system: supporting cells and conducting cells. The supporting cells consist of the glia, epedenymal cells lining cells the ventricles, the meningeal coverings of the brain, the circulating blood cells, and the endothelial lining cells of the blood vessels. The conducting cells, or neurons, form the circuitry within the brain and spinal cord and their axons can be as short as a few microns or as long as one meter. The supporting cells are constantly being replaced, but the majority of conducting cells/neurons, once formed, remain throughout our lives.

Any investigation of the structure of the nervous system is complicated by the fact that no single stain demonstrates all details of a neuron or of the glia. Instead, many techniques are used for microscopic examination of the nervous system. But before nerve tissues can be examined, they must be preserved (fixed). Neutral-buffered formalin is the most commonly used fixative in light microscopy.

Golgi Neuronal Method (Figs. 3-1 and 3-2)

The shapes of neurons and glia can best be seen by means of the Golgi neuronal method. Brain slices 3 to 5 mm thick (either fixed or unfixed, normal or abnormal, from vertebrates or invertebrates) are encrusted with heavy metals (usually dichromate or mercury) and then immersed in silver nitrate. With the Golgi method, only about 1 in every 70 cells stains completely and reveals the axon, soma, dendrites, and dendritic spines in full detail. With either a camera lucida or the modern technique of image analysis, one is able to fully reconstruct the cell and determine the morphology of normal or diseased cells.

With the Golgi neuronal method some of the most elegant cells have been identified, including the Purkinje cell of the cerebellum, the pyramidal cell of the cerebrum *(Fig 3-1)*, the stellate cells of the cerebrum *(Fig 3-3)* and the mitral cell of the olfactory bulb. A disadvantage is that it does not

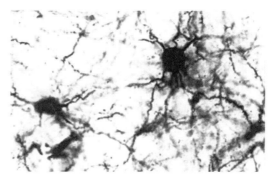

Figure 3-1. Golgi type II cells in the motor cortex of the rat. (Golgi-Cox stain, <X>450.)

reveal details of the neuron=s internal structure. However, the most pronounced organelle in the soma, the rough endoplasmic reticulum, or Nissl substance, is demonstrable with basophilic dyes.

The Neuron

The basic functional unit of the nervous system is the neuron. The neuron doctrine (postulated by Waldeyer in 1891, described the neuron as having one axon, which is efferent, and one or more dendrites, which are afferent. It was also noted that nerve cells are contiguous, not continuous, and all other elements of the nervous system are there to feed, protect, and support the neurons.

Although muscle cells can also conduct electric impulses, only neurons, when arranged in networks and provided with adequate informational input, can respond in many ways to a stimulus. Probably the neuron's most important feature is that each is unique. If one is damaged or destroyed, no other nerve cell can provide a precise or complete replacement. Fortunately, though, the nervous system was designed with considerable redundancy, so it takes a significant injury to incapacitate the individual (as in Alzheimer's disease).

Neuronal parts and their functions are shown in Figure 3-2 and Table 3-1.

Neurons in the adult nervous system are either pseudounipolar, bipolar, or multipolar.

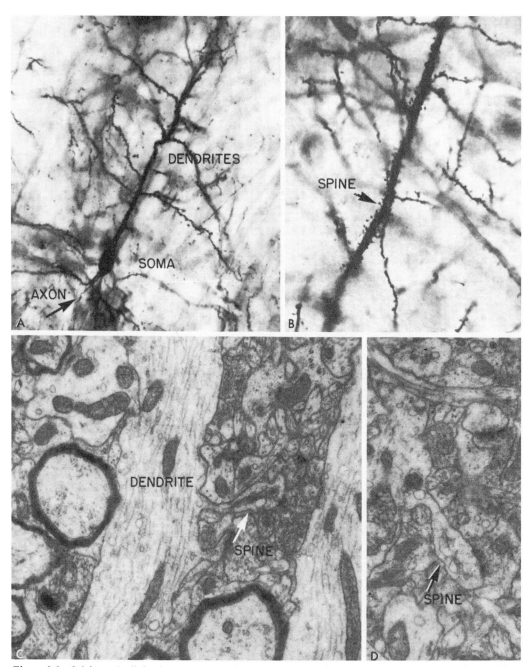

Figure 3-2. Golgi type I cells in the motor cortex of a rat. A, shows entire cell--soma, axon, and dendrite (Golgi rapid stain, <X>100). B demonstrates dendritic spines (Golgi rapid stain, <X>350). C and D are electron micrographs of dendritic spines (<X>30,000).

True unipolar cells are found in the invertebrate nervous system (Fig 1-1). In the mammalian central nervous system pseudounipolar cells (fig 1-1) are found in the sensory ganglia of the spinal cord (dorsal root ganglia) and cranial nerves and in the mesencephalic nucleus of cranial nerve V, as a single process acts as the axon and the dendrite. *Bipolar neurons* (fig 1-1), which are sensory in function, are found in the rods and cones of the retina, in the olfactory neuroepithelial cells, in the olfactory mucosa at the upper end of the nasal passages, and in the vestibular and auditory

NEUROCYTOLOGY

TABLE 3-1. PARTS OF A NEURON

Soma	The neuron=s trophic center, containing the nucleus, nucleolus, and many organelles. The majority of inhibitory synapses are found on its surface.
Dendrites	Continuation of the soma; has many neurotubules and majority of synapses on its surface; type I neurons have dendritic spines
Axon	Conducts action potentials to other neurons via the synapse. Ranges from a few millimeters to a meter in length; In CNS covered by myelin, an insulator.
Synapse	The site where an axon connects to the dendrites, soma or axon of another neuron. Consists of a presynaptic part containing neurotransmitters and postsynaptic portion with membrane receptors separated by a narrow cleft.

receptors of the inner ear. *Multipolar neurons(Fig 1-1)* are found throughout the central nervous system and in the sympathetic ganglia of the peripheral nervous system. They convey both sensory and motor impulses. Multipolar neurons vary greatly in size and in the complexity of their axonal and dendritic fields.

Dendrites

The dendritic zone receives input from many different sources. The action potential originates at the site of origin of the axon and is transmitted down the axon in an all-or-nothing fashion to the synapse, where the impulse is transmitted to the dendritic zone of the next neuron on the chain.

Dendrites have numerous processes that increase the neuron=s receptive area. The majority of synapses on a nerve cell are located on the dendrite surface. With the electron microscope the largest dendrites can be identified by the presence of parallel rows of neurotubules, which may help in the passive transport of the action potential *(Fig. 3-5, 3-12)*. The dendrites in many neurons are also studded with small membrane extensions, the dendritic spines.

Soma

The soma (perikaryon, or cell body) of the neuron varies greatly in form and size. Unipolar cells have circular cell bodies; bipolar cells have ovoid cell bodies; multipolar cells have polygonal cell bodies.

Golgi Type I and II Neurons

Neurons can also be grouped by axon length: those with long axons are called Golgi type I (or pyramidal) cells; those with short axons are called Golgi type II (or stellate) cells (Gray, 1959).

Golgi type I axons are projectional (Fig. 3-2). They form the tracts and commissures in the central nervous system, as well as the axons of the peripheral nervous system. In Figure 3-6 #1, a cerebral pyramidal cell (Golgi type I) with a long axon is compared in figure 3-6 #2, with a stellate (Golgi type II) cell with a short axon.

The *Golgi type I cell (Fig 3-1)* has an apical and basal dendrite, each of which has secondary, tertiary, and quaternary branches, with smaller branches arising from each of these branches that extend into all planes. Spines are absent from the initial segment of the apical and basal dendrite of pyramidal neurons, but they become numerous farther along the dendritic branches. The axons of pyramidal neurons run long distances within the cortex, but they may also exit from the cor-

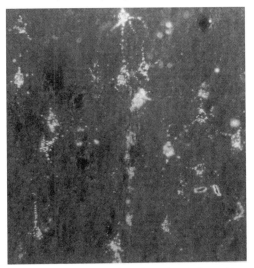

Figure 3-3. Demonstration of axoplasmic flow with the horseradish peroxidase method. A, Pyramidal cells in the sensory cortex after injection in the opposite hemisphere. B, Pyramidal cell in motor cortex labeled after injection into gyrate hemisphere. (Dark field, <X>350.)

tex and distribute to the subcortical nuclei.

The *Golgi type II cell* has a small axonal field and dendrites (fig 3-3). The axon usually extends only a short distance within the cerebral cortex (0.3 to 5 mm). Golgi type II cells have fewer dendritic spines than type I cells. Spines, which are common to many neurons, are small knob-shaped structures approximately 1 to 3 microns in diameter (Fig. 3-1). Their importance stems from the fact that they greatly expand the dendrite=s receptive synaptic surface.

Neuronal Cytoplasmic Organelles

Organelles found in the cytoplasm allow each neuron to function (Fig. 3-9) In these eukaryotic cells the organelles tend to be compartmentalized and include the nucleus, polyribosomes, rough endoplasmic reticulum, smooth endoplasmic reticulum, mitochondria, and inclusions (Fig. 3-10). Most neuronal cytoplasm is formed in the organelles of the soma and flows into the other processes. Newly synthesized macromolecules are transported to other parts of the nerve cell, either in membrane-bound vesicles or as protein particles. As long as the somas with a majority of its organelles are intact, the nerve cell can live. Thus it is the *trophic* center of the neuron. Separation of a process from the soma produces death of that process.

Nucleus. The large ovoid nucleus is found in the center of the cell body (Figs. 3-7, 3-8 and 3-9. Within the nucleus there is usually only a single spherical nucleolus, which stains strongly for RNA. The DNA, which can be demonstrated by staining the neuron by the Feulgen method or fluorescent markers, appears dispersed (heterochromatic) in mature neurons. (These cells are very active in metabolizing protein; consequently, the DNA is dispersed.)

In females, the nucleus also contains a perinuclear accessory body, called the *Barr body* (Fig.3-9). The Barr body is an example of the inactivation and condensation of one of the two female sex, or X, chromosomes (Barr and Bertram, 1949). The process of inactivation of one of the X chromosomes is often called *lyonization*, after the cytogenetist who discovered it, Mary Lyons.

Recently much progress has been made in the localization of genes associated with neurologic processes, e.g., Huntington's Chorea, Down=s syndrome.

Endoplasmic Reticulum. The largest membrane in the eukaryotic nerve cell is the endoplasmic reticulum (ER). It consists of a rough endoplasmic reticulum, which is the site of ribosome and protein synthesis, and the smooth endoplasmic reticulum, which is the site of the synthesis and metabolism of fatty acids and phospholipids.

Rough Endoplasmic Reticulum (Figs. 3-10, 3-11 and 3-12). The rough endoplasmic reticulum, or Nissl substance, is the chromidial substance found in the cell body. It can be demonstrated by using a light microscope and basic dyes, such as methylene blue, cresyl violet, and toluidine blue. The appearance and amount vary from cell to cell. With electron microscopy, cisterns containing parallel rows of interconnecting rough endoplasmic reticulum are revealed (Fig. 3-11). Ribosomes (clusters of ribosomal RNA) are attached to the outer surfaces of the membranes and consist of a large and a small RNA-protein subunit. Protein synthesis begins when there is a combination of initiation factors, messenger RNA (mRNA) and transfer RNA (tRNA) with the small subunit. This is then followed by the presence of an elongation factor, which then starts the peptide chain to grow.

The Nissl substance is most concentrated in the soma and adjacent parts of the dendrite (Fig. 3-12A). It is, however, also found throughout the dendrite (Fig. 3-12B). Before the electron microscope it was always presumed that the axon hillock was devoid of Nissl substance, but it has

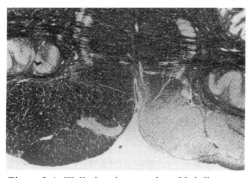

Figure 3-4. Wallerian degeneration. Medullary pyramids in a human some months after an infarct in the motor-sensory strip. Left side is normal; note the absence of myelin on right side. (Weigert myelin sheath stain, <X>80.)

now been shown that there are polyribosomes in this region.

Smooth Endoplasmic Reticulum (Fig. 3-10). All nerve cells have some smooth endoplasmic reticulum, but in neurosecretory cells in the hypothalamus, the smooth endoplasmic reticulum is greatly enlarged. The smooth endoplasmic reticulum consists of GERL--Golgi apparatus, endoplasmic reticulum, and lysosomes--which work together to synthesize, modify, or even degrade secretory proteins. The Golgi apparatus is found in all cells and is visible by a light microscope with osmium and silver stains as an irregular network in a perinuclear location. In electron micrographs, the Golgi apparatus consists of stacks of flattened smooth-surface membranes called saccules.

The protein secretion from the Nissl substance is transferred to the Golgi apparatus where a carbohydrate component is added to the protein. The product is released in a secretory vesicle.

Lysosomes (Figs. 3-10, and 3-11). Lysosomes are common in the cell body, appear as dense bodies, and function as centers of degradation. They are membrane-bound, vary in size from 0.35 to 3.0 microns in diameter, and commonly contain small granules. Lysosomes contain acidic hydrolytic enzymes (4.8 pH) that are capable of breaking down proteins, DNA, RNA,

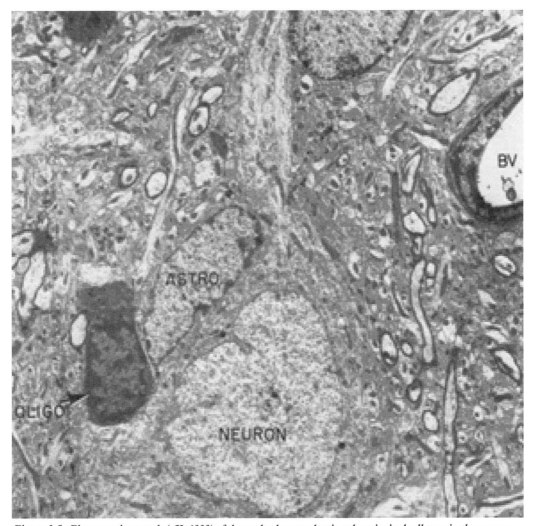

Figure 3-5. Electron micrograph (<X>6000) of the cerebral cortex showing the principal cell types in the nervous system: neuron, astrocyte (astro), oligodendrocyte (oligo), and a blood vessel (BV)

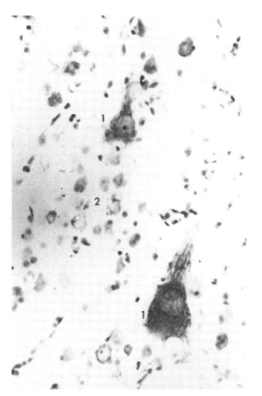

Figure 3-6 Motor cortex of the chimpanzee, demonstrating a pyramidal neuron. 1, Golgi type I cells (neurons with long axons). 2, Golgi type II cells (with short axons). (Nissl stain, <x> 1500.)

and certain carbohydrates. The lysosomes help digest macromolecular polymers into subunits (de Duve and Wattiaux, 1966). They are necessary for the degradation of older portions of membranes as newer ones are formed and also help in the elimination of deleterious toxins from the nerve cells.

Tay-Sachs disease illustrates the importance of the lysosome in the normal function of the nerve cells. When a specific lysosomal hydrolase is missing, B-N-hexosaminidase A, the degradation of the ganglioside G_{M3} is stopped, and Tay-Sachs results. Other instances of lysosomal enzymatic defects can also result in lysosomal storage diseases in the brain and spinal cord (Greenfield, 1993).

Peroxisomes. These small membrane-limited organelles are similar in appearance to lysosomes but have different functions. Peroxisomes contain enzymes that break down fatty acids, amino acids, and the enzyme catalase, which degrades the deleterious hydrogen peroxide formed by many reactions.

Mitochondria (Figs. 3-10, 3-11, and 3-12). Mitochondria are the principal site of adenosine triphosphate (ATP) production in the cell. These organelles, found throughout the neuron, are the third largest organelles after the nucleus and endoplasmic reticulum. They are rod-shaped and vary from 0.35 to 10 microns in length and 0.35 to 0.5 microns in diameter. Mitochondria can be demonstrated in light microscopy, but details of their structure are best seen in electron micrographs.

The wall of a mitochondrion consists of two layers--an outer and inner membrane. The outer membrane contains pores that render the membrane soluble to proteins with molecular weights of up to 10,000. The inner membrane is less permeable and has folds called cristae that project into the center of the mitochondrial matrix. The interior of the mitochondrion is filled with a fluid denser than cytoplasm. Cations and mitochondrial DNA have been demonstrated in the mitochondrial matrix. Mitochondrial DNA is derived from the mother. An intriguing study links this mitochondrial DNA to a common human female ancestor, Lucy, who lived in Africa over 300,000 years ago.

On the inner membrane are found enzymes that provide much of the energy required for the nerve cell. These respiratory enzymes (flavoproteins and cytochromes) catalyze the addition of a phosphate group to adenosine diphosphate (ADP), forming ATP. ATP is broken down in

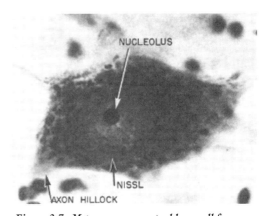

Figure 3-7. Motor neuron; ventral horn cell from a human cervical spinal cord, demonstrating the Nissl substance (rough endoplasmic reticulum), axon hillock, and nucleolus. (Nissl stain, <X>600.)

the cytoplasm to ADP, providing the energy required for cellular metabolic functions. In the cytoplasm are found enzymes that break down glucose into pyruvic and acetoacetic acid. These substances are taken into the mitochondrial matrix and participate in the Krebs citric-acid cycle, which allows the mitochondria to metabolize amino acids and fatty acids.

Centrosomes. Centrioles within the centrosome are seen in the immature dividing neuroblast as well as the adult neuron. However, since mature neurons are incapable of dividing, the function of centrosomes there is not clear.

Inclusions. Substances stored in a cell include pigments, glycogen, and lipid droplets. Pigment granules (melanin) are common in certain parts of the brain, particularly the substantia nigra, locus ceruleus, and reticular formation. In humans, lipochrome pigment (lipofuscin) is found in most cells *(Fig. 3-11A)*. The amount appears to increase with age. Lipofuscin consists of pigment combined with fatty material and probably is a metabolic by-product of lysosomal activity that is not readily disposable. It is commonly referred to as the "wear-and-tear" pigment.

Glycogen (Fig. 3-11B). Glycogen is a polymer made up of D-glucose monomers and is commonly seen in electron micrographs of nerve cells and glia. Glycogen appears in the form of electron-dense rosettes, which are much larger than the RNA rosettes. It is a local source of energy.

Lipid Droplets (Fig. 3-11). Lipid droplets are also seen in the soma. They represent a local store of energy as well as a source of carbon chains for membrane formation.

Neurosecretory Granules. Neurons in the supraoptic and paraventricular nuclei of the hypothalamus form neurosecretory material (Bodian, 1963 and 1966; Palay, 1957; Scharrer, 1966). The axons of these cells form the hypothalamic-hypophyseal tract, which runs through the median eminence, down the infundibular stalk to the neurohypophysis (pars nervosa), where the axons end in close proximity to the endothelial cells. The secretory granules are 130 to 150 millimicrons in diameter and are found in the tract *(Fig. 3-13)*. They increase in size as one approaches the endothelial end of the axons.

The protein in the secretory granules is made in the Nissl substance; the granules are formed in the Golgi apparatus and transported by the axons of the hypothalamic-hypophyseal tract to the infundibulum, where they are stored in the neural lobe. The sites of storage are called Herring bodies (Fig. 3-14B). Interruption of the hypothalamic-hypophyseal tract produces diabetes insipidus.

Neuronal Cytoskeleton

In silver-stained sections examined in a light microscope, a *neurofibrillary* network can be seen in the neurons (Fig. 3-14). Electron micrographs can distinguish microtubules, 3 to 3 nm in diameter, and neurofilaments, 1 nm in diameter. It appears that fixation produces clumping of the tubules and filaments into the fibrillar network seen in light micrographs.

Neurons in common with other eukaryotic cells contain a cytoskeleton that maintains it shape. This cytoskeleton consists of at least three types of fibers:

1. Microtubules 30 nm in diameter
3. Microfilaments 7 nm in diameter
3. Intermediate filaments 10 nm in diameter.

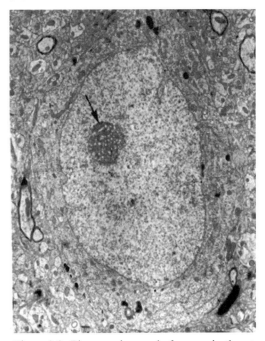

Figure 3-8. Electron micrograph of neuron in the rat sensory cortex, demonstrating the nucleus and nucleolus (arrow). (<X>10,000.)

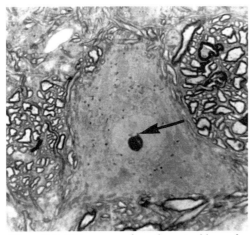

Figure 3-9. Motor neuron from the ventral horn of a female squirrel monkey. Note the nucleus, nucleolus, and accessory body of Barr (arrow). (One-micron epoxy section, <X>1400.)

If the plasma membrane and organelle membrane are removed, the cytoskeleton is seen to consist of actin microfilaments, tubulin-containing microtubules, and criss-crossing intermediate filaments.

Neurotubules (microtubules) predominate in dendrites and in the axon hillock, whereas microfilaments are sparse in dendrites and most numerous in axons (Fig. 3-15). Microtubules and intermediate filaments are found throughout the axon.

The microtubules help to transport membrane-bound vesicals, protein, and other macromolecules. This orthograde transport, or anterograde axonal transport, is the means whereby these molecules formed in the soma are transported down the axon into the axonal telodendria.

The individual microtubules in the nervous system are 10 to 35 nm in length and together form the cytoskeleton. The intermediate filaments are associated with the microtubules. The wall of the microtubule consists of a helical array of repeating tubulin subunits containing the A and B tubulin molecule. The microtubule wall consists of globular subunits 4 to 5 nm in diameter; the subunits are arranged in 13 protofilaments that encircle and run parallel to the long axis of the tubule. Each microtubule also has a defined polarity. Associated with the microtubules are protein motors, kinesins and dyneins, which when combined with cAMP may well be the mechanism of transport in the central nervous system (Brady, 1985; Vale et al., 1985 and 1987). The products transported down the microtubules probably move like an inch worm and not like a train on a track. During mitosis, microtubules disassemble and reassemble; however, a permanent cytoskeleton lattice of microtubules and intermediate filaments in the neuron is somehow maintained. It is not yet known how long each microtubule exists, but there is evidence of a constant turnover.

Neruofibrillar tangles are bundles of abnormal filaments within a neuron. They are helical filaments, that are different from normal cytoskeletal proteins and they contain the tau protein a microtubule binding protein (MAP) that is a normal component in neurons. In Alzheimer's disease there are accumulations of abnormally phosphorylated and aggregated forms the microtubule binding protein tau. These large aggregates form the tangles that can be physical barriers to transport, may interfere with normal neuronal functions, and are probably toxic. Mutations in the human tau gene are found in autosomal domininant neuronal degenerative disorders isolated to chromosome. 17. These familial disorders are characterized by extensive neurofibrillar pathology and are often called "taupathies" (Hutton 2000). Functions of neuronic organelles are listed in Table 3-3.

Axon and Axon Origin

The axon contains some elongate mitochondria, many filaments oriented parallel to the long axis of the axon *(Figs. 3-15, and 3-16)*, and some tubules. In contrast, a dendrite contains a few filaments and any tubules, all arranged parallel to the long axis of the dendrite *(Fig. 3-5)*. Polyribosomes are present, but the highly organized, rough endoplasmic reticulum is absent.

Axon Hillock. The axon hillock is a slender process that usually arises from a cone-shaped region on the perikaryon (Fig. 3-16). This region includes filaments, stacks of tubules, and polyribosomes (Fig. 3-16B). The initial segment of the axon, arising from the axon hillock, is covered by dense material that functions as an insulator membrane at the hillock is covered by an electron dense material *(Fig. 3-16)*.

Myelin. In the nervous system axons may be

TABLE 3-3. FUNCTIONS OF DYNAMIC ORGANELLES IN THE NEURON

Microtubules	Provide the structural basis for transport, and axoplasmic flow. Found throughout the neuron; part of the neuronal cytoskeleton.
Microfilaments	Form much of the cytoskeleton of the entire neuron.
Nissl substance (rough endoplasmic reticulum)	Protein manufacturing unit in the nerve cell. The most commonly stained organelle with basophilic dyes; very sensitive to cell injury.
Golgi bodies	Form the lipophilic portion of all the membranes in the neuron.
Nucleus	Chromatin is dispersed as nerve cells are very active metabolically. Eukaryotic in the adult nerve cell; important in all normal cell functions. Demonstrable abnormalities in many diseases, including trisomies, Alzheimer=s, Huntington=s, and Parkinson=s.
Nucleolus	Contains the messenger RNA that is activated by chromatin.

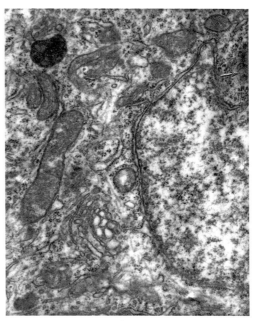

Figure 3-10. Electron micrograph of a small pyramidal neuron in the rat cerebral cortex, demonstrating the following organelles: Golgi apparatus, mitochondria, lysosome, and Nissl substance. Note the nuclear pore (arrow). (<X>60,000.)

myelinated or unmyelinated. Myelin is formed by a supporting cell, which in the central nervous system is called the oligodendrocyte and in the peripheral nervous system, the Schwann cell. The immature Schwann cells and oligodendrocyte have on their surface the myelin-associated glycoprotein that binds to the adjacent axon and may well be the trigger that leads to myelin formation. Thus, the myelin sheath is not a part of the neuron; it is only a covering for the axon. Myelin consists of segments approximately 0.5 to 3 mm in length. Between these segments are the nodes of Ranvier. The axon, however, is continuous at the nodes, and axon collaterals can leave at the nodes. The myelin membrane like all membranes contains phospholipid bi-layers (Fig. 3-17). In the central nervous system myelin includes the following proteins:

    Proteolipid protein (50%)
    Myelin basic protein (40%)
    Myelin-associated glycoprotein (1%)
    3,3-cyclic nucleotide (4%)

The oligodendrocytic process forms the myelin sheaths by wrapping around the axon. The space between the axonal plasma membrane and the forming myelin is reduced until most of the exoplasmic and cytoplasmic space is finally forced out. The result is a compact stack of membranes. The myelin sheath is from 3 to 100 membranes thick and acts as an insulator by preventing the transfer of ions from the axonal cytoplasm into the extracellular space.

Myelin Sheath. Myelin sheaths are in contact with the axon. In light microscopy they appear as discontinuous tubes 0.5 to 3 mm in length, interrupted at the node *(Fig. 3-29)*. The axon is devoid of myelin at the site of origin (the nodes) and at the axonal telodendria. At the site of origin, the axon is covered by an electron-dense membrane, and at the site of the synaptic telodendria the various axonal endings are isolated from one another by astrocytic processes.

In electron micrographs each myelin lamella actually consists of two-unit membranes with the entire lamella being 130 to 180 Å thick (Fig. 3-17). Myelin is thus seen to consist of a series of light and dark lines. The dark line, called the major dense line, represents the apposition of the

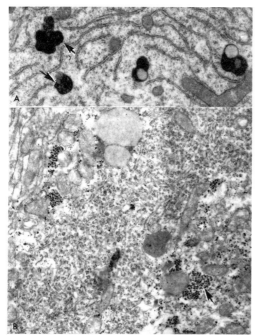

Figure 3-11. Electron micrographs showing inclusions. A, Lipofuscin. B, Glycogen, lipid, and Nissl substance. (<X>30,000.)

inner surface of the unit membranes. The less dense line, called the interperiod line, represents the approximation of the outer surfaces of adjacent myelin membranes.

Only at the node of Ranvier is the axonal plasma membrane in communication with the extracellular space. The influx of Na+ at each node causes the action potential to move rapidly down the axon by jumping from node to node (see Chapter 5-Part II).

Myelination. The process of myelination has been followed with the electron microscope. An axon starts with just a covering formed by the plasma membrane of either the Schwann cell or the oligodendrocyte. More and more layers are added until myelination is complete. One theory is that myelin is laid down by the processes of the Schwann cell twisting around the axon (Geren, 1956; Robertson, 1955); this indeed occurs in the peripheral nervous system. In the central nervous system each oligodendrocyte enwraps many axons, and they also appear to twist around the axons as they myelinate.

The sequence of myelination has been studied in great detail; it begins in the spinal cord, moves into the brain stem, and finally ends up

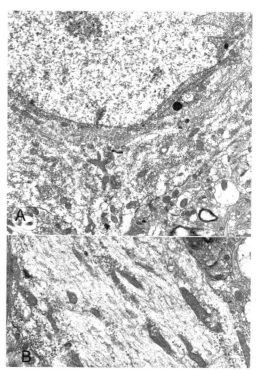

Figure 3-12. Electron micrographs of a pyramidal neuron in the rat cerebral cortex. A, Soma and nucleus. B, Dendrite. Note the large amount of Nissl substance in the soma, but the dendrites have less Nissl substance and many microtubules. (<X>35,000.)

with the diencephalon and cerebrum last. A delay in myelination can result from many factors, including genetic and nutritional ones, and is usually very harmful to the fetus.

Peripheral Nervous System

In the peripheral nervous system, there is usually only one Schwann cell for each length or internode of myelin. In the central nervous system each oligodendrocyte may form and maintain myelin sheaths on 30 to 60 axons.

The unmyelinated axons in the peripheral nervous system are found in the cytoplasm of the Schwann cell. There can be as many as 13 unmyelinated axons in one Schwann cell. The unmyelinated axons in the central nervous system are usually found in small bundles without any special covering.

Axoplasmic Flow (Fig. 3-18).

With the protein manufacturing apparatus present only in the soma, and to a lesser degree in the dendrites, a mechanism must exist to

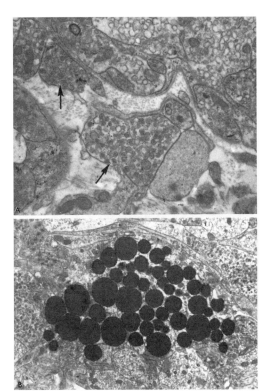

Figure 3-13. Electron micrograph of a rat neurohypophysis. A shows neurosecretory granules in the axoplasm of fibers of the hypothalamo-hypophyseal tract (<X>30,000). B demonstrate a Herring body, a storage site of neurosecretory material. (<X>8,000.)

Figure 3-16. Cytoskeleton. Neurofibrillary stain of a ventral horn cell in the cat spinal cord, showing neurofibrillary network in soma and dendrites (A) and in the axons (B). (<X>400).

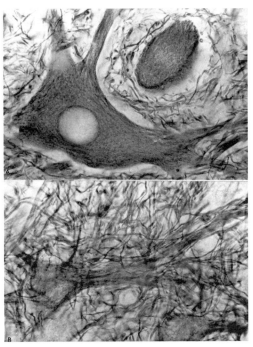

Figure 3-14. Cytoskeleton. Neurofibrillar stain of a ventral horn cell in the cervical spinal cord of the cat, showing neurofibrillar network in soma and dendrites(A): and in axons(B). X400

transport proteins and other molecules from the soma, down the axon, and into the presynaptic side. Weiss and Hisko (1948) demonstrated by tieing off a peripheral nerve, which caused swelling proximal to the tie, that material flows from the soma, or trophic center, into the axon and ultimately to the axon terminal. The development of techniques that follow this axoplasmic flow has revolutionized the study of circuitry within the central nervous system. The ability to map this circuitry accurately has given all neuroscientists a better understanding of the integrative mechanisms in the brain. There are many compounds now available to follow circuitry in the brain and they include horseradish peroxidase, wheat germ agluttin, tetanus toxin, fluorescent molecules and radiolabeled compounds.

This mechanism of transport is not diffusion but rather retrograde axonal transport associated with the microtubule network that exists throughout the nerve cell. The rate of flow varies depends upon the product being transported and ranges from more than 300 mm/day to less than 1 mm a day. The main direction of the flow is anterograde, from the cell body into the axon and synapse. There is also a very active retrograde flow from the synaptic region back to the cell body that may be a source for recycling many of the substances found at the synaptic ending.

The particles that move the fastest consist of small vesicles of the secretory and synaptic vesicles, and the slowest group is the cytoskeletal components. Mitochondria are transported down from the cell body at an intermediate rate. The retrograde flow from the synaptic telodendria back into the soma, returns any excess of material for degradation or reprocessing. The retrograde flow permits any excess proteins or amino acids to recycle. It also permits products synthesized or released at the axonal cleft to be absorbed and then transported back to the cell body, where they can affect the basic function of the cell--the signaling process.

Fast axonal transport is associated with the microtubules. The slower components including membrane associated proteins (MAPS) are transported inside the microtubules, but the mitochondria actually descend the axonal cytoplasm (Table 3-3).

TABLE 3-3. RATE OF AXONAL TRANSPORT OF CELLULAR STRUCTURES (DATA FROM GRAFTSTEIN AND FORMAN, 1980; MCQUARRIE, 1988; WUJEK AND LASEK, 1983.)

Transport	Rate (mm/day)	Cellular Structure
Fast	300<->400	Vesicles, smooth endoplasmic reticulum, and granules
Intermediate	50	--Mitochondria
	15	--Filament proteins
Slow component B	3<->4	Actin, fodrin, enolase, CPK, calmodulin, and clathrin
Slow component A	0.3<->1	Neurofilament protein, tubulin, and MAPS

Peripheral Versus Central Nerve Structure

Peripheral Nerve Structures. The structure of a peripheral nerve is different from that of fiber bundles in the central nervous system (Table 3-4). Peripheral nerves consist of many axons held together in a fascicle by connective tissue of mesodermal origin *(Fig. 3-19)*. The outer layer that covers the nerve trunks and fills between the individual fascicle is called the *epineurium*. It consists of connective tissue cells, collagen, and some fat cells. Each of the fascicles is wrapped in a dense layer of connective tissue, which is called the *perineurium*. Strands of collagen, fibroblasts, and other cells that run between the individual nerve fibers are called the *endoneurium*. The term endoneurium is also applied to the delicate trabeculum surrounding each nerve fiber, which is also called the sheath of Key-Retzius. The peripheral nerve fiber or axon is engulfed in the Schwann cytoplasm (the neurilemmal sheath), which also forms the myelin sheath. Large blood vessels are found in the epineurium and perineurium and capillaries are seen in the endoneurium.

TABLE 3-4. CONTENTS OF A PERIPHERAL NERVE BUNDLE

Epineurium	The outer layer, covering the nerve trunks and filling between the individual fascicles consists of connective tissue cells, collagen, and some fat cells.
Perineurium	A connective tissue layer that surrounds the nerve fascicles
Endoneurium (sheath of Key-Retzius)	Strands of collagen and fibroblasts between individual axons. Endoneurium also refers to the delicate trabeculum surrounding each nerve fiber.
Sheath of Schwann	Engulfs each individual axon and forms myelin.

Fibers can vary in diameter from less than 0.5 microns to 33 microns (see Chapter 5 Part II). Axons can be classified by size and function into three major groups (Table 3-5).

Central Nervous Structure. Axons in the central nervous system also vary in size (5 to 33 microns) and in length (0.5 mm to 1 m), but these axons cannot be separated into functional categories based on axonal diameter. The axons in the central nervous system run in groups called tracts that are enwrapped by the processes of fibrous astrocytes. However, no specific cov-

TABLE 3-5. FUNCTIONAL COMPONENTS OF PERIPHERAL NERVES

Fiber Type	Description	Diameter (microns)	Conduction Speed (meters per second)
Type A	Myelinated somatic afferent and efferent fibers.	1<->33	5<->130
Type B	Myelinated efferent preganglionic autonomic fibers.	1<->3	3<->15
Type C	Unmyelinated afferent or efferent fibers, pain fibers, and postganglionic sympathetics fibers.	0.3<->1.3	0.5<->3

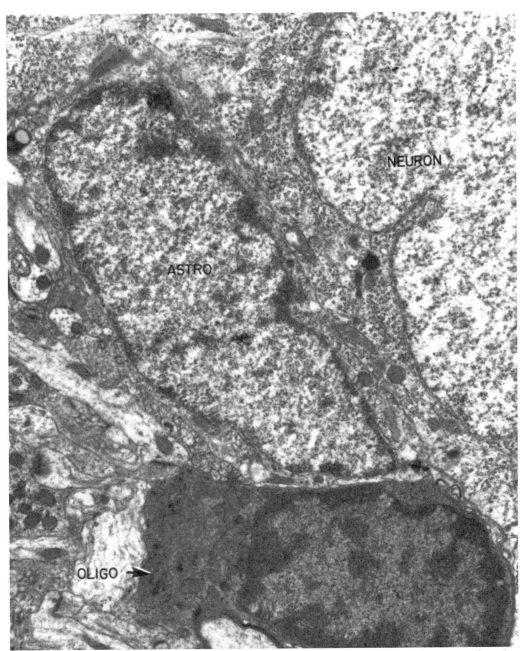

Figure 3-15. Electron micrograph of the rat cerebral cortex, demonstrating the difference between a myelinated axon and a dendrite. Note the axon has numerous microfilaments while the dendrite has numerous microtubules. (<X>30,000.)

ering corresponds to the sheath of Schwann in the peripheral nervous system. Each tract has a distinct function in the nervous system.

The axonal arborization of neurons is not as elaborate as that of the dendrite, but it can be very extensive. Although the surface area of many dendrites may total more than that of an axon, some axons run very long distances. For example, from the type I pyramidal cell in the cerebral cortex to the ventral horn cells in the sacral spinal cord, or from a ventral horn cell in the spinal cord, down a peripheral nerve to a foot muscle.

Action Potentials. The action potential is

generated at or near the axon hillock and is then propagated down the axon as an all-or-none phenomenon to the synapse (see Chapter 5 - Part II). Several axon terminals that come from many different sources are found on any neuron. Unlike the axon, the dendrite does not respond in an all-or-none fashion like the axon. Instead, each nerve impulse at a given site on the dendrite produces a change in the electrical activity. The sum total of all these electrical changes result in a variation in the membrane potential in the neuron either below or above the firing threshold.

Synapse: Synapses can be seen at the light microscopic level *(Fig 3-20)*, however to identify all the components of a synapses the electron microscope must be use. At the electron microscopic level the synapse consists of the axonal ending, which forms the presynaptic side, and the dendritic zone, which forms the postsynaptic side *(Fig. 3-21)*. Collectively, the pre- and postsynaptic sides and the intervening synaptic cleft are called the synapse.

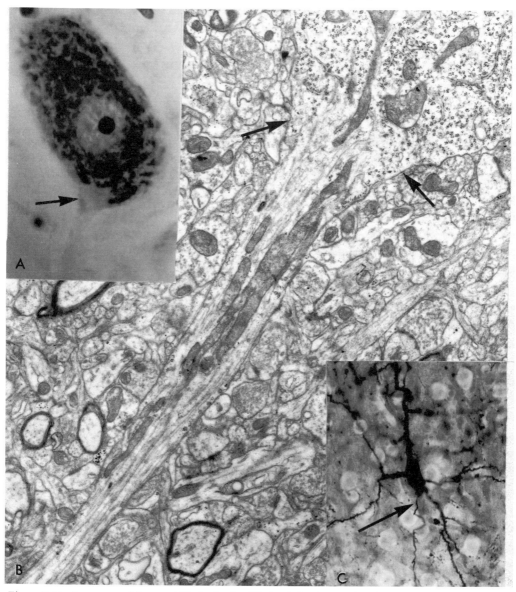

Figure 3-16. Appearance of the axon hillock: A, after Nissl staining (<X>400), B, in an electron micrograph (<X>15,000), and C, after Golgi rapid staining (<X>350).

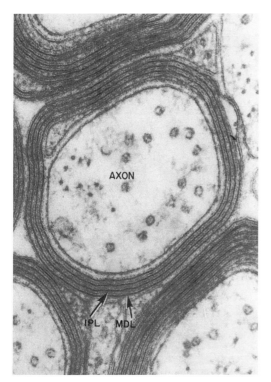

Figure 3-17. Electron micrograph of myelin sheath from the optic nerve of the mouse demonstrating repeating units of the myelin sheath, consisting of a series of light and dark lines. The dark line, called the major dense line (MDL), represents the apposition of the inner surface of the unit membranes. The less dense line, called the interperiod line (IPL), represents the approximation of the outer surfaces of adjacent myelin membranes. (<X>67,000.) (Courtesy of Alan Peters, Department of anatomy, Boston University School of Medicine)

At the synapse the electrical impulse from one cell is transmitted to another. Synapses vary in size from the large endings on motor neurons (1 to 3 microns) to smaller synapses on the granule and stellate cells of the cortex and cerebellum (less than 0.5 microns). Synapses primarily occur between the axon of one cell and the dendrite of another cell. Synapses are usually located on the dendritic spins but are also seen on the soma and rarely between axons. At the synapse the axon arborizes and forms several synaptic bulbs that are attached to the plasma membrane of the opposing neuron by intersynaptic filaments (Fig. 3-21).

Structure. Synapses can be identified by light microscopy, but the electron microscope

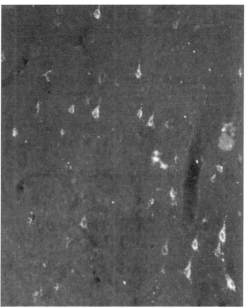

Figure 3-18. Demonstration of the cells of origin of callosal axons with the horseradish peroxidase<->diaminobenzidine reaction method in layers III, and V of the somatosensory cortex. The horseradish peroxidase was injected into the contralateral cortex 34 hours previously. The predominant cell type is pyramidal. (Dark field, <X>100.)

has revealed many new details in synaptic structure (Bodian, 1970; Colonnier, 1969; Gray, 1959; Palay, 1967). In an electron micrograph, the presynaptic or axonal side of the synapse contains mitochondria and many synaptic vesicles (Fig. 3-21). Synaptic vesicles are concentrated near the presynaptic surface with some vesicles actually seen fusing with a membrane (Fig. 3-21), illustrating that this site releases neurotransmitters. Neurofilaments are usually absent on the presynaptic side. Pre- and postsynaptic membranes are electron-dense and are separated by a 30 to 40 nm space, the synaptic cleft, which is continuous with the extracellular space of the central nervous system.

Synaptic Types. Two types of synapses, electrical and chemical, differ in location and appearance. Most of the synapses in the mammalian central nervous system are chemical.

Electrical synapses are connected by membrane bridges, gap junction connections, which permit the electric impulse to pass directly from one cell to the other. Electric synapses have almost no delay and little chance of misfiring.

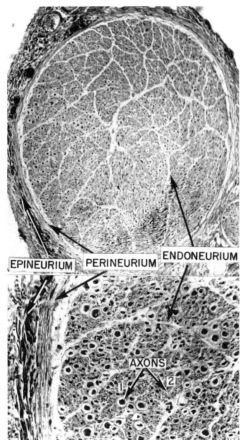

Figure 3-19. Peripheral nerve of a cat. A shows wrapping of the nerve trunk, the epineurium; each nerve fascicle is surrounded by the perineurium while each nerve fiber is embedded in endoneurium (Bodian stain, <X>100). In B, 1 demonstrates a large myelinated axon; 3 points to a small myelinated and unmyelinated axon (Bodian stain, <X>350).

These synapses are seen in many fish.

Chemical synapses have a presynaptic side, containing vesicles and a gap, and the postsynaptic side with membrane receptors. The neurotransmitter released by the action potential is exocytosed and diffuses across the synaptic cleft and binds to the specific receptor on the postsynaptic membrane.

Synapses are either excitatory or inhibitory. Synapses that depolarize the membrane potential (make it more positive) are *excitatory*. Synapses that hyperpolarize the membrane potential (make it more negative) are *inhibitory*.

Excitatory synapses in the central nervous system are asymmetrical, having a prominent

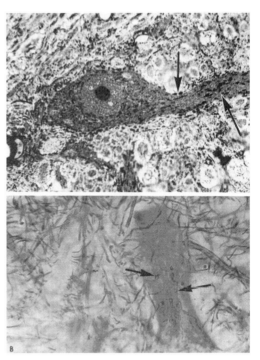

Figure 3-20. Silver stain of a one-micron plastic embedded section. A demonstrates synaptic boutons on neurons in the reticular formation. B demonstrates boutons on ventral horn cells. (<X>400.)

postsynaptic bush with presynaptic vesicles *(Fig 3-21)*. This type of synapse is most commonly seen on dendrites. Glutamate has been identified in excitatory synapses. At the excitatory synapse there is a change in permeability that leads to depolarization of the postsynaptic membrane and which can lead to the generation of an action potential. Glutamate has been identified in excitatory synapses.

Inhibitory synapses in the central nervous system are symmetrical with thickened membranes on the pre- and postsynaptic side and vesicles only on the presynaptic side. GABA has been identified in the inhibitory synapses. At an inhibitory synapse the neurotransmitter binds to the receptor membrane, which changes the permeability and tends to block the formation of the action potential. Synapses on the soma are symmetrical and they are considered inhibitory.

Throughout much of the central nervous system, spines are found on the dendrites. These dendritic spines are bulbous, with a long neck connecting to the dendrite. Many axon terminals are located on the spines. In the cerebral cortex, a spine apparatus is found within the spine, which seems to function like a capacitor, charging and then discharging when its current load is exceeded (see Fig. 3-2, 3-15).

Synaptic Vesicles. The synaptic vesicles differ in size and shape and may be agranular, spherical, flattened, or round with a dense core. The method of fixation for electron micrographs affects the shape of a vesicle. Bodian (1970) has shown that osmium fixation produces only spheroidal vesicles. Aldehyde followed by osmium produces spheroidal and flattened vesicles. The shape of flattened vesicles may also be modified by washing the tissue in buffer or placing the tissue directly from the aldehyde into the osmium. The spheroidal vesicles retain their

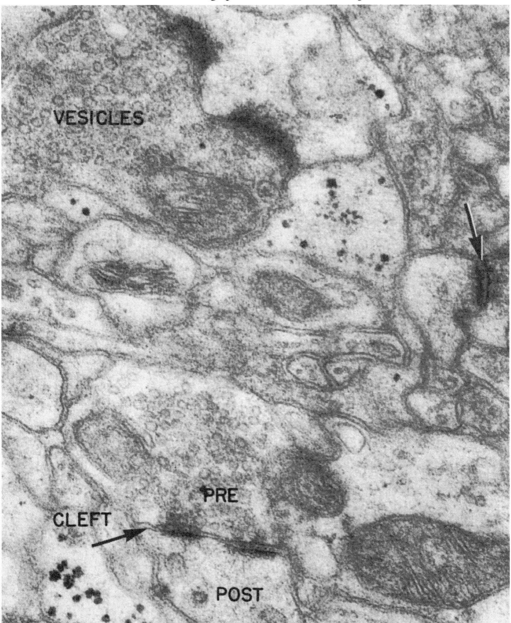

Figure 3-21. Synapse in the sensory cortex of the rat demonstrating agranular synaptic vesicles (300 to 400 Å) in the presynaptic axonal side. Note the electron-dense synaptic membranes and the intersynaptic filaments in the synaptic cleft. Electron micrograph (<X>65,000.)

shape regardless of any manipulation.

There are four basic categories of synaptic vesicles (Palay, 1967), as described in Table 3-6.

Treating the Type 1 vesicles with high osmolality aldehyde fixatives produces the commonly seen spherical vesicles and also a group of flattened, ovoid, or disc-shaped vesicles. Flattened vesicles are known to be inhibitory synapses, and the spherical vesicles are assumed to be excitatory synapses.

Synaptic Transmission. Current evidence in the mammalian central nervous system suggests that synaptic transmission is primarily a chemically and not an electrically mediated phenomenon, based on the presence of:

1. A 30 to 40 nm cleft
2. Synaptic vesicles
3. Appreciable synaptic delay due to absorbance of the chemical onto the postsynaptic receptor site.

In contrast electrical synapses have cytoplasmic bridges that interconnect the pre- and postsynaptic membranes resulting in a minimal synaptic delay as transmission is ionic rather than by release of chemical from a vesicle.

Neurotransmitters

Many compounds have been identified as neurotransmitters. These substances are found in synaptic vesicles on the presynaptic side. Introduction of the compound into the synaptic cleft produces the same change in the resting membrane potential as stimulation of the presynaptic axon; the compound is rapidly degraded, and the membrane potential returns to the resting state.

The neurotransmitters are either amino acids, derived from amino acids, or small neuropeptides. The classic neurotransmitters in the central nervous system include acetylcholine, epinephrine, norepinephrine, serotonin, glycine, glutamate, dopamine, and GABA. At certain synaptic sites the following compounds may also function as modulators (usually a slower transmitter) form of neurotransmission: adenosine, histamine, octopamine, B-alanine, ATP, and taurine. Many of the neuropeptides, such as substance P, vasoactive peptide, peptide Y, and somatostatin are also active in neurotransmission or neuromodulation. Catecholamines and 5-hydroxytryptamine are transmitters linked to synaptic transmission in the central nervous system. Noradrenaline is a transmitter at the preganglionic synapses.

Many steroids and hormones have also been linked to synaptic transmission. It is still uncertain whether these compounds play a direct role in nervous transmission or if they are just related by their importance to the ongoing functions of the entire nervous system (Table 3-7).

At excitatory synapses the following com-

TABLE 3-6. CATEGORIES OF SYNAPTIC VESICLES

Type	Diameter	Locations
1. Spheroidal or flattened with a clear center; most common type (Figs. 3-33 and 3-34).	30 to 40 nm	At neuromuscular junction and throughout central nervous system.
2. Spheroidal with 38 nm electron-dense granule in the center	40 to 80 nm	Found in autonomic endings in the intestines, vas deferens, and pineal body; contains catecholamines.
3. Spheroidal with a 50 nm electron-dense granule in the center	80 to 90 nm	Found at preganglionic sympathetic synapses, at neuromuscular junctions in smooth muscle, and in part of the hypothalamus, basal nuclei, brain stem, and cerebellum; catecholamines present in vesicles.
4. Spheroidal with a large droplet that nearly fills the vesicle (Fig. 3-15).	130 to 300 nm	Characteristic of nerve endings in the hypothalamus and neurohypophysis; also found in the soma, axons, and presynaptic endings of nerve cells of the hypothalamic-hypophyseal tract; vesicles contain vasopressin and oxytocin.

TABLE 3-7. LOCATION AND FUNCTION OF NEUROTRANSMITTERS

Agent	Location	Function
L-Glutamine	Excitatory neurons	Excitation
GABA	Inhibitory neurons	Inhibition (fast/slow)
Acetylcholine	Motor neurons, basal forebrain, and midbrain and pontine tegmentum	Excitation and modulation
Monoamines -- Norepinephrine -- Serotonin -- Histamine	Brain stem and hypothalamus --Locus ceruleus --Raphe nuclei --Hypothalamus	Modulation
Neuropeptides	Limbic, hypothalamus, autonomics, and pain pathways	Modulation

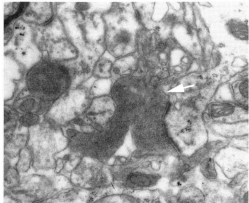

Figure 3-22. Degenerating synaptic ending in the sensory cortex of the adult rat, showing dense vesicles and filaments replacing the normal appearance seen in Figure 3-21. Electron micrograph <X> 115,000

pounds have been found: acetylcholine, norepinephrine, dopamine, serotonin, glutamate, and aspartate. Inhibitory neurotransmitters include GABA, histamine, neurotensin, and angiotensin.

Acetylcholine, the best documented transmitter in the peripheral nervous system, has been isolated in synaptic vesicles. Acetylcholine esterase has been found throughout the central and peripheral nervous systems and at postganglionic sympathetic endings.

Effectors and Receptors

Each peripheral nerve, whether sensory, motor, or secretory, terminates by arborizing in a peripheral structure (Fig.1-3).

Effectors

The motor nerves of the somatic nervous system end in skeletal muscles and form the *motor end plates* (see Chapter 6). Nerve endings in smooth and cardiac muscle and in glands resemble the synaptic endings in the central nervous system. *Visceral motor endings* are found on muscles in arterioles (vasomotor), muscles in hair follicles (pilomotor), and sweat glands (sudomotor).

Sensory Receptors. A stereogram of the skin is shown in Figure 1-3. Table 3-8 lists the mechanicoreceptors in the body.

Sensory Endings (Fig 1-3). Sensory endings, found throughout the body, subserve pain, touch, temperature, vibration, pressure, heat, and cold in the skin, muscles, and viscera as well as the specialized somatic and visceral sensations of taste, smell, vision, audition, and balance. Visceral sensory receptors are similar to somatic sensory receptors associated with the somatic nervous system, except that they are located in the viscera and their accessory organs.

Free Nerve Endings (Fig. 1-3). Free nerve

TABLE 3-8. MECHANICORECEPTORS

Modality	Receptors
Light touch and vibration	Encapsulated endings-- Meissner=s and Pacinian corpuscles
Proprioception	Muscle spindles and Golgi tendon organs in joints
Pain and temperature	Free nerve endings

endings are formed by sensory fibers and arborize in various tissues, including the stratified epithelium, muscles, tendons, connective tissues, mucosa, and serous membranes in joints. They are considered to be pain receptors because they are found in tissues where pain is the primary sensation, such as tooth pulp, dentin, and the cornea. Crude touch may also be subserved by these receptors.

Free nerve endings are also found in terminal networks around the disc-shaped tactile cells of Merkel and around the hair follicles in the dermal sheath and outer root sheath. These structures appear to subserve touch.

Encapsulated Sensory Endings (Fig. 1-3). In these endings, the nerve is surrounded by a specialized connective tissue capsule of varying thickness. Encapsulated endings include Meissner=s and pacinian corpuscles, muscle and tendon spindles, the cylindrical end bulb of Krause, and the end bulb of Golgi-Mason.

Meissner's Corpuscles(Fig 1-3). Meissner=s (tactile) corpuscles are presumed to subserve touch. They are elliptical and may have from one to five myelinated nerve fibers arborizing in their lamellated capsule. These end organs are found in dermal papillae, being most numerous in the fingertips, soles, palms, lips, glans penis, and clitoris.

Pacinian Corpuscles (fig 1-3). These corpuscles resemble a sliced onion. Many concentric layers built upon the centrally placed axon. They are found throughout subcutaneous tissue and are especially numerous in the hand, foot, mammary glands, clitoris, and penis. Pacinian corpuscles are pressure-sensitive receptors. Herbst=s corpuscles are similar to pacinian corpuscles but smaller.

End Bulbs of Krausse and Golgi-Mason(fig 1-3). These endings are found throughout the body and contain a single, extensively ramified axon within the matrix. Many variants of this structure have been identified. This organ is presumed to record changes in heat and cold.

Muscle and Tendon Spindles (fig 7-20). The muscle spindles and annulospiral endings (see Chapter 5), as well as the tendon spindles, transmit information concerning muscular activity and tendon stretching to the central nervous system.

Remember that all of the sensory endings form the primary neurons in the sensory system. Their cell bodies are located in the spinal dorsal root ganglia or cranial nerve ganglia, and their axons enter the central nervous system. The motor or effector axons represent the lower motor neuron or final neuron in the motor system.

Supporting Cells of the Central Nervous System

The central nervous system has billions of neurons, but the number of supporting cells exceeds them by a factor of five or six. Supporting cells form a structural matrix and play a vital role in transporting gases, water, electrolytes, and metabolites from blood vessels to the neural parenchyma and in removing waste products from the neuron. In contrast to the neuron, the supporting cells in the adult central nervous system normally undergo mitotic division. The supporting cells are divided into macroglia and microglia.

The *macroglia,* which include astrocytes, oligodendrocytes, and ependyma, are the supporting cells or neuroglia (nerve glue) of the central nervous system (fig 3-23). *Schwann cells, satellite cells,* and *fibroblasts* are supporting cells of the peripheral nervous system. Mesodermal *microglia* cells include the perivascular cells and any white blood cells found within the parenchyma of the central nervous system. Functions of the different supporting cells in the nervous system are summarized in Table 3-9.

Astrocytes (Figs. 3-23, 3-24, and 3-25)

Astrocytes are of two types: fibrous (most common in white matter) or protoplasmic (most

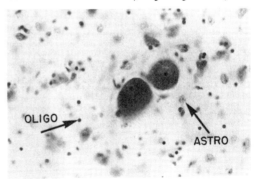

Figure 3-23. Appearance of neuron, astrocyte (astro), and oligodendrocyte (oligo). Nissl stain.(<X>300.)

NEUROCYTOLOGY 3-21

TABLE 3-9. FUNCTIONS OF SUPPORTING CELLS

Cell Type	Functions
Astrocytes --Fibrous type (white matter) --Protoplasmic type (gray matter)	Major supporting cells in the brain, forming microenvironment for neurons; act as phagocyte; isolate synapses, enwrap blood vessels, and form membranes on brain=s inner and outer surface.
Oligodendrocytes	Form and maintain myelin
Ependymal cells	Ciliated lining cells of the ventricular system
Endothelial cells	Lining cells of blood vessels in the brain that form blood-brain barrier.
Microglia (pericytes)	Supporting cells and multipotential cells found in the basement membrane of blood vessels and within brain parenchyma
Mononuclear cells	White cells from the circulation that readily enter the brain (lymphocytes, monocytes, and macrophages) and function as sentinels for the immune system

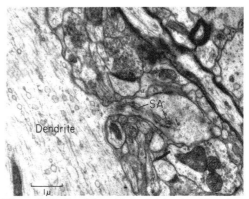

Figure 3-24. Dendrite with dendritic spine. The spines greatly increase the surface area of many neurons. Note the many microtubules in the dendrite and the neck of eh dendritic spin containing a spine apparatus. The3re are several synapses on the spine. Electorn micrograph <X> 80,000.

common in gray matter). All astrocytes are larger and less dense than the oligodendrocytes. The astrocytes form a complete membrane on the external surface of the brain called the external glial limiting membrane, which surrounds all blood vessels, fuses with the ependymal processes, and isolates neuronal processes.

In light micrographs astrocytes appear as pale cells with little or no detail in the cytoplasm. The nuclei are smaller than those of a neuron but larger and less dense than those of an oligodendrocyte *(Fig. 3-24)*. Electron micrographs demonstrate that fibrous astrocytes have many filaments, which in places appear to fill the cytoplasm. There are few microtubules, and the processes appear pale. The nuclei of these cells have some condensed chromatin adjacent to the nuclear membrane. Glycogen is also common in astrocytic processes. Protoplasmic astrocytes have nuclei that are a little darker than those of a neuron. They resemble fibrous astrocytes except that they have just a few filaments.

Astrocytes not only form the skeleton of the central nervous system but also tend to segregate

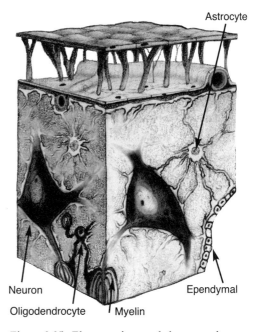

Figure 3-25. Electron micrograph demonstrating appearance of the neuron, astrocyte (astro), and oligodendrocyte (oligo). (<X> 18,000.)

synapses and help form the blood-brain barrier by enwrapping brain capillaries (outer and inner limiting membranes). If the brain is damaged by infarction, for example, astrocytes proliferate and form scars.

Astrocytes and adjacent neurons form the microenvironment of the nervous system. There are approximately 100 to 1,000 astrocytes per neuron, depending on the size of the neuron. In areas with high glutamate concentration, such as the cerebral cortex, or areas with high dopamine content, such as the basal nuclei, the glia take on the chemical characteristics of the adjacent neuron. Glia are involved in neuronal functions because they absorb transmitters and modulators in their environment and often release them back into the synapse.

When central nervous system diseases affect only astrocytes, the reaction is called primary astrogliosis. More commonly, the disease process affects nerve cells primarily, which is called secondary astrogliosis. When an injury occurs to the nerve cell in the central nervous system without concomitant injury to blood vesicles and glia, the nerve cells are phagocytized and the astrocytes proliferate and replace the neurons, forming a glial scar (replacement gliosis). In the case of a more severe injury to the nervous system, such as an infarct that damages glia, nerve cells, and blood vessels, the astrocytes proliferate along the wall of the injury, and the dead neurons and glia are phagocytized, leaving only a cavity lined with meninges. In all other organs there are enough fibroblasts to proliferate and form a scar, but in the central nervous system there are only a few fibroblasts, so cavitation is a common sequela to extensive destruction.

Oligodendrocytes

In light micrographs the oligodendrocyte has a small darkly stained nucleus surrounded by a thin ring of cytoplasm (Fig. 3-23). In electron micrographs oligodendrocytes are dense cells with many microtubules and few neurofilaments (Figs. *3-25 and 3-26*). Dense clumps of rough endoplasmic reticulum and clusters of polyribosomes are seen in the cytoplasm, which is denser but scantier than that in neurons. The nucleus tends to be located toward one pole of the cell; the nuclear chromatin tends to be heavily clumped. In electron micrographs oligodendrocytes can be distinguished from astrocytes because they have a darker cytoplasm and nucleus, few if any filaments, and more heavily condensed chromatin (Fig. 3-26).

The role of the oligodendrocyte is to form and maintain myelin (although they may also be responsible for breaking down myelin in multi-

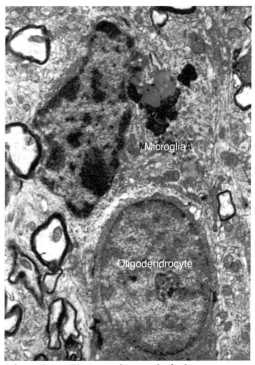

Figure 3-26. Electron micrograph of a human cerebral cortex demonstrating differences in the density of the DNA in the nuclei of oligodendrocytes (oligo) and microglia. (<X>30,000.)

ple sclerosis). The formation of myelin is under genetic control. Oligodendrocytes are usually seen in close proximity to astrocytes and neurons, and all three cell types are important in forming and maintaining myelin.

Endothelial Cells

Endothelial cells form the lining of the capillaries in the central nervous system(fig 3-25). They are of mesodermal origin and bound together by tight junctions. Their tight junctions and apinocytosis provide the basis of the blood-brain barrier (see below).

Mononuclear Cells

Mononuclear cells--lymphocytes, monocytes, and histiocytes--are found in the central nervous system, where they seem to act as phagocytes, breaking down myelin and neurons. Myelin destruction always triggers intense macrophage reaction within 48 hours, followed by infiltration of monocytes first and then lymphocytes. Note that astrocytes have also been shown to engulf degenerating myelin sheaths, axonal processes, and degenerating synapses.

The central nervous system was once considered an Aimmunologically privileged site@ because:

1. No specific lymph drainage from the central nervous system alerts the immune system of infection.

2. Neurons and glia do not express the major histocompatibility complex.

3. The major cell for stimulating the immune response (leukocyte dendritic cells) is not normally present in the disease-free nervous system.

However, recent studies have shown that there is a regular immune surveillance of the central nervous system, which is sufficient to control many viral infections (Sedgwick and Dorries, 1991). It is now known that immune cells regularly enter the brain through the capillaries and that macrophages infected with human immunodeficiency virus (HIV), for example, can infect the brain directly, the so-called Trojan-horse phenomenon (Haase, 1986; Price et al., 1988).

Microglia

Neurons, astrocytes, and oligodendrocytes are ectodermal in origin, but microglial cells are mesodermal in origin (Fig. 3-26). The ovoid microglia cells are the smallest of the supporting cells and are divided into two categories: (1) those that form the perivascular cells and (3) the resting microglial cells in the brain parenchyma. Microglial cells originate from monocytes that enter the brain (Table 3-10).

Pericytes are found in relation to capillaries but external to the endothelial cells and enwrapped in the basal lamina. In electron micrographs they are not as electron-dense as oligodendrocytes and lack the neurofilaments of the astrocyte and the tubules of the oligodendrocyte. The cytoplasm is denser than that of astrocytes and contains fat droplets and laminar dense bodies. The granular endoplasmic reticulum consists of long stringy cisterns. Microglia are considered multipotential cells because with the proper stimulus they can become macrophages (Vaughn and Peters, 1968). The pericyte contains actin, and this cell may well be important in controlling the channels entering the endothelial cells (Herman and Jacobson, 1988).

During early development, *monocytes* enter

TABLE 3-10. TYPES OF MICROGLIA CELLS

Cell Type	Function
Monocytes	Enters brain during early development and is the stem cell of microglia.
Pericytes (perivascular cell)	Found inside the brain in the basement membrane of the blood vessel; can act as a macrophage.
Amoeboid microglia	Transitional form leads to resting microglia.
Resting microglia (ramified)	Down-regulated from amoeboid microglia; probably the sentinels in the brain that raise the alarm for invasive diseases.
Activated microglia	Upregulated resting cell changes into partially activated macrophage with MHC class I.
Reactive microglia	Fully activated macrophage with MHC class II and phagocytic properties.
Giant multinucleated cells	Forms from fusion of reactive cells; associated with viral brain infections and Creutzfeld-Jakob disease; hallmark of AIDS dementia.

the brain and, after formation of the blood-brain barrier, become trapped (Davis et al., 1994; Ling and Wong, 1993). The monocytes pass through an intermediate phase of development, the *amoeboid microglia*, which evolve into a down-regulated resting form, the *ramified microglia*. These *resting microglia* are found throughout the central nervous system and may well be the sentinels that alert the immune system to disease in the brain. With the appeearance of any central nervous system disease (e.g., multiple sclerosis, stroke, trauma, or tumors), the resting microglia are upregulated and become *activated microglial cell*. The factor or gene that upregulates or down-regulates these cells is currently unknown. Once the disease process has been resolved, the activated microglia can revert to resting microglial cells.

The activated microglia cell is a partially activated macrophage containing the CR3 complex and class I major histocompatibility complex (MHC). The active microglia then become a reactive microglial cell, which is a fully active macrophage containing class II MHC and phagocytic activity. These cells are very active during all major disease states in the brain. Activated microglial cells can also evolve into giant multinucleated cells by the fusion of reactive cells. They are seen in viral infections and are considered the hallmark of AIDS dementia. Also called gitter cells, *giant multinucleated cells* are often found in patients with Creutzfeldt-Jakob disease (Fig. 3-31), a disease caused by proteinaceous infectious particles, or prions.

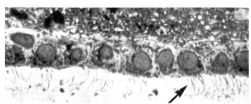

Figure 3-27. Ependymal lining cells in the third ventricle of a rat. Note prominent cilia (arrows) in this one-micron thick plastic-stained section. (<X>1,400.)

Ependymal Cells (Fig. 3-27)

Ependymal cells line all parts of the ventricular system. They are cuboidal, ciliated, and contain filaments and other organelles. The processes of these cells extend in the central nervous system and fuse with astrocytic processes to form the *inner limiting glial membrane*. Highly modified ependymal cells are found attached to the blood vessels in the roof of the body of the lateral ventricles, the inferior horn of the lateral ventricles, and the third and fourth ventricles. There they form the choroid plexus, which secretes much of the cerebrospinal fluid (Fig. 3-35). Ependymal cells originate from the germinal cells lining the embryonic ventricle, but they soon stop differentiating and stay at the lumen on the developing ventricles.

Satellite Cells (Fig. 3-28)

Satellite cells, which are found only in the peripheral nervous system among sensory and sympathetic ganglia, originate from neural crest cells. Many satellite cells envelop a ganglion cell. Functionally, they are similar to the astrocytes, although they look more like oligodendrocytes.

Schwann Cells

Schwann cells are ectodermal in origin (neur-

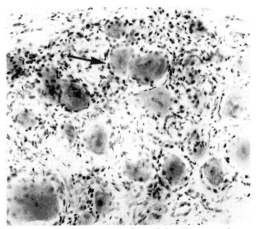

Figure 3-28. Sensory ganglion of the rhesus monkey, demonstrating pseudounipolar cells surrounded by their satellite cells (arrow). (<X>300.)

al crest) in the peripheral nervous system and function like oligodendrocytes, forming the myelin and neurilemmal sheath. In addition, the unmyelinated axons are embedded in their cytoplasm. Schwann cell cytoplasm stops before the nodes of Ranvier (fig 3-33), leaving spaces between the node and Schwann cells. In an injured nerve Schwann cells can form tubes that penetrate the scar and permit regeneration of the peripheral axons. Nerve growth factor is important to proliferation of the Schwann cells.

Neural Crest Cells

These cells originate embryologically as neuroectodermal cells on either side of the dorsal crest of the developing neural tube but soon drop dorsolaterally to the evolving spinal cord area. Neural crest cells migrate out to form the following: dorsal root ganglion cells, satellite cells, autonomic ganglion cells, Schwann cells of the peripheral nervous system, chromaffin cells of the adrenal medulla, and pigment cells of the integument.

Response of Nervous System to Injury

Degeneration

Neuronal death or atrophy may result from trauma, circulatory insufficiency (strokes), tumors, infections, metabolic insufficiency, developmental defects, and degenerative and heredodegenerative diseases. These neuropathologic processes produce a range of responses in the neurons and glia (Ramon y Cajal, 1938; Young, 1943). In this section, the neuronal response to

NEUROCYTOLOGY 3-25

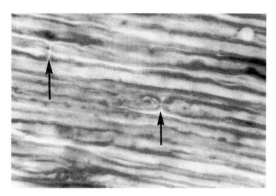

Figure 3-29. Longitudinal section of a peripheral nerve fixed in osmium, demonstrating the nodes of Ranvier (arrows). (<X>1,000.)

injury will be examined in the cell body and axon.

Retrograde Changes in the Cell Body (Fig. 3-30). Section of the axon obr direct injury to the dendrites or cell body produces the following series of responses in the soma:

1. The nucleus, cell body, and nucleoli swell. The nucleus is displaced from the center of the cell body and may even lie adjacent to the plasma membrane of the neuron.

2. A slow dissolution of Nissl substance starts centrally and proceeds peripherally, until only the most peripherally placed Nissl substance is left intact (which is probably essential to the protein metabolism of the rest of the cell). This dissolution of the Nissl substance (ribosomal RNA), called chromatolysis, allows the protein-manufacturing processes to be mobilized to help the neuron survive the injury. The mRNA then begins the manufacturing of membrane that is transported down the intact tubules into the growing axonal ending (growth cone).

3. All other organelles in the cell body and dendrites also respond to the injury. The mitochondria swell, and the smooth endoplasmic

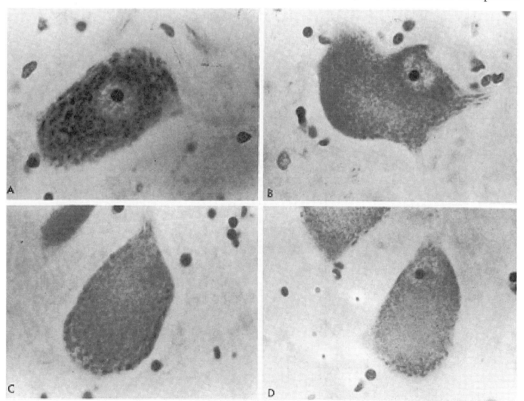

Figure 3-30. Ventral horn cells in the human lumbar spinal cord. A, Normal. B to D, Wallerian retrograde chromatolytic changes in ventral horn cells following injury to the peripheral nerve. B, Chromatolytic neuron with eccentric nucleus and some dissolution of the Nissl substance. C, Chromatolytic neurons, showing a peripheral ring of Nissl substance (peripheral chromatolysis). D, Chromatolytic neuron, showing eccentric nucleus and only a peripheral ring of Nissl substance. (Nissl stain, <X>400.)

reticulum proliferates to help in the formation of new plasma membrane and new myelin.

These responses represent the increased energy requirements of the nerve cell and the need to form plasma membrane during the regenerative process. If the cell survives the injury, all organelles return to normal: the nucleus returns to the center of the cell body, and the soma returns to its pretraumatic size. If the injury is too extensive, the neuron atrophies or dies. The responses of neuronal soma to injury (chromatolysis) can be summarized in three steps:

1. Swelling of nucleus, nucleolus, and cytoplasm with nucleus becoming eccentric as a direct response to injury of the axon or dendrite. Nissl substance appears to dissolve.

2 Proliferation of metabolic processes in the nucleus including mRNA occurs. Endoplasmic reticulum and mitochondria starts manufacturing membranes and increasing the energy available in the cell.

3. With successful recovery, the cell returns to normal size. If seriously injured, the cell becomes atrophic or may be phagocytized.

Atrophic Change. In atrophic change, the nerve cell is too severely damaged to repair itself. Consequently, the cell body shrinks and becomes smaller. This response is similar to the response of a nerve cell to insufficient blood supply, which produces an ischemic neuron. If necrosis occurs, the neuron cannot survive. The Nissl substance begins to disperse, and after 7 days the nucleus becomes dark and the cytoplasm eosinophilic. Within a few days, these cells are phagocytized.

Wallerian Degeneration. When an axon is sectioned, the distal part that is separated from the trophic center (cell body) degenerates, a process called *wallerian,* or *anterograde, degeneration.* At the same time, the cell body undergoes a process called *axonal,* or *retrograde, degeneration.* If the cell body remains intact, the proximal portion begins to regenerate. The distal stump is usually viable for a few days, but its degeneration begins within 13 hours of injury. The axon starts to degenerate before the myelin sheath. In 4 to 7 days, the axon appears beaded and is beginning to be phagocytized by macrophages, which enter from the circulatory system (Fig. 3-4). Fragments of degenerating axons and myelin are broken down in digestion chambers (Figs. 3-31

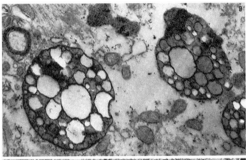

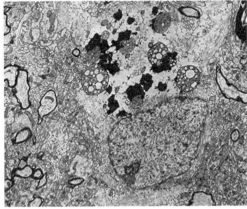

Figure 3-31. A and B, Electron micrograph of a reactive astrocyte in the cerebral cortex of a person with Jakob-Creutzfeldt disease. Note the prominent digestion vacuoles shown in higher power in B. (A, <X>8,000; B, <X> 35,000.)

and 3-32), and it may take several months before all of the fragments are ingested. In the proximal portion degenerative changes are noted back to the first unaffected node. As the myelin degenerates, it is broken up into smaller pieces that can be ingested more easily *(Figs. 3-32 and 3-33).*

REGENERATION

Peripheral Nerve Regeneration

Within a few days after section, the proximal part (attached to a functional neuronal soma) of the nerve starts regrowing. Nerve growth factor is produced after injury to the axon, and it promotes the axonal sprouting. If the wound is clean, e.g., a stab sound, sewing the nerve ends together can dramatically increase the rate of recovery in the affected limb. The regenerating nerves may cross the scar within several weeks (Fig. 3-34). The crossing is helped by the Schwann cells and fibroblasts, which proliferate from the proximal end of the nerve. The Schwann cells form new basement membrane

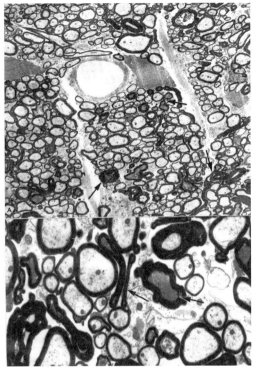

Figure 3-32. Electron micrograph of degenerating axons in several fascicles of the medullary pyramid of a rat 15 days after a cortical lesion. A, Arrows point to degenerating axons (<X>8,000). B, Detail of degenerating axons (<X>30,000). Note the collapsed axons and dense axoplasm and that many axons were unaffected by the lesion.

and provide tubes through which the regenerating axons can grow.

In certain peripheral nervous system diseases only segmental degeneration occurs. One example is diphtheria: the myelin sheath degenerates but the axon remains intact. Phagocytes break down the myelin, and Schwann cells rapidly reform myelin.

The rate of movement of the slow component of axoplasmic flow probably accounts for the rate of axonal regrowth, which is limited to about 1 mm a day. Slow components of the axoplasmic flow (-Scb) carry actin, fodrin, calmodulin, clathrina, and glycolytic enzymes that form the network of microtubules, intermediate filaments, and the axolemma, which limit the rate of daily axonal regeneration, although functional recovery may be a little faster (McQuarrie and Grafstein, 1983; Wujek and Lasek, 1983; McQuarrie, 1988; Kandel and Schwartz 2000).

As the regenerating axon grows, the axonal end sprouts many little processes. If one axonal sprout penetrates the scar, the other sprouts degenerate, and the axon follows the path established by the penetrating sprout. If an axon reaches one of the tubes formed by the Schwann cells, it grows quickly and after crossing the scar descends the distal stump at a rate of approximately 1 mm/day (Jacobson and Guth, 1965; Guth and Jacobson, 1966).

When the motor end plate is reached, a delay occurs while the axon reinnervates the muscle and reestablishes function. At this stage the average rate of functional regeneration is 1 to 3 mm a day.

Only a small percentage of the nerves reach the effectors or receptors. The basal laminae helps direct the regenerating nerve to the motor end plate. If a sensory fiber innervates a motor end plate, it remains nonfunctional and probably degenerates, and the cell body atrophies. A sensory fiber that reaches a sensory receptor may become functional, even if the receptor is the wrong one. For example, after nerve regeneration some patients complain that rubbing or pressing the skin produces pain. In these cases it would appear that fibers sensitive to pain have reached a tactile or pressure-sensitive receptor. A

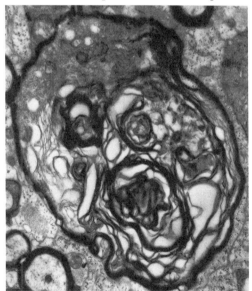

Figure 3-33. Electron micrograph of a degenerating myelin sheath in the medullary pyramid of a rat 30 days after a cortical lesion. Note the unraveling and vacuolization of the myelin. (<X>30,000.)

motor fiber may also reinnervate the wrong motor end plate, as when a flexor axon innervates an extensor. In such a case, the patient has to relearn how to use the muscle.

Muscle that is denervated assists the regenerating axons by expressing molecules that influence the regenerating axons. Some of the molecules are concentrated in the synaptic basal lamina of the muscle. Other molecules are upregulated following denervation and help in attracting and reestablishing the synapse in the muscle. These upregulated molecules include: growth factors (IGF-3 and FGF-5), acetylcholinesterase (AChE), agrin, laminin, s-laminin, fibronectin, collagen a3, and the adhesion molecules N-CAM and N-cadherin (Hall and Patterson, 1993; Horner and Gage 3000).

Successful nerve regeneration also depends on an adequate blood supply. For example, in a large gun shot wound, nerves attempt to regenerate but may not succeed. A summary of the sequence of regeneration in the peripheral nervous system can be broken down into seven steps:

1 The peripheral nerve is sectioned by an injury.

2 The axon dies back to the first unaffected node of Ranvier, with the myelin and distal axon beginning to degenerate within 34 hours.

3 At the site of injury, the axon and myelin degenerates to form a scar. Phagocytosis begins within 48 hours.

4 Axons separated from the cell body degenerate. With an adequate blood supply the portion of the axon still connected to the intact cell body begins regenerating by sprouting.

5 Within 73 hours, Schwann cells begin to proliferate and form basement membranes and hollow tubes. Nerve growth factor is also formed and released, which further encourages sprouting.

6 From each of the severed axons, sprouts attempt to penetrate the scar. After one sprout successfully grows through the scar the other sprouts die. Nerves take a month or more to grow through the scar.

7 Once an axon penetrates the scar, it grows at 1 mm/day; about a third of the severed axons actually reinnervate muscle and skin.

Central Nerve Regeneration

After an injury, axons in the central nervous system regenerate, but there seems to be no equivalent to the Schwann cell because oligodendrocytes and astrocytes do not form tubes to penetrate the scar. Instead, they form a scar that is nearly impenetrable. Even if the axons penetrate the scar, they have no means of reaching the neuron to which they were originally connected.Horner and Gage (2000)have reviewed the question of how to regenerate the damaged central nervous system in the brain that is inherelty very plastic. The y have noted that it is not the failure of neruoaln regeneration , but it is rather a feature of the dmaged environment; and it is now possible to reintroduce the factors preent in the developing nervous system tha produced this wonderful organ. The gene responsible for needed growth factors is probably missing or inactivated in adult tissue. Recently a brain-derived neurotrophic factor has been identified, which may eventually help in finding a way to guide the axon (Goodman, 1994).

Animal studies have shown that neurons have considerable *plasticity*. That is, if some axons in a region die off, bordering unaffected axons will sprout and form new synapses over many months, filling in where the synapses were and resulting in major functional reorganization. This reorganization may eventually produce some recovery of function.

Stem Cells. In the Adult Brain neuronal stem cells have been identified in the adult brain and spinal cord. These cells under the right conditions may well be activated and help to reverse the effects of lesions in the CNS (Kornack & Rackic 1999). After the implantation of immature neurons (neuroblasts) in regions affected by certain diseases (e.g., the corpus striatum of patients with Parkinson's disease), there has been some recovery (Sladek and Gash, 1984; Gage et al., 1991). In Parkinsonian patinet's the age of the individual receiving the transported cells seems to effect the outcome with younger patients(less than 50 years of age) more likely to show some improvement. Regenerating axons in the central nervous system may also grow and form a nonfunctional neuronal ball, or neuroma. Similar long-lasting neuromas may form in the peripheral nervous system and may be a source of pain. In time, the dead neurons and axons are phagocytized by the glia and macrophages, but the astrocytic scar remains.

Factors that promote neuronal survival and axon outgrowth (e.g. brain derived neurotrophic factor-BDNF) have been identified and the focus is now on getting these cells to produce axons to grow into the injured areas and then to grow through into the uninjured area.

Nerve Growth Factors. The first nerve growth factor was isolated by Levi-Montalcini and Angeletti in 1968, but only recently have biotechnology techniques been able to produce these factors in large quantities. Attempts have been made to help regeneration in the central nervous system, for instance, by placing Teflon tubes on Schwann cells through the scarred portion of the spinal cord in the hope that the nerves would follow these channels. However, even though nerves do grow down these channels, no functional recovery occurs. Further information is needed to understand better the chemical nature of scar tissue that retards axonal regeneration.

With the identification of neurotrophic factor (netrins 1 and 3) that produces axonal growth (Serifini et al., 1994) and with the studies of programmed cell death beginning to identify genes that may be responsible for premature neuron death (Oppenheim, 1991), we may be entering an era of brain research that offers great promise to help patients with neurodegenerative diseases, Huntington=s, Parkinson=s and Alzheimer=s.

Glial Response to Injury

Neuronal death triggers an influx of phagocytic cells from the blood stream and the microglia proliferate and break down the dying neurons.

Necrosis. Within a few days of an ischemic attack with infarction, neutrophils are seen at the site of injury. Shortly thereafter, microglial cells and histocytes are seen in the region of the dying cells. Since the blood-brain barrier is usually compromised, monocytes may now migrate into the parenchyma of the central nervous system in greater numbers and assist in phagocytosis.

The time it takes for the complete removal of injured cells depends on the size of the lesion.

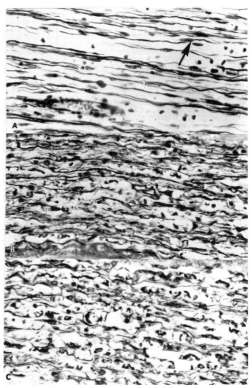

Figure 3-34. Sciatic nerve of a rat, demonstrating the appearance of regenerating nerves 3 weeks after a crushing injury. A, At the site of the crushing lesion, the nerves look normal with regenerating axons grown past the site of the injury; arrow indicates Schwann cell nucleus. B, About 30 mm distal to the site of injury, note some regenerating axons and still many degenerating axon fragments. C, About 35 mm distal to the crush site the transition from regenerating axons to only degenerating axons is shown. (Bodian silver stain, <X>500.)

Large infarcts may take several years before phagocytosis is complete. If the lesion is huge, such as a large infarct in the precentral gyrus, a cavity lined by astrocytic scar will form. In small lesions the neurons are phagocytized glia proliferate, a process called replacement gliosis. In organs with numerous fibroblasts, necrotic areas are soon filled with proliferating fibroblasts, but in the central nervous system there are few fibroblasts for this, and the astrocytes do not proliferate in sufficient numbers.

Blood-Brain Barrier

In the *peripheral nervous system* the endothelial cells are fenestrated and very active in pinocytosis: these two factors increase the ability of compounds to readily enter the nerves. Also, molecules move through peripheral endothelial cells by fluid phase, or receptor-mediated, endocytosis. *Fluid phase endocytosis* is relatively non-specific; the endothelial cells engulf molecules and then internalize them by vesicular endocytosis. In *receptor-mediated endocytosis*, a ligand first binds to a membrane receptor on one side of the cell. After binding to the ligand the complex is internalized into a vesicle and transported across the cell, the ligand is usually released.

In the *central nervous system* endothelial cells that line the capillaries and the choroid plexus are joined together by tight junctions, zonula occludens(fig 3-25). The capillaries are not perforated, and the endothelial cells show very little pinocytosis or receptor-mediated endocytosis (Brightman, 1988). This endothelial lining is called the blood-brain barrier because it is very selective to certain large molecules and dyes and limits the entry of other substances, including amino acids, water, glucose, and electrolytes into the brain parenchyma.

Plasma in blood vessels is separated from the central nervous system tissue by the endothelial lining of the blood vessel and the basement membrane. Pericytes (perivascular) cells are also found within the basement membrane of the capillary wall. The extracellular space of the central nervous system lies external to the basement membrane. All vascular branches within the central nervous system are surrounded by a thin covering formed by astrocytic processes *(Fig. 3-25)*. However, the astrocytic processes do not fuse with the endothelial lining of the blood vessel or with the processes of other cells, so they have minimal effect on limiting the entry of solutes into the brain parenchyma. Thus the extracellular space can be entered once the materials pass through the endothelium.

The intravenous perfusion of various dye compounds (trypan blue, Evans blue, proflavin HCI, and horseradish peroxidase) demonstrates passage through the blood-brain barrier. These dyes demonstrate that the blood-brain barrier is leaky in certain midline regions of the third and fourth ventricle, the circumventricular organs, and in the choroid plexus and locus ceruleus (Brightman, 1989; Dempsey and Wislocki, 1955; Wislocki and Leduc, 1953). The cir-

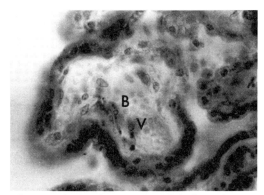

Figure 3-35. Site of cerebrospinal fluid formation: the choroid plexus in the fourth ventricle. Note the blood vessel (BV) in the center and the cuboidal epithelial cells (arrow) on the outside of the vessel. (<X>300.)

cumventricular organs include: pituitary, median eminence, organum vasculosum, subfornical organ, subcommissural organ, pineal gland, and the area postrema of the fourth ventricle. These open connections between the brain and the ventricular system permit neuropeptides from the hypothalamus, midbrain, and pituitary to enter the cerebrospinal fluid and to be widely distributed in the brain and spinal cord, thus forming an alternate pathway in the neuroendocrine system.

Large molecules (such as ferritin and horseradish perioxidase) injected invascularly do not pass through the endothelium; instead, they fill the extracellular spaces between the glia and neurons and do not reenter the blood vessels (Reese and Karnovsky, 1968). Studies have shown that the blood-brain barrier is impermeable to certain large molecules including proteins, but substances such as small lipid-soluable compounds, including alcohol and anesthetics, gasses, water, glucose, electrolytes (NA+, K+, and CL-), and amino acids, can pass from the plasma into the intracellular space (inside neurons and glia) or into the extracellular space between neurons and glia.

Acute lesions of the central nervous system, including those caused by infections, usually increase the permeability of the barrier and alter the concentrations of water, electrolytes, and protein. In viral diseases the infected leukocytes (macrophages) more easily penetrate into the brain by passing between the normally tight junctions in the endothelial cells, which is one way that HIV enters the brain directly from the blood. Tumors within the central nervous system produce growth factors that cause blood vessels to sprout. These new blood vessels have immature tight junctions that are also quite leaky. The leakiness of the capillaries within tumors has been exploited with some success to deliver chemotherapeutic agents specifically to the tumor (Neuwelt and Dahlborg, 1989). There have also been attempts to interfere with the formation of the blood vessel growth factors as a way to starve tumors.

Stress has been shown to open the blood-brain barrier by activating the hypothalamic-hypophyseal-adreanal axia and releasing CRH (Esposito et al. 2001). Acute lesions of the central nervous system including those caused by infections usually increase the permeability of the barrier and alter the concentrations of water, electrolytes, and protein. In some viral diseases, for example, infected leukocytes (macrophages) more easily penetrate directly into the brain by passing between the normally tight junctions in the endothelial cells. This is one way HIV enters the brain from the blood. Also, central nervous system tumors produce growth factors that cause blood vessels to sprout. These new capillaries have immature tight junctions that are also quite leaky and have been studied with some success as a way to deliver chemotherapeutic agents specifically to the tumor (Neuwelt and Dahlborg, 1989).

Extracellular Space

Between the cells in the central nervous system is the extracellular space, measuring between 30 and 40 nm and filled with cerebrospinal fluid(CSF) and other solutes. The CSF is formed primarily by the choriod plexus in the lateral ventricle, IIIrd ventricle, and IV ventricle (Fig 3-35). The amount of extracellular space in the brain is still a matter of controversy. Some solutes can readily pass from the blood plasma through the endothelial lining into the extracellular space, and the solutes present in this space (whether deleterious or not) affect the functions of the central nervous system. A portion of the cerebrospinal fluid appears to be formed by the diffusion of extracellular fluid. Cerebrospinal fluid may also be reabsorbed after temporary storage in the extracellular space. Fat-soluble compounds that readily pass through the blood-brain barrier can enter the extracellular space and may be useful in resolving infections in the central nervous system or in improving the function of certain brain cells.

CHAPTER 4
Neuroembryology and Congenital Malformations

Introduction

The brain undergoes a series of incredible changes in utero as it changes from a flat plate into a tube and then evolves into the convoluted cerebral hemisphere. This complex evolution results in several structures that start very close together (particularly the fornix, stria terminalis, corpus callosum and anterior commissure) moving quite a distance apart.

The in utero development of the central nervous system can be divided into three periods that roughly correspond to the first, the second, and the third trimester of gestation (Table 4-1).

TABLE 4-1. IN UTERO BRAIN DEVELOPMENT

First Trimester	Establishment of basic organization of the CNS beginning as the neural plate, the neural plate sinks inward to form the neural groove. The neural groove continues to deepen until the dorsal lips of the meet and fuse, forming the neural tube, the beginning of the hollow dorsally placed central nervous system.
Second Trimester	Tubular like nervous system is changed by the massive waves of migration of neuroblasts from the germinal cell lining producing an immature model of the adult nervous system with the beginnings of much of the circuitry in place.
Third Trimester	Neuronal processes mature, and the gyri enlarge, myelination begins, and the basic circuitry continues to form preparing the infant for functioning outside the womb.

Newborns, through genetic inheritance, can make about 40 different sounds, and distinguish some sounds as being either safe or dangerous. They can also perceive certain shapes and usually have enough myelin for an organized grasp and suckling reflex.

I. Formation of the Central Nevous System

The first evidence of the differentiation of the nervous system appears at the end of the second week of gestation. At that time a "heaping up" of ectodermal cells can be noted at the caudal end of the embryonic disc. This accumulation of cells is known as the primitive streak *(Fig. 4-1A)*. As development continues, surface ectodermal cells involute through the primitive streak and migrate laterally, forming the intraembryonic mesoderm, interposed between the surface ectoderm and subjacent endoderm. As the intraembryonic mesoderm forms, another thickening of ectodermal cells called Hensen's node occurs at the cephalic end of the primitive streak (Fig. 4-1A). Cells migrate inward through this zone and extend between the surface ectoderm and subjacent endoderm, forming the notochord plate. The notochord is the primary "inductor" for the thickening of the ectoderm overlying and on either side of the notochord to form the neural plate, that is the origin of the central nervous system. The many factors in the notochord that underlie the formation of the neural tube and the transformation of embryonic cells into adult cells are currently being identified. The notochord ultimately degenerates as the body of the vertebrae appears and it persists as the nucleus pulposi of the intervertebral disc.

On either side of the neural plate is a thin strip of ectoderm known as the neural crest. As development proceeds, the neural plate sinks inward to form the neural groove *(Figs. 4-1B and 4-2)*. At the beginning of the forth week this midline groove continues to deepen until the dorsal lips of the groove meet and fuse, forming the neural tube, neurulation *(Figures 4C and 4-2)*. The adjacent neural crest separates from the overlying ectoderm and comes to lie lateral to the neural tube (Fig. 4-2). The central nervous system is derived from the neural tube; the neural crest yields many parts of the peripheral nervous system. The remaining ectoderm becomes the epidermis of the embryo (somatic ectoderm).

Before the formation of the neural tube the paraxial mesoderm begins to form the somites. Fusion of the dorsal margins of the neural groove, forming the neural tube, begins on

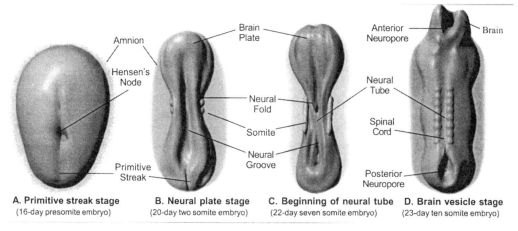

Figure 4-1. Dorsal view of human embryo. A, Primitive streak stage (16-day presomite embryo). B, Neural plate stage (20-day two-somite embryo). C, Beginning of neural tubes (22-day seven-somite embryo). D, Brain vesicles stage (23-day ten-somite embryo).

about the twenty-second day in the region of somites 4 to 6 *(Fig. 4-1C)*. Fusion continues in both cranial and caudal directions, and by the twenty-fifth day, only the cranial and caudal ends of the neural tube remain open (called the anterior and posterior neuropores respectively). The lumen in the neural tube will become the ventricular system of the central nervous system.

After delineation of the neural and somatic ectoderm, the neural crest cells appear as an almost continuous column of cells along the dorsal surface, crest, of the neural tube from the mesencephalic level through all spinal cord levels *(Fig. 4-2)*. The neural crest cells are the primordium of the sensory ganglia, the cranial nerves and the spinal ganglia, the neurilemmal cells, satellite cells, the autonomic ganglia, and probably the pia-arachnoid membrane. Shortly after the closure of the neural tube, processes from the neural crest cells enter the spinal cord or brain stem and form the sensory roots of the spinal and cranial nerves.

Histogenesis

The walls of the neural tube throughout the nervous system consist of germinal epithelial cells that form a pseudostratified epithelium. After closure of the neural tube, another cell type is found external to the germinal cell layer. This new cell is the neuroblast (or immature neuron), these cells originate from the mitotic divisions of the germinal cell and form a new layer, the mantle or intermediate layer, and external to the mantle layer is seen a cell-free zone, the marginal layer. Subjacent to the germinal layer the subventricular region develops.

The glia are instrumental in the final formation of the central nervous system. They migrate out from the subventricular layer prior to the neuroblasts and form processes that reach the submeningeal surface and remain connected to the ventricular surface. The neurons migrate out

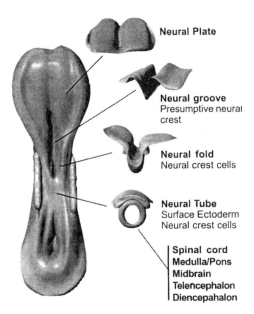

Figure 4-2. Coronal section showing differentiation of neural ectoderm in the neural plate stage, the neural groove stage, the neural fold stage, and the neural tube stage.

into the region formed by the glia.

Initially the neuroblasts are connected to the lumen by an elongate process, a transient dendrite. As they migrate into the subjacent mantle layer they lose this process and become apolar neuroblasts. Shortly thereafter two new processes appear; one becomes the axon, the other the dendrite. The axonal process elongates and more dendritic branches appear, forming a multipolar neuroblast.

Some neurons form only a single process; e.g., these are the sensory ganglion of the spinal cord and cranial nerves, and the mesencephalic nucleus of cranial nerve V. Some neurons remain bipolar, e.g., the olfactory receptor cells, rods, and cones of the retina, the bipolar cells in the retina, and the receptor cells in the vestibular and cochlear ganglion.

It should be noted that the germinal cells also form the supporting cells in the central system *(Fig 4-3)*. Throughout the central nervous system neurons migrate into the intermediate (mantle) zone to become the definitive structures of the spinal cord, much of the brain stem, and the basal nuclei. Sidman and Miale (1959) using tritiated thymidine identified this migration. In these regions the cells in the mantle layer become the gray matter, while their axons form the marginal zone or white matter. By contrast, the cerebral cortex and cerebellar cortex are formed by the migration of neuroblasts through

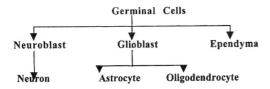

Figure 4-3 Evolution of Germinal Cells.

the mantle zone into the marginal zone.

Repair of Damaged Nervous System.

Since immature cells from fetuses are not as easily rejected, it has been proposed to recover neuroblasts from the brains of aborted human fetuses. These fetal stem cells from human embryos would be implanted in adults replacing damaged neurons in patients with Huntington's chorea and Parkinson's disease. This line of transplant research holds great promise not only for the nervous system but also for pancreas, adrenal, pituitary.

The implantation of fetal stem cells seems especially promising, but there is much public controversy over the use of fetal or aborted material. Recently adult stem cells have been harvested from surgical specimens and at postmortem examinations; this approach also shows much promise.

Principles of Differentiation within the CNS:

The following are certain rules that can be applied to the differentiation within the CNS (after M. Jacobson 1978):

1. .Spatial and temporal gradients of proliferation exist in the neuroepithelial germinal zone.

2. Neighboring regions have separate origins and are different cytoarchitectonically.

3. In an orderly sequence large neurons are produced first, then intermediate-sized neurons, and finally small neurons.

4. Glial cells tend to originate before neurons in any particular region of the brain.

5. Phylogenetically older parts of the brain are predisposed to arise earlier in ontogenesis.

Growth Cone Guidance.

Ramon y Cajal at the turn of the last century identified the growth cones that form connections in the developing nervous system. Since that time neurobiologists have been searching for the mechanisms that permit the formation of these synapses. Several mechanisms have been proposed including cell adhesion, repulsion and chemoattraction. Recently a diffusible factor that promotes outgrowth of axons in vitro has been found in the floor plate of the embryonic spinal cord (Serfanin et al, 1994; Kennedy, 1994)). Two membrane-associated proteins were purified from the embryonic chick brain, netrin-1 and netrin-2, and each protein promotes growth cone outgrowth. Cloning cDNA encoding the netrins showed that they were homologous to UNC-6 a laminin-related protein (found in the extracellular matrix) required for migration of cells and axons in the nematode Coenorhabditis elegans.

This homology suggests that growth cones in the vertebrate and nematode respond to a sim-

TABLE 4-2 LISTS THE FATE OF THE LAYERS IN THE NEURAL TUBE:

Marginal zone	-at the surface covered by the presumptive pia contains pathways in the adult spinal cord and neurons in the cerebrum and cerebellum
Intermediate or mantle zone	-becomes gray matter of the adult spinal cord, brain stem and diencephalon, and contains white matter of the cerebrum and cerebellum
Subventricular zone	-forms the macroglia (astrocytes, Oligodendrocytes).
Ventricular lining zone	-early on has germinal cells, that form neuroblasts; -in mature brain becomes ependymal lining of ventricles.

ilar molecule. Further studies with UNC-6 and the netrins have shown that these molecules may be guidance and targeting signals rather than just attractive versus repulsive molecules (Goodman 1994). We must expect that further investigations of these molecules will lead to a better understanding of how the nervous system actually forms and connects. An understanding of how the nervous system forms connections which may ultimately lead to better methods to repair damage in the brain.

Programmed Cell Death.

During the development of both the central and peripheral nervous system, the number of immature neurons formed is actually double that that will ultimately survive (Cowan et al '83). This apparent excess in formation of neurons is a mechanism whereby the size of the neuronal pool is ultimately matched to the amount of tissue actually innervated (Kane 1995, Lee '93, Majno & Joris 1995).

Neuronal Death

1. *Necrosis*. Neuronal death is necrosis and it is unplanned cell death characterized by cellular edema and destruction of the neuronal organelles and membranes. Necrosis is commonly accompanied by the inflammatory response. This occurs usually following an injury (e.g. stroke, trauma, and tumor).

2. *Programmed cell death*. This type of degeneration is normal and planned, called apoptosis (a Greek term referring to the seasonal falling off of leaves) begins with condensation and degeneration of the chromatin, blebs of the plasma membrane, shrunken cytoplasm with the organelles intact (Oppenheim '91, Schwartzman & Cidlowski '93, Taylor '93). Caspases (cysteine-dependent, aspartate-specific proteases) are endoproteases that are integral to the disassembly of the cell (Cohen 1997; Cortese 2001). These enzymes exist normally in an inactive form, but once they are activated they set off a proteolytic cascade that spreads and destroys the cell. In both of these events glia and macrophages phagocytose the nerve cells, but in apoptosis the nerve cells die quietly without any inflammatory response or major activation of the phagocytes. With necrosis, there is an inflammatory response with phagocytes activated, while in apoptosis there is no injury or interruption of the blood supply and yet the cell begins to degenerates with the cell bodies appearing pyknotic.

Researchers have become more aware of this "natural" type of cell death and have also called it programmed cell death, biological/physiological cell death, or even cell suicide. Many areas in the vertebrates and invertebrates nervous system are now known to have apoptosis. It is also well established that programmed cell death occurs in most other tissues and organs and is an important means to shape the final form of the body. There have also been studies with hormones and other compounds that have decreased the number of immature neurons dying by apoptosis.

With the recognition that apoptosis is a normally occurring phenomenon, neurobiologists using molecular probes have identified genes that regulate this process (Kane 1995). The finding of genes that control these phenomena has also permitted researches to questions whether diseases such as Huntington's chorea, Parkinson's disease and even Alzheimer's are a consequence of this same gene pool being activated later in life to produce the resultant neuronal death. This promising research may well lead to gene therapy which permits treatment of these presumed programmed cell death caused disease by turning off the offending genes.

Development of Blood Vessels in the Brain.

The blood supply to the developing brain begins by the invasion of solid cords of mesodermal endothelial cells, called Wilson's Buds, which over a period of several days develop a lumen and a basement membrane. These arteries develop in the pharyngeal arches during the fourth week and are derived from the third pair of aortic arches and dorsal aortae and by 6 weeks there are well-established internal and external carotid vessels. These buds of growing endothelial cells are especially prominent as the intermediate and marginal layers evolve. The blood supply evolves in direct relationship to the maturation of the CNS and as a consequence of the many factors released into the developing parenchyma of the CNS. As the astrocytes form within the CNS, some of their end-feet surround the basement membrane of the brain capillaries and begin to form a blood-brain barrier. The actual tight junctions between the endothelial cells that delineate the definitive blood-brain barrier are not complete until just before birth.

Ventricular System

When the neural folds fuse, a neural tube is formed. At the rostral end of the nervous system two lateral dilations appear: the primordia of the cerebral vesicles. These regions will grow rapidly, extending anteriorly, inferiorly, and posteriorly as they form the ventricles in the cerebral hemispheres.

Lateral Ventricles. The cavity in each cerebral vesicle is called a lateral ventricle. As the cerebral hemispheres grow by continued outward migration of nerve cells, the ventricles take up less and less space in the brain and are modified into a C-shaped structure with a spur extending occipitally. Each lateral ventricle in its final form is divided into an anterior horn, body, inferior horn, and posterior horn.

Foramen of Monro. Early in development, the lumen in the paired telencephalic vesicles is continuous with that of the neural tube. As the telencephalon differentiates into cerebral cortex and basal nuclei, the ventricles take up relatively less space. A narrow channel, called the interventricular foramen of Monro, persists between the cerebral ventricles and the third ventricle (Fig 1-21).

Third Ventricle. The centrally placed cavity in the diencephalon -- the third ventricle -- narrows as the parenchyma of the diencephalon forms.

Cerebral Aqueduct. In the midbrain the tectum and tegmentum expand greatly and constrict the lumen so that only the narrow cerebral aqueduct remains.

Fourth Ventricle. In the metencephalon (cerebellum and pons) and myelencephalon (medulla), the tegmentum undergoes a series of flexures and expands laterally, and is overgrown by the cerebellum. As a result, the cavity in the myelencephalon and metencephalon called the fourth ventricle assumes a rhomboid-shaped. The fourth ventricle is continuous with the narrow cerebral aqueduct above and the constricted spinal canal below.

Spinal Canal. In the spinal cord, the expanding gray and white matter nearly obliterates the ventricular cavity so that only the small spinal canal remains.

Formation of Peripheral Nervous System

Peripheral structures are innervated from nuclei in the brain stem (by the cranial nerves) and from nuclei in the spinal cord (by the spinal nerves). The twelve cranial-nerve nuclei innervate skin, glands, viscera, striated muscles, and special sense organs (eye and ear) in the head and neck. The spinal nerves innervate skin, glands, and striated muscles in the upper and lower extremities and in the thorax, abdomen, and pelvis.

Spinal Cord Differentiation

The origin of the spinal cord can be traced to the twentieth day. The neural plate then has two distinct regions: caudally, a single elongate cylindrical region, which will become the spinal cord; and, cranially, a shorter broader region, which will become the brain (Fig. 4-1B).

Until the third month the spinal cord fills the vertebral column. From then on the cartilage and bone grow faster than the central nervous system, and by birth the coccygeal end of the spinal cord lies at the level of the third lumbar vertebra. In an adult it is found at the level of the first lumbar vertebra.

At the end of the somite period (days 30 to 35), mitotic divisions in the spinal cord region produce thickened walls and a thin roof and floor; with almost complete obliteration of the

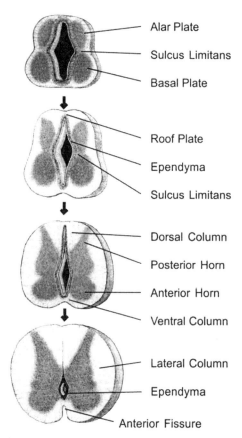

Figure 4-4. Schematic representation of the development of the spinal cord. (After Hamilton, W.J. Boyd, J.D., and Mossman, H.W.: Human Embryology. Cambridge, England, Heffer and Sons, 1964).

TABLE 4-3 STRUCTURE INNERVATED BY CRANIAL NERVES

STRUCTURE INNERVATED BY CRANIAL NERVES	CRANIAL NERVE
1. Muscles of somite origin	XII, VI, IV, and most of III.
2. Muscles and skin of pharyngeal arches	V, VII, IX, X and XI.
3. Preganglionic parasympathetic	IX, X, VII, V, and a small part of III.
4. Special Sensory cranial nerves	I, II, VII, VIII, IX, and X.

central canal *(Fig. 4-4)*. The ventral portion of the lateral walls develops and expands earlier than the dorsal region. The ventral portion is called the basal plate and the dorsal portion is called the alar plate. Processes from cells in the mantle layer enter the external-most region of the cord, which is called the marginal zone.

Gray matter. Four columns appear in each half of the gray matter of the spinal cord: two motors and two sensory (Table 4-3). The somatic afferent and efferent column innervates skin and muscle related to the somites, while the visceral afferent and visceral efferent columns innervate the mucosa, glands and smooth muscles in the visceral structures. The somatic efferent cells are the motor neurons in the ventral horn of the spinal cord; the somatic afferent cells are the neurons in the spinal ganglia. The visceral efferent cells consist of sympathetic and parasympathetic neurons

Axons of developing motor neurons in the basal plate enter the adjoining somites, while sensory roots derived from the adjacent neural crest zone enter the dorsal or alar plate. In the spinal cord (Fig. 4-3) the plate will become the dorsal or sensory associative horn, and the basal plate will become the ventral or motor horn. In the medulla *(Fig. 4-5)*, pons (Fig. 3-6), and midbrain (Fig. 4-7) the alar and basal plates are restricted to the floor of the developing ventricular system. The remaining bulk of these zones consist of nuclei and tracts that are related to the head, neck, and special senses.

In the spinal cord the marginal zone also increases in width as axons that form the ascending and descending tract enter (Fig. 4-4). The formation of the dorsal and ventral horns delimits the dorsal, ventral, and lateral columns or funiculi in the marginal zone. An intermediate horn develops dorsal and lateral to the anterior horn in the thoracic and lumbar levels of the spinal cord. Cells in this horn are the preganglionic sympathetic neurons. The postganglionic visceral neurons are in the paravertebral sympathetic chain or in ganglia associated with the visceral organs. The preganglionic parasympathetic innervation originates from the motor neurons in cranial nerves III, VII, IX and X and the sacral spinal cord and innervates ganglia that connect the appropriate organ or gland. Again, these ganglia differentiate very close to the appropriate structure. In the dorsal or sensory horn of the spinal cord there are sensory neurons that pro-

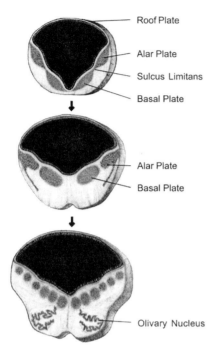

Figure 4-5. Schematic representation of the development of the medulla. (After Hamilton, W.J. Boyd, J.D., and Mossman, H.W.: Human Embryology. Cambridge, England, Heffer & Sons, 1964).

vide connections throughout the spinal cord or brain stem.

The segmental arrangement of the central nervous system stops at the spinomedullary junction, which marks the site of transition from the spinal cord to the brain stem. The brain stem and cerebrum are suprasegmental with no evidence of segmentation present. These regions of the central nervous system are specialized to control structures in the head and neck as well as exerting control over the entire body.

During development the spinal cord and the somites lie close together, so the distance the axon must travel to reach its appropriate muscle is very short. Undoubtedly, chemical factors help direct the axon to the correct muscle. The sensory fibers also have only a short distance to run from the ganglia into skin or muscle and their axon into the central nervous system; again, probably by chemical mediation they establish the connection with the correct dorsal horn cells.

Brain Differentiation

The brain begins development as a broad zone at the upper end of the developing central nervous system with three distinct vesicles- the prosencephalon, the mesencephalon, and the rhombencephalon.

Rhombencephalon (Hind Brain)

The rhombencephalon differentiates into the metencephalon *(Fig.4-6)* and myelencephalon (fig 4-5). This region is further divided by their relationship to the ventricular system. That portion forming the floor of the ventricle is called the tegmentum, while that forming the roof of the ventricle is called the tectum. In the spinal cord the basal plate was ventral and the alar plate dorsal. In the medulla, pons, and midbrain both basal and alar plates are restricted to that portion of the tegmentum that forms the ventricular floor. The alar or sensory region is lateral, while the motor region is medial. The sulcus limitans, on the ventricular floor, separates the medial motor region from the lateral sensory region.

In the spinal cord the neurons are in the gray

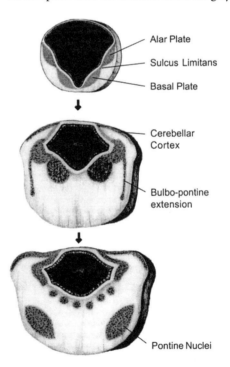

Figure 4-6. Schematic representation of the development of the pons. (After Hamilton, W.J., Boyd, J.D., and Mossman, H.W.: Human Embryology. Cambridge, England, Heffer and Sons, 1964).

matter, in the center of the spinal cord while the fiber tracts are on the outside. In the brain stem motor nuclei migrate to a medial position and sensory nuclei to lateral. The fiber tracts are on the external and medial surface; some tracts actually in the center of the brain stem.

The dorsal portion of the alar plate in the metencephalon (pons) forms the primordium of the cerebellum. This region, the rhombic lip, thickens and the cerebellum continues to expand laterally and then medially so that pair of lateral regions, the cerebellar hemispheres, and a medial zone, the vermis, is now formed. The ependymal lining and pia form the roof of the lower medulla, the tela choroidea.

On the anterior surface of the medulla, pons, and midbrain is the basilar region, which include the more recently phylogenetically added descending cortical fibers. In the medulla these fibers consist of the pyramids, and, in the pons, consists of the middle cerebellar peduncle and pyramids, and in the midbrain they consist of the cerebral peduncles. Other tracts and nuclei form the bulk of the tegmentum of the pons, medulla, and midbrain.

The ventricular lumen in pontine and medullary levels is called the fourth ventricle. The neural tube in these zones changes from a tube to a rhomboid-shaped fossa as a result of the broadening of the tegmentum and the overgrowth of the cerebellum.

The cerebellum and deep cerebellar nuclei are formed by the migration of cells (Fujita et. al., 1966). In the cerebellum there are two distinct germinal centers:

1) a region in the roof of the IV ventricle gives origin to the germinal Purkinje and Golgi II cells and glia which migrate into the mantle layer of the cerebellum, and

2) The external granule zone that forms beneath the pia of the cerebellum, the external granule zone. This region gives origin to the granule cells, stellate, and basket cells and some of the glia.

The Purkinje cell layer is the first to form. A little later cells migrate inward from the more external layers to form the granular layer. In the adult the molecular layer has only a few cells, but the Purkinje layer and granule cell layer have many neurons. It appears that as the primitive neurons are migrating into their definitive place, they form many of the connections necessary for their specialized functions.

The factors which control the migration of cells into the cerebellum and cerebrum and which provide the connections that are necessary for the proper function of these structures are only now being identified. (These factors produce some of the following changes- the glia and neurons migrate and elongate at certain critical times and is supplied with an adequate blood supply that permits all this to happen).

Mesencephalon

The mesencephalon *(fig 4-7)* begins as a wide tube with a conspicuous floor (tegmentum) and a thin roof (tectum). The addition of fibers and nuclei in the mantle zone of the tegmentum, concomitant with the formation of the superior and inferior colliculi in the tectal area, produces a narrow cerebral aqueduct with a large tegmentum and tectum.

The tectum consists of neurons that receive

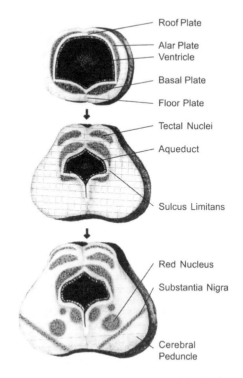

Figure 4-7. Schematic representation of the development of the midbrain. (After Hamilton, W.J. Boyd, J.D., and Mossman, H.W.: Human Embryology. Cambridge, England, Heffer and Sons, 1964).

input from the cranial nerves that relay auditory (inferior colliculus) and visual information (superior colliculus). The cerebral peduncles form the anterior surface or basis of the midbrain, with the tegmentum located between the basis and tectum. The cerebral peduncles consist of fibers descending from the cerebrum to the brain stem and spinal cord.

Prosencephalon

Three primary vesicles of brain regions are seen as the neural tube begins to close; prosencephalon (forebrain), Mesencephalon(midbrain) and rhombencephalon (hindbrain. The prosencephalon *(Fig. 4-9)* begins to differentiate at about 30 days as a dilation of the cranial end of the neural tube (Fig. 4-2C, and D) this dilation evolves into a right and left dilation. Early in development optic vesicle, which will become the eye, appear as a lateral diverticulum of the anterior portion of the prosencephalon. In front of and above the optic stalk a pair of cerebral vesicles form. The cerebral vesicles expand superiorly, anteriorly, and posteriorly.

Diencephalon.

The diencephalon originates from the caudal portion of the prosencephalon (Fig. 4-2D) and represents the thickened lateral wall of the primordium of the third ventricle. The dorsal half of the diencephalon will become the thalamus and epithalamus, while the floor differentiates into the hypothalamus and neurohypophysis. The alar and basal plates are no longer found at diencephalic levels although functionally the basal plate may well be represented as the hypothalamus. Fibers to and from the telencephalon will form the lateral limit of the diencephalon. These fibers are the internal capsule.

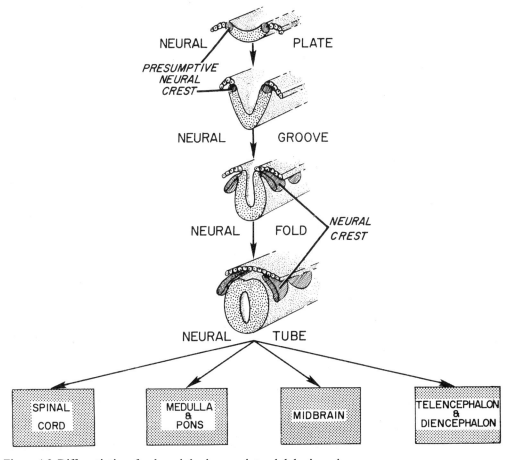

Figure 4-8. Differentiation of embryonic brain areas into adult brain regions.

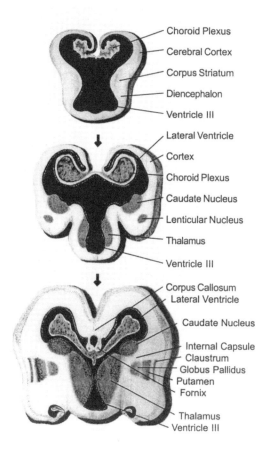

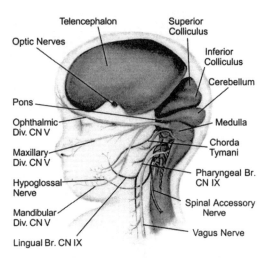

Figure 4-9. Schematic representation of the development of the telencephalon and diencephalon. (After Hamilton, W.J., Boyd, J.D., and Mossman, H.W.: Human Embryology. Cambridge, England, Heffer and Sons, 1964).

Figure 4-10. Lateral surface of 11-week human embryo demonstrating the cranial and upper spinal nerves. (After Patten, B.M.: Human Embryology, New York, McGraw-Hill, 1953).

Cranial Nerves

The cranial nerves form connections to muscles, skin, glands, and blood vessels in the head and neck. A detailed discussion of the individual cranial nerves can be found in Chapter 11. In this section, the cranial nerves Table 4-3 and Figure 4-10 will be categorized on the basis of the embryonic derived structures they innervate:

1. Cranial Nerve Innervation for Muscles of Somite Origin

Nerve III (oculomotor) is a purely motor nerve that innervates the inferior oblique, medial, superior, inferior recti muscles and levator palpebrae superior muscles. The extrinsic eye muscles develop from mesenchyme in the orbital region.

Nerve IV (trochlear) is a purely motor nerve and innervates the superior oblique muscle of the eye. This is the only cranial with rootlets that leave the central nervous system from the posterior surface of the brain.

Nerve VI (abducens) is a purely motor nerve and innervates the lateral rectus muscle of the eye.

Nerve XII (hypoglossal) is a purely motor nerve and innervates the extrinsic and intrinsic muscles in the tongue. Occipital myotomes migrate ventrally and form the tongue musculature that is innervated by nerve XII. The epithelium and general connective tissue of the tongue arises by the fusion of pharyngeal endoderm and branchial mesoderm that are innervated by nerves VII, IX and X. -

2. Cranial Nerves Innervating Muscles (Skeletal) and Skin in the Pharyngeal Arches.

Cranial nerve innervation of the pharyngeal arches is shown in Table 4-4. After somites begin to form, five ectodermal grooves appear lateral to the embryonic pharynx and caudal to the somatoderm, or primitive mouth. The grooves are separated from one another by elevations, which gradually elongate, fuse with the opposite side, and extend laterally and anteriorly around the pharynx, forming arches. In lower vertebrates these arches form the gills, while in vertebrates without gills they are called pharyngeal arches. Many of the muscles and bones in the face and neck originate from these arches.

Mammals have a total of six arches:

1. First and second arches named respectively the mandibular and hyoid and the lower four unnamed (the sixth arch is rudimentary).

a. The trigeminal nerve innervates the first arch - muscles of mastication form here;

b. The facial nerve innervates the second arch - the muscles of facial expression form here;

c. The glossopharyngeal nerve innervates the third arch - the stylopharyngeus forms in the third and,

c. The vagus nerve innervates the fourth, fifth and sixth arches- the laryngeal forms here.

Nerve V (trigeminal) is a mixed nerve (motor and sensory) that innervates the first or mandibular arch and also contains a portion (ophthalmic) that distributes to the skin up to the vertex of the skull. The motor division of this nerve innervates the muscles of mastication derived from the first arch. The sensory division innervates the skin on the face and forehead.

Nerve VII (facial) is a mixed nerve that distributes to the second pharyngeal or hyoid arch. The motor division of this nerve innervates the muscles of facial expression. The sensory division innervates the taste buds on the anterior two-thirds of the tongue.

Nerve IX (glossopharyngeal) is a mixed nerve that innervates structures associated with the third arch. The motor division of this nerve innervates the stylopharyngeus muscles.

Nerve X (vagus) is a mixed nerve that innervates structures associated with the fourth and sixth arches (the fifth arch regresses). As well as providing afferent and efferent components to the heart, lungs, and much of the gastrointestinal system.

Nerve XI (accessory) is purely motor and innervates the sternomastoid and trapezius muscles, which originate from somatic and branchial mesenchyme in the cervical region.

3. Preganglionic Parasympathetic Innervation to Smooth Muscle.

Nerve III innervates the ciliary ganglion; postganglionic fibers then pass to the sphincter of the iris, which constricts the pupil in bright light.

Nerve VII provides innervation to the submaxillary and lacrimal ganglia associated with the following glands: submaxillary, sublingual, nasal, and lacrimal.

Nerve IX provides innervation to the parotid gland via the otic ganglion.

Nerve X provides innervation to the heart, lungs, and gastrointestinal system up to the transverse colon and the axons from the vagus synapse upon ganglia associated with these organs.

4. Cranial Nerves Associated with the Special Senses.

Nerve I (olfactory) cell bodies are located in the nasal placode in the upper fifth of each nasal cavity. The neuroblasts in the olfactory epithelium differentiate into olfactory epithelial cells. The axons of these cells grow toward the cerebral hemisphere, pierce the roof of the nasal cavity and enter the region of the hemisphere that will become the olfactory bulb. The primary fibers synapse in the olfactory bulb. The secondary cell bodies are located in the olfactory bulb and form the olfactory tracts.

Nerve II (optic) originates from tertiary ganglion neurons in the retina. These axons are really a tract in the central nervous system and do not constitute a peripheral nerve. The axons synapse in the lateral geniculate nucleus of the thalamus. (Students may also want to review the details of the formation of the eye from the optic vesicle and lens vesicles in a standard embryology test.

Nerve VIII (vestibulocochlear) primary cell bodies are located in the cochlea (auditory division) and vestibule (vestibular division) of the inner ear. (Students may also want to review the details of the formation of the inner ear from the otic placode in a standard embryology textbook).

Nerves VII, IX, and X provide innervation to the taste buds on the tongue and pharynx.

Telencephalon

The Telencephalon, part of the forebrain, consists of two regions a cortical zone that becomes the cerebral hemispheres and a deeply placed nuclear region the corpus striatum that becomes the caudate, putamen, and globus pallidus.

The telencephalon begins at 10 weeks as paired small smooth telencephalic vesicles. By birth a large, convoluted cerebral cortex is present. The first prominent features seen in the hemisphere are the formation of the primary sulci or depressions.

Primary Sulci.

The first of these sulci, the primary sulci, appear at about 19 weeks; they are initially the lateral or Sylvian sulcus, on the lateral surface *(Fig. 4-11)*, and the parieto-occipital sulcus, on the medial surface *(Fig. 4-12)*. By the 24th week, the Sylvian sulcus is more pronounced on the lateral surface and the calcarine sulcus is prominent on the medial surface becomes prominent (Fig. 4-11). Shortly thereafter the other primary sulci—the central, callosal, and hippocampal—appear.

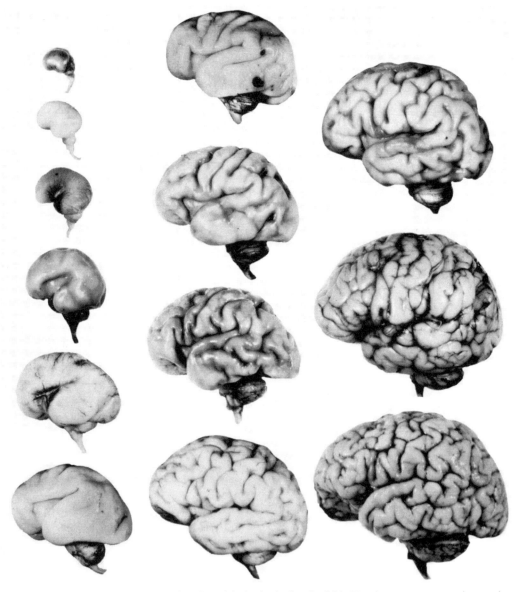

Figure 4-11. Development of the lateral surface of the brain during fetal life. Numbers represent gestation age in weeks. (From Larroche, J. Cl: The Development of the Central Nervous System During Intrauterine Life. In Falkner, F. (ed.): Human Development, Philadelphia, W.B. Saunders, 1966, p. 258).

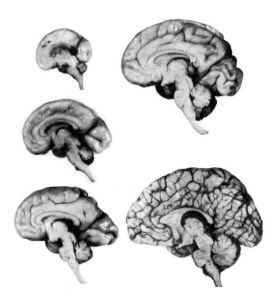

Figure 4-12. Development of the medial surface of the brain during fetal life. Numbers represent gestation age in weeks (From Larroche, J. Cl.: The Development of the Central Nervous System During Intrauterine Life. In Falkner, F. (ed.): Human Development. Philadelphia, W.B. Saunders, 1966, p. 259).

After about 30 weeks the primary sulci are formed, and there is a rapid increase in the secondary sulci and gyri. By term, formation of the gyri and sulci is complete. After birth the gyri increase in bulk as the neurons continue to differentiate.

The telencephalic vesicle rapidly changes shape as the frontal lobe forms and pushes anteriorly, the temporal lobe grows inferiorly, and the occipital and parietal regions expand. The olfactory bulb forms at the base of the frontal lobe by an evagination of cells that quickly obliterates any trace of the ventricle in the olfactory bulb.

The sulci begin as rather shallow depressions. However, as the neurons continue to migrate in from the ventricular surface, more and more cells are added and the gyri become larger and the sulci become deeper. In the term and adult brain more than 70% of the cortical surface is hidden from view, deep in the banks of the sulci. Therefore, when we speak of the total surface of the cerebral hemispheres we must include not only the superficial cortex but also the hidden sulcal cortex.

The brain at birth weighs 300 to 400g; by 1 year it weighs 1000 grams. The brain continues to grow and by late childhood reaches its final weight of approximately 1500 g.

Development of Cerebral Cortex

The development of certain areas is of special importance in understanding the final form of the brain. As the lateral sulcus forms, the cortex above and below the lateral fissure overgrows the cortex deep in the lateral/Sylvian fissure. This cortex, the insula, soon disappears from view and is covered over by this overgrowth of frontal, parietal, and temporal cortex.

Formation of the cerebral cortex begins around the seventh week and is completed in the first year of life. Developing from the inside to out, the first layer to form is layer VI and the last to form is layer II (tritiated thymidine studies- Angevine and Sidman, 1961; Berry and Rogers, 1965). The pyramidal cells in the cerebral cortex appear to mature prior to the stellate cells (Altman, 1966). Radial glial fibers form connections from the luminal surface to the pial surface that helps the primitive neurons migrating later into the more superficial layers to traverse the increased cell densities in the deeper layers. The formation of the cerebrum from inside to outside permits much of the circuitry to form and also permits the columnar arrangement in the cortex to evolve.

The development of the six-layered cerebral cortex starts with the migration of neuroblasts into the deepest layer (VI-IV) first, then the cells into the external layers (I-III) last- "inside-out". Tritiated thymidine studies have been used on rodents and primates for labeling nerve cells in the germinal layer of the ventricle on their "birth days" and following them into the cerebrum and identifying their final destination (Sidman and Feder 1959; Sidman and Miale, 1959; Rakic, 1972).

Commissures. Fiber bundles, called commissures *(Fig. 4-13 A, B, C, and D)* interconnect the two cerebral hemispheres. The anterior commissure interconnects the olfactory bulbs and anterior parts of the temporal lobe. The huge corpus callosum interconnects the remaining portions of the hemisphere. The hippocampal commissure interconnects the hippocampus and adjacent portions of the hippocampal gyrus. These commissures form in the lamina terminalis

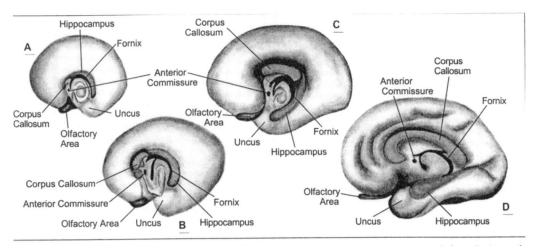

Figure 4-13. Development of the fornix and commissures of the cerebral hemispheres. A, 3-month fetus. B, 4-month fetus. C, Fetus at beginning of 5th month. D, Fetus at end of 7th month. (After Keibel, F., and Mall, F.P.: Manual of Human Embryology, Philadelphia, Lippincott, 1910-12).

that marks the site of the closure of the anterior neuropore. The ventral surface of the hemispheres at this place fuse, and the fibers can then cross and interconnect the two hemispheres.

Initially, the anterior commissure, corpus callosum, and fornix (which connects the hippocampus and hypothalamus) are close to one another *(Fig. 4-13A)*, but as the brain grows, the corpus callosum expands anteriorly, inferiorly, superiorly, and caudally with only the rostral tip remaining attached to the lamina terminalis.[1]

PRENATAL DEVELOPMENT OF CEREBRAL CORTEX:

The developing cerebral hemispheres initially show the three regions common to the neural tube, ventricular, intermediate and marginal zones. Then the subventricular zone appears. Cells of the intermediate zone migrate into the marginal zone and form the cortical layers. So that the gray matter is on the periphery and the while matter, medullary center, is located centrally.

The initial progenitor neurons first appear in the matrix/germinal area a proliferative cell layer around the ventricle at approximately 33 - 42 days of development and form a structure known as the preplate (the primordial plexiform layer). Later formed neurons divide the preplate into 2 layers: the marginal zone (future layer 1) and the subplate (the future neurons of layer 6 and the white matter). Between the marginal zone and the subplate a transient layer the cortical plate is formed beginning approximately at embryonic day 52.

Neuronal Migration. The migration of neurons is facilitated by the presence of special primitive radial glial cells that extend between the ventricular surface and the pial surface. These glial cells essentially serve as guide wires for the migrating neurons. This radial or vertical orientation provides the basis for the columnar organization that is a characteristic of cerebral cortex. The initial migrating neurons travel through the entire marginal layer and form a temporary organizing layer in what will eventually become the relatively acellular molecular layer. This organizing layer apparently secretes an extracellular protein: Reelin. (Reelin is named after the mutant mouse –reeler–which has ataxic reeling movements due to the absence of this protein and in which there is total disorganization of the usual laminar pattern). When the successive waves of migrating neurons encounter the organizing layer, the Reelin provides a signal to detach, from the glial guide wires. The initial cells to detach form the neurons of layer 6. The apical dendrites of these pyramidal cells in layers 6 and subsequently layer 5 will extend to the pial surface.

---

[1]*The hippocampus anteriorly is disrupted by the formation of the corpus callosum and a small dorsal hippocampus or indusium griseum remains on the dorsal surface of the corpus callosum.*

Subsequent waves of neurons detach at successively higher levels. In addition to the putative role of the protein Reelin, other factors also influence this process of migration. The radial guided migration is blocked by NMDA–antagonists suggesting a role for the transmitter glutamate. In addition calcium ion channels may also be involved in migration. Although the previous discussion has emphasized the radial movement of neurons it is also evident that some neurons may migrate in a direction perpendicular to the radial glial fibers. Thus there are horizontal connections in addition to the vertical organization. The process of neurogenesis and migration is completed in the human fetus by approximately week 24. The subsequent step in the process of cortical development is a removal of those neurons in the molecular layer that served as the initial organizing layer by the process of apoptosis. The neurons in the ventricular zone, the subventricular zone and the white matter (which have failed to reach the cortical plate forming layers 2-6) also die off by this same process. The radial glial cells lose their long fibers and differentiate into astrocytes.

The mature cerebral cortex is laminated and consists of the following six layers:

Layers 1, 2 and 3- supragranular phylogenetically newest layers and last layers to form

Layer 4 - granular layer,

Layers 5 and 6 - infragranular layers phylogenetically are first layers to form

II. CHANGES IN CORTICAL ARCHITECTURE AS A FUNCTION OF POSTNATAL AGE

During postnatal development the brain increases significantly in weight from an average of 375 to 400 grams at birth to 1000 grams at 1 year, to a maximum of 1350 to 1410 grams at age 15 years (males). This increased growth does not affect all areas of the brain to an equal extent. Thus this growth involves a rapid expansion of the convexities of the frontal, parietal, and temporal lobes. This results in a completion of the covering of the insular lobe by the surrounding operculum and a relative displacement of the parahippocampal gyrus and occipital lobe. In addition, the cortical surface area is in a sense being increased by the deepening of sulci, with increasing complexity of the primary and secondary fissures. This process continues after the second year with the development of tertiary fissures.

Myelin formation is necessary for the functioning of the nervous system. The following table lists the chronological sequence of myelination in the brain Table 4-4. and by implication the beginning of functions in many of these same regions.

At birth most of the intrinsic fiber systems of the spinal cord and brain stem have myelinated. The long corticospinal tracts and corpus callosum begin to myelinate during the first postnatal months. The frontopontine, temperopontine, and thalamocortical projections myelinate during the second postnatal month.

Myelination within the cerebral hemisphere and within the cerebral cortex occurs at different rates in different areas paralleling the changes in psychomotor development. Flechsig demonstrated that the sequence of myelination in cerebral cortex followed a logical pattern (Fig.17-21).

Although myelination of the cerebral cortex begins during the early postnatal months, it is evident that this process is not completed at this time (Fig. 17-22). The initial myelination begins in the vertically oriented fibers of the infragranu-

TABLE 4-4. MYELINATION OF BRAIN REGIONS.

BRAIN REGION	WHEN MYELINATES
Spinal Cord	Last trimester
Brain Stem	Last Trimester
Fronto, parieto, tempero, occipito-pontine fibers	First- 2 month postnatal (PN)
Primordial Cortical Fields: Motor –Sensory Strip	PN 1-2 Month
Intermediate Cortical Fields: Association areas adjacent to primary motor and sensory cortex - Premotor and Superior Parietal Lobule-	PN 2-4 Month
Terminal Cortical fields: Association or integration areas - Prefrontal, Inferior parietal, posterior temporal cortex	PN 4th month

lar layers and then extends into the supragranular layers. The myelination of horizontal plexuses occurs at a later stage. Myelination continues to increase beyond 15 years of age even until age 60 with a gradual increase in density of the myelinated fibers particularly in the horizontal plexuses. It is important to note that even in the adult not all cortical areas will reach the same density of myelinated fibers since the underlying pattern of distribution and density of pyramidal cells and their processes will differ significantly among different cortical areas.

Neuronal Maturation. In addition to changes in the degree of myelination, more basic changes in the relationship between nerve cells are occurring. Thus the neocortex has received its full complement of neurons by the end of the 6th fetal month. During post natal development as shown in silver stains there is a progressive growth of the apical and basilar dendritic arborizations. The studies of Conel [25] have demonstrated that the more primitive allocortex (e.g. hippocampus) attains its final structural form at an earlier stage than the neocortex. The process of maturation however involves more than a progressive growth and elaboration of dendrites and synapses. In addition, a re-modeling of neuronal connections occurs in large part based on neuronal activity. In the early perinatal period, there may be an excessive projection of axons to their targets. Maturation in part consists of the elimination of these aberrant collateral axon branches. In the primate there are estimated to be 3 to 4 times as many axons within the corpus callosum at birth compared to the adult state. In this process the NMDA receptor may have a considerable role in both neocortex and hippocampus, with the same cascade of the events that occur in long-term potentiation also responsible for the remodeling and synaptic plasticity. Abnormalities in development or sensory/environmental deprivation may retard the structural and functional development of these processes.

III. ABNORMAL DEVELOPMENT

The complexity of development of the nervous system means that there is a great chance for things to go wrong. In fact, more malformations occur in the nervous system than in any other organ system. Of the 0.5% of newborn infants with a major malfunction, roughly 60% affect the nervous system.

Many defects in the brain and spinal cord are evident at birth (Table 4-5). Some defects may be limited to only the nervous system, but others may include overlying ectodermal and mesodermal structures (bone, muscle, and connective tissue). Many of the malformations are due to genetic abnormalities. Other categories of defects can be acquired from the maternal environment: infections (syphilis, rubella, AIDS), drugs (including alcohol, cocaine etc.), ionizing radiation, environmental toxicants, metabolic disease and poor nutrition. The most serious defects are fatal, but even the less severe abnormalities may significantly impairment function. The following discussion covers many of the major groups of malformations of the peripheral and central nervous system. More complete lists can be found in pathology textbooks.

Malformations Resulting from Abnormalities in Growth and Migration with incomplete development of the brain.

Heterotopias. Displaced islands of gray matter appear in the ventricular walls or white matter due to incomplete migration of neurons and this

TABLE 4-5. CONGENITAL NEUROLOGIC MALFORMATIONS

CATEGORY	RESULTANT DEFICITS
1. Abnormalities in growth and migration	Heterotopias; anencephaly; holoprosencephaly; lissencephaly; micro-, macro-, or polygyria; porencephaly; schizencephaly; agenesis; fetal alcohol syndrome
2. Chromosomal trisomy and translocation	Down's, Edward's, and Patau's syndromes
3. Defective fusion	Spina bifida, cranial bifida, Arnold-Chiari malformation
4. Abnormalities with excessive growth of ectodermal and mesodermal tissue.	Tuberous sclerosis, neurofibromatosis (von Recklinghausens=Disease), Sturge-Weber syndrome,
5. Abnormalities in ventricular system	Syringomyelia, syringobulbia, hydrocephalus

defect in neuronal migration may be seen in many of the following abnormalities. In this category of malformation, the neurons or glia have failed to migrate, and consequently the brain development is incomplete. In the normal development of the cerebrum the neurons which form the deeper cortical layers arrive first and then cells destined for outer layers appear. In this abnormality the later waves of migration stop. The axons from the first cells to arrive begin to extend throughout the brain and actually form a barrier that interferes with the next wave of migration. The newly arriving neurons, which were destined for the superficial layers, instead now form a disorganized mass of neurons deep to the cortex.

The disruption of the normal pattern of cerebral cortical development results in an imbalance in the equilibrium of excitation-inhibition. Seizures often result. The routine use of MRI for the evaluation of patients with epilepsy, has demonstrated the high frequency of migration disorders particularly in the pediatric and young adult age group. Depending on the severity of the disorder, the neurological examination and intellectual development may be normal. Areas of microgyria, macrogyria or polygyria may be encountered. For some migration disorders, a genetic basis has been established as summarized in Figure 4-18 derived from Walsh [23]. Various seizure types are present, but neurological development, examination and intelligence are normal. (17-20). A recent review by summarizes other genetic disorders with cerebral cortical changes.

Genetically Linked Migration Disorders

1) *Type 1 X-linked dominant lissencephaly.* Most of the neurons never reach their expected destination but instead are arrested in the subcortical white matter and the subventricular proliferative zone. The cortex is smooth, since development never proceeds to the stage of development of sulci and gyri that is necessary to accommodate the large numbers of cortical neurons that would otherwise be present in a normally developed cerebral cortex. In this mutation, the cortex per se is thin with the lamination pattern consisting of a rudimentary molecular layer 1 and a thin layer 5/6. Since males have only one X chromosome this type of lissencephaly occurs in males. Severe psychomotor retardation and seizures are the result of this process. Since females have 2 X chromosomes the effects of a single mutant X chromosome, are less severe. The double cortex (subcortical band heterotopia, or laminar heterotopia or diffuse cortical dysplasia) syndrome is the result . There is only a partial migration defect. Some neurons arrest in the subcortical white matter but sufficient neurons reach the intended targets so that a normal 6-layered laminar pattern is present. Cognitive function fully develops but seizures are present.

2) *Miller–Dieker syndrome.* This different genetic type of lissencephaly occurs in the autosomal recessive with linkage to locus 17p13.3.Infantile spasms and mixed seizures are associated with motor and mental retardation and a characteristic facies.

3) *Bilateral nodular periventricular heterotopia* is another X-linked dominant, [24] a relatively continuous series nodules composed of neurons is present in the ependymal layer and subventricular zone, the former proliferative zone for neurons during development. The cortical lamination pattern is normal. In these cases,

4) *Cortical microdysgenesias* is a subtle form of neuronal migration disorder. It has been described in many patients with idiopathic epilepsy: absence, juvenile myoclonic epilepsy and primary generalized tonic-clonic. This disorder reflects events at a late stage of prenatal cortical development. Neurological examination and psychomotor development are normal, although subtle changes in cognitive function may be present.

In the following malformations, neurons or glia have failed to migrate, and consequently the brain's development is incomplete. In the normal development of the cerebrum the neurons that form the deeper cortical layers arrive first and then cells destined for outer layers appear. In this abnormality the later waves of migration stop. The axons from the first cells to arrive begin to extend throughout the brain and actually form a barrier that interferes with the next wave of migration. The newly arriving neurons, which were destined for the superficial layers, instead now form a disorganized mass of neurons deep to the cortex. The following case is an example of abnormal migration.

Case 4-1: Bilateral subcortical band heterotopias: double cortex SEE Details on CD

This 24 year old right handed mother of 2 children and home health aide was admitted to the neurology service after a recurrence of seizures with a flurry of 4 seizures on the day of admission. Seizures had begun at age 11 and would occur every 2 months. She described the seizures as beginning with an involuntary driving of the head and eyes to the right "as though she was going to look over her right shoulder." At this point she might experience a sensation of fear. She was aware that involuntary movements of both legs would then occur as well as clonic or tonic movements of the right upper extremity. On some occasions, she would then lose consciousness and a secondarily generalized tonic-clonic seizure would then be witnessed. She had been initially treated with anticonvulsants: phenytoin and phenobarbital, without control of seizures. Carbamazepine was then utilized but produced side effects. She had discontinued all medications 14 months prior to admission.

Family history: negative for neurological or seizure disorders.

Past history: head trauma right frontal with short period of unconsciousness at age 9 years. She had a tubal ligation and bilateral carpal tunnel surgery, 5 years prior to admission.

Neurological examination: mental status, cranial nerves, motor system, reflexes and sensory system were all normal.

Clinical diagnosis: Focal seizures (partial epilepsy) originating left frontal eye field with subsequent spread to left supplementary motor cortex and amygdala (or possibly cingulate gyrus).

Laboratory data:

1. *Electroencephalogram*: normal
2. *MRI*: bilateral triangular shaped foci of heterotopic gray matter were present in the frontal white matter extending from the superolateral aspect of the frontal horns to the gray white junction superiorly (see MRI Fig 4-1.).

Subsequent course: Anticonvulsant medications were re-adjusted. Valproic acid was added to the carbamazepine. Compliance continued to be a problem. She continued to have at least 1 generalized tonic-clonic and possibly several simple partial seizures per year over the next 4 years.

Comment: The description of the initial clinical phenomena and the subsequent evolution suggest initial discharge in the left frontal eye field (area 8) with subsequent spread to the (left) supplementary motor cortex (an area where there is bilateral representation of the lower extremities) .In contrast to the primary motor cortex where simple clonic movements occur on stimulation, complex postures and bilateral leg movements occur on supplementary motor cortex stimulation .The tonic posture or movement of the right arm would also be consistent with such a pattern of spread. The clonic movements of the upper extremity would be consistent with additional spread to the left motor cortex. The sensation of fear could have reflected spread to the limbic system possible by means of the adjacent cingulate gyrus eventually involving the amygdala. The actual pathology in these areas of frontal lobe consisted of heterotopias of gray matter, (termed double cortex or subcortical band heterotopia) a long-standing migration disorder. With the increasing use of the MRI to investigate patients with seizure disorders, there has been increasing recognition of such migration disorders. This is an X chromosome linked disorder. In the female with 2 X chromosomes, this is a relatively benign disorder. In contrast in the male with only 1 X chromosome, the effects are much more severe, most neurons never reach their expected destination but instead are arrested in the subcortical white matter. The result is type 1 X linked lissencephaly. The cortex is smooth (lissencephalic) since development never proceeds to the stage of sulci and gyri. Sulci and gyri develop normally in the process of accommodating the large number of neurons that would under normal circumstances reach the cortical layers. Such males have severe psychomotor retardation as well as seizures.

Anencephaly. Anencephaly is a lethal condition in which skull, cerebral hemisphere, diencephalon, and midbrain can be absent. The notochord does not induce the anterior neuroectoderm and the anterior neuropore does not close. The cerebral cortex and upper brainstem are represented by a tangle of meninges, glia and vessels at the base of the skull. The amount of brain tissue absent can vary from only the cerebrum to all of the diencephalon and telencephalon. Part of

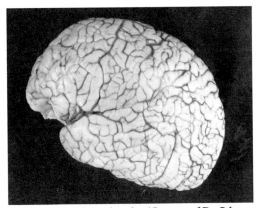

Figure 4-14. Micropolygyria. (Courtesy of Dr. John Hills, New England Medical Center Hospitals).

the brainstem, and remnants of higher structures are usually present. The pituitary gland and other endocrine glands are small or absent. The eyes and ears are well developed, but the optic nerves are usually absent. These children are usually born alive, but do not survive the first few days. This category of almost complete absence of the brain has been used extensively for transplants of the normally formed heart, lung or kidney into infants in need of organ donors.

In this condition the brain has failed to form two distinct cerebral hemispheres. Instead, there is a single partially differentiated hemisphere. The patient has a single ventricle, representing both lateral ventricles and the third ventricle. There are grades of severity of the malformation. In some cases of holoprosencephaly the children had trisomy 13y. In some cases these children have had only one eye, cyclopia, and a proboscis like nose with only a single nostril.

Lissencephaly (Agyri-Pachygyri). The brain has failed to form sulci and gyri and corresponds to a 12-week embryo. There are usually also dysmorphic facial features. In some cases a few gyri do form. Neuronal migration has been arrested at this stage due to a chromosomal defect, ischemia, or infection. The studies of Reiner et al (1993) have identified a specific gene deletion at chromosome 17 region p13.3, the LIS-1 gene.

Micropolygyria (Fig. 4-14). The gyri are more numerous, smaller, and more poorly developed than normal.

Macrogyria. The gyri are broader and less numerous than in the normal brain.

Microencephaly. This is another abnormal defect in which development of the brain is rudimentary and the individual has a low-grade intelligence.

Porencephaly. There are symmetrical cavities in the cortex due to the absence of cortex and white matter in these sites.

Schizencephaly (Fig 4-15). There is an incomplete development of the cerebral hemispheres with the presence of abnormal gyri. The patient is usually retarded.

Agenesis of Corpus Callosum (Fig 4-16). There is a complete or partial absence of the corpus callosum and septum pellucidum (See also Fig 21-1)).

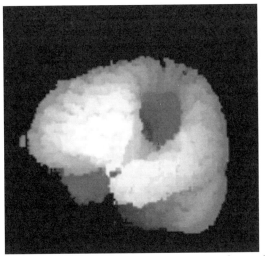
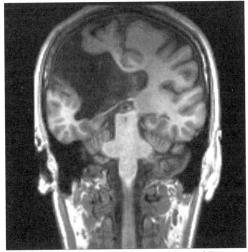

Figure 4-15. Schizencephaly. A. 3D reconstruction and B coronal view. MRI, T2 (Courtesy of Dr. Val Runge, University of Kentucky Medical Center).

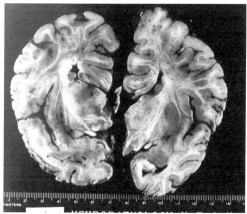

Figure 4-16. Agenesis of corpus callosum. (Courtesy of Dr. John Hills, New England Medical Center Hospitals).

The following case Case 4-2 is an example of Agenesis of the Corpus Callosum.

Case 4-2. Agenesis of corpus callosum: Patient of Dr. Sandra Horowitz)

This 20-year-old female warehouse employee was referred for re-evaluation of generalized convulsive seizures that had developed at age 3 years. She had been seizure free for 5 years. She had mild developmental delays. Full scale Wechsler Adult Intelligence score IQ was 82, with a verbal score of 88 and a performance score of 77(all were within low average range). MRI (Fig.4-2 on CD) demonstrated complete agenesis of the corpus callosum. With associated alterations of gyral and sulcal patterns. Midline sagittal section T 1 weighted.

Holoprosencephaly. Only a single forebrain has formed. This deficit is also commonly accompanied by a deficit in medial facial development *(Fig 4-17)*.

Figure 4-3 on CD. This 10month old white male had severe spastic diplegia (a form of "cerebral palsy") and severe developmental delays. CT scan demonstrated a fusion of the two lateral ventricles and no frontal horns . A large cystic area extended superiorly and posteriorly.

Cerebellar Agenesis. Portions of the cerebellum, deep cerebellar nuclei, and even the pons are either absent or malformed. In some instances portions of the basal ganglia and brain stem and spinal cord may also be malformed or absent.

The cerebellar cortex forms differently than the cerebrum by migration of primitive cells from the rhombic lip of the brainstem. These cells form a layer called the external granular cell layer. From this layer, cells then migrate inward to form the internal granular cell layer.

Environmentally Induced Migration Disorder. Fetal Alcohol Syndrome. Fetal alcohol syndrome (FAS), mothers drinking alcohol during pregnancy cause one of the more common nongenetic causes of mental retardation (one out of 300 to 2000 live births, or up to 2%). There are other environmentally generated disorders but the FAS has the clearest etiology. The current recommendation is for pregnant women to avoid all alcohol, especially during the first trimester, or to cut back to no more than two drinks per day. Pregnant alcoholics are at high risk for premature delivery.

In FAS, pre- and postnatal growth retarda-

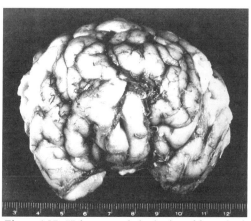
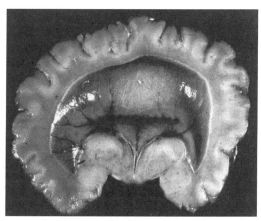

Figure 4-17. Holoprosencephaly. A, view of frontal poles, B, coronal section. Note the absence of any differentiation of the hemispheres. (Courtesy of Dr. John Hills, New England Medical Center Hospitals).

tion occurs along with cardiovascular, limb, and craniofacial abnormalities, such as a short palpebral fissure, thin vermillion border, and smooth philtrum. Children often have impaired fine and gross movements and developmental deficits. Microcephaly, heterotopias, and lissencephaly have been reported (Greenfield, 1992; Diminski and Kalens, 1992). Related to the problem of FAS is the huge rise in "cocaine babies," which have similar deficits that require lifetime care.

Malformations Resulting from Chromosomal Trisomy and Translocation

Through advances in molecular biology, more and more genetic defects are being localized to specific genes. For example, in translocation syndromes like trisomy 21 (Down's syndrome), one chromosomal segment is transferred to a nonhomologous chromosome; this is called a Robertsonian translocation and most commonly affects chromosomes 13, 14, 15, 18, 21, and 22. *Figure 4-19* shows the clinical features of trisomy 21. Other defects, such as trisomy 18 (Edwards' syndrome) and trisomy 13 (Patau's syndrome), have similar neurologic findings.

Most children with Down's syndrome are moderately mentally retarded and of short stature. They have hypoplastic faces with short noses, small ears with prominent antihelices, and prominent epicanthal folds (which perhaps suggested the now outdated term mongolism). Other defects include a protruding lower lip, a fissured and thickened tongue, and a simian crease on the palm. A predisposition to leukemia and congenital heart defects often prove fatal. Children with Down's syndrome constitute about 5% of children in institutions for the mentally retarded.

The brains of these children show little gross abnormality. The frontal and occipital poles are rounder than normal and the superior temporal gyrus is thinner. Conventional stains show the brain to be normal histologically, but the Golgi stain, which specifically impregnates axons and dendrites, shows sparser dendritic arborizations, fewer dendritic spines, and more anomalous spines (Marin-Papilla, 1972; Purpura, 1974; Becker et al., 1986).

Patients who reach middle age may develop a degenerative disease of the nervous system called Alzheimer's disease (see chapter on dementia).

Malformations Resulting from Defective Fusion of Dorsal Structures

Spinal Bifida. In spinal bifida, the most common malformation in this category, the arches and dorsal spines of the vertebrae are absent. Often the bony deficit alone is present. The spinal cord, however, may be malformed either at one level or at many levels. In some instances the spinal cord, nerve root, and meninges have herniated through the midline defect in the skin and bone -- meningomyelocele *(Fig. 4-18)*. In other instances only the meninges have herniated through the midline defect, the sac is called a meningocele.

Cranial Bifida. Cranial bifida is less common the spinal bifida and usually occurs only in the suboccipital region. The cranial bones either fails to fuse or do not form. The abnormality may only be restricted to the underlying portion of the cerebrum or it may be accompanied by a herniation of meninges -- meningocele -- or

Figure 4-18. Meningomyelocele at sacral level of the spinal cord. Arrow refers to sac, meninges, neural tissue, and overlying skin. (Courtesy of Dr. John Hills, New England Medical Center Hospitals).

meninges and brain tissue -- encephalocele. Cranial bifida is commonly associated with hydrocephalus and extensive brain damage.

Arnold-Chiari Malformation. In this condition there is elongation and displacement of the brain stem and a portion of the cerebellum through the foramen magnum into the cervical region of the vertebral canal. Hydrocephalus, spina bifida with meningocele, or meningomyelocele may also be associated with this abnormality, which may be fatal or produce neurological symptoms due to compression of the cervical roots and the overcrowding of the neural tissue in the posterior fossa (Fig. 15-7).

Malformations Characterized by Excessive Growth of Ectodermal and Mesodermal Tissue-affecting skin, nervous system and other tissues.

These neurocutaneous disorders are referred to as phacomatosis (mother spot). They are all characterized by hereditary transmission, involvement of multiple organs, and a slow evolution of the disorder during the childhood and adolescent years. Due to maldevelopment, benign tumors are formed-referred to as hamartomas (body defect tumor). There is moreover a tendency to malignant transformation.

Tuberous Sclerosis. The triad of seizures, mental retardation and a skin disorder called "adenoma sebaceum" characterizes this autosomal dominant disorder. The skin disorder involves the formation of nodules over the face-composed of angiofibromas. Within the nervous system, the nodules are usually present in a periventricular subependymal location composed of glioblast and neuroblasts. Other nodules may be present in the cerebral cortex. Giant neurons and astrocytes may be present. As these nodules (tubers) expand the ventricular system may be blocked. Rarely malignant transformation occurs. The abnormal gene has been isolated to chromosome 9. In some manner this gene location may control an inhibitory factor. When defective, excessive growth occurs.

Neurofibromatosis (von Recklinghausen's Disease).

Two types of neurofibromatosis have been described. Type I is related to a gene locus on chromosome 17. This autosomal dominant disorder is characterized by the proliferation of cells derived from the neural crest. The cells identified in this defect include Schwann cells, melanocytes and endoneurial fibroblasts. The presence of melanocytes produces multiple areas of skin with hyperpigmented (cafe au lait spots). Multiple tumors form in the cutaneous and subcutaneous tissues. The subcutaneous lesions may be discrete tumors and may be attached to a nerve or more diffuse-described as plexiform neuromas. The latter may produce considerable distortion of face and body. Tumors may involve peripheral, spinal or cranial nerves. Tumors of glial and meningeal origin also occur within the central nervous system.

Type II is characterized by bilateral acoustic neuromas and has been identified with a gene locus on chromosome 22. (Case history examples and additional discussion will be found in Chapters 9, 15, and 27.)

Cutaneous Angiomatosis with associated malformations of the central nervous system.

Various disorders are included in this category, often some type of vascular anomaly affecting skin (a vascular nevus), is associated with a malformation in the affected areas of the nervous system. Hemangiomas of the spinal cord may be associated with a congenital vascular nevus in the corresponding dermatome.

Sturge-Weber Syndrome. Patients with this abnormality have a large, deep port red wine nevus over much of the forehead corresponding to the distribution of the ophthalmic branch of the trigeminal nerve. The vascular malformation involves the meninges overlying the parieto-occipital cortex. In this cortical region there is calcification of the second and third layers with neuronal degeneration and proliferation of reactive glial. Atrophy occurs in the cerebrum. The specific genetic basis of this syndrome is uncertain.

Malformations Resulting from Abnormalities in the Ventricular System

Syringomyelia. In this condition the spinal canal is abnormally large. The cavity may enlarge to destroy parts or all of the gray matter and adjacent white matter producing a characteristic "cape like deficit" in pain sensation (Chapter 9).

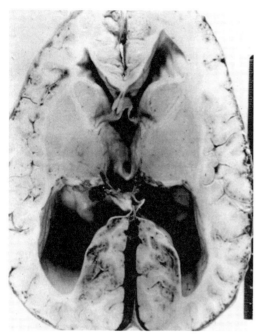

Figure 4-19. Noncommunicating hydrocephalus, Dandy-Walker syndrome. (Courtesy of Dr. John Hills, New England Medical Center Hospitals).

Syringobulbia. In this condition there are abnormal slit like cavities extending from the fourth ventricle into the subjacent tegmentum of the medulla (most common) and pons disrupting gray and white matter with accompanying cranial nerve and long tract signs (Chapter 9).

Hydrocephalus. Any abnormality in absorption of cerebrospinal fluid that produces increased cerebrospinal fluid pressure and dilation of the ventricular system is called hydrocephalus. Blockage of the flow in the ventricular system is called noncommunicating hydrocephalus, while blockage of the cerebrospinal fluid in the subarachnoid space is called communicating hydrocephalus.

Noncommunicating hydrocephalus may be caused by a CSF blockage at the interventricular foramen, the cerebral aqueduct, or in the fourth ventricle, as well as by a malformation of any portion of the ventricular system. The passageways connecting the fourth ventricle to the subarachnoid space (foramina of Luschka and Magendie) may be blocked or ill formed (as in the Dandy-Walker malformation, *Fig. 4-19)*.

Communicating hydrocephalus is caused by the inability of fluid to pass from the subarachnoid space into the venous channels (via the arachnoid granulations) or the malformation of the arachnoid villi, produces excess fluid in the subarachnoid space with resultant pressure on the central nervous system. Obliteration of the subarachnoid cisterns or of the subarachnoid channels may produce this syndrome.

Hydrocephalus can produce thinning out of the bones of the skull with a prominent forehead and atrophy of the cerebral cortex and white matter, compression of the basal ganglia and diencephalon, and herniation of the brain into the foramen magnum. Prior to closure of the sutures such pressure will result in an increase in the size of the head. Depending on the severity of the brain damage the infant may die or survive with mental retardation, spasticity, ataxia, and other defects.

CHAPTER 5
Physiology

PART I. CELL PHYSIOLOGY

The cell is the basic building block of the nervous system and indeed, the whole body. On our way to an understanding of the nervous system, a working knowledge of the capacities of individual cells will be most helpful. The cell membrane (Chapter 3) delimits the cell and is the boundary between the sodium and chloride rich extracellular fluid and the potassium and protein containing intracellular fluid.

The Cell Membrane

The membrane is composed of lipids, particularly lecithin, cholesterol, cephalin and sphingomyelin, arranged as a bimolecular leaflet *(Fig. 5-1)* with the non-polar, water insoluble, ends in the center and the polar, water soluble ends facing the aqueous solutions of the extra and intra cellular fluid compartments. The outside is covered by a thin layer of protein; the total about 5-5 nm thick. Similar bimolecular leaflets can be formed in the laboratory. Such a membrane would easily pass substances (such as ethanol) which were soluble enough in water to reach it and also soluble enough in lipid to pass through it. The membrane would be permeable to such a substance. Highly water-soluble substances such as glucose and ions such as sodium or potassium would not pass through; the membrane would be impermeable to them.

Such a membrane would also be impermeable to water, yet we can easily demonstrate water movement into and out of cells by observing volume changes and eventual bursting of cells. Water apparently moves through the cell membrane in protein-lined pores or channels 0.3 to 0.5 nm in effective diameter. The changing cross sections of a single muscle fiber shown in *Fig. 5-2* are but one example of water movement. Even though we can study water movement easily by observing volume changes, we have made very little progress in characterizing the pores or channels through which it moves. Finkelstein (1987) summarizes both the theory and reality.

The force moving water into the cell and stretching its membrane comes from a difference in the concentration of water on the two sides of the membrane and a tendency for any substance to diffuse from a region of higher to a region of lower concentration. Concentration of water, you say? Certainly!

Pure water contains 55.56 moles H_2O per liter and is 55.56 molar. As substances are dissolved in the water they replace and displace H_2O molecules, the number of moles of water per liter is reduced. When red blood cells are placed in distilled water, the concentration of water outside the cells is much higher than inside where the water is diluted by ions and proteins. As a consequence of this difference in concentration, water moves into the cell. The cell swells and because the inward force is so great, it bursts. This is not just a laboratory curiosity, the person who aspirates fresh water into their lungs when drowning will burst many red blood cells in the pulmonary circulation and liberate potassium into the blood heading back to the heart.

We should digress to develop an intuitive approach and a few simple equations describing the forces acting on the water molecules as they move through the cell membrane. Any text of physical chemistry will provide a more rigorous approach to the derivation of the same equations.

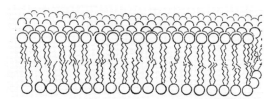

FIG 5-1. A drawing of the phospholipids bilayer that comprises the bulk of plasma membranes. The lipid soluble tails face each other while the more water-soluble heads face the extracellular fluid and the cytoplasm. Most membranes are covered with a thin protein skin.

Energy

We know intuitively that a reservoir full of water contains a certain amount of energy; it has

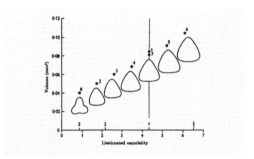

Figure 5-2. The influence of osmotic strength of the bathing solution on the cross-section area and volume of an isolated single muscle fiber. (From Blinks, J.: J. Physiol., 177: 52, 1965.)

the potential to do some work as it flows from the reservoir. Whether this work is the frictional heat of water falling on rock or the rotation of the turbine of an electric generator is largely a measure of the cunning of the engineer. We know, however, that the energy of the reservoir is conserved; it goes somewhere and is turned into some kind of work.

As water flows from a reservoir, the distance the waterfalls are the potential of the system to do work. The amount of work done is also dependent on the size or capacity of the reservoir. The total extractable work, the energy of the system, is proportional to the distance (i.e., potential) the water can fall and the amount (i.e., capacity) of water available to fall.

Similarly, the total energy of any system is equal to the product of a potential factor and a capacity factor. Several examples are given in the following table:

Type of Energy	Potential (or Intrinsic) Factor	Capacity (or Extrinsic) Factor
Gravitational	Height	Weight
Expansion	Pressure	Volume
Electrical	Voltage	Current
Heat	Temperature	Caloric Content

It is the potential factor that determines whether work will proceed at all. We can get work from two reservoirs of water only if they are at different heights. Heat will flow between two bodies only if they are of different temperatures. The idea of a difference in potential is crucial to an understanding of the direction in which a system can go. Of its own accord, a stone can only roll downhill. If we find rocks moving uphill we infer that some form of energy is (or has been) expended.

Chemical potential. Cells exist in aqueous solutions. Consequently, it is the energy of aqueous systems and they're potential for doing work that is of interest to us in the study of cell physiology. The potential factor, called the chemical potential, is analogous to all the other potential factors, such as height, temperature, pressure, or voltage. The capacity factor is the number of moles involved in the reaction.

Metallic sodium reacts with water to form sodium hydroxide, hydrogen, and heat. The chemical potential of pure sodium is greater than that of sodium hydroxide and hydrogen. Consequently, the reaction proceeds spontaneously and energy (heat) is liberated. The reaction proceeds from a state of higher chemical potential to a state of lower chemical potential; all spontaneous reactions do. The reactants have a higher chemical potential than do the products. The total amount of energy liberated, as you might expect, depends on the number of moles of sodium that react.

The chemical potential (μ_a) of a simple, uncharged reactant (a) is, to a first approximation, a very simple function of concentration.

$$\mu_a = \mu°(a) + RT \ln C_a$$

Where
  R = the gas constant
  T = absolute temperature of the solution
  C_a = molar concentration of the reactant
  $\mu°(a)$ = standard state chemical potential

$\mu°(a)$ is the chemical potential of a 1 molar solution of the reactant at 273°K (0°C) and 760 mm Hg pressure. It enters the equation because of the units chosen for concentration. Note that when C_a = 1 mole, $\mu = \mu°$, because ln 1 = 0.

Because we are usually interested in the difference in u between two solutions on either side (i and o) of a membrane, $\mu°$ drops out and $\mu_i - \mu_o$ = RT ln (C_i/C_o).

In the study of the movement of a substance through cell membranes, an important factor is the difference, if any, in chemical potential between the substance inside the cell and the substance in the outside solution. Unless this is known, only conjectures can be made about the nature of the process by which the substance enters the cell. If the substance is moving from a region of high chemical potential to a region of low chemical potential, a permeable membrane is all that is required. If the substance is moving from a region of low chemical potential to one of high chemical potential, we know immediately that energy must be expended, an active transport.

To sum up, the chemical potential is, as the name implies, a potential function. We expect that energy will flow from a phase of greater chemical potential to a phase of lower chemical potential. This principle is the cornerstone of thermodynamics. The maximum work we can get from a system is equal to the product of the difference in chemical potential times the number of moles involved.

If, however, the chemical potential of a substance is equal in two phases, we do not expect a net flow of the substance. Two solutions are, by definition, in equilibrium when the chemical potentials of all the substances in the two solutions are equal. In the case of a semi-permeable membrane, only permeable substances are considered when determining equilibrium across the membrane. Water is usually in equilibrium across the membrane.

Diffusion

The flow of mass is one form of work in which energy in the form of chemical potential can be expended; in which a difference in chemical potential can be put to work. All of you have seen the classic demonstration of a crystal of potassium permanganate dropped into a cylinder of water. The purple color gradually spreads throughout the cylinder. The color change indicates that permanganate ions are spreading through the water by the mechanism of diffusion. The permanganate ions are moving from a region where they have a high chemical potential (high concentration) to a region where they have a low chemical potential (low concentration).

Diffusion is an irreversible process and is described by the equation

moles/time = $-DA\, C_i/X$

where

t = time A = area available for diffusion

D = constant C_i/X = concentration gradient

The number of moles of glucose per minute that will flow from a solution of 2 molar glucose to a solution of 1 molar glucose is dependent on the area of contact between the two solutions. Clearly, the greater the area exposed, the greater the flow. M. H. Jacobs (1967) discuss specific solutions to the diffusion equation for various boundary conditions (physical situations).

Linear Diffusion. When a substance is diffusing in a long tube, the rate of diffusion depends on the square of the distance. After diffusion has progressed a while, the concentration at any point along the tube will be proportional to the square of the distance from the original interface.

An example of linear diffusion is shown in the following table. One million sugar molecules were concentrated in a very fine layer at the bottom of a graduated cylinder. One hour later the distribution was:

Distance	Number of Molecules
0-1 mm	553,000
1-2 mm	319,000
2-3 mm	157,000
3-5 mm	55,000
5-5 mm	13,000
5-6 mm	2,000
6-7 mm	170
Over 7 mm	20

As you can see, almost half the molecules have not moved 1 mm from the starting position. A handy rule of thumb for simple electrolytes in water is that it will take 1 millisecond for a substance to diffuse 1 µm.

Diffusion into a Cylinder. The second case concerns the diffusion of nitrogen into a cylindrical muscle fiber. Consider, for example, that pure nitrogen is suddenly bubbled into the fluid surrounding a muscle fiber. How long will it take before the concentration reaches 90 percent of its final value at the center of the cell? That depends on the diameter of the cell. Values for both 50 percent and 90 percent of the value at the surface for cylindrical cells of varying diameter are given in the following table:

Diameter	Time (sec) for 50 percent Saturation	Time (sec) for 90 percent Saturation
20μm	0.008	0.055
50μm	0.035	0.177
100μm	0.213	1.11
200μm	0.852	5.55
1mm	21.3	111
2mm	852	5500

Since the situation for other small molecules, such as O_2 or Ca^{++}, is similar, it can readily be seen that large cells are at a considerable disadvantage since they must obtain all their nutrients and excrete all their waste products by diffusion. The time it takes the large cell to move even small quantities of these substances is so long that they cannot metabolize at a reasonable rate. For that reason large (200 μm) cells are very rare and cells which move substances rapidly, such as red blood cells, are small (7μm).

Furthermore, calculations such as these show that the distance between a cell and the nearest capillary must be fairly short to provide rapid diffusion.

Water Movement

The energy to drive water into cells and change their volume comes then from differences in chemical potential of water between the solutions inside and outside the cell and the speed is described by the diffusion equation. The single muscle fiber shown in Fig. 5-2 maintained its normal volume when placed in a 120 mM NaCl solution. The chemical potential of water is equal on both sides of the red cell membrane. When placed in an 80 mM NaCl solution, the fiber swelled. The chemical potential of water is slightly greater in the 80 mM NaCl solution; it is diluted less. Water will move down the chemical potential gradient, swell the cells and dilute the cell contents until the chemical potential of water on both sides of the membrane is the same, an equilibrium reached and a new volume established.

Osmolarity.

We describe the degree of dilution of water by the number of particles diluting it and express this in milliosmoles. Human plasma osmolarity is 285-295 mosm/l and is made up of Na^+, 150mM; K^+, 3mM; Ca^{++} (2x5mM=10mosm); Cl^-, 100mM; HCO_3^-, 25mM; with the remainder from serum proteins. Because the osmolarity of a solution is a function of the number of particles, osmolarity belongs to the much larger and well-studied group of colligative properties of water such as vapor-pressure lowering, freezing point depression and boiling point elevation. The standard, commercial osmometer really measures freezing point depression, but is calibrated in milliosmoles. We describe solutions that maintain cell volume as isotonic, such as isotonic saline for injection. Hypertonic solutions contain more particles, particularly particles that cannot enter cells, and cause cells to shrink. As the cell shrinks due to water movement down its chemical potential gradient, the concentration of the cell contents increases until the chemical potential of water is equal on both sides of the membrane. Hypotonic solutions contain fewer diluting particles and cause cells to swell.

Osmotic pressure refers to the ability of physical pressure to alter the chemical potential of water and hence balance the chemical potential of H_2O by pressure rather than concentration. Plants and bacteria have cell walls that allow the creation of a pressure within the cell, indeed it is the only way a cell can survive in distilled or fresh water. Animals have no cell wall and are unable to maintain pressure within individual cells.

The adult brain, however, is encased in the nonexpendable skull and does maintain a small physical pressure, the cerebrospinal fluid pressure, normally 70-80 mm H_2O or 5-7 mmHg. A large number of conditions, including trauma, cause the brain to swell and hence increase intracranial pressure and decrease consciousness. Intracranial pressure can be reduced by adding particles to the plasma which cannot cross capillaries or get into brain cells. *Fig. 5-3* is an example where isosorbide was added to plasma, serum osmolarity was increased and water moved from brain cells into plasma. Intracranial pressure was reduced dramatically and the patient's condition improved. Mannitol is more commonly used for reducing intracranial pressure.

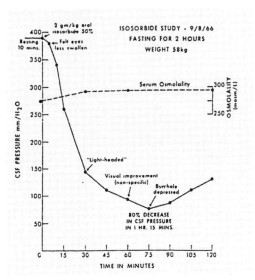

Figure 5-3. The effect of increased serum osmolarity (dashed line) on intracranial pressure (solid line). (From Wise, B.L., and Mathis, J.L.: J. Neurosurgery. 28:125, 1968.)

Potassium Channels

Cells contain a high concentration of potassium (150 mM) while the extracellular fluid contains a low (3 mM) concentration; there is a significant difference in the concentration of potassium across the cell membrane. The lipid bilayer membrane discussed earlier would not allow potassium to move across the membrane because potassium ions are insoluble in lipid. Soon after adding radioactive potassium to the extracellular space, however, the cells contain radioactive potassium and when these cells are returned to a non-radioactive solution, the radioactivity leaves the cells; the cell membrane is clearly permeable to potassium ions.

Potassium ions cross the cell membrane via specialized protein channels in the cell membrane *(Fig. 5-4)*. These potassium channels and all channels that have been studied are constantly opening completely and after an apparently random interval close abruptly. The channel opens a door, some other force, concentration difference, for example, provide the energy necessary for the ion to move. Channels allow only a single ion species to traverse an open pore.

Recording from such a channel *(Fig 5-4A)* shows it opening and closing rapidly. This particular channel was recorded from a small patch of membrane attached to a pipette by mild suction and pulled off the cell so that the inside surface of the membrane is exposed to the bathing solution (here 150 mM K+), a so-called inside out patch.

These channels are quite selective for potassium. In *Fig. 5-4B* the solution against the inside of the membrane was changed to one containing 150 mM Rb+ and the number of ions flowing through the channel (the current) dropped dramatically. On the basis of whole cell recording, the selective permeability series is K+>Rb+>NH5+>>Na+ or Li+. High ionic selectivity is characteristic of most channels.

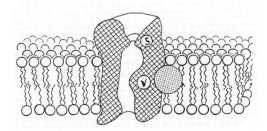

Figure 5-4. A cartoon of a voltage gated potassium channel inserted in a lipid bilayer. S represents the selectivity gate or filter that discriminates between potassium and other monovalent ions. The dotted sphere represents the voltage sensor and V the voltage modulated gate.

Potassium equilibrium voltage. Most cells contain several thousand such potassium channels on their surface; indeed, it is very difficult with a 0.1 μm pipette to find a place with only one channel. On a statistical basis, then, a large number of these channels are open at any one time and provide a sizable pathway for potassium ions to move down their concentration gradient. Yet the potassium ion concentration of cells remains constant and the potassium flux leaving the cells is quite small. Why with a sizable pathway and an equally sizable concentration gradient (150 mM > 3 mM) doesn't potassium rush out of the cell *(Fig 5-5)*?

Remember for a minute that potassium ions are positively charged and the equal number of negative charges inside the cell is predominantly on proteins that do not move around and certainly cannot leave the cell. As a positively charged potassium ion moves into the transmembrane channel, down its concentration gradient, it leaves behind a negative charge; the inside of the cell becomes ever so slightly negative. As more positively charged potassium ions start down their concentration gradient through the potassium channels of the membrane, the cell becomes increasingly more negative because of the negative charges on the protein, until a stable voltage of 0.1 volt, inside negative, is maintained. As the inside of the cell becomes slightly negative, a second force, electrostatic attraction, begins to work on the potassium ion acting to hold it within the cell *(Fig 5-5)*. An equilibrium is rapidly established between the concentration force tending to move potassium down its concentration gradient, out of the cell, and the electrostatic force attracting the positively charged potassium ions back into the negatively charged cell. Only about .005% of the intracellular potassium moves into the membrane to establish this voltage, called the resting membrane voltage.

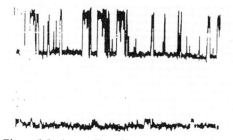

Figure 5-5. A recording of ions passing through a single potassium channel. In A fluid facing the cytoplasm surface contained K+ and many ions traversed the channel each time it opened. When Rb+ was substituted for K+ (B) the current was greatly reduced even though the channel opened as often as before. Patch clamp data adopted from Findley, J. Physiol. 350:179-195, 1985.

Because this is an equilibrium, once the work of putting the potassium-protein complex into the cell is completed, the voltage continues without the expenditure of energy. All nerve and muscle cells maintain such a negative voltage. It is derived from the potassium ion gradient, the potassium permeability of the membrane, and an intracellular, impermeable anion.

Nernst equation. In addition to concentration, a second factor, electric field, also influences the chemical potential of an ion (b):

$$-b = -_o + RT \ln C_b + zF$$

where z = valence, F = Faraday and = the absolute electric field.

Once an electric field is established, the electrochemical potentials of the potassium ions inside and outside the cell become equal. When only one ion is permeable, in this case potassium, the potassium equilibrium voltage (V_K) is described quantitatively by the Nernst equation; an equilibrium between the electrical force and the chemical concentration force each acting on potassium ions

$$zFV_K = - RT \ln K^+_i/K^+_o \text{ or}$$

$$V_K = - RT/zF \ln K^+_i/K^+_o$$

where V is the voltage (o- i) across the membrane, K+i and K+o are the potassium concentrations, R & F are constants, T the absolute temperature and z the valence of the ion. For potassium ions at 20°C, this equation becomes

$$V_K = -58 \log 150mM/3 mM = -97mV$$

At 37°C the constant is -61.

V_K is the potassium equilibrium voltage, at this voltage and concentration ratio, K^+ is in equilibrium across the cell membrane. The Nernst equation can be solved for any permeable ions such as H+, Cl- or for Na+ if only that ion was permeable; these are equilibrium voltages for single permeable ions.

Resting membrane voltage. Actual voltages across the membrane of a single muscle fiber at a varying K_o are shown in *Fig. 5-6* and are well described by the Nernst equation that assumes that potassium is the only permeant ion; the resting membrane voltage closely approximates the potassium equilibrium voltage. The potassium ion is in equilibrium across the resting membrane. There is no net force acting on it. The potassium equilibrium voltage is the voltage across the membrane, which brings the electrochemical potentials of potassium, inside, and out, into equilibrium.

We find electrical potentials across the membranes of most types of cells, and most of these potentials can be described to a first approximation by the Nernst equation for potassium and range from -80mV to -100 mV.

When discussing changes in the actual mem-

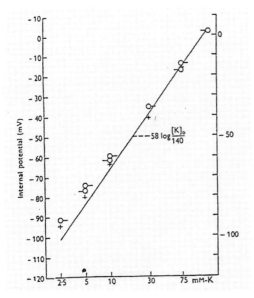

Figure 5-6. The relationship between the external potassium concentration and the resting membrane voltage fitted with the Nernst equation. (From Hodgkin, AL; and Horowicz, P.: J. Physiol., 158:127-160, 1959).

brane voltage of a cell, we use the term depolarize to denote a movement to more positive voltages (i.e. -90 to -60 mV) and hyperpolarize to denote movement to more negative voltages (i.e. -90 to -110 mV). We use the convention that the outside solution is zero millivolts.

Several Permeable Ions

  Sodium channels. The preceding discussion of the resting membrane voltage as a potassium equilibrium voltage assumed that the membrane was permeable only to potassium ions; it contained only potassium selective channels. Real cell membranes don't quite reach that degree of perfection. They are also permeable to sodium ions; radioactive sodium ions enter and leave cells slowly. Some of the Na+ enters via K channels; they are only 1000/1 selective for K, and some Na+ enters through the Na channels we will consider in the next chapter. That sodium ions enter should be no surprise, the Na+ concentration gradient (120 mM/20 mM) is into the cell and the negative interior voltage of the cell is very inviting to a positively charged ion. Indeed, the sodium equilibrium voltage is:

V_{Na} = -58 log 20 mM/120 mM = +55 mV

The sodium equilibrium voltage is positive because the Na+ concentration gradient is into the cell. Only a voltage of +55 mV inside the cell would bring Na+ into electrochemical equilibrium across the membrane if Na+ were the only permeant ion. In a cell with a resting membrane voltage of -90 mV, sodium ions, if permeable to the membrane, are 135 mV away from equilibrium, a significant electrochemical potential difference.

  The membrane is about 100 times more permeable to potassium ions than to sodium ions. Each time a sodium ion manages to sneak into the cell, the cell becomes a little less negative. A potassium ion is now liberated from the constraints of the electrostatic field and escapes from the cell. In the steady state, the membrane voltage is a little less negative than that predicted by the potassium equilibrium. Inspection of Fig. 5-6 shows that at Ko of 2.5 and 5 the experimental voltage points are less negative than predicted by the Nernst line.

  The Goldman Equation. This is a useful equation describing the membrane voltage when more than one ion is permeable:

$$V(mv) = -58 \log \frac{P_K K_i + P_{Na} N_{ao}}{P_K K_o + P_{Na} N_{ao}}$$

Where PK and PNa are the membrane permeabilities to potassium and sodium. When PNa = 0, it becomes the Nernst Equation for potassium. When two (or more) ions are permeable, each ion influences the final membrane voltage in proportion to its ability to cross the membrane. The equation can be rewritten as

$$V = +58 \log \frac{K_o + P_{Na}/P_K N_{ao}}{K_i + P_{Na}/P_K N_{ai}}$$

  Because both the ratio PNa/PK (0.01) and Nai (20) are small, the product is very small in relationship to the internal potassium (150), that term can be dropped and the equation inverted to become:

$$V = 58 \log \frac{K_o + P_{Na}/P_{Na} N_{ao}}{K_i}$$

  This equation is plotted in *Fig. 5-7* together with membrane potential data. The addition of the $P_{Na}/P_K N_{ao}$ term adds greatly to the fit to experimental data at low potassium concentrations and very negative membrane voltages. All membranes are at least slightly permeable to more than one ion and the Goldman equation is very useful in describing the observed membrane voltage.

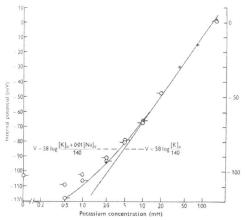

Figure 5-7. The resting membrane voltage for frog single muscle fibers in solutions of varying potassium concentration fitted by the Goldman equation. The normal values for frog muscle are Na_o = 120 mM/L, Na_i = 20 mM/L, K_i = 150 mM/L, and K_o = 2.5 mM/L. (From Hodgkin, A.L., and Horowicz, P.: J. Physiol., 158:127-160, 1959).

Active Transport

Just as it was clear that sodium ions were moving down an electrochemical gradient into the cell, any sodium ion movement out of the cell must be up an electrochemical gradient and must require energy; an active transport with ATP as the energy source. The active transport of sodium out of cells has been extensively studied (Stein, 1988) and the protein responsible isolated. Under most circumstances 3Na+ move out for every 2 K+ moving in so there is a mild imbalance in charge. Other ionic pathways, usually the potassium channel we have already discussed, make up the difference in charge (moving a third K+ into the cell) and there is usually no component of membrane voltage attributable to active transport. If these other pathways are blocked, a large hyperpolarization (more negative voltage) is associated with active transport.

All cells leak a little sodium in and potassium out and active transport keeps the internal potassium high and sodium low. About 10% of our resting metabolic rate is consumed with the Na/K active transport. Cold and the digitalis glycosides block Na/K active transport. When red blood cells are stored in the cold they gain sodium and loose potassium.

Control of potassium Channels

At resting membrane voltages, one group of potassium channels are opening and closing as previously described to provide the significant potassium permeability of the resting cell membrane and the energy source for the resting membrane voltage. This might suggest these were static channels, the probability of opening and closing never changing; nothing could be farther from reality. Potassium channels come under at least three types of control: membrane voltage, internal calcium and external neurotransmitters. They are the most diverse group of channels yet studied.

Voltage Control. Some channels open rapidly but transiently where the cell is depolarized; because of their role in the action potential. A second type of voltage sensitive potassium channels is usually open but closes a few seconds after depolarization. These are the channels that provide the resting potassium permeability in many cells.

To understand the closing behavior when the cell is depolarized, remember that any membrane voltage less negative than V_K will result in an outward electrochemical gradient for K^+ ions and if potassium channels are open, K+ ions will flow out of the cell, usually in exchange for Na+ ions. The cell must then expend energy to actively transport the Na+ out and K+ back in. These K channels decrease their probability of opening whenever the membrane voltage is less negative than V_K, specifically to prevent potassium loss. These channels likewise increase their probability of opening when the cell becomes more negative than VK as it might when a 3Na/2K active transport is working. This channel passes K^+ much more easily when the potassium electrochemical gradient is into rather than out of the cell. We call these channels inward rectifying K channels and they probably are the major source of open potassium channels at the resting membrane voltage of many cell types.

These channels are also capable of great mischief whenever the external (plasma) potassium concentration falls as it often does in diuretic therapy. As external K falls from 2.5 to 1 mM (*Fig. 5-8*) V_K changes from -101 to -125 mV while the actual membrane voltage hyperpolarizes from -95 to only -105 mV. The difference between V_K and membrane voltage increases from 7 to 20 mV, a potassium electrochemical gradient out of the cell. This increasing outward gradient decreases the probability that the inward rectifying K channels will be open and consequently reduces the total potassium permeability of the membrane. As potassium permeability of the membrane decreases, the sodium permeability, which has not changed, exerts a stronger, depolarizing, force on the membrane voltage leading to the sudden depolarization shown in *Fig 5-8*. The ratio P_{Na}/P_K increases from 0.03 to 0.15. The abrupt change in P_{Na}/PK permeability ratio gives the Na ion and the Na equilibrium potential (+55 mV) a greater "voice" in the final membrane voltage. Such a sudden fall in resting membrane voltage in the heart can cause cardiac stand still.

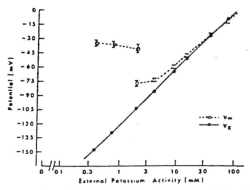

Figure 5-8. The relationship between the logarithm of the external potassium activity (concentration) and the resting potential of heart cells. Note the abrupt depolarization of approximately 2 mM K_o^+ caused by the closing of potassium conserving channels. Their partial closure increases the role Na+ permeability plays in determining the resting membrane voltage. Adapted from data in Shen, S.S. et al., Circ. Res. 57:692-700, 1980.

The Goldman equation suggests the extreme membrane voltages are V_K (-100 mV) when only K is permeable and V_{Na} (+55 mV), when only Na is permeable. Any membrane voltage in between can be achieved by a judicious ratio of P_{Na} and P_K.

An analogy that may help you is to consider the final temperature of water coming out of a faucet. The extremes are only hot water running and only cold water running and any temperature in between can be achieved by the appropriate mixing of flow rates of hot and cold water.

Ca_i Controlled. Internal Ca as showed in *Fig. 5-9* controls another group of potassium channels. These records are from a single isolated potassium channel incorporated into an artificial bilayer so that the solution on both sides of the channel can be controlled. As the calcium concentration increases from 3 _M to 95 _M, the percent open time increases until the channel is open almost all of the time.

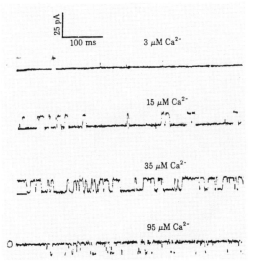

Figure 5-9. Records of a patch clamp of a single calcium activated potassium channel. As the calcium concentration on the cytoplasmic surface increases the channel opens more frequently and, on average, remains open longer until at 95 M Ca the channel is almost always open. From Latorre, R., C. Vergora and C. Hidalgo. Proc. Nat. Acad. Sci. 79:805-809, 1982.

When these Ca_i controlled potassium channels are incorporated into a cell, any increase in intercellular C^{a++} opens these channels and increases the potassium permeability. Increasing potassium permeability will hyperpolarize the membrane and move the membrane voltage toward the potassium equilibrium voltage of approximately -100 mV. This hyperpolarization will stabilize the membrane and tend to inhibit spontaneous or rhythmic activity. We will come upon many examples of these channels as we explore the cells of the nervous system.

Chemically Controlled K Channels. A third major type of potassium selective channel is controlled by chemicals applied to the exterior of the cell; particularly those released by nearby nerves. The classic example is the slowing of the heart by acetylcholine released from the vagus nerve. The acetylcholine does not act directly on the channel but through an intermediate G protein step *(Fig.5-10)*. Acetylcholine binds to the external surface of the receptor and activates the reaction of the G protein with GTP and Mg. The activated subunit binds to a nearby G gated K channel and increases its probability of opening. Increased potassium permeability moves the resting potential (-70 mV) of the cell closer to the potassium equilibrium voltage (-100 mV). As we shall discuss in the next chapter, this makes the cell less excitable and lowers the heart rate.

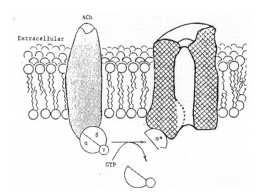

Figure 5-10. A cartoon of a G protein gated, potassium selective channel. Ligand binding (for example, Acetylcholine) on the extracellular receptor site of the transmembrane receptor complex catalyzes the intracellular reaction of GTP with a G protein. Ultimately the a* subunit binds with an intracellular receptor site of the potassium channel. Occupation of this site increases the probability the K channel will open and remain open longer. The active a* subunit is deactivated by hydrolysis of the bound GTP. Adapted from AM. Brown and L. Birnbaumer. Am. J. Physiol. 255: H501-H510, 1988.

PART II. NERVE PHYSIOLOGY

All excitable cells such as nerve and muscle cells maintain a small (100 mV) voltage across their surface membrane. This voltage id generated across the semipermeable cell membrane utilizing the difference in potassium concentration inside (-150mM) and outside the (-3mM) the cell. *Axons* are long cylinders carrying a stereotyped electrical message, an action potential, from cell body to synaptic region of the nerve cell. Like most cells, the axon maintains a negative resting membrane voltage *(Fig. 5-11)* derived from the potassium ion gradient (Ch. 5-I). Whenever the axon is stimulated (Fig. 5-11), in this case by a brief electrical shock, the voltage across the membrane becomes positive and within 1 msec returns to the resting membrane voltage. This sudden reversal of polarity (electrical sign) is called an *action potential* and his constant throughout the axon.

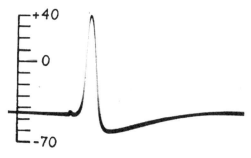

Figure 5-11. An intracellular recording of an action potential from a squid giant axon showing (left to right) resting membrane voltage (-45 mV), stimulus and action potential. The electrical signal is m respect to the bathing solution which is considered to be at zero potential. (From Hodgkin and Huxley, A.F.: J. Physiol. 104:176, 1954.

The Action Potential

Many types of stimuli lead to the generation of action potentials but they all depolarize the axon membrane to a critical voltage threshold. The action potential is propagated unchanged down the axon and is remarkably similar wherever it is measured in the nervous system. Since the action potential is a stereotyped event (constant duration and magnitude), information must be carried down the axon in the form of the interval between action potentials, action potential frequency. For example, more intense sensory inputs are transmitted as higher action potential frequencies.

Ionic Mechanism. Each patch of axonal membrane participates actively in the propagation of the action potential; each section of membrane acts as a repeater station. During the action potential, the ease of ionic passage through the membrane (membrane conductance) increases (*Fig. 5-12*). The positive voltage of the action potential is derived from the sodium equilibrium voltage (*Fig. 5-13*). As the sodium concentration of the external solution is reduced, the peak voltage falls.

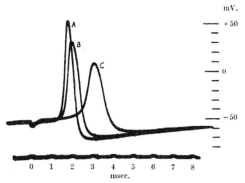

Figure 5-13. Action potentials recorded across the membrane of a squid giant axon whose axoplasm has been squeezed out and replaced with an artificial solution. In record A the solution is isotonic potassium sulfate, 600 mM K^t. In record B the internal solution is 3/4 K^t, 1/4 Nat, 450 mM K^t. In record C the internal solution is 1/2 K^t, 1/2t. The Na in the seawater is 460 mM. (From: Baker, Hodgkin and Shaw: Nature, 190: 885, 1961).

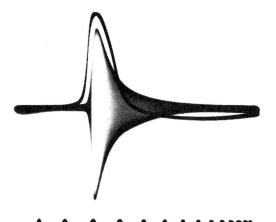

Figure 5-12. The change in the ease of passage of ions (impedance) through the membrane during an action potential, Two superimposed records are shown. The thin line is an action potential of a squid giant axon and the continuous curve is the impedance change. (From Cole, K.S., and Curtis,HJ, J. Gen. Physiol. 22:649,1938.)

Sodium ions are suddenly and briefly able to move down their concentration gradient, through open sodium channels in the membrane, carrying excess positive charge into the axon, and swinging the voltage across the membrane positive. Sodium channels open rapidly in response to a stimulus, allow Na to move into the cell, and then close quickly. Many types of stimuli lead to the generation of action potentials but they all depolarize the axon membrane to a critical voltage, *threshold*, which opens the sodium channels and generates an action potential (Fig. 5-4). Some force, external to the patch of axon, does work upon this patch of axon to depolarize the axon from the resting membrane voltage (-65 mV in this case) to the threshold voltage of -55mV. At threshold, the sodium channels open transiently and the sodium equilibrium voltage briefly dominates the membrane. In the terminology of the Goldman Equation, the ratio of Pna/ Pk becomes about 20. Sodium channels are voltage gated; they change their properties from mostly closed to mostly open in response to depolarization of the membrane to the threshold voltage.

Sodiums Channels

Channels consist of long stretches of a hydrophobic helix cylinder spanning the membrane. Channels can be opened by a number of. stimuli. Sodium channels are an excellent example of a channel opend by transmembrane voltage; a voltage gated channel. The individual records in *Figure 5-13A* of a patch of membrane containing a single sodium channel show the channel usually opens in response to a depolarizing stimulus. Rarely (top trace) the channel opens for 10-15 msec, sometimes for 5 msec. and often very briefly. Axons have about 100 sodium channels/1m2, occupying about 1% of the surface area. It is the average response of many hundreds of sodium channels that is responsible for the action potential we record. The average can be obtained by summing either many sodium channels reacting to a single stimulus or one sodium channel reacting to many stimuli. Figure 5-13B is the average of 64 responses to the same depolarizing stimulus and shows a sudden increase in the probability sodium channels will be open immediately following

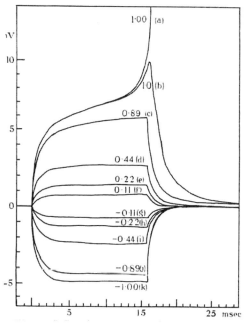

Figure 5-14. The response of a crab axon of 80-im diameters to current passage. The resting membrane potential was -65 mV. Depolarization is shown as the upward deflection. The numbers beside each trace give the current strength relative to the threshold current. The upstroke seen in (a) is the beginning of an action potential similar to the one seen in Figure 5-1. (From Hodgkin, A.L., and Rushton, W.A.H.: Proc. Roy. Soc., B133: 444,1946.).

tial of sodium ions inside the cell; sodium ions are nearing electrochemical equilibrium across the membrane. The experiment shown in Figure 5-I3A were done with a stimulus to +50 mV, no current (Nat ions) would flow through the open channel because sodium would be in equilibrium across the membrane. Were the sodium channels to remain open, membrane voltage would remain at +40 mV. The sodium channels do not remain open for very long, as we saw in Figure 5-13 SB, so the membrane voltage returns to its resting (-80 mV) level. The closing of the sodium channels is called *inactivation* and will be discussed later in this chapter.

Sodium Selectivity.

These channels are highly sodium selective with permeability ratios: Pli/P_{na}= 1, PK/P_{Na} = 0.08 and Pca/PNa = 0.016 which shows Na s Na permeability decreasing with increasing crystal radius. This suggests it is the unhydrated ion that transits the channel. The Na^+ selectivity filter is on the outside of the channel *(Fig. 5-16)* and is the site of attachment of a number of naturally occurring paralyzing toxins including tetrodotoxin (TTX). These toxins bind to the outside of the channel (they are ineffective when injected into the axon) and block the channel, thereby preventing Na^+ ions from entering and altering the voltage across the membrane.

depolarization below threshold. The probability of the channel being open falls rapidly, within a few milliseconds, and then remains low, but not zero, as long as the depolarization continues. The same pattern pertains to the hundreds of sodium channels of any patch of axon membrane. Open sodium channels permit sodium ions to move down their concentration gradient, into the cell, and carry excess positive charge into the cell. The peak rate of influx is 150 Na^+/msec/channel. The first few sodium ions entering displace the potassium ions in the membrane and allow them to move down their concentration gradient, out of the cell. The rest of the sodium ions bring excess positive charge through the membrane, into the axon, and cause the voltage across the cell membrane to become positive. As the voltage across the membrane approaches the sodium equilibrium voltage (+50 mV) the number of sodium ions entering slows because of the increased electrochemical poten-

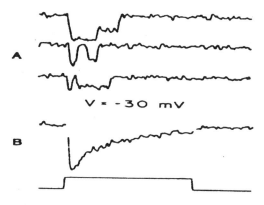

Figure 15. Opening of a Na channel in response to a depolarization from –lOO mV to -25 mV for 30 msec. Notice the variability in response. If the channel opens, it usually does so soon after the depolarization. B. The summed response (on a different time scale) of 64 consecutive depolarization's like the ones above. The average response is a rapid opening followed by closing.

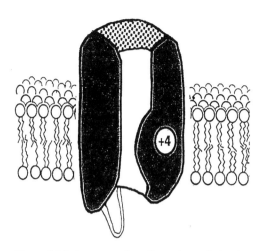

Figure 5-16. A cartoon of a sodium channel at rest.

Gating Currents. The axon and hence the sodium channel is activated by an externally induced voltage change across the membrane from the resting membrane voltage (-80 to -100 mV) to the threshold voltage (-55 mV), a depolarization. This voltage change does work on a charged particle or dipole within the channel structure. Because of its position within the membrane the dipole can sense voltage across the membrane. Dipole movement within a changing voltage field would require no energy beyond the externally induced voltage charge. As the dipole moves it carries a very small, transient current, *a gating current*, shown in Figure 5-17. In Figure 5-17, all ionic currents are blocked by a combination of substitution of ions impermeable to the axon membrane and blockage of the sodium channel by TTX. Within 10 msec. of depolarization, a small outward current is recorded, the gating current. By 500 msec. all the + charges have moved to the outside of the membrane and the current signal ceases. Four positive charges move outward per channel. In figure 5-17B, TTX was not added but only 5% of the usual Na was present. Following the early outward gating current, a larger inward current (Nat ions) flows through the now open Na Channels. After a few msec. the Na current declines to zero, the channel inactivates.

Threshold.

Individual Na channels open with 50% probability from -60 to -40 mV, yet threshold for a patch of axon is remarkably constant at -50 mV (Fig. 5-17) because the process is regenerative.

Once a few channels open, Na+ rushes in and further depolarizes the patch of axon which in turn opens more channels, which allows more Na to enter, which in turn opens more channels; positive feedback, a cascade. Once some external force depolarizes the patch of membrane to threshold, an internal process takes over which requires only the energy in the Na ionic gradient.

Inactivation. Sodium channels close quite soon after they open without receiving or producing any external signal. Indeed, as we have shown in Figures 5-15 and 5-17 the channel closes even when the membrane remains depolarized and the charges that activate the voltage gate remain on the Figure 5-17A. Although not shown in Figure 5-17, repolarization of the membrane gives a transient, inward (down) current of the same size and duration as the initial gating current. The gate that opens in response to depolarization is not the one that closes to block further flow of Na+. We call the gate responsible for the early termination of Na + flow through the Na channels the *inactivation gate*.

The inactivation gate seems to close, and block further Na+ entry, based on the number of Na+ ions entering through the channel. We haven't the vaguest idea of how it does it but can describe its actions, particularly in whole nerve, very well. The gate acts to limit the amount of Na+ entering the axon and is the major cause of the brief duration of the action potential.

Once the inactivation gate is closed, it does not open immediately upon repolarization. The rate at which these gates open, remove inactivation, is a function of both time, on the msec. scale, and membrane voltage. Until the membrane is repolarized to -50 mV, the rate of recovery is virtually zero; the inactivation gates remain closed and no further action potentials can be generated. The patient paralyzed with familial periodic paralysis (Chapter 5, problem 2) has depolarized the muscle membrane below -50 and cannot generate further action potentials. The heart cell, which suddenly depolarized from -65 to -40 mV, would fire a single action potential upon depolarization but the inactivation gate, once closed, would not reopen until the membrane repolarized to -55 mV. Consequently no further action potentials would propagate across the* ventricle and no further contraction would occur.

As the membrane is further repolarized *(Fig. 5-18)*, the rate of resetting the gate increases, as does the final percent of channels open. Even at -80 mV when the rate is near maximum, the gate takes a couple of msec. to completely reopen and reset the channel.

The inactivation gate can be enzymatically removed by briefly applying pronase to the inside of the axon. Then depolarization results in a prolonged flow of Na~ ions through the open sodium channel. Because pronase is only effective when applied on the inside of the axon, we believe the inactivation gate is on the inside of the Na channel (Fig. 5-16).

In *Fig 5-19* each cartoon is identified on the voltage trace below. At the second stimulus (arrow) the reset. Inactivation gate is still closed and even though the voltage sensitive gate opens, Na+ cannot enter and further depolarize the cell; the channel is refractory.

The Action Potential Cycle.

The cartoon in Figure 5-19 illustrates the

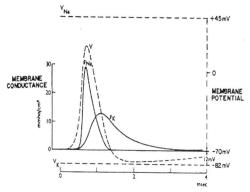

Figure 5-18. The solid line shows both the rate of removal of inactivation of a group of sodium channels and the final % of these channels activated, as a function of membrane voltage. Removal of inactivation is slow and incomplete at -50 mV, rapid and complete at -70 mV. The dashed line shows the shift in the relationship to more negative voltages induced by local anesthetics. During this time, the refractory period, a second stimulus will not result in a second action potential. Membrane voltage also determines the probability of resetting the inactivation gate. At -60 mV, for example, only 40% of the inactivation gates ever reset.

cycle an average sodium channel completes in response to a brief depolarization. The membrane voltage trace is shown below. The stimulus (arrow) does external work on the resting channel (A) to bring the transmembrane voltage to threshold (Th) and opens the voltage gate (A> B). Na~ rushes down its electrochemical gradient and depolarizes the axon (B). The inactivation gate closes (BA> C) at about the time the voltage peak is reached and the membrane voltage slowly returns to the resting voltage. The voltage gate closes quite rapidly (C > D) upon repolarization but the inactivation gate takes its own sweet time reopening (D>A). A second stimulus (arrow) delivered when the inactivation gate is still closed does not result in an action potential; the axon (or this channel) is refractory. Only after a couple of msec, when the inactive gate has reset (opened), does the channel return to its original state (A) and can once again be stimulated to generate an action potential (third arrow).

Toxins and Local Anesthetics.

A number of naturally occurring toxins paralyze by blocking sodium channels. Conveniently for the scientist, these toxins block by different mechanisms and bind firmly enough to be useful

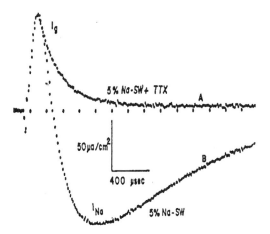

Figure 5-17. Sodium current and gating current recorded from a voltage-clamped squid giant axon. The axon was in a seawater solution containing only 5% of the normal sodium concentration (replaced by tris ions), and was internally perfused with cesium fluoride. Both trace A and trace B are the sum of the current following ten positive and ten negative steps of 90 mv amplitude from a holding potential of -70 mV. Tetrodotoxin (TTX) elirninates I but leaves I (gating current) unaffected. Adapted from C.M.Armstrong, The Physiologist. 18: 93-98, 1975

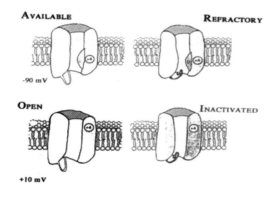

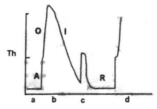

Figure 5-19. Cartoons of a Na channel during an action potential cycle. At A, the channel is resting and primed the voltage gate is closed the inactivation gate open. Following an external stimulus (A> B) the voltage gate opens and Na+ rushes through the channel carrying the interior voltage to + 20 mV. B. At about the peak of the action potential the Inactivation gate closes (B> C) Na+ flow through the channel ceases and the voltage drifts back to the resting membrane voltage. Immediately upon repolarization (C> D) the voltage gate closes (DJ and a few msec. later the inactivation gate opens (1)> A)

for anatomic localization and biochemical tagging. Tetrodotoxin (TTX) and Saxitoxin bind to the Na recognition site, the selectivity gate and block Na currents (Fig.5-17). Another group including Veratridine and Aconitine act as pronase does to remove the inactivation gate and cause long lasting opening of the Na channel. The toxins of North African scorpions and sea anemones slow inactivation. A fourth group of toxins from American scorpions modify Na channel activation and provide a probe for the voltage-activated gate.

Local anesthetics block nerve conduction by binding to the cytoplasmic end of Na channels and prevent the Na channels opening in response to depolarization *(Fig. 5-20)*. The early inward sodium current is blocked by 0.1% procaine.

These local anesthetics act by altering the voltage at which recovery from inactivation occurs, Figure 5-18, dashed line. At a low concentration of local anesthetic, removal of inactivation might not begin until -60 my and never is more than 20% complete. At higher concentrations, removal of inactivation would not occur until -120 mV, more negative than the cells' resting membrane voltage, so recovery never occurs. In axons, we use high concentrations to block action potential conduction. In the heart we use lower concentrations to delay recovery from inactivation, particularly in damaged ventricular cells with low resting membrane voltages which may fire action the voltage of the resting region, a little further down the axon, is close to the potassium equilibrium voltage. Ions will flow between these two regions of unequal voltage. When the ionic flow crosses the membrane of the resting region, it depolarizes the membrane to threshold, the sodium permeability increases, and resulting in further depolarization, and an action potential is generated in this previously resting region. This process continues on in each thin region of the axon. As can be seen in Figure 5-13, there is a short time delay between the peak of the action potential at A and B. The conduction velocity is obtained by dividing the distance, X, by the time, t. The conduction velocity of nonmyelinated nerves is roughly proportional to the square of the diameter and varies from 0.3 meters! sec for a 0.7 im axon to 25 meters/sec for a 500 im squid giant axon.

Salutatory Conduction.

In unmyelinated nerves, increased conduction velocity is achieved by increasing the diameter of the axon. Nature has found another way to

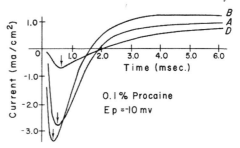

Figure 5-20. The effect of 0.1% procaine at pH 7.9 on the membrane currents, particularly the inward (down) Na current following a depolarizing step.

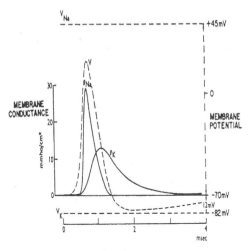

Fig 5-21. Membrane Conductance

achieve this; by increasing the distance between active regions along the nerve fiber. The time, t, between successive peaks of the action potential is little affected by the distance between the active regions. Consequently, the conduction velocity will increase as the distance between active regions increases. The myelin sheath around many nerves prevents ions from moving across the membrane, and therefore only the spaces between successive myelin sheaths, the nodes of Ranvier, are available for ionic movement. Consequently, the action potential jumps from node to node, about 1 mm at a jump. Conduction velocities in myelinated nerves run from 10 meters/sec for a 2 im fiber to 100 meters/sec for a 20 im nerve.

Refractory Period.

Immediately after an action potential, the axon is incapable of carrying a second action potential because the inactivation gates are still closed. This period lasts for about the duration of the action potential, 1 to 2 msec. This absolute refractory period is followed by a relative refractory period during which a supra-normal stimulus is necessary to evoke an action potential. This period lasts for another 2 to 3 msec. These two refractory periods put an upper limit of the nerve's frequency response at about 200 per second.

Active Transport.

As was previously discussed, sodium is very far from electrochemical equilibrium across the resting axon membrane. A small efflux of sodium can be measured; sodium ions are moving up their electrochemical gradient. This process must require energy since work is being done. This conclusion is confirmed by the almost complete cessation of sodium efflux when the metabolic process .of the axon are shut off by poisons such as cyanide or DNP, as seen in Figure 5-14, while injection of ATP will increase the sodium efflux for a short time. The ATP is broken down while providing the energy for sodium transport.

The efflux of three sodium ions is coupled with the influx of two potassium ions. A third potassium ion enters via the resting potassium channel. When potassium is removed from around the axon, the sodium efflux falls to very low values.

In a few systems the sodium is actively transported in conjunction with an anion, thereby preserving electroneutrality. In other systems only the sodium is actively transported and the accompanying ion is "dragged" through the membrane by an increased potential across the membrane.

In nerve, active transport is a relatively slow, continuous process that increases slightly after a large number of action potentials. Active transport maintains the ionic gradients on an hour-to-hour basis. It is not necessary to pump out the sodium that enters during a single action potential before a second action potential will fire. In fact, the average nerve will conduct thousands of action potentials while metabolically poisoned.

Dendrites and Cell Body

In addition to the axon with its rapid conduction mechanism, nerve cells have extensive dendrites and a cell body. These are covered with synapses and serve as the information gathering and integration portion of the nerve cell. The dendrites and cell body fluctuate in voltage. If the cell body depolarizes to -55 mV, an action potential fires at the first node of Ranvier and propagates along the axon. Most dendrites and cell bodies do not contain voltage gated Na channels so cannot fire an action potential. An anterior horn cell of the spinal cord may receive upwards of 5,000 different incoming synaptic connections. Some of these always cause firing of the axon by depolarizing it to threshold, most

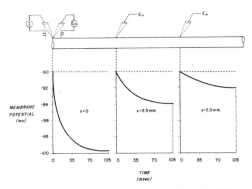

Figure 5-22. Decremental conduction in a dendrite.

excitatory synapses acting individually only depolarize towards threshold. Other synapses increase potassium permeability (Pk) and hyperpolarize the cell body, inhibiting the possibility of the axon firing an action potential.

Decremental Conduction. Dendrites and cell bodies communicate via continuously variable voltage fluctuations that become smaller with increasing distance, decremental conduction. These properties are similar to those of long undersea cables, so are often referred to as cable properties and are illustrated in *Figure 5-22*.

Current is passed into the left hand end of the dendrite to hyperpolarize the dendrite. The voltage of a nearby patch of membrane alters rapidly. When the voltage is measure 2.5 mm away from the current passing electrode, the induced voltage is reduced and the voltage changes more slowly. Voltage changes are more rapid for large structures and very slow in the thin dendrites. Consequently, synapses on the very ends of a dendritic tree have less of an influence on cell body voltage, and hence axon firing, than do synapses on the cell body itself. At each synaptic region the dendritic membrane contains a variety of receptors linked to ionic channels, such as the potassium channel we discussed in Part I of this chapter. We will discuss transmitter-receptor interaction in both chapters 6 and 7.

Calciom Channe;ls

Calcium channels are widely distributed in excitable cells. Ca channels have two important potencies: Ca entry carries charge to move the membrane voltage toward VCa, + 50 mV and the CA^{++} that enters the cell activates a wide range of reactions once it is within the cell.

Activation of the contractile proteins, are a few of the reactions activated by Ca entry. Most cells maintain very low Ca++, 1iM, while Ca++ =1-5 mM; consequently Vca is greater than +100 mV. Unlike the sodium ion that carries charge across the membrane and then becomes a nuisance that needs to be actively transported out of the cell, calcium has a dual role, both as a depolarizing charge carrier and as an intracellular messenger, a universal provocateur. Ca ions carry an electrical message on their way through the membrane and then activate a vast array of intracellular processes. Contraction, for example, is controlled by intracellular free calcium, some of which enters through calcium channels in the membrane. Calcium channels are intriguing because they exist in a number of types and a large number of highly specific and clinically useful blockers have been discovered.

Transient Calcium Channels.

Three major types of voltage gated calcium channels have been described. Transient (T type) channels have properties similar to sodium channels. *Figure 5-23* shows the major characteristics.

The peak Ca current (the number of Ca ions) passing through the channel is small, and it is open 40-60 msec. Threshold (Fig. 5-23B) is at -50 mV (solid line, open squares) and recovery from inactivation (solid line, close squares) begins at -60 mV and reaches maximum at -80 mV, the resting membrane voltage of these cells. These channels often coexist with Na channels and are co activated with the Na channel. One role of these channels is in cells that fire bursts of 15-20 action potentials. Ca enters the cell with each action potential; Ca++ builds up and eventually turns on enough Ca activated K channels to block further firing, even though the initial stimulus remains.

T channels by themselves carry enough charge to be the sole source of action potentials in many tissues. The rise time is not so rapid and the duration is longer than Na channel action potentials.

Long Lasting Calcium Channels.

L type channels *(Fig. 5-23)* scarcely inactivate even after several seconds and carry much larger peak Ca ion influxes. Threshold (Fig. 5-23B) (dotted line, open circles), however, is quite positive, -10 mV; some other voltage sensitive process, such as Na channels or T type Ca channels, must depolarize the cell to this threshold. Once opened, these channels inactivate slowly. Once inactivated, recovery (dotted line, closed circles) occurs at membrane voltages well positive to the resting membrane voltage. The major function of these channels is to bring Ca into the cell to raise Ca and serve as an intracellular messenger. These channels are very sensitive to the dihydropyridine drugs including nifeedine. An intermediate, N type channel has a threshold at -20 mV and has an open duration of 100-200 msec. with a peak Ca influx intermediate between T and L type channels. They are usually found in the presynaptic region of axons. The Ca^{++} that enters triggers neurotransmitter release.

Ligand gated Calcium Channels.

Channels of this type have been demonstrated in a number of tissues including CNS and smooth muscle. Various neurotransmitters. are responsible for both increasing and decreasing the number of open calcium channels. Opening these channels depolarizes the cell toward the threshold for activation of voltage gated Na and Ca channels and brings C+ into the cell to act as a second messenger.

Regulation.

L type calcium channels can be recorded in whole cell preparations quite easily by the patch clamp method

(Fig. 5-24). When the patch is pulled off the cell and the inside of the membrane exposed, L type channel activity quickly disappears (Fig 5-24, at 5-7 mm). Activity does not reappear until the catalytic subunit of cAMP protein kinase is added, together with Mg ATP, to the cytoplasmic side of the membrane. This enzyme system phosphorylates the inner side of the L type Ca channel to reestablish voltage gating (Fig 5-24 at 25-35 min). The initial loss of voltage sensitivity was the result of dephosphorylation, probably catalyzed by calcineurin, a calcium and calmodulin-dependent phosphatase. Here is a system in continuous flux; one system to activate channels by phosphorylation, another system to inactivate channels by dephosphozylation; a regulatory system operating on the scale of minutes. It regulates the number of channels capable ~of opening in response to depolarization, allowing Ca^{++} to transit the membrane and bringing in positive charge to depolarize the membrane and Ca^{++} to activate any number of intracellular systems.

The number of activated channels is controlled by a large number of neurotransmitter and hormones acting via receptors proteins in the membrane to modulate c AMP and hence channel phosphorylation. The number of Ca channels in the heart ventricle is increased by epinephrine, the action potential is more positive and prolonged and more CA enters the cell. Increased CA results in a more forceful contraction. The number of systems and the subtlety of their action are expanding too rapidly to make listing them all useful. Patch Clamp

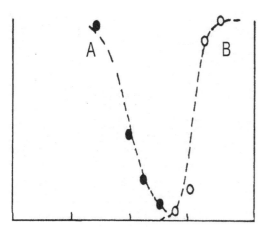

Figure 5-23. A. Ca flux (current) through a fully excited T type or L type Ca Channel. B. T type Ca channels (, solid curves) are quite similar to Na channels; threshold is near -50 mV, they open briefly and then inactivate. Recovery from inactivation begins at -50 mV and is rapid and complete at -80 mV. L types channels have a threshold close to zero (, dotted curves) but once opened remain open for seconds and allow large amounts of Ca to enter the cell. Recovery from inactivation is a voltage much less negative than the resting membrane voltage. (Adapted from Fox, A.P., Nowyck, M.C. & Tsien, R.W. J. Physiol 44; 149-172).

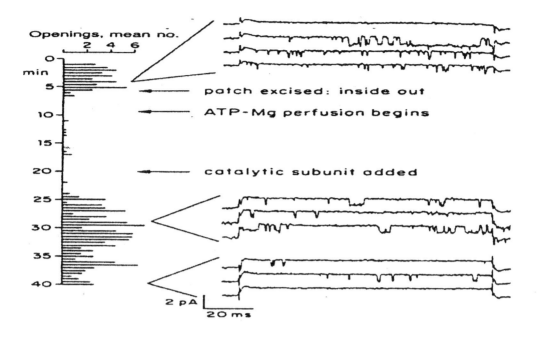

Fig 5-24. The inner surface of the channel is exposed to an artificial cytoplasm; the site is dephosphorylated and channel activity cease. Channel activity is regained when the phosphorylation enzyme system is added. (From Armstrong and Eckert, Proc.Natl. Acad. Sci. USA 84: 2518-2522, 1987)

CHAPTER 6
Skeletal Muscle and Nerve-Muscle Junction

GROSS STRUCTURE AND FUNCTION.

Skeletal muscles are the major ending of the efferent branch of the central nervous system. We work our will upon the outside world through these muscles. Skeletal muscles occupy about 80 % of the total weight of the body. They use about 6 % of the resting oxygen consumption to maintain ionic gradients; after strenuous exercise they may use as much as 70 % of the oxygen consumption. Each anatomic muscle is delimited by strong fascial sheets and has a characteristic origin and insertion. The basic unit of the muscle is the muscle fiber or cell that runs from one end of the muscle to the other end and has a diameter of 50 to 100 μm.

Motor Units.

Anatomic muscles are subdivided into bundles of many fibers (Fig. 6-1). Each muscle fiber is innervated by only one motor nerve fiber. Groups of muscle fibers are delimited functionally by their nervous innervation. As the *Fig. 6-1* shows, the motor nerve branches and innervates a number of muscle fibers. The muscle fibers that are innervated by a single motor nerve are called a motor unit or group. All these fibers act in the same manner since a single nerve controls them. The number of muscle fibers in a motor unit varies from 300 to 400 in the gastrocnemius (calf) muscle to 4 to 6 in the extraocular muscles. In general, the size of the muscle group is proportioned to the delicacy of the required movement. The extraocular muscles make very small, fine adjustments; the gastrocnemius muscle, coarse, powerful movements.

When the motor nerve is stimulated, an action potential travels down the axon until it reaches the end-plate region, where it releases a chemical transmitter, acetylcholine. The acetylcholine diffuses to a specialized portion of the muscle surface, the motor end plate, and initiates a second action potential on the surface membrane of the muscle fiber. We will discuss the motor endplate in greater detail later in the chapter.

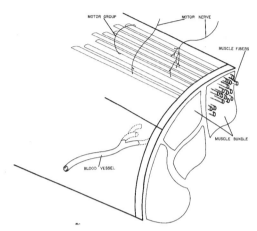

Figure 6-1. The organization of muscle fibers into structural units, muscle bundles and functional units, motor groups.

Contraction.

The action potential travels along the muscle surface from the end-plate region with a conduction velocity of about 1 meter/sec. The response to a single stimulus, either to the motor nerve or to the muscle surface, is called a twitch. Muscle activity usually occurs in response to a series of action potentials, a partial or complete tetanus as illustrated in *Figure 6-2B,C, and D*. It can be seen that increasing the frequency of stimulation to a muscle increases the tension generated. Since motor units are in parallel, their tension is additive. The total tension a muscle produces is primarily a function of the number of motor units activated.

Sarcomeres and Filaments.
When we study the structure of the muscle fiber, the mechanism of contraction becomes clearer. *Figure 6-3, A-D* shows the structure of a 100-μm diameter muscle fiber and its component 1 μm myofibrils. The banded pattern (D) is clearly seen in the light microscope in either single living fibers or fixed and stained material. A dark A band alternates with a light I band. The I band is bisected by a thin, dark Z disk. The basic contractile unit is a length of myofibril from Z line to Z line

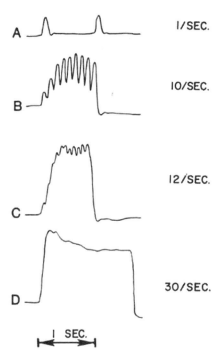

Figure 6-2 Isometric tension in response to stimuli of constant voltage and a varying frequency. Note that the total tension is greater when the frequency of stimulation is increased until a maximum (tetanus) tension is reached. Record from a human flexor carpi radialis muscle in situ. The subject's arm was held to a table with adhesive tape; stimulation was via a carbon electrode over the muscle mass in the upper forearm and a large ECG electrode at the wrist. The tension transducer was in contact with the styloid process on the wrist below the base of the thumb. Both wrist and finger flexors can be stimulated by this method.

called a sarcomere. After isolation, this unit, 1 μm in diameter and 2.5 μm long, will still contract.

At higher magnification, in the electron microscope *(Fig 6-3E)*, it is clear that the bulk of the muscle structure is made up of two types of filaments. The larger of these filaments, the thick filament, is 10 nm in diameter and 1.5 μm long and is located entirely within the A band. Indeed, all of the properties of the A band can be attributed to these filaments. The thin filaments are 4 nm in diameter and 1.0 μm long and run from the Z line various distances into the A band. During contraction the thin filaments are pulled past the thick filaments to reduce sarcomere length.

Excitation Contraction Coupling

Muscle activation begins when an action potential spreads over the surface and then into the depth of each fiber. Calcium released from intracellular structures allows the thick and thin filaments to interact, produce tension, and shorten. Contraction ceases when CA++ is transported into the same intracellular structures.

Reticular Structures. Skeletal muscle has an enlarged and specialized reticular network that is shown in *Figure 6-4*. It can be subdivided into two portions: the transverse tubules or T system and sarcoplasmic reticulum (SR). The T system is a tubular network that is continuous across the whole fiber and contains extracellular fluid. If a perfect cross section were cut across the fiber, the T system would look like a chicken wire fence with the fibrils running through the holes in the wire. The sarcoplasmic reticulum wraps around the myofibrils like the bun around a hot dog.

The (T) system conducts the surface action potential rapidly inward to initiate contraction. When small patches of the surface membrane are stimulated to induce local contraction, the sensitivity of an area depends on its location with respect to the T system. The most sensitive location varies; it is at the Z line in the frog and at the A-I junction in the lizard and many mammals (arrows, Fig. 6-4). Inward conduction is an active, Na+ dependent process in the tubular wall, probably much like the surface action potential. Depolarization of the transverse tubular system generates a charge = movement signal which precedes Ca^{++} release from the SR. This charge = movement signal has many similarities to the gating current of the axon.

The basic ionic mechanisms underlying the muscle action potential are quite similar to those in nerve, a regenerative increase in sodium permeability (to depolarize) which quickly inactivates, followed by an increase in potassium permeability (to repolarize). In skeletal muscle there is also a large, but unchanging, chloride permeability which participates in the repolarization phase. Chloride is in equilibrium across the membrane at a resting potential of -90 mV.

The surface area of the T system gives the 'surface' action potential a slow velocity (1 m/sec). Most of the potassium channels are in the T = tubule membrane so the potassium efflux during the falling phase of the action potential is into the lumen of the T system. After several action potentials, K⁺ builds up in the T lumen and begins to depolarize the fiber. In normal muscles this depolarization is not large and is buffered by the chloride conductance in the surface membrane.

Myotonia. In muscle fibers from myotonic goats the potassium buildup in the T tubules causes a significant depolarization and action potentials continue to fire after stimulation ceases. Much the same result is obtained when nor-

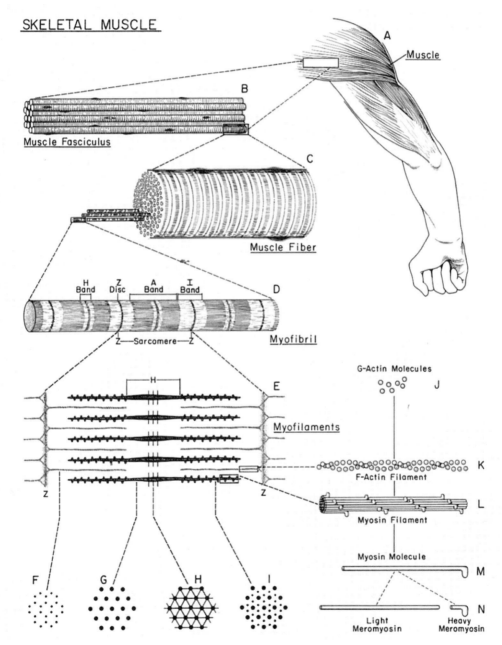

Figure 6-3 Diagram of the organization of skeletal muscle from the gross to the molecular level. F,G,H, and I are cross sections at the levels indicated. (Drawing by Sylvia Colard Keene from Bloom and Fawcett: A Textbook of Histology. Philadelphia, W.B. Saunders, 1968.)

mal fibers are placed in chloride-free solution.

T-SR Coupling.

The structure of the T system - terminal cisternae junction is shown in *Figure 6-5A*. The T

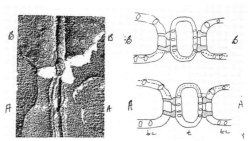

Figure 6-5. A. The T system - terminal cisternae junction. A freeze fracture study showing the T system running vertically flanked by terminal cisternae. B. A cross sectional drawing showing the fracture planes at AA and BB. Fracture lines follow the hydrophobic (center) line of the membrane. The foot processes of the SR are stippled. (Electron micrograph courtesy of Clara Franzini-Armstrong, Ph.D., University of Pennsylvania, Philadelphia, PA)

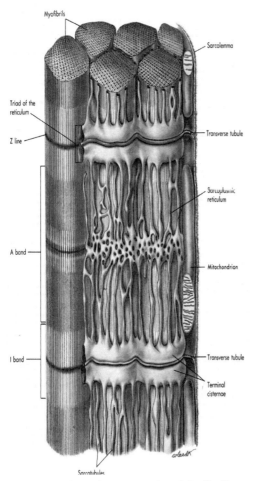

Figure 6-4 Schematic representation of the distribution of the sarcoplasmic reticulum around the myofibrils of skeletal muscle. The longitudinal sarcotubules are confluent with transverse elements called the terminal cisternae. A slender transverse tubule (T tubule) extending inward from the sarcolemma is flanked by two terminal cisternae to form the so-called triads of the reticulum. The location of these with respect to the cross-banded pattern of the myofibrils varies from species to species. In frog muscle, depicted here, the triads are at the Z line. In mammalian muscle there are two to each sarcomere, located at the A-l junctions. (Modified after L. Peachey, from Fawcett, D.W., and McNutt, S.: J. Cell Biol., 25:209, 1965. Drawn by Sylvia Colard Keene.)

tubule runs vertically in the center. The T system is ovoid and the portion adjacent to the terminal cisternae contains prominent particles.

The terminal cisternae (tc) of the sarcoplasmic reticulum shows a rich array of structures which appear as either pits or particles depending upon the plane of fracture. At the T-tc junction are "feet" which create a distinct morphological gap. The terminal cisternae are not do not participate in any of the electrical events of the muscle fiber indicating there is no ionic connection between the two. The feet are the site of Ca^{++} release from the SR that initiates contraction. After Ca^{++} is released from the terminal cisternae, the Ca^{++} concentration in the sarcoplasm rapidly builds up as shown in Figure 6-6, well in advance of tension generation.

The ultimate result of the depolarization of the T system is Ca^{++} release from the terminal cisternae. How depolarization induced charge movement in the T system walls induces release of Ca^{++} from the SR is far from clear. One of the author's suspects there is a chemical transmission and Ca^{++} is the first messenger. Other authors suggest a molecular rod that spans the foot process that controls Ca release from the SR.

Dantroline sodium (Dantrium) is a muscle-relaxing drug that acts directly upon the muscle fiber. One-half hour after a clinical dose, the twitch tension is half it normal value; recovery occurs within 24 hours. Dantrolene sodium acts to reduce the amount of Ca^{++} released per action potential, largely by blocking release of Ca^{++} from the terminal cisternae.

The large particles elsewhere in the SR membrane are the site where Ca++ from the sarcoplasm is taken back into the SR bringing about relaxation.

MOLECULAR ARCHITECTURE OF CONTRACTION

Muscle Proteins. Four major protein types have been extracted from skeletal muscle: actin, myosin, tropomyosin and troponin. The first two are the major protein constituents; the others are a relatively small fraction of the protein. Myosin has a molecular weight of 420 KD and contains two quite different regions; a long double helical tail and a dual headed structure containing both the ATPase and actin binding activities. Figure 6-3, L-N shows the packing of the myosin molecules within the thick filaments of the band. The myosin molecule can be readily divided into two subunits - heavy and light meromyosin. The heavy meromyosin retains the adenosinetriphosphatase (ATPase) and actin binding activity of the intact myosin. This fragment forms the bridges between this and the filaments. Light meromyosin rods aggregate to form a tension bearing rod, the thick filament.

Actin has a molecular weight of 60kD and readily forms long chains of a fibrous protein (Fig. 6-3,J-K). Two chains of F-actin, wound around each other, form the thin filaments along with tropomyosin and troponin. These two regulatory proteins alter the ATPase activity of the actin-myosin complex so that calcium ions are required for ATP breakdown and consequently for muscle activity.

The immediate energy source for contraction is ATP and muscle converts chemical to mechanical energy with an efficiency of about 50%. Intact muscle contains relatively large quantities of creatine phosphate, an energy storage protein.

Ca^{++}-Troponin Interaction. Free Ca^{++} binds to troponin on the thin filaments causing the troponin-tropomyosin to 'roll-back' from the active site of actin. Actin and myosin can now interact, bridges can form and tension is generated. The interaction continues for as long as the Ca^{++} remains elevated. In the twitch shown in *Figure 6-6*, Ca^{++} is rapidly taken up by the sarcoplasmic reticulum and tension declines. During tetanic stimulation, Ca++ remains elevated. After the surface membrane and reticulum have been removed by chemical treatment, tension generation is related to Ca++ concentration *(Fig. 6-7)*. Tension is maintained for as long as Ca++ is present.

Filament Interaction.

A muscle fiber shortens when thin filaments slide past thick filaments. In this manner the distance between individual Z lines (sarcomere length) decreases. The decrease in muscle length is proportional to the product of the decrease per sarcomere and the number of sarcomeres per muscle.

Cross sections through the array of thick and thin filaments are shown in Figure 6-8, F-I. Each set of filaments is arranged in a basically hexagonal array. In the region of overlap, section I, each of the thin filaments has three thick filaments and each neighboring thick filament has six thin filaments. Contact between the filaments is made by cross bridges *(Fig. 6-8)*. These bridges stick out from the thick filaments and are arranged in a six-fold helix, like the treads on a spiral staircase. In this case, the stairs make a complete revolution in six steps.

Cross-Bridges.

It is the interaction between the two filament

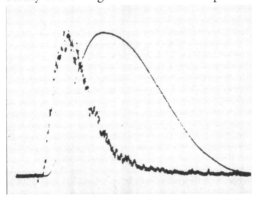

Figure 6-6. The broken line shows the average light emission from Ca++ sensitive protein aequorin inside a muscle fiber while the solid line is the twitch tension. Aequorin, a protein extracted from a luminescent jellyfish, emits light in proportion to Ca++ concentration. The light emission, hence the increasing Ca concentration within the fiber, proceeds the tension generation (solid line). From Blinks, JR, R. Rudel and SR Taylor, J. Physiol. 277. 291-323, 1978.

arrays that produces tension; the bridges *(Fig. 6-8)* of the thick filaments "hook on" to and move the thin filaments. The three-dimensional structure of the heavy meromyosin (Fig 6-3) which contains both the actin and ATP binding sites has recently been worked out. There is a prominent cleft near the actin-binding site: when it is closed *(Fig 6-9A)*, actin-myosin binding is strong, when it is open (Fig 6-9B) binding is weak. A lateral pocket contains the ATP binding site; closing of the ATP binding pocket imparts a curvature to the head and a 5 nm movement of the actin binding site along the actin chain (Fig 6-9 B>C). Fig 6-9 shows a molecular model of bridge interaction consisting of actin-myosin dissociation with ATP binding (A>B), curving of the bridge with P hydrolysis to move the myosin head 5 nm

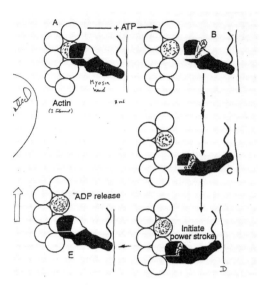

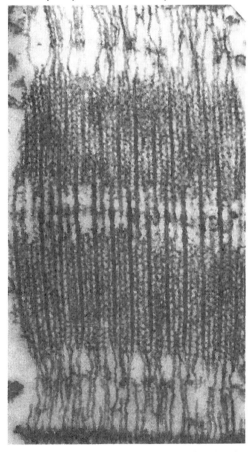

Figure 6-7. A view of the central region of the band at 600,000 x magnification. The projections from the thick filaments, the bridges, are the site of interaction between the thick and thin filaments. (From Huxley, H.E.: J. Biophys. Biochem. Cytol., 3:631, 1957.

Figure 6-8. The power cycle of the bridge unit of myosin. In A the cleft between the two domains of the myosin head is closed and the myosin head is attached to an actin monomer (speckled for identification). The ATP binding pocket is open. As ATP begins binding, B, the cleft opens and the myosin head unbinds from actin. As ATP binding is completed and P hydrolysis occurs, the ATP pocket closes and the myosin head changes curvature, C, to move 5 nm along the actin chains, to bring it into alignment with the next actin monomer. As the P leaves, D, the cleft closes and the head attaches to the next actin monomer. As ADP is released, E, the pocket opens, the myosin head straightens forcing the actin chain up by 5 nm, the power stroke. Adapted from Rayment et al, Science 261:50-65, 1993.

along the actin chain (B>C), head reattachment (D) and finally the power stroke as ADP dissociates (D>E).

Length - Tension: Relations. Tension production, then, should be related to the number of bridges connected to thin filaments, to the degree of overlap of actin and myosin filaments, and 2 to the sarcomere spacing. When resting muscle is fixed at various sarcomere spacing, the length of each set of filaments remains constant, as does their diameter *(Fig. 6-10)*. The amount of overlap between the thick and thin filaments varies in direct proportion to sarcomere length, but the A bandwidth remains constant during contraction. Experiments with contracting isolated fibrils also show the bandwidth with the I band decreases in width with decreasing sarcomere spacing. When the sarcomere spacing is just greater than the sum of the lengths of the and I band filaments (3.6 µm) there should be no overlap, no bridge interaction, and consequently no tension production, Figure 6-9B. As the sar-

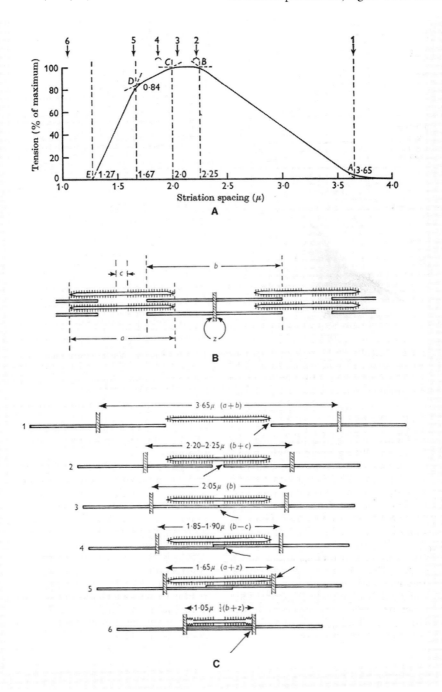

Figure 6-9. The length tension curve (A) and electron micrographs (B,C, and D) of skeletal muscle showing the overlap of the sliding filaments. When the filaments barely overlap (B), tension is low. When all bridges are attached (C), tension is maximum. When the thin filaments overlap (D), tension declines. (Data from AL Gordon, AF, Huxley and F.J. Julian. J. Physiol. 184:170-192, 1966. Electron micrographs courtesy of Brenda Eisenberg, Ph.D., University of Illinois College of Medicine, Chicago, IL.)

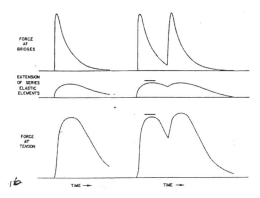

Figure 6-10. Calculated values for bridge tension, length of series elastic component, and tension. Note especially that a second stimulus can capitalize upon extension of the series elastic element and thereby increase the tension at the tendon. F.J. Julian, Harvard Medical School, kindly supplied (these records. For further details of the computation, see Julian, F.J.: Biophysical J. 9:547, 1969.)

comere spacing decreases, the number of bridges increases and so does the tension. The number of bridges and the tension increase until all of the bridges are attached. There are no bridges in the center of the band, as can be seen in Figure 6-8.

Further shortening gives no increase in tension since there is no more bridges to interact. In fact, further shortening leads to decreased tension since one thin filament must force its way past its opposite filament, probably disturbing the bridge filament interaction. Further shortening also requires compression of the band filaments and most of the tension that is produced goes to compress the thick filaments.

Force-velocity Relation.

We have concentrated so far on the generation of tension at constant length; to do useful work the muscle must shorten. When a muscle shortens against a very light load it shortens very quickly. As the load increases, the velocity

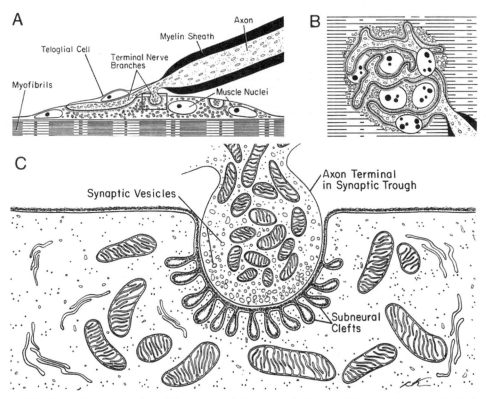

Figure 6-11. Schematic representations of the motor end plate as seen by light and electron microscopy. A, End plate as seen in histological sections in the long axis of the muscle fiber. B, As seen in surface view with the light microscope. C, As seen in an electron micrograph of an area such as that in the rectangle on A (After R. Couteaux. From Bloom and Fawcett: A Textbook of Histology. Philadelphia, W.B. Saunders, 1968.)

decreases. A force-velocity curve is shown in *Figure 6-10*. A human forearm was pulling against increasingly heavy weights. The same curve can be obtained with an isolated muscle; maximum tension and maximum speed will vary, however, from muscle to muscle.

The force-velocity curve is described by the equation

$(P+a)(V+b)=(P_o+a)b$, where P is the force, P_o is the isometric tetanic force, V is the velocity of shortening, and a and b are constants. The constants, a and b, can be fitted uniquely and are a major goal for theoretical models of muscle function.

Active State.

When the bridges from the thick filaments interact with the thin filaments, they develop tension 3 to 5 msec after the action potential runs along the surface. Tension cannot, however, be measured in the tendon for 10 to 20 msec following stimulation. What causes the delay? There is a great deal of elastic material in series with the tension-generating element, some indeed in the bridges themselves. The tension produced by the bridges must first stretch the elastic material before it can be transmitted to the tendon and the load. Consider for a moment a person pulling a stretchy nylon rope that is attached to a large rock. At first the rope stretches only after a delay does it transmit the full force of his pull. Remember that a pull of a given force will extend the rope to a characteristic length and no further, but the stretching will take time.

Let us return to the case of the muscle. Following a single stimulus, the bridges generate their maximum tension for a very short time *(Fig. 6-10)*. This short phase corresponds to the active state in A.V. Hills model. The tension is mainly expended in stretching the series elasticity; about one-half of the bridge tension appears as tendon tension. If a second stimulus follows soon after the first, the renewed bridge tension pulls on partially extended elastic elements and consequently must do less work on them to transmit tension to the tendon; more of the bridge tension is applied to the tendon. If a third stimulus follows the first two, the work the bridges do on the series elastic elements is further reduced; consequently, the work done and tension produced on the tendon is much greater during a tetanus (Fig. 6-10D) than during a twitch.

NERVE-MUSCLE JUNCTION

The Endplate

To return to the control of skeletal muscle contraction, the motor nerve enters the muscle, branches, and forms a very close junction, a synapse, with the center of each muscle fiber. The axon (presynaptic element) and muscle (postsynaptic element) remain two distinctly separate cells. The structure of the region is shown in Figure 6-11. Transmission between the axon and muscle is by means of the chemical, acetylcholine (ACh).

Acetylcholine. To be considered a transmitter, a substance must meet four criteria: (a) it must be effective at the postsynaptic surface, (b) it must be liberated by the presynaptic surface in response to and in proportion to nerve stimulation, (c) there must be an enzyme system for destroying or reabsorbing liberated transmitter, and (d) there must be an enzyme system for synthesizing the transmitter.

Each of these criteria has been met by acetylcholine at the nerve-muscle junction. The idea of a chemical transmitter between nerve and muscle came from experiments by Otto Loewi in 1921 that demonstrated the slowing of the heart when the vagus nerve was stimulated. He noticed that Ringer's solution that had dripped over a slowed heart caused a second heart to slow as well. The only link between the two hearts was the Ringer's solution. Vagal stimulation liberated acetylcholine into the solution that affected the second heart. Acetylcholine can be collected in from small veins in stimulated muscles.

Acetylcholine is effective in stimulating contraction if it is injected very close to the muscle. When acetylcholine is microinjected onto an excised muscle, it elicits contraction only when injected at the end-plate region. The rest of a normal muscle does not have ACh receptors. Both biochemical and histochemical studies show the presence of a cleaving enzyme, acetylcholine esterase at the myoneuronal junction. It is present in both the postsynaptic surfaces in the subneural folds and on the presynaptic side. It is also present in fairly high concentration in the blood that explains the general ineffectiveness of acetylcholine when injected intra-arterially.

Another enzyme, choline acetylase, reconstitutes the acetylcholine, into the synaptic vesicles. The enzymes that produce the vesicles are produced in the nerve cell body; the number of vesicles decreases rapidly after section of the motor nerve. When an action potential enters the presynaptic terminal, the accompanying depolarization opens voltage gated Ca channels. Increasing intracellular Ca++ is essential for neurotransmitter release, catalyzing the fusion of synaptic vesicles with the surface membrane to release ACh into the synaptic cleft *(Fig 6-12)*.

Acetylcholine acts on the postsynaptic membrane and destroys its selective permeability to all small ions. The membrane has an equilibrium voltage of zero millivolts. This voltage is, of course, never reached, but would be if acetylcholine were continuously applied to the end plate. The high concentration of acetylcholine esterase prevents such an event. The depolarization brought about by the usual amount of acetylcholine causes a current to flow to the surrounding muscle fiber surface. As the current flows through the peripheral membrane, it depolarizes that membrane and sets up a propagated action potential. The end-plate region itself is not electrically excitable; current must flow to the surrounding area which area contains classi-cal sodium channels that must be repolarized before it can fire again.

Acetylcholine Activated Channels. The receptor and channel are contained in a single transmembrane protein aggregate (Fig 6-12) made up of 5 subunits each of which is thought to contain three alpha helical columns with a 0.7 nm channel down the center. The receptors on the alpha subunits are some 6 nm from the gated ionic channel that is on the cytoplasmic end of the channel.

Single channel recordings *(Fig. 6-13)* show the channel to be either open or closed. ACh concentration alters the % time the channel is open. When a microelectrode is inserted into the end-plate region, the usual resting membrane potential is recorded. When the motor nerve is

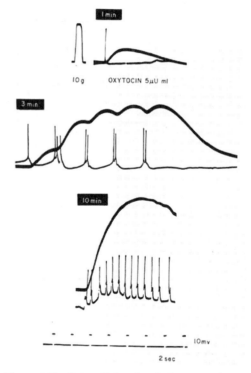

Figure 6-13. Intracellular electrical events at the end-plate region (A), and (B), 2 mm away. The upper traces are at high gain (3.6 mv) and slow time scale, (47 msec). In (A) the miniature end-plate potentials (minis) are clearly shown, whereas in (B) they are almost nonexistent. The two bottom traces are at low gain, (50 mv) and fast time scale, (2 msec). Note the end-plate potential at the leading edge of the action potential in A. (From Fatt, P., and Katz, G.: J. Physiol., 117:109, 1952.)

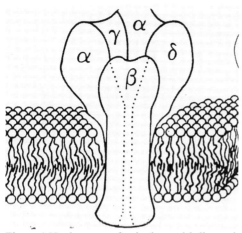

Figure 6-12. A cartoon of a single acetylcholine activated channel. Each channel has two ACh binding sites; one on the cytoplasmic side of each alpha subunit. Miles, K. and R.L. Huganir, Molecular Neurobiology 2: 91-124, 1988.

stimulated, the microelectrode records an action potential as shown in the lower traces of Figure 6-13. Careful examination of the left hand action potential will reveal a hump on the leading edge of the action potential. This hump is the part of the end-plate potential caused by the release of acetylcholine. It is a local change in the end-plate region; no end-plate potential is seen 2-mm away (B). The end-plate potential can be seen more clearly when curarine is applied to reduce the effectiveness of the acetylcholine and to produce a subthreshold depolarization.

Small depolarizations are recorded at times when the motor nerve is quiet. They originate in the end-plate region and can be recorded only very close to it. These miniature end-plate potentials ("minis") are quantized *(Fig 6-13)*. They represent the release of several vesicles of ACh. The amplitude of these potentials increases as more vesicles are released. Each pocket or vesicle releases some 10^{-17} moles of acetylcholine that results in a depolarization of 0.4 mV. An action potential arriving at the presynaptic terminal causes an increase in the rate of vesicle release so that about 100 vesicles discharge their acetylcholine over a very short interval; this amount of acetylcholine is sufficient to initiate an action potential on the muscle surface.

MECHANISMS OF DRUGS ACTING ON NERVE-MUSCLE JUNCTION

Drugs that Compete for Receptor Site. D-tubocurarine competes for the same postsynaptic site, as does acetylcholine; it binds more strongly to the site but does not cause a permeability change. Neuromuscular transmission is blocked because the acetylcholine cannot get to the sites. Gallamine (Flaxedil) is a synthetic d-tubocurarine. Several snake neurotoxins, including bungarotoxin, also bind very strongly and block the receptor. Hexamethonium and tetramethylammonium (TEA) also block by competitive antagonism. These drugs are slowly broken down and paralysis wears off.

Continued application of acetylcholine, and continued depolarization, does not result in continued skeletal muscle activity because the action potential mechanism must be reset after the first action potential. Any drug that produces a continual depolarization at the end plate produces a depolarization block. Nerve-muscle block and muscle relaxation should be achieved by infusing large quantities of acetylcholine but this method is very inefficient because the acetylcholine is broken down rapidly. Succinylcholine binds to the postsynaptic surface and causes a permeability change. Since it is only slowly broken down, it produces a long-lasting depolarization. It depolarizes the muscle membrane, which e contracts once and then relaxes. Decamethonium also acts in this manner.

Drugs that block the nerve-muscle junction are often used in conjunction with anesthetics to achieve muscle relaxation during surgery. It should always be borne in mind that the respiratory muscle would be blocked so that the patient must be artificially ventilated.

Drugs that prolong the action of acetylcholine. Blocking the esterase activity o that the action of the released acetylcholine is prolonged can also block the end plate. The anticholinesterase, such as neostigmine, edrophonium (Tensilon), eserine, and diisopropyl fluorophosphate (DFP), all combine with acetylcholine esterase and prevent it from cleaving acetylcholine, so ACh continues to depolarize the end-plate region and to block transmission. Many insecticides block acetylcholinesterase.

Agents that block the release of acetylcholine from the presynaptic terminal will be discussed below. The various disorders affecting the neuromuscular junction are discussed in great detail below.

EFFECTS OF MOTOR NERVE ON SKELETAL MUSCLE

From the discussion so far, one might conclude that the motor nerve supplies only the stimulus to contract; this is far from the case. When the motor nerve is cut, the muscle rapidly atrophies. Its volume and strength of contraction decrease – and all of the fibers decrease in size, as does the total muscle bulk. After three months, the muscle bulk may have decreased to as little as 25 % of its original bulk. This atrophy is not due to muscle disuse alone since disuse, such as that caused by immobilization, will cause muscle to atrophy to 3/4 of its original size in the same 3-month period. An intact motor nerve is necessary for continued survival of the muscle fiber. It would appear that the continued release of acetylcholine, or possibly some other transmitter, from the nerve, is the agent responsible for this trophic influence.

Denervation Sensitivity. When the muscle is denervated, the entire surface of the muscle fiber slowly becomes sensitive to acetylcholine. After several weeks, application of ACh anywhere on the surface results in an action potential and contraction. Patients who have had a motor nerve severed show great sensitivity to injected ACh for this reason.

Reinnervation. If the cut ends of a motor nerve are rejoined, sprouts from the central end will grow down the tube left by the degenerated axon. If the distance between the cut and the muscle is short and the cut ends are well aligned, the nerve will make contact with the muscle. A new end plate will form, and muscle function will be restored. Passive exercise of the muscle during regrowth is helpful to prevent disuse atrophy and contractures.

Slow and Fast Muscles.

There are two sorts of skeletal muscle: red and white. The best known example of this difference is the white and dark meat of a chicken; the same qualitative differences appear in mammals.

Most muscles, particularly in mammals, are not pure red or white muscle but are made up of a mixture of fibers *(Fig 6-14)*. The fibers in red muscles are called Type I. They are rich in mitochondria but have a relatively low myosin ATPase activity. They generate ATP from glucose as it is used. Enzymes of oxidative metabolism cause the red color. Type II fibers found in white muscles have a very active myosin ATPase activity and are relatively lacking in mitochondria. Type II fibers which are responsive for quick, phasic contractions depend upon an anaerobic, glycolytic metabolism; they produce large amounts of lactic acid.

Red muscle is characterized by a long twitch time, and concomitantly a low frequency of stimulation will result in tetanus. These muscles are primarily antigravity or postural muscles; their movements are characterized by long-sustained contractions. The white muscles, on the other hand, have short twitch times and are used for quick phasic movements.

The speed of contraction is determined by the pattern of activity in the motor nerve. When nerves from fast and slow muscles are crossed, the fast muscle slows down and the slow muscle speeds up *(Fig 6-14)*. When a limb is immobilized, the activity in motor nerves to a slow muscle, which is usually intense, decreases. After several weeks the speed of contraction has increased markedly. It is not clear how the pattern of activity alters the biochemical control mechanisms and the contractile properties of a muscle. Some authors suggest a second transmitter released in very small quantities from motor nerves is responsible for modulating which proteins are expressed from the muscle genome.

DISEASES OF MUSCLE

Most classifications divide these disorders into two groups: inherited and acquired (Morgan & Hughes 1992).

Muscular Dystrophies

The major inherited disease is the muscular dystrophies that are characterized by a primary degeneration of skeletal muscle. The slow but progressive destruction of muscle results in a progressive weakness initially affecting the proximal muscles.

Duchenne's Muscular Dystrophy. The most common varieties of muscular dystrophy are X linked almost all cases occur in males (rarely females with Turner's syndrome or X-chromosome translocation may be affected). Duchenne originally described the most common type within this group in 1868. The typical patient with

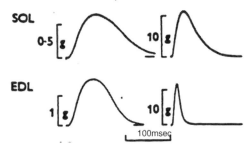

Figure 6-14. Twitch time in a fast, extensor digitorum longus (EDL) and a slow, soleus (SOL) muscle. At birth (left) the difference in twitch time is not striking, yet 5 weeks later (right) the difference is very pronounced. (From Close, R.: J. Physiol., 180:542, 1965.)

Duchenne's muscular dystrophy (DMD) is a boy who has delayed walking. At some point between the ages of 2 and 5 years, the patient is noted to be clumsy and slow in exercises and games. It is soon evident that proximal pelvic girdle lower extremity weakness is present. In rising from the floor the patient uses his upper extremities and hands placed on the thighs to force the body in to an erect position (Gower's sign). Thus counteracting the weakness at hips and pelvis. At this point, marked enlargement of the calf muscles is present (pseudo hypertrophy). With progression of the disease, increasing atrophy of muscle and increasing weakness occurs. By age 12, the patient is confined to a wheel chair. Death occurs late in the teens or early twenties from respiratory complications and/or cardiac failure (heart muscle is involved). The blood creatine kinase level while the patient is still ambulatory is at least 40 times the upper limit of normal. Levels may be 300-400 times normal. Electromyogram (EMG) demonstrates myopathic features. The muscle biopsy taken early in the course of the disease before severe atrophy is present demonstrates characteristic myopathic features including: 1) wide spread necrosis and phagocytosis of muscle fibers 2) regeneration of muscle fibers, 3) marked variation in fiber size and 4) large rounded "hypercontracted hyalinized" fibers. *(Fig. 6-15)*. In late stages replacement of muscle by proliferation of endomysial connective tissue and fat occurs.

DMD is the most common lethal, X linked disease with an estimated incidence of 1 in 4000 live male births (see Moser 1984). The maternal carrier can be identified based on family history and an elevated creatine kinase level in the carrier. A small % of carriers also has mild weakness or EMG or muscle biopsy changes. There is however a high frequency of isolated cases, 30-50% suggesting a high incidence of new mutants or mutant maternal carriers.

In families at risk, prenatal diagnosis is possible (Darras et al. 1987, and see below).

Becker's muscular dystrophy. Becker's muscular dystrophy described in 1955 is a less common (1 in 20,000 male births) and less severe form of x-linked disorders. Age of onset is later; and the rate of progression is slower. The patient is still ambulatory at age 15. Life expectancy is only slightly reduced. Creatine Kinase levels are markedly increased. The EMG demonstrates myopathic features. The muscle biopsy indicates features that are similar but less marked than those noted above.

Recent major advances have been made in our understanding of the genetics and molecular biology of these disorders (Arahata et al, 1989, Hoffman et al. 1988, see also review of Rowland 1988). The precise locus has been identified: the short arm of the X chromosome at the region designated as Xp 21. The specific affected gene has been isolated and characterized. DNA analysis has shown that both DMD and BMD affect the same gene (allelic). Portions of the coding sequence of the gene have been used to produce polyclonal antisera directed against the normal muscle protein produce of the normal gene. The specific protein, dystrophin, is a normal component of the plasma membrane, transverse tubule system of the normal muscle fiber. In patients with Duchenne's muscular dystrophy, the muscle contained less than <3% of the amounts of this protein found in control patients. In patients with Becker's muscular dystrophy, the dystrophin was normal but the size of the protein (molecular weight) was abnormal. Patients with an intermediate clinical course had results bridging these two disorders. Patients with other types of neuromuscular disorders had normal dystrophin.

Thus a defective gene at this specific site may result in a total defect in a failure to produce this muscle protein (DMD) in the production of a

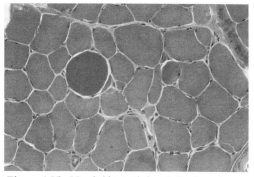

Figure 6-15. Muscle histopathology I: Duchenne's Muscular dystrophy Marked variation in fiber size is present with a large dense hypercontracted hyalinized fiber (Compare to 6-21 and 8-21). H&E x 63- Courtesy of Dr. Tom Smith, U.Mass Medical Center.

muscle protein of abnormal size (BMD) or in various combinations of these deficits. Specific therapy has not yet been achieved. Corticosteroids may produce minor improvement (Brown 1989).

The non X-linked muscular dystrophies are less common.

Facioscapulohumeral (FSH) dystrophy is an autosomal dominant with the gene linked to chromosome 40. The incidence is 0.5-5.0/100,000 persons. The phenotypic expression is variable. Some cases begin in childhood, and have a poor prognosis. Most begin in the late teens, or early twenties with a slow rate of progression. Often the specific age of onset is difficult to identify. The name of the disease reflects the prominent early features of bilateral facial weakness and proximal upper extremity involvement with weakness of shoulder abduction and winging of the scapula. Subsequently lower extremities are involved. In most cases little disability of significant degree occurs before the thirties or forties. Life span is relatively well preserved. Serum creatine kinase is borderline or only mildly elevated reflecting the slow rate of muscle breakdown. The EMG and muscle biopsy reflects the myopathic features.

Limb girdle dystrophies. The limb girdle dystrophies are heterogenous autosomal recessive disorders and some are autosomal dominants. Some begin in the upper extremities at shoulder and scapula, some in the lower extremes at pelvic girdle and knees. Progression is slow and severe disability usually not present until the thirties or even the fifties. Serum creatine kinase is moderately elevated. EMG demonstrates myopathic features; muscle biopsy demonstrates dystrophic features. From a clinical standpoint, there is considerable overlap with cases of indolent polymyositis, motor neuron disease (spinal muscular atrophy), endocrine myopathies, and carriers of the DMD gene with minor clinical manifestations.

The following case history illustrates this type of case:

Case History # 6-1.

This 21-year-old white male college graduate was referred for evaluation of bilateral leg weakness. The patient was delayed in walking until age 2. During grammar school, he was the slowest runner in the class and could not keep up with his peers. In high school he first noted minor weakness in climbing stairs. By college he had trouble in climbing one flight. By his senior year, he had difficulty descending stairs.

Past History and Family History not remarkable.

General Physical Examination: Unremarkable

Neurological Examination: The relevant findings included an absence of deep tendon reflexes at triceps and radial periosteal and trace reflexes at biceps. Proximal weakness was present with intact distal strength and normal muscle bulk.

Motor system:

At hip[1] strength was 3/5; at shoulders, 4/5. Gower's sign was present in attempting to stand from a recumbent position.

Gait was waddling - consistent with proximal weakness at hips.

Deep tendon reflexes[2] were absent at triceps and radial periosteal and trace at biceps. Quadriceps and Achilles reflexes were normal.

b. *Plantar responses* were flexor.

5) *Sensation*: All modalities were normal.

Lab:

1) *Erythrocyte sedimentation rate (ESR)*, thyroid studies, and electrolytes were all normal.

---

[1]*The primary grading system for strength is that suggested by the Medical Research Council (MRC) during World War II: 0=no contractions; 1= flicker or trace of contraction; 2=active movement with gravity eliminated; 3=active movement against gravity; 4=active movement against gravity and resistance; 5=normal power. The scale is not a linear function- and because of the wide range included in grades 4 and 5 many examiners will grade 4(-), 4, 4(+) and 5(-).*

---

[2]*Deep tendon reflexes are graded as follows: 0=absent; trace=minimally present; 1=hyperactive; 2=normal; 3= hyperactive-brisk; 4=hyperactive-unsustained clonus; 4+ sustained clonus. As above minor gradations maybe superimposed such as 2+, or 3+.*

2) *Muscle enzymes,* CK, SGOT, LDH and aldolase were all mildly elevated.

3) *Motor and sensory nerve conduction velocities* were all normal.

4) *EMG* - demonstrated myopathic features with decreased amplitude and duration of motor units with a full interference pattern on volitional effort.

5) *Muscle biopsy* (left deltoid) reported significant dystrophic features:

a. Marked variation in muscle fiber size in a random distribution.

b. Central position of subsarcolemmal nuclei.

c. Significant increase in endomysial connective tissue.

Subsequent Course.

Follow Up one year late indicated no progression. Two years later, minor progression was noted with a minor decrease in hand strength. Over the years slow progression occurred. The patient reported 17 years after his initial evaluation that he was still able to walk without assistance. Climbing stairs was a problem. Face and hands were not involved. Additional discussion of these less common dystrophies van be found in Padberg 1993.

Muscular dystrophies with predominant involvement of cranial nerves. Cranial nerves may be affected during the course of the more common muscular dystrophies. There have however been several families described with adult onset. Diseases beginning with predominant and relative selective involvement of cranial nerves - usually with autosomal dominant pattern of inheritance: chronic progressive ophthalmoplegia and oculopharyngeal dystrophy.

Myotonic Dystrophy: This is the most common inherited form of muscular dystrophy affecting adults. It is an autosomal dominant disorder with an estimated incidence of 1:8000 in which cranial nerve and distal limb involvement are prominent[3]. Multiple organ systems are involved with variable expression. The non muscular striated manifestations include cataracts, baldness, mental subnormality, gonadal atrophy, other endocrine disturbances, low plasma IgG, glucose intolerance, smooth muscle autonomic involvement and cardiac conduction defects. The latter may lead to syncope and sudden death. Although cases can present in the neonatal period, the most common ages of onset are late adolescence or early adult life.

The early symptoms relate to complaints of gait difficulty and clumsiness. At this point, the more specific diagnosis may not be apparent to the non-neurological observer. (The author once saw 6 patients in a 8-month period at an Army Basic Training base with this diagnosis, referred with the more general complaints). Examination however will demonstrate a distal limb weakness. Percussion of the thenar hand muscles or of the tongue will often demonstrate a myotonic reaction. (prolonged contraction delayed relaxation). Myotonia can also be demonstrated by asking the patient to squeeze the examiner fingers and to then rapidly open the hands. Myotonia of the eyelids may be noted in the continued retraction of eyelids and a lid lag after prolonged upgaze. With progression of the disease, a significant weakness and atrophy of muscles occur in muscles innervated by the facial, mandibular supplied muscles, and the accessory nerve to the sternocleidomastoid muscles occurs with the development of characteristic "myopathic facies." Most patients are disabled and unable to walk by the thirties or forties. Life expectancy is reduced because of cardiac and pulmonary complications.

The creatine kinase level is usually normal. The EMG demonstrates myopathic features plus the prolonged myotonic discharges. When audio amplified the EMG myotonic discharge sounds like a "dive bomber". The muscle biopsy shows characteristic features of chains of central nuclei, ringed fibers and selective atrophy of type 1 fibers.

Myotonic dystrophy should be easily distinguished from syndromes in which myotonic occurs in relative isolation: Myotonic congenita Thomsen's disease and an autosomal recessive

---

[3]*Recent studies have demonstrated that the disorder is related to an increased number of cytosine-thymidine-guanine trinucleotide repeats in the region of the protein kinase gene located on the long arm of chromosome 19. A DNA probe that detects directly the mutation is available (Shelbourne et al 1993, Ptacek et al 1993).*

form (Becker). The myotonic syndromes are characterized by an abnormal increase in membrane excitability with persistent runs of action potential in the surface membrane. Abnormal low chloride conductance has been implicated in some types of myotonia: myotonia congenita in humans, congenital myotonia of the goat (due to mutagens) and myotonia following ingestion of aromatic carboxylic acids.

Subsequent K^+ accumulation in the transverse tubular lumen leads to excessive depolarization and the persistent firing of action potentials. Similar result occur when normal muscle fiber are placed in a chloride free solution

However, this mechanism is apparently not the explanation for the myotonia of myotonic dystrophy, where intracellular sodium concentration is elevated apparently because of abnormalities in the regulation of sodium channels. The underlying molecular basis of myotonic dystrophy has recently been discovered. An unstable DNA fragment occurs on chromosome 19q 13 related to an expanded CAG trinucleotide repeat at the end of a region encoding a protein kinase. The greater the number of repeats, the more sever the disease (Wang et. al 1994). Ptacek et provides a more complete review of the physiology of the myotonic disorders. al (1993).

Congenital myopathies.

(Refer to Fardeau 1982). These are relatively rare disorders of muscle which primarily presents with hypotonic weakness in childhood: "The floppy infant" and delay in motor developmental milestones. The specific name assigned to each of these syndromes is based on the unique features on muscle biopsy and the special histochemical stains. The first type described by Shy and Magee in 1956 is so named because type I fibers lack mitochondria in the central core which instead contains disorganized tightly packed myofilaments. In nemaline myopathy there are peripheral collections of rod shaped bodies - possibly derived from the Z bands. In centronuclearmyopathy there are chains of central muscles - surrounded by a zone devoid of myofibrillar - ATPase activity.

Metabolic Myopathies: (Refer to Rowland et al 1986 and DiMauro 1985)

Muscle contraction utilizes ATP (adenosome triphosphate) which is converted from ADP (Adenosine diphosphate) by the action of CK (Creatine kinase).

Moderate exercise is fueled by aerobic conditions with glycogen as the main fuel source. After 5-10 minutes blood glucose is utilized. Subsequently fatty acids are utilized. After 4 hours of exercise lipids and amino acids are utilized.

In high intensity exercise anaerobic glycogen breakdown and glycolysis generate additional fuel. At rest lipids are the predominant energy source.

The student is already aware that multiple enzymes are involved in the events of aerobic and anaerobic carbohydrate metabolism and in lipid fatty acid metabolism. The metabolic myopathies are relatively rare inborn errors of metabolism, which involve these enzymes. The specific defect may involve the lysosomal and cytosolic enzymes: The glycogen storage diseases - or the enzymes related to the mitochondrial respiratory chain for aerobic energy production: The mitochondrial myopathies.

Depending on the specific deficit - a) there may be multiple system involvement with persistent weakness, b) selective muscle involvement with persistent weakness or c) exercise intolerance with easy fatigue and muscle cramps but with little persistent weakness.

The first hereditary myopathy in which a specific enzyme defects was identified McArdle's Disease falls into this last category. The enzyme muscle phosphorylase is absent as demonstrated on special stains for this enzyme applied to the muscle biopsy. When the patient exercise glycogen can not be broken down to pyruvate and lactic acid and the expected rise in blood lactic acid in ischemic limb exercise fails to occur. As exercise continues, fatigue and painful cramps occur. If the patient persists, muscle breakdown will occur with a significant rise in blood and urine levels of myoglobin. Significant acute renal tubular impairment is a complication of any acute condition in which rapid breakdown of muscle occurs producing myoglobinuria.

Periodic Disorders of Muscles:
Familial Periodic Paralysis[4]

Other rare inherited (autosomal dominant) disorder of muscle that may also produce acute transient weakness. These are not related to energy metabolism but are characterized instead by episodic failure of muscle membrane excitability. The attacks are often associated with marked alterations of serum potassium. There are hypokalemic and hyperkalemia (and possibly normokalenic forms). In the hypokalemic form, at the onset of the attack a significant movement of potassium and sodium ions into skeletal muscle occurs with a decrease in serum potassium. Attacks may be induced by the administration of insulin and glucose or of sodium chloride, by alcohol, stress, cold exposure, or by rest after exercise. In the hyperkalemic form attacks may be precipitated by fasting, cold, stress or by rest after exercise. In all types mild persistent proximal weakness may develop later in the disease course. Muscle biopsy may demonstrate vascular changes and tubular aggregates.

In both forms the changes in serum levels of potassium are of such magnitude as to produce EKG changes. In the hypokalemic form serum K may be depressed to 2-3 mEg/Liter. In the hyperkalemic form the lead may be as high as 7-8 mEgl/Liter. (Normal 3.9-5.0 mEg/liter).

The following case history illustrates many of these points. **Case History 6-2.** Patient of Dr. Thomas Twitchell.

This 30-yr. old white housewife had the onset of episodes of night paralysis at age 9 years. These occurred several times the month but with the administration of potassium supplementation, attacks decreased to 1-2 times per year. At age 32, she remarried and was under increased stress. One year later she began to have frequent episodes of nighttime paralysis again. She could relate her attacks to high carbohydrate meals and cold water exposure as well as stress.

Shortly before the onset of these attacks she had noted a mild persistent proximal lower extremity weakness which was non-progressive but did result in difficulty in climbing steps.

---

[4]*The relationship of the periodic paralysis disorders to the myotonic disorders and the more specific genetics and physiology are discussed in Ptacek et al 1993.*

Past History: - Negative
Family History :- Negative
General Physical Examination: - Negative
Neurological examination:.
1) *Mental status* - Normal
2) *Cranial Nerves* Normal
3) *Motor system*
a) No atrophy
b) Strength: Intact except: Hip flexors 4/5, Triceps 4/5
Deep tendon reflexes: present but relatively quiet at quadriceps and Achilles
5) *Plantar responses* flexor
6) *Sensation:* Normal
Laboratory data:
1) *Sed. rate and thyroid functions* normal
2) *Serum potassium on admission* 3.9 mEq/liter (normal 3.9-5.0 mEq/liter)
3) *Creatine phosphokinase* 12.8/12.0
4) *EMG* - Normal
5) *Nerve conduction studies* normal.
Additional studies

With a baseline potassium of 4.7 mEq/liters, the patient was given 38 grams of glucose solution by mouth plus 5% dextrose in water intravenously plus 15 units of regular insulin. Within 15 minutes after insulin administration, the patient had increased weakness in all four extremities. By 30 minutes, the potassium level had fallen to 2.1 mEq/liter. At 60 minutes, the patient had total paralysis of all four limbs. (Quadriplegia). and no deep tendon reflexes could be obtained. She was flaccid unable to lift her head from the bed. However, she was conscious she could speak, and breath and all cranial nerves were intact. EKG did show the typical features of low serum potassium (flat T. waves). She was given oral potassium and within 2 hours had returned to her baseline state. She was subsequently treated with potassium supplements plus acetazolamide a drug which affects sodium potassium transport. Studies by Tacek and associates 1994 have demonstrated a dihydropyridine receptor mutation produces a calcium channel disorder resulting in hypokalemias. A mutation in the sodium channel produces the hyperkalemic form of periodic paralysis.

These cases of primary periodic paralysis must be distinguished from the more common secondary varieties which occur in relationship to the electrolyte alterations of renal disease, adrenal cortical disease, effects of diuretics, cathartics and in gastrointestinal disorders with severe diarrhea.

Acquired Disorders of Muscle

The major categories are as follows:

1) *metabolic:* hypokalemic/hyperkalemic myoglobinuria and alcoholic. (see above). Drugs such as colchicine used in the treatment of gout may also produce a myopathy.

2) *Endocrine:* Hyperthyroid (Thyrotoxic myopathy) hypothyroid (Myxedema Myopathy) and corticosteroid. The corticosteroid myopathy occurs not only in patients with Cushing's disease but also in many patients receiving long term corticosteroid therapy. Typically the weakness begin in the pelvic girdle and proximal muscles of the lower extremities.

3) *Trauma:* Closed muscle compartment compression due to trauma or a deep unconscious state produced by drugs and alcohol may produce considerable rhabdomyolysis of an acute nature (see Owen et al 1979).

4) *Inflammatory.* (Refer to Dalakas 1991).

Inflammation of muscle may occur in a) viral disease such as influenza on Coxsackie, b) bacterial infections by staphylococcus aureus or c) in parasitic infections due to toxoplasmosis, cysticercosis, and trichinosis that are of the major importance in some areas of the world.

The major considerations in this section are those cases where the polymyositis occurs on autoimmune bases.

In some cases, (idiopathic, polymyositis and dermatomyositis) the autoimmune disease is restricted to muscle (and skin). The cases cover a wide age range. The onset and source may be acute, subacute or chronic. The chronic gradually progressive pattern is the most common. Cranial nerves are usually not involved although difficulty in swallowing may be present due to posterior pharyngeal striated muscle involvement. Proximal muscles are involved predominantly in dermatomyositis. There is more diffuse involvement in polymyositis but with a proximal predominance (limb girdle and neck flexors). In dermatomyositis there is edema and bluish (heliotrope) discoloration of the eyelids, with a scaly red rash on the face (in a butterfly pattern), shoulders, upper chest, back and extensor sur-

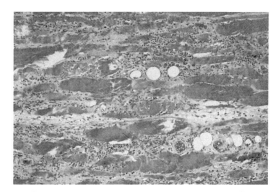

Figure 6-16. Muscle histopathology II: Polymyositis wide spread necrosis of muscle fibers is present with extensive infiltration by mononuclear inflammatory cells H&Ex25. (Compare to 6-15). Courtesy of Dr. Tom Smith, Neuropathology, U. Mass. Medical Center.

faces of the limbs, Sedimentation rate and muscle enzymes are significantly increased. Muscle biopsy indicates segmental muscle necrosis, regeneration and inflammatory infiltrates, *Figs 6-16*. EMG indicates not only myopathic features but also the presence of fibrillations.

There is an increased incidence of underlying neoplasm in adults with dermatomyositis and polymyositis (particularly in older adults). Common sites are lung and ovary. Overall incidence in all cases of dermatomyositis and polymyositis -is 9% (Sigurgeirsson et al 1992).

The term overlap polymyositis refers to patients who develop polymyositis within the context of already existing autoimmune connective tissue diseases: systemic lupus erythematosus (SLE), rheumatoid arthritis, periarteritis nodosa, systemic sclerosis and Sjögren's syndrome. It has been estimated that 5-15% of all patients with these diseases will manifest polymyositis (Isenberg, 1984).

Most patients (60-70%) with polymyositis and dermatomyositis have a significant response to corticosteroid therapy. Non responders may require more definitive immunosuppressive therapy. Approximately one third of non-responders have another variety of myositis inclusion body myositis. Such cases have a male predominate relatively normal CPK and greater distal involvement (see discussion Dalakas 1991). Case 6-3 demonstrates dermatomyositis complicating ovarian malignancy.

Case History 6-3.

This 69-yr. old white female had partial resection of ovarian carcinoma in Dec. 1973 and received radiotherapy and chemotherapy in January and February 1974. Two weeks prior to admission she had developed progressive weakness of all four extremities with proximal more than distal involvement. Because of difficulty swallowing, she had been on a liquid diet for two weeks prior to admission. General Physical Examination: 1) There was superficial redness; swelling involving the skin of both upper extremities 2) Liver was enlarged.

Neurological examination

1) *Mental Status:* Intact
2) *Cranial Nerves:* Intact except absent gag reflex
3) *Motor System:* Weakness of all four extremities proximal greater than distal.
4) *Reflexes:* Deep tendon reflexes were depressed in the upper extremities and absent in the lower extremities
5) *Sensory System:* Intact.

Laboratory Data

1) *Erythrocyte sedimentation rate* increased to 50 mm per hour.
2) *Muscle enzymes* (SGOT LDH, CPK) all significantly increased with creatine phosphokinase elevated to 201/12 units.
3) *Antinuclear antibody titer* was not significantly elevated (test for SLE)
4) *EMG:* Abnormal - consistent with polymyositis.

Hospital Course:

The patient was treated with high dosage corticosteroids. - (Prednisone). Despite a significant improvement in muscle enzymes (CPK 201>9.9, SGOT.295>102 LDH 320 > 102), the weakness continued to progress. During the last week of life, she had three episodes of aspiration and increasing respiratory distress expiring 3-29-74.

Comment. This patient received a diagnosis of dermatomyositis complicating ovarian malignancy. In such patients deep tendon reflexes may be totally absent. Changes may be present on sensory or cranial nerve examination, suggesting a peripheral neuropathy. Peripheral nerve changes reflect - a) paraneoplastic effect of malignancy or b) the effects of chemotherapy.

The continued progression of the weakness in this patient is not unusual. The paraneoplastic polymyositis syndrome does not have the favorable response to corticosteroids found in the idiopathic syndrome.

DISEASE OF THE NEUROMUSCULAR JUNCTION

Diseases affecting the neuromuscular junction may be classified as to location: postsynaptic or presynaptic (see Newsome-Davis 1992, Engel 1984, Drachman 1986).

Postsynaptic Disorders-Myasthenia Gravis

The major primary disease due to neuromuscular junction pathology in this country and western Europe is myasthenia gravis. The prevalence is 5-7.5 per 100,000. This is a disease characterized by fluctuating weakness of voluntary muscle worse on exercise, improved by rest and by administration of anticholinesterase drugs. This is primarily a disease of adults with onset between ages 10 and 70. However neonatal and congenital cases are recognized. The peak age of onset is between ages 20 and 30. Under age 40, females are affected more often than men in a ratio of 70:30. After age 50, there is a male predominance of 60:40. Ten % of patients have an associated tumor of the thymus gland (thymoma). These patients tend to be older with a male predominance.

The symptoms usually first appear in the muscles supplied by the cranial nerves. The extraocular muscles for control of eye and eyelid movement are first affected in 60% of cases and are affected at some stage in 90% of patients. In some cases, with a relatively benign course, the symptoms may remain confined to the eye muscles, ocular myasthenia. The patient has a ptosis (closure or droop) of one or both eye lids, often most apparent on sustained upward gaze. Double vision (diplopia) results from a weakness of one or more extraocular muscles. The early occurrence of diplopia may be a reflection of the precise synchronization required between the two eyes in movement under normal conditions. A minimal weakness of one medial or lateral rectus muscle would disrupt such a precise synchronization. In the extensive series of Grob et al with 1487 patients followed 1940-1985) 14% of patients continued to have localized ocular myasthenia.

Generalized myasthenia refers to progression to involve other cranial nerves and the extremity and trunk muscles. The patient will have difficulty swallowing (dysphagia), speaking (dysarthrias) and chewing (mandibular nerve supplied muscles). In addition a bilateral weakness of facial muscles may provide a suggestive myasthenic faces. The patient may support head or jaw with the hand to compensate for weakness. As the disease progresses, more generalized involvement of proximal limb muscles develop.

In some patients there is significant involvement of the muscles of respiration. In the series of Grob maximum level of weakness was reached during first year in 55% first 3 years in 70% and first 5 years in 85%. After 3-5 years, the disease stabilizes.

The prognosis depends on the age of the patient, the duration of the localized form of the disease and on the associated diseases: (thymoma, thyroid disorders other autoimmune disorders).

Myasthenia was once associated with a significant mortality due to respiratory insufficiency and infection but this mortality has now been markedly reduced. In the series of Gob, et al. for patients with maximum weakness in the 1940-1957 era - mortality was 31%, rate of remission 10%. In 1958-1966 era, the mortality was 15% rate of remission 10%. In the 1966-1985 era, mortality was 7% remission rate was 11%.

The diagnosis is readily apparent from a clinical standpoint: Fluctuating weakness a) no sensory finding b) Intact reflexes and no long tract findings. However there are several confirmatory diagnostic tests. Jolly in 1895 first described the electrical test that carries his name. Repetitive stimulation of a motor nerve result in a rapid decrease in the amplitude of muscle contraction "decremental response". Jolly demonstrated that the apparently fatigued muscle would still respond to direct galvanic stimulation.

In modern clinical neurophysiology the motor nerve is stimulated at a rate of 3-5 per second and the reduction in amplitude of the muscle action potential is measured. (Abnormal is greater than 10% reduction). An agent which blocks acetylcholine esterase, is then administered (edrophonium or neostigmine). This increases the duration of available acetylcholine at the neuromuscular junction (edrophonium or neostigmine) and the decremental response is reversed.

The edrophonium (Tensilon) test can be carried out in the outpatient office. This agent administered intravenously will often transiently reverse or reduce the signs of weakness. Clinical effect occurs in 30-60 seconds and lasts 4-5 min. Since related cholinergic agents (anticholinesterases) play a major role in therapy the test can also be performed to determine whether the patient is receiving too little on too much of the therapeutic agent.

A very specific test is the measurement of the blood level of the acetylcholine receptor antibody positive in 85-90% (see below).

Recently major advances have been made in our understanding of the underlying pathogenesis and pathophysiology of this disorder. This constitutes a major accomplishment for neurobiology that has been translated into improved results as regards therapy of the disease.

The role of the thymus appears critical from several standpoints. 1) The thymus has a central role in immunology. 2) The thymic hyperplasia involves the presence in the medulla of many germinal centers and surrounding T cell areas. Such germinal centers are rare in non-myasthenic thymus. 3) The thymus contains muscle like cells, and ACH receptors can be demonstrated when these cells are placed in tissue culture. Drachman 1978 has suggested that this receptor bearing muscle cells may be particularly vulnerable to immune attack. Some alteration of these cells by lymphocytes (perhaps triggered by viral infection of thymus etc.) could initiate any acute immune response directed against ACH receptors.

The essentials of therapy in myasthenia gravis follow from the neurobiology of the disease.

1) Initial therapy consists of anticholinesterase drugs. The main drug used is pyridostigmine (Mestinon) which has a somewhat longer duration of action than neostigmine. For some patients this will suffice particularly if only local ocular myasthenia is present and if this responds well.

2) Generalized myasthenia will almost always

require more definitive therapy directed at the immune systems.

a) The most definitive procedure is total thymectomy. As discussed by Rowland, (1987) there is now a consensus that all adults with generalized myasthenia should have this procedure relatively early in the course of the disease. In the "maximum thymectomy" series of 72 non-thymoma patients previously with moderate to severe generalized myasthenia reported by Younger et al. (1987), 46% of patients were in complete remission (on no medication). 33% were asymptomatic on 60-240 mg of pyridostigmine daily and 10% were asymptomatic on steroids. Approximately 90% then were in complete remission or asymptomatic. An additional 6% were improved. At least 1-4 years were required to see the maximum response to therapy. There was in significant decline in the titre of acetylcholine receptor antibody. Although this was not a matched control series, the results are in striking contrast to the contemporary results of non surgical treatment of moderate to severe generalized myasthenia (see Grob. 1987). Other major centers have confirmed these results. Schumm et al. (1985) has also recommended the procedure in patients with pure ocular myasthenia, if no spontaneous remission and no satisfactory response to cholinesterase inhibitors occurs in a 6-month period. In their series, thymectomy prevented the subsequent development of generalized myasthenia; none of 18 patients progressed over two years from ocular to generalized myasthenia.

Thymectomy was originally performed primarily to remove thymomas in patients who coincidentally also had myasthenia gravis; some improvement in the myasthenia also occurred (Blalock et al. 1939). Considerable experience in subsequent years has demonstrated a poorer response of the myasthenia in such cases of thymoma to thymectomy.

b) Other types of immunosuppression may also be employed in the pre or postoperative period or in those patients unable to tolerate thymectomy. These include corticosteroids, plasmapheresis (plasma exchange) and cytotoxic drugs (azathioprine).

Myasthenia Gravis

Adams and Victor (1989) have summarized the major historical landmarks in the study of myasthenia. Welch in 1877 and Erb in 1875 recognized the lack of any pathology in brain stem to explain the cranial nerve motor findings. The electrical studies of Jolly in 1895 suggested what has come to be recognized subsequently as the location of pathology: the neuromuscular junction. The term's pseudoparalysis as well as myasthenia gravis was originated by Jolly since no pathology was present at autopsy. He also recommended the use of an anticholinesterase physostigmine. Walker in 1934 noted the similarity of some of the signs to those produced by the poison curare and began the use of physostigmine. Buzzard in 1905 in a detailed clinical pathological analysis of 5 cases described the lesions in thymus; (Thymic hyperplasia which now has been noted in over 80% of patients) and the minor lymphocytic collections in muscle. Buzzard proposed that an autotoxic agent could produce all of these findings. He also noted the association of the disease with thyrotoxic cases. Simpson in 1960 proposed an autoimmune basis for the disease because of the increased incidence of other putative autoimmune disease: thyroiditis, lupus erythematosus and rheumatoid arthritis.

Electron microscopic studies by Zack et al. 1962 and by Engel and Sarta 1971, Engel et al 1976 had demonstrated significant changes in the postsynaptic region with shallow postsynaptic folds, a widened synaptic cleft. In contrast the presynaptic area was normal as regards the number and size of presynaptic vesicles. Fambrough, Drachman and Satyamurt; (1973) then demonstrated a marked (3 fold) reduction in acetylcholine receptors per neuromuscular junction in motor point nerve biopsies of myasthenic patients. Compared to control subjects, there were a decreased number of binding sites for a radioactive labeled snake poison, alpha bungarotoxin which can be purified from the venom of the cobra and krait. This toxin binds in an irreversible manner to the acetylcholine receptor. In 1973, Patrick and Lindstrom demonstrated that repeated immunization of rabbits with acetylcholine receptor protein derived from the electric organs of eels reproduced the disease. These animals develop all of the clinical and electrical features of myasthenia gravis. Antibodies to the ACH receptor protein could be identified attached to the ACH receptor. Moreover, normal animals receiving these antibodies also developed myasthenia.

Subsequently Lindstrom et al (1976) and several other groups reported a radioimmunoassay for acetylcholine receptor antibody. This is now a standard laboratory test for myasthenia gravis with a 90% detection rate in the serum of myasthenic patients. (The serum level does not correspond to the severity of the disease). Engel (1984) were able to demonstrate immune complexes at the postsynaptic junction in biopsies from myasthenic patients. When the immunoglobulin derived from serum of myasthenic patients is administered to mice, the characteristic myasthenic syndrome develops in the recipient animal (Toyka, et al. 1974). The syndrome of neonatal myasthenia also suggests a serum or plasma transmissible factor. 15% of children born to myasthenic mothers will manifest clinical signs of weakness lasting several weeks due to passive transfer of anti ACH receptor antibodies across the placenta. Circulating ACH receptor antibodies and electrophysiological findings can be demonstrated in these infants and even in additional neonates who do not manifest clinical weakness.

Subsequent studies reviewed by Drachman (1986) suggest several mechanisms for antibody action: 1) there is a significant increase in the rate of degradation of ACh receptors, 2) the antibody actually blocks the binding sites of the acetyl choline receptor and, 3) there is a complement mediated damage to and subsequent change in the geometry of the junction with a reduction in efficiency of transmission.

Cells cultured from patients with myasthenia gravis but not from control subjects can synthesize acetylcholine receptor antibody in vitro. The highest levels of production are obtained from thymic cells of patients with medullary thymic hyperplasia. Peripheral blood lymphocyte lymph nodes and bone marrow also produce the ACH receptor antibody.

Case History 6-4. NECH.

This 44-year-old, white, married, machine operator for several years had noted a general sensation of fatigue in his arms and legs at the end of a day. Approximately 6 to 7 weeks prior to admission, on Dec 7, 1975 the patient began to have more significant difficulty with a marked increase in the degree of weakness of the arms and legs. At the same time, the patient noted significant slurring of words, difficulty in swallowing, and drooping of the lids. All of these symptoms were transient; they were not present in the morning; they were clearly precipitated by exercise. For example, although the patient would initially have strong chewing movements, as soon as he began to chew for a short period he would develop fatigue of jaw muscles. The degree of ptosis was sufficient to result in difficulty in driving.

The patient had been receiving neostigmine 15 mg every 6 hours. He reported that 30 mg every 6 hours produced significant diarrhea and hypersalivation. In addition this dose exacerbated swallowing and chewing problems.

Past history: The patient had a long-standing problem of marked obesity.

Neurological examination: Four hours after neostigmine 15 mg.

1. *Mental status*: intact
2. *Cranial nerves*:

a. A significant bilateral ptosis of the eyelids was present. The degree of ptosis was markedly increased by exercise when complete closure of the left eyelid occurred.

b. On repetitive upward gaze, a bilateral weakness of superior rectus developed.

c. There was a bilateral facial weakness, worse on exercise, more marked on the right than on the left.

d. Jaw movements (opening, closing, and lateral movements) were weak. The degree of weakness was increased by exercise.

e. Lateral tongue movements, particularly on sustained pressure, became weak.

3. *Motor system*:

a. There was a significant weakness in shoulder abductors, elbows flexor and extensors, and handgrip. The degree of weakness was markedly increased by repetitive exercise.

4. *Reflexes:*

a. Deep tendon reflexes were symmetrical.

b. Plantar responses were flexor.

5. Sensory system: All modalities were intact.

Laboratory data:

1. Chest x-rays and tomogram studies of thyroid function were normal. Subsequent tests for lupus erythematosus were negative.

2. A tensilon edrophonium test was performed. This demonstrated an almost immediate eye-opening effect with the disappearance of the bilateral ptosis. The ptosis, however, had reappeared 3 to 4 minutes after injection of 10 mg of the agent.

Subsequent course: The patient was treated with pyridostigmine and stabilized on a dosage of 90 mg 5 times per day (approximately every four hours). On the dosage his ocular bulbar and generalized weakness stabilized until June 1973 when increasing generalized weakness developed. The patient had gradually increased his dosage to 180 mg every three hours. Respiratory distress increased and the patient was readmitted in extreme distress in the superior position. Tidal volume was zero. He had marked bilateral ptosis and diffuse skeletal muscle weakness. Analysis suggested the patient was in "cholinergic crisis." He was intubated, ventilated and received no medication for 60 hours. Repeat tensilon test now demonstrated marked improvement in respiratory effort and in skeletal weakness. Pyridostigmine was reinstituted and based on edrophonium tests, the patient was stabilized on 90 mg every 3 hours. A tracheostomy was performed and thoracic surgery consulted regarding thymectomy. The surgeons considered that the marked obesity constituted too great an operative risk for surgical thymectomy. The patient instead received radiation of the anterior mediastinum, 3000 rads over one month. At the time of discharge tidal volume was now 3000 cc. The patient did well with no additional problems over the next three years. An exacerbation occurred in 1976.

He was subsequently treated with long term prednisone therapy eventually being maintained an alternate day therapy. On March 26 1983, he had a respiratory arrest followed by cardiac arrest. Despite resuscitation he died after a one-month period of anoxic encephalopathy.

Other post synaptic syndromes: pharmacologic and toxic: (see also discussion above)

1) Acetylcholinesterase inhibitors: The usual action of acetylcholine esterase at the neuromuscular junction is to rapidly terminate the action of ACH. This allows high rates of transmission across the synapse. If acetylcholine esterase is inhibited, acetylcholine remains on the receptor and in the synaptic space.

There are two classes of acetylcholine esterase inhibitors:

a) reversible: (carbamates), and

b) relatively "irreversible" (organophosphates)

a) The reversible agents include those anti ACH esterase drugs used in the treatment of myasthenia gravis. An increased availability of acetylcholine is of value in counteracting the weakness due to decreased numbers of receptors in this syndrome. Excess amounts of such drugs produce excessive amounts of acetylcholine at the receptors as noted in the above case history resulting in increased weakness referred to as "cholinergic crisis" The dose of the drug must be reduced. In the management of such patients the edrophonium ("Tensilon") test is of value.

b) The relatively irreversible agents (organophosphates) constitute a much more serious problem - not only does a significant acute cholinergic effect occur but there is damage to the postsynaptic membrane and delayed effects on peripheral nerve. Senanagake and Karolliedde 1987 distinguished three syndromes. 1) The acute cholinergic syndrome is accompanied by autonomic effects fasciculations, and at times central effects (coma) 2) The intermediate syndrome occurring in 10% of patients after recovery from the acute syndrome 24-96 hours after the poisoning - identical to an acute myasthenic syndrome with respiratory cranial nerve and proximal limb muscle involvement. If the patient survives the respiratory problems, recovery may occur in 5-18 days. 3) A delayed distal motor neuropathy occurs in some patients at 2-5 weeks after exposure.

The organophosphates are the neurotoxic agents of chemical warfare. They are also extensively employed as pesticides. As discussed by Davis (1987) this is a major problem in the developing countries where the sale of such chemicals is often unregulated, the chemicals are often repackaged in unlabeled containers. The original containers are often reused. The storage and use are often unsupervised and the population is often unable to read any warning label.

The magnitude of the problem is staggering. In Sri Lanka alone between 1978 and 1980, there were 80,000 patients admitted with pesticide poisoning and 6083 of these patients died. Seventy five % of all cases were due to organophosphorous. In Sri Lanka, the majority of complex cases were the result of ingestion with suicidal intent.

2) Agents that bind to the acetylcholine receptor and thus block access of acetylcholine to the receptor.

a) *d-tubocurarine* - competes for the receptor in a reversible manner. Initially a South American Indian arrow poison - now used in anesthesia.

b) *Repolarizing agents:* biquartenary ammonium salts. These resemble acetylcholine. They are agonists that initially produce depolarization and then block the site. They are utilized to produce muscle relaxation during general anesthesia and during mechanical respiration.

c) *Snake toxins*: derived from the venom of the cobra and krait. These snakes are common in India, Sri Lanka, Southeast Asia and Taiwan. The neurotoxic effect produced is often an acute myasthenic syndrome with respiratory paralysis. As discussed by Watt et al. (1986) and by Drachman (1986) for some of the toxins, - alpha cobra toxin, the binding is reversible and can be treated with anticholinesterase. For other toxins in the group, alpha bungarotoxin from the Taiwan Krait, the binding is less reversible and the effects of anticholinesterase less consistent. Some snake venoms also contain toxic agents that act at the presynaptic region (B. bungarotoxin of the krait).

1) *Eaton-Lambert (or Lambert-Eaton) Syndrome:* The patient usually an adult has predominantly proximal limb weakness which may be progressive but which fluctuates. In contrast to myasthenia, a) bulbar ocular and respiratory muscle involvement is not common. b) Exercise improves the weakness. c) Repetitive stimulation of the motor nerve produces an incremental response in the muscle action potential.

Most patients with this disorder have an underlying malignancy, most often a small cell carcinoma of the lung. The neuromuscular syndrome however may precede the appearance of the tumor by months or years. In some patients (1/3) a malignancy may not be identified, but there may be other evidence of autoimmune disease.

The underlying pathophysiology relates to a decrease in the release of acetylcholine quanta from the nerve terminal (Not only at the neuromuscular junction but also at cholinergic nerve endings in the autonomic nervous system). As discussed by Newsom-Davis (1992), this defect in release relates to immunoglobulin g antibody that binds to presynaptic Ca channels and prevents their voltage gated opening. As the action potential enters the presynaptic region, fewer Ca channels open, less Ca++ enters to catalyze the binding of vesicles to the surface membrane. Hence less ACII is released. The disorder may be transferred passively (via plasma from patients) to mice. The patients are improved by plasma exchange or by drugs that increase acetylcholine release. Treatment of the underlying malignancy may produce some improvement.

2) *Botulism*: The toxin of the bacteria Clostridium Botulinum produces an acute syndrome of paralysis involving extraocular muscles, cranial nerves muscles of respiration and then progressively descending to involve limbs. Cholinergic autonomic junction is also involved. Thus the pupil which is spared in myasthenia often is involved (dilated). Anaerobic bacteria that may contaminate improperly canned food or preserved meat produce the neurotoxin. Clusters of cases bring this problem to alteration the attention of public health officials. Many cases are fatal. In infants and some adults, the bacteria itself may be present in the gastrointestinal tract or in a wound producing the toxin. (See Bartlett 1986). The toxin binds rapidly to cholinergics nerve endings and blocks the quantal release of acetylcholine. Once the toxin enters the nerve ending, the available therapeutic antitoxin if administered has little action. The flaccid paralysis then is often long lasting (see Chia et al. 1986).

3) *Antibiotics:* The aminoglycoside antibiotics: neomycin, streptomycin, and kanamycin act to interfere with the quantal release of acetylcholine. In most patients this presents no signif-

icant clinical problem. When blood levels are high due to high dosage or renal failure and when the patient has myasthenia or a subclinical myasthenia. There may be the presentation of an acute paralysis. Most often there is an associated factor of anesthesia and the patient fails to regain normal ventilation after anesthesia has been discontinued. The effects may also be seen with another type of antibiotics.

4) *Spider venom:* Black widow. This toxin produces rapid release of quanta of acetylcholine storage vesicles. The clinical syndrome is characterized by severe muscle contraction followed by paralysis.

CHAPTER 7
Spinal Cord: Structure and Function

Since the spinal cord is the best-understood and least complex of the major elements of the central nervous system, it is appropriate to begin the discussion of the CNS with it. The spinal cord has two fairly distinct functions. It conducts action potentials to and from the brain and it relates to "its" segment.

The spinal cord is segmented, one segment per vertebra, and in the less complex nervous system of fish, the region of the body relating to each segment is a cylindrical band the width of the vertebra. Into this spinal cord segment flows all of the sensory information from the body segment: information on pain, temperature, and position of the muscles, touch, and vibration. The axons that innervate the muscles of that body segment have their nuclei in the spinal cord segment.

In the human there is only a general relationship between the sensory segments and muscles innervated by the same segment of the cord. The size and shape of the regions vary considerably, but the principle remains the same - that of relating, in sensory reception and motor control, to a delimited region of the body. At each segmental level, incoming sensory information connects, in a stereotyped manner, to motor outflow. The knee jerk reflex and the reflexive withdrawal from pain are examples of segmental spinal cord activity.

In addition to relating these functions to specific regions, the spinal cord processes incoming sensory information and sends some of it on to the brain. This is a second and basically distinct function of the spinal cord: conduction to and from the brain. These two functions are anatomically separated. Inspection of a cross section of spinal cord shows white columns surrounding a butterfly-shaped interior gray column *(Fig. 7-1)*. The myelin sheaths of the thousands of axons running up and down in the white matter, the dorsal, lateral and ventral funiculi, produce the white color. The thousands of cell bodies in the center give that region a gray appearance. We speak, then, of gray matter, which contains cell bodies and synapses relating to the sensory and motor activities of the segment, and of white matter, which contains axons running to and from the brain.

Each spinal cord segment has four nerve roots attached to it *(Fig. 7-1)*. Each root is made up of many nerve bundles entering the body of the cord almost continuously. The two anterior roots, which carry axons out of the spinal cord to innervate muscles, are known as motor roots. The two posterior roots carry sensory information into the spinal cord. Since man stands vertically, we refer to posterior and anterior; in animals, dorsal and ventral. The cell bodies of all of the sensory fibers are in the posterior root ganglion. Damage to the right anterior root will paralyze motor units on the right side of the body. Damage to the left posterior root will interrupt all modalities of sensation from a small region on the left side of the body. The roots join just before they leave the spinal canal, through the intravertebral foramen, and form a single spinal nerve.

GROSS ANATOMY

The gross anatomy of the spinal cord is illustrated in *Figure 7-2*. The cord runs through the spinal canal of each vertebra, as is shown in

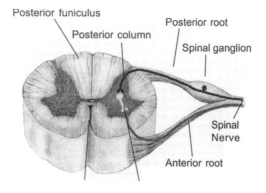

Figure 7-1. The posterior and anterior roots in relation to the gray and white matter of the spinal cord.

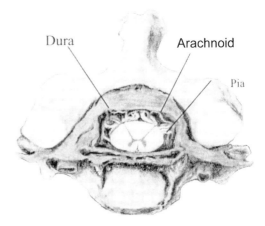

Figure 7-3. The relationships of the vertebral column, the meninges and spinal cord.

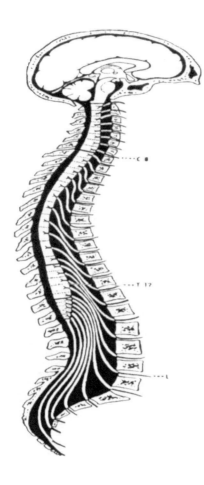

Figure 7-2. The lateral aspect of the spinal cord exposed within the vertebral canal. The spinous processes and the laminae of the vertebrae have been removed and the dura mater has been opened longitudinally. (From Clemente, C. (Ed) Gray's Anatomy, Philadelphia, Lea & Febiger, 1985.)

Figure 7-3, and is protected by a number of layers of meninges. The pia mater is closely adherent to the cord. Cerebrospinal fluid occupies the subarachnoid space. The arachnoid lies snugly against the thick, tough dura. The spinal dura forms a protective tube beginning at the dura of the skull and tapering to a point in the region of the sacral vertebrae. Between the dura and the vertebrae lies the epidural space, usually filled with fat. The denticulate ligaments anchor the spinal cord to the dura.

The spinal cord and, more importantly, the spinal roots, are named in relation to the vertebral column. In humans there are 8 cervical, 12 thoracic, 5 lumbar, and 5 sacral spinal cord segments. They are shown in relation to the vertebral column in Figure 7-2. There are 7 cervical vertebrae; the C8 in Figure 7-2 is the eighth cervical spinal nerve that arises from the anterior and posterior roots of the eighth cervical spinal cord segment. The eighth cervical nerve emerges below the seventh cervical vertebrae and above the first thoracic vertebrae. The spinal nerves are always named for the spinal cord segment of origin. The intervertebral disk and foramen (Fig 7-5) are normally named by the vertebrae above and below, i.e., L4, 5 disk; L5, S1 foramen.

During development and growth, particularly intrauterine growth, the vertebral column grows faster than the spinal cord it contains so that, in the adult, the spinal cord does not extend the length of the vertebral column (Fig 7-2). As a result, the spinal roots run interiorly before they leave the spinal canal. The spinal cord ends at the first or second lumbar vertebra, and below this level the spinal canal contains only spinal roots. This mass of roots reminded early anatomists of a horse's tail, so it was named cauda equina. The nerve roots run through the subarachnoid space, which is filled with cerebrospinal fluid. When a needle is inserted through the space between the fourth and fifth lumbar vertebra *(Fig 7-4)*, the point penetrates the dura and pushes aside the spinal roots. If the needle were inserted above L2 there would be danger of damaging the spinal cord. Samples of cerebrospinal fluid can be taken through this

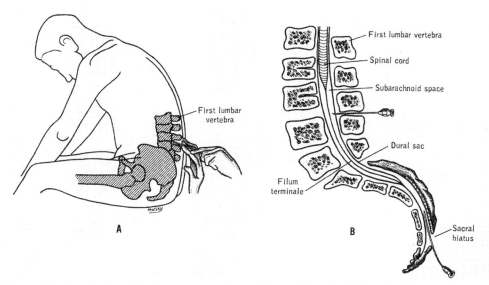

Figure 7-4. The technique of lumbar puncture. (From House and Pansky: A Functional Approach to Neuroanatomy. New York, McGraw-Hill, 1960.

needle; pressure can be measured or anesthetic agents or radio-opaque dyes (as in Fig 7-4) injected.

The anterior and posterior roots join to form a spinal nerve (Fig 7-1) just before leaving the vertebral canal through a recess in the posterior process of each vertebra as shown in *Figure 7-5*. The canal, the intervertebral neural foramina, is only a little larger than the nerve so that any swelling of the nerve or diminution of the diameter of the canal will pinch the nerve. It is not uncommon for the intervertebral disc to rupture posteriorly or laterally and press on a spinal nerve as it exits. The very intense pain this pressing causes is referred to the area of the skin where the nerve began. Frequent sites where the nerve roots are pinched are the intervertebral foramina between L2 and S1, which give rise to the sciatic nerve. The pinched nerve gives the patient the impression that a hot knife is being dragged along the posterior aspect of his calf.

Cross-sections.

The relative amounts of white and gray matter vary with the spinal cord level *(Fig 7-6)*. White matter decreases in bulk as the sections are further from the brain. Motor tracts from the brain leave the white matter to enter the gray matter and synapse with motor neurons. Sensory fibers entering the cord, section by section, also form the white matter.

A section of the cervical cord contains sensory fibers running up to the brain from the thoracic, lumbar, and sacral sections. It also contains motor fibers running down to innervate motor axons of the thoracic, lumbar, and sacral sections. The white matter of a section of the lumbar cord, however, only contains fibers to and from the

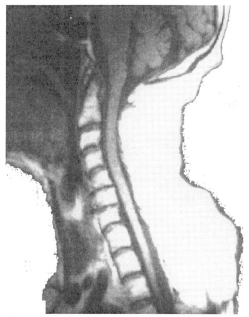

Figure 7-5. An MRI of a normal spinal cord in the spinal canal. Sagittal section.

lumbar segments inferior to it and the sacral segments. The white matter columns have the shape of thin pyramids with their bases at the foramen magnum and their tips at the last sacral section.

The situation in the gray matter is quite different. Since the gray matter innervates a segment, the size of the gray matter (the number of cells it contains) is related to the complexity of the segment. The hand, for example, is innervated by cervical segments 6, 7 and 8 (C6, C7 and C8) and the first thoracic segment (T1). The hand has the highest concentration of sensory receptors of any region of the body. All of these receptors send their axons into C6 to T1 where they synapse, thus increasing the girth of the posterior gray matter. The muscles of the hand can carry out very fine and intricate movements. They are innervated by nerves having their cell bodies in the gray matter of C6-T1. Such movements require many motor nerves and many cell bodies in the anterior gray matter. Consequently, the cord is enlarged at C6-T1. The situation is much the same in the lumbar region. Sensory input from and motor output to the leg is complex and the gray matter is large. The thoracic and sacral segments, on the other hand, have very small gray matter areas since they innervate only a few muscles and receive relatively uncomplicated sensory messages.

Segments can be recognized, then, by the amount of gray and white matter relative to the whole cross section. In the lower cervical segments the section is large, oval, and white matter and gray matter are nearly equal. The thoracic segments have much more white matter than gray matter and the shape of the gray matter, a thin H, is very characteristic. The lumbar segments have more gray matter than white matter, and the sacral segments are very small and have much more gray matter than white matter. The segments can also be recognized by the shapes of the gray matter. A careful review of Figure 7-6 and Spinal cord Figures 1-4 in descriptive atlas should provide a basis for recognizing the segmental levels of spinal cord sections. The area of the body which sends sensory fibers into a given spinal cord segment is called a dermatome. These have varying shapes and sizes. Figure 7-7 may help you to remember the general area

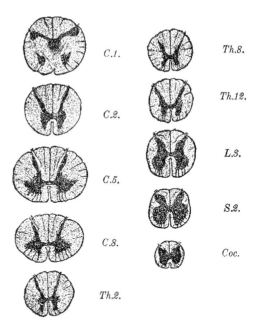

Figure 7-6. Typical spinal cord cross sections. (From Gross, C.M. (ed.): Gray's Anatomy. Philadelphia, Lea and Febiger.

served by the cord segments. The muscles underlying these areas have essentially the same innervation. More detailed areas are shown in chapter 8.

SEGMENTAL FUNCTION

Anterior horn Cells

The final effector cell of the spinal cord, the anterior horn cell *(Fig. 7-8)* is probably the best place to begin a discussion of the segmental function of spinal cord segments. The extensive dendritic tree allows upwards of 20-50,000 individual synaptic areas or knobs while the large cell body may have another 1-2,000 synaptic knobs. Incoming axons may have more than one synaptic connection and hence exert greater control. Each anterior horn cell gives rise to one large (8 to 12 _m) axon, called an alpha motor neuron, which innervates a motor group or unit made up of 10 to 200 individual muscle fibers. There is only one motor end plate on each muscle. Therefore, for a motor group to contract, the anterior horn cell on the proximal end of the axon must fire an action potential. The muscle group is chained to its anterior horn cell. This anatomical relationship is often called the final common pathway. Activation of a specific anterior horn cell precedes activity of a motor group. There are many ways to activate this anterior horn cell, many thousand axons synapse upon it and its extensive dendritic tree.

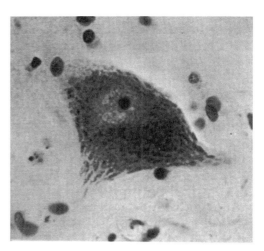

Figure 7-8. *A motor neuron from the cervical spinal cord of a human with prominent Nissl substance (rough endoplasmic reticulum), AAThionin Stain <X 75>.*

Neurons in the anterior horn can be divided into medial, and lateral groups *(Fig. 7-9)*.

1. The medial nuclear division is divided into posterior medial and anterior medial groups. The posterior medial nucleus is most prominent in the cervical and lumbar enlargement. The medial nucleus innervates the muscles of the axial skeleton.

2. The lateral nuclear division innervates the appendicular musculature. In the thoracic region the intercostal and associated muscles are innervated by this region. In the cervical and lumbar enlargement, these nuclei become especially prominent and are divided into individual columns of nuclei (Fig. 7-9) and these nuclei columns include anterior, anterior lateral, accessory lateral, posterior lateral and retroposterior lateral. These nuclei are represented functionally so that from medial to lateral one passes from midline spine, to trunk, upper and lower limb girdle, upper leg and arm and lower leg and arm to hand and foot. The most lateral nuclear groups innervate the muscles in the hand and foot.

3. The preganglionic autonomic nuclei lie in the intermediolateral column, a prominent lateral triangle in the gray matter (Fig 7-6; Th.2). There are two distinct groupings. The intermediolateral nucleus from C8 to L2 is the origin of the preganglionic sympathetic neurons which synapse again either in the sympathetic trunks or

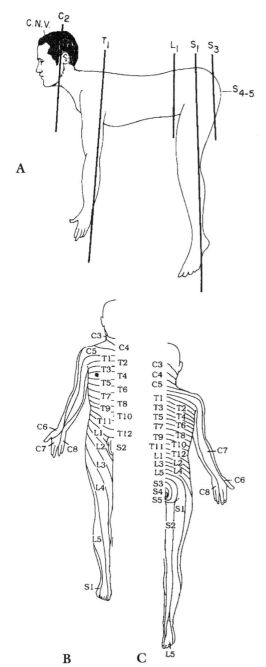

Figure 7-7. Key dermatome boundaries in man. A. Anatomical Position. B-anterior surface and C-posterior surface. From Zimmerman J and Jacobson S, Gross Anatomy, Little Brown 1990.

in remote ganglia. The sacral nucleus in S2 to S4 is the origin of preganglionic parasympathetic fibers that run to several ganglions in the walls of the pelvic organs.

REFLEXES OF A SINGLE MUSCLE

All of the sensory axons entering the posterior roots are bipolar cells having their cell bodies in the posterior root ganglion. The axons can be divided into many classes by the sensory modality they carry. As far as we know, each axon carries information about only one modality, such as pain. The intensity of the pain or other sensation is coded as action potential frequency. Later in this chapter, the various modalities will be dis-

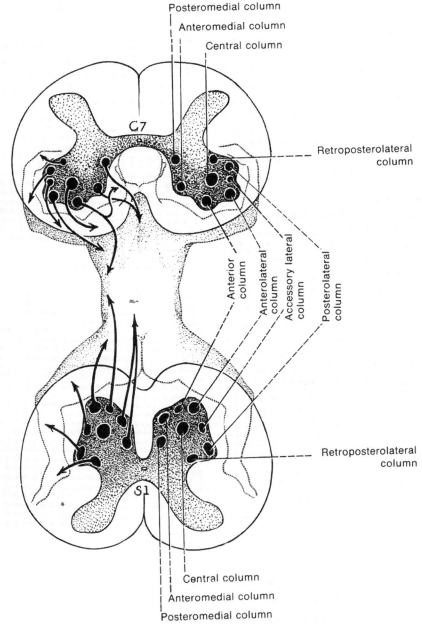

Figure 7-9. Functional localization within the anterior horns. (From Bossy: Atlas of Neuroanatomy. Philadelphia, W.B. Saunders, 1970.)

cussed in detail. For the present, two modalities will suffice for examples: muscle stretch and pain. Pain is carried by very small axons, many of them unmyelinated, which have high thresholds and slow conduction velocities. The stretching of muscles, on the other hand, leads to a barrage of action potentials in large, heavily myelinated nerve fibers. These axons have low thresholds and rapid conduction velocities.

Stretch Reflexes. The classic stretch reflex is the knee jerk, produced by tapping the patellar tendon. This simple involuntary response is such a familiar part of the physical examination and has been used so frequently in medical humor that precise description seems unnecessary. This simple test can give the astute examiner many clues to the function and dysfunction of the nervous system.

The knee jerk is described quantitatively by *Figure 7-10*, which is taken from the early work of the great English physiologist Sir Charles Sherrington. As the figure indicates, after a brief stretch, the muscle contracts with a delay of about 10 msec. It can readily be shown that this is a response mediated by the spinal cord. When the nerve to the muscle is cut, the response is abolished. When the spinal cord is severed from the rest of the nervous system, the response remains.

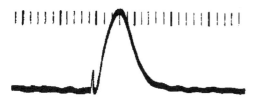

Figure 7-10. A brief stretch reflex recorded from the quadriceps tendon. The initial small, sharp response is the quick stretch which initiated the large response. The time marks are 20 msec apart. (From Ballif, et al.: Proc. Roy. Soc., B98, 589, 1925.)

The pathway of this response is very simple *(Fig. 7-11)*. The axon from the stretch receptor runs into the posterior horn, and while it branches many times in the gray matter, it eventually synapses directly on the cell body of the anterior horn cells of the muscle stretched. The stretch reflex is often referred to as a monosynaptic reflex. The only muscle which contract is the one stretched. The reflex continues for as long as the muscle is stretched *(Fig. 7-12)*.

There are stretch reflexes in all muscles, but they are much stronger in the antigravity muscles, particularly the leg extensors. In addition to the familiar knee jerk from the quadriceps group, brisk stretch reflexes can be elicited from the ankle (the Achilles tendon), the jaw, and the biceps and triceps muscles in the arm.

Although the reflex loop is completed within a few segments of spinal cord, the magnitude of the response can be drastically altered by input from other levels. For example, grasping the hands together and pulling will greatly enhance the knee-jerk response (reinforcement). As will be discussed later, the magnitude or briskness of the response can give many clues to pathological processes.

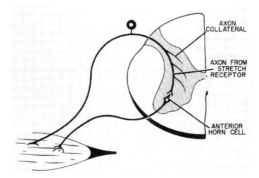

Figure 7-11. The pathway for the monosynaptic stretch reflex.

Reflexive Response to Pain. The majority of spinal cord responses have more flexibility than the monosynaptic response, as can be seen in the reflex withdrawal from pain. This is one of the most important of the protective reflexes of the spinal cord. The flexors are activated and the injured limb withdrawn. Figure 7-12 shows the responses of the ankle flexor muscle, the tibialis anterior, to stimulation of the nerve branches leading to the skin - branches which contain predominantly pain fibers. It can immediately be seen that the combined response is considerably greater than the sum of the two. This augmentation of response is known as facilitation and is one of the most basic integrative responses of the nervous system. The reaction is graded in response to the intensity of the stimulus and the area stimulated.

The mechanism of facilitation is shown in *Figure 7-13*. Stimulation of nerve A excites two anterior horn cells to threshold. Two more are excited in a subthreshold manner; they do not fire an action potential. Nerve B stimulates two totally different anterior horn cells to threshold and the same two to just below threshold. Simultaneous activation of nerves A and B brings all six cells to threshold and elicits a greater tension from the muscle.

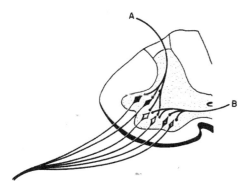

Figure 7-13. A mechanism of summation. Stimulation of nerve A fires the solid anterior horn cells and excites the clear cells. Stimulation of nerve B excites the lined cells to fire and excites the clear cells. Stimulation of both nerves fires all of the cells.

RECIPROCAL INNERVATION OF A JOINT

When the biceps femoralis muscle contracts, the opposing muscle, the quadriceps, is stretched because both muscles are connected to the tibia. The quadriceps group has a strong stretch reflex which is not elicited when the biceps femoralis contracts. When the quadriceps tendon is stretched, stretch reflex tension develops *(Fig. 7-14)*. The tension is maintained until the biceps tendon is stretched; then the tension abruptly drops. This diminution or abolition is called inhibition. Inhibition is, together with facilitation, the keystone of the integrative action of the spinal cord and probably the entire nervous system.

The pathway for the inhibitory response is shown in *Figure 7-15*. A collateral (branch) of the stretch receptor nerve from the biceps muscle runs to an interneuron, which in turn synapses on the anterior horn cells of the quadriceps. As far as we know, all inhibitory responses are carried out through at least one interneuron. The endings of one nerve, such as the stretch receptor nerve, all have the same effect, either facilitation or inhibition. For the nerve pathway to exhibit the other type of function, an interneuron must intercede.

The pathway for the inhibition shown in *Figure 7-14*. Trains of action potentials from the stretch receptors of biceps enters the anterior root and stimulate biceps anterior horn cells and via an interneuron, inhibits the quadriceps muscle.

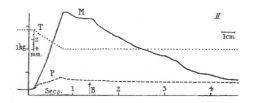

Figure 7-14. Inhibition of a stretch reflex (M) in an extensor muscle, quadriceps, by stretch, at B, of a flexor muscle, biceps. T is the applied stretch to the extensor tendon that initiated the stretch reflex. (From Liddell and Sherrington: Proc. Roy. Soc., B97, 267, 1925.)

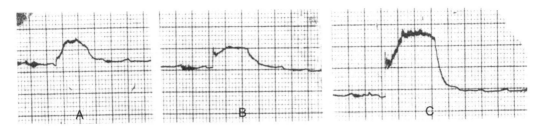

Figure 7-12. Tension response in the tibialis anterior to stimulation of the skin of (A) the ipsilateral foot, (B) ipsilateral calf, or (C) both. Stimulation of both sites results in a tension response greater than the sum of the individual responses: summation. Spinal rabbit. (From unpublished experiments of B.A. Curtis, M.C. Fleming, and E.M. Marcus.)

SPINAL CORD: STRUCTURE AND FUNCTION 7-9

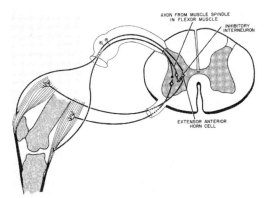

Figure 7-15. The pathway for the inhibition shown in Figure 7-14. Trains of action potentials from the stretch receptor of biceps enters the anterior root and stimulate biceps anterior horn cells and, via an interneuron, inhibits the quadriceps anterior horn cells.

MEMBRANE BASIS OF INTEGRATION

When a microelectrode is inserted into the anterior horn of the spinal cord, the tip, with luck, eventually penetrates an anterior horn cell. This penetration is signaled by the abrupt jump of voltage between the tip and the extracellular fluid from 0 to -70 mV. This intracellular work was pioneered by Sir John Eccles and his collaborators and is discussed at length in his book, *The Synapse*. In this experiment the anterior roots are cut to prevent action potentials, generated by the stimulator, from traveling up the motor axons antidromically (backwards). Stimulation, then, mimics only sensory input. Low voltage stimulation activates only the larger, stretch receptor axons. The extracellular recording from the posterior root gives a signal proportional to the number of axons stimulated. After the cell is penetrated it must be identified; it is almost certain to be an anterior horn cell because of their large size. Other cell bodies and glia cells are so small that penetration is very unlikely. The destination of its axon is determined by using the stretch reflex pathway. Each muscle nerve is stimulated in turn until a short latency, all or nothing; spike (action potential) response is obtained. This is the equivalent of a monosynaptic stretch reflex. A spike response recorded from an anterior horn cell is the counterpart of muscle tension.

Excitatory Postsynaptic Potential, EPSP.
When a flexor anterior horn cell has been located, the record in Figure 7-16 is obtained by stimulating a cutaneous nerve to mimic pain. When the nerve is stimulated at high voltage, many small pain fibers are stimulated. The response in the anterior horn cell can be the firing of an action potential, as in the lowest record. This response is the electrically recorded counterpart of the reflexive withdrawal from a painful stimulus, as shown in Figure 7-13. When the stimulus intensity is reduced, upper records; intensity is reached when no action potential is produced, yet the cell body still responds. The response is a subthreshold depolarization of the anterior horn cell called an excitatory postsynaptic potential, abbreviated EPSP. This response brings the anterior horn cell closer to threshold, that is, closer to firing an action potential that would cause contraction of a motor group. The EPSP is localized to the cell body and dendrites. The axon is unaffected. In contrast to the action potential, it is a graded response.

The EPSPs can add up until threshold is reached and an action potential is fired as is shown in Figure 7-17. Note that the second EPSP in A is larger (greater depolarization) than the first. The effect of the first had not yet "worn off" and the second could add on top of it. In B the second EPSP follows sooner and adds up to a threshold depolarization, initiating an action potential.

This is the basic mechanism of facilitation. The two classic types of facilitation, temporal and spatial, are the same in terms of a membrane response: the EPSP's add up. Spatial response refers to the EPSP's generated as a result of the activation of two different nerves as Figure 7-13. Temporal summation, as illustrated in Figure 7-20, means summation in time; its membrane basis depends upon the time course of the EPSP. The membrane repolarizes slowly following the initial rapid depolarization of an EPSP and a second EPSP can add upon the first. Frequently this "added boost" will bring the cell membrane to the threshold for firing an action potential in the axon.

Delay.
There is a delay of approximately 0.8 msec between the arrival of an action potential from a stretch receptor axon and response of an anterior

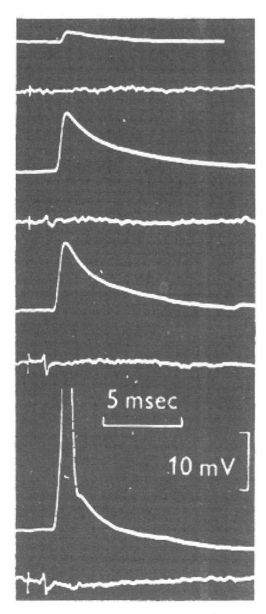

Figure 7-16. EPSP's: The upper trace of each pair is an intracellular record from a biceps anterior horn cell. The lower trace of each pair is the action current from the posterior root and is proportional to the number of sensory axons firing. As the number of axons firing increases, the depolarization (an EPSP) increases until threshold for the anterior horn cell is reached and an action potential fires. The bottom pair of records is the intracellular version of a stretch reflex. The rest of the action potential of the lower record is off the scale. The average resting potential is -70 mV. (From Brock, Coombs, and Eccles: J. Physiol., 117, 431, 1952.)

horn cell. This delay and other features strongly support chemical transmission at the synapses on the anterior horn cell. The identity of the transmitter is still not entirely clear although there is a great deal of evidence to suggest that L-glutamate and L-aspartate are the transmitters for many of the excitatory synapses in the spinal cord.

These transmitters destroy the selective permeability of the postsynaptic membrane, just as acetylcholine does at the myoneural junction. The equilibrium potential is then zero millivolts. The membrane never reaches zero under the influence of excitatory transmitters since they are rapidly removed. At the anterior horn cell membrane, however, unlike the myoneural junction, the amount of excitatory transmitter released is not always sufficient to depolarize the cell body to threshold.

Inhibitory Postsynaptic Potential, IPSP. When the nerve leading to the quadriceps muscle is stimulated, the membrane of a flexor (an antagonist) anterior horn cell hyperpolarizes, as shown in the recordings in *Figure 7-18*. This hyperpolarizing response called an inhibitory postsynaptic potential (IPSP), moves the membrane further from threshold, making it more

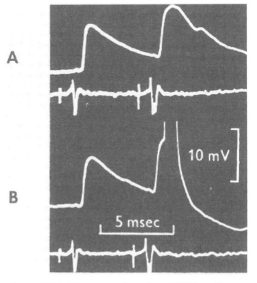

Figure 7-17. The summation of two EPSP's to fire an action potential. In B the second EPSP followed sooner and added upon the first to achieve a threshold depolarization, initiating an action potential. (From Brock, Coombs, and Eccles: J. Physiol., 117, 431, 1952.)

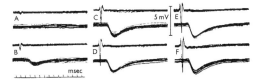

Figure 7-18. The response (an IPSP) of a biceps anterior horn cell to stimulation of the quadriceps nerve. The upper trace of each pair is the action current from the posterior root; the lower is the intracellular record. The greater the number of sensory axons firing the more negative and longer lasting is the IPSP. (From Coombs, Eccles, and Fatt: J. Physiol., 130, 396, 1955.)

difficult to fire an action potential and is a postsynaptic inhibition. The IPSP is the membrane equivalent of inhibition. As noted earlier, inhibition is effected through an interneuron (Fig. 7-14).

The transmitters substances responsible for the IPSP include GABA and glycine. They specifically increase the chloride conductance of the anterior horn cell, typically for hundreds of msec. The membrane potential at rest is -74 mV, indicating there is considerable sodium conductance in relationship to potassium and chloride conductances. The calculated values for potassium, chloride, and sodium equilibrium voltages are: VK = -90 mV, VCl = -80 mV and VNa = +45 mV. A large increase in chloride conductance will result in a membrane potential of -80 mV and this is the equilibrium potential at the height of the IPSP.

Summation of IPSP's and EPSP's can occur as shown in *Figure 7-19*. The microelectrode is recording from an anterior horn cell innervating biceps (a flexor). When the stretch receptor nerve from the biceps is stimulated (Fig. 7-19A), an action potential fires; this is the stretch reflex. When the nerve to the quadriceps (an opposing extensor) is stimulated (Fig. 7-19B), a large IPSP is recorded; this is the inhibition of a stretch reflex of an opposing muscle. When the two are stimulated 45 msec apart (Fig. 7-19C), the cell fires. When the delay was reduced, the membrane depolarizes but does not reach threshold (Fig. 7-19D).

SYNAPTIC MECHANISMS

The dendritic tree and cell bodies of spinal cord neurons contain a varied collection of receptor mediated ionic channels that are sum-

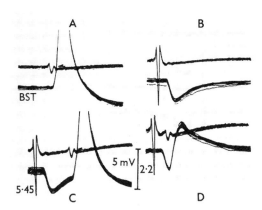

Figure 7-19. Summation of an IPSP and an EPSP. The lower trace of each pair is an intracellular recording from a biceps anterior horn cell. The upper trace is the action current in the posterior root. In A the nerve to the biceps was stimulated and an action potential evoked; this is the intracellular equivalent of a stretch reflex. In B the nerve to quadriceps was stimulated and an IPSP was evoked. In C the two were stimulated 45 msec apart and a spike resulted. In D the two were stimulated 2.2 msec apart and no action potential resulted; the IPSP generated by the quadriceps stimulus inhibited the biceps anterior horn cell. In C the hyperpolarization effect of the IPSP had declined by the time the second nerve was stimulated. (From Coombs, Eccles, and Fatt: J. Physiol., 130, 396, 1955.)

marized in Figure 7-19 along with some of the neurotransmitters implicated in their activation. All of these ionic channels act on a time scale of milliseconds to seconds.

The fast EPSP channel (A) is very similar to the ACh channel at the myoneuronal endplate. It opens to all small ions and depolarizes rapidly and decisively. The somewhat slower IPSP Cl- channel (B) is equally important. Selective increase in PK (C) usually occurs after action potential firing in the axon and is thought to be mediated by increasing Cai, possibly Ca entering during the very positive voltages of the action potential. All of these mechanisms depend on increasing ionic permeability (decreasing cell resistance) and to rapidly changing intracellular ionic concentrations which must be restored by energy expenditure.

The next two mechanisms (D,E) rely on decreases in permeability and are more energy efficient.

The active transport of unequal numbers of

ionic changes can result in changes in membrane voltage; 3Na/2K and 3Na/Ca are examples. The relative importance of this electrogenic mechanism is not clear.

Intermediate term changes in neuron baseline characteristics (more or less excitable) are produced by altering the number of active channels on the cell surface. For example, in response to noxious stimuli 5 HT is co-released from sensory a nerve which increases the C-AMP level of the interneuron causing rapid inactivation by dephosphorylation of potassium channels. Whenever an action potential enters the presynaptic region, the depolarization is prolonged because the normal increase in PK that hastens repolarization is lacking. Consequently, the voltage gated Ca channels remain open longer allowing greater Ca entry and hence greater transmitter release. This augmented response may last for several hours after 5 HT liberation. The key to specificity is both the type of receptor and the enzyme chain-second messenger system linked to it. Receptors for Epinephrine & Norepinephrine increase the number of phosphorylated, hence active, Ca channels.

Enhancement of synaptic transmission lasting one or more days occurs by increasing the amount of neurotransmitter released from each synaptic vesicle. Increase in synaptic vesicle content presumably reflects changes in gene expression that could last a very long time and might be the synaptic basis of long-term memory. This mechanism alters the importance of existing synapses. A synaptic junction that previously needed the cooperative effort of 10 other synaptic junctions to depolarize the cell body to threshold can now do so alone.

The only other major variants seen when recording from anterior horn cells are long-lasting (30 to 40 msec) responses. These are thought to involve many interneurons possibly arranged in a circular path so that an action potential "chases its tail" around the circuit for several revolutions, stimulating the anterior horn cells on each revolution. They may also be produced by long acting neurotransmitters co-released from the same axons that release the shorter acting neurotransmitters.

Slow Potentials. Action potentials in axons making synaptic contact with the dendrites or cell body of an anterior horn cell are expressed within the anterior horn cell as depolarizations (EPSP's, a very few of which will, by themselves, fire the cell) and hyperpolarization (IPSP's, which inhibit firing). The membrane potential of the cell is constantly changing under the influence of these two types of input. Whenever the membrane potential reaches the firing level of the axon (-55 mV), action potentials are generated with a frequency proportional to the depolarization past threshold. For the purposes of transmission, depolarization, more positive than threshold, is coded as frequency. When these action potentials bombard another cell they are once again decoded into changes in the membrane potential of that cell body.

It is not clear why the EPSP's and IPSP's recorded from the cell body are of different amplitude. This may be due to differences in the amount of transmitter released or may be due to the location of the synapse on the cell. Since the dendrites, like the cell body, conduct potential disturbances in a decremental, cable fashion, synapses close to the cell body will have a greater influence on the membrane potential.

It is possible to show inhibition of firing of anterior horn cells without any change in the resting membrane potential; no IPSP precedes the inhibition. The inhibition is thought to take place presynaptically, either at a nearby unidentified interneuron or more probably, in the junction region between the axon and the dendrite. We will discuss this type of inhibition later in this chapter.

STRETCH RECEPTORS

The Muscle Spindle. In the previous discussion of spinal reflexes, reference to stretch receptors was very general. The major dynamic stretch receptor is the muscle spindle. This is a bundle of modified muscle fibers that lie parallel to the rest of the muscle fibers. Its structure is shown in Figure 7-20; it is composed basically of three to five small muscle fibers each containing a specialized, nonstriated region in their center. The striated ends of the muscle fibers can contract and are innervated by small, gamma motor neurons. These gamma motor neurons have cell bodies in the anterior horn, just as do the large alpha fibers that innervate the bulk of the muscle.

flower-spray endings on either side of the annulospiral ending. These axons rise to the ipsilateral cerebellum, carrying information on "unconscious position sense". They apparently play no part in the stretch reflex.

The basic function of the stretch receptor is to fire when the muscle is stretched. The response of the annulospiral ending is of short latency. The frequency of firing is at first high; it then slows down, but never adapts completely.

The Gamma System. The function of the intrafusal muscle fibers is more difficult to understand. The first function is to keep the muscle spindle tight as the extrafusal fibers contract; Figure 7-26 shows this. At the rest length (A) the spindle is tight. Any further stretch will set up a volley of action potentials in the Ia sensory nerve that will lead to a stretch reflex; contraction of the extrafusal fibers. When the extrafusal fibers contract because of firing in the alpha motor neuron alone, the spindle goes slack (B) and is no longer responsive to stretch. Indeed the muscle would have to be pulled out slightly further than the rest position for the spindle to react. To rectify this situation, the gamma motor neuron fires, thus contracting the intrafusal fiber, and the spindle is tight again (C). Through the mechanism of the gamma motor neurons the sensitivity of the stretch reflex is maintained throughout the entire range of the limb movement. For example, the knee-jerk reflex can be elicited in many positions of the lower leg.

Another function of the intrafusal fibers is to change the sensitivity of the annulospiral ending (Fig. 7-21). It can easily be seen, in the last column, that the response to 20 gm tension or an equivalent stretch can be drastically altered by activity of the intrafusal fiber. Apparently the annulospiral ending reacts to stretch or deformation whether it is from without, as in stretch, or from within, as in intrafusal muscle fiber activity.

Probably the most important function of the intrafusal fibers is to modulate contraction of the extrafusal fibers via the stretch reflex. It is clear from Figure 7-21 that activation of the gamma fibers and subsequent contraction of the intrafusal fibers sets up a response in the Ia sensory fibers which is indistinguishable from the response to stretch (compare response in the upper right and lower left in Figure 7-22. This

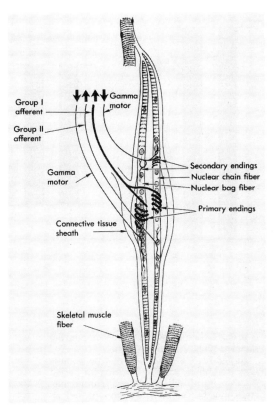

Figure 7-20. A muscle spindle. (From Gardner: Fundamentals of Neurology. Philadelphia, W.B. Saunders, 1968.)

The small muscle fibers in the spindle are called intrafusal fibers, and the large muscle fibers that make up the bulk of the muscle are referred to as extrafusal fibers. The extrafusal fibers are primarily innervated by large (8-12 μ) axons. The pattern is not exclusive and there is some dual innervation; alpha motor fibers to both intra and extrafusal fibers.

There are two sensory nerves which take origin from the unstriated center region of the muscle spindle. The largest, (12 μ) classified Ia, comes from the center of the sensory region. The unmyelinated ends of the nerve wrap around each of the muscle fibers and are called primary or annulospiral endings. From these endings arise the action potentials that stimulate the stretch reflexes. This axon gives off many types of collateral within the gray matter of the cord that then travel up and down the cord for several segments.

The second ending gives off a smaller nerve, classified IIa, from specialized, secondary or

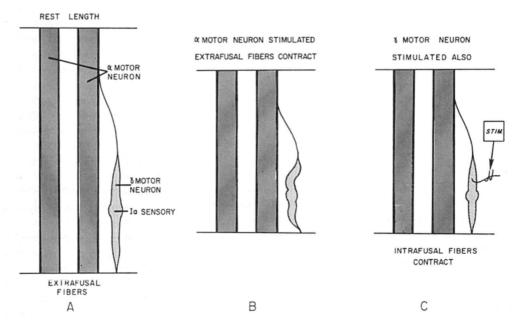

Figure 7-21. The effect of gamma activation of intrafusal muscle fibers upon maintenance of "tone" in the muscle spindle. When just the extrafusal fibers contract (B) the spindle becomes slack and unresponsive to small stretches. When both alpha and gamma systems are activated, the spindle is once again taut and responsive to small stretches.

activation of the Ia sensory fiber leads to contraction of the extrafusal fibers. This is probably best shown by the accompanying sketches (Figs. 7-22).

The length of A is the equilibrium point for the stretch reflex when there is no gamma activity; stretch of this muscle would lead to extrafusal contraction. At A the gamma fiber is suddenly activated. During the period B, the intrafusal fibers are contracting, stretching the annulospiral ending and altering the sensitivity of the primary stretch receptor to the dashed line. The Ia ending increases its rate of firing and causes reflexive shortening of the extrafusal fibers and shortening of the muscle as a whole, as is happening in C. This process continues in D until point E is reached. At E the muscle as a whole has contracted sufficiently to decrease the stress on the center region of the muscle spindle and reduce the rate of firing of the Ia fiber to threshold for the stretch reflex.

The shortening and subsequent maintenance of muscle length and, consequently, of the joint angle was brought about by a constant gamma activation (Fig 7-24). The rate of firing of the gamma system, initiated by the brain, has not changed during the entire time. The new position (E) is not affected by the load the muscle had to move. If the load was light, the new position was reached quickly; if the load was heavy the new position was reached slowly. In any event, the new position was reached and maintained without further judgment from the motor centers of the brain.

This loop will provide good length control if movement is very slow. If movement is faster, this simple type of loop will begin oscillating because of nerve conduction and synaptic delays. As the muscle is contracting, its length at t=0 is

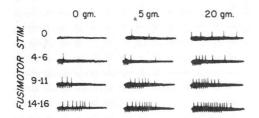

Figure 7-22. The response of an Ia fiber from an annulospiral ending to stretch and gamma activity. Notice that the number of action potentials fired increases with stretch (weight) and gamma activation. The downward deflections are stimulus artifacts of gamma stimulation. (From Kuffler, et al.: J. Neurophysiol, 14, 29, 1951.

SPINAL CORD: STRUCTURE AND FUNCTION

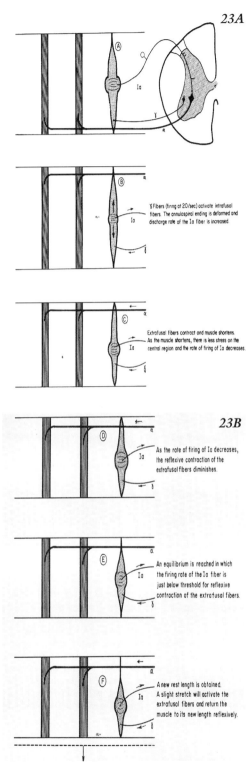

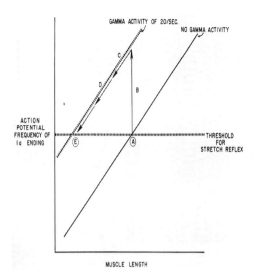

Figure 7-24. The effect of gamma activation upon maintenance of "tone" in the muscle spindle.

sent up the sensory nerve and arrives at the spinal cord at t+15 msec. After a 5 msec delay the rate of firing of the anterior horn cell varies. This variation arrives via the motor nerve at the muscle in another 15 msec. Thirty-five msec have elapsed, the muscle is no longer at its original length and the "correction" is no longer correct. This will lead to a cycle of over corrections around the intended length: oscillations.

The oscillations can be rectified if the control system is given information on the rate of change of position, velocity. With this information the control system in the spinal cord can figure out where the muscle will be at the end of the delay time and make corrections accordingly. Careful inspection of Figure 7-21, particularly the right hand column, will show that the firing rate at the beginning of movement is faster: there is velocity information in the signal from the stretch receptor.

The difficulty with the pure gamma stimulation of muscle activity theory is that recordings such as are shown in Figure 7-21 show that the alpha fibers are activated at about the same time as the gamma fibers, not 30-40 msec later as the theory predicts. Other evidence suggests that muscle movement probably gets started by direct alpha stimulation and is later reinforced by reflexive gamma stimulation. The termination point may well be found and maintained mainly through the gamma system.

Figure 7-23A. & 23B The role of the gamma system in setting a new muscle length. (Refer to text for discussion.)

Let us digress a moment and consider the simplest type of muscle movement - opposing the thumb to the palm. Cells in the anterior horn are activated by fibers from the large descending corticospinal tract, a direct pathway, of which more will be said later. The intensity of stimulation and the number of motor groups activated are determined by the motor cortex. The motor cortex basically sets the tension that is to be developed. The distance moved is a function of the resistance to movement and the duration of the stimulus. The extent of movement is visually controlled; when the thumb reaches the desired position, the motor cortex stops stimulating the anterior horn cells.

In contrast to this type of movement, which requires constant attention if the desired position is to be achieved, many of our movements, such as walking, require little attention beyond the decision to walk along the sidewalk. Most of us can even chew gum at the same time! These movements are probably carried out through mediation of the gamma system, which allows the brain to set a desired position and then forget about it. Compare, then a corticospinal system, which produces a force, and a second system, the gamma system, which produces a new position. Although it is not completely clear at this time, it would appear that the gamma anterior horn cells are innervated by descending motor tracts other than the corticospinal. This system is controlled by neurons in the brain stem; the reticular formation, the vestibular nucleus and the red nucleus.

Golgi Tendon Organs. Golgi tendon organs are a second type of stretch receptor found in the tendinous insertion of muscles. It contains no muscle fiber system. This receptor is in series with the muscle fibers and signals tension via Ib sensory fibers. It is much more sensitive to tension generated by the muscle than to stretch of the muscle. Activation of the sensory fibers from the tendon organ causes an inhibition of the contraction of the muscle. Among the functions of the tendon organ is to protect the insertion of the muscle from too great a stress which might tear the insertion from the bone. Tension information from the tendon organ is transmitted to higher centers via both dorsal and ventral spinocerebellar tracts.

INTERNEURONS

Only a very small portions of the cells of the gray matter of the spinal cord are anterior horn cells; the vast majority are interneurons - small cells with short dendrites and axons. These cells interconnect incoming sensory axons and descending spinal cord tracts with each other and with anterior horn cells.

One such interneuron is the Renshaw cell that must be studied in a somewhat roundabout fashion. After the posterior (sensory) roots have been cut, stimulation of a muscle nerve (for example, to the gastrocnemius) will result in action potentials traveling back up the motor nerve (antidromic) and entering the spinal cord. An extracellular electrode records a burst of high frequency action potentials from the Renshaw cell in response to a single volley of action potentials *(Fig. 7-26)*. Electrophoresing dye from a recording electrode has localized these cells. After a Renshaw cell has been characterized physiologically by its response to antidromic stimulation, dye is deposited at the tip of the electrode by passing current through the microelectrode. An example of this widely used technique is shown in Figure 7-27. When many such dots are placed together on a spinal cord cross section (Fig. 7-27) they all appear in the most anterior portion of the anterior horn. This is the same region in which the axons from the anterior horn cells branch.

The firing of anterior horn cells activates the Renshaw cell with acetylcholine as the neurotransmitter. Studies of a Renshaw cell show it to

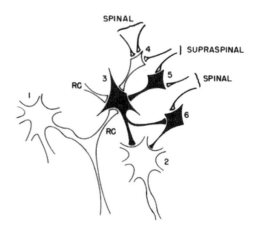

Figure 7-25. Major classes of input to Renshaw cell.

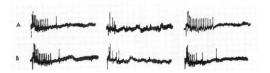

Figure 7-26. Renshaw cell activity, left hand responses are produced by stimuli to ventral roots conducted antidromically. This activity can be inhibited (right hand responses) by (A) a squeeze of the ipsilateral toe and (B) by stimulus to the contralateral biceps-semitendinous nerve. (From Wilson, Talbot and Kato: J. Neurophysiol., 27, 1063, 1964.)

be activated by antidromic stimulation of any of a great many motor nerve fibers; even from different muscles. These muscles, however, usually belong to a single functional grouping, such as knee flexors or ankle extensors. The effect of Renshaw cell activation upon other anterior horn cells is inhibitory; it depresses or totally inhibits firing. This is shown diagrammatically in Figure 7-25. Activity in anterior horn cell 1 will inhibit, via the Renshaw cell, anterior horn cell 2. This inhibition is usually to opposing muscle groups. The Renshaw cell itself can be inhibited by a variety of pathways, such as squeezing the ipsilateral toes and stimulating the muscle nerve on the contralateral side (Fig. 7-26). Note that in each case the response to the antidromic stimulation is much reduced.

Since the Renshaw cell is inhibitory to anterior horn cells, inhibiting the Renshaw cell will remove inhibition from the anterior horn cell 2 in Figures 7-26, 27, 28. The effect of inhibition of an inhibitory cell is called disinhibition. This doesn't mean that anterior horn cell 2 will fire, since that requires a facilitatory stimulus, but it does mean that no inhibition is being applied via that particular Renshaw circuit. This phenomenon of disinhibition is quite common in the nervous system.

POSTERIOR HORN

Laminar Organization. The posterior horn contains the entry of the sensory fibers and their rich synaptic connections. Many names have been proposed for the regions of the posterior and intermediate horns to describe their anatomical and physiological variations. In order to clear up this confusion about the terminology of the organization of the gray matter of the cord, Rexed proposed a laminar organization of the cat spinal cord which has since been extended to all primates. The gray matter is divided into nine layers with a thin tenth region surrounding the central canal, *Figure 7-28*.

Lamina 1 forms the cap of the dorsal horn and is penetrated by many fibers. It includes the nucleus posterior marginalis. Many intersegmental pathways arise from this layer.

Lamina 2 corresponds to the substantia gelatinosa. This nucleus extends the entire length of the cord and is most prominent in the cervical and lumbar levels. Cells in this lamina also form intersegmental connections.

Lamina 3 is the broad zone containing many myelinated axons and receives many synapses from the dorsal root fibers.

Lamina 4 is the largest zone and consists primarily of the nucleus proprius of the dorsal horn. This nucleus is conspicuous in all levels.

Lamina 5 extends across the neck of the dorsal horn and in all but the thoracic region is divided into medial and lateral portions. The lateral portion consists of the reticular nucleus that is most conspicuous in the cervical levels. Corticospinal and posterior root synapses have been identified in this lamina.

Lamina 6 is a wide zone most prominent in the cervical and lumbar enlargements. In these levels, it is divided into medial and lateral zones. Terminals from the posterior roots end in the

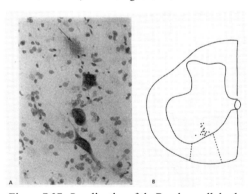

Figure 7-27. Localization of the Renshaw cells by the method of dye electrophoresis from the recording electrode. A, The dye spot among the anterior horn cells. The dye spot, arrow, is an azure blue while the cells are purple. B, The location of a number of dye spots superimposed upon a tracing of a lumbar spinal cord segment. (From Thomas and Wilson: Nature, 206:211, 1965.)

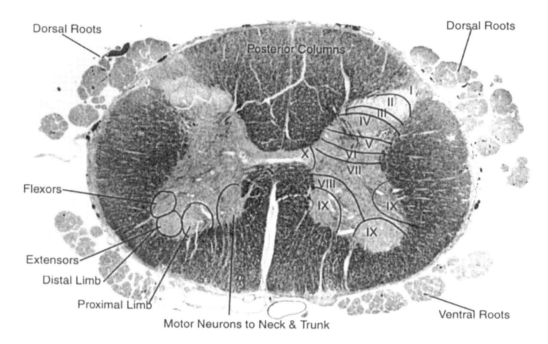

Figure 7-28. Rexed's lamination pattern of the spinal gray matter on the right and the location of ventral horn cells on the left, Lumbar Section. Myelin Stain,

medial region while descending fiber tracts project to the lateral zone.

Lamina 7 includes most of the intermediate region of the gray matter in the spinal cord. In this lamina are found the intermediolateral and intermediomedial nuclei. The nucleus dorsalis or nucleus spinal cerebellar of Clark is obvious in C8 through L2 levels. The axons arising from this nucleus form the posterior spinocerebellar tract. In the cervical and lumbar enlargements, this lamina includes many of the internuncials and the cell bodies of gamma efferent neurons. Axons from posterior roots, cerebral cortex and other systems end in lamina 7. Cells in lamina 7 form tracts that project to higher levels including cerebellum and thalamus. In the thoracic and sacral regions axons also leave this lamina to form preganglionic autonomic connections.

Lamina 8 in the cervical and lumbar enlargement is confined to the medial part of the anterior horn. Many of the axons from these nerve cells form commissural fibers in the anterior white column. Axons from descending pathways originating in the brain stem terminate here.

Lamina 9 includes the largest cell bodies in the spinal cord, the horn cells. The axons of these cells (alpha motor neurons) form much of the ventral rootlets that supply the extrafusal muscle fibers. The medial group innervates the muscles of the axial skeleton while the lateral group innervates the muscles of the appendicular skeleton.

Lamina 10 includes the commissural axons.

Posterior root fibers. Sensory nerve fibers enter the spinal cord gray matter (Figure 7-28) by a medial bundle of large, heavily myelinated fibers and a lateral bundle consisting of thinly myelinated and unmyelinated axons. The heavily myelinated medial bundles convey information from the large dorsal root ganglion cells subserving encapsulated somatic receptors (muscle spindles, pacinian corpuscles, Meissner's corpuscles), carrying information on touch, position and vibratory senses. Upon entering the spinal cord, the axons divide into ascending and descending processes, both giving off collaterals. Many of these collaterals end in the segments above or below the level of entry. Branches from the larger fibers enter the ipsilateral posterior funiculus and ascend to the medulla. The large myelinated axons also have numerous terminals in Clark's nucleus in lamina 7. Collaterals from posterior

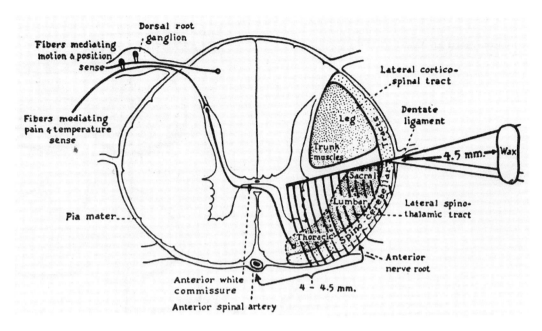

Figure 7-29. Diagram illustrating a chordotomy. The cross section of the spinal cord shows the lamination of the spinothalamic tract, the position of the pyramidal tract in relation to it, and the presence of other tracts in the lower quadrant. A piece of bone wax is mounted 4.5 mm. from the tip of the knife as a depth gauge. Heavy curved lines in the ventral quadrant indicate the sweep of the knife. Note that a desire to spare the lateral corticospinal tract would result in sparing the sacral dermatomes. (From Kahn and Rand: J. Neurosurg. 9:611-619, 1952.)

root fibers also enter Lamina 7 and 9 ending on internuncial neurons and anterior horn cells for reflex activity.

The smaller, lateral bundle of thin axons (A delta & C fibers) conveys information from free nerve endings, tactile receptors and other unencapsulated receptors. These fibers, conveying impulses of tissue damage, temperature and light touch sensation, enter Lissauer's tract (fasciculus dorsal lateralis) (Figures 7-28, 7-34). Axons from Lissauer's tract enter laminas I, II and III. Axons from this lamina ascend and descend in Lissauer's tract to reenter the same lamina,

NOCICEPTION AND PAIN

Any stimulus, such as heat, trauma or pressure, which produces tissue damage or irritation, is a nociceptive stimulus. Nociceptive stimuli are the afferent arm of many reflexes; locally, a red wheal develops around a cut, withdrawal of a limb from a hot pipe is a spinal cord reflex, while tachycardia from an electric shock to the finger is a brain stem autonomic reflex. Only when this nociceptive information reaches the thalamus and cerebral cortex should we talk about pain. Pain perception by these higher centers triggers affective responses and suffering behaviors. The pain experience varies enormously from person to person and the circumstances may alter the response. Contrast the pain of your thumb being hit with a hammer when you are: fixing your sail boat, doing a "chore" around the house or holding the nail for someone else.

As a gross oversimplification we perceive pain in two ways. As pricking, itching or sharp and easily localizable, such as a razor blade cut or a mosquito bite. This type of pain is usually short lasting, up to a day or so, and is usually tolerated. In contrast, pain may be described as dull, aching or burning and is poorly localized. This type of pain is longer lasting (rheumatoid arthritis) or repetitive (menstrual cramps) and is often poorly tolerated, frequently coloring the persons entire view of life.

Receptors. Tissue damage releases a variety of typical intracellular substances including K^+, H^+, bradykinin, as well as specialized compounds such as: serotonin from blood platelets,

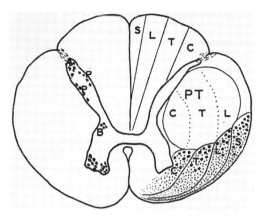

Figure 7-30. The lamination pattern of the major tracts of the spinal cord. (From Walker: Arch Neurol. Psychiat. (Chicago), 43:284, 1940.)

substance P from nerve terminals and histamine from mast cells. All of these substances directly stimulate the free endings of small, myelinated A delta fibers (1-5 μ diameter) and smaller, non-myelinated C fibers (0.25-1.5 μ diameter). A second group of released substances including prostaglandin precursors and leukotrienes act as sensitizes. Many of these small fibers have collaterals ending in regions containing neurotransmitter vesicles that apparently amplify nociceptive stimuli by releasing substance P when the free nerve ending is stimulated.

In addition to these relatively non-specific receptors, the upper end of the range of stimuli to temperature and pressure receptors generates nociceptive responses. For example, heating a patch of skin to 40°C with a heat lamp gives a comfortable, warm sensation. Heating to 47°C generates a painful, but tolerable experience. Higher temperatures are perceived as unbearable pain.

Cold is a divers sensation and recently a cold channel has been identified (McKeeney, Neuhauser & Julius 2002). Cold and warmth are sensed by thermoreceptor proteins on the free nerve endings of the somatosensory neurons; one channel, vanilloid receptor subtype 1 (VR1) os activated by temperatures above 43° C, and by vanilloid compounds including capsaicin the component of hot chili peppers. The other vanilloid receptor type 1 (VRL-1) is sensitive to temperatures above 50°C. The cation channel that is methanol activated (CMR1-cold methanol type 1) is activated by temperatures of 8-30°C and is a member of the transient receptor potential family of ion channels with VR1 and VRL-1 also members of this family.

These nerve fibers carrying nociceptive information join peripheral nerves, segregate into spinal nerves and once inside the spinal dura, join posterior roots. Their cell bodies are in the posterior root ganglia while the axon continues and enters the spinal cord in the lateral bundle. Nociceptive information entering on one posterior root projects for 2-3 segments up and down the cord in lamina I-III. From centuries old clinical observation we know all nociceptive information reaching consciousness, had hence causing pain, crosses the neuro-axis close to the segment of entry and rises in the contralateral white column.

Modulation of Pain Transmission. Transmission of pain information through the chain of neurons in the posterior root is influenced by many factors, some of them originating within the segment, some from higher centers. Most of us, after cutting or bruising ourselves, rub the surrounding area and obtain relief of the pain for as long as we rub. An animal that has been hurt will lick the wound, presumably to obtain relief of pain. This subjective phenomenon is frequently spoken of as counter irritation; stimulation of the large, myelinated touch fibers reduces the magnitude of transmission of pain sensation through the posterior horn. This modulation takes place in the substantia gelatinosa, lamina II, within the "Gate" proposed by Wall and Melzak.

Several clinical observations point to the importance of large fiber inhibition of nociceptive transmission, even in the absence of apparent nociceptive stimulation. Perhaps the most painful of these condition is avulsion, traumatic tearing out, of posterior roots. Surgical severing of posterior roots, instead of giving pain relief, often results in intolerable pain. In both cases large fiber input is lost. In contrast, destruction by a laser beam of the lateral dorsal root entry zone (for small fibers) is often highly successful in blocking the flow of nociceptive information up the neuro-axis. The large fiber input is largely spared because they enter in the medial zone.

Severed peripheral nerves often generate itching or burning sensations, causalgia, which

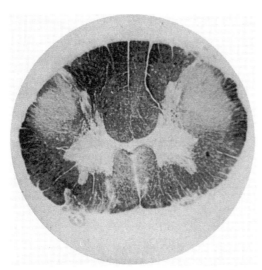

Figure 7-31. (A) The location of the corticospinal tracts as shown by degeneration caused by a lesion in the internal capsule. (From Wechsler: Clinical Neurology. Philadelphia, W.B. Saunders, 1963.) (B) The collateral branches of a corticospinal axon originating in the monkey motor cortex and innervating ulnar nerve motor neurons. The collaterals extend up and down the spinal cord for 2-3 mm (1 segment). From Futami, Brain Res. 164:279-284, 1979.

can be alleviated by stimulating the large diameter axons central to the cut. The large diameter axons have low thresholds so can be stimulated without stimulating the smaller, high threshold nociceptive fibers. Activity in the large fibers suppresses background nociceptive inflow. TENS, Trans Epithelial Nerve Stimulation, is often effective in preventing, or reducing, peripherally generated nociception from becoming pain; effective in closing the gate.

As mentioned earlier, branches of many incoming large axons enter the posterior column. Stimulation of the posterior column sends antidromically-conducted action potentials into the posterior horn and often reduces the pain experience.

Damaged small axons are prone to developing alpha2 adrenergic receptors, so the axon becomes sensitive to norepinephrine liberated by the peripheral sympathetic nerves. Pharmacological blocking or surgical severing sympathetic outflow often reduces pain.

The major excitatory neurotransmitter in the pain system is substance P that is found in abundance in the posterior horn, particularly lamina I and II. The mechanism of postsynaptic action is a reduction in potassium permeability leading to depolarization. The major inhibitory transmitters in the pain pathway are the enkephalin/endorphin series of peptides released from cells entirely within the posterior horn, mostly lamina II. Their action is closely mimicked by the opioid peptides. At least three modes of action have been shown, although all studies are complicated by the very small size of the cell bodies in the substantia gelatinosa (lamina II). An increase in potassium permeability has been shown which hyperpolarizes the cell and acts as an IPSP. Many studies suggest these compounds also compete with the excitatory transmitters of the region, substance P and glycine, for postsynaptic sites. Other studies suggest reduction of Ca++ influx into the presynaptic terminal in response to incoming depolarization and consequent reduction of the amount of excitatory transmitter released. Other studies implicate presynaptic inhibition (below).

Injection of opioids into the lumbar CSF often reduces nociceptive transmission and reduces the pain experience. The effect is blocked by the universal narcotic blocker, naloxone. Successful acupuncture increases endorphins in the lumbar CSF.

Descending axons in both the anterolateral and posterior white columns influence transmission across this chain of synapses.
Supraspinal control of these enkephalin/endorphin neurons is from the brain stem, particularly from the locus ceruleus (Norepinephrine) and Raphe nuclei (5HT), both in the medulla. These areas in turn are modulated by the periaqueductal gray of the midbrain that also contains an inhibitory endorphin/enkephalin system. Stimulation of these structures results in profound anesthesia (electroanesthesia).

While the descending pathways are not clear, we can suppress the flow of nociceptive information through the posterior horn by positive thinking, for example, all "pain" studies are bedeviled by a 40% placebo effect. Yes, the third, post surgical morphine injection can be replaced with saline and still "work" half the time.

Presynaptic Inhibition. In contrast to the cells in the anterior horn, the major mechanism of inhibition in the posterior horn is presynaptic.

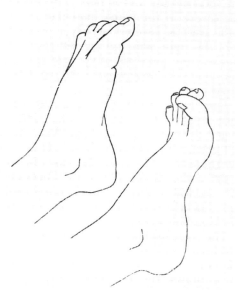

Figure 7-32. The Babinski response. Upper, The normal adult response to stimulation of the lateral plantar surface of the foot. Lower, The normal infant and abnormal adult response.

After a volley of action potentials enter the posterior root, prolonged depolarization can be measured in that and adjoining posterior root axons as well as in the substantia gelatinosa. All available evidence suggests that the next cell in the transmission pathway is not depolarized. If an action potential(s) enters the endplate region of cell II, a reduced amount of excitatory transmitter is released because of the existing depolarization. The reduced amount of transmitter released by cell II causes a smaller EPSP in cell III (panel C); in this example cell III does not reach threshold. Presynaptic inhibition appears to be particularly useful in the sensory system because specific inputs to cell III can be blocked; while other inputs (such as cell IV) have not lost their influence.

Projection fibers. After the extensive processing just described, large diameter, myelinated axons with cell bodies in lamina I and V cross the neuro-axis near the central canal (Fig 7-28) to join the anterior-lateral white matter. These fibers are described later in this chapter as the Spinothalamic tract as they project to the brainstem and ultimately the cerebral cortex.

TRACTS

The white matter of the cord contains axons which run up and down the spinal cord connecting segment to segment and the segments to the brain. (The white matter is divided into three regions that are delimited by the presence of the dorsal roots that separate the posterior from the lateral funiculi and the ventral roots that separates the lateral from the anterior funiculi Fig. 7-28).

The posterior funiculus consists primarily of the posterior columns. The lateral funiculus is a solid mass of myelinated nerve fibers containing many tracts with the ascending fibers on the outside and the descending fibers closer to the gray

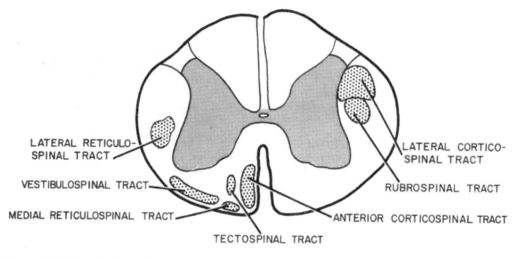

Figure 7-33. The major descending tracts in man.

SPINAL CORD: STRUCTURE AND FUNCTION 7-23

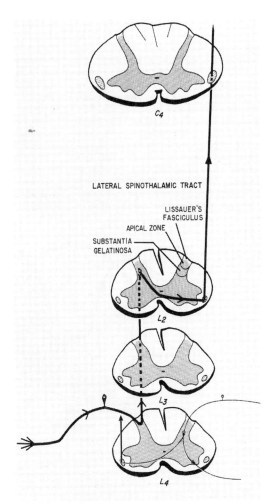

Figure 7-34. *The spinothalamic tract/anterolateral system. Incoming fibers that are activated by tissue damaging stimuli may rise ipsilaterally for up to 3 spinal cord segments (as the fiber entering on the left) before synapsing and crossing the neuro-axis and entering the spinothalamic tract. The fiber entering on the right synapses at the level of entry, one axon crosses the neuro-axis and rises, the other axon enters the anterior horn to participate in local reflexes, such as withdrawal from a hot surface. The major projections of the spinothalamic tract are: midline medulla (o), periaqueductal grey of the midbrain, and two nuclear groups of the thalamus, the ventral posterior lateral nucleus which projects in turn to the post central gyrus, as well as the interlaminar nucleus which projects widely.*

matter. The only way that the location of individual tracts can be determined is to examine sections after an injury in the nervous system. The anterior funiculus likewise carries many ascending and descending tracts. The location of the tracts must be determined by observing any degeneration caused by discrete lesions.

In this section we will provide an overview of the pathways in the spinal cord. A detailed discussion of the individual sensory and motor pathways are included in the diencephalic chapter 17. The discussion in this section will focus on topics relevant to only the spinal cord.

DESCENDING TRACTS IN the spinal cord.

Corticospinal Tracts. Commands for voluntary movement travel from the brain and through the spinal cord in the corticospinal tract. This tract has its origin in the cerebral cortex, most prominently from the motor and premotor cortex of the frontal lobe(see Chapter 17). At the junction between the medulla and the spinal cord (the level of the foramen magnum) most of the fibers cross the neuroaxis, decussate, and move laterally and posteriorly to form the lateral corticospinal tract (Fig. 7-36).

The location of this pathway in the human is determined by analyzing cord sections obtained at autopsy from cases where cortical destruction has occurred in the motor areas of the precentral gyrus or in the internal capsule and axons degenerated following death of the cell body (Fig 7-31).

The Lateral Corticospinal Tract is the major tract for voluntary control of skeletal muscle. Destruction of this tract leads to paralysis of skeletal muscle and the loss of voluntary movement. This paralysis is usually total distally, in the hand, and somewhat less severe in the trunk musculature. If the right tract is severed at C1, a paralysis of both the right arm and leg will result, Hemiplegia, a paralysis of the arm and leg on the same side. Monoplegia is the paralysis of a single limb.

ASCENDING SENSORY TRACTS

All ascending sensory systems have three types of neurons;

1) Primary sensory neuron in dorsal root ganglion attached to each segment of the spinal cord

2) Second order neuron in dorsal horn of spinal cord or gracile and cuneate nuclei of medulla.

The axons of the second order sensory neuron cross the neuro-axis, become contralateral, and form ascending pathways within the spinal cord and brainstem that terminate on third order neurons.

3) Third order neurons in the thalamus. Third order neurons project to the ipsilateral sensory area of the cerebral cortex.

Tactile Discrimination-Posterior columns. Fibers in the posterior fasciculus, also known as the dorsal column, *(Fig. 7-35)*, are the major, if not exclusive, pathway for signals conveying joint position sense, tactile localization, 2 point discrimination, and vibratory sensation. Fibers conveying touch sensation rise both in the posterior column and several other fasciculi. The fibers in this region consist primarily of heavily myelinated dorsal root ganglion fibers. Upon entering the posterior fasciculus via the medial entry root zone, these fibers divide into ascending and descending branches.

Ascending fibers from lower levels (sacral levels) lie medially while those from upper levels lie laterally. In uppermost thoracic and all cervical levels, the posterior column is divided into the medially placed fasciculus gracilis that includes fibers from the lower extremity and lower thorax and the laterally placed fasciculus cuneatus which include fibers from the neck, upper extremity and upper thorax. The axons in the dorsal column are primary uncrossed axons and continue with-

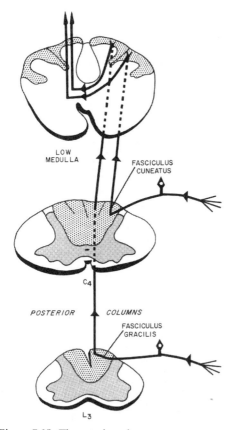

Figure 7-35. The posterior columns

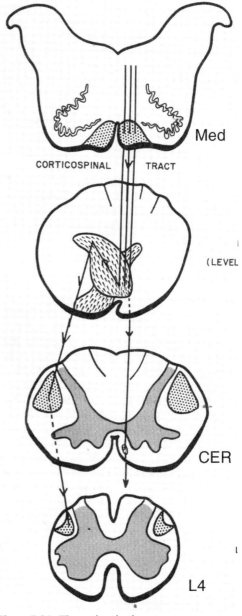

Figure 7-36. The corticospinal tracts.

out a synapse ipsilaterally up to the lower medulla where the secondary neurons begin. The axon of the secondary neuron crosses to the contralateral side and forms the medial lemniscus that ascends through pons and midbrain to the ventral posterior lateral nucleus of the thalamus. The third order neuron in the thalamus sends their axons ipsilaterally by the posterior limb of the internal capsule onto the postcentral gyrus of the cortex.

Lesions in this pathway usually diminish tactile localization, 2-point discrimination and vibratory sensation. There is also loss of ability to appreciate weight differences. Position and movement sense is also affected. These deficits are most pronounced in the fingers and extremities than in the thorax or abdomen. These

TABLE 7-1: MAJOR PATHWAYS IN THE SPINAL CORD

Pathway	Origin	Termination	Function
Corticospinal Contralateral	Motor Cortex	Lamina 7 and 9 in spinal cord	Voluntary movement of limbs
Rubrospinal Contralateral	magnocellular portion of red nucleus in midbrain	Lamina 7 and 9	Involuntary Support of movement. Facilitates flexor and inhibits extensor motor neurons, particularly arm flexors
Tectospinal Contralateral	Deep layers of superior colliculus	Lamina 7 & 9 of cervical spinal cord	Support corticospinal pathway
Vestibulospinal Uncrossed	Lateral vestibular nucleus of Cranial nerve VIII	All levels of cord lamina 7&8	Coordination of eye and neck movements
Reticulospinal	Nucleus reticularis gigantiocellularis in the medulla and from nucleus reticularis pontus caudalis and oralis of the pons	Lamina 7 & 8 of all levels of cord but especially cervical and lumbar enlargements	nfluence gamma motor system Facilitory to extensor motor neurons.
MLF	Cranial nerves iii, v,vi, vii & VIII, XI	Cervical cord	Coordination of eye and neck movements
Descending Autonomic Pathways Bilateral	Hypothalamus and brain stem	Sympathetic to lamina 7 in C8-L2 Parasympathetic to Sacral nucleus of S2-S4	
Spino-spinal System.	All spinal cord levels intersegmental and dorsal root fibers	Lamina	involved in intersegmental connections or taking part in the 2-neuron reflex arch.
Posterior spinal cerebellar some of largest myelinated axons origin of group IA and IB afferent fibers. Ipsilateral	Clark's nucleus in lamina 7 from C2-L2	Ipsilateral vermis of cerebellum	unconscious tactile and proprioceptor information. Encapsulated(Golgi tendon organs, stretch receptors and muscle spindles
Anterior spinocerebellar Contralateral	lamina 5, 6 & 7 from most levels	Contralateral vermis of cerebellum	unconscious tactile proprioception
Cuneocerebellar Ipsilateral	accessory cuneate nucleus in low medulla	anterior lobe and also go into the pyramis and uvula of the cerebellum.	unconscious tactile and proprioceptive information for upper extremity and neck

deficits also produce poorly coordinated movements; posterior column ataxia.

The Anterolateral pathway. Fibers carrying information on pain and temperature sense from the body all rise in the contralateral spinothalamic tract *(Fig.7-34)*. Compression, intrinsic disease or deliberate section all result in anesthesia of the contralateral body beginning 3 segments below the level of disruption.

Pain and Temperature. Primary Cell Bodies-Cutaneous receptors for pain and temperature send axons to small and medium sized dorsal root ganglion cells. These axons enter the spinal cord via the lateral aspects of the dorsal root entry zone. Most of these fibers enter Lissauer's tract and branch extensively (over 2 or 3 segments on either side of the segment of entry) before entering the posterior horn, lamina I, II and III (Fig. 7-28).

The secondary neurons arise primarily from cell bodies in lamina I and V that give origin to large axons which cross in the anterior white commissure and ascend in the contralateral anterior-lateral funiculus. As discussed earlier in the chapter (Nociception and Pain) many factors modulate information transfer across the posterior horn; between incoming nociceptive fibers and outgoing spinothalamic fibers.

There is a somatotopic arrangement in the spinothalamic tract (Fig. 7-34) with the most lateral and external fibers representing sacral levels while the most intermediate and anterior fibers representing cervical levels. Pain fibers are located anteriorly to the more posteriorly placed temperature fibers.

Perhaps as many as half the fibers in the tract, often called the spinoreticular tract, end in the brainstem, while many of the rest send collaterals into the brainstem on their way to the thalamus (Chapter 15). There are endings in two major nuclei of the thalamus: the ventral posterior medial nucleus and the intralaminar nuclei. (see Chapter 15)

Recordings from individual spinothalamic neurons reveal four major functional categories, all contralateral.

1) low threshold units, activated only by gentle stimuli, for example, by stroking hairs on the arm.

2) wide dynamic range units, activated by many types of stimuli of both high and low intensity with response graded by intensity.

3) high threshold units, activated only by nociceptive stimuli.

4) thermosensitive units, with high action potential frequency signaling nociception.

All of these units have small receptive fields peripherally and larger fields toward the midline. The posterior horn seems to tease apart what stimulus from what part of the body. Recording from the thalamic endings of these fibers also shows this separation of function as well as adding a wake up function- ouch! A unilateral section of this tract (Fig. 7-29) for relief of pain produces a complete absence of pain and temperature from the opposite side of the body lasting for 6-9 months, but pain sensation slowly returns. Nociceptive information probably rises in Lissauer's tract until it is above the cut and then crosses.

There are several "pain" responses to nociceptive stimuli. A direct spinothalamic pathway to the contralateral ventral posterior medial nucleus of the thalamus with third order projection to the postcentral gyrus probably mediates sharp, localizable pain. Stimulation of the post central gyrus, however, rarely generates the sensation of pain. Pain is rarely reported by patients during epileptic (cortical) seizures. Apparently the thalamus tells us what (pain) and the post central gyrus tells us where. We test this pathway with the light prick of a pin and expect the patient can tell us, or point to, the location of the pin prick.

The dull throbbing quality of the pain probably ascends by a multi-synaptic pathway via brain stem synapses to the midbrain and then to interlaminar thalamic nuclei with much wider cortical projection including the limbic system. The limbic system (Chapter 22) has much to do with our outlook upon life. The thalamus alone can signal poorly localized pain to consciousness.

In addition to the spinal thalamic and spinal reticular pathways, there are other pathways that may convey pain and thermal information to the thalamus. The cervicothalamic pathway originates from the lateral cervical nucleus in the lateral column at level C1 and C2 and projects to the brain stem and thalamus. Another is the spinotectal tract running from the spinal cord to

deep

Upper and Lower Motor Neurons. Causes of muscle weakness or paralysis can be grouped functionally into two categories. If the difficulty is located in the corticospinal or other descending motor tract the problem is called an upper motor neuron lesion. If the problem is in the anterior horn cell, its axon, or its motor group, the problem is called a lower motor neuron lesion.

Weakness, hyperreflexia, the Babinski sign, and a type of increased resistance to passive movement at a joint, spasticity, characterize the upper motor neuron syndrome. Passively moving or rotating the ankle, knee, hip, shoulder, elbow, or wrist joints easily demonstrates the increased tone. In the normal individual, the joints all move easily. When spasticity is present, the joint is easy to move for a short interval, then resistance to movement increases rapidly; upon further pressure, the resistance suddenly gives way. This latter phenomenon is often referred to as a clasp-knife reflex. In the leg, spasticity is greatest in the extensors, whereas in the arm the flexors are more affected.

Upper Motor Neuron Dysfunction	vs	Lower Motor Neuron Dysfunction
Spasticity		Flaccidity
Hyyperreflexia		Hyporeflexia
Babinski Sign		Fasciculation
Very little atrophy		Severe muscle atrophy

Descending motor tracts normally exert an inhibitory effect on spinal cord reflexes. Hyperreflexia is a sure sign of decreased corticospinal function. Decreased function of the corticospinal tract also effects a curious reflex of the foot, Babinski's sign. If a moderately sharp object such as a key, is drawn over the lateral boundary of the sole of the foot, the toes of an adult flex (curl). In very young children and in adults with corticospinal tract destruction, the toes extend and fan out as shown in Figure 7-22.

Weaknesses, loss of reflexes, extreme muscle wasting, and a flaccid tone, hyporeflexia, to the muscles, characterize the lower motor neuron syndrome. The reasons for the weakness and muscle wasting are discussed in Chapter 6 and relate to the loss of the anterior horn cell or the motor axon. The loss of reflexes and the flaccid tone of the muscles relate to the loss of the motor side of the stretch reflex pathway. Often, it is possible to observe spontaneous twitching of the muscle, fasciculayers of the superior colliculus. It also conveys pain and thermal information with some synaptic interruption before reaching the thalamus.

Other Spinal Pathways. We have just listed the major ascending pathways. There are also pathways from the spinal cord to reticular formation of the brain stem to the vestibular nuclei of the medulla and pons, the inferior olive of the medulla and to nuclei and pons and midbrain.

(See Chapter on functional localization in the Brain Stem and the Major Pathways –Chapter 15).

CHAPTER 8
A Survey of Diseases of Peripheral Nerve, and Nerve Root

INTRODUCTION:

In chapter 2, we have already considered, the differential features of disease of muscle compared to disease of peripheral nerve compared to disease of the nerve root compared to central nervous system disease. Diseases of peripheral nerve and nerve root both represent, lower motor neuron disorders. In contrast to disorders of muscle, both involve, motor and sensory features. As we will discuss in greater degree, peripheral nerve disorders, may involve specific nerves in a local manner (mononeuropathies) or peripheral nerve in a more generalized manner (polyneuropathies). Mononeuropathies must be distinguished from radiculopathies. This distinction can be made based on the pattern of motor deficit, the pattern of deep tendon stretch reflex deficit, the pattern of sensory deficit and the distribution of pain. Figure 8-1 compares radicular (dermatomal or segmental) sensory innervations to the superficial sensory innervations of peripheral nerve.

PERIPHERAL NERVE DISEASE.

Diseases involving the peripheral nerves have a combination of motor and sensory symptoms and signs. Lower motor neuron findings are present with a flaccid type weakness and atrophy. Sensory symptoms and findings (involving to a variable degree all modalities of sensation) are present within the same distribution as the motor findings. These patients, however, do not show evidence of damage to the long fiber systems involved in transmitting sensory and motor information within the central nervous system.

In general, mental status is intact. Deep tendon reflexes and superficial reflexes (response to plantar stimulation) are absent within the distribution of the involved peripheral nerves. Fasciculations may be present within the distribution of the involved peripheral nerves. Damage to the sympathetic fibers, traversing the peripheral nerves, may result in alterations in sweating and skin temperatures. Nerve conduction studies demonstrate reduction in velocity (if the basic process involves loss of myelin, demyelination) or in the amplitude of the motor or sensory action potential (if the basic process involves predominately a loss of axons). In mononeuropathies the specific site of damage (conduction block) may be demonstrated. The EMG will demonstrate abnormal spontaneous activity: fibrillations (onset 10-25 days after the axonal damage) positive sharp waves and fasciculations.

As will be discussed later in greater detail, diseases of peripheral nerves are essentially of two types: (a) localized mononeuropathies involving a single peripheral nerve, often due to trauma or compression or less often, occlusion of blood supply. (b) Symmetrical polyneuropathies - usually distal and usually due to a metabolic disturbance involving many nerves. The polyneuropathies reflect many of the systemic, toxic and metabolic disease considered in pathology. As a general rule any patient with diabetes mellitus or with chronic alcoholism or receiving chemotherapy is likely to present symptoms or signs (often subclinical) of a polyneuropathy. In many cases we are not able to establish a specific cause for the polyneuropathy. We may not have sufficient information about environmental or industrial exposure. In some cases we have insufficient information regarding the family history.

In a mononeuropathy, the weakness and sensory signs and symptoms are clearly within the distribution of a specific plexus peripheral nerve, e.g., sciatic, radial, median, or ulnar. *(Fig. 8-1)*. Common sites of compression include the radial nerve at the radial (or spiral) groove of the humerus, the ulnar nerve at the

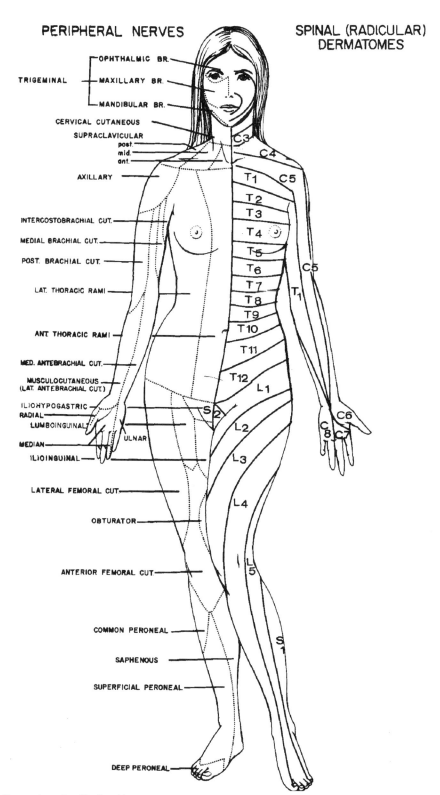

Figure 8-1. Comparison of radicular (dermatome or segmental) and peripheral nerve innervation.

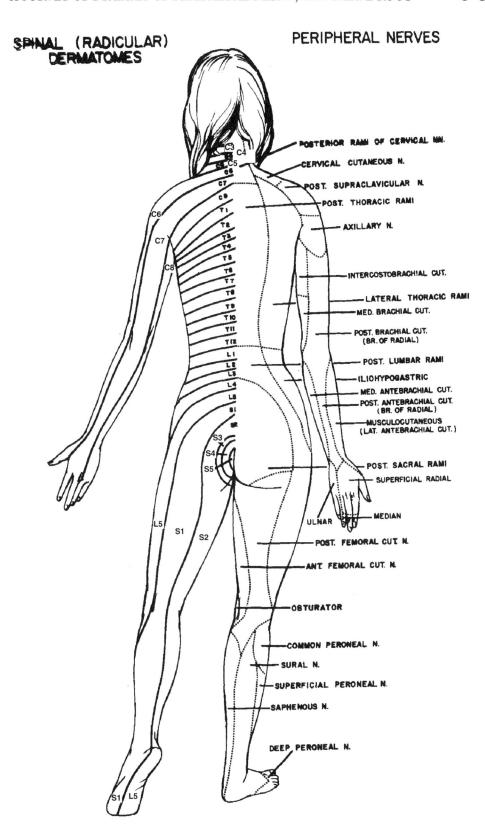

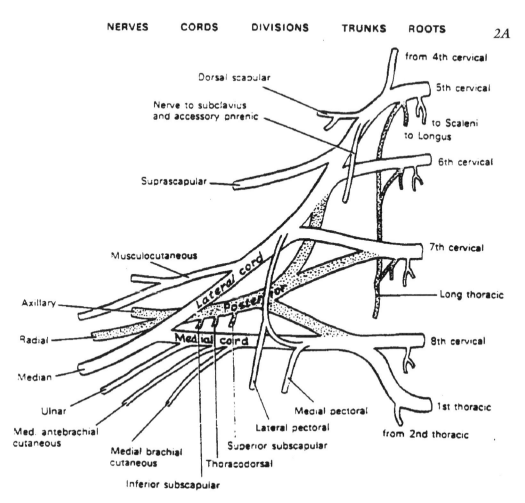

Figure 8-2. A) Plan of the Brachial Plexus: Note origin of median nerve from both medial and lateral cord, of ulnar nerve only from medial cord with radicular origin, from C8, T1, of radial nerve from posterior cord only. (From Clemente, C. Gray's Anatomy. Philadelphia. Lea and Febiger. 30th Ed. 1985, p. 1205).

olecranon process of the elbow, the median nerve in the carpal tunnel, the brachial plexus at the thoracic outlet, and the peroneal nerve as it leaves the popliteal fossa and curves around the head of the fibula.

A specialized form of mononeuropathy: mononeuropathy multiplex reflects multifocal involvement of peripheral nerve for example in inflammatory diseases of blood vessels.

In addition, in neurofibromatosis (Von Recklinghausen's disease), multiple peripheral nerves and nerve roots may be involved by tumors arising from the Schwann cell and mesodermal components. This disorder will be discussed in relation to the spinal cord.

MONONEUROPATHIES: UPPER EXTREMITY

BRACHIAL PLEXUS –[C5, 6, 7, 8 T1- (Fig.8-2, 8-3)]

Patients frequently present with a complaint of transient distal sensory symptoms and weakness involving the upper extremity occurring in relationship to sleeping posture or position of the arm. The symptoms are often bilateral. Rarely the symptoms are persistent rather than intermittent. These symptoms are referred to as the **thoracic outlet syndrome or the neurovascular syndrome of the thoracic outlet.** Both the brachial plexus and the subclavian artery pass through a relatively narrow

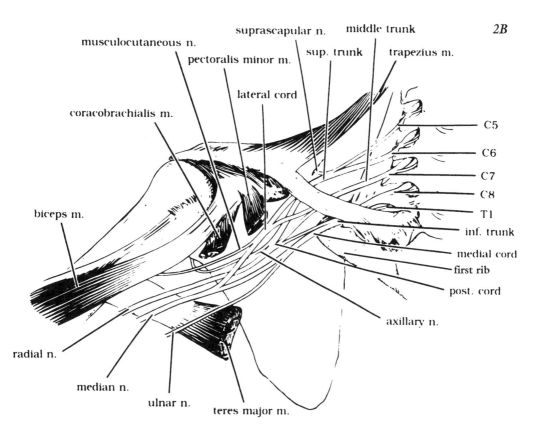

Figure 8-2. B) Brachial plexus. Relationship to the neural foramina the first rib and the clavicle. (From Zimmerman J. and Jacobson S. Anatomy. Boston. Little Brown 1989 p.176.

area with clavicle anteriorly and the first rib posteriorly *(Fig.8-2)*. **Alterations in posture may produce compression of the brachial artery - producing a tingling pins and needles sensations ("paresthesias") involving all fingers and a distal weakness.** Often specific maneuvers will demonstrate the intermittent compression of the brachial artery. **Neural syndromes of the outlet tend to involve primarily the ulnar nerve, or the lower roots (C8-T1).** At times a cervical rib or tight band may further compress the narrow outlet.

Upper Plexus: In addition to the relatively benign thoracic outlet syndrome, often trauma, traction downwards as in birth injuries (Erb's) or brachial neuritis may involve the upper plexus roots C5-6 *(Fig.8-2)* resulting in weakness of muscles about shoulder and flexors at elbow. Brachial neuritis presumably has a post infectious or immunologic or familial etiology. The specifics are often unclear. Brachial neuritis is often associated with considerable pain in the arm. Radiotherapy to the axilla for breast carcinoma may also involve the upper plexus in a relatively nonpainful syndrome.

Lower Plexus: Malignant infiltration of the plexus from tumors at the apex or upper lobe of the lung ("Pancoast tumor") often involves the lower (C7, 8,T1) half of the plexus producing severe pain. The related sympathetic plexus is often involved producing a Horner's syndrome. Upward traction on the plexus may occur at birth (Klumpke's) or in children who suddenly are pulled up by the arm producing damage to the lower half of the plexus.

Case 8-1 provides an example of a brachial plexopathy and Horner's syndrome in a patient subsequently found to have an apical carcinoma of the lung.

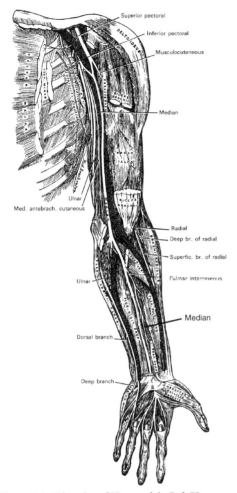

Figure 8-3. Dissection of Nerves of the Left Upper Extremity from anteriorly. Note exposed position of the ulnar nerve in the olecranon groove at the elbow, position of median nerve at the carpal tunnel at the wrist and relationship of the brachial artery to the brachial plexus and clavicle. See also fig 8-4. (From Clemente, C. Gray's Anatomy 30th Ed. Philadelphia. Lea and Febiger 1985, p. 1214).

Case 8-1 This 57 yr. old right-handed white male with a 30-year history of heavy cigarette smoking (2 packs per day) had a 2-year history of progressive pain in the left upper extremity beginning with mild pain at the left elbow, then a lack of sensation in the left ulnar distribution. Eight months prior to admission, the left hand had become swollen and painful. Severe pain on motion at the shoulder also developed. Six months prior to evaluation drooping of the left eyelid was first noted followed by a decrease in sweating on the left side of face and body. Three months prior to admission the patient had the onset of weakness of the left arm. The patient soon noted.

Physical examination demonstrated swelling, redness, change in temperature and sweating and limitation of motion of the left arm, plus distension and firmness of the left supraclavicular space. Orthostatic hypotension was present

Neurological examination: *Symphabetic System:* Seen as a full Horner's syndrome on the left. *Motor System:* atrophy in the median (thenar eminence) and ulnar (first interosseus) distribution, weakness in left arm most marked in the median and radial distribution. *Reflexes:* Decreased deep tendon reflexes left arm *Sensory System:* Decreased pain in the left C7, C 8 dermatomes.

Clinical diagnosis: Brachial plexopathy and Horner's syndrome due to apical lung tumor (Pancoast tumor).

Laboratory data: *Computerized tomography of the lower neck and upper chest* after venous injection of the contrast material into the left arm indicated: near occlusion of the left subclavian vein at the level of the first rib and a necrotic mass in the left lower neck posterior to the thymus and carotid sheath and invading the muscular structures of the prevertebral region and arising from or extending to the left apex of the lung. *Nerve conduction and EMG studies* indicated a severe axonal lesion of the lower half of the brachial plexus with a marked involvement of median and ulnar nerves and less of the radial nerve. *Biopsy of the left supraclavicular mass* revealed adenocarcinoma presumably of pulmonary origin.

Subsequent course: The patient received radiotherapy (3000 rads) to the left lung apex and the left supraclavicular area with a decrease of pain and of swelling in the arm. The patient expired 1.5 years after radiotherapy and 3.5 years after onset of symptoms.

Often a brachial neuritis involves only a limited portion of the brachial plexus. Limited involvement of the long thoracic nerve (C5, 6, 7) is relatively frequent.

Case 8-2 presented on CD ROM provides an example of a neuropathy of the long thoracic nerve occurring in a woman 1 day after delivery.

Ulnar nerve (C8, T1): This nerve is often exposed to trauma as it passes through the groove behind the medial epicondyle at the elbow *(Fig.8-3)*. Each of us has struck our "funny bone" at some time and we are all acquainted with the ulnar distribution of positive sensory symptoms, ring and 5th finger. Repeated trauma to the nerve or fractures at the elbow may produce more persistent symptoms including atrophy of the hypothenar and interossei muscle and weakness of digiti quinti abductor, ring and fifth finger flexors, the interossei, and the thumb adductor.

Median nerve (C5, C6, C7, C8, T1): The major syndrome of the median nerve relates to compression at the carpal tunnel by the overlying transverse carpal connective tissue. *(Fig. 8-3, 8-4)*. Pain and paresthesia extend into the median nerve supplied fingers predominantly index and middle, with less invovement of thumb and median side ring finger when severe. There is significant weakness and atrophy involving the median supplied thenar muscles of the hand:-abductor pollices brevis, opponens and the finger flexors. A positive Tinel's sign is present on percussion over the carpal tunnel. Carpal tunnel syndrome may complicate any process where edema or swelling occurs at wrist or hand: rheumatoid arthritis (as in case 8-3), trauma, myxedema (hypothyroid state) or acromegaly (a state associated with excessive production of pituitary growth hormone-see Chapter 16).

Case 8-3 This 83 yr. old right-handed white widow and retired shoe factory worker with a 20-30 year history of rheumatoid arthritis, in relationship to an exacerbation of her joint symptoms, had the onset of tingling paresthesia in the left median distribution of hand and fingers most prominent in the middle finger, less in index and ring and minor in thumb. There was pain in the hand particularly or making a fist.

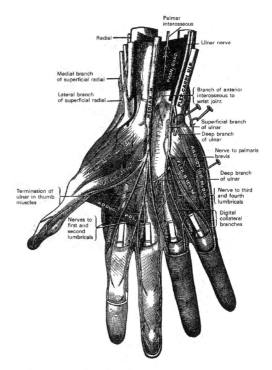

Figure 8-4 Dissection of Nerves and Muscles of the palm of the right hand.(From Clemente, C: Gray's Anatomy 30th Ed. 1985, p. 1219 after Testut).

Physical examination demonstrated mild inflammatory synovitis in the hands, with mild soft tissue swelling in the wrists, ankles, proximal interphalangeal joints, and metacarpal phalangeal joints.

Neurological examination demonstrated a minor weakness in left hand grip and left thumb opponens and a marked tenderness over the left carpal tunnel area with a positive Tinel's sign on palpation or percussion of the median nerve in the carpal tunnel (tingling paresthesias extending from the carpal tunnel into the median nerve supplied fingers).

Clinical diagnosis: Compression of median nerve at carpal tunnel.

Laboratory data:Nerve conduction studies indicated severe delays in both sensory and motor conduction for median nerve at carpal tunnel on the left with mild findings on the right.

Subsequent course: Despite the use of nonsteroidal anti-inflammatory agents, re-evaluation 2 weeks later indicated additional pro-

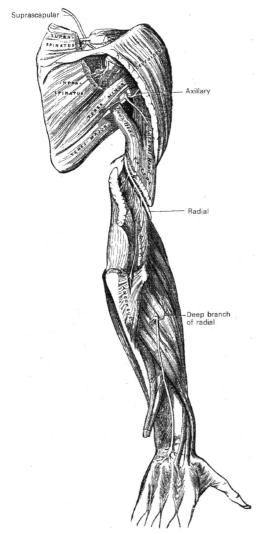

Figure 8-5 Dissection of the nerves of the right upper extremity - from posteriorly. Note the radial course of the radial nerve providing innervation to the triceps muscle, and to the extensors of wrist and finger (From Clemente, C.: Gray's Anatomy. 30th Ed. 1986, p. 1220.)

gression: greater weakness in grip and pain and touch sensation now decreased bilaterally in the median nerve distribution. Despite the use of local steroid injections, reevaluation one month later indicated persistence of symptoms, with atrophy present in the thenar eminence on the left. Carpal tunnel surgery on the left hand was performed with improvement of symptoms. When symptoms developed in the right hand, similar surgery was performed with relief.

Radial nerve (C6, 7, 8): The major syndrome of this nerve relates to injury at the spiral groove of the humerus (Fig.8-5) related to fractures or to so called "Saturday Night Palsy" (pressure effects on arm draped over a hard surface are noted due to excessive alcohol intake so that the paresthesias fail to awaken the patient and the compression therefore continues). The characteristic weakness involves the wrist and finger extensor as well as the long abductor of the thumb.

The brachioradialis muscle is usually involved. Lesions at this level however usually spare the triceps muscle also supplied by radial nerve. A wrist drop is characteristic. A cock up wrist splint is utilized to maintain the hand in a physiological position. With time recovery will occur.

MONONEUROPATHIES : LOWER EXTREMITY

LUMBAR SACRAL PLEXUS [L1-S3 (FIG.8-6, 8-7)].

The lumbar plexus is composed of nerve roots L1 - L4, and the sacral plexus of nerve roots L4-S3. Both may be involved by malignancies within the pelvis. Both may also be commonly involved in the painful diabetic mononeuropathy that occurs on a vascular basis.

Neuropathies may commonly involve the following branches of the lumbar plexus.

1. Lateral femoral cutaneous nerve of the thigh (L2, L3). This nerve supplies the anterior-lateral aspect of the thigh. It enters the thigh by passing between the two points of attachment of the lateral aspect of the inguinal ligament to the anterior superior iliac spine (Fig.8-8). There is sensitivity of the supplied cutaneous area to contact from any clothing or repetitive tactile stimulation. Tingling paresthetica are present. At times burning painful paresthesia is present; thus, the name "meralgia paresthesia". The common causes are weight gain or weight loss. The following underlying diseases are frequent: obesity, pregnancy, dia-

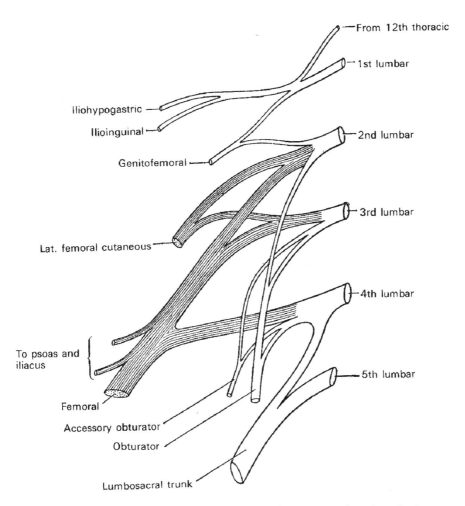

Figure 8-6. Plan of the Lumbar Plexus. The anterior divisions of L2, L3, L4 unite to form the obturator nerve. The posterior divisions of L2, L3, L4 (shaded) unite to form the femoral nerve. Twigs from L2, and L3 forms the lateral femoral cutaneous nerve. (From Clemente, C.: Gray's Anatomy, 30th Edit. 1985 p. 1226).

betes and infrequently pelvic malignancy.

2. Obturator nerve (L3, L4), *(Fig.8-8, 8-9,8-10)* **this nerve supplies the adductors of the thigh.** The major causes of injury relate to difficult deliveries: pressure from fetal head or forceps.

3. Femoral nerve (L2, 3,4), *(Fig-8-8, 8-10).* This nerve supplies the quadriceps muscle necessary for extension at the knee and the iliacus and psoas muscles necessary for hip flexion. Diabetic mononeuropathy on a vascular basis is the most common cause of this neuropathy. Hematomas into the iliac muscle may compress the nerve. Pelvic malignancies and pelvic surgery may injure the nerve.

SCIATIC NERVE AND DIVISIONS

Neuropathies may commonly involve the sciatic nerve, (the major trunk originating from the sacral plexus) or the two major divisions of the sciatic nerve: the common peroneal and the posterior tibial nerves. The two divisions are bound together from the plexus to just above the popliteal fossa. They are separate nerves below that point. The other branches of the sacral plexus are a) the gluteal nerves to the gluteus medius and gluteus maximus; b) the pudendal nerve (S2, 3,4) to the perineal muscle of the anal sphincter.

Sciatic nerve: (L4, L5, S1, S2), *(Fig.8-11).* This nerve supplies the hamstring muscles and

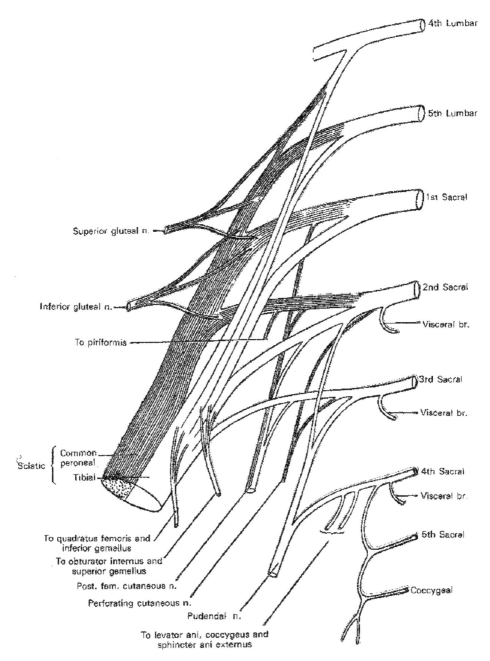

Figure 8-7 Plan of the Sacral Plexus and Coccygeal Plexus: Posterior divisions are depicted as striated. Anterior divisions are unshaded. Note origin of common peroneal nerve from posterior divisions L4-S2 and of tibial nerve from anterior divisions (L4-S3). The two nerves are joined together to form the sciatic nerve (From Clemente, C: Gray's Anatomy, 30th Ed. 1985, p. 1235).

all muscles below the knee. Common causes of injury are a) fractures of pelvis or femur, b) gunshot wounds of the buttock; c) Injections of medications into the buttock - e.g. penicillin. d) Diabetes mellitus mononeuropathy.

Peroneal Nerve: (L4, L5, S1), *(Fig.8-10, 8-11)*: This nerve has a) superficial branch: cutaneous and muscular supply to the everter of the foot (peroneal muscles) and b) deep peroneal or anterior tibial branch supplying the

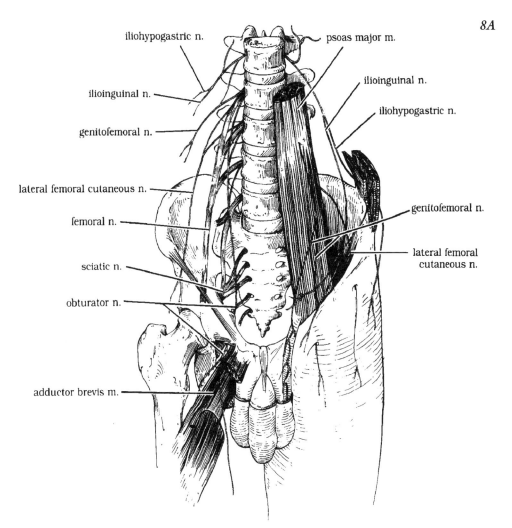

Figure 8-8. A) Dissection of the lumbar and sacral plexuses. demonstrating their relationship to the vertebral foramina and the inguinal ligament and the sciatic notch. (From Zimmerman, J.and Jacobson, S. Anatomy 1989 p.108.

ankle and toe dorsi flexor muscles and sensation to the dorsum of foot and lateral aspect of the calf. This nerve passes around the head of the tibula from the popliteal fossa to the anterior aspect of the leg. In this location, the nerve is relatively superficial and very subject to pressure from leg crossing (when seated in tight coats or tight boots) or operating or obstetrical room stirrups. The major manifestation of compression is a foot drop or unstable ankle (due to weakness of ankle dorsiflexors and everter).

Tibial Nerve: (L4, L5, S1, S2): *(Fig.8-11)*

This nerve supplies the posterior calf muscles. The gastrocnemius and soleus muscles, (the plantar flexor and then continues as the posterior tibial nerve to supply intrinsic foot muscles. It also supplies sensation to the plantar surface. The posterior tibial nerve passes behind the medial malleolus in a relatively superficial location and enters the tarsal tunnel. In this location, the nerve may be palpated and is subject to compression. Entrapment also occurs due to disease in the tendon sheaths that accompany the nerve in the tunnel.

The following case 8-4 provides an example of a mononeuropathy involving the lumbar -

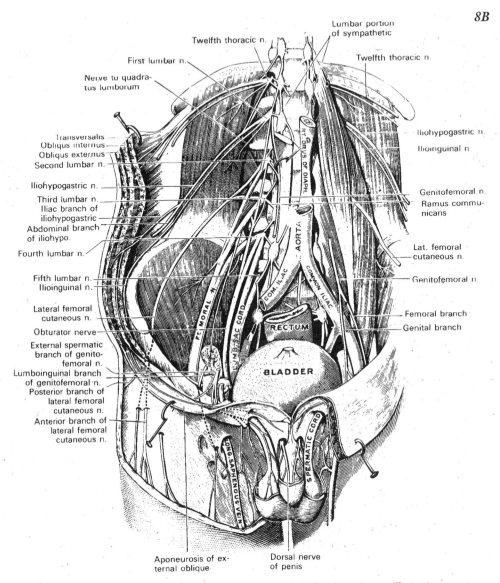

Deep and superficial dissection of the lumbar plexus. (Testut.)

Figure 8-8. B) *Dissection of the Lumbar Plexus. Note relationship of the plexus to the psoas muscle (dissected away on right side of abdomen) iliacus muscle and pelvic structures. Note the course of the lateral femoral cutaneous, femoral and obturator nerves and possible sites of compression. (From Clemente, C: Gray's Anatomy 30th Ed. 1985 p. 1228).*

sacral plexus in a diabetic patient who also had a peripheral neuropathy.

Case 8-4: This 75 yr. old right-handed white widow with a one-year history of non-insulin dependent diabetes mellitus had the sudden onset of weakness in the right leg, most prominent at hip and knee, three weeks prior to evaluation. Shortly thereafter, she had the onset of a severe toothache like pain in the lateral and anterior surface of the right leg from hip to knee. No other symptoms were present.

Neurological examination: *Motor System:* Weakness was present in the proximal muscles of the right lower extremity. *Reflexes:* Deep tendon stretch reflex was absent at the right Achilles and patellar compared to left. *Sensory*

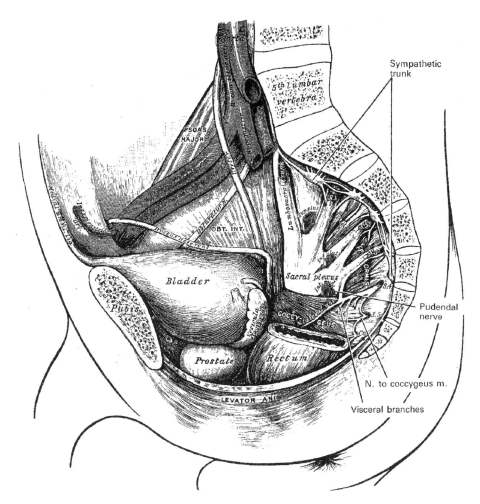

Figure 8-9. Dissection of the Sacral and Pudendal Plexuses. Sagittal View. Note the relationship of the plexus to the visceral structures. Note also the course of the obturator nerve (from the lumbar plexus) as it passes forward to enter the obturator canal and foramen. See also Fig. 8-6B (From Clemente, C.: Gray's Anatomy, 30th Ed. 1985, p. 1233).

System: Pain sensation was decreased in a symmetrical manner over the toes and feet and two thirds of the distance up the calves. However there was also a focal deficit in pain sensation over the right third lumbar dermatome. Vibration was decreased at toes compared to ankles compared to knees in a bilateral manner.. Significant tenderness was present over both the sciatic nerve at the sciatic notch and the femoral nerve in the femoral canal of the anterior thigh.

Clinical diagnosis: 1. Acute diabetic mononeuropathy: lumbar and to a lesser degree sacral plexopathy; 2. Diabetic distal peripheral neuropathy.

Laboratory data: *nerve conduction studies* indicated slowed conduction over the left sural nerve, absence of conduction right sural nerve and both peroneal nerves - leg to ankle. The *EMG studies* demonstrated acute denervation in sampled muscles supplied by the right femoral nerve: rectus femoris, vastus medial and iliopsoas plus muscles supplied by major components of the sciatic nerve gastrocnemius and peroneus longus. *Myelogram and CT scan* of the pelvis were normal.

Subsequent Course: The patient improved but then 9 months later has an acute exacerbation in the right leg and one year later had a similar process involving the left leg.

8-14 CHAPTER 8

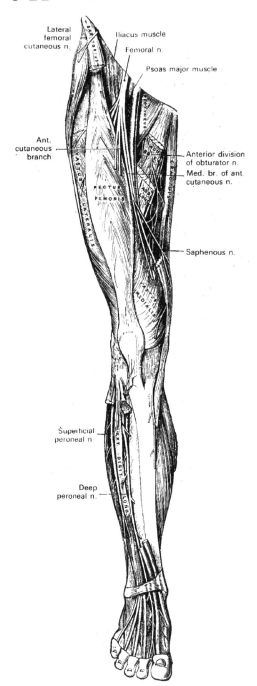

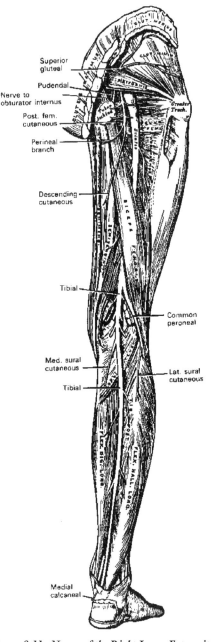

Figure 8-10. Nerves of the Right Lower Limb Anterior View: Note the emergence of the femoral, obturator and lateral femoral cutaneous nerves into the upper anterior thigh. The relatively superficial course of the common peroneal nerve as it passes around the head of the fibula from posterior - (popliteal) location to anterior calf is best seen in Fig. 8-11. (From Clemente, C.: Gray's Anatomy, 30th Ed. 1985, p. 1231).

Figure 8-11. Nerves of the Right Lower Extremity Posterior View: The medial and lateral sural cutaneous nerves have been shifted in position by the dissection. Note a) The relationship of the sciatic nerve to possible sites of intramuscular injection into the buttock. b) The relationship of the tibial and peroneal nerves to the popliteal fossa, c) The relatively exposed position of the peroneal nerve in relationship to lateral calf head of fibula (not fully dissected), d) Relatively superficial position of the tibial nerve behind medial malleolus of ankle. (From Clemente, C.: Gray's Anatomy, 30th Ed. 1985, p. 1238)

Over the subsequent 9 months significant improvement occurred so that she was able to ambulate with a cane.

POLYNEUROPATHIES:

Polyneuropathies may be classified from several standpoints:

1. Predominantly motor vs. predominantly sensory vs. predominantly autonomic vs mixed.

2. Acute vs. subacute vs chronic

3. The site of pathology: a) axonal - "dying back phenomena" -distal degeneration as in metabolic diseases or arsenic intoxication. b) Segmental demyelination with the preservation of axons - as in the immunological post infectious polyneuropathy or some connective tissue disorders or chronic lead poisoning.

4. Etiology: Major categories.

a) Infection - leprosy.

b) Inflammation: Vasculitis - periarteritis.

c) Immune disorders.

d) Intoxications: Arsenic, lead, anti neoplastic agents, industrial chemical exposures)

e) Metabolic disorders: nutritional deficiencies of B vitamins, diabetes mellitus, and uremia, critical care unit neuropathies

f) Degenerations: hereditary.

g) Tumors: neurofibrobromas, schwannomas.

From a practical standpoint, the usual approach is to combine the several types of classification.

1. First classify as to acute, subacute or chronic.

2. Then classify as to predominantly motor or sensory or autonomic. The majority are mixed.

3. If possible classify as to axonal, or demyelinating based on EMG /nerve studies and in some cases; on nerve biopsy. (A survey of peripheral nerve histopathology is presented in *Fig.8-12--8-16*)

4. Then classify as to etiology.

Acute progressive motor neuropathy with variable sensory features.

The Guillain Barré Syndrome provides the most common example of an acute rapidly progressive polyneuropathy. (Acute idiopathic polyneuritis, acute inflammatory demyelinating poly infectious polyneuropathy, Landry-Guillain-Barré Disease).

This is a rapidly progressive predominantly motor peripheral neuropathy that evolves over 10-14 days and affects legs, arms and in many cases cranial nerves. In some cases there is progression to involve the muscles of respiration. Sensory symptoms are present but in most cases sensory findings are minor compared to the motor findings. In severe cases, tracheotomy and mechanical respiratory assistance may be required. Prior to the era of modern respiratory assistance, the death rate was as high as 25%.

The essential pathology consists of perivascular lymphocytic infiltrates, perivenous segmental demyelination. Nerve roots as well as peripheral and cranial nerves are involved. In more severe cases, axons as well as myelin are involved with secondary Wallerian degeneration. In very severe cases, with involvement of the proximal nerve or nerve root, the motor neuron may also be damaged and destroyed. The cerebrospinal fluid is usually without a cellular reaction but does shown an increase in protein that peaks several weeks after onset. In most cases, slowing of motor nerve conduction will be demonstrated, consistent with a demyelinating peripheral neuropathy. Acute motor axonal neuropathy is a variant in which motor axonal degeneration occurs with only minimal demyelination and inflammation. In the Miller Fisher variant or syndrome, a gait ataxia, areflexia, and ophthalmoplegia are present but limb weakness is absent and nerve conductions are normal, although CSF protein is increased. There are also sensory and autonomic variants.

The underlying pathogenesis of the syndrome is now apparent. *In 60-70% of patients, a viral respiratory or gastrointestinal infection has occurred 1-2 weeks prior to the onset of the neurological symptoms.* Other patients have experienced infectious mononucleosis, viral hepatitis, other viral illnesses, surgical proce-

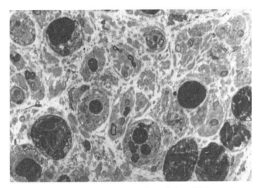

Figure 8-12. Histopathology of Peripheral Nerve I Wallerian Degeneration. Considerable necrosis and degeneration of axons, myelin, and Schwann cells has occurred. The field is dominated by large macrophages containing remnant of axons, myelin, fat globules etc. ("Digestion Chambers"). Toluidine Blue x 100 (approx.). Courtesy of Dr. Tom Smith, Neuropathology, U. Mass. Medical School.

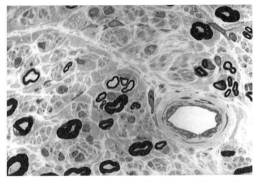

Figure 8-13. Histopathology of Peripheral Nerve II. Chronic Axonal neuropathy. A marked loss of myelinated fibers has occurred. However these are clusters of axons surrounded by a thin myelin sheaths suggesting that regeneration of axons has occurred. Toluidine Blue X 100. Courtesy of Dr. Tom Smith,

dures or immunizations - (old type of serum antibodies vaccine, swine influenza vaccine of 1976). The acute motor axonal neuropathy variant appears to follow infection with the gastrointestinal agent: Campylobacter jejuni.

The pathological and clinical manifestations of the disease are considered to reflect an immunological reaction directed at peripheral nerves. A similar syndrome - experimental allergic neuritis has been produced in rabbits and other laboratory animals 2 weeks after immunization with homogenized peripheral nerve (Waksman and Adams 1955). Sera from patients with the syndrome react with multiple antigens in peripheral nerve. The levels of IgM antibodies against peripheral nerve-myelin correlate with the disease course. Sera from some patients with Guillain Barré Syndrome will produce demyelination in neuronal tissue culture (McFarlin in Asbury and Gibbs 1990; Keski in Dyck 1993).

In more severe cases, plasmapheresis is now employed. In the cases treated by the Guillain Barré Study Group (1985) within two weeks of onset, there was a definite reduction in the length of time for hospitalization, for mechanical respiration and for resumption of walking. More recent studies have suggested that intravenous immune globulin-G maybe even more effective (for a review of current concepts see Ropper 1992). Glucocorticoid steroids usually have no significant effect although massive intravenous therapy (1000mg /day of methylprednisolone) in uncontrolled trials has been reported to have a possible effect.

The following case history 8-5 provides an example of a mild case of this syndrome.

Case 8-5: Approximately two weeks after influenza infection, this 71-year-old right-handed male awoke with tingling of the plantar surfaces of his feet, unsteadiness and weakness at the knees. Over the next 10 days, he experienced a gradual decrease in lower extremity power but no progression of the sensory symptoms. He denied low back pain, bowel or bladder symptoms.

Neurological examination: *Motor Systems:* Weakness was present in the lower extremities both proximal and distal slightly greater on right than left. *Reflexes:* A minimal dysmetria was present in the upper extremities. All deep tendon stretch reflexes were absent in upper and lower limbs except for a trace triceps. Plantars demonstrated no response to stimulation. *Sensory System:* There was a mild decrease in pain and temperature sensation over toes. Perianal sensation was normal. To a greater degree, position and vibratory sensation were decreased in the toes.

Clinical diagnosis: Guillain Barré syndrome: acute post infectious polyneuropathy.

Laboratory data: CBC, glucose, renal functions, liver functions, electrolytes, protein immunoelectrophoresis and ESR (40mm/hr) and MRI of the spine were all normal. CSF: protein was increased to 206 mg%, with 0 white blood cells and a glucose of 76mg%. Serological tests and cultures were negative. EMG/nerve conduction studies demonstrated slowing of motor nerve conduction velocity. - consistent with a demyelinating neuropathy.

Course: the patient's condition stabilized and he was transferred to a rehabilitation facility.

Other causes of the syndrome of acute progressive motor neuropathy:

1) *The exotoxin produced by the bacillus, c. Diphtheria:* pharyngeal and laryngeal muscles are affected 1-2 weeks after the pharyngeal infection followed by a general polyneuropathy: arms then legs after 4-5 weeks.

2) *Acute intermittent porphyria* is inherited as an autosomal dominant trait. A metabolic defect in the liver results in increased production of and high urine levels of porphobilinogen and its precursor delta aminolevulinic acid (involved in hemoglobin metabolism). The severe motor neuropathy may be accompanied by abdominal pain or psychosis or convulsions. Anticonvulsants, barbiturates, sulfa drugs and estrogens may trigger attacks.

3) *Toxic polyneuropathies:*

a) Triorthocresylphosphate :the so-called Jamaican Ginger polyneuropathy seen during the prohibition era reflected contamination of bootleg alcohol by this agent). b) Thallium salts.

All of these acute polyneuropathies as well as many subacute and chronic polyneuropathies are characterized by an acellular cerebrospinal fluid that often has an increased protein content.

Acute polyneuropathies are distinguished from subacute and chronic polyneuropathies. There is no clear-cut distinction, subacute may evolve into chronic over years.

Subacute Polyneuropathies:

The common causes are toxic and metabolic producing a distal symmetrical sensory, motor neuropathy often with painful feet (dysesthesia and hyperesthesia)

1) *Nutritional deficiencies of B vitamins:* thiamine, pyridoxal phosphate, and folic acid, B12, (often multiple) are frequently encountered. a) In association with alcoholism b) in chronic elderly intensive care patients –"critical illness polyneuropathy of Bolton et. al., 1993 c) in the "dry" or neuropathic form of beri beri seen in prisoner of war camps, in famine, or in underdeveloped parts of the world.

2) *Heavy metal intoxication)* Arsenic - distal sensory motor. A more acute form associated with acute arsenic intoxication also occurs .b) Lead - predominantly motor with greater involvement of the upper extremities (wrist drops due to radial nerve involvement).

3) *Hexacarbon industrial solvents* used in glues and plastic production n-hexane and methyl-n-butyl ketone (see Spenser et al 1975).

4) *Other industrial agents:* acrylamide, trichloroethylene.

5) *Medication induced*: a) Isoniazid (INH) for treatment of tuberculosis -produces a pyridoxine deficiency. b) Nitrofuradontins used for the treatment of bladder infection) Anti-neoplastic agents including: cis-platinum and vincristine d) Thalidomide no longer available once marketed as a tranquilizer e) Drugs used to prevent seizures (anticonvulsants)-phenytoin is the most prominent in this group - usually the syndrome is subclinical.

6) *Uremic polyneuropathy* occurs in 60-70% patients with chronic renal failure. The syndrome clears slowly with renal transplantation.

Chronic polyneuropathy: Common causes are as follows:

1) *Diabetes mellitus:* distal symmetrical predominantly sensory - axonal - peripheral neuropathy (see case 4 above) etiology is not clear but may involve accumulation of glucose, fructose and sorbitol in nerves). Autonomic neuropathies or painful distal neuropathies may occur. Successful pancreatic transplantation may halt progression of the disease (Kennedy et al 1993).

2) *Remote effects of malignancy:* carcinoma of lung, ovary, and breast. Mixed sensory motor or motor or selective sensory.

3) *Complications of connective tissue disorder:* periarteritis, rheumatoid arthritis. In some cases this is a complication of HIV infection.

4) *Leprosy:* the most common world wide infectious cause of peripheral neuropathy

5) *Chronic inflammatory demyelinating polyneuropathy (CIDP) Subacute-chronic "Guillain Barré Syndrome"* progressive or relapsing remitting form of the more common acute idiopathic polyneuropathy. These disorders occur on an immunological basis; components of peripheral nerve are the targets of the immune system. These neuropathies are often very responsive to the administration of steroids such as prednisone or drugs that modify the immune system (immunoglobulin G) or to plasmapheresis. (Refer to Dyck, et al 1981, and Dalakas et al. 1981). Some of these cases are associated with HIV infection. (See Cornblatt, 1981).

6) *Abnormalities of plasma proteins usually a monoclonal abnormality of immunoglobulins* (multiple myeloma, macroglobulinemia, cryoglobulinemia and benign monoclonal gammopathy).

7) *Hereditary peripheral neuropathies.*
A. Those disorders in which a specific metabolic defect has been identified. Several may be cited.

1) *Familial amyloid polyneuropathy autosomal dominant (Andrade type).* Amyloid accumulates in blood vessel walls and in the endoneurium. An often-painful syndrome begins in the adult with sensory and autonomic symptoms and slowly progresses to a full polyneuropathy over 10-15 years.

2) *Refsum Disease.* Hereditary ataxic polyneuritis (Autosomal recessive). Phytanic acid (a fatty acid) accumulates in blood and nerves. The symptoms begin in late childhood or adolescence. In addition to a peripheral neuropathy, cerebellar ataxia, a degeneration of the retina (Retinitis pigmentosa) neurogenic deafness and a degeneration of cardiac muscle (cardiomyopathy) are present in most patients.

3) *Bassen Kornsweig Syndrome* - a rare autosomal recessive syndrome which begins in infancy - the metabolic defect is a deficiency of beta lipoprotein and of cholesterol. As above, peripheral nerve, cerebellum, heart, and retina are all affected. In addition - fatty stools, (steatorrhea) retarded growth and an abnormal appearance of red blood cells (acanthocytosis) is present.

4) *Familial dysautonomia (Riley-Day Disease).* A deficiency of serum dopamine B hydroxylase, the enzyme that converts dopamine to norepinephrine results in severe abnormalities of the sympathetic autonomic nervous system. In addition, small myelinated and unmyelinated fibers are involved producing a selective loss of pain and temperature sensibilities.

5) *Porphyria* in contrast to the previous processes produces an acute syndrome (see above).

B. *Mixed sensory motor inherited polyneuropathies without a clearly defined metabolic abnormality.* In the series of 205 patients referred to the Mayo Clinic (Dyck et al 1981) with chronic polyneuropathies of unknown cause, 86 or 42% were found to have an inherited disorder of this type.

The most common disorder in this group is peroneal muscular atrophy -Charcot Marie Tooth Disease (CMT). This disease is inherited as an autosomal dominant. Distal motor involvement is prominent. Sensory fibers are less involved. There are often associated malformations of the feet pes cavus (high arches). Relatives may manifest minimal, minor or partial forms of the diseases which are sometimes asymptomatic - so called "forms fruste". The age of onset is usually in the second or third decade, occasionally later. More recent studies have suggested at least two major forms of the disease referred to as hereditary motor sensory neuropathy I & II, (HMSN I & II) and several minor forms (HMSN III-VII). (Refer to Asbury 1992, and Harding and Thomas 1980).

HMSN Type I is characterized by demyelination with the hypertrophic onion bulb changes that are seen when demyelination and remyelination has occurred *(Fig. 8-14)*. The nerves may also be grossly enlarged. A marked reduction occurs in nerve conduction velocity. *Three subtypes have been identified: CMT 1A, CMT 1B and CMT 1C.* All of these subtypes are transmitted as an autosomal dominant and have slow conduction implying a demyelinating disorder. Type CMT 1A accounts for 60% of all hereditary peripheral neuropathies and is linked to a point mutation occurring on chromosome 17 at the PMP 22 gene segment. This gene encodes myelin protein and appears to be duplicated (Roa et al 1993). Type CMT 1B has been linked to chromosome 1.

HMSN Type II (CMT 2) has a somewhat later age of onset. Nerve conduction velocity is normal; there is no histologic evidence of demyelination. The underlying pathology is considered neuronal. Most subtypes 2A-2D have autosomal dominant inheritance

HMSN Type III -Dejerine Sotta's neuropathy - (CMT-3) is a less common but more severe recessive or autosomal dominant disorder beginning in the first decade of life with both sensory and motor features. The disorder may map to 1q or 17 p or to other sites. The essential pathologic change is demyelination with subsequent remyelination to produce the hypertrophic onion bulb formations *(Fig 8-15). Nerves are grossly enlarged to palpation;* progression is more rapid and disability greater.

Case 8-6 presented on CD ROM provides an example of a patient with autosomal dominant Charcot Marie Tooth disease and slow nerve conduction consistent with type 1A.

8) *Unknown Cause:* Even when the patient has been fully investigated in a specialized center, a specific etiology may still not be found. In the Mayo Clinic series (Dyck et al 1981) previously cited, 24% of patients remained in the unknown etiology category.

DISORDERS OF THE NERVE ROOT: RADICULOPATHY:

The major problem affecting the nerve root is compression by a ruptured disk or by osteophytes projecting into the neural foramen from the degenerative process affecting the disks (spondylosis).

These processes primarily involve the cervical and lumbar areas. These problems are among the most frequently encountered in the neurological office. In the cervical area the C 5-C 6 and the C6-C7 interspaces are most frequently involved with compression of the C6 or C 7 nerve roots nerve roots Less often the C5 or C 8 nerve roots are involved. The relationship of the disc space to the nerve root and to the spinal cord is demonstrated in *Figure 8-17*.

In the lumbar area the L 4—5 (L5 nerve root) and L5-S1 (S1 or L5 nerve root) interspaces are most frequently involved. Less often the L3-4 or L2-3 disc interspaces are involved with involvement of the L4 or L3 nerve roots.

In the diagnostic analysis of a radiculopathy the following features should be considered:

1) The acute onset of pain in the limb in a radicular distribution (Fig.8-1) with associated pain in the neck or lumbar spine.

2) In general the radicular pain is out of the proportion to the neck or back pain.

3) The radicular pain is described a sharp, shooting, burning, or electric shock or tooth ache like pain. The pain is exacerbated by maneuvers that suddenly increase pressure within the CSF space, for example coughing, sneezing or straining at stool.

4) Radicular sensory symptoms plus or minus radicular sensory findings may be pre-

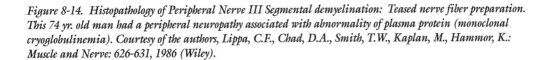

Figure 8-14. Histopathology of Peripheral Nerve III Segmental demyelination: Teased nerve fiber preparation. This 74 yr. old man had a peripheral neuropathy associated with abnormality of plasma protein (monoclonal cryoglobulinemia). Courtesy of the authors, Lippa, C.F., Chad, D.A., Smith, T.W., Kaplan, M., Hammor, K.: Muscle and Nerve: 626-631, 1986 (Wiley).

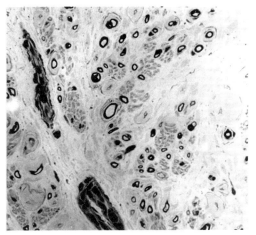

Figure 8-15. Histopathology of Peripheral Nerve IV Hypertrophic "onion bulb" neuropathy. Repeated episodes of segmental demyelination and remyelination may result in finely myelinated axons surrounded by whorls of overlapping intertwined processes of Schwann cells. In addition, this microscopic field demonstrates a significant decrease in total numbers of fibers and a significant increase in connective tissue. Toluidine Blue X63 (Approx.). Courtesy of Dr. Tom Smith.

sent

5) Selective weakness of muscles supplied by a specific cervical or lumbar nerve root may be present.

6) Selective depression of a deep tendon stretch reflex supplied by a specific cervical or lumbar nerve root may be present.

7) In restricted lateral disk herniations in the cervical area, no long tract sensory or motor or reflex findings should be present that is there is no evidence for an upper motor neuron lesion. In restricted lateral disk herniations in the lumbar area no bladder symptoms should be present.

8) With midline herniations in the cervical area, signs of spinal cord compression may be present and are considered below in chapter 9.

9) With midline herniations in the lumbar area, urinary retention and a disturbance of sexual functions may be present due to involvement of the cauda equina.

The usual distribution of hypalgesia and of radicular pain in the upper extremity is demonstrated in *figure 8-18*. See also figure 2-1 for the lower extremity

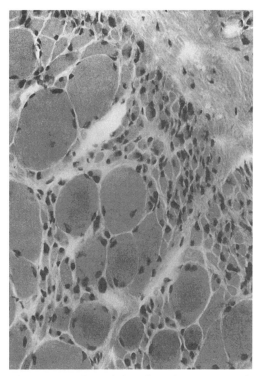

Figure 8-16. Histopathology of Muscle III Neurogenic Atrophy in a patient with Charcot Marie Tooth Disease, Grouped atrophy is evident. H&E X 63 (Approx.) Courtesy of Dr. Tom Smith.

In considering sensory symptoms as compared to sensory findings, when there is involvement of a single nerve root, it is important to note that sensory symptoms (radicular pain and paresthesias) may be present although the examination demonstrates no actual sensory deficit as regard pain or touch sensation. This is a reflection of the fact that there is an overlap of radicular sensory fields. Thus one half of the dermatome of C-7 is also supplied by C-6 and the other one half by C-8 *(Fig.8-19a)*. The total sensory field then for a single nerve root is therefore more extensive than demonstrated in the standard sensory diagrams *(Fig.8-19b)*.

The following case history (8-7) provides an example of lumbar root compression secondary to a herniated disk.

Case 8-7: This 37 year old right-handed white female day care worker in relationship to heavy lifting had the acute onset of persistent pain in the lumbosacral area shooting as a sharp

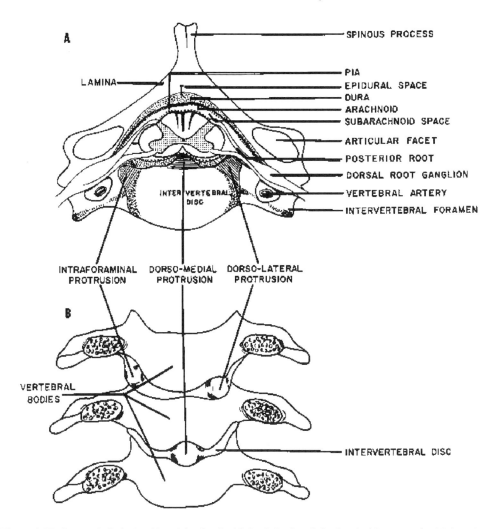

Figure 8-17. Anatomic Relationships of the Cervical Spinal Cord and the Cervical Intervertebral Discs. Lateral and midline disc protrusions are indicated. (Modified after Frykholm, Acta. Chir. Scand. 101:345, 1951).

pain into the left buttock. One month later, she developed numbness left posterior thigh and left posterior calf to the ankle. Three months later, she had the acute onset of severe pain in the lumbar area now extending to the posterior thigh and subsequently to the posterior calf as a shooting pain. Tingling paresthesias now extended into the small toes. The pain in the leg was triggered by coughing or straining at stool. Despite a prolonged period of bed rest, pain medication and non-steroidal anti-inflammatory agents, no improvement occurred.

Neurological Examination (6 months after onset of symptoms): *Motor System:* Significant weakness was present on the left at toe extensors, ankle dorsiflexors, everter and inverters with minor atrophy in the calf. *Reflexes:* The Achilles deep tendon stretch reflex was absent on the left compared to active (3) right Achilles and both patellar reflexes. *Sensory System:* Pain sensation was markedly decreased on the left side over the L5 and S1 dermatomes and to a considerable degree over S2-5. On straight leg rising, pain was present in both leg and back at 45-50 degrees on the left. *Maneuvers:* On palpation, there was marked tenderness over the left sacroiliac area and over the left sciatic nerve at the sciatic notch and over the left posterior tibial nerve behind the medial malleolus of the ankle. The patient was in severe pain sitting, standing or recumbent.

Clinical Diagnosis: Rupture of disk with involvement of nerve roots L5, S1 and probably cauda equina.

Laboratory data: MRI, *(Fig.8-20)* demonstrated a massive rupture of the disk on the left at the L4-L5 level with inferior extension and marked compression of the cauda equina.

Subsequent course: The patient eventually agreed to removal of the ruptured disk, 5 weeks after the initial neurological evaluation. She had some relief of pain but over the next 3 years she continued to have a foot drop with significant weakness at left ankle and toes, an absent left Achilles reflex, and sensory deficits over the left L 5 distribution.

Most patients (75%) with a more limited lumbar radiculopathy respond to a period of strict bed rest for 3 to 5 days on a firm mattress and bed board, nonsteroidal anti inflammatory agents, heat or cold and possibly the temporary use subsequently of a back support. Those who fail this therapy may require local steroid injection or surgical therapy. Only if surgical therapy is a consideration should MRI studies be obtained.

Benign non-malignant back pain alone, either acute or chronic is an extremely common complaint Usually it is benign in nature. However a complete history and general physical examination is indicated to identify immediate precipitating events. In addition, the history should survey for, any past history of malignancy, or change in bowel habits, urinary pattern or menses etc. A rectal and or pelvic examination is indicated particularly when there is no clear-cut history of trauma. In the middle aged or older male patient, appropriate laboratory studies to rule out prostatic malignancy are indicated (prostatic specific antigen/antibody-PSA). If such a survey is negative and no neurological symptoms or signs are present, then neurological consultation and lumbar MRI scan are not indicated. Plain films of the lumbar spine will usually suffice to rule out orthopedic disease. Such benign back pain does not require a course of bed rest. Non

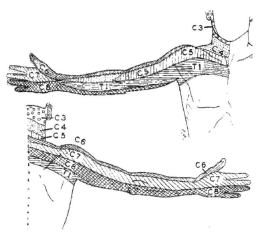

Figure 8-18 Dermatome charts of the upper extremity in man outlined by the pattern of hypalgesia, following rupture of an intervertebral disk. From Keegan, J.J., and Garrett, F.D.: Anat. Rec., 102: 417, 1948 (Wiley).

steroidal anti-inflammatory agents, modification of posture, heat or cold and weight loss if indicated are appropriate measures. Activities at work may have to be temporarily modified, use of a support belt or corset may be indicated.

For cervical radiculopathy, at home cervical traction, a cervical collar, cervical, pillow and, nonsteroidal anti-inflammatory agents are usually effective. Most (75%) patients will usually respond to these measures. When these measures fail and radicular symptoms and findings persist, surgical therapy may be considered. At such a time MRI scans is appropriate.

The following case history 8-8 demonstrates the effect of a ruptured cervical disk producing compression of a cervical nerve root.

Case 8-8: This 45 year-old right-handed married white female educational coordinator 2 weeks prior to evaluation had the acute onset pain in the neck radiating into left arm with pain and tingling in the index and middle finger of the left hand. The pain would shoot into the arm is she coughed or sneezed or strained to move her bowels. 4 months previously the patient had experienced pain in the neck extending to the left shoulder area but that symptom had cleared. In the last several days prior to consultation the patient had

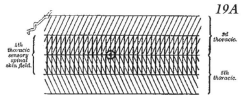

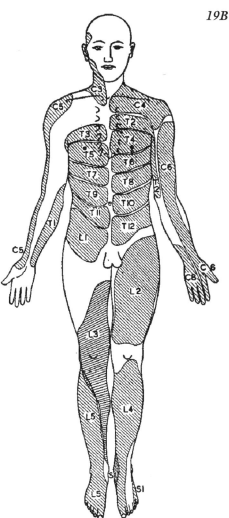

Figure 8-19 A) Method of remaining sensibility to demonstrate the sensory skin field of a nerve root. Sherrington sectioned three roots above and three below the intact root to be studied. In this diagram the overlap of the third and fifth thoracic spinal roots is demonstrated. (From Ransom, S., and Clark, S.: The Anatomy of the Nervous System, 10th Edition. Philadelphia, W. B. Saunders, 1959, p. 129).
B) Dermatomes in man demonstrated by the method of remaining sensibility. (From Lewis, T.: Pain. New York, Macmillan, 1942, p. 20, after Foerster).

developed twitching of biceps and triceps muscles. She had no leg or bladder symptoms.

Neurological examination: The left triceps deep tendon stretch reflex was absent or reversed. Pain sensation was decreased over the left index and middle fingers. Neck motion was limited for extension and rotation and there was tenderness over the spinous processes of C 7 and T1 and over the left supraclavicular area.

Clinical diagnoses: Cervical 7 radiculopathy secondary to lateral rupture of disk.

Treatment: The use of cervical traction, cervical collar, nonsteroidal anti inflammatory agents, various pain and anti muscle spasm agents, epidural injection and various measures in physical at therapy for 4 weeks failed to produce any relief. Pain was now predominantly in the arm and her examination now demonstrated additionally significant weakness at the left triceps muscle. MRI *(fig 8-21)* demonstrated a lateral disk rupture at the C 6 -7 interspace. The patient underwent a left sided laminectomy at that level with removal of ruptured disk material. She had a resolution of symptoms except for a minor residual tingling, a decrease in pain sensation over the index finger, and a mild decrease of the left triceps deep tendon stretch reflex.

Schwannomas arising from the nerve root: These tumors arise from Schwann cells of the peripheral nerve or nerve root. When the nerve root is involved the tumor may be present within the bony canal and/or external to the bony canal. The tumor may enlarge the neural foramen. Depending on the size and location of the tumor the nerve root alone may be involved *(Fig.8-22)* or the spinal cord may also be involved (refer to chapter 9). When the tumor arises from a nerve root within the cauda equina, multiple nerve roots may be involved. (Fig.8-23).

Tabes Dorsalis: this late complication of syphilis involving the nervous system is a result of the infectious agent, the spirochete involving the posterior (dorsal) root and to some extent the dorsal root ganglion producing secondary degeneration in the posterior columns.

Disease of the dorsal root ganglion: The

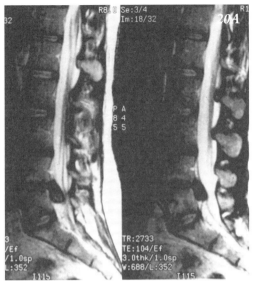

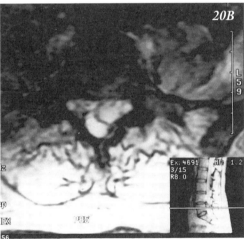

Figure 8-20: Lateral protrusion of a Lumbar disc: Case 8-7: lumbar radiculopathy MRI A) sagittal sections B) transverse section.

major disease involving the dorsal root ganglia is infection by the herpes zoster/varicella virus. In actuality this is a reactivation of a virus remaining dormant in the dorsal root ganglion after previous varicella (chickenpox) infection. Diseases that alter the immune system such as infection, e.g. HIV, or lymphomas may activate the virus and in some cases generalized zoster may occur. Ten-25% of patients with generalized lymphoma will develop H.Zoster. In addition local irritation of a nerve root or a primary neoplastic process arising in organs closely related to the dermatome may activate the virus. The clinical disorder begins with sharp lancinating pain in a radicular distribution. Within 3 or 4 days, the involved dermatome demonstrates a vesicular eruption. The thoracic dermatomes are most frequently involved. In approximately 20% of cases, cranial nerve root sensory ganglion are involved most often cranial nerves 5 (primarily the ophthalmic division) and 7. Although the dorsal root ganglion is primarily involved, occasionally the anterior root or horns are involved producing atrophy and weakness. Rarely, the spinal cord is involved, producing a clinical myelitis. The anti viral agent, acyclovir is effective in reducing the course of disease. *Post herpetic neuralgia* is the term applied to the persistent severe pain occurring in 20% of patients one month after the rash has healed. The occurrence of this chronic pain syndrome is more frequent in elderly patients; 75% of patients with H.Zoster who are over the age of 70 years continue to have severe pain one month after healing of the cutaneous lesions. Treatment with lidocaine skin patches, the tricyclic antidepressant nortriptyline or the anticonvulsant gabapentin or oral opioids may produce a moderate reduction in pain in some of these patients. A recent study has demonstrated the effect of intrathecal methylprednisolone in patients who were still refractory at one year (Kotani et al 2000).

The herpes simplex virus may also remain dormant in the sensory root ganglion and may be activated to produce radicular pain. With the type 2 virus (vaginalis), the sacral segments are often involved, and urinary retention may occur.

A SURVEY OF DISEASES OF PERIPHERAL NERVE, AND NERVE ROOT 8-25

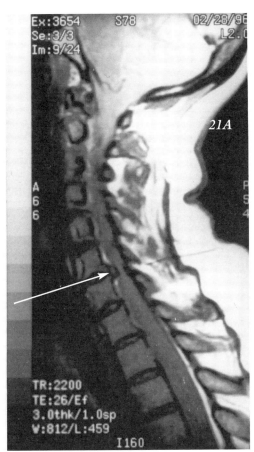

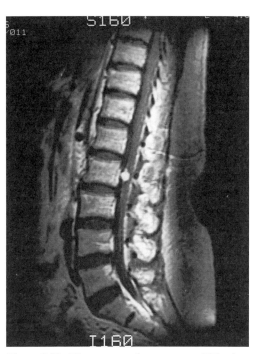

Figure 8-22: Schwannoma of nerve root at L2 level. MRI. This 53-year-old female had an 18-month history of progressive low back pain that began to radiate to the right foot with numbness of the lateral aspect of the right foot and numbness of left thigh. Examination demonstrated only decreased pain sensation of the left anterior thigh and lateral aspect of the right foot. An MRI of the lumbar spine demonstrated a circumscribed Schwannoma at the lumbar-2 level, subsequently removed by Dr. Alex Danylevich.

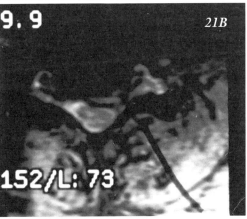

Figure 8-21: Lateral protrusion of a Cervical disc: Case 8-8: Cervical radiculopathy. MRI A) sagittal section B) transverse section.

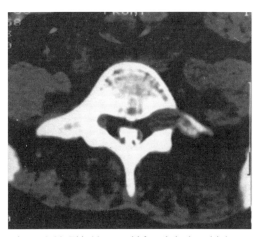

Figure 8-23. This 23-year old female had multiple neurofibromas. In this CT scan a Schwannoma at the lumbar level is demonstrated.

CHAPTER 9
Spinal Cord: Clinical Considerations

Disease affecting the spinal cord is a frequent cause of chronic neurological disability. Often, young adults are involved as in injuries due to war, motor vehicle accidents, motorcycle accidents, diving and skiing accident. Special units have been developed for the care, and rehabilitation of the unfortunate victims of these injuries. Early recognition and management of a potential spinal cord compromise is essential.

It is important to differentiate extrinsic (compressive) and intrinsic diseases of the spinal cord.

Continued spinal cord compression will produce irreversible damage. For extrinsic diseases, surgical therapy is then indicated as early as possible.

In general, extrinsic compressive lesions (Table 9-1) manifest a number of local segmental features at the level of the lesion as well as long tract findings below the level of the lesion.

Acute spinal cord compressions or transactions: The phenomena of spinal shock and the pattern of reflex recovery: After the acute and sudden loss of supra segmental control, a temporary depression of segmental reflex activities occurs. Deep tendon stretch reflexes are depressed. Cutaneous and autonomic reflexes may also be affected. The excitability of alpha and gamma motor neurons and of interneurons is depressed. This is more prominent in man than in species such as the cat or dog related to the progressive encephalization of function that has occurred in the primates. In the frog, the duration is fleeting; in the cat, spinal shock lasts a matter of minutes or hours. In the monkey duration is a matter of days or weeks. In humans, the duration is usually of several weeks. In the cat or dog spinal shock follows damage to the vestibulospinal and ventral reticular spinal tracts. In primates, spinal shock depends more on damage to the cortical spinal tracts. It is important to note that the depth and duration of spinal shock depends on the rapidity (acuity) and completeness of spinal cord section. Slowly evolving chronic spinal cord compressions are not char-

TABLE 9-1. CONSEQUENCES OF EXTRINSIC COMPRESSION OF SPINAL CORD

Symptoms and signs	Anatomical correlation
1. Radicular pain at level of compression.	Posterior root and ganglion
2. Radicular sensory symptoms at level of compression	Posterior root and/or posterior horn
3. Segmental atrophy, weakness and fasciculations, at level of compression	Segmental lower motor neuron findings due to involvement of the anterior horn cells and anterior roots.
4. Segmental depression of deep tendon stretch reflexes	Involvement of the anterior root and anterior horn or of posterior root and or horn. The monosynaptic reflex arc has been interrupted as regards its afferent or efferent component
5. Long tract findings below the level of the lesion:	a. Spastic (UMN) weakness, increased deep tendon reflexes & Babinski sign (extensor plantar response) due to damage to descending pyramidal tracts. b. Spastic neurogenic bladder due to bilateral involvement of UMN control with release of the segmental bladder stretch reflexes. c. Deficits in position, vibration, and other proprioceptive and light touch sensation due to posterior column damage. d. Deficits in pain and temperature sensation due to lateral spinothalamic pathway damage.

acterized by spinal shock. Fever or infection will prolong the duration of spinal shock or may result in a recurrence of spinal shock once recovery has occurred. Immediately after an acute transection, the patient will demonstrate a depression of all the deep tendon stretch reflexes and to a lesser degree the flexion reflexes. The limbs will be flaccid and paralyzed. The bladder will be flaccid that is hypotonic or atonic due to depression of the bladder stretch reflex. Within a short period of time (1 - 5 weeks), the flexion reflexes including the sign of Babinski will recover and remain prepotent. The flexion reflex, (the withdrawal of the foot and leg on painful stimulation) is a polysynaptic reflex. The afferent component of the reflex arc is mediated by group II and III myelinated fibers and group IV unmyelinated fibers, from a variety of receptors in skin, muscle, and joints. Initially in the human, the flexion reflex may be manifested as a mass reflex. When fully developed, this mass reflex may consist of generalized reflex flexor spasms of the muscles of the trunk and lower extremities accompanied by sweating, emptying of the bladder and in males penile erection and ejaculation. Initially, the receptive field for triggering this response is quite extensive, involving any tactile or nociceptive stimulus to the foot, leg, abdomen or thorax. With the passage of time, the mass reflex becomes less prominent and local sign develops. The more specific reflex movement then depends on the more specific location of the stimulus. For example, a painful stimulus to the outer border results in flexion, inversion, and adduction as opposed to a similar stimulus to the inner border of the foot that results in flexion, eversion, and abduction. The mass reflex may return if fever or infection occurs. Over a number of weeks to months depending on the completeness of transection (longer if complete), stretch reflexes with other extensor reflexes and postures will return. The deep tendon stretch reflexes will then become hyperactive. There will be a spastic rather than a flaccid weakness. Bladder emptying and bowel evacuations return 3 to 4 weeks after injury. The bladder will subsequently become hypertonic that is spastic with small capacity. The extensor capabilities may consist of crossed extension, (the contralateral lower limb extends when the ipsilateral will lower limb is responding to a painful stimulus by flexion). Sweating below the level of transection may not return for 3-4 months. In the cat and dog-alternate flexion/extension, alternate stepping and in high cervical preparations inter limb reflexes may occur (the upper limb movements are appropriately triggered by lower limb responses). With marked extensor tone, a partial capacity for brief periods of standing may return. Clearly there is a phylogenetic factor; such extensor capabilities are more limited in the spinal human. In comparison the capabilities of the frog or chicken with a high cervical transection are well described in common terminology.

The Brown Sequard hemisection syndrome (Table 9-2): Transactions of the lateral one-half of the spinal cord or lateral compressions of the spinal cord for example by tumors such as Schwannomas, meningiomas or metastases will produce a classical syndrome which reflects the anatomy of the spinal cord and of the spinal cord pathways.

The progressive subacute or chronic anterior midline compression syndrome: The spinal cord is gradually compressed backwards against the bony canal elements. Based on the clinical and experimental studies of Tarlov, the following sequence is often noted.

Initially the symptoms and signs will be relevant to posterior columns followed by symptoms and signs of lateral column involvement and finally by symptoms and signs relevant to the lateral spinothalamic system. From a clinical standpoint, the initial symptoms may consist of a numbness of the lower extremities and a sensory ataxia. An ascending spastic weakness of the lower extremities will then develop followed by an ascending deficit in pain and temperature sensation. Such a pain and temperature deficit will initially involve the sacral segments then lumbar segments followed by the thoracic segments and finally the lower cervical segments. This sequence reflects the lamination pattern of the lateral spinal thalamic system in which sacral segments are most extrin-

TABLE 9-2 CLINICAL SIGNS IN BROWN SEQUARD SYNDROME

Deficit	Cause of Clinical Deficit
A spastic weakness below and ipsilateral to the side of transection/damage	Injury to the lateral column/cortical spinal system.
Deep tendon reflexes increased below and ipsilateral to the side of transection/damage	Injury to lateral column/cortical spinal system.
The sign of Babinski will also be present ipsilateral to the side of transection/damage	Injury to pyramidal tract in the lateral column.
Ipsilateral loss of vibration, position sense and all other proprioceptive sensory modalities below the level of transection/compression	Injury to posterior column
A contralateral loss of pain and temperature sensation beginning 1 or 2 segments below the level of the transection/compression	Involvement of the lateral spinothalamic tract.
An ipsilateral band of weakness atrophy and fasciculations may be present at the level of transection	Damage to the anterior root and or the anterior horn.
Ipsilateral depression of deep tendon stretch reflexes at the level of transection/compression	Damage to the afferent or efferent components involved in the monosynaptic stretch reflex arc.
An ipsilateral band of loss of pain, temperature and touch sensation may be present at the level of transection	Damage to the posterior root and posterior horn

sic and cervical segments most intrinsic. As a general rule then, extrinsic lesions do not have sacral sparing. The earlier involvement of posterior and lateral columns may reflect the sequence of structural compression or the greater vulnerability of the heavily myelinated fibers.

It is important then to realize that the apparent upper level of sensory level for pain and temperature may not be the actual level of compression. When the patient is seen early in the course of spinal cord compression the apparent level of sensory deficit will be considerably lower than the actual level of compression.

It is important to recognize that slowly evolving extrinsic lesions (such as intradural meningiomas and neurofibromas) despite considerable distortion of spinal cord may produce only minimal symptoms and have a good response to surgical therapy. In contrast, rapidly evolving lesions such as metastatic epidural tumors, more often present a relatively acute or subacute course and often have a poor response to therapy. At times such metastatic lesions are complicated by vascular compromise.

SPECIFIC EXTRINSIC LESIONS OF THE SPINAL CORD.

Fracture dislocations: *(Fig.9-1)* In civilian life, most injuries to the cord do not represent actual lacerations of the spinal cord but are a result of crush injuries caused by fracture dislocations in the cervical or thoracic area. Vascular effects often complicate such crush injuries. Infarction (related to compression of anterior spinal artery and of veins) and hemorrhage may occur often in a central cord location (the latter is referred to as hematomyelia). After resorption of a hemorrhage, a cystic cavity may remain. Falls transmit force to the thoracic vertebrae; the resulting compression fractures may collapse the vertebrae and caused a sharp angulation of the axis of bony support. A similar compression and collapse; may also occur from other non-traumatic causes such a tuberculosis of the spine (Pott's disease), metastatic carcinoma and osteoporosis. Injuries to the cervical spine resulting from fracture dislocation are common in auto accident, and in diving accidents. In rheumatoid arthritis, the odontoid is commonly fractured and dislocated. It is important to note that fractures of the vertebrae may occur without a marked degree of dislocation. In such cases great care must be taken in moving these patients. Thus, in fractures of the cervical spine, the patient should remain supine with the neck immobilized to avoid flexion and extension during transportation. (Utilize a makeshift collar or sandbags.) Once the patient arrives at a treatment center,

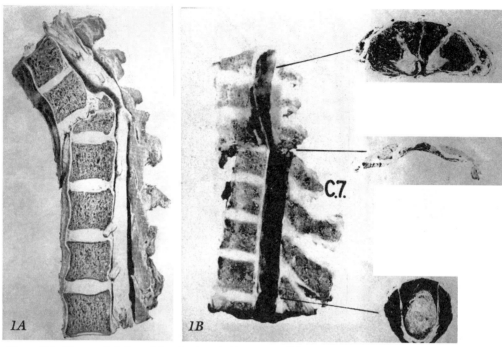

Figure 9-1. Traumatic lesions of the spinal cord: A) Fracture dislocation of thoracic vertebra has crushed and almost transected the thoracic spinal cord. B) A fracture dislocation of the cervical spine (due to an auto accident 2 weeks prior to death) has produced an almost complete transection of the spinal cord at the C7 level. (From Blackwood, W., Dodds, T.C., and Somerville, J.C.: Atlas of Neuropathology, 2nd Edition, Baltimore, Williams and Wilkins, 1964, p. 147).

additional measures such as traction or spinal fusion must be considered. The use of very high dosage corticosteroids may have some value in reducing the edema of the spinal cord.

Cervical disk disease: Acute ruptures and chronic cervical spondylosis with cervical spinal cord compression: This is probably the most common cause of cervical spinal cord compression. Acute lateral or midline rupture of disk material may occur in the cervical area related to trauma or sudden coughing or straining. The term cervical spondylosis refers to a more chronic process. The disk degenerates with increasing age, the disk material begins to bulge, the disk space narrows, and the secondary formation of osteophytes (bony spurs) occur. The osteophytes may project laterally into the neural foramina producing nerve root compression and/or may be located in a midline location producing spinal cord compression *(Fig. 8-17)*. Not all bulging disks and osteophytes compress the spinal cord. In some patients, the vertebral canal is sufficiently wide that no compromise of the spinal cord occurs. In other patients, the bony canal in relatively narrow (spinal stenosis) and compromise occurs. The pattern of ascending and descending degeneration found at autopsy following spinal cord compression in a patient with cervical spondylosis is demonstrated in *Figure 9-2*. This 64 year old male had a 6 year history of progressive spastic paraparesis with atrophy and fasciculations at C4-C8, bilateral corticospinal and posterior column findings in the lower extremities and a deficit in pain sensation below the T 10 level. Softening of the spinal cord was present at the cervical 7 levels. The following case illustrates the syndrome of spinal cord compression secondary to cervical disk disease.

Case 9-1: This 34-year-old female after 2 weeks of prolonged hyperextension of her neck while painting walls and ceilings developed a transient 6-hour period of weakness in both her legs. She then developed pain in the right side of her neck, her right shoulder and right upper arm, followed by weakness in the right leg. Shortly thereafter she noted intermittent prickly sensations (paresthesias) in both lower extremities. And she was unable to walk.

Neurologic exam: *Motor:* weakness was present in the right upper extremity most prominent at the triceps and wrist extensors (4/5) with a lesser degree of weakness at deltoid, pectorals, wrist flexors and finger abductors plus a minor degree of weakness at right ankle dorsiflexion (4.5/5). Gait was ataxic. *Reflexes:* Bilateral ankle clonus was present with bilateral Babinski signs. *Sensation:* Vibratory sensation was decreased in the right lower extremity at toes ankle and knee with associated proprioceptive deficits. Pain sensation was decreased in the left lower extremity from the toes to the groin. The entire buttock and the perianal area were also involved in this deficit.

Clinical diagnosis: Partial Brown-Sequard syndrome due to acute ruptured cervical disk involving the right side of cervical cord.

Laboratory data: *CT myelogram* demonstrated a probable extruded disk at the C5-C6 level to the right of the midline displacing the spinal cord posteriorly and to the left *(Fig.9-3)*.

Cerebral spinal fluid protein was significantly elevated to 141 mg % (normal is less than 45mg %), consistent with a spinal cord block.

Subsequent course: She was unwilling to have an MRI or any neurosurgical procedure undertaken to remove the disc. With the use of a cervical collar, she had had rapid resolution of her symptoms and continued to do well over the next 8 years. Refer to CD ROM for additional details.

The MRI scan remains the study of choice in patients with cervical spine pathology as demonstrated in *Figure 9-4*. This 31-year-old male jumped off a truck carrying an 80-pound bag on his shoulder. The following morning, he awoke with severe pain throughout the spine on coughing. And intermittent tingling of both hands for approximately 4 months Nine months later, he still had an equivocal plantar response and MRI demonstrated a ruptured disc at the C5-6 interspace.

Metastatic carcinoma: May involve the vertebrae without producing spinal cord or nerve root compression. However continued growth of the tumor or vertebral collapse may lead to spinal cord compression. The tumor may spread to the epidural space from the vertebral involvement producing spinal cord compression. Vertebral collapse may also produce spinal cord compression from bony elements of the vertebrae. A metastatic tumor may also appear in the epidural space without direct involvement of the vertebrae. In such cases, tumor within the thoracic or abdominal cavity may have spread through the neural foramina into the epidural space. Lymphomas in the thoracic or abdominal cavity in a paravertebral

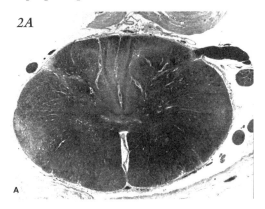

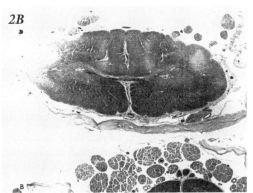

Figure 9-2. Ascending and Descending Degeneration: Cervical Spondylosis with Spinal Cord Compression Myelin stains. A) Upper cervical spinal cord above level of compression: ascending degeneration predominantly in posterior columns. B. Lower cervical cord just below area of softening: descending degeneration predominantly in lateral columns. (Courtesy of Dr. Jose Segarra, Boston Veterans Administration Hospital).

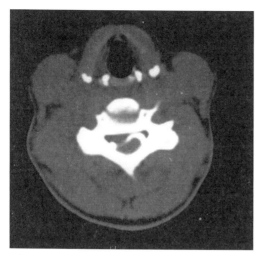

Figure 9-3. Cervical Disc Rupture: Brown Sequard Syndrome, Case History # 9-1: CT Scan of cervical cord post metrizamide myelography. Refer to text.

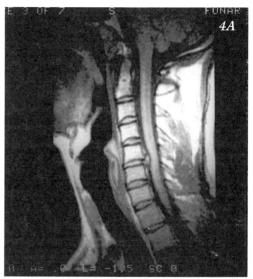

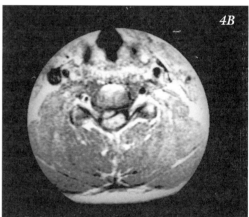

Figure 9-4 Acute Rupture of cervical disc midline and right lateral at C5-C6. A) MRI T1 weighted sagittal section of cervical spinal cord slightly to right of midline demonstrating a ruptured intervertebral disc at C5-C6 interspace. B) Transverse sections at C5-C6 interspace refer to B for correlation of levels- slice- 9 and 10. Cross-section at approximately C5, C6 demonstrating an extra dural mass to the right of the midline displacing the spinal cord posteriorly and to the left.

location often spread in this manner. Spread of metastatic tumor may also occur via the venous plexus or via other hematogenous routes. The majority of metastatic tumors involve the thoracic and lumbar areas. The majority of cases originate from primary disease in the lung, breast, prostate and kidney. Based on the autopsy studies of Barron in 1959, 5% of all systemic cancers will eventually develop spinal epidural spread. A significant proportion of patents presenting with epidural spinal cord compression secondary to metastatic disease may not have been previously known to have a malignancy. In a series from a large general hospital (The London Hospital) in 1982, 47% of the patients were in this category. 14% of patients never had the primary location determined. **Case 9 - 2** demonstrates the consequences of metastatic carcinoma producing spinal cord compression.

Case 9-2: This 55 year-old white housewife first noted pain in the thoracic-right scapular area, 5 months prior to admission Three months prior to admission, a progressive weakness in both lower extremities developed resulting in a bed ridden state in which she was unable to move her legs or even to wiggle her toes .She had also noted a progressive pins and needles sensation involving both lower extremities. At the same time she developed difficulty with the control of bowel movements and urination. At age 40, 15 years prior to this admission, she had undergone a left radical mastectomy for an infiltrating carcinoma of the breast with regional lymph node involvement.

Neurological examination: *Motor system:* a marked flaccid weakness in both lower extremities with retention of only a flicker of flexion at the left hip. *Reflexes:* Plantar responses were extensor bilaterally (bilateral sign of Babinski). *Sensory system:* Position sense was absent at toes, ankles, and knees and impaired at the hip bilaterally. Vibratory sensation was absent below the iliac crests bilaterally. Pain sensation was absent from the toes through the T 6-T 7-dermatome level bilaterally with no evidence of sacral sparing. Tenderness to percussion was present over the midthoracic vertebrae (T4 and T5 spinous processes)

Clinical diagnoses: Spinal cord compression at T4-T5 vertebral level most likely secondary to metastatic epidural tumor.

Laboratory data: *XRay Studies:* The patient had multiple metastatic lesions in lungs, T4, T5 vertebrae and head of femur. An emergency *myelogram* demonstrated a complete block to the flow of contrast agent at the T5 level due to an extradural lesion displacing the spinal cord to the left.

Subsequent Course: An emergency thoracic T3-T4 laminectomy demonstrated adenocarcinoma presumably metastatic from breast in the vertebral processes, laminae and epidural space displacing the spinal cord. Following removal of tumor from epidural space and subsequent radiotherapy, movement in the lower extremities had returned to 30% of normal and pain sensation had returned to the lower extremities to a moderate degree.

In this case, a myelogram was performed, at the present time; the most appropriate diagnostic procedure would be an MRI scan. However in circumstances where that study cannot be performed, on an emergency basis then an emergency myelogram (or myelogram-CT) is appropriate. The presence of persistent mid thoracic - scapular back pain in a patient with a past history of breast malignancy should always prompt a search for metastatic lesions in vertebrae. In any case, the earliest development of motor or sensory symptoms in a patient with a past history of malignancy and back pain should lead to prompt neurological evaluation and investigation. (The present standard of investigation would include appropriate MRI study of the spine, which has the advantage of allowing studies not only of the specific area of compression but also of the spinal cord above and below the area of compression. Thus other epidural lesions may be identified). The best treatment of metastatic spinal cord compression is a high index of suspicion and early recognition to prevent the results seen in this case.

Acute epidural abscess: The dura mater of the spinal cord is separated from the periosteum of the surrounding bony canal by a narrow space containing fatty tissue (the epidural space). This space may be infected by an infection in adjacent tissues for example skin (furuncles) or bone (osteomyelitis involving the vertebrae). Alternatively, hematogenous spread of infection from distant sources may occur. The usual organism is staphylococcus aureus .The usual site is midthoracic. The epidural abscess is rare compared to the epidural tumors just considered and rare compared to meningitis and brain abscess. Nevertheless early recognition is necessary. Prompt drainage, decompressive laminectomy and antibiotic therapy are essential to avoid the poor prognosis of continued spinal cord compression.

The early symptoms of severe back pain and fever are followed by radicular pain or weakness and then the acute or subacute development of the signs of spinal cord compression. Approximately half the cases evolve over days to 2 weeks; half evolve over a period greater than 2 weeks. These latter cases are usually secondary to vertebral osteomyelitis. If treatment is delayed, functional transection of the spinal cord is the usual outcome related both to the direct effects of compression and the indirect effects of compression on the blood supply and the involvement of the blood vessel walls by the infectious process. The diagnosis is established by the appropriate clinical history, an MRI study or myelogram-demonstrating blockage of the subarachnoid space by an epidural mass and subsequent aspiration of pus from the epidural space.

Tuberculous involvement of the vertebrae with secondary spinal cord compression: Tuberculosis was once a common disease in urban centers of the United States and Europe. In other parts of the world for example, India, tuberculosis remains a common disease. Tuberculous involvement of joints, tuberculous arthritis, is a complication of untreated tuberculosis in children. A very common site of involvement is the dorsal spine. The process often appears to begin in the disc space and then involves the adjacent vertebrae. The disease process, tuberculous spondylitis (or Pott's disease), results in destruction of the body of the vertebrae and of the intervertebral disc. Collapse of the vertebrae and severe angulation of the bony canal occurs. In addition the chronic infection may spread into the epidural space as a local mass of infection or may accumulate as a mass under the ligament posterior to the vertebral bodies. All of these factors may contribute to spinal cord compression. Diagnosis can now be established by MRI scan. Treatment consists of drainage of any mass pockets of infection compressing the spinal cord, anti tuberculous chemotherapy, immobilization and possibly fusion of the spine.

Intradural - extra medullary spinal cord tumors (meningiomas and Schwannomas): In many earlier operative series this category of extrinsic tumor internal to the dura but external to the substance of the spinal cord, accounted for the largest percentage of spinal cord tumors. Such series accumulated prior to the advent of the MRI tended to underestimate epidural metastatic tumors because such patients were usually not admitted to specialized neurological-neurosurgical units. For example among 567 cases of spinal cord tumors collected from the literature by Merritt, in 1967, 59% were intradural extra medullary, 25% were extradural and 11% were intramedullary. The extradural tumors have already been considered in relationship to metastatic spread of carcinoma and lymphoma. The intramedullary tumors will be considered later under intrinsic lesions: gliomas and ependymomas. Essentially 2 types of tumor constitute in a relatively equal proportion all of the intradural lesions: meningiomas and Schwannomas (neuromas). Both types are benign in the sense that they are not locally invasive and do not spread to distant sites.

Meningiomas: These tumors arise from the arachnoidal cell clusters. Since the more common location is the cerebral hemisphere a more complete discussion of histologic types will be found in a later chapter. The typical gross appearance of this tumor in relation to the spinal cord is shown in *Figure 9-5*.

These tumors occur most frequently in middle-aged females (many meningiomas have estrogen receptors). The most frequent location is the thoracic portion of the spinal cord,

Figure 9-5. Gross appearance of meningioma compressing the ventrolateral aspect of thoracic spinal cord. (From Russell, D.S., and Rubinstein, L.J.: Pathology of Tumors of the Nervous System, 2nd Edition. Baltimore, Williams and Wilkins, 1963, p. 45).

in part because the thoracic segments constitute the longest extent of the spinal cord. Other areas however are not immune. In the upper cervical area these tumors may arise in relation to the foramen magnum producing both lower brain stem and upper cervical cord symptoms. The symptomatology of a typical meningioma arising in the thoracic spinal cord area might include thoracic back pain that radiated around to the interior chest in coughing and straining at stool. Ascending sensory upper motor neuro symptoms in the lower extremities might evolve slowly over 6 to 12 months. Examination might demonstrate the findings of the bilateral transverse or Brown Sequard hemisection discussed above. In addition local tenderness would be present over the spinous processes at the level of involvement. Note that local tenderness over the spinous processes is common in extradural and intradural extra medullary tumors. MRI scan (or CT- myelogram if MRI is not available) would reveal an intradural extramedullary tumor with partial or complete block to the flow of CSF. CSF protein would be increased. These tumors require early neurosurgical intervention to avoid progression to a chronic paraplegic state. With the slow growth of the tumor, the capacity for recovery of function despite considerable distortion of the spinal cord is usually quite surprising once the compression has been relieved.

Schwannomas: These benign tumors arise from the cells of the nerve sheath. These tumors as noted above may arise in relationship to peripheral nerve, cranial nerves or nerve root. As discussed above those that arise in relation to nerve root with in the bony canal may compress nerve root and/or the spinal cord depending on location and size of the tumor. The symptoms and signs may be very similar to those of a meningioma. Because these tumors arise from the nerve root radicular pain is often more common than in meningiomas. The use of MRI to image Schwannomas is demonstrated in Figure 9-6. This 43 year old male had several months of pain in the cervical area that radiated into the anterior chest on coughing plus tingling from this area to the sacral area on sneezing. He had a marked decrease in all deep tendon reflexes in the right upper extremity. and percussion tenderness over the C6-7 spinous processes. All symptoms resolve following removal of the tumor. Note that the tumors may attain considerable size and produce only minimal spinal cord symptoms. As with all spinal cord lesions, the preferred neuroimaging technique is the MRI scan.

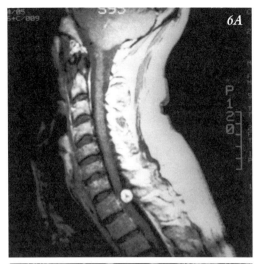

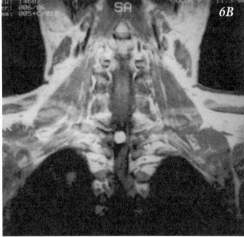

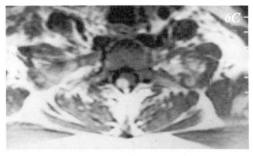

Figure 9-6. Schwannoma with compression of spinal cord. MRI with contrast enhancement. A) Sagittal section-midline. B) Coronal section between posterior arches and dorsal surface of spinal cord. C) Transverse section-T1-vertebral level.

Case history 9-3 presented on the CD ROM demonstrates the course of an Schwannoma compressing spinal cord.

Neurofibromatosis Type I (von

Recklinghausen's disease): This is an autosomal dominant inherited disease characterized by multiple peripheral neurofibromas (tumors composed of Schwann cells, collagen and reticulin fibers). There are also various anomalies of the skin (multiple > 6 cafe au lait spots and cutaneous neurofibromas), iris (Lisch nodules –pigmented hamartomas) and skeletal system. Large plexiform neurofibromas may produce marked facial deformities. Patients may also manifest meningiomas or intrinsic gliomas involving the optic nerve, brain or spinal cord but such central lesions are less common than in type II neurofibromatosis considered below. This form of neurofibromatosis is also referred to as peripheral neurofibromatosis; and occurs once in 3,000 live births. The mutated NF1 gene has been localized to chromosome 17q. The CT scan and MRI appearance of Schwannomas involving nerve roots has been presented in chapter 8.

Neurofibromatosis Type II: This type also known as central neurofibromatosis is also an autosomal dominant but is rare compared to neurofibromatosis type 1 occurring 1 in 50,000 births. The mutated NF 2 gene has been localized to chromosome 22. In contrast to type 1, peripheral manifestations are uncommon. Skin and bone lesions do not occur. Instead the patients have multiple types of tumors of the central nervous system. Most patients eventually develop bilateral acoustic neuromas (actually vestibular Schwannomas). This problem will be discussed in greater detail in the brain stem chapter.

INTRINSIC DISORDERS OF THE SPINAL CORD

Three general categories of intrinsic disease must be distinguished:

1. *Local disease affecting one or more adjacent segments:* a). Infarcts produced by occlusion of the anterior spinal artery b) transverse myelitis c) intramedullary spinal cord tumors: gliomas and ependymomas d) syringomyelia

2. *System diseases affecting one or more neuronal or fiber systems:* a) motor neuron disease, b) tabes dorsalis c) combined system disease d) spinal cerebellar degeneration

3. *Multifocal disorders:* various levels of the nervous system are affected. At any given level, a specific lesion is not restricted to a specific neuronal or fiber system: a) multiple sclerosis, b) other demyelinating disorders c) collagen-vascular disease lupus erythematosus and vasculitis.

SPECIFIC INTRINSIC SYNDROMES

LOCAL DISORDERS:

Vascular disease: anterior spinal artery occlusion: Primary vascular disease affecting the spinal cord is not common although a secondary vascular component may be present in many of the spinal cord compression syndrome discussed above. Any understanding of the clinical syndrome found in vascular disease of the spinal cord is dependent on knowledge of the anatomy of the arterial supply of the spinal cord *(Fig.9-7)*. The major artery of the spinal cord is the anterior spinal artery that is located in a midline position at the anterior median fissure. This single thin midline vessel has a bilateral origin from the intracranial portions of each vertebral artery. The arterial flow as this vessel descends is dependent on additional supply from the radicular arteries. These radicular arteries are derived from the cervical portion of the vertebral artery particularly at C3 and C 5 levels and the inferior thyroid artery at the C6 level. Additional radicular arteries originate from the aorta as the intercostal thoracic lumbar and sacral arteries. Most of the radicular arteries of the thoracic area do not contribute a significant supply. However, the middle thoracic artery usually at T 7 and the artery of the lumbar enlargement, (the artery of Adamkiewicz) which usually arises between T10 and L2 are of particular importance. There are then border zones of blood supply in the rostral-caudal axis between the cervical and lumbar segments of major supply. The actual border zones of circulation are at segments T4 and L1.

When the transverse anatomy is considered:

1. Anterior spinal artery: This midline vessel supplies the anterior and lateral columns and almost all of the gray matter except for the posterior horns.

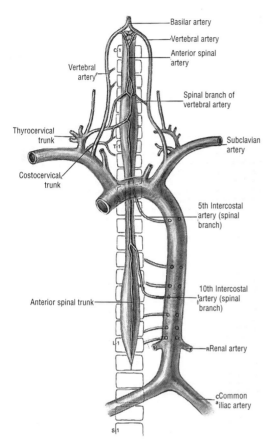

Figure 9-7. Vascular Anatomy of the spinal cord. Anterior spinal artery: the major radicular (segmental) blood supply is diagrammed. From Clemente, C., Gray's Anatomy, 30th Ed, Philadelphia, Lea and Febiger, 1988.

2. Posterior spinal arteries: These paired vessels supply the posterior horns and the posterior columns, each derived from the intracranial segment of a vertebral artery. These posterior spinal arteries also receive contributions from the posterior branches of the radicular arteries.

3. Anastomotic Vessels. At each level, coronal arteries at the periphery of the spinal cord connect the anterior spinal artery and the posterior spinal arteries. These anastomotic vessels at the periphery also serve to interconnect the blood supply of adjacent segments. A transverse border zone or watershed must also exist in the central gray matter where penetrating branches of the anterior spinal artery meet the penetrating branches of the posterior spinal arteries.

Since the largest area of the spinal cord is supplied by a single anterior spinal artery, it is not surprising that most vascular disease involving the spinal cord presents as the syndrome of the anterior spinal artery that is infarction of the territory supplied by this vessel *(Fig.9-8)*. This 71-year-old male had the acute onset of a flaccid paraplegia with an absence of deep tendon reflexes in the lower extremities and a pain sensory level at T7-8. No recovery occurred.

Actual occlusion of this vessel is rare. The usual cause of the syndrome relates to diseases of the aorta or to surgical procedures involving the heart, aorta, or related vessels. Clamping of the upper aorta for the surgical treatment of an aneurysm of the aorta may result in a decrease in blood flow in the critical intercostal branches supplying the spinal cord. A dissecting aneurysm of the aorta (a tear within the wall of the vessel with blood under high pressure dissecting down the media of the wall of the blood vessel) often results in the occlusion of intercostal and other vessels arising from the aorta. Other arteries arising from the arch of aorta such as vertebral and carotids may also be occluded. In other instances, the occlusion of intercostal branches particularly at T 10 may occur during surgical procedures on the kidney

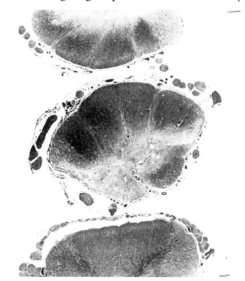

Figure 9-8. Infarction of spinal cord in distribution of anterior spinal artery secondary to occlusion of radicular artery at T6. Courtesy of Dr. Jose Segarra

or lumbar sympathetic ganglia.

The resultant neurological syndrome is manifested by the acute onset of a paraplegia which is usually flaccid due to spinal shock and a bilateral deficit in pain and temperature below the level of lesion with preservation of vibratory and position sense. If the area of infarction involves the lumbar sacral area the anterior horn cells will also be destroyed. The legs will then remain flaccid, muscle atrophy will develop, deep tendon stretch reflexes will never recover and spasticity will never develop. The prognosis then for a significant degree of recovery is usually poor. Theoretically, if the thoracic spinal cord were infarcted, with preservation of the lumbar and sacral segments, deep tendon reflexes could recover and spasticity without atrophy in the lower extremities could develop after several days.

In distinguishing other acute processes producing paraplegia, the rapidity of onset and the anatomical pattern of infarction seen in occlusion of the anterior spinal artery should be considered. Other processes such as trauma or hemorrhage into the spinal cord (hematomyelia) from rupture of a malformed blood vessel may produce an acute syndrome. The history, and findings as well as the MRI scan and the CSF examination would be distinguishing factors. Spinal arteriovenous malformations may also produce a progressive syndrome with episodes of acute exacerbation related to infarction and hemorrhage, but the anatomical pattern will differ from the acute anterior spinal artery syndrome. Multiple sclerosis may also produce a relatively acute onset of symptoms but the anatomical pattern will differ and lesions will usually be present elsewhere in the nervous system. Acute transverse myelitis and the spinal cord compressions discussed above do not produce the anatomical pattern of the anterior spinal artery syndrome.

The syndrome of acute or subacute transverse myelitis: When all other causes of an acute or subacute myelopathy have been excluded, there remain a group of patients without evidence of an extrinsic compressive or intrinsic vascular lesion. Usually the upper or mid thoracic spinal cord is involved. In the 1978 series of Ropper and Poskanzer, (in which all patients had myelography, since this was prior to the modern era of neuroimaging) 138 patients presented from 1955 to 1975 with an acute myelopathy or myelitis. In this overall group, 82 (59%) had an anatomical mass lesion usually an epidural metastatic tumor, four (3%) had a dissecting aortic aneurysm and presumably an anterior spinal artery syndrome, and 52(38%) remained in the transverse myelitis category.

Of the 52 patients in this residual transverse myelitis category, 33% had a prior viral illness and presumably could have had a post infectious myelitis. 6% had cancer of the lung, ovary, or prostate but without spinal cord compression and presumably had a remote effect of malignancy. 13% of patients eventually developed multiple sclerosis with the transverse myelitis as the presenting syndrome. The remaining 48% had no specific possible etiology.

In 21% of these patients the appearance of symptoms was acute with mid thoracic back pain, weakness of legs, sensory deficit below the level of lesion and urinary retention evolving over less than 1 to 12 hours. In 69% of patient's the symptoms progressed in an ascending manner over 14 days and then stabilized with the findings of a spastic paralysis and spastic bladder. In 10% of patients, the course was a stuttering progression over 10- 28 days. The progressive patients had a better prognosis than those with the acute onset. In the era prior to MRI scans, myelograms were usually normal but occasional patients were found to have mild swelling of the spinal cord.

In cases of transverse myelitis, pathological examination of the spinal cord reveals an acute or subacute necrotic process involving a number of segments. In other cases, areas of demyelination are present.

The MRI is able in most cases to image these pathologic changes. In mild cases, the MRI may be normal.

Case 9-4 presented on the CD ROM and in *Figure 9-9A* illustrates the diagnostic dilemma provided by such cases of transverse

SPINAL CORD: CLINICAL CONSIDERATIONS 9-13

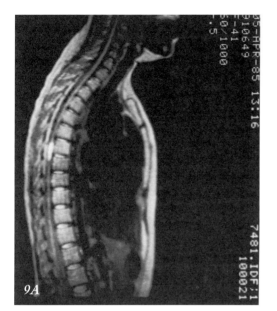

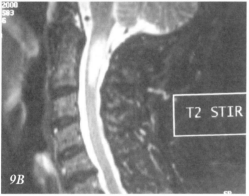

Figure 9-9. Transverse myelitis: A) Case 9-4 MRI (T1) (CASE ON CD ROM) This 42 year old female had the sudden onset of back pain at T6-8, followed by tingling and weakness in the lower extremities plus urinary retention evolving over 12 hours and then improving. B) This 54 year old female with a prior history of non Hodgkin,s lymphoma developed a progressive quadriparesis over 6 days ,and then rapidly improved possibly related to high dosage corticosteroids.

myelitis A 42 year old woman had the relatively sudden onset of back pain at T6-T8 and a tingling in the lower extremities. Over the next 12 hours, she developed weakness in the lower extremities and urinary retention. She then had improvement over the next 24 hours. A CT myelogram indicated a widened spinal cord at T7-8. Eventually clinical and MRI improvement occurred over several months.

Devic's syndrome (neuro myelitis optica) is a variant of transverse myelitis in which a relatively acute transverse myelitis is associated with an acute unilateral or bilateral optic neuritis. Some of these cases represent a variant of multiple sclerosis; others a variant of a post infectious myelitis, or rarely a paraneoplastic syndrome.

Subacute –chronic HTLV associated progressive myelopathy HAM/ tropical spastic paraparesis: This is a myelopathy produced by the retrovirus HTLV that causes human T cell leukemia. A chronic meningoencephalomyelitis is associated with demyelinating lesions in posterior and lateral columns. A sensory level may be present in the thoracic area. In tropical areas, the incidence may be as high as 128/100,000. The risk factors for transmission of the disease include many of the factors found with HIV infection: sexual transmission, intravenous drug use and blood transfusions. The CSF and MRI findings are similar to multiple sclerosis (which actually has a lower incidence in tropical as opposed to temperate areas). Differentiation may be made by a determination of antibody levels in serum and CSF.

Intrinsic spinal cord tumors: Rarely metastatic tumors may be found within the substance of the spinal cord. However most intramedullary tumors are intrinsic arising from the glial (astrocytic or ependymal) components. The glioma arising from the astrocytic series of cells is the most frequent intrinsic tumor of the spinal cord found in the cervical-thoracic spinal cord. Most spinal cord ependymomas arise at the lower end of the spinal cord in the filum terminale. Intrinsic tumors of the spinal cord are rare compared to intrinsic tumors of the cerebral hemispheres and compared to extrinsic tumors of the spinal cord.

Ependymomas are often relatively localized and are to some extent discrete tumors. Astrocytomas on the other hand often tend to infiltrate the surrounding tissue without any particular limiting border. Several or many adjacent segments may be thus involved by the tumor. In some children, the tumor does appear to have limiting borders and thus may be shelled out at surgery. A cystic component is often present and the cyst may be drained at surgery. The overall diameter of the spinal cord is increased and this may be visualized at myelography or on MRI scans. The increased mass of the spinal cord produces pressure on the pedicles of the vertebrae. Plain x-ray films of the spine may demonstrate the resultant increased interpeduncular distance. From a histological standpoint the spinal cord astrocytoma is usually of a uniform appearance with a relatively low to moderate grade of malignancy (grade 1-2). This is reflected in the usual clinical course that extends over a period of several years with a 5-year survival after surgery of up to 90% of patients. Contrast this to the astrocytoma that infiltrates the cerebral hemispheres which has a much more malignant histological grade of 3-4, and a much shorter survival.

A typical case is that of a young adult or child who presents with a several year history of a slowly progressive lower motor neuron weakness and atrophy of 1 extremity involving multiple segments. Sensory symptoms in that extremity might develop at the same time or subsequently with or without radicular pain. As the process continues upper motor neuron weakness of the ipsilateral leg might be followed by a similar involvement of the opposite leg. Subsequently a slowly progressive lower motor neuron weakness might develop in the opposite upper extremity. Depending on the pattern of infiltration, long tract sensory findings might be present or absent. Bladder function would usually be well preserved. Very little local tenderness would be present over the vertebrae in contrast to extrinsic tumors. The changes in plain spine x-rays, myelography/CT and MRI scan have been noted above. The patient would improve following laminectomy, drainage of any cystic component and in some cases removal of tumor in cases (usually pediatric) where the tumor could be discreetly separated from the surrounding tissue. Radiotherapy would also produce considerable improvement.

Syringomyelia: this disease usually affects the cervical portion of the spinal cord but may extend into thoracic and lumbar segments. The basic pathology involves the formation of an irregular cavity (syrinx) in a central or paracentral location *(Fig.9-10)*. The cavity is surrounded by a border of gliosis, (an area of proliferation of astrocytes with the production of glial fibers. The cavity is usually ventral to and distinct from the central canal that is lined by ependymal cells .As the disease progresses, the syrinx may however extend into the central canal. The location of the syrinx is critical for understanding the early symptoms and signs that develop. The syrinx initially involves those pain and temperature fibers crossing the midline in the anterior white, commissure. Thus the initial symptoms and signs relate to a selective loss of pain and temperature sensation in the cervical segments producing a cape like deficit *(Fig.9-11)*. Since, initially touch, vibratory and proprioceptive sensation are preserved, this is referred to as a dissociated sensory deficit. The patient often reports painless burns and painless trauma to the hands and arms. Although these symptoms and findings are usually bilateral, occasionally because of the irregular shape of the syrinx the symptoms and signs may be predominantly unilateral. The subsequent course of syringomyelia is variable since the syrinx often varies in its shape and pattern of expansion. The anterior horns however are often subsequently involved resulting in local atrophy fasciculations and local flaccid paralysis of hand or upper extremity muscles with a segmental loss of deep tendon stretch reflexes. As the syrinx continues to expand, lateral and at times posterior columns may be involved with resultant long tract motor and sensory findings. Although the process is most

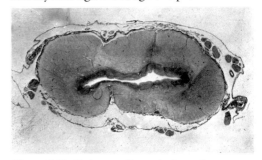

Figure 9-10. Syringomyelia producing enlargement of the cervical spinal cord, destroying the anterior white commissure and the anterior horn. A dense border of gliosis is evident around the cavity. (Holzer stain for glia). (Courtesy of Dr. E. Ross).

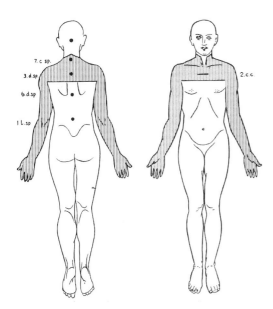

Figure 9-11. Syringomyelia with dissociated sensory loss. The cape-like distribution of a selective pain and temperature deficit with intact touch sensation over the upper extremities is demonstrated.

prominent in the cervical spinal cord, the syrinx may extend into the thoracic and lumbar spinal cord. In addition, the process may extend into the medulla as syringobulbia. In such cases, the syrinx is usually present as a slit like cavity extending in a ventral lateral direction from the floor of the 4th ventricle (the continuation of the central canal) into the dorsal lateral medullary tegmentum. In this location, many of the symptoms and signs relevant to the medulla will be similar to those of the lateral medullary infarct to be discussed in the brain stem chapter. The symptoms and signs of syringobulbia are often predominately unilateral although bilateral lesions may occur with involvement of the nucleus ambiguous. The slit like cavity may be difficult to visualize on MRI scans of the brain.

The precise etiology of syringomyelia is variable, and the underlying pathophysiology is often not clear.

In *type I syringomyelia*, there is an abnormality, usually developmental at the cervicomedullary junction. The Chiari or Arnold Chiari malformation is a frequent association. Two types of the Chiari Malformation are recognized: type I and type II. In *type I Chiari malformations*, the cerebellar tonsils extend down through the foramen magnum compressing the cervical medullary junction. In *type II Chiari malformations*, there is caudal displacement of the lower end of the medulla and 4th ventricle as well as of the cerebellar tonsils through the foramen magnum. As a result, the brain stem is elongated and distorted with an abnormal bend. Other abnormalities at this junction may include the Dandy Walker malformation in which the 4th ventricle is markedly dilated due to a closure or failure of development of the foramina of Magendie and Luschka. In other cases a meningioma may be present at the junction or inflammation of the meninges (arachnoiditis) may have occurred. In all of these abnormalities at this junction, a dilatation of the central canal (hydromyelia) usually occurs, and the syrinx appears to develop from this dilated central canal. In some of these cases hydrocephalus is also present.

In *type II syringomyelia*, no specific etiology is present. These cases referred to as idiopathic constitute approximately 60% of all cases.

In *type III syringomyelia*, the syrinx appears to relate to other disease of the cervical spinal cord.

In approximately 8 - 16% of patients with syringomyelia, there is an associated intramedullary tumor.

Syringomyelia may also follow severe spinal cord trauma or following spinal cord compression, developing late after the event from areas of necrosis (myelomalacia).

Diagnosis can be made based on the clinical findings with confirmation by MRI studies of the spine and brain.

The following case 9-5 provides an example of syringomyelia.

Case 9 - 5: This 33 year-old female credit union manager had a 3-year history of intermittent pain and paresthesias extending from the left cervical area to the left arm and hand predominantly involving the middle finger. In the last 3 months, weakness of the left arm and had developed and paresthesias of the left arm had become more continuous.

Neurological examination: *Motor system:*

A mild weakness was present at the left triceps muscle. *Reflexes:* There was depression of all deep tendon stretch reflexes in the left upper extremity. *Sensory system:* There was a selective cape like decrease of pain and temperature sensation in the left upper extremity (shoulder and arm). In addition, there was a selective decrease in pain sensation over the T 3-T5 dermatomes on the right.

Clinical Diagnosis: syringomyelia

Laboratory data: *MRI scan* of the spinal cord demonstrated an extensive irregular cavity extending from C2 to T 9 with a major enlargement of the spinal cord at the T 4 vertebral level *(Fig.9-12).*

MRI scan of the head demonstrated a type 1 Arnold Chiari malformation with displacement of the cerebellum below the foramen magnum *(Fig.9-13).*

Subsequent course: Dr. Alex Danylevich performed a laminectomy at T4 and shunted the large cavity into the subarachnoid space.

Figure 9-14 and **Case 9-6** (presented on the CD ROM) represent a more complex and more advanced example of syringomyelia and syringobulbia.

SYSTEM DISEASES OF THE SPINAL CORD:

These are diseases in which there is a relatively selective involvement of particular neuronal cell groups and/or their fibers. At times, several related cell or fiber systems are involved. System diseases are not of a uniform etiology. Nutritional deficiencies, infections and degenerative disorders are found in this category.

SPECIFIC SYSTEM DISEASES:

Disorders of the anterior horn cell:

Acute anterior poliomyelitis: This disease is caused by a filterable virus that invades the central nervous system. The virus spreads from the gastrointestinal tract by means of a viremia or by spreading up the axis cylinders of the autonomic nerve fibers. On reaching the central nervous system, the virus then involves preferentially the large motor neuron of the spinal cord and brain stem resulting in damage, degeneration, or death of these neurons and

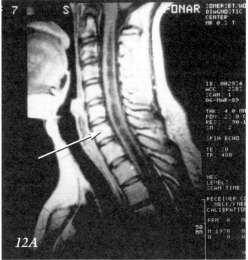
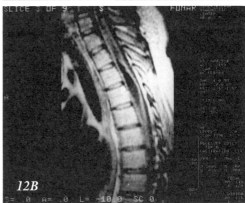

Figure 9-12. Syringomyelia. Case 9-5 MRI Scan. A) Cervical 2-Thoracic 5: Sagittal Section 10mm to left of midline-the arrow is on the cervical 7 vertebral body. B) Cervical 6- Thoracic 9: Sagittal section 10 mm right of the midline. The point of widest enlargement of the spinal cord is opposite the T4 vertebral body.

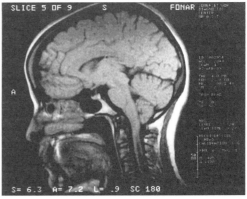

Figure 9-13: Arnold Chiari malformation associated with syringomyelia Case 9-5 MRI (See text)

SPINAL CORD: CLINICAL CONSIDERATIONS 9-17

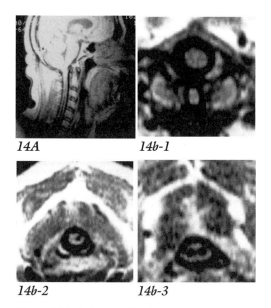

Figure 9-14 Syringomyelia. Case History #9-6. (CASE ON CD ROM). MRI Scan. This 60 year old male had clinical findings of syringomyelia and syringobulbia. A) Midline Sagittal Section demonstrating a central cavity beginning at the cervical two segment and extending well into the upper thoracic segments. B) Transverse Sections demonstrating the irregular nature of the central cavity. 1) Cervical 1-level, no definite cavity. 2) Cervical 2 Level-cavity is predominately right sided extending into right posterior column, 3) Cervical 4 Level-bilateral cavities.

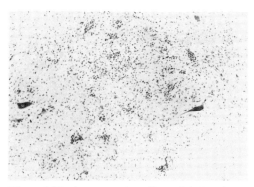

Figure 9-15. Acute anterior poliomyelitis. Lumbar spinal cord, with a loss of large anterior horn motor neurons and replacement by clusters of mononuclear cells. Cresyl violet stain (X100) (Courtesy of Dr. Jose Segarra).

inflammatory changes in the surrounding tissue *(Fig.9-15)*. Neurons in the posterior horn and interneurons of the spinal cord are to a lesser degree often involved in the acute stage.

The correlated clinical phenomena consist of a prodromal period of fever, malaise, headache plus gastrointestinal and upper respiratory symptoms. This is followed in some patients, by a stage of meningeal irritation and then in some patients, by the development of a flaccid paralysis. This involves in an irregular manner, many of the muscles of the extremities and trunk (the spinal form) or the muscles supplied by the bulbar motor nuclei (the bulbar form). Muscles affected in the bulbar form include the facial, palatal, pharyngeal and tongue. In the acute phase cramps are often a significant symptom. Respiratory paralysis may occur in severe cases because of the involvement of the motor nuclei of the diaphragm and intercostal muscles or from damage to the respiratory centers in the medulla. The spinal fluid findings are typical of aseptic meningitis with a predominantly lymphocytic response; however very early in the disease polymorphonuclear cells may predominate. The majority of patients with anterior poliomyelitis make a complete or significant recovery. A significant proportion may actually be labeled as nonparalytic cases because a paralysis fails to develop even in the acute state. However an estimated 20-30% of all cases have a significant residual deficit or disability: a flaccid paralysis with atrophy, fasciculations, and loss of deep tendon stretch reflexes. In many cases, one limb, often a lower extremity is predominantly involved. In bulbar cases, many patients remain respirator dependent. We may presume that in the majority of cases many neurons invaded by the virus manifest a transient dysfunction but are not destroyed.

Recent interest has centered on the development of additional weakness and atrophy many years after the acute episode called the post polio syndrome. The underlying pathological basis for this progression is not entirely clear. The muscles involved however appear to be those previously affected in a clinical or subclinical manner with or without residual paralysis. Electromyography in many patients after the acute phase of poliomyelitis demonstrates changes in many muscles that were not clinically affected by the acute disease.

With the development of the Salk vaccine (formalin inactivated virulent strains) and the Sabin (attenuated live virus) vaccine, infections with the poliomyelitis virus are now extremely rare in the United States and other developed countries. Worldwide eradication is now possible but economic factors and warfare has prevented such eradication in many parts of Africa, and Asia. In countries where the natural disease (due to the "wild," naturally occurring virus) has been eliminated rare cases are still encountered as a result of infection from immunization with the live attenuated virus. In addition, adults who have not been immunized may be infected by exposure to infants or children who have just been immunized with the live attenuated virus.

Case 9-7 presented on the CD ROM presents an example of poliomyelitis in an adult who had not been immunized.

Spinal muscular atrophies: This is a group of genetic disorders that involve the anterior horn cell, without involvement of the corticospinal or sensory systems. All are autosomal recessive. Four types are identified based on age of onset. The later the age of onset, the slower the progression. The earlier the age of onset, the more likely that bulbar motor neuron involvement and respiratory failure will be present in addition to the findings of anterior horn cell involvement. The incidence of the infantile and juvenile forms of the disease is estimated as 1 in 6,000 - 20,000 births. The major disease to be differentiated is usually a disease of muscle i.e. a myopathic process. EMG studies, muscle biopsy and genetic studies are useful in establishing the diagnosis. Types are based on age of onset.

Type I, infantile form (Werdnig-Hoffman Disease): onset: last trimester of pregnancy-6 months: floppy.

Type II (Intermediate/arrested form of spinal muscular atrophy):onset <18 months: delayed motor dates.

Type III Juvenile form (Kugelberg-Welander Disease):onset 5-15 years: proximal limb weakness.

Type IV, Adult onset spinal muscular atrophy: onset>20years:very slowly progressing proximal limb weakness. Note overlap with progressive muscular atrophy form of amyotrophic lateral sclerosis discussed below.

Underlying molecular basis: types I, II, III have all been mapped to chromosome 5 at region 5q11.2--5q13.3 where a survival motor neuron (SMN) gene is located. Deletions in exons on this gene occur in these patients. The genetic defect responsible for the type IV, adult onset cases is unknown.

X-Linked Recessive Bulbospinal Neuronopathy (Kennedy's Disease): This adult onset disease affects males over the age of 30 years. Degeneration of motor neurons in spinal cord and brain stem occurs. There are minor sensory findings. The disease is rare compared to the progressive muscle atrophy form of amyotrophic lateral sclerosis to be considered below but somewhat more common than the adult onset type of spinal muscular atrophy. The importance of this disorder relates to the underlying molecular basis: an abnormal increase in the trinucleotide cytosine –adenine-guanine (CAG) repeats in the region of the androgen receptor gene on the X chromosome. Many of these patients will have gynecomastia; some will have testicular atrophy. In normal individuals, there are 17-26 repeats. In the patients, there are 40-65 repeats. A similar expansion in the number of CAG trinucleotide repeats occurs in other diseases involving different aspects of motor function and affecting different chromosomes: Huntington's disease and the spinocerebellar degenerations. The CAG trinucleotide repeats encodes poly glutamine. With excessive function, protein aggregation occurs in the motor neuron nucleus resulting in degeneration. As in these other disorder,s the higher the number of repeats, the earlier the age of onset.

Disorders of the motor system affecting both the lower motor neuron and the upper motor neuron

Amyotrophic lateral sclerosis (motor neuron or motor system disease, ALS): this is a relatively common degenerative disease of unknown etiology affecting predominantly adults of 40 to 60 years of age. Prevalence is estimated at 4-10 per 100,000. Approximately 10% of cases are inherited usually as an autosomal dominant rarely as an autosomal recessive. In 15-20 % of the autosomal dominant families, a mutation has been mapped to the gene locus for superoxide dismutase on chromosome 21. This enzyme is involved in the detoxification of the free radical superoxide to hydrogen peroxide. Whether injury from free radicals is responsible for the neuronal degeneration in sporadic cases is uncertain.

Clustering of cases of ALS has been noted among the Chamorros on Guam, on the Kii peninsula of Japan and in western New Guinea. In these clusters, there is an overlap with cases of Parkinson - dementia complex. In these areas, the prevalence of ALS was formerly 100 times that of sporadic cases in other parts of the world when studied in the 1950's. With improvement in nutrition and in water supplies, the incidence has decreased. Thus possible toxic factors have been suggested, although a possible predisposition may also be present in these populations. As regards the etiology in most sporadic cases of ALS, the hypothesis of glutamate excitotoxicity has gained prominence in recent years. Glutamate serves as the major excitatory transmitter in the central nervous system. Excessive long-term ingestion of the chick pea (Lathyrus sativus) that contains a neurotoxin, a glutamate receptor agonist may produce a neurological syndrome (lathyrism) in which selective damage to upper motor neurons in the motor cortex occurs. As we will discuss later in relationship to dementia, an epidemic of food poisoning related to contaminated mussels occurred in the Canadian Maritime Provinces in the early 1980s with manifestations of motor neuron disease and dementia. Domoic acid, a potent glutamate receptor agonist was found to be the specific neurotoxin. Related to glutamate excitotoxicity is the role of the subsequent influx of excessive amounts of intracellular calcium. Intracellular calcium binding proteins protect against the effects of excessive intracellular calcium. Those neurons that degenerate in ALS such as the large cortical pyramidal cells the bulbar motor neurons and the alpha motor neurons of the spinal cord have low activity of these calcium-binding proteins. In contrast, those neurons which are not affected in ALS; the oculomotor nuclei, the sensory neurons, the cerebellar Purkinje cells and the nucleus relevant for the bladder (Onuf's nucleus) have a high level of immunoreactivity for calcium binding proteins.

An additional hypothesis as to etiology, concerns possible failure of action of nerve growth factors. Table 9-3 outlines the clinical features and the neuropatholic correlation in ALS.

The majority of cases referred to as *classical ALS* eventually have involvement of all of the systems listed in Table 9-3. However the disease may begin with involvement of the bulbar motor neuron or the spinal cord lower motor neuron or the upper motor neuron. The onset may be very asymmetrical and limited to one limb. The terms applied to those cases with only one of the systems involved are presented in table 9-3.

The diagnosis in those cases with a classical ALS syndrome can be established on the basis of the clinical symptoms and signs. Those cases with a progressive bulbar symptomatology may require MRI studies to exclude other brain stem pathology such as a brain stem glioma. Those cases with a progressive cortical spinal syndrome will require MRI study of brain and cervical spinal cord to exclude other treatable conditions such as parasagittal or foramen magnum meningioma. Those cases with a progressive spinal muscular atrophy syndrome will require nerve conduction studies and electromyography to exclude motor neuropathies. These latter patients may also require studies for immunological and toxic causes.

Electromyography in anterior horn cell disease will show evidence of both denervation and reinnervation. Some of the muscle fibers of a motor unit that have lost their innervation may be reinnervated by axonal sprouting from surviving anterior horn cells. Muscle biopsy will demonstrate grouped motor unit atrophy (as in figure 8-16) Thus a group of muscle fibers all innervated by the same anterior horn cell will be atrophic where as a neighboring group of muscle fibers innervated by an intact anterior horn cell will be well preserved. MRI studies of the brain particularly in patients with prominent upper motor symptoms and signs will demonstrate atrophy in the motor and premotor areas.

Overall median duration of disease ranges from 23 - 52 months. 20 to 25% of all patients live longer than 5 years and 8 -16% of patients survive beyond 10 years. The prognosis of patients with ALS depends on the type of disease and on the initial manifestations as indicated in Table 9-4.

Note that questions have been raised been raised as to whether cases of primary lateral sclerosis and progressive muscular atrophy represent diseases that are distinct from classical ALS. Cases that remain as pure anterior horn cell involvement longer than 36 months are usually classified as PMA. The age of onset is usually younger as well. Overall younger patients have a better prognosis.

The following **case history 9-8** demonstrates the full clinical extent of a classical case of ALS.

Case 9-8: This 66 year-old married white male merchant, 9 months prior to evaluation had the insidious onset of a progressive weakness and atrophy involving the lower extremities and subsequently a similar but lesser involvement of the upper extremities. Three-month prior to evaluation, the patient had the onset of thickness of speech, a difficulty in swallowing solids and to a lesser degree liquids. At the same time stiffness in both lower extremities developed. There had been no sensory symptoms, no urinary symptoms, and no change in mental status. There had been a weight loss of 30 lb.

Neurological examination: *Cranial nerves:* V: the jaw jerk was hyperactive (deep tendon stretch reflex). VII: A bilateral peripheral paralysis was present involving the upper and lower face with a paucity of facial expression. IX, X: Although a gag reflex was present, elevation of the uvula was poor. The voice was hoarse and speech was of low volume. XII. Ridges of atrophy and fasciculations were present along the lateral borders of the tongue. *Motor system:* Widespread muscular atrophy, fasciculations and weakness were present in all four extremities. *Reflexes:* The deep tendon stretch reflexes at the biceps, triceps, and patellar were hyperactive at 3 +. However the right Achilles reflex was 2 + and left was decreased at 0 to 1. The plantar responses were both extensor (bilateral sign of Babinski). *Sensory system:* Intact.

Clinical Diagnosis: amyotrophic lateral sclerosis (fully developed-classical type).

Laboratory data: *The EMG studies* muscle and biopsy were consistent with the diagnosis.

Subsequent course: The patient experienced additional difficulty in swallowing. He expired 3 months after the above evaluation, approximately 1 year after the onset of his disease.

Other diseases with selective involvement of the cortical spinal tracts:

TABLE 9-3. CORRELATION OF CLINICAL SIGNS AND NEUROPATHOLOGY IN ALS AND RELATED DISORDERS

Clinical Sign	Location of Neuropathology	If this system only is involved
1. atrophy, fasciculations, flaccid weakness and loss of deep tendon stretch reflexes (DTR's)	Alpha motor neurons of anterior horn	Progressive muscular atrophy (PMA)
2. bulbar palsy (CN 5,7,9,10,11,12)	Bulbar motor neurons of medulla and pons	Progressive bulbar palsy
3. upper motor neuron findings of spastic weakness, increased DTR's and sign of Babinski	Degeneration of corticospinal tracts secondary to loss of large/giant pyramidal cells in motor cortex (Fig.9-16)	Primary lateral sclerosis (PLS)
4. Pseudobulbar findings of increased jaw jerk and gag plus pseudobulbar speech.	Bilateral degeneration of corticobulbar tracts secondary to loss of large/giant pyramidal cells in motor cortex	Progressive pseudobulbar palsy

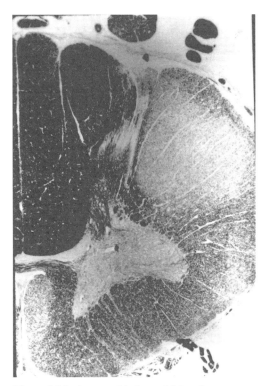

Figure 9-16. *Amyotrophic Lateral Sclerosis. Degeneration of the corticospinal tracts is demonstrated in this myelin stain. (Courtesy of Dr. Emanuel R. Ross).*

TABLE 9-4: SURVIVAL RELATED TO ONSET OR TYPE (DERIVED FROM MATSUMOTO ET AL, 1998 AND MACKAY, 1963)

Classical ALS	Duration (mean)	Relatively Pure forms	Duration
OVERALL	36 months		
Bulbar onset	17 months*	Pogressive bulbar palsy	17 months
Corticobulbar onset	24 months		
Spinal motor neuron onset	33 months	Progressive muscular atrophy (PMA)	159 months (13 years)
Corticospinal onset	36months	Primary lateral sclerosis (PLS)	224 months (19 years)**

* Worst prognosis **Best prognosis

In addition to primary lateral sclerosis, another very slowly progressive disorder has been identified: hereditary (or familial) spastic paraplegia. The prevalence of this disease may be as high as 1 in 10,000. In 70-85% of cases the transmission of the disease occurs as an autosomal dominant. Genetic studies have indicated linkage to different chromosomes for different families: 14q or 15q or 2 p. Less often an autosomal recessive pattern is present. In most cases, onset of disease occurs in childhood or adolescence, less often, after age 35. Initially the lower extremities are primarily involved, with delays in motor development or clumsiness in walking or athletic activity. As the disease progresses the spastic paraparesis becomes evident. As in many of these degenerative disorders a deformity of the feet, pes cavus may occur. Later in the disease, the upper extremities may be involved. In addition a minor decrease in vibratory sensation in the lower extremities may be noted, and pathological examination of the nervous system demonstrates a minor degeneration of the posterior columns in addition to the marked degeneration of the corticospinal tracts.

Degeneration of the posterior columns secondary to disease of the posterior root:

Tabes dorsalis *(Fig.9-17)*: this is a late complication of the sexually transmitted disease syphilis caused by the spirochete, Treponema pallidum. Three stages of the disease are recognized: Primary, secondary, and tertiary. A lesion of abraded skin and or mucous membranes that begin as a painless papule, which then subsequently ulcerates to form the chancre, characterizes the primary stage. The chan-

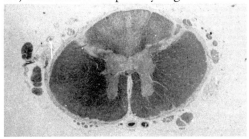

Figure 9-17. *Tabes Dorsalis. Ascending degeneration of the posterior columns in a 67-year-old white male with tabes dorsalis, abdominal crises, luetic aortitis, and luetic optic neuritis. Myelin stain (Courtesy of Dr. Jose Segarra).*

cre may involve the genitalia or lips or oral cavity or anus. The *primary lesion* develops after an incubation period of approximately 20 days, and may persist for an additional 14 -40 days. There is an associated painless lymphadenopathy. Primary lesions involving the vagina or cervix or anus may go unrecognized and in other locations may be mistaken for other transient diseases. In approximately a one-third of these patients asymptomatic spread to the nervous system may be documented by a cerebral spinal fluid examination. The *secondary stage of disease* begins at a variable interval of 2 - 12 weeks after contact. It is characterized by a generalized maculopapular rash which may be confused with other infectious exanthems and constitutional symptoms such as fever fatigue and generalized lymphadenopathy. The symptoms may be so mild that the patient has little awareness of this stage of the disease. Again approximately one-third of patients would demonstrate a meningeal infection were spinal fluid examination to be performed. However only 1 to 2% of patients in the secondary stage actually have symptoms of meningitis. After a variable period following infection of months to 30 years, the late or tertiary manifestations involving the nervous system and the cardiovascular system appeared in 28% of all patients in the pre antibiotic era. Symptomatic neurosyphilis developed in 7-9% of all patients in that era. Prior to the antibiotic era, tertiary neurosyphilis constituted a major neurological problem. A resurgence of the disease may occur in the future. The various types of tertiary neurosyphilis are listed in Table 9-5. Note that the primary process in almost all these syndromes relates to the late effects of meningeal infection and inflammation.

The Clinical Symptoms and Signs of Tabes Dorsalis are listed in table 9-6.

Case 9-9 provides an example of tabes dorsalis.

Case 9-9: This 40 year old male had a 4-year history of severe lancinating type pains in either leg worsened by exposure to cold weather or hot water. This was accompanied by a progressive deterioration of gait, which was particularly impaired in the dark. He could not determine where his feet were placed. He would often go 18-20 hours without voiding.

Neurological examination: *Cranial nerves:* pupils were small, did not respond to light but did respond to accommodation. *Motor system:* A gross ataxia of gait and stance was present worse with eyes closed (positive Romberg sign). *Reflexes:* Deep tendon reflexes were absent in the lower extremities. *Sensory system:* Position and vibratory sensation was absent at toes and ankles. Pain sensation was decreased over the nose and nipples.

TABLE 9-5: THE SYNDROMES OF TERTIARY NEUROSYPHILIS (DERIVED FROM MERRITT, ADAMS &SOLOMON, 1946)

Type and % of all tertiary neurosyphilis (predominant syndrome)	Location of Pathology	Delay in onset after infection
Acute meningitis	Meninges plus cranial neuropathies	2 months to 26 years usually <1 year
Meningo-vascular (16%)	Meninges and arteries in subarachnoid space (arteritis with infarcts in brain and spinal cord)	Several months to 20 years, average of 7 years
General paresis (12%)	Invasion of cerebral cortex by the spirochete with dementia and personality changes (see chapter 30)	20 years
Tabes dorsalis (30%)	Inflammation and infection of posterior roots and secondary degeneration of posterior columns plus (see table 9-6 below)	10-25 years.
Mixed usually general paresis and tabes	Combined cerebral cortex and posterior roots plus (see table 9-6 below	10-25 years
Chronic granulomas (gummas)	Skin, bones, CNS Rare	
Asymptomatic (31%)	Asymptomatic meningitis	Usually 1 year

Clinical diagnosis: Tabes dorsalis

Laboratory data: The diagnosis was confirmed by positive serological tests for syphilis in serum and CSF.

The diagnosis of tabes dorsalis is dependent on the recognition of the clinical pattern of disease. It may not be possible to obtain a reliable history of the primary or secondary lesions. The clinical diagnosis is confirmed by serological tests of the serum and by cerebrospinal fluid examination demonstrating a positive serological test, some mild increase in mononuclear cells, a mild elevation of protein and an increase in gamma globulin. The serological test are of 2 types: 1) the nonspecific reagin antibody tests such as the RPR and the VDRL. 2) The specific treponemal antibody test such as the fluorescent treponemal antibody absorption test (FTA ABS). As regards the nonspecific antibody tests, a positive serum test alone indicates only previous infection without necessarily indicating neurosyphilis. Moreover a false positive test may occur in other febrile illnesses and other immunological disorders. A false negative nonspecific test in the serum may occur in up to 30% of patients with chronic neurosyphilis. The specific antibody tests on serum are positive in almost all cases of neurosyphilis. The positive nonspecific test of the spinal fluid when contamination of the spinal fluid by blood has been ruled out is diagnostic of neurosyphilis.

Penicillin is the treatment of choice of neurosyphilis. The current recommendations of the U.S. public health service are that the patient receive 24,000,000 units a day intravenously for at least 14 days. This should eliminate the activity of the organism. However many of these symptoms and signs will persist since the chronic damage to nerve roots and nervous system from the meningeal or the meningoencephalitic inflammation or vascular components has already occurred. A reexamination of the spinal fluid at 6 months may continue to show a weakly positive serological test for syphilis but all other abnormalities should clear. The serum nonspecific serological tests may continue to remain weakly positive.

TABLE 9-6: TABES DORSALIS: CORRELATION OF CLINICAL FINDINGS AND LOCATION OF NEUROPATHOLOGY

Clinical Sign	Location of Pathology
1. Loss of conscious proprioception: involving the lower extremities	Degeneration of heavily myelinated fibers entering the posterior columns
2. Loss of unconscious proprioceptive information.	Loss of medium diameter fibers, in the posterior root and possibly posterior horn on their way to neurons in the nucleus dorsalis (Clarke's column)
3. Unsteadiness of stance and gait: sensory ataxia. Positive Romberg sign	Loss of proprioceptive information in posterior columns from lower extremities to cerebellum and cerebrum
4. Loss of the deep tendon stretch reflexes manifested as an absence of the patellar and Achilles reflexes.	Loss of the heavily myelinated Ia fibers in the posterior root.
5. Bladder dysfunction - hypotonic flaccid dilated bladder occurs, atonic neurogenic bladder	Loss of the afferent fibers in the posterior roots of S2, S3 and S4 conveying stretch information from detrusor muscle of the bladder (interferes with reflex contractions of the detrusor muscle and subsequent emptying of the bladder)
6. Fleeting sharp pains in the legs the back, the body or face so-called ightening lpains. At times the viscera may also be involved.	Damage to small diameter fibers conveying pain sensation, in the posterior roots
7. Loss of sensation in the feet and subsequent trauma may result in trophic ulcers	Damage to sensory fibers from the leg with degenerative changes in the joints (Charcot's joints)
8. Abnormalities of the pupillary response to light (the Argyll Robertson pupil). Pupil is small (miotic), irregular, fails to respond to light and to sympathetic stimulation. Pupil does accommodate.	Lesion in pretectal region of upper midbrain or possibly in ciliary ganglion

Combined degeneration of the posterior and lateral columns:

Subacute combined degeneration: Combined system disease due to vitamin B12 deficiency: The combination of posterior and lateral column degeneration may be seen in several diseases involving the spinal cord. Of these various causes, the most important characteristic syndrome is that of subacute combined degeneration secondary to vitamin B12 (cobalamine) deficiency. In most cases the basic defect is a lack of intrinsic factor, the enzyme produced by the parietal cells of the gastric mucosa, the same cells that produce gastric acid. In most cases, this occurs on a genetic or immunological basis or as a result of a previous gastrectomy. Intrinsic factor is necessary for the absorption of vitamin B12 that occurs in the distal ileum. In a few cases, the deficiency of vitamin B12 does not indicate a deficiency of intrinsic factor but rather reflects an overall severe nutritional deficiency (eg. a diet deficient in this vitamin) or represents a failure of the absorption of vitamin B12. Such a malabsorption syndrome may occur with a tropical or non-tropical sprue. In other instances a blind loop has occurred and bacterial growth has resulted in consumption of vitamin B-12. The presence in the small intestine of the fish tapeworm d. Latum may also result in the consumption of vitamin B12.

Two active forms of cobalamine have been identified in humans. Methylcobalamin functions as the cofactor for the conversion of homocysteine to methionine. When the reaction fails to occur, folate metabolism is impaired and as a result there is a defect in DNA synthesis. As a result, megaloblastic maturation fails to occur and the characteristic megaloblastic picture of pernicious anemia results. The methionine that results from this conversion is also utilized for the production of choline and choline containing phospholipids. When defective, the formation and maintenance of myelin may be impaired. The second form of cobalamine, adenosylcobalamin is necessary for the conversion of methylmalonyl - coenzyme A to succinyl -coenzyme A. As a result, abnormal fatty acids may be synthesized and incorporated into neuronal lipids. Although the precise metabolic defect responsible for the neurological syndromes is not entirely clear, the first reaction is considered to be the more likely explanation.

Approximately 30 to 70% of all patients with pernicious anemia develop neurological symptoms and signs. Even when severe anemia is present the neurological state may be relatively normal and the explanation is unclear. On the other hand, severe signs of neurological involvement may be present in the absence of anemia. In some instances folic acid may been present in the absence of vitamin B12. Folic acid will correct the anemia but will not affect the progression of the neurological syndromes.

The neurological syndromes relate to the degeneration of myelin and subsequently of axons.

1. The major site of involvement is the spinal cord *(Fig. 9-18)*. Initially the process is most severe in the heavily myelinated fibers of the posterior columns. This results in paresthesias in the extremities and a sensory ataxia accompanied by a + Romberg sign a

2. At the same time, the same process affecting the heavily myelinated sensory fibers also affects peripheral nerves resulting in paresthesias and a loss of deep tendon reflexes.

3. Subsequently the heavily myelinated fibers of the lateral columns are involved predominantly affecting the lateral cortical spinal tracts. Bilateral Babinski signs will be found. Hyperactive deep tendon reflexes are not likely

Figure 9-18 Combined Systems Disease. Degeneration of posterior and lateral columns is demonstrated in this myelin stain of cervical spinal cord. (Courtesy of Dr. Emanuel R. Ross).

because of the effects of the peripheral neuropathy.

4. Less often the cerebral hemisphere white matter may be involved. Rarely for unknown reasons this site predominates producing a progressive dementia.

5. Occasionally the optic nerve is involved producing an optic neuropathy.

Diagnosis: Clinical syndrome + low B12 level (<200pg/ml) + Schilling test demonstrating low absorption and subsequent excretion (<8%) of labeled vitamin B 12 in urine with improvement after combined B12 and intrinsic factor.

Treatment: Monthly injections (IM) of 1000mcg of vitamin B12, after loading during the first month, 1000mcg daily X 5 days, then weekly X 3 weeks.

Case 9-10 demonstrates the symptoms and signs of a typical case of combined system disease.

Case 9-10: This 48 year-old house painter had a 16-week history of progressive impairment of gait mainly unsteadiness and imbalance worse in the dark than in the light and tingling paresthesias of all his toes. Twelve weeks prior to admission, tingling began in his fingers. Weakness had not been a major complaint although he had noted shortly before admission some sense of heaviness in his legs on climbing steps.

Neurological examination: *Motor system:* walking a tandem gait with eyes closed was difficult. The Romberg test was positive. *Reflexes:* Achilles deep tendon stretch reflexes were absent. Bilateral Babinski sign was present. *Sensory system:* vibratory sensation was absent at toes, ankles and knees. Position sense was decreased at toes. There was a minimal decrease in pain and touch sensation over the toes.

Clinical diagnoses: Combined system disease.

Laboratory data: consistent with the diagnosis. Serum B12 level: none detected. Schilling test demonstrated 1% excretion of radioactive B12 in the urine which increased to 12% with intrinsic factor.

Subsequent course: The patient was treated with vitamin B12 injections with improvement in symptoms.

Spinal forms of spinal cerebellar degeneration: Spinocerebellar degenerations represent overlapping groups of degenerative diseases that are usually of unknown etiology. In some cases and families the predominant pathology involves peripheral nerve; in others the cerebellum, in others, the cerebellum and brain stem, in others, the cerebellum, brain stem and basal ganglia. In the type to be considered in the section, the predominant involvement is of the spinal cord. However as will be evident, peripheral nerve is also often involved during the course of these cases.

Friedreich's Ataxia: In most series of hereditary ataxia, this type accounts for 50% of all cases. It is the most common of the autosomal recessive type and is certainly the most common of the early onset hereditary ataxias. The prevalence of this disease in Europe and the United States is 1-2 cases per hundred thousand. The initial symptom: an ataxia of gait appears in childhood or adolescence although rarely symptoms may appear in the young adult years. Subsequently, pyramidal tract findings and dysarthria may appear. Approximately 50% of the patients have a distal wasting suggesting a peripheral nerve involvement. Most patients have a distal loss of deep tendon reflexes and a distal sensory neuropathy involving large diameter fibers. Deformities of the foot including pes cavus *(Fig.9-19)* are common. Deformities of the vertebral spine such as scoliosis (curvature) or kyphoscoliosis are also common. In some cases, blindness due to optic nerve involvement and deafness due to 8th cranial nerve involvement is prominent. The majority of patients have abnormal electrocardiograms. Cardiac arrhythmias and ventricular enlargement/hypertrophy are common due to myocardial muscle and conduction system involvement. Ten % of patients has overt diabetes mellitus. The disease slowly progresses with a mean age of death in the 30s usually due to the cardiac involvement and to the bed ridden state.

Pathological examination *(Fig.9-20)* demonstrates a loss or degeneration of axons

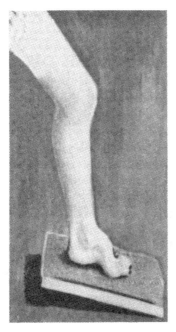

Figure 9-19. Friedreich's Ataxia: Pes Cavus. A similar abnormality of the foot may occur in other disease states. (From Wechsler, I.: Clinical Neurology, 9th Edition. Philadelphia, W.B. Saunders, 1963, p. 123).

that is most severe in the posterior columns of the cervical spinal cord and is accompanied by a significant loss of cells in the posterior root ganglion. The corticospinal and spinocerebellar (lower extremities) and cuneocerebellar (upper extremities) pathways also demonstrate degeneration of axons. There is also a loss of neurons in Clark's column, the cells of origin of the spinocerebellar pathway. In some cases, there is a minor loss of Purkinje cells in the

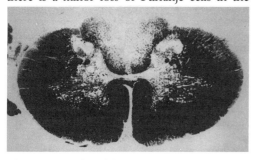

Figure 9-20. Friedreich's Ataxia. Degeneration of posterior columns and, to a lesser extent, of lateral columns is demonstrated in this myelin stain of cervical spinal cord. (From Wechsler, I.: Clinical Neurology, 9th Edition, Philadelphia, W.B. Saunders, 1963, p. 28)

superior vermis of the cerebellum, the termination of the spinocerebellar pathway. There is usually a mild loss on neurons in the dentate nuclei of the cerebellum and of the inferior olivary nuclei of the medulla which projects to the cerebellum. A more significant loss of neurons is found in the nuclei for cranial nerves 8, 10 and 12. Mild degenerative changes may be noted in the optic nerves. In contrast the peripheral nerves demonstrate a more significant distal loss of large myelinated axons.

From the standpoint of clinical pathological correlation, the ataxia is primarily a sensory ataxia due to the involvement of posterior columns, dorsal root ganglia, peripheral nerve and a lesser degree, the spinocerebellar pathways. However some cerebellar component is also present. The dysarthria reflects involvement of cranial nerves 10 and 12 in addition to some possible cerebellar component.

The underlying genetic basis of the disease is an unstable expansion of a trinucleotide repeat GAA (guanine adenine adenine) which maps to the chromosome locus 9 q 13 . On normal chromosomes, the number of GAA repeat ranges from 7 to 22. In contrast, 96% of patients with Friedreich's ataxia have both alleles expanded to 100 to 2,000 repeats. Patients with a larger number of repeats have an earlier age of onset and a more severe form of the disease. They are also more likely to have a cardiomyopathy. The trinucleotide expansion apparently results in a decrease in a protein which has been labeled frataxin and which may function as a mitochondrial iron transporter. The buildup of untransported iron could then result in defects of free radical regulation and oxidative metabolism.

Case history 9-11 presented on the CD ROM demonstrates the typical history and findings in a case of Friedreich's ataxia followed over a long period of time.

MULTIFOCAL DISORDERS AFFECTING THE SPINAL CORD:

Two categories of pathology produce multifocal symptoms and signs affecting central nervous system disease: demyelinating disease and vascular disease particularly small vessel disease such as vasculitis. When the spinal cord is considered the first variety occurs frequently; the second is encountered less frequently.

Demyelinating diseases: In this category of disease we mean a pathological process in which there is a primary destruction of normally formed myelin. We have already indicated other disorders in which damage to myelin in system disorders is but one step in the destruction of the axon. In addition, destruction of myelin and axons may occur as a consequence of vascular infarction. There is another rare group of disorders, the leukodystrophies occurring in childhood, adolescence or early adult life, in which an extensive diffuse loss of myelin occurs. The loss however involves the destruction of defectively formed myelin.

Multiple sclerosis: Among the primary demyelinating diseases, the most common variety is multiple sclerosis also referred to as disseminated sclerosis. This is a relatively common disease of unknown etiology primarily affecting young or middle-aged adults with most patient's first demonstrating symptoms between the ages of 18 and 50. The frequency of disease varies based on geographical origin. Those who reside in the northern temperate zone for the first 18 years of their lives are at greatest risk as opposed to those who have spent those years in a more tropical climate Within a given geographic area whites are at higher risk than blacks. Approximately 1 in 1,000 individuals of northern European origin who have resided in temperate climates during those early critical years will develop multiple sclerosis at some point during their lifetime. An autoimmune disorder of the central nervous system has been suggested and therapy has been directed at this suspected etiology. As with many autoimmune disorders, frequency is higher in females than in males, (ratio of 1.6/1.0). As discussed recently by Noseworthy (1999), while it is unlikely that a viral invasion of the nervous system directly causes the disease, it is possible that molecular mimicry-the antigenic similarity between viral organisms and neural tissue - may trigger the autoimmune reaction. Viral infections may also trigger exacerbations of the disease. Genetic and familial clustering has also been noted.

The disease is characterized by the dissemination of the pathological process in time and space. Thus there is a diagnostic criterion that various lesions be acquired at different times. At any given moment, the various lesions than will be at a different stage of development. It is also a diagnostic criterion that multiple levels of the central neural axis are affected. At a given level the lesion cuts across various fiber systems; it is not a system disease.

The pathological picture varies with the stage of the disease. The initial areas of involvement are perivenous with infiltration of mononuclear cells; plasma cells and lymphocytes .A slight degeneration of oligodendrocytes may be noted. There is a destruction of myelin. There is than a stage of infiltration by macrophages in which the destroyed myelin is removed. There is then a stage of astrocytic proliferation with the production of glial fibers. These areas of gliosis result in the firm sclerotic gray appearance of old lesions and thus the name multiple sclerosis *(Fig.9-21)*. In some early cases, re myelination by oligodendrocytes may occur. As the disease progresses actual destruction of the axons may occur as demonstrated at pathological examination and by MRI scans. Perivascular infiltration of plasma cells and lymphocytes is the probable source of production of the increased amounts of immunoglobulin -G in the cerebrospinal fluid of most patients with active multiple sclerosis. Among all patients with multiple sclerosis 40 - 60% will have such an elevation. Immuno-electrophoresis of the CSF will demonstrate a more specific abnormal population of oligoclonal bands within the IgG in 75 to 85% of patients. A moderate increase in the number of lymphocytes will also be found in the CSF during acute exacerbations These CSF finding however are not specific for multiple sclerosis, also being present in neurosyphilis, and post infectious encephalomyelitis.

Cases are classified according to the anatomical level involved in the initial acute stage for example spinal, cerebellar- brain stem, optic nerve, cerebral hemisphere. A more

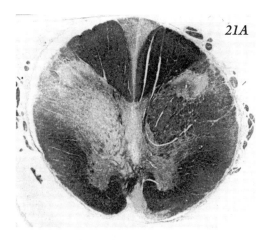

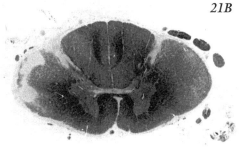

Figure 9-21. Multiple sclerosis affecting the spinal cord. A) A 58-year-old male with a long history of progressive multiple sclerosis. At postmortem examination multiple plaques of variable age were present in the nervous system. (Courtesy of Dr. Jose Segarra). (8-26A) B) This 44-year-old male 20 years previously had experienced a single acute episode of possible multiple sclerosis involving spinal cord and was asymptomatic in the interim until his acute death from an unrelated cause. (Courtesy of Dr. Jose Segarra).

important classification takes account of the pattern of evolution of the disease for example: relapsing- remitting, primary progressive or secondary progressive. Some patients have a benign disease with only 1 or 2 episodes with a relatively complete remission. Overall 85% of patients began with a relapsing /remitting course and 15% with a primary progressive course.

Diagnosis of multiple sclerosis is based on the clinical history and examination. When patients have clear-cut evidence of lesions disseminated in time and space they are referred to as manifesting definite multiple sclerosis. When dissemination in time or in space is present the label of probable multiple sclerosis has been employed. When patients of appropriate age present with the initial findings of syndromes that are seen commonly in multiple sclerosis such as transverse myelitis, or optic neuritis, or the brain stem syndrome of internuclear ophthalmoplegia or an acute cerebellar syndrome they are labeled as possible multiple sclerosis.

The main ancillary technique for diagnosis is the MRI scans. The MRI scan has simplified the diagnosis since many cases of possible or probable now have been shifted into the definite category at first presentation. The visual evoked potential may also be of value in providing evidence of an optic nerve lesion in patients with involvement of other areas of the nervous system.

The following **case 9-12** is an example of multiple sclerosis beginning with an episode affecting spinal cord, but in which the MRI scans demonstrated not only the suspected spinal cord lesion but also, cerebral lesions.

Case 9-12: This 32 year old female registered nurse had a 6 month history of tingling paresthesias in the lower extremities beginning in the toes on the right foot and then the left and gradually spreading to the rib margin on the right and slightly lower on the left. Three months later she developed tingling of the ring and 5th fingers bilaterally. Six weeks prior to evaluation, a positive Lhermitte's sign developed: flexion of the neck produced electric shock sensations that extended down from the buttocks into lower extremities.

Neurological examination confirmed the positive Lhermitte's sign. There was no local cervical tenderness. *Reflexes:* The left plantar response was extensor; the right equivocal. *Sensory system:* Pain and cold were decreased up to the L1-L2 vertebrae posteriorly (but possibly at times to the rib margin on the right), and anteriorly up to D7 dermatome on the right and to D9 dermatome on the left. Vibratory sensation was bilaterally decreased at toes, ankles, and knees with a greater defect on left than right.

Clinical diagnosis: Cervical myelopathy. While the Lhermitte's sign could be seen in multiple sclerosis, this sign also could be produced by compression of posterior columns.

Laboratory data: *MRI of the cervical spine (Fig.9-22)* demonstrated a large area of demyelinating at C2 involving the right lateral column and to the lesser degree the left posterior column. *MRI brain (Fig 9-23)* demonstrated multiple demyelinating lesions in both cerebral hemispheres.

CSF studies were consistent with the diagnosis in terms of an increased count of 18 lymphocytes plus a markedly elevated Immunoglobulin G index). Two oligoclonal bands were present (normal = 0-1).
Subsequent course: The patient received a 5-day course of high dose (1000mg/day) intravenous methylprednisolone with some improvement in the sensory symptoms in the legs. She had subsequent episodes affecting spinal cord and brain stem and was treated with beta interferon.

Case history 9-13 presented on the CD ROM includes several episodes indicating clear-cut clinical involvement of the spinal cord in a patient with multiple sclerosis followed over a number of years. In addition several episodes relevant to brain stem are documented. The history of the patient's brother with a single episode of probable transverse myelitis is also presented.

Present clinical and MRI studies indicate that most patients with relapsing/remitting multiple sclerosis do have an increasing lesion load resulting eventually in symptoms or signs that do not resolve. The present approach then is to utilize early in the disease course agents that modify the immune system. Agents such as the beta interferons decrease relapses and secondary progression. The effects are evident from both a clinical and MRI standpoint. Many would now advocate beginning such therapy after the first episode in a patient who had clear-cut MRI evidence of multiple lesions such as case 9-12.

Acute exacerbations are usually treated with high-dose intravenous corticosteroids with the patient receiving 1,000 mg a day of methylprednisolone for 5-7 days. This shortens the course of the exacerbation with the possibility of less residual disability. There is no clear-cut therapy for primary progressive multiple sclerosis although various major immunological therapies have been investigated.

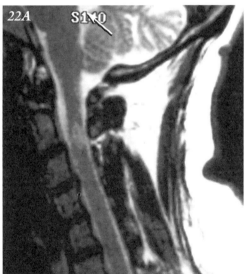
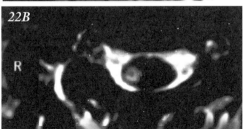

Figure 9-22: Multiple sclerosis involving the spinal cord. Case 9-12 see text. MRI A) Sagittal section. B) transverse section.

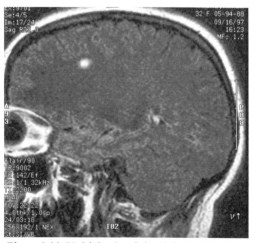

Figure 9-23: Multiple sclerosis involving brain. Case 9-12. MRI. See text.

CHAPTER 10
Case History Problem Solving: Part I
Spinal Cord, Nerve Root, Perepheral Nerve and Muscle

LESION DIAGRAMS: For each of the following diagrams, indicate the structures involved, the clinical symptoms and signs that would be present and the most likely pathology. That is name the disease or syndrome. Diagrams 1-10 follow.

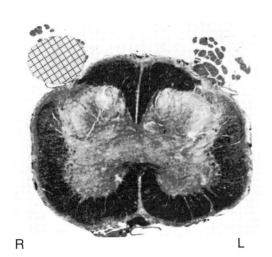

Figure 10-1

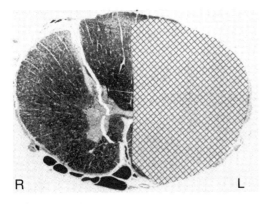

Figure 10-2

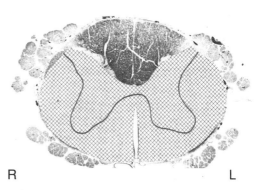

Figure 10-3

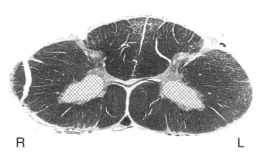

Figure 10-4

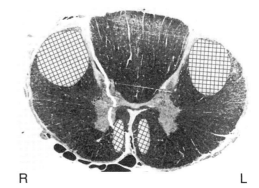

Figure 10-5

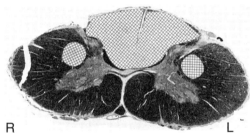

R   L
Figure 10-7

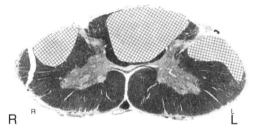

R   L
Figure 10-9

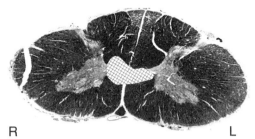

R   L
Figure 10-8

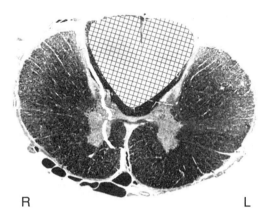

R   L
Figure 10-10

CASE HISTORY PROBLEM SOLVING PART I - SPINAL CORD, NERVE ROOT, PERIPHERAL NERVE AND MUSCLE

Each of the following case histories deals with disease at the level of muscle peripheral nerve root and spinal cord. For each case relevant to spinal cord, be prepared to draw a diagram of the lesion indicating the appropriate spinal cord level, the location of the lesion, and the nature of the pathologic process. If disease of the spinal cord is present, decide whether the process is intrinsic or extrinsic and whether it involves a single-level lesion or is a system disease.

CASE 10-1: A 48-year-old white male, while lifting a heavy object, approximately two months prior to neurological consultation, had the sudden onset of pain in the posterior cervical area and radiating into the right shoulder, down the posterior aspect of the arm into the elbow. Coughing or straining at stool resulted in a shooting, burning electric shock-type pain in the above distribution, also extending at times into the index and middle fingers. In association with the pain, he experienced tingling Pins-and-needles paresthesias in the right upper extremity in a distribution similar to that noted above. During this period the patient also noted minor weakness of the right hand. The lower extremities and bladder were not involved. When symptoms persisted, neurological consultation was obtained.

NEUROLOGIC EXAMINATION

Mental status: Intact.
Cranial nerves: II-XII intact.
Motor system:

    a. There was significant weakness of the right triceps (50% of normal) with minimal weakness of the right wrist extensor and the finger abductors of the right hand. There was no weakness of the lower extremities.

    b. Gait was intact.

    c. No definite atrophy was present. Rare fasciculations were noted in the right triceps muscles.

Reflexes:
   a. Deep tendon reflexes
      Biceps: right, 2+; left, 2+
      Triceps: right, 0; left, 2+
      Brachioradials: right, 1 to 2+; left, 2+
      Patellar: right, 2+; left, 2+
      Achilles: right, 2+; left, 2+
   b. Plantar responses:
   Right flexor, left flexor.

Sensory System:
Minor decrease in pain sensation was present over the right middle finger. Pain, touch, position, and vibratory sensation were otherwise intact.

Neck: There was limitation of neck motion in all directions due to pain. Pressure over the lower cervical spinous processes produced a radiation of pain onto the right upper extremity. There was also local tenderness on pressure over the right supraclavicular area.

QUESTIONS

1. Is the spinal cord directly involved by this lesion? If so, indicate the level. If not, cite evidence against such involvements. Are the anterior roots involved? If so, indicate the level. If not, cite evidence against such involvement. Are the dorsal roots involved? If so, indicate the level. If not, cite evidence against such involvement.
2. What is the localizing significance of the relatively selective weakness of the right triceps muscle?
3. What is the explanation for the occasional fasciculations noted in the right triceps muscle?
4. What is the localizing significance of the selective depression of the triceps deep tendon reflex? Review also the segmental levels involved in the biceps, radial, patellar and Achilles deep tendon reflexes.
5. What is the localizing significance of the distribution of pain and numbness experienced by the patient?
6. What is the localizing significance of the restricted pain deficit in the right hand?
7. Speculate concerning the pathology involved in this case.
8. Why did coughing and straining at stool produce an exacerbation of the pain?
9. How would you manage this problem?
10. Which imaging studies should be obtained and when should these studies be obtained?

SUBSEQUENT COURSE

The patient had a reduction in pain and sensory symptoms following the use of cervical traction. Strength improved and the triceps deep tendon reflex returned.

CASE 10-2: This 58-year-old white housewife presented with a 7-month history of throbbing midthoracic back pain, which at time would radiate to the anterior chest and was aggravated by coughing or by straining at stool. Two months prior to admission, numbness of the right foot was noted, which gradually spread up the leg so that within a 6-week period, a level just below the breast anteriorly and just below the scapula posteriorly was involved. Six weeks prior to admission, a similar ascending numbness of the left foot developed. One month prior to admission, weakness of both lower extremities was noted, the right being weaker than the left. For 2 months, difficulty in control of urination had been present.

GENERAL PHYSICAL EXAMINATION:
Normal.

NEUROLOGIC EXAMINATION:
Mental Status: Intact
Cranial Nerves: Intact
Motor System:
   a. Bilateral decrease in strength was present at hip, knees and ankles.
   b. Gait was shuffling, possibly spastic.
Reflexes:
   a. Deep tendon reflexes were active in the upper extremities, possibly related to a significant degree of anxiety. The patellar reflexes were markedly hyperactive (4+).
   b. The plantar response was markedly extensor on the left and borderline on the

right.

Sensory System:
   a. Pain and temperature sensation were decreased bilaterally below the xiphoid sternum anteriorly and the T6 spinous process posteriorly. The degree of impairment was greater on the right than the left side.
   b. Position sensation was absent at the left toes.
   c. Vibratory sensation was absent at the left toes and decreased at the left ankle and knee.

Vertebral percussion tenderness was present over the T3-T5 spinous processes.

LABORATORY DATA:

X-rays: Thoracic spine X-ray was negative. Chest X-ray was negative.

QUESTIONS

1. What is the significance of this selective combination of signs and symptoms:
   a. The degree of impairment of pain and temperature sensation was greater on the right side than the left.
   b. Position and vibratory sensation was absent at left toes but intact on the right.
   c. The plantar response was markedly extensor on the left but borderline on the right.
2. Where is the lesion located? Be specific and indicate basis for your conclusions.
3. Is the pathology intrinsic or extrinsic?
4. Provide a differential diagnosis as to the nature of the pathology and defend your conclusions.
5. What diagnostic studies would you request? When would you obtain those studies?
6. What therapy would you recommend? What type of consultation would you request? When would you obtain this consultation? What results would you expect from therapy?

CASE 10-3: This 16 year old right-handed white female high school student, 7 days prior to admission awoke with a bilateral sensation of numbness and tingling of teeth and gums. Three days prior to admission, she awoke with numbness of the hands and the plantar surfaces of the feet. On the day prior to admission, she began to experience diffuse weakness of all 4 limbs and of her face. This worsened over 24 hours.

PAST HISTORY:
unremarkable, she did smoke 5-10 cigarettes per day.

GENERAL PHYSICAL EXAMINATION:
Blood pressure slightly elevated (140/90), with elevated pulse of 116. Temperature 97.6 dig. (Oral). Respiration 20. There was mild erythema of the pharynx, but no lymphadenopathy.

NEUROLOGICAL EXAMINATION:
Mental status and cranial nerves were normal.
1. *Motor system:* diffuse weakness: right upper extremity 3+-4+/5, left upper 5-/5.
Symmetric weakness in lower extremities in the following muscles: iliopsoas 3/5, hamstrings 4/5, gastrocnemius 4/5, anterior tibials 4/5, extensor hallucis longus 4/5.
2. *Reflexes:* Deep tendon stretch reflexes were everywhere absent. Plantar responses were flexor.
3. *Sensory system:* intact except for a minimal decrease in pinprick over the soles of the feet.

LABORATORY DATA:
1. *CSF:* day of admission: tube #1; 1800 RBC's, 5 lymphocytes; tube #4: 3 RBC's, 1 lymphocyte. Protein normal: 36 mg%, glucose normal: 70.
   CSF: one week after admission: protein: elevated to 92 mg%, no significant cells were present
2. *WBC:* slightly elevated: 12,600
3. *ESR*, and *ANA* were normal.
4. *Monospot and hepatitis A IgM antibody* were positive.

QUESTIONS:
1. Localize the lesion. Is this a problem of spinal cord, peripheral nerve or muscle?

2. How do you interpret the absence of deep tendon reflexes?
3. How do you interpret the CSF findings?
4. Now be more specific as to the type of pathology –assign a clinical diagnosis.
5. What would EMG/nerve conduction studies demonstrate?
5. What would a nerve biopsy demonstrate?
6. How would you manage this case in terms of treatment if clinical findings improved? Note that findings may already have been improving: the bilateral facial weakness of which she had complained was no longer present.
7. What measures should be undertaken if weakness was progressing and respiration was compromised?

CASE 10-4: This 47-year-old white male, inspector of small parts entered the hospital complaining of weakness and numbness in both legs. Five months prior to admission, the patient had noted a burning-type pain extending over the right forearm. Soon thereafter, he became aware of a gradually increasing numbness, pins-and-needles sensation involving his legs; first the right and then the left. This was followed by a weakness of both lower extremities, which increased in severity. Urgency of urination, occasional bowel incontinence, and increasing impotence were also noted.

NEUROLOGIC EXAMINATION:

Mental Status: Intact.
Cranial Nerves: Intact.
Motor System:
    a. Strength was intact in the upper extremities but decreased in the lower extremities to 50% of normal. The involvement of the right leg was greater than that of the left leg.
    b. A mild degree of spasticity was present at the knees and ankles, with a spastic gait.
    c. No atrophy or fasciculations were present.
Reflexes:
    a. Deep tendon stretch reflexes at triceps were decreased bilaterally (0 to 1+) compared to biceps and radial periosteal (brachioradialis), which were active (2 to 3+). The patellar and Achilles deep tendon reflexes were hyperactive (3+ to 4+).
    b. *Superficial reflexes:* bilateral extensor plantar responses were present (bilateral Babinski signs).
Abdominal reflexes were decreased bilaterally.
Sensory System:
    a. Position sense was defective for gross movements of the great toes bilaterally.
    b. Vibratory sensation was markedly decreased below the level of the T7 spinous process.
    c. Pain and light-touch sensation were markedly decreased below the level of the umbilicus. Sacral segments were involved. There was also a poorly defined band of decreased pain sensation over the upper thorax. Pain sensation was somewhat greater on the left than on the right.

LABORATORY DATA:

1. *Glucose tolerance test* revealed a diabetic-type curve, with a fasting blood sugar of 122, 30 minute sample of 205, 60 minute sample of 218, and 150 minute sample of 142 mg/100 ml.
2. *Lumbar puncture* revealed a partial dynamic block with the head in extension. Cerebrospinal fluid protein was elevated to 80 mg/100 ml (normal 45 mg/100 ml).

QUESTIONS

1. Indicate the level of the lesion in this case and the structures involved in the pathology.
2. Assuming that the pain in the upper extremities early in the course of the disease had some localizing significance, why was the sensory level for pain present only up to the level of the umbilicus (T10) (that is, pain sensation was absent or decreased below the umbilicus)?
3. Was the pathology in this case intrinsic or extrinsic (compressive) to the spinal cord? Cite the evidence for your conclusions.
4. What diagnostic procedures would you undertake prior to surgery?

5. Granted that the patient had, among other findings, posterior and lateral column signs, why is the diagnosis of combined system disease unlikely in this case?

HOSPITAL COURSE

Surgery was performed by Doctor Samuel Brendler. Examination two weeks after surgery revealed some return of pain sensation over the thorax, abdomen and lower extremities. Walking had improved. Evaluation 2 months after surgery indicated continued improvement as regards gait. Position sense had returned to the lower extremities, but vibration sense was still absent.

CASE 10-5: This 73-year-old retired executive was referred for evaluation of paresthesias (pins-and-needles sensation and numbness) involving all four extremities. Approximately 8 months previously, while hospitalized for gallbladder surgery, the patient developed a glove-and-stocking distribution of paresthesias involving all four extremities in a symmetric manner. The level of sensory symptoms gradually ascended over the ensuing weeks and months; in the lower extremities, as far as the perineum; in the upper extremities, to the level of the elbows. In the several months prior to admission, the patient had also noted increasing unsteadiness of gait, primarily when attempting to walk in the dark.

Family History There were several relatives with pernicious anemia but no relatives with neurologic disease.

NEUROLOGIC EXAMINATION:

Mental Status: Intact.
Cranial Nerves: Intact.
Motor System:
    a. Strength was intact.
    b. Gait was broad-based, with unsteadiness on the turns. The degree of ataxia was increased by eye closure.
    c. The patient with eyes open and standing on a narrow base was relatively steady. When his eyes were closed, a significant swaying was apparent (Romberg test positive).
    d. Cerebellar tests, such as bringing the finger-to-the-nose, when performed with eyes open, were not remarkable.
Reflexes:
    a. Deep tendon reflexes were everywhere absent.
    b. Plantar responses were extensor bilaterally (bilateral Babinski signs).
Sensory System:
    a. Vibratory sensation was absent at the toes and ankles and markedly decreased at the knees and iliac crests in a symmetric manner. There was also a significant decrease at the fingers, with a lesser decrease at the wrists and elbows.
    b. Joint position sense was defective for fine-amplitude movements at the toes but elsewhere was intact.
    c. Pain and light-touch sensation were intact.

QUESTIONS

1. **This patient pres**ents a common neurologic syndrome. Indicate the site of pathology. Present a differential diagnosis and indicate the most likely pathology.
2. Does this patient have a single-level lesion or a system disease?
3. Discuss the diagnostic significance of the sensory symptoms and findings.
4. Explain the absence of deep tendon reflexes.
5. What is the significance of a "positive Romberg sign"?
6. Indicate the significance of bilateral extensor plantar responses (Sign of Babinski).
7. Outline what additional tests you would perform to confirm the diagnosis. Indicate the normal values for those tests.
8. What treatment would you undertake? (Be specific)

CASE 10-6: This 21-year-old single white female was admitted to the hospital because of progressive difficulty in walking. Nearly four years before admission, the patient first noted a relatively rapid onset of weakness in the left hand. Three months later, she noted a throbbing pain in the neck that radiated into the left arm and left hand and was accompanied by some numbness (paresthesias) of the fingers of the left hand. Weakness of the left upper extremity slowly progressed and was accompanied by atrophy. Four months before admission, a more rapid progression of left arm symptoms was noted, and weakness of both lower extremities began, initially on the left and to a greater extent than the right. Just before admission, weakness of the right upper extremity was noted. Bladder symptoms were not present.

NEUROLOGIC EXAMINATION:

Mental Status: Intact.
Cranial Nerves: Intact.
Motor System:

　　a. Severe atrophy of all muscles of the left upper extremity was present, with flexion (claw-hand deformity) contracture. Fasciculations were present in almost all of the left upper extremity muscle groups. Mild atrophy of the intrinsic muscles of the right hand was also noted.

　　b. There was weakness of the muscles of the left upper extremity, most marked distally but also involving the shoulder girdle, with a lesser weakness of the muscles of the right upper extremity. Both lower extremities were weak, the left more so than the right.

　　c. Spasticity was present bilaterally on passive movement at the knees and ankles.

Reflexes:

　　a. Deep tendon reflexes were depressed in the left upper extremity. The biceps reflex was absent: the triceps and brachial radialis (radial periosteal) reflexes were hypoactive (1+). In the right upper extremity the biceps reflex was depressed (1+), whereas the triceps was 2+. In the lower extremities, the patellar and Achilles were hyperactive to a marked degree (4+).

　　b. *Superficial reflexes:* Plantar responses were extensor bilaterally. Abdominal reflexes were absent.

Sensory System:

　　a. Pain and temperature sensation were intact.

　　b. There was a slight distal decrease in vibratory sensation at the toes.

Cranium and Vertebral Column:
No local tenderness or abnormalities were present.

LABORATORY DATA:

1. *Cervical spine X-rays* showed bony changes: widening of the spinal canal n anteroposterior and lateral diameters was noted.
2. *Thoracic spine X-rays* showed that kyphoscoliosis was present.
3. *Cerebrospinal fluid* protein was increased to 150 mg/100 ml.

QUESTIONS

1. Where is the pathology located (be specific)?
2. Present a differential diagnosis. In this case you will need to consider the radiology findings.
3. What is the most likely pathology?
4. What is your diagnostic approach to this patient?
5. Outline your therapeutic approach to this patient.
6. What therapeutic results are expected?

Case 10-7: This 39-year-old white married airplane mechanic had the gradual onset of muscle weakness approximately 18 months prior to admission. This involved both arms and legs, with greater involvement of the proximal muscles than of the distal muscles. Gradual progression occurred so that 14 months before admission, significant difficulty in walking was experienced. Shortly thereafter, the patient experienced difficulty in swallowing, but this problem improved following his hospitalization the previous year. Weakness of extremity musculature also improved following the use of cortisone. The patient at no time had difficulty in voiding. He had no sensory symptoms; he had had no actual pain or tenderness in the muscles.

NEUROLOGICAL EXAMINATION:

Mental Status: Intact.
Cranial Nerves: Intact.
Motor System: The patient had severe weakness and atrophy of muscles (both proximal and distal) of all four extremities but with clearly more marked involvement of the proximal musculature. The patient was unable to lift his arms above the shoulder and unable to lift his legs off the bed. No fasciculations were present.
Reflexes:
    a. Deep tendon stretch reflexes were everywhere absent except at the ankle where normal 2+ reflexes were found.
    b. Plantar responses were flexor.
Sensory System: No abnormalities were present.

LABORATORY DATA:

1. *Erythrocyte sedimentation rate* was elevated to 45 mm/hr.
2. *Electromyogram and muscle biopsy* were consistent with the clinical diagnosis.

SUBSEQUENT COURSE:

Following treatment with prednisone in high dosage, the patient had a gradual improvement in strength.

QUESTIONS:

1. Does this patient have a problem localized to spinal cord, nerve root, peripheral nerve or muscle?
2. Present a differential diagnosis of this problem and indicate the most likely diagnosis.
3. Outline the expected findings on electromyography and muscle biopsy.
4. Would any other laboratory data be of help in establishing the diagnosis?

CASE 10-8: This 37-year-old white male was evaluated by Doctor Sandra Horowitz for a diagnosis of upper extremity disability. His neurologic history had begun, 20 years previously with weakness in both upper extremities. An extensive laminectomy had been performed 19 years earlier.

NEUROLOGIC EXAMINATION:

1. Severe flaccid weakness in voluntary movement in the upper extremities, particularly at the shoulders.
2. Spasticity was present on passive motion of the lower extremities.
3. The gait was also ataxic.
4. The deep tendon reflexes were absent in the upper extremities but hyperactive in the lower extremities at patella and Achilles. The plantar responses were extensor bilaterally.
5. Sensation for pain and temperature was selectively decreased over the shoulder and arms. An MRI scan of the cervical spine was performed.

QUESTIONS:

1. What is the significance of the selective decrease in pain and temperature over the shoulders and arms?
2. What would you expect the MRI to demonstrate?
3. Select the appropriate MRI from the group of illustrations below that best corresponds to this case.
4. What therapeutic approaches are possible?

CASE HISTORY PROBLEM SOLVING PART I: SPINAL CORD   10-9

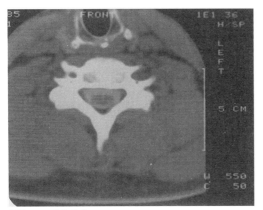

Figure 10-1A

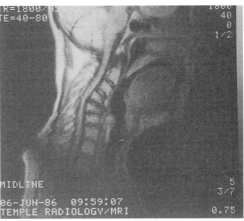

Figure 10-2A

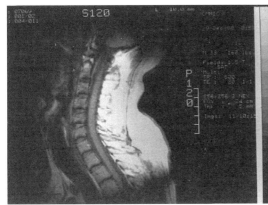

Figure 10-3A

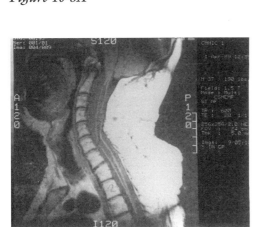

Figure 10-4A

CHAPTER 11
Functional Anatomy of the Brain Stem

I. INTRODUCTION

This chapter presents an overview of the anatomy of the brain stem and describes certain functional centers. The brain stem consists of medulla, pons and midbrain *(Fig 11-1 and 11-2)*. We will first start at the medullospinal junction and then discuss several levels of the medulla, pons, and midbrain. The discussion of each level will include the important functions, nuclei and tracts found in that level. The medulla, pons, and midbrain are subdivided by their relationship to the ventricular system. One division, the tectum, forms the roof and walls or the posterior surface; the other division, the tegmentum, forms the floor or anterior surface. The basis forms the anterior surface of the tegmentum.

cord. Table 11-1 lists the general contents of each region of the brain stem.

TABLE 11-1. REGIONS IN THE BRAIN STEM

REGION	CONTENTS
Tegmental-anterior floor of fourth ventricle and cerebral aqueduct	Cranial nerve nuclei and all ascending and descending Tracts except for tracts in basis
Basis-anterior surface of tegmentum	Corticospinal, corticonuclear, And corticopontine tracts
Tectum-posterior surface of fourth ventricle and aqueduct	Cerebellum for medulla and pons; superior and inferior colliculi of midbrain

TEGMENTUM

The tegmentum of the brain stem (medulla, pons, and midbrain) is continuous *(Fig 11-3)*, and it can be divided functionally into five zones: ventricular, lateral, medial, central and basilar. This is done to help the student understand the basic plan of the brain stem. Each half of the brain stem should be considered organized like an apple with the core being the

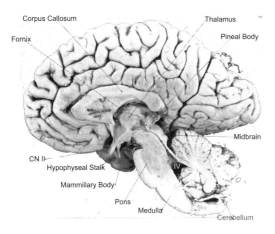

Figure 11-1. Gross View of the Brain Stem in Sagittal Section of the brain.

The tectum in medullary, pontine, and midbrain levels consist of regions that have highly specialized functions related to the special senses and movement. The tegmentum contains cranial nerve nuclei, the reticular formation, and tracts that interconnect higher and lower centers as well as tracts that interconnect the brain stem with other portions of the central nervous system. The basis consists of the descending fibers from the cerebral cortex to the brain stem, cerebellum, and spinal

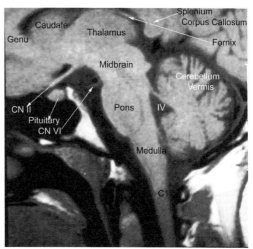

Figure 11-2. The regions of the brain stem. Sagittal MRI-T1.

reticular formation and the pulp formed by the four surrounding zones *(Figures 11-3 A,B,C, Table 11-2)*.

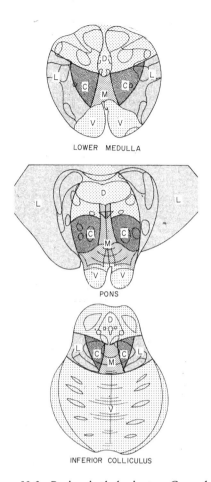

Figure 11-3. *Regions in the brain stem, Coronal sections; A- medulla, B-pons, C-midbrain.*

II. FUNCTIONAL LOCALIZATION IN CORONAL SECTIONS OF THE BRAIN STEM

The previous sections on the brain stem have discussed gross anatomy and cranial nerves. In this section, we will discuss the functional anatomy of the brain stem by examining coronal sections at representative levels through the medulla, pons and midbrain. Each level will contain the following illustrations:

A - a myelinated section with the nuclei and tracts labeled on the outline below.

B - an MRI at a similar level (Please remember that the MRI is from a living patient while the brain sections were obtained from postmortem examination)

TABLE 11-2 THE FIVE ZONES IN THE TEGMENTUM

Zone	Gray Matter	White Matter
Ventricular (Dorsal)	medulla:hypoglossal & vagal trigones, solitary nucleus vestibular & cochlear nerves & nuclei pons: median eminence with facial colliculus midbrain: CN II & III	all levels – descending autonomics medulla – solitary tract
Lateral	inferior olive medulla & pons - descending nucleus of V	all levels spinothalamic,& Rubrospinal. medulla- spinocerebellars trigeminal afferents inferior cerebellar peduncle pons - lateral lemniscus midbrain - lateral lemniscus & medial lemniscus
Medial	reticular formation raphe nuclei	all levels – MLF & tectospinal. medulla – medial lemniscus
Central (Core)	reticular formation central & lateral nuclei. medulla- ambiguous pons - motor nucleus of V, facial nucleus, parapontine nucleus. midbrain - red nucleus & substantia nigra	ascending & descending reticular tracts, descending autonomics
Basilar	Medulla; arcuate nucleus pons: pontine gray	medulla- pyramid pons-inferior cerebellar peduncle midbrain- cerebral peduncle

Although the discussion in this chapter of each level is confined to the coronal section under examination, it should be noted that most nuclei and all tracts extend through more than just one level. Therefore, when reading this section the student should keep in mind the origin and destination of structures identified in each of these sections.

It is especially important that the student understand the function and clinical importance of the anatomical structures they are

TABLE 11-3. GRAY & WHITE MATTER EQUIVALENTS IN SPINAL AND BRAIN STEM STRUCTURES

SPINAL CORD REGION	EQUIVALENT BRAIN STEM REGION
Dorsal (sensory) horn	Ventricular zone, lateral region with sensory cranial nerves VII, IX, X; lateral zone cranial nerve V
Ventral (Motor) Horn	Ventricular Zone, medial Motor Cranial Nerves - III-VI, X-XII Reticular Formation with motor nuclei of Cranial Nerves V, VII, IX & X
Intermediate Zone	Reticular formation and associated nuclei
Intermediolateral Cell Column (Sympathetic) T1-T11,L1 & L2 ; Parasympathetics S2-S4.	Parasympathetic Nuclei with Cranial Nerves III, VII, IX & X.
Lateral Column = Corticospinal/crossed, rubrospinal, Spinothalamic	Basilar Zone= corticospinal/not crossed Lateral Zone – spinothalamic & rubrospinal
Posterior Column = uncrossed Gracile & cuneate fasciculi	Medial Lemniscus = crossed tactile information
Anterior Column = MLF, vestibulospinal, Reticulospinal, spinothalamic	Medial Zone - MLF, tectospinal reticular formation reticulospinal lateral zone vestibulospinal

studying. The cranial nerve nuclei and many pathways (e.g., corticospinal, corticonuclear, and spinothalamic) in the brain stem are clinically important since any functional abnormality in these systems help to identify where the disease process is occurring in the CNS.

DIFFERENCES BETWEEN THE SPINAL CORD AND BRAIN STEM

The interior gray matter and exterior white matter of the spinal cord gradually changes to a mixture of gray and white matter in the brain stem. Table 11-3 identifies the nuclei and tracts in the spinal cord and lists their functional equivalents in the brain stem. The student should be aware that the development of the cerebellum, midbrain, and cerebrum has greatly affected the neural contents of the brain stem and has lead to the formation of many nuclei and tracts (for example the following nuclei - inferior olive of the medulla, pontine gray of the pons, red nucleus and substantia nigra of the midbrain, and the corticonuclear and corticopontine tracts).

MEDULLA
Blood supply-Vertebral arteries and its branches

The first level we have included is at the sensory and motor decussation *(Fig 11-4)*. This section has mostly spinal cord elements, the ventral horns and the three white matter columns; however, one can also note brain stem elements--the gracile and cuneate nuclei. This region is the transitional zone and the region is the size and shape of the spinal cord. As we move up the brain stem, we come to the widened medulla with the opening of the fourth ventricle.

In the brain stem tegmentum the gray and white matter appear intermingled; however, there is a basic organization of the tracts in the brain stem zones, as shown in Table 11-4.

TABLE 11-4. MAJOR TRACTS IN TEGMENTAL ZONES

Ventricular zone –	solitary tract and autonomic tracts
Medial zone –	MLF, tectospinal, medial lemniscus
Lateral zone –	spinothalamics, rubrospinal, cerebellar peduncles, and lateral lemniscus. Medial lemniscus enters this zone in pons and midbrain.
Central core (reticular formation)-	descending autonomics and central tegmental tract.
Basilar zone –	corticospinal and corticonuclear.

MEDULLA

Gross Landmarks in the Medulla. The tegmentum of the medulla has two distinct levels, a narrow lower portion which is similar to the spinal cord and an broader upper portion which includes the landmarks that are a hallmark of this level, the medullary pyramids and the inferior olivary prominence.

LEVEL: SPINOMEDULLARY JUNCTION WITH MOTOR DECUSSATION *(Fig. 11-4)*

Gross Features. This level resembles the spinal cord as the dorsal columns - the funiculus gracilis and the funiculus cuneatus are conspicuous; the trigeminal funiculus is more laterally placed.

On the anterior surface of the spinal cord, the medullary pyramids (corticospinal tracts) are evident *(Fig 11-4A)*. Note that this descending pathway on the right side has first crossed (decussated) with the fibers, shifting from the anterior surface of the medulla, then entered the lateral funiculus of the spinal cord. The narrow spinal canal is present in the gray matter above the pyramidal decussation..

Motor Cranial Nerve Nuclei (Fig 11-4A). Only the cranial portion of nerve XI is present at the lateral margin of the reticular formation. It innervates the Sternocleidomastoids and trapezius muscles that rotate the head and elevate the shoulders, respectively.

Sensory Cranial Nerve Nuclei (Fig 11-4A). The primary axons convey pain and temperature from the head and neck and are located in the spinal tract of the fifth cranial nerve; these fibers have descended from the pons. The second order axons originate from the underlying descending nucleus and their axons cross and then ascend contralaterally to the ventral posterior medial nucleus of the thalamus. The nucleus lateral to the pyramidal tract is the inferior reticular nucleus, a portion of the reticular formation.

White Matter (fig 11-4A). At this level the tracts are still in the same positions as in the spinal cord. Tactile discrimination (fine touch, pressure, vibration sensation, and two-point discrimination) and proprioception from the extremities, thorax, abdomen, pelvis, and neck are carried via the dorsal columns. The nuclei can now be seen that are the 2° neurons in this pathway, the gracile and cuneate nuclei. The gracile and cuneate tracts (fasciculi) have reached their maximum bulk with the addition of the last of the fibers from the uppermost cervical levels. The somatotopic arrangement of the tactile fibers in the posterior column are as follows: the most medial fibers are sacral, then come the lumbar, thoracic, and most laterally, cervical fibers.

Spinothalamics/anterolateral column. Pain and temperature from the extremities, abdomen, thorax, pelvis, and neck are carried by the lateral spinothalamic tract located at the lateral surface of the medulla. Light touch from the extremities, thorax, abdomen, pelvis, and neck is carried via the anterior spinothalamic tract, which is seen on the surface of the medulla just posterior to the corticospinal tract. In the spinothalamic pathways sacral fibers are on the outside while cervical fibers are on the inside. The dorsal and ventral spinocerebellar tracts are found in the lateral funiculus.

The rubrospinal and tectospinal tracts, which are important in supporting voluntary motor movements, are found in the lateral funiculus and near the midline in the medullary and pontine levels).

LEVEL: LOWER MEDULLA AT SENSORY DECUSSATION *(FIG. 11-5)*

Gross Features. At this level the fourth ventricle narrows. The funiculus gracilis and the funiculus cuneatus are conspicuous on the posterior surface of the spinal canal, while the medullary pyramids are prominent on the anterior surface *Fig 11-5A and B.*

Ventricular Zone.

Motor Cranial Nerve Nuclei (fig 11-5A). This section contains the inferior extent of the hypoglossal nerve (XII) and the dorsal motor nucleus of the vagus nerve (X). The hypoglossal nucleus innervates the intrinsic and extrinsic musculature of the tongue, while the dorsal motor nucleus of the vagus nerve provides parasympathetic preganglionic innervation of the viscera. This level marks the superior extent of the cranial portion of the eleventh cranial nerve in the ambiguous nucleus.

FUNCTIONAL ANATOMY OF THE BRAIN STEM

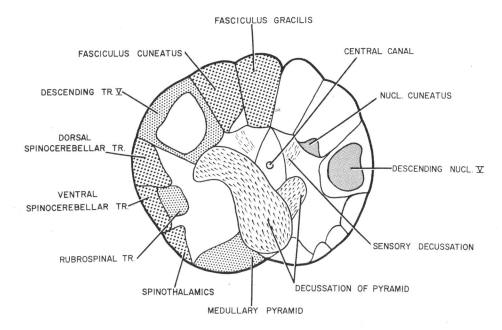

Figure 11-4. Brain Stem at transition level between cervical spinal cord and medullar-Sensory Decussation. Coronal

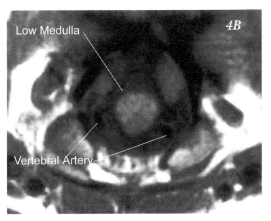

Lateral Zone.

Sensory Cranial Nerve Nuclei (Fig 11-5A). Pain and temperature from the head are conveyed by the descending nucleus and tract of nerve V that is prominent anterior to the cuneate nucleus. Note that myelinated ipsilateral primary axons are on the outside of the second order neurons while the second order

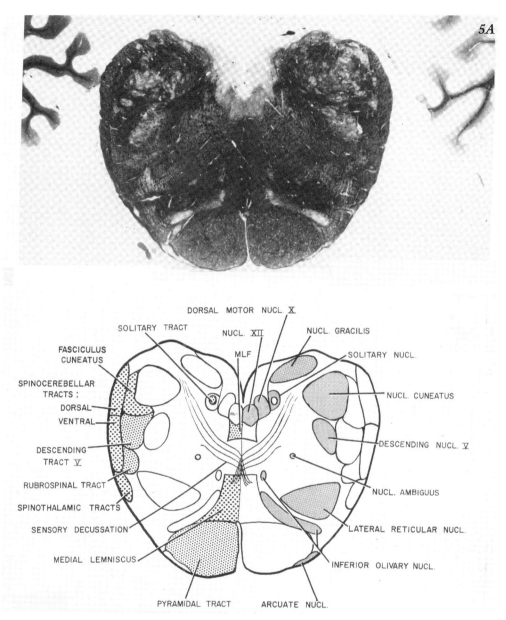

Figure 11-5. Brain Stem at lower medullary level. Coronal

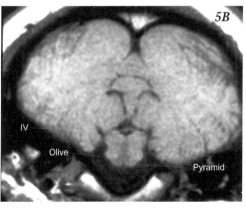

neurons are leaving the inner surface of the nucleus, crossing and entering the medial lemniscus, forming the trigeminothalamic (quintothalamic) tract.

White Matter (Fig 11-5A). Pain and temperature from the extremities, thorax, abdomen, pelvis, and neck are carried via the lateral spinothalamic tract. The spinothalamics are located in the lateral funiculus throughout the spinal cord and brain stem! The anterior spinothalamic tract is adjacent to the lateral spinothalamic tract.

The tract carrying unconscious proprioception from the upper extremity originates from the external cuneate nucleus and its fibers enter the cerebellum through the posterior spinocerebellar tract. The function of this nucleus is similar to that of Clark's column in the spinal cord.

The vestibulospinal tract is found internal to the spinothalamic tracts. The dorsal and ventral spinocerebellar tracts are seen on the surface of the medulla covering the spinal tract of nerve V and the spinothalamic tracts.

The rubrospinal tract is an important afferent relay to the alpha and gamma neurons in the spinal cord. It originates from the red nucleus in the tegmentum of the midbrain, crosses the midline, and descends. It is found in the lateral funiculus of the medulla internal to the spinocerebellar tract and anterior to spinal nerve V throughout the pons and medulla. It is important in postural reflexes. The tectospinal tract is phylogenetically an old tract, being the equivalent of the corticospinal tract in non-mammalian vertebrates. It originates from the deep layers of the superior colliculus and to some extent from the inferior colliculus. It is seen anterior to the medial longitudinal fasciculus throughout the medulla, pons, and midbrain and is important in coordinating eye movements and body position.

Medial Zone.

The medial lemniscus is the largest pathway in the medial zone. However, as we progress superiorly in the brain stem these fibers migrate laterally to finally enter the lateral zone in the midbrain. This migration is necessary for the medial lemniscus to be correctly positioned as it enters the thalamus.

Tactile discrimination and proprioception are conveyed from the extremities, thorax, abdomen, pelvis, and neck by the axons in the fasciculus gracilis and cuneatus. The nucleus gracilis and the nucleus cuneatus form the second order neurons in this pathway and are conspicuous in this section. Internal arcuate fibers are seen leaving the inner surface of the gracile and cuneate nuclei, curving around the ventricular gray, crossing the midline (sensory decussation), and accumulating behind the pyramid and beginning the formation of one of the major ascending highways, the medial lemniscus. The serotonin containing raphe nucleus of the reticular formation is found in this zone throughout the brain stem.

The medial longitudinal fasciculus is located in the midline of the tegmentum just below the hypoglossal nucleus and above the tectospinal and medial lemniscal pathways. In all levels of the brain stem it will be in this subventricular position.

Central Zone

The central core of the medulla, pons, and midbrain consists of the reticular formation, which is important for many vital reflex activities and for the level of attentiveness.

Basilar Zone

This level contains the corticospinal fibers just before they cross in the medullospinal junction. Note that the pyramidal system is named for the passage of the corticospinal fibers through this pyramidal shaped region respectively *(Fig 11-5 A-B)*. One of the important concepts from this section is that axons that form the volitional motor pathway, pyramidal system, originate in the cerebral cortex. These tracts are located in the basis of the brain stem and include the corticospinal and corticonuclear fibers that run in the medullary pyramid. Motor system fibers that do not run in the medullary pyramid are considered extrapyramidal (e.g., cerebellar peduncles, rubrospinal tectospinal, etc.) The location of the major structures in the tegmentum of the medulla is listed below in Table 11-5.

LEVEL: INFERIOR OLIVE OF THE MEDULLA *(FIG. 11-6)*

Gross Features. At this level the medullary tegmentum expands laterally with the prominent inferior olive located behind the pyramids. The floor of the fourth ventricle con-

TABLE 11-5. MAJOR TRACTS AND NUCLEI IN MEDULLARY TEGMENTAL ZONES

Ventricular Zone- Cranial nerve nuclei of VIII, X & XII (motor nuclei medial; sensory lateral) solitary nucleus and tract.

Lateral Zone- Cranial nerve rootlets exit in relation to the inferior olive with the XII nerve anterior to the olive and the rootlets of IX, X and XI posterior to the olive. Rootlets for VIII exit in the cerebellopontine angle.

Tracts: Inferior cerebellar peduncle, spinothalamic, rubrospinal; nuclei: external cuneate, descending nuclei and tract of V.

Medial Zone- Nuclei - raphe or medial reticular. Tracts = MLF, tectospinal, medial lemniscus

Central Zone- Nuclei of reticular formation, ambiguous nuclei of IX-XI, inferior olivary complex, Tract: central tegmental tract, the major ascending reticular pathway

Medullary Basis – Tracts contains corticospinal and corticonuclear tracts in medullary pyramid.

---

tains the median eminence and the vestibular and cochlear tubercles. The median eminence consists of aqueductal gray, while the vestibular tubercle is formed by the medial vestibular nucleus and the descending root and nucleus of cranial nerve VIII. The dorsal cochlear nucleus forms the cochlear tubercle. The ventricle at this level is at its widest extent. *(Fig. 6A and 6B)* The sulcus limitans in the ventricular floor separates the medially placed motor cranial nuclei from the laterally placed sensory cranial nuclei.

Ventricular Zone

Motor Cranial Nerve Nuclei (Fig 11-6A). This level marks the superior extent of nerve XII and the dorsal motor nucleus of nerve X. The ambiguous nucleus in the lateral margin of the reticular formation contains cell bodies innervating the pharynx and larynx (nerve X).

Lateral Zone
Sensory Cranial Nerve Nuclei.
   a. Trigeminal. Pain and temperature are conveyed from the face by nerve V. The descending nucleus and tract of cranial nerve V are nearly obliterated by the olivocerebellar fibers.

   b. Solitary Tract & Nucleus. Taste and visceral sensations are found in the solitary nucleus and tract in the tegmental gray below the medial vestibular nucleus. Fibers from cranial nerves VII, IX, and X ascend and descend in this tract carrying general sensations from the viscera (nerve X) and gustatory sensations from the taste buds in the tongue (nerves VII and IX) and epiglottis (nerve X).

   c. Vestibular. The medial and descending vestibular nuclei are present with first order axons *(Fig. 11-6B)* found in the descending root of cranial nerve VIII, interstitial to the descending nucleus. Fibers may be seen running from the vestibular nuclei into the medial longitudinal fasciculus.

   d. Auditory. The dorsal and ventral cochlear nuclei are second order nuclei in the auditory pathway and receive many terminals from the cochlear portion of nerve VIII seen within its borders.

Cerebellar. The conspicuous inferior olivary nucleus in the anterior portion of the medullary tegmentum consists of the large main nucleus and the medial and dorsal accessory nuclei. The entire complex is important in supplying information to the cerebellum. Removal of the olive in animals produces contralateral increase in tone and rigidity in the extremities, with concomitant uncoordinated movements. The olive connects to the contralateral cerebellar hemispheres via the inferior cerebellar peduncle. The climbing fibers found on the dendrites of Purkinje's cells in the cerebellum originate in the inferior olive. The olive also has strong connections with the red nucleus and receives input from the spinal cord, cerebellum, red nucleus, intralaminar nuclei, and basal ganglia. The central core of this section, as in other levels, consists of neurons of the reticular formation.

Medial Zone (Fig 11-6 A). The serotinergic raphe nuclei of the reticular formation are found in this zone. The position of the MLF, tectospinal and medial lemniscus corresponds to that in the previous level.

FUNCTIONAL ANATOMY OF THE BRAIN STEM

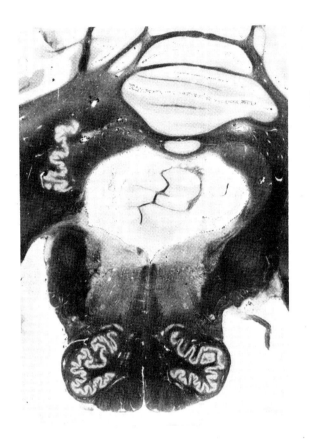

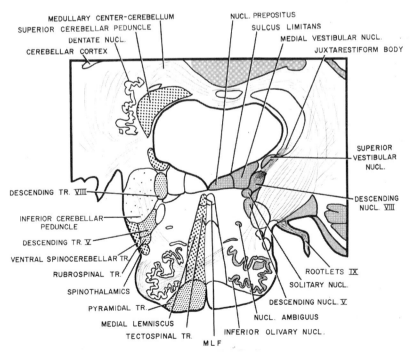

Figure 11-6. *Brain Stem at mid-medullary level. Coronal*

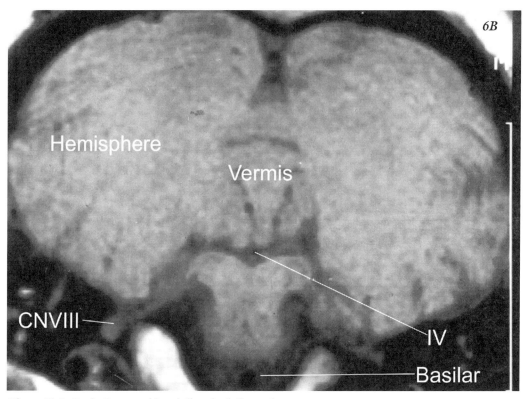

Figure 11-6. Brain Stem at mid-medullary level. Coronal

Basilar Zone (11-6A&B). The medullary pyramids are prominent at this level. The rootlets of cranial nerve XII exit lateral to the pyramid.

PONS
Blood Supply - Basilar Artery and its branches

Gross Landmarks. The hallmark of pontine levels is the prominent basis ponti that is present on the anterior surface. The cerebellum forms the posterior surface. The tegmentum takes up much less of these levels due to the massive enlarlement of the pontine basis. With the cerebellum removed the posterior surface of the tegmentum demonstrates the three-cerebellar peduncles (see Fig. 1-15) inferior cerebellar peduncle, middle cerebellar peduncle, and superior cerebellar peduncle - posterior surface and the fourth ventricle is now opened.

LEVEL: LOWER PONS AT LEVEL OF FACIAL NERVE AND FACIAL COLLICULUS (FIG. 11-7) VENTRICLE EQUIVALENT LEVEL MRI.

Gross Features. The bulk of this section consists of the middle cerebellar peduncle (brachium pontis) and the cerebellum. In *Figure 11-7* the bulk of the cerebellum has been removed. The fourth ventricle narrows as it nears the cerebral aqueduct. The medullary pyramids are present at the anterior surface. Note that the inferior olive is no longer present.

VENTRICULAR ZONE
Motor Cranial Nerve Nuclei (Fig 11-7B).

1. The nucleus of nerve VII is conspicuous at the lateral margin of the tegmentum. This nucleus innervates the muscles of facial expression. In the pons, in close proximity to the ventricular floor, the rootlet of this nucleus swings around the medial side of the nucleus of nerve VI forming the internal genu of nerve

VII. These fibers then pass lateral to the nucleus of nerve VII and exit from the substance of the pons on the anterior surface in the cerebellopontine angle.

2. The nucleus of nerve VI innervates the lateral rectus muscle of the eye. The rootlet leaves the anterior surface of the nucleus and has the longest intracerebral path of any nerve root. The fibers finally exit on the anterior surface of the brain stem near the midline at the pontomedullary junction. The close proximity of the nucleus of nerve VI to the rootlets of nerve VII demonstrates why any involvement of the nucleus of nerve VI usually produces a concomitant alteration in function of nerve VII. The nucleus of nerve VI and the internal genu of the rootlets of nerve VII form the prominent facial colliculus on the floor of the fourth ventricle.

Cerebellar Pathways (fig 11-7 B). At this level the middle and inferior cerebellar peduncles form the lateral walls of the ventricle as well as the bulk of the cerebellar medullary center. The ventral spinocerebellar tract is seen lateral to the superior cerebellar peduncle that it enters and then follows back into the cerebellum. The superior cerebellar peduncle consists primarily of axons carrying impulses from the dentate nucleus of the cerebellum to the red nucleus and the ventral lateral nucleus in the thalamus. This tract is also called the dentatorubrothalamic tract. The tractus uncinatus connects the deep cerebellar nuclei with the vestibular nuclei and reticular formation bilaterally.

Lateral Zone

Sensory Cranial Nerve Nuclei (fig 11-7B). Pain and temperature are conveyed from the head. At pontine levels the descending nucleus of nerve V is small, while the tract is large. The superior vestibular nucleus is seen at the lateral margin of the ventricle with primary vestibular fibers present in its substance. Auditory sensation at this level is related to the superior olive that is seen inferior to the motor nucleus of nerve VII. The superior olive is one of the secondary nuclei in the auditory pathway. The auditory fibers are seen accumulating inferior to the superior olive and cutting through the medial lemniscus to form the lateral lemniscus.

White Matter (fig 11-7B). At this pontine level, some of the ascending tracts are starting to move more laterally. The medial lemniscus will shift laterally and approach the rubrospinal and spinothalamic tracts.

Tactile discrimination and proprioception from the limbs, thorax, pelvis, abdomen, and neck are conveyed by fibers in the medial lemniscus, which is seen posterior to the corticospinal tract.

Pain and temperature from the head are conveyed by the secondary trigeminothalamic tracts, which are found in the medial lemniscus and ascend to the nucleus ventralis posteromedialis in the thalamus.

Pain and temperature from the extremities, thorax, abdomen, pelvis, and neck are carried in the lateral spinothalamic tract, which is near the anterior surface of the pons, close to the medial lemniscus.

The anterior spinothalamic tract that is mixed in with the lateral spinothalamic tract carries light touch.

The gustatory fibers from the tongue are found in the solitary tract. The secondary fibers ascend bilaterally in the medial lemniscus. Fibers carrying visceral sensations also synapse in the solitary nucleus.

Medial Zone.

The raphe or the medial reticular nuclei are present in this level. The medial lemniscus at this level contains secondary fibers from the dorsal columns and the spinothalamic, trigeminothalamic, and solitary tracts. The medial longitudinal fasciculus and tectospinal tracts are still present near the midline in the floor of the fourth ventricle with the medial longitudinal fasciculus conspicuous under the floor and the tectospinal tract below it.

Central Zone (Fig 11-7 A&B). In the center of the pontine tegmentum, the main efferent ascending tract of the reticular system, the central tegmental tract, occupies the bulk of

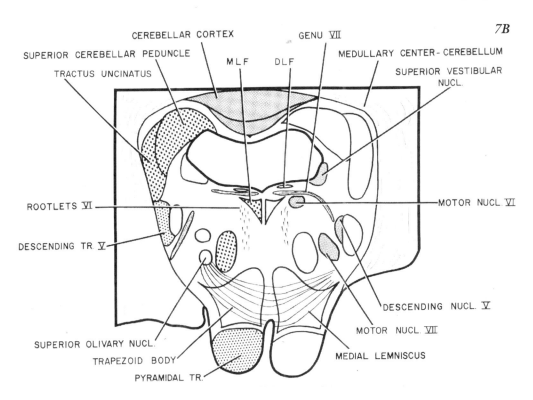

Figure 11-7. Brain stem at pontine level of Cranial Nerve VII. Coronal

the reticular formation.

Internal to the nucleus of VI is the parapontine reticular nucleus that coordinates eye movements. Fibers descend the opposite

TABLE 11-6. MAJOR CONTENTS OF TEGMENTAL ZONES OF THE PONS

Ventricular Zone: Nuclei = Median eminence with facial colliculus (nucleus of VI), superior vestibular nucleus of VIII.
Lateral Zone: Nuclei-main sensory of V, descending of V; Tracts – spinal of V. Tracts = spinothalamic and rubrospinal.
Central Zone: Nuclei =reticular formation, facial nucleus, parapontine reticular nucleus, superior olive and lateral lemniscus. Tracts = crossing of lateral lemniscal nucleus of cochlea.
Medial Zone: Nuclei = Raphe nuclei of the reticular formation. Tracts = MLF, tectospinal, crossing auditory fibers, medial lemniscus.
Basilar Zone: Middle cerebellar peduncle with pontine gray that forms the bulk of the pons at these levels. Tracts = Corticospinal and corticobulbar.

frontal eye fields, synapse here and then connect via the MLF to the nuclei of IV and III.

Basilar Zone (fig 11-7 B). The descending corticospinal and corticonuclear pathways are nearly obscured by the fibers of the middle cerebellar peduncle and the pontine gray. Table 11-6 reviews the contents of the pontine tegmentum.

LEVEL : UPPER PONS AT THE MOTOR AND MAIN SENSORY NUCLEI OF NERVE V (Fig. 11-8)

Gross Landmarks. The bulk of this section consists of the middle cerebellar peduncle and cerebellar cortex. The fourth ventricle begins to narrow as it nears the cerebral aqueduct.

Ventricular Zone. In this level there are no motor cranial nerve nuclei on the floor of the ventricle. With the nuclei in the ventricular floor being for visceral functions.

Lateral Zone

Motor Cranial Nerve Nuclei *(Fig 11-8A)*. *The rootlet of CN V and motor nucleus of* nerve V is conspicuous bilaterally in the reticular formation, medial to the sensory nucleus and the entrance of the root of nerve V. Each of these nuclei provides innervation to the ipsilateral muscles of mastication.

Sensory Cranial Nerve Nuclei (fig 11-8A). Tactile discrimination from the face is carried by the trigeminal root into the main sensory nucleus of nerve V, seen lateral to the intrapontine root of nerve V. The secondary fibers originate from this nucleus and ascend crossed and uncrossed. The crossed fibers run adjacent to the medial lemniscus (ventral trigeminothalamic), while the uncrossed fibers run in the dorsal margin of the reticular formation (dorsal trigeminothalamic). Both fiber pathways terminate in the ventral posterior medial nucleus of the thalamus.

Proprioception from the majority of the muscles in the head is carried in by the mesencephalic root and ends in the mesencephalic nucleus of nerve V, located lateral to the walls of the fourth ventricle in the upper pontine and midbrain levels. Note cranial nerves V & VII in the MRI 11-8B. These neurons are the only primary cell bodies (dorsal root ganglion equivalent) in the CNS. These axons synapse on the Chief Nucleus of V. The jaw jerk is a monosynaptic stretch reflex. The receptor is the mesencephalic nucleus of nerve V and the effector is the motor nucleus of nerve V.

The superior olive is one of the nuclei in the auditory system. Auditory fibers are seen running through the superior olive or inferior to it cutting through the medial lemniscus and crossing the midline, forming the trapezoid body. The fibers are then found lateral to the medial lemniscus and are known as the lateral lemniscus. The pontine nuclei are conspicuous in the basilar portion of the pons.

White Matter (Fig 11-8A). At this level the medial lemniscus has moved from the midline to a more lateral position, and many of the ascending sensory systems are now either in the medial lemniscus or adjacent to it.

Medial Zone. The raphe nuclei of the reticular formation are evident in this level. The medial longitudinal fasciculus is larger in this

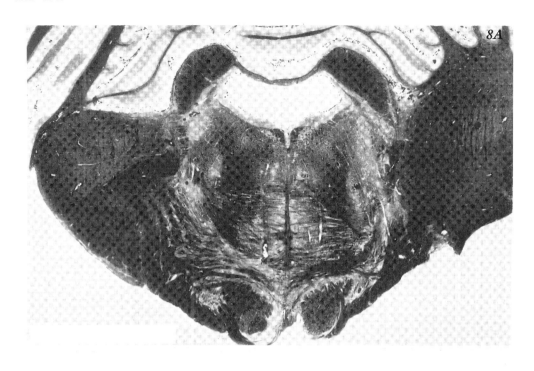
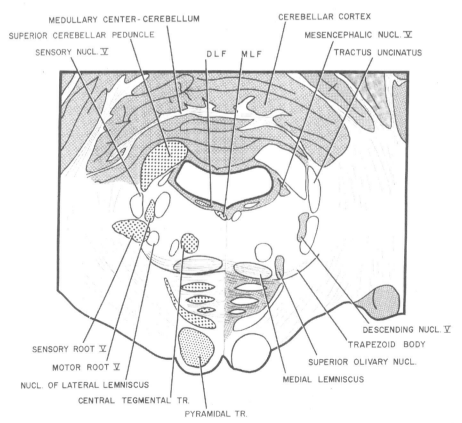

Figure 11-8. Brain stem at pontine level of Cranial Nerve V. Coronal

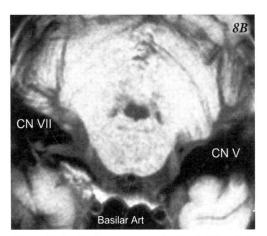

Figure 11-8. Brain stem at pontine level of Cranial Nerve V. Coronal

section between the motor nuclei of cranial nerves III and VI because it contains many vestibular and cerebellar fibers necessary for accurate eye movements. The tectospinal fibers always lie below the MLF.

Central Zone. In the reticular formation the central tegmental tract is conspicuous.

Basilar Zone. The corticospinal tract is found in the pontine gray and white matter.

MIDBRAIN
Blood Supply-Basilar Artery and Posterior Cerebral Arteries

Gross Landmarks. On the anterior surface of the midbrain we find the principal landmark of this level, the cerebral peduncles with the third cranial nerve exiting from the medial surface of the cerebral peduncles in the interpeduncular fossa. On the posterior surface of the midbrain we find the corpora quadragemini, the superior and inferior colliculi. The fourth cranial nerve exits from the posterior surface of the midbrain and the junction with the pons..

In a coronal section through the midbrain one finds the red nucleus in the tegmentum of the midbrain. *(Fig. 11-9B).* Also in the inferior levels in the tegmentum just before the red nucleus one finds the crossing of the superior cerebellar peduncle.

LEVEL: INFERIOR COLLICULUS AND PONTINE BASIS (Fig. 11-9)

In this section the midbrain forms the roof and floor of the narrow cerebral aqueduct, while the pons makes up the basilar portion. The roof is the inferior colliculus, an important nucleus in the auditory pathway.

TECTUM

Inferior colliculus (Fig 11-9 A&B). The gray matter of the inferior colliculus forms the tectum at this level. The inferior colliculus is divided into three nuclei, a large central nucleus, and a thin dorsal nucleus the paracentral or cortical nucleus, and an external nucleus. The central and cortical nucleus functions as a relay center for the cochlea. The external nucleus functions for the acusticomotor reflexes through the tectospinal pathway. The brachium of the inferior colliculus carries auditory information onto the medial geniculate nucleus of the diencephalon. The tectospinal pathway descends from this region and terminates on motor nuclei of the lower cranial nerves and upper cervical ventral horn cells.

TEGMENTUM

Ventricular Zone *(Fig 11-9 A&B).* The periaqueductal gray is conspicuous in midbrain levels. Descending and ascending tracts associated with the visceral brain are found here. In the midbrain, the lower half of the periaqueductal gray and some other nuclei in the tegmentum are part of the midbrain limbic area. This zone is important in our level of attentiveness. Bilateral lesions to the periaqueductal gray and midbrain tegmentum usually produce comatose patients.

Motor Cranial Nerve Nuclei *(fig 11-9 A).* The nucleus of cranial nerve IV is seen indenting the medial longitudinal fasciculus in lower regions of the midbrain. The axons of this nucleus pass posteriorly (unique for a cranial nerve) and exit the brain at the midbrain-pontine junction where the fibers then proceed anteriorly to reach the superior oblique muscle through the superior orbital fissure.

Sensory Cranial Nerve Nuclei *(fig 11-9A).* Proprioception is conveyed from the muscles in the head and neck. The mesencephalic nucleus of nerve V is located at the

11-16 CHAPTER 11

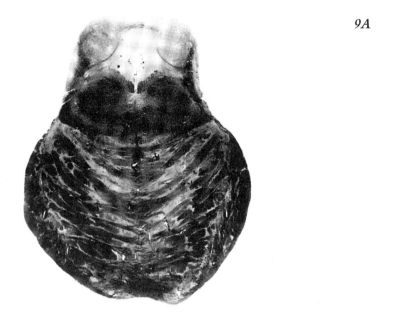

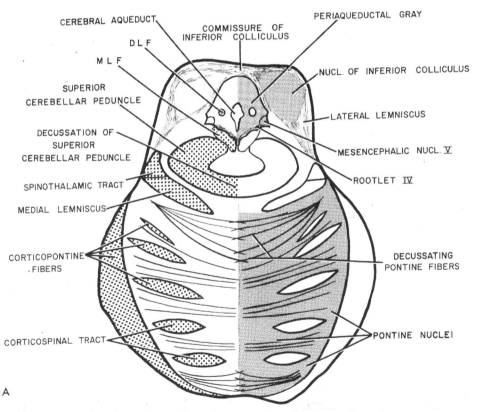

Figure 11-9 Brain stem at inferior collicular level. Coronal

FUNCTIONAL ANATOMY OF THE BRAIN STEM 11-17

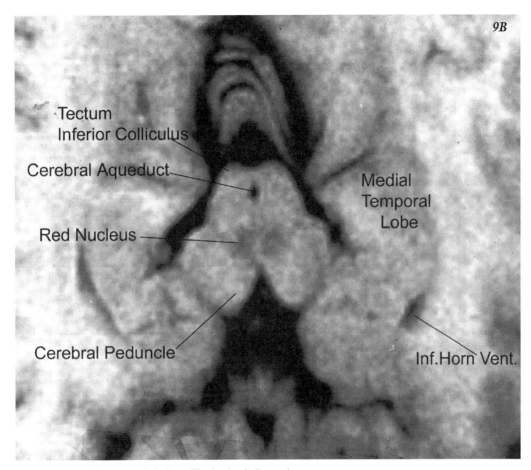

Figure 11-9 Brain stem at inferior collicular level. Coronal

lateral margin of the periaqueductal gray. Remember these are the only primary sensory neurons in the CNS; their axons join the medial lemniscus and ascend to the ventral posterior medial nucleus in the thalamus. This nucleus controls the superior oblique muscle of the eye.

Lateral Zone *(Fig 11-9A)*. At this level the medial lemniscus and the other major ascending tracts are stretched over the lateral surface of the tegmentum of the midbrain. The spinothalamic and trigeminothalamic fibers are found lateral to the medial lemniscus while the auditory fibers are located above them. Thus the following sensory modalities for the entire body are located here: tactile discrimination, proprioception, pain and temperature, gustatory and visceral sensations.

Auditory Fibers *(fig 11-9A)*. The lateral lemniscus is seen superior to the medial lemniscus. These secondary auditory fibers are now entering the inferior colliculus. Many of these fibers synapse and then continue up into the medial geniculate as the brachium of the inferior colliculus.

Medial Zone *(fig 11- 9 A)*. The raphe nuclei of the reticular formation are very conspicuous in these levels. The medial longitudinal fasciculus is prominent in the floor of the ventricle as it nears cranial nerve III; the fibers located within it are important in coordinating ocular movements.

Central Zone *(fig 11- 9 A)*. This zone in the lower tegmentum of the midbrain is nearly completely filled with the crossing fibers of the superior cerebellar peduncle. Ultimately, all these fibers decussate and continue up to the red nucleus where some fibers synapse, but the majorities of these axons bypass the red nucleus and terminate on the ventral lateral thalamic nuclei.

Basilar Zone *(fig 11-9-A)*. The bulk of this region is taken up by the pontine gray and descending fibers from the cerebrum to the pontine gray, the corticopontine system. The corticospinal and corticobulbar fibers are also present here.

LEVEL: SUPERIOR COLLICULUS AND PONTINE BASIS (Fig. 11-10)

TECTUM

Superior Colliculus (Fig 11-10). The superior colliculi form the rostral portion of the midbrain tectum. The parenchyma of the superior colliculus is organized in layers. There are four layers containing gray and white matter from inside out: stratum zonale, stratum cinereum, and stratum opticum and stratum lemnisci. The superficial layers of the superior colliculus receive their input from the retina and visual cortex with the contralateral upper quadrant medially and the contralateral lower quadrants laterally. In contrast the deeper layers receive their input from polysensory sources including the cerebellum, inferior colliculus, spinal cord, reticular formation gracile, cuneate, trigeminal nuclei, and visual regions. A portion of the tectospinal pathway originates at this level and descends in the lower brain stem to terminate on the lower cranial nerves and ventral horn cells of the spinal cord. (Remember the tectospinal and tectobulbar fibers terminate on the same motor nuclei as the corticonuclear and corticospinal pathways). The zone of the tectum just above the superior colliculus, the pretectal zone is important in light reflexes.

TEGMENTUM

Gross Features. In this section the roof and floor are formed by the midbrain, while the basilar portion is made up of peduncles and the pons. The roof is the superior colliculus, an important station in the visual pathway. The cerebral aqueduct forms the narrow ventricular lumen. *(Fig. 11-10B)*. The superior colliculus is a laminated structure important in relating eye movements and body position. The tectospinal tract originates from its deepest layer and provides connections onto certain cranial and spinal neurons.

Ventricular Zone

Sensory Cranial Nerve Nuclei *(fig 11-10 A)*. Proprioception is conveyed from the muscles in the head and neck by the mesencephalic nucleus of nerve V, located at the lateral margin of the periaqueductal gray in the superior collicular levels as well as in the upper pontine and inferior collicular levels.

Motor Cranial Nerve Nuclei *(fig 11-10 a)*. Cranial nerve III is visible in the floor of the cerebral aqueduct, adjacent to the medial longitudinal fasciculus. This nerve supplies the medial, inferior, and superior rectus muscles and the inferior oblique and superior levator muscles of the eyelid. The Edinger-Westphal nucleus of nerve III is also present; it provides preganglionic parasympathetic innervation to the constrictor muscle of the pupil via the ciliary ganglion.

Lateral Zone.

White Matter *(Fig 11-10A)*. Medial Lemniscus. At this level in close proximity to the medial lemniscus includes most of the ascending sensory fibers (Table 11-7). This tract is stretched out over the inferior and lateral surface of the tegmentum, with the fibers that mediate pain and temperature from the extremities, thorax, abdomen, and pelvis located at its superior extent. The trigeminal fibers form its middle portion, and the fibers that mediate tactile, proprioceptive, and visceral sensations are placed medially.

Anterolateral Column.

Lateral Spinothalamic. Pain and temperature from the limbs, thorax, and pelvis are found in this pathway.

Anterior spinothalamic. Light touch fibers from the limbs, abdomen, and neck are seen in close proximity to the lateral spinothalamic pathway.

Auditory Pathway. The fibers from the

FUNCTIONAL ANATOMY OF THE BRAIN STEM

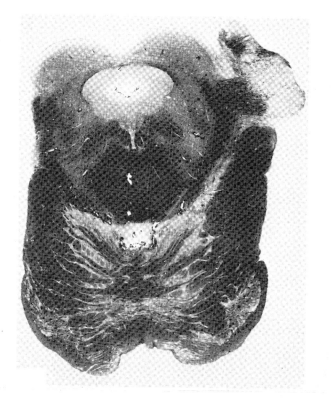

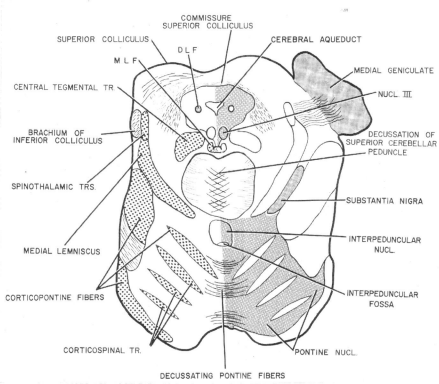

Figure 11-10. Brain stem at superior collicular level. Coronal

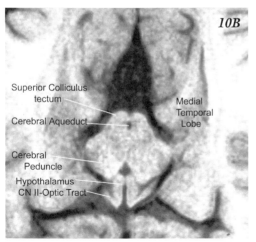

Figure 11-10. Brain stem at superior collicular level. Coronal

inferior colliculus (called at this level the brachium of the inferior colliculus) are seen at the inferior surface of the superior colliculus. On the left side they enter the medial geniculate nucleus of the metathalamus at which point they reach their final subcortical center. There is a complete discussion of the auditory and vestibular pathways in the Special Sensory Section found on the CD ROM.

Cerebellar Fibers. The superior cerebellar peduncle continues to cross in the tegmentum of the midbrain; just above this level many of these fibers synapse in the red nucleus, while others will continue to the ventral lateral nucleus in the thalamus.

Central Zone *(fig 11-10A)*. The central tegmental tract is conspicuous in the midbrain tegmentum in the reticular formation. The red nucleus is found in the medial edge of the reticular formation. In the human it consists of a small magnocellular region and a large parvocellular region. Fibers from all cerebellar nuclei, but especially the dentate, synapse here.

Corticorubral fibers run bilaterally from the motor cortex synapse here also terminate here in a somatotopic relationship. Fibers from the red nucleus terminate directly or indirectly in the cerebellum and on many of the motor nuclei in the brain stem and spinal cord, which also receive input from the corticospinal and corticonuclear pathways. The red nucleus has minimal if any projections in the human onto the thalamus, although it most likely influences the cerebellar input onto the thalamus.

The substantia nigra is also located in this level just behind the cerebral peduncle. Its functions are discussed in the relationship between the basal ganglia and movement (see Chapter 18).

Medial Zone *(fig 11-10 A)*.

The serotinergic raphe nuclei of the reticular formation are present in this level. The medial longitudinal fasciculus in the floor of the cerebral aqueduct is conspicuous and connected across the midline. The oculomotor complex indents the MLF and receives ascending fibers from cranial nerves VI and VIII through the MLF.

The tectospinal fibers at this level are primarily from the superior colliculus.

Basilar Zone *(Fig 11-10 A&B)*

Cerebral Peduncles. At this level, the frontopontine fibers, which occupy the most medial part of the cerebral peduncles, enter the pons. *Figure 11-11* is a stained section through the cerebral peduncles demonstrating the relationship between the peduncles, and the III cranial nerve. The corticospinal fibers and many of the corticobulbar fibers are still in the peduncle. In Figure 11-10B, a coronal section through the brain, the relationship of the cerebral peduncle to the medial temporal lobe demonstrates that the cerebral peduncles are adjacent to the medial temporal lobe. This may be significant if herniation of the temporal lobe produces compression of the cerebral peduncle and cranial nerve III with the accompanying cranial nerve and upper motor neuron signs. Table 11-7 lists the significant structures in the tegmental zones of the midbrain.

III. FUNCTIONAL CENTERS IN THE BRAIN STEM

Now that the anatomical features of the brain stem have been discussed, it is appropriate to identify some important functional centers located in the brain stem and diencephalon. During this discussion, it will become evident that the cranial nerves are the pivotal point for many of these activities.

A. RETICULAR FORMATION

The central core of the medulla, pons, and midbrain tegmentum consists of the reticular formation. Upon microscopic examination, this region is seen to consist of groupings of neurons separated by a meshwork of medullated fibers. With a Golgi neuronal stain the Golgi Type I cells with long axons, neurons are shown with their dendrites extending

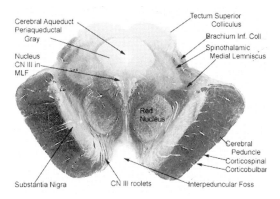

Figure 11-11. Brain stem at level of cerebral peduncles and III cranial nerve. Weil stain. Coronal section

transversely and the axons bifurcating into ascending and descending branches, which run throughout the system. Each neuron receives input from at least 1000 neurons, and each neuron connects to as many as 10,000 neurons in the reticular formation.

The reticular formation of the brain stem blends inferiorly into lamina VII of the cord and while superiorly it is continuous with the hypothalamus and dorsal thalamus of the diencephalon. Many nuclei have been identified in the reticular formation.

Functionally, the nuclei can be divided into cerebellar and noncerebellar nuclei. (Table 11-8). The cerebellar portion of the reticular formation includes primarily the lateral and paramedian nuclei of the medulla, and the tegmental nucleus of the pons.

The noncerebellar portion nuclei are divided anatomically into three columns: the raphe,

TABLE 11-7. MAJOR CONTENTS OF TEGMENTAL ZONES IN MIDBRAIN

Ventricular Zone: Cranial nerve nuclei are Nuclei: IV (Inferior collicular level), and III (Superior Collicular Level).

The parasympathetic nucleus of Edinger-Westphal, associated with cranial nerve III, to the pupillary constrictor is also present. The interstitial nucleus of Cajal, and the nucleus of Darschewtiz important pupillary reflexes are also found here.

Tracts= MLF and descending autonomics and ascending reticular fibers.

Lateral Zone: Tracts = Medial lemniscus, lateral lemniscus, spinothalamics and brachium of inferior colliculus, with a confluens of the major ascending tracts.

Central Zone: Nuclei = Reticular formation with red nucleus and origin of rubrospinal tract. Substantia nigra present just posterior to the cerebral peduncle.

Tracts = central tegmental and origin of rubrospinal

Medial Zone: Nuclei = Raphe nuclei which forms the serotinergic pathway.

Tracts - MLF, with tectospinal and decussation of superior cerebellar peduncle and rubrospinal tract. The tectospinal and rubrospinal tracts originate in the midbrain.

Basilar Zone. Tracts = Cerebral peduncle containing from medial to lateral frontopontine, corticonuclear, corticospinal, parietopontine, temperopontine, and occipitopontines.

central, and lateral groupings. Most of the nuclei in the reticular formation consist of large cells with ascending and descending axons. Functionally, the reticular formation in the medulla is especially important since the descending reticulospinal fibers and much of the ascending reticular system originate there.

Descending Reticulospinal System. The *nucleus reticularis gigantocellularis* is found at

TABLE 11- 8 : FUNCTIONAL GROUPINGS OF NUCLEI IN THE RETICULAR FORMATION

Non-Cerebellar	Cerebellar Related
Raphe, central, and lateral in tegmentum of brain stem	Lateral and paramedian nuclei of medulla and tegmental nucleus of the pons

the rostral medullary levels, dorsal and medial to the inferior olive. This nucleus gives origin to much of the lateral reticulospinal tract, which is primarily an ipsilateral tract, running in the lateral funiculus of the spinal cord in all levels and terminating on internuncial neurons. The axons from the nucleus reticularis pontis oralis and caudalis form much of the medial reticulospinal tract, which runs in the anterior funiculus in all levels and terminates on internuncial neurons.

The *lateral reticular nucleus* is located in the lateral margin of the reticular formation dorsal to the inferior olive, while the ventral reticular nucleus is found in the caudal end of the medulla, dorsal to the inferior olive. These two nuclei, in conjunction with the nuclei in the pons and midbrain, form much of the ascending reticular fibers in the central tegmental tract distributing to neurons in the thalamus (intralaminar and reticular), hypothalamus, and corpus striatum. The paramedian reticular nucleus is found near the midline at mid-olivary levels, dorsal to the inferior olive, and provides direct input into the anterior lobe of the cerebellar vermis.

Input to Reticular Formation. The reticular formation receives information via the ascending spinal tracts (spinotectal, spinoreticular, spinothalamics, and spinocerebellar), from the brain stem itself (olivoreticular, cerebelloreticular, and vestibulospinal), from the cerebral hemispheres (corticoreticular), from the basal ganglia and hypothalamus. The cranial nerves are another important source, especially nerves I, II, V, VII, VIII, and X for sensory information that appears to project most heavily onto the central nuclei. The hypothalamus and striatum also project to the reticular system via the dorsal longitudinal fasciculus on the floor of the cerebral aqueduct and fourth ventricle and the more diffuse descending fiber systems in the core of the reticular formation.

Output. The medial lemniscus is a specific point-to-point relay system with few synapses (a closed system), while the fiber tracts of the reticular system are a multisynaptic non-specific system (an open system). (Table 11-9).

The central tegmental tract is the principal fiber tract of the reticular formation. Its descending portions are located in the medial tegmentum, and its ascending portions are located in the lateral tegmentum. The ascending system projects to the thalamic intralaminar and reticular nuclei, hypothalamus, basal ganglia, substantia nigra, and red nucleus. The descending system synapses via the reticulospinal tract onto interneurons that mediate their effects through alpha and gamma motor neurons in the spinal cord and via autonomic pathways and cranial nerves onto visceral neurons.

Ascending Reticular System. This fiber system is the structural and functional substrate for maintaining consciousness. It also receives proprioceptive, tactile, thermal, visual, auditory, and nociceptive information via the spinal and cranial nerves. Many sensations (cutaneous, nociceptive, and erotic) activate the system. The reticular system is functionally important in controlling our "posture" by its reflex relationship to the position of our body in space and in controlling our internal milieu by maintaining the stability of our viscera. The sensory information ascends via the central tegmental tract into the limbic-midbrain area from which information can be more directly passed into the thalamus and hypothalamus. This midbrain to diencephalon

TABLE 11-9: MAJOR PATHWAYS OF THE RETICULAR FORMATION

Tract	Reticular Origin	Termination of Pathway
Lateral Reticulospinal	Gigantocellularis	spinal cord & autonomic interneurons
Medial Reticulospinal	Pontis oralis & caudalis	spinal cord & autonomic interneurons
Central tegmental	Lateral & ventral	halamic-intralaminart & reticular, hypothalamus, corpus striatum, substanita nigra

to telencephalon circuit seems especially important in determining our level of consciousness and motivation.

The reticular formation is also important in controlling posture and orientation in space. Stimulation of the caudal medulla inhibits the knee jerk. Laterally, stimulation facilitates the knee jerk.

The lateral part of the reticular formation is the receptor area, while the medial portion is the effector zone and the origin of the central tegmental and reticulospinal tracts. Many of the functions vital to the maintenance of the organism are found in the medulla.

Neurochemically Defined Nuclei in the Reticular Formation Effecting Consciousness.

1. Cholinergic Nuclei. – a. located in the dorsal tegmentum of the pons and midbrain, in the mesopontine nuclei, and in the basal forebrain region and b project diffusely to the cerebral cortex through the thalamus and have a modulating influence on cerebral cortical activity and wakefulness.

2. **Monoamine Nuclei.** In the reticular formation we find cells containing norepinephrine and serotonin.

a. The norepepinephrine containing cells are found in the locus ceruleus (blue staining) in the upper pons and midbrain and they project widely upon nuclei in the spinal cord, brain stem, thalamus, hypothalamus and corpus striatum which are important for maintaining attention and wakefulness.

b. Serotinergic nuclei. These nuclei are found in the raphe of the medial tegmental zone in the medulla, pons and midbrain. The nuclei in the pons and medulla project onto the spinal cord and brain stem while the nuclei in the upper pons and midbrain project onto to the thalamus, hypothalamus, corpus striatum and cerebral cortex. The serotinergic system facilitates sleep. An area outside of the reticular formation, the histamine containing area of the posterior hypothalamus, is also important in maintaining wakefulness.

B. RESPIRATION CENTERS

Respiration is under the control of neurons in the respiratory center in the upper medulla and pons. Sensory fibers from the lungs ascend via cranial nerve X and enter the solitary tract and proceed onto the neurons in the reticular formation in the medulla and pons. Specific regions in the medulla control either inspiration or expiration. The medullary respiratory center itself is responsive to the carbon dioxide content in the blood. Increased carbon dioxide produces increased respiration and decreased carbon dioxide produces decreased respiration.

In the pons the pneumotaxic centers are related to the frequency of the respiratory response. These centers play onto the medullary respiratory center to determine the respiratory output. The axons from cells in the medullary reticular center descend to the appropriate spinal cord levels to innervate the diaphragm and the intercostal and associated muscles of respiration. The vagus nerve itself provides preganglionic innervation of the trachea, bronchi, and lungs.

The respiratory response is also modified by cortical and hypothalamic control; i.e., emotionally stimulated individuals breathe more rapidly because of the hypothalamic influence on the brain stem and spinal cord. The actual respiration occurs with the intercostal muscles, the accessory muscles, and the lungs expanding and contracting in concert with associated vascular changes (autonomic nervous system).

C. CARDIOVASCULAR CENTERS

The *carotid body* (innervated by sensory fibers of cranial nerve IX) is found distal to the bifurcation of the common carotid artery, and the aortic body, innervated by the sensory fibers of nerve X, is found near the origin of the subclavian arteries. These carry information on blood pressure into the solitary tract and then into the medullary respiratory centers. These sites are sensitive to oxygen-carbon dioxide pressure; when oxygen is reduced, ventilation increases; when oxygen increases, ventilation decreases. Neural control over the tone of the arterioles is exercised through vasoconstrictor fibers (and sometimes vasodilator fibers). A vasomotor center in the medulla extends from the midpontine to the upper medullary levels. In the lateral reticular formation of the medulla, stimulation elicits an increase in vasoconstriction and an increase in heart rate--pressor center. In the lower medulla, stimulation of the depressor center produces a decrease in vasoconstriction and a decrease in heart rate. The carotid body (cranial nerve IX) and the aortic body (cranial nerve X) contain specialized pressure receptors stimulated by an increase in the size of these blood vessels. This information runs via nerves IX and X to the medulla and depresses activity. Most sensory nerves, cranial and spinal, contain some nerve fibers, which when stimulated, cause a rise in arterial pressure by exciting the pressor region and inhibiting the depressor centers. Higher centers also influence respiratory and vascular centers in the medulla and alter blood flow, depending on the psychic state; i.e., mental activity decreases peripheral blood (constricts vessels), while emotional states can produce increased blood flow, blocking vasodilation.

D. DEGLUTITION

The act of swallowing starts volitionally but is completed by reflex activity.

1. The food is first masticated (motor nucleus of nerve V) and reduced to smaller particles, called the bolus, which is lubricated by saliva from the salivary glands (nerves VII and IX).

2. The bolus is propelled through the oral pharyngeal opening when the tongue is elevated against the soft palate (nerve XII), and the facial pillars are relaxed.

3. The oral pharyngeal pillars close and the superior and middle pharyngeal constrictor muscles (ambiguous nucleus of nerve X) force the bolus along into the laryngeal pharynx where the pharyngeal walls contract, closing the opening superiorly.

4. At the same time, the tracheal opening is closed by the epiglottis and glottis, as the larynx, trachea, and pharynx move up. This movement is noted externally as the bobbing of the thyroid eminence (Adam's apple).

5. Finally the inferior constrictor muscle contracts, pushing the bolus into the esophagus where peristaltic waves and gravity carry it through the esophagus.

Cranial nerves V, IX, X, and XII perform the motor part of the activity, while nerves V, VII, IX and X form the sensory function.

E. VOMITING

Vomiting is produced by many stimuli and is usually a reflex activity. The vomitus is composed of the gastric contents.

1. Caused commonly by irritation of the oropharynx, gastrointestinal mucosa, and genitourinary and semicircular canals. Stimulation of the vestibular system including the semicircular canals and the nerve itself may also produce vomiting.

2. The afferent nerves travel via the vagus and glossopharyngeal tracts into the solitary tract. After synapsing on the dorsal motor nucleus of cranial nerve X, the information is conveyed via the vagus nerve to the stomach. Impulses are also passed down to the cervical and thoracic level onto the ventral horn cells that control the simultaneous contraction of the intercostal, diaphragmatic, and abdominal musculature.

3. Nausea and excessive salivation precede the deep inspiration associated with retching. The glottis is closed and the nasal passages are sealed off. The descent of the diaphragm and the contraction of the abdominal muscles exert the pressure that causes the stomach to contract in a direction contrary to normal peristalsis and forces vomitus through the relaxed cardia of the stomach and into the esophagus.

F. EMETIC CENTER

The area postrema is found in the caudal end of the fourth ventricle above the obex of the medulla. This area is very vascular and contains many venous sinuses. The blood-brain barrier is lacking in this region. The vagal, glossopharyngeal, and hypoglossal nuclei are strongly interconnected with the area postrema. This region is very sensitive to changes in pressure or to drugs whose passage into this area is not inhibited by any blood-brain barrier.

G. COUGHING.

Irritation of the lining of the larynx or trachea produces coughing. The stimulus is picked up by free nerve endings associated with the internal laryngeal branch of cranial nerve X, which carries it up into the solitary tract, following the same pathway as in vomiting, except that the muscles contract alternately rather than simultaneously with the intercostal muscle contracting suddenly.

IV. LOCALIZATION OF DISEASE PROCESSES IN THE BRAIN STEM

Pathological processes affecting the brain stem can usually be well localized from an anatomical standpoint. This is a direct consequence of the fact that these disease processes in general involve a particular combination of cranial nerves (as in acoustic neuroma at the cerebellar pontine angle) or produce a particular combination of ipsilateral cranial nerve dysfunction and contralateral long tract findings (as in the midbrain infarction of Weber's syndrome which results in a combination of ipsilateral third nerve dysfunction and contralateral pyramidal tract findings).

Moreover, when anatomical-pathological correlation is considered, the brain stem, of all sites in the central nervous system, provides the best example of how the anatomical pattern of involvement may allow the prediction of the actual pathological processes. The student has, of course, already encountered several examples of this phenomenon at the level of the spinal cord. Thus, the clinical finding of a selective dissociated loss of pain and temperature in a cape-like distribution over the shoulders and upper extremities implies an anatomical process in a pericentral location involving the anterior commissure. One could readily deduce that an intrinsic pathological process was present and, moreover, could assume that in most cases the pathological process was that of syringomyelia.

At the level of the brain stem such examples of close correlation occur frequently. Thus, the combined progressive involvement of the ipsilateral cranial nerve VIII (auditory and vestibular), nerve VII (facial), and nerve V (trigeminal), manifested by deafness, tinnitus (a buzzing or ringing sound in the ear), dizziness, peripheral facial paralysis, and unilateral facial numbness, not only indicates the location of the pathology (the cerebellar pontine angle) but also suggests the actual nature of the pathology: acoustic neuroma (tumor arising from Schwann cells of the sheath of nerve VIII). While other pathological lesions may occur at this site, they are much less common.

There are several possible approaches to the problem of localization of disease processes in the brain stem. We will first consider a number of guidelines for localization. We will then consider in greater detail the particular anatomical syndromes and the various types of pathology in terms of specific extrinsic and intrinsic syndromes. When speaking of the anatomical syndromes of the brain stem in clinical neurology, we will limit our considerations to the medulla oblongata, pons, and midbrain.

A. GUIDELINES FOR LOCALIZING BRAIN STEM DISEASE

1. Mental status is not directly involved (except that lesions involving the tegmentum of the upper brain stem – reticular formation etc, may alter level of consciousness).

2. No muscle atrophy is presented, except that relevant to local involvement of cranial nerves.

3. Limb weakness if present involves central control and spasticity.

4. Deep tendon reflexes are increased in a bilateral or unilateral manner. A unilateral or bilateral sign of Babinski maybe present.

B. EFFECTS OF INTERRUPTING LONG MOTOR PATHWAYS- CORTICOSPINAL AND CORTICOBULBAR PATHWAYS

1. The pyramidal (corticospinal) tracts - decussate low in the medulla.

- Damage to the pyramidal tract above this decussation will produce contralateral upper motor neuron signs. Within the brain stem, the left and right pyramidal tracts are situated closest together at the level of the pyramidal decussation and are most widely separated at the level of the upper midbrain-internal capsule. Lesions in the medulla will often produce bilateral effects; in the midbrain, unilateral or bilateral effects.

Throughout their extent in the brain stem, the corticospinal tracts are located in a ventral (or basilar) location. Lesions limited to the dorsal (tegmental or tectal) portions of the brain stem then are less likely to produce signs of pyramidal tract involvement.

2. Corticobulbar fibers - descend to the cranial nerve motor nuclei in close proximity to the corticospinal tracts.

- lesions here produce a supranuclear lesion, as regards motor control of the cranial nerves, below the lesion.
- for each cranial nerve motor nucleus (V, VII, and XII, ambiguous) decussate, separately, somewhat above the level of that motor nucleus.
- In humans (compared to the cat) direct corticobulbar fibers can be traced to the trigeminal, facial, hypoglossal, and spinal accessory motor neurons.
- Corticobulbar control of the cranial nerve motor nuclei is bilateral to the laryngeal, pharyngeal, palatal, and upper facial musculature.
- The motor neurons supplying the lower half of the face and the genioglossus muscle of the tongue receive predominantly unilateral corticobulbar fibers.
- The effects of a unilateral corticobulbar lesion then will be limited to an upper motor neuron type weakness of the contralateral lower facial muscles (termed supranuclear or central) and a minor weakness of the tongue (with slight deviation of the tongue on attempted midline protrusion to the side contralateral to the lesion).
- Bilateral damage to the corticobulbar fibers will produce an upper motor neuron type weakness of the muscles of the pharynx, larynx, tongue, face (upper and lower muscles), and jaw. The resulting syndrome, termed a pseudobulbar palsy, will have the qualities which the student has come to associate with an upper motor neuron lesion: impairment of voluntary control, weakness without atrophy, and a release of segmental brain stem reflex activity from higher control. Thus, the jaw jerk, a stretch reflex, will be hyperactive. Since the muscles and brain stem motor nuclei are also involved in emotional expression (consider laughing, crying, the facial expression of rage, and so forth), bilateral corticobulbar lesions will result in a loss of higher control and a release of these motor components of emotional expression.

C. CEREBELLAR DYSFUNCTION

1. Cerebellar symptomatology may be noted frequently in diseases affecting the brain stem and may provide information as to localization. On the other hand, expanding lesions of the cerebellum may secondarily compress the brain stem. Diseases affecting the cerebellum will be considered later in greater detail (Chapter 19), here we will simply indicate certain general rules at this point.

1. Diseases affecting the midline cerebellum (vermis) or its fiber systems result in a disturbance of equilibrium (sense of balance) and an ataxia (unsteadiness) of the trunk. This may be evident on sitting, standing, or walking. If an ataxia in walking occurs, we speak of an ataxia of gait.

2. Diseases affecting the lateral aspects of the cerebellum or of the fiber systems related to the lateral cerebellum produce lateralized symptoms affecting the limbs. Thus, an unsteadiness (ataxia of dysmetria) of arm and leg movements will be noted. In addition, a characteristic tremor will be present--intention tremor, in which oscillations of movement perpendicular to the line of movement occur (in the finger to nose and heel to shin tests). The lateralized unsteadiness of the lower extremity will result in an impairment of gait. The ataxia of gait will have, however, a lateral-

ized quality in the sense that the patient will tend to fall or deviate to a particular side.

3. In general, pathology affecting the cerebellar hemisphere produces symptoms ipsilateral to the side of involvement. This reflects the fact that the dorsal spinocerebellar and cuneocerebellar pathways are essentially uncrossed. Moreover, the major outflow from the cerebellar hemisphere, the dentatorubrothalamic pathway of the superior cerebellar peduncle, decussates in the upper pons before reaching the rubral and ventral lateral thalamic areas and then onto the motor cortex. The major efferent pathways from the motor cortex cross again in the pyramidal decussation. Because of this double decussation, data then from the right cerebellar hemisphere will eventually influence the anterior horn cells of the right arm and leg.

Lesions of Cerebellar Peduncles:

Inferior cerebellar peduncle and the adjacent spinocerebellar and cuneocerebellar pathways in the lateral medulla will produce ipsilateral symptoms.

- Superior cerebellar peduncle below its decussation will produce ipsilateral symptoms; above the decussation, contralateral symptoms.

4. Whether lateralized cerebellar symptoms reflect intrinsic disease of the brain stem or disease of the cerebellum will depend on the associated signs and symptoms. Thus, the ipsilateral intention tremor and ataxia seen in the lateral medullary infarction (see below chapter 16) and due to damage of the restiform body and the adjacent spinocerebellar and cuneocerebellar pathways is clearly associated with the signs and symptoms referable to the involvement of the adjacent intrinsic structures: the ipsilateral descending spinal tract of the fifth nerve; the lateral spinothalamic pathway (producing contralateral deficits in pain and temperature), the ipsilateral Horner's syndrome, and the ipsilateral nucleus ambiguous or vagal dysfunction producing hoarseness, defective gag reflex, and dysarthria. On the other hand, the ipsilateral intention tremor and ataxia due to an abscess or tumor (e.g., cystic astrocytoma) of the lateral cerebellum will not have this association but rather will be associated with the signs and symptoms of increased intracranial pressure. Common midline lesions of the cerebellum secondarily compressing the brain stem are tumors such as medulloblastomas and hemangioblastomas. The ependymoma although arising from the floor of the fourth ventricle is sometimes included in the midline cerebellar syndrome category.

D. DECEREBRATE RIGIDITY

Decerebrate rigidity is a consequence of a pathological transection of the brain stem between the vestibular nuclei and the midbrain. It is characterized by;

- a marked increase in extensor tone due to the release of the extensor facilitory area in the reticular formation in the tegmentum of the pons and midbrain and a marked increase in stretch reflexes,

- level of transection is at a point between the vestibular nuclei and the red nuclei; in general, damage has occurred at a midbrain or upper pontine level,

- pyramidal tracts and lesions of the pyramidal tracts are not involved in the phenomena. Refer to Chapter 18 for further discussion

E. INTERRUPTION OF LONG SENSORY SYSTEMS-MEDIAL LEMNISCUS AND SPINOTHALAMICS

Effects of Lesions: The spinothalamic tracts are on the lattermost margin of the brain stem-therefore in the medulla they are separate from the medial lemniscus. An intrinsic paramedian lesion is likely to produce bilateral involvement of the medial lemniscus. Since the medial lemniscus is quite separate from the lateral spinothalamic pathway, selective involvement of contralateral position and vibratory sensation, as opposed to pain and temperature sensation, may occur.

- At the level of the midbrain and thalamus, such selective involvement is likely. However, selective involvement of positional vibration as opposed to pain and temperature

F. CRANIAL NERVE DYSFUNCTION

1. Motor Cranial Nerves

a. The effects of a lesion involving the motor nuclei of a cranial nerve or of the motor fibers of a cranial nerve will be ipsilateral to the lesion. The effects will be those of a lower motor lesion: atrophy and weakness, and a loss of segmental reflex activity. With regard to the facial nerve, the effects are often referred to as peripheral. In a local lesion of the brain stem, it is the segmental ipsilateral motor findings, in association with the contralateral corticospinal and corticobulbar findings below the level of the lesion, to specific levels of the brain stem which allow for localization (termed hemiplegia alternans or alternating hemiplegia).

b. The nuclei of the somatic motor cranial nerves (III, IV, VI, and XII) are located close to the midline. These nuclei tend to be involved, then, by paramedian lesions rather than by lateral lesions. Intrinsic lesions in a paramedian location tend to produce, early in their course, a bilateral involvement of these nuclei. These nuclei are also all located in a relatively dorsal position. These nuclei would be involved early by a lesion in a dorsal or tegmental location. Except for cranial nerve IV, the fibers of these cranial nerves all exit in a relatively paramedian ventral location.

Sensory Cranial Lesions

2. Cranial Nerve V Lesions:

a. In the main sensory root of the trigeminal nerve, in its course within or external to the midpons, will produce ipsilateral deficits in pain, temperature, and touch over the face.

b. In the descending spinal tract of the trigeminal nerve or of its associated nucleus will produce ipsilateral deficit in pain and temperature sensation over the face with sparing of facial touch sensation. These structures are situated close to the lateral spinothalamic tract, the combination of contralateral pain and temperature deficit over the body and extremities with ipsilateral pain and temperature deficit over the face is frequently found as a consequence of lesions which involve the dorsal lateral tegmental portion of the medulla, e.g., infarction of the territory of the posterior inferior cerebellar artery which supplies this sector. A similar combination may be found in the infarction of the dorsolateral tegmentum of the caudal pons: the territory of the anterior inferior cerebellar artery.

c. The descending spinal tract (analogous to and, in a sense, the direct continuation of Lissauer's tract) and associated nucleus (analogous to the substantia gelatinosa), descends into the upper cervical spinal cord. Although there is considerable overlap the primary trigeminal fibers synapse on the nucleus of the descending spinal nucleus at the following level:

-mandibular division fibers at a lower pontine--upper medullary level;

- the maxillary division at a medullary level,
- and the ophthalmic division fibers at a lower medullary and upper cervical cord level.

There is overlap at the upper cervical cord level between the upper cervical segment pain fibers and the ophthalmic division pain fibers. It is not surprising, then, that pain originating in the C2 and C3 roots or segments of the spinal cord may sometimes be referred to the ophthalmic division of the face (orbit and forehead), e.g. the pain of cervical 2 and 3-occipital neuralgia is often referred to the orbit.

d. The secondary pain fibers decussate, join the ventral secondary trigeminal tract (also labeled trigeminothalamic or quintothalamic tract) which, at a medullary level, is located just lateral to the medial lemniscus. At a pontine and midbrain location, the secondary trigeminal tract is just lateral to and essentially continuous with the medial lemniscus and adjacent to the lateral spinothalamic path. A lesion that involves this tract would then usually produce contralateral deficits in pain and temperature over a variable portion of the face in association with a contralateral deficit in pain and temperature and/or position and vibration were the affected arm, leg, and trunk.

G. CLINICAL SIGNS AFTER LESIONS IN THE TEGMENTUM OF THE BRAIN STEM.

Although all nuclei and tracts in the brain stem are important to the proper function of the CNS, lesions in only certain tracts and nuclei produce abnormal responses detectable at the bedside and important for diagnosis and treatment of disease. Remember injury to any cranial nerve usually produces a defect that helps you to localize just where in the brain stem the lesion is located. (See also Chapters 15 and 16.) Table 11-11 provides only a brief overview of the major deficits seen after lesions in one of the five zones of the tegmentum.

1. Disease involving the brain stem does not directly affect most aspects of mental status such as memory, abstract reasoning, comprehension, and the ability to perform calculation--functions usually associated with the cerebral cortex. Brain stem lesions may interfere with the ability to test these functions in the sense that a patient with a transection of the upper pons or midbrain will usually be unable to convey information from the cerebral cortex to the cranial nerve and spinal cord motor neurons so that expression of the cortical functions cannot occur. The patient is akinetic and mute. Lesions involving the reticular formation at the diencephalic and mesencephalic junction will interfere with mental status in the sense that the alertness and arousal aspects of consciousness are defective. The patient then remains in a comatose state, either failing to arouse from this state when stimulated or arousing only for brief periods. For additional discussion refer to chapter on consciousness.

2. Pathological processes that involve the brain stem may have effects moreover which are not simply limited to the brain stem. Thus, occlusion of the basilar artery may not only produce infarction (ischemic damage) of the brain stem but also may produce infarction of the territory supplied by the posterior cerebral arteries. This includes the calcarine cortex, the medial and inferior aspects of the temporal lobe, and the diencephalon. The medial aspects of the temporal lobe (the hippocampus) and the medial thalamic areas apparently provide the anatomical substrate for recent memory (the learning of new information).

3. Space-occupying lesions of the brain stem or posterior fossa may also secondarily interfere with mental status through a blockade of the ventricular system. With blockade of the fourth ventricle or aqueduct (or for that matter the third ventricle), cerebrospinal fluid pressure increases and is transmitted to the lateral ventricle. The lateral ventricles dilate and since the skull is a rigid container (after closure of the sutures), the white matter and gray matter of the cerebral hemisphere are subjected to

TABLE 11-11. CLINICAL DEFECTS FROM LESIONS IN THE TEGMENTAL ZONES OF THE BRAIN STEM.

Ventricular Zone: Interruption of motor cranial nerves III, IV, VI, X, and XII produce lower motor neuron signs. Interruption of sensory nuclei of VIII produces dizziness and deficits in hearing while injury to the solitary nucleus produces deficits in taste
Lateral Zone: Interruption of the spinothalamic tract produces contralateral decrease in pain and temperature from the body. Injury to the rootlet or chief sensory or descending nucleus of CN V produces sensory finding of ipsilateral deficits in pain, temperature, and touch from the head and ipsilateral weakness in chewing.
Medial zone: Interruption of the medial lemniscus, depending on the level of the lesion, produces loss of touch, pain, and temperature from the head or body; injury to the medial longitudinal fasciculus produces deficits in coordination of eye movements.
Central zone: Lesions here produce abnormalities in the functions associated with cranial nerve nuclei of VII, IX, X and XI. The descending parasympathetic and sympathetic fibers can also be interrupted affecting the pupil (Horner's) and sweat glands (See Chapter 11 on Cranial Nerves).
Basilar zone: Destruction of the corticospinal tract produces an upper motor neuron lesion with loss of volitional movement in the contralateral body muscles; interruption of the corticonuclear tract produces weakness of volitional movement of the muscles in the contralateral head and neck.

compression with an alteration in mental status. The patient frequently presents a drowsy appearance and when roused he often manifests a diffuse defect in memory, comprehension, and abstract reasoning. With relief of the obstruction to the flow of cerebrospinal fluid, a rapid reversal in this state should occur. An increase in intracranial pressure, particularly when sudden, is often associated with headache, nausea, and vomiting.

More recent diagnostic approaches have involved earlier diagnosis by CT or MRI scan (MRI is the preferred imaging approach). The surgical approach may now include excision via a supracerebellar or transtentorial approach to the tumor utilizing the operating microscope (see Stein 1979).

V. BRAIN STEM AND EYE MOVEMENTS

Cortical Control of Eye Movements. Conjugate eye movements originate in the frontal and occipital lobes, however the pathways that coordinate these movements are found in the pons and midbrain where they are commonly affected. See Chapter 17.

Subcortical Control of Eye Movements

1. MLF Syndrome

a. superior division. Assume lesion above the abducens nucleus, when right lateral gaze is attempted:

- the right eye will fully abduct but the left eye will fail to adduct fully,
- -the left medial rectus, however, will usually be otherwise intact for adduction in other movements, such as convergence,
- horizontal nystagmus, whether occurring spontaneously or induced by caloric stimulation, will be most prominent in the abducting eye.

b. Inferior division of the medial longitudinal fasciculus-Dolls Eye Phenomena. This pathway can be utilized in the comatose patient (but not in the conscious patient) to elicit eye movements and in a sense to test the integrity of the medial longitudinal fasciculus. The movements induced by head turning (the oculocephalic reflex) are commonly referred to as the dolls' head eye or "Dolls eye" phenomenon and serve to center fixation regardless of head positions..

- Flexion of the neck leads to upward deviation of the eyes; extension of the neck to downward deviation of the eyes. This reflex is not dependent on the presence of the cerebral cortex. Although some have postulated a vestibular basis for the reflex, it is clear that the reflex may be obtained in patients in whom the ocular response to caloric vestibular stimulation can no longer be obtained. (A possible role of the nucleus prepositus hypoglossi - located below the floor of the ventricle has been postulated).

2. *Convergence.* From a theoretical standpoint, selective damage to the nucleus of Perlia could occur without involvement of the nuclei supplying the medial rectus muscles. Such a selective lesion, however, would not be common in view of the fact that the entire oculomotor complex occupies a relatively small area. An intrinsic lesion would be implied. A relative defect in convergence may be commonly encountered in any circumstance in which consciousness has been impaired. A supranuclear pathway from occipitoparietal cortex has been postulated.

3. *Effect of Cranial Nerve Lesions on Eye Movements.*

a. Third Cranial Nerve symptoms. The signs may be frequently noted in mass lesions that involve the cerebral hemisphere. Thus laterally placed, space-occupying lesions such as collections of blood in the epidural or subdural space, brain tumor, or abscess in the temporal lobe may all displace the temporal lobe medially and downward. The medial aspect of the temporal lobe (uncus and hippocampal gyrus) is then forced through the tentorial opening compressing the third cranial nerve- herniation of temporal lobe. As compression continues, the fibers to the levator palpebrae and then to the other extraocular muscles are involved, resulting eventually in a complete third nerve paralysis.

b. Sixth Cranial Nerve symptoms. A small paramedian infarction in the tegmentum of the lower pons may produce the combination of ipsilateral peripheral facial palsy and ipsilateral lateral rectus palsy, due to involvement of the abducens nucleus and of the facial nerve fibers in the genu about the abducens nucleus. As we have already indicated, such a lesion would also produce a paralysis of conjugate lateral gaze, due to involvement of the "lateral gaze center." This combination of cranial nerve findings may be associated with a contralateral hemiplegia due to a more extensive paramedian infarction (the Foville pontine syndrome; see Wolf 1972).

Disease processes in the posterior cranial fossa, producing an extraocular palsy limited to the lateral rectus muscle, may affect the sixth cranial nerve due to its long extramedullary course. A bilateral involvement is frequently seen in any situation where an increase in intracranial pressure has occurred. A unilateral paralysis is also frequently found in locally invasive disease involving the base of the skull (nasopharyngeal carcinoma). Such a lesion would progress to involve other cranial nerves at the base of the skull.

4. Combined unilateral involvement of all extraocular muscles indicates disease extrinsic to the brain stem. These findings originate from growths within the orbit, compression at the superior orbital fissure, or thromboses of sinus due to spread of infection from face or orbit within the cavernous sinus may involve all three nerves (III, IV, and VI). In addition the ophthalmic division of the trigeminal nerve will be involved with the cavernous sinus syndrome. At times, the adjacent maxillary division is involved as well. Bilateral involvement of the extraocular muscles may occur in myasthenia gravis or in familial degenerative disease involving the extraocular muscles in the process of muscular dystrophy (progressive ophthalmoplegia and oculopharyngeal dystrophy). In these diseases the extraocular muscles are selectively involved; the pupillary reactions remain intact.

5. Pupillary reactions.

Light Reflex. Fibers from the retinal ganglion cells pass through the optic nerve and tract. Rather than terminating in the lateral geniculate, these fibers diverge from the optic tract just rostral to the lateral geniculate and enter the pretectal region at the level of the posterior commissure. Fibers then pass to the Edinger-Westphal nucleus of the same side (for the direct pupillary response).

Accommodation and Pupillary constriction. Accommodation (the shift in gaze from a distant to near object) and to convergence depend on fibers in the optic tract reaching the occipital-calcarine cortex via the lateral geniculate. Fibers from the occipital association cortex then reach the superior colliculus and the nucleus of Edinger-Westphal. The lesion in the pretectal area that has been postulated for tabes dorsalis then would affect pupillary response to light but not to accommodation. Involvement of descending sympathetic fibers in this area could be postulated to explain the fact that in tabes dorsalis the pupil is relatively small and constricted in the resting state. Pseudotabetic pupil may occur in other diseases, e.g., diabetic peripheral neuropathy.

CASES EFFECTING EYE MOVEMENTS

Case 11-1. (Full case on CD).

This 75-year-old white male had the onset of diplopia in March. Examination in April revealed only a variable possible lag in movement of the left medial rectus muscle. On readmission in June there was now a clearly defined left abducens nerve paralysis and nystagmus on right lateral gaze. A left peripheral facial paresis was now also noted. Cerebrospinal fluid examination was negative.

He was readmitted in July because of progression of the left facial paralysis. On examination he was unable to close the left eye and he was drooling from the left side of the mouth. He was also complaining of diplopia on gaze to either left or right side. Examination of eye movements now demonstrated severe problems:

1) On attempted gaze to the left; a total paralysis of left CN VI plus a lesser paresis of

TABLE 11-12. DYSFUNCTIONS IN THE EYE PRODUCED BY SUBCORTICAL LESIONS

DYSFUNCTION IN EYE MOVEMENTS	CLINICAL DEFICITS
Role of MLF - Moves the eyes in a conjugate, coordinated manner as though yoked together.	-Syndrome of MLF is usually bilateral syndrome due to close paramedian location of MLF fiber systems of the two sides. Usually intrinsic disease: multiple sclerosis, vascular disease with infarction, or an infiltrating glioma
- Ascending fibers, the vestibular nuclei, the pontine center for lateral gaze (see below), the abducens nucleus, and the oculomotor nuclei. - Descending fibers; above nuclei & proprioceptive information from cervical cord	-Dolls Eyes Phenomena (Oculocephalic Reflexes) Side to side rotation results in contraversive conjugate eye deviation.
Lateral Gaze Center in pons	Section of the abducens nerve root results in only an ipsilateral lateral rectus paralysis, while destruction of the abducens nucleus results in a long-lasting ipsilateral paralysis of conjugate lateral gaze with ipsilateral lateral rectus palsy.
Disturbance of Conjugate Gaze-. Paralysis of upward gaze is often accompanied by a paralysis of downward gaze and by a paralysis of pupillary response to light.	Damage to the pretectal area, the Edinger-Westphal nucleus (in rostral position in the oculomotor complex) or to the fibers running between these two areas.
Convergence- fixation on approaching visual objects on the fovea and corresponding retinal points of the two eyes	Damage to nucleus of Perlia, in midline component of third nerve complex
DYSFUNCTION OF CRANIAL NERVES: LEADING TO ABNORMAL EYE MOVEMENTS	
-Third Cranial Nerve-pupillary dilation & controls medial, superior, & lateral rectus, superior oblique and sup. Levator palpebrae. Lesion produces weakness in eye movements and diplopia	-Compression initial sign, paralysis of the pupillary constriction, resulting in a fixed dilated pupil ("blown pupil"), as constrictor fibers are most peripheral among the third nerve fibers.
-Sixth Cranial Nerve to lateral rectus and also coordinates connections to CN III via MLF lesions in pons usually also affects cranial nerve VII.	A small paramedian infarction in the tegmentum of the pons produces peripheral facial palsy and ipsilateral lateral rectus palsy, and also produces a paralysis of conjugate lateral gaze, due to involvement of the "lateral gaze center." A more extensive paramedian infarction produces hemiplegia (the Foville pontine syndrome; see Wolf 1972).
PUPILLARY DYSFUNCTION PRODUCED BY DISEASE EXTRINSIC TO BRAIN STEM	
Disturbance of Pupillary Reactions Cranial Nerve III Contains fibers that constrict Pupil	Compression of Cranial Nerve III produces dilations of pupil, as only sympathetic fibers are still unaffected.
-Tabetic Pupil (tertiary neurosyphilis)	Selective impairment of the response to light, with preservation of the response to accommodation. (See diseases of the spinal cord).
-Adie's syndrome	A tonic pupil (once dilated is slow to respond to light, once constricted is slow to dilate in the dark) benign with absence of deep tendon reflexes in the lower extremities.
-Sympathetic Pupil –Constricted Pupil	Injury to thoracic cord, superior cervical sympathetic ganglia or interruption of sympathetic fibers on aorta or in brainstem

conjugate lateral gaze as regards the right eye.

2) On gaze to the right a) full abduction of the right eye, b) some adduction of the left eye but this was incomplete, and c) coarse nystagmus of the abducting right eye.

3) In primary gaze, the left eye was slightly medial.

4) Convergence was intact for both eyes.

5) Up gaze was intact but minor vertical nystagmus was present.

Conclusions were:

1) Left CN VI paralysis.

2) Left lateral gaze center involved

3) Left medial longitudinal fasciculus involved.

The remainder of the neurological examination was normal.

Clinical Diagnosis: Tumor in pons at the level of the facial colliculus.

CASE HISTORY 11-2.

This 16-year-old, white male in November was first noted to be lethargic. Gradually from November to March there was progressively poor performance in school with apathy and somnolence. In December the patient had the onset of diplopia with difficulty in reading. In March of the next year he began to sleep throughout the day and could be awakened only with difficulty. Headaches on arising, with nausea and vomiting had been noted. In April evaluation at the Buffalo General Hospital had revealed that the pupils failed to respond to light but did respond to accommodation. Upward gaze was intermittently defective, and conjugate lateral gaze was intact. Findings improved to some degree over a 3-day period.

On the day of admission, an episode of urinary incontinence occurred.

Neurological Examination:

1. *Mental status:* All areas were relatively intact except for a slowness of response, emotional lability, and immaturity of behavior and questions.

2. *Cranial nerves:*

a. The right eye was deviated outward; the patient was unable to converge. There was a paralysis of upward and downward gaze. Bilateral lid ptosis was present. However, spasm of the eyelids was easily stimulated. Pupils were sluggish in response to light. The right pupil was slightly larger than the left. Horizontal nystagmus was present.

b. The jaw jerk was hyperactive.

c. Guttural and lingual sounds were slurred in a pattern that was consistent with pseudobulbar palsy.

d. Hearing was decreased bilaterally.

3. Motor system:

a. Strength was intact.

b. Spasticity was present on passive motion in the lower extremities and possibly in the upper extremities.

c. There was slowness in alternating hand movements. A fixed facies was present.

d. A bilateral intention tremor was present on finger-to-nose testing.

e. Gait was broad-based with truncal ataxia.

4. *Reflexes:*

a. Deep tendon reflexes were hyperactive bilaterally with ankle clonus.

b. Plantar responses were extensor bilaterally (bilateral sign of Babinski).

Clinical Diagnosis: Possible pineal region tumor.

CHAPTER 12
The Cranial Nerves

INTRODUCTION

The cranial nerves originate from a diverse group of nuclei and ganglia and innervate primarily the skin and muscles of the shoulder, head and neck. The structures innervated by the cranial nerves are especially important for many of the communications skills of humans including facial expression, language, sight, smell, taste and hearing. Another major function of cranial nerves III, VII, IX and X is to provide the parasympathetic innervation to the eyes, heart and GI system in the body; that is one of the major survival mechanisms for our species.

Components of the Cranial Nerves

In the spinal cord, the neurons are arranged in continuous columns of cells, each of which is connected to a structure with a specific function. This grouping of neurons with similar anatomic and physiologic functions is called a nerve component. In the spinal cord, the neurons Innervate general structures such as skin skeletal muscles, blood vessels, glands, and viscera. They do not innervate the special sense or-gans: the eye, the ear, or the taste buds.

In the brain stem, cranial nerve nuclei that innervate structures of a similar embryonic origin are found in approximately the same position throughout the medulla, pons, and midbrain. These nuclei are arranged in columns, near the ventricular floor or in the reticular formation *(Fig 12-l)*.

All the motor and sensory nuclei in the spinal cord are of a general nature and are designated in the nerve component scheme of Herrick (1948) as follows:

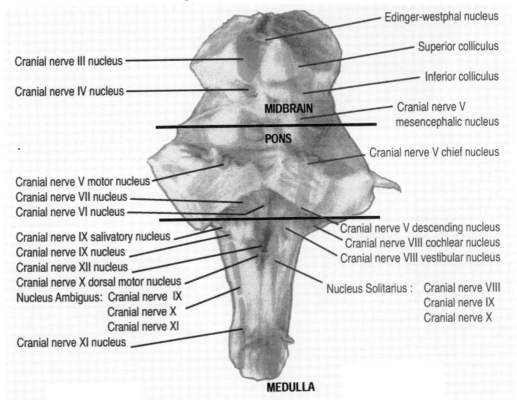

Figure 12-1. Posterior surface of brain stem showing cranial nerve nuclei.

TABLE 12-1. COMPONENTS OF THE CRANIAL NERVES

Function	Cranial Nerve
1. Motor to skeletal muscles of somite origin	Nerve III (midbrain), IV (midbrain), VI (pons), and XII (medulla). Brain Stem Tegmental Zone = ventricular
2. Motor to the parasympathetic (See Fig 11-13)	Edinger-Westphal Nucleus (midbrain), superior salivatory nuclei nerve VII (pons), inferior salivatory nucleus of nerve IX (medulla), dorsal motor nuclei of nerve X (medulla). Brain Stem Tegmental Zone = ventricular.
3. Motor to skeletal muscles of pharyngeal and visceral origin.	Motor nucleuses of nerves V, VII, IX, and X (pons and medulla). Brain Stem Tegmental Zone = ventricular
4. Sensory—visceral and gustatory sensation.	Solitary nucleus. Primary cell bodies are located in the sensory ganglia of nerves VII, IX, and X. General sensation from the viscera as well as the taste. Zone = ventricular.
5. Sensory—cutaneous and proprioceptive from skin and muscles in the head and neck	Trigeminal nuclei--mesencephalic nucleus (midbrain), chief sensory nucleus (midpontine), and descending nucleus of nerve V (lower pons, upper cervical level). General sensations carried in by nerves III, IV, VI, VII, XI, and XII synapse in the descending nucleus. Zone = lateral
6. Special sensory	Olfaction I, vision II, hearing and balance VIII

Motor (efferent)
  To skeletal muscle, CN III -VII, IX-XII
  To smooth muscle & glands, CN III, IX & X
  - To cardiac muscle, CN X

Sensory (afferent)
  - From cutaneous and proprioceptive receptors V, VII, IX & X
  - From viscera, CNIX & X

In addition to these general categories in nerves of the spinal cord, we find in the cranial nerves, neurons innervating the special sensory receptors (ear, eye, taste buds) or muscles in the face, pharynx, and larynx, which originated from pharyngeal arches. These special categories are present only in cranial nerves.

The discussion that follows will review the functions and clinical disorders associated with each of the 12 cranial nerves. At the end of this chapter we have included Table 12-5 that summarizes the cranial nerve syndromes produced by mass lesion.

I. GENERAL FUNCTIONS AND CLINICAL DISORDERS OF EACH CRANIAL NERVE

CRANIAL NERVE I - OLFACTORY

The olfactory nerve (*Fig. 12-2*) originates from receptor cells in the nasal mucosa. The unmyelinated nerve fibers combine into about 20 bundles, pierce the cribriform plate, and end in the glomerular layer of the olfactory bulb (See Chapter 22 for a discussion on chemoreception). These fibers have an especially important input to the amygdala of the temporal lobe. This is the only cranial nerve associated solely with the telencephalon.

Clinical Disorders. Defects in smell or Anosmia.

The most frequent causes of unilateral or bilateral Anosmia are:

1. Disease of nasal mucosa producing swelling and preventing olfactory stimuli from reaching the olfactory receptors.

2. Head trauma – resulting in shear-effects tearing the filaments of olfactory receptor cells passing through the cribriform plate.

3. Less common causes of unilateral Anosmia is an olfactory groove meningioma

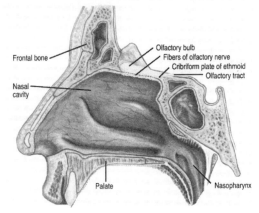

Figure 12-2. The Olfactory Nerve – Cranial nerve I

TABLE 12-2. CRANIAL NERVE COMPONENTS

CRANIAL NERVE	LOCATION OF CELL BODIES	FUNCTION
I Olfactory	Neuroepithelial cells in nasal cavity	Olfaction
II Optic-central pathways discussed in Chapter 21.	Ganglion cells in retina	Vision
III Oculomotor	-Oculomotor nucleus in tegmentum of upper midbrain. -Edinger-Westphal nucleus in teg-mentum of upper midbrain – preganglionic to ciliary ganglion.	Eye movements, levator palpekal supenored, superior medial and inferior rectus, inferior oblique Pupillary constriction and accommodation of lens for near vision through ciliary ganglion
IV Trochlear	Trochlear nucleus in tegmentum of lower midbrain	Eye movements (contralateral), superior oblique muscle
V. Trigeminal	Primary trigeminal ganglion, Secondary Mesencephalic (midbrain), chief sensory (Pons), and descending spiral nuclei (pons, Medulla, upper cervical levels) Motor nucleus in pons	Cutaneous and proprioceptive sensations from skin and muscles in face, orbit, nose, mouth, forehead, teeth, paranasal sinuses, meninges, and anterior two-thirds of tongue. Muscles of mastication
VI. Abducens	Abducens nucleus in tegmentum of pons	Eye movements, ipsilateral, involving lateral rectus muscle
VII Facial	Primary geniculate ganglion: Secondary Nucleus solitarius -Motor nucleus in lateral margin of pons -Superior salivatory nucleus in pons -preganglionic parasympathetic to ganglia associated with these glands	Gustatory sensations from taste buds in anterior two-thirds of tongue Muscles of facial expression and platysma; extrinsic and intrinsic ear muscles, stapedius muscle. Glands of nose and palate and the lacrimal, submaxillary and sublingual glands through pterygopalatine and submandibular ganglia
VIII. Vestibulo-acoustic. (Central pathways discussed in chapter 15)	Primary vestibular & cochlear ganglion, temporal bone; Secondary cochlear nuclei in medulla Secondary vestibular nuclei in medulla and pons	Audition Equilibrium, coordination, orientation in space
IX Glossopharyngeal	Primary inferior ganglion; Secondary Nucleus solitarius Inferior salivatory nucleus in medulla Nucleus ambiguous Preganglionic to otic ganglion	Interoceptive palate and posterior one-third of tongue, carotid body; Gustatory sensations from taste buds in posterior one-third of tongue Secretions from parotid gland through otic ganglion Swallowing, stylopharyngeus mm Visceral sensations
X Vagus	Primary Superior inferior ganglion; Dorsal motor nucleus – preganglionic (Parasympathetic innervation) Nucleus ambiguous in medulla	Visceral sensations from pharynx, larynx aortic body and thorax and abdomen Gustatory Sensation from taste buds in epiglottis and pharynx Skeletal muscles in pharynx & larynx Smooth muscles in heart, blood vessels, trachea, bronchi, esophagus stomach, intestine to lower colon
XI Spinal Accessory	Cranial Portion Ambiguous nucleus in medulla Spinal portion: C2 C3 & 4	Innervation Sternomastoid Trapezius and mm of pharynx, larynx, except cricothyroids
XII Hypoglossal	Hypoglossal nucleus in medulla	Innervation of intrinsic and extrinsic muscles of the tongue except palatoglossal MM

compressing olfactory bulb and tract.

4. The normal process of aging and degenerative CNS disorders, including Alzheimer's disease, Huntington's disease, and Parkinson's disease.

CRANIAL NERVE II – OPTIC

The optic nerve *(Fig. 12-3)* because it is invested with glial cells not Schwann cells is actually a tract of the central nervous system. The optic nerve originates from the axons of the ganglion cells (tertiary neurons) of the retina, which are ensheathed with myelin as they leave the neural retina. These axons form the optic nerve, which exits the orbit via the optic foramen chiasm, and tract and terminate in the optic thalamus (lateral geniculate nucleus) and superior colliculus.

Clinical disorders

A Unilateral loss of vision is produced by:
1. Retinal disease as a consequence of, vascular: transient – carotid occlusive disease
2. Central retinal artery thrombosis – sudden and persistent
3. Degeneration – retinitis pigmentosa
4. Optic nerve: optic neuropathy:
5. Optic neuritis:
6. Demyelination as manifestation of multiple sclerosis
7. Ischemic optic neuropathy
8. Toxic affects, e.g., methanol usually bilateral
12. Deficiency states: e.g., Thiamin and B12: usually bilateral
10. Compression: by masses - inner third sphenoid wing or olfactory groove meningioma
   - Glioma of optic nerve
   - Long-standing papilledema
   - Pituitary masses (usually involve optic chiasm producing a bitemporal hemianopia

CRANIAL NERVE III—OCULOMOTOR

The Oculomotor nucleus *(Fig 12-4)* is a pure motor nucleus and is found in the tegmentum of the midbrain in sections containing the superior colliculus. This nucleus

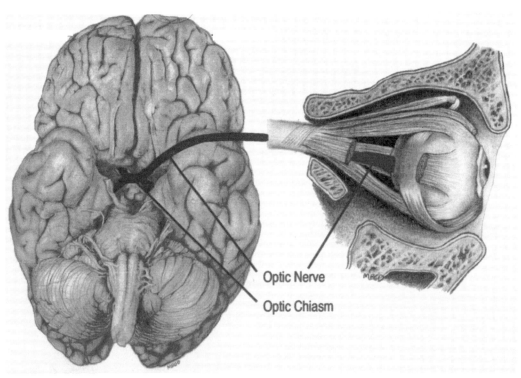

Figure 12-3. The Optic Nerve – cranial nerve II

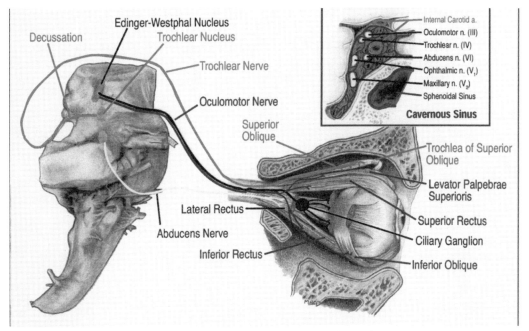

Figure 12-4. The Oculomotor, Trochlear, and Abducens nerves (respectively cranial nerves III, IV and VI), and the location of their nuclei in the pons and midbrain of the brain stem.

indents the medial longitudinal fasciculus. The Oculomotor complex can be differentiated into a somatic and visceral portion.

The somatic portion consists of the lateral nuclear complex and a single central nucleus, the Edinger-Westphal. The individual muscles are represented in the lateral nuclear complex, but data on the exact location and extent of crossed and uncrossed innervation of the eye muscles is incomplete.

The visceral portion of the Edinger-Westphal nucleus in the floor of the cerebral aqueduct provides the preganglionic parasympathetic fibers to the ciliary ganglion in the orbit. The postganglionic motor fibers pass to the muscles of accommodation and pupillary constriction, permitting, dilation of the pupil and an increase in the thickness of the lens for accommodation (focus on objects within 6 inches of the eye).

The somatic fibers from the nuclear complex of nerve III pass anteriorly through the tegmentum of the midbrain. Some fibers running through the red nucleus and others passing lateral or medial to it, finally entering the medial margin of the of the cerebral peduncle and reuniting in the interpeduncular fossa to form nerve III. The III nerve emerges from the midbrain superior to the pons between the superior cerebellar artery and the posterior cerebral artery. It then penetrates the dura, enters the cavernous sinus where it lies lateral to the internal carotid artery and enters the orbit via the superior orbital fissure *(Fig. 12-4)* and separates into superior and inferior divisions. The superior division innervates the lavatory palpebra superioris and the superior rectus muscles, while the inferior division supplies the medial and inferior rectus and inferior oblique muscles and sends roots from the Edinger--Westphal nucleus to the ciliary ganglion. The superior rectus muscle elevates the eye-ball and turns it upward and inward. The medial rectus muscle adducts the eyeball. The inferior rectus muscle depresses the eyeball and adducts it to some degree, turning the eye downward and inward. The major action of the inferior oblique muscle occurs when the eyeball is adducted; contraction of the inferior oblique muscle both elevates and rotates the eye. The superior levator palpebra controls the eyelid.

Clinical Disorders.

1. Complete paralysis of nerve III produces ptosis of the lid, paralysis of the medial and upward gaze, weakness in the downward gaze, and dilation of the pupil. The eyeball deviates laterally and slightly downward: the dilated pupil does not react to light or accommodation. Paralysis of the intrinsic muscles (the ciliary sphincter, or pupil) is called *internal ophthalmoplegia,* while paralysis of the extraocular muscles (recti and obliques) is called *external ophthalmoplegia.*

2. Causes of cranial nerve III dysfunction:

- Aneurysm of posterior communication internal carotid artery junction

- Lesions within the cavernous sinus - SEE CDROM Case 12-9

-Herniation of temporal lobe

-Tumors – meningioma of tuberculum sellae or sphenoid wing (Compression involves both pupillary and extraocular muscles).

-Diabetic cranial neuropathy (vascular) – pupil may be spared, at time of extraocular involvement. However, diabetes mellitus may be associated with a pseudotabetic pupil

The following case is an example of involvement of the occulomotor nerve 2nd to an aneurysm.

Case 12-1. This 50-year-old right-handed, white male factory foramen was referred for evaluation of ptosis and diplopia involving the left eye. Approximately two weeks prior to admission, the patient developed a bifrontal headache; during the week prior to his evaluation the headache had become a left sided aching pain. It increased in intensity during the two days prior to admission, and was present as a constant pain interfering with sleep. If the patient were to cough he had additional pain in the left eye. On the day of admission the headache became much more severe, it was now the worst headache he had ever experienced.

In retrospect, the patient reported that lights had been brighter in the left eye for approximately one week.

At 3:00 PM on the day of admission the patient noted the sudden onset of diplopia, which was more marked on horizontal gaze to the right and much less marked on horizontal gaze to the left. At approximately the same time he noted the rapid onset of ptosis involving the left lid.

General physical examination: There was moderate resistance to flexion of the neck.

Neurological examination:

1. The left pupil was fully dilated to approximately 7mm. There was no response to light or accommodation.

2. Total ptosis of the left eyelid was present.

3. No medial movement of the left eye was present; there was no upward movement of the left eye possible. The patient had minimal downward gaze of the left eye. He had full lateral movement of the left eye. Movements of the right eye were full.

Clinical diagnosis: Subarachnoid hemorrhage secondary to an aneurysm at the junction of the posterior communicating and internal carotid arteries.

The occurrence on the day of admission, of the worst headache ever experienced by the patient suggested the possibility of a subarachnoid hemorrhage. Most patients with an acute subarachnoid hemorrhage have an underlying saccular aneurysm.

There are three major locations for single intracranial aneurysms:

1) the junction of the posterior communicating and the internal carotid arteries,

2) the junction of the anterior communicating and anterior cerebral arteries and

3) the bifurcation of the middle cerebral artery in the Sylvian fissure.

The posterior communicating artery is adjacent to and runs parallel to the 3rd cranial nerve. With hemorrhage, compression of (or bleeding into) the nerve symptoms will develop relevant to cranial nerve 3. Thus in patients with prodromal symptoms, a third nerve syndrome allows the prediction of this specific location of the aneurysm. Because of the arrangement of the fibers within this nerve, the initial symptoms of compression often involve the pupillary fibers. In retrospect, the early symptoms of lights being brighter in the left eye, one week before the ptosis and diplopia developed would be consistent with a pupil that was unable to constrict in response to light. We will discuss in greater detail in a later chapter, that one can predict the middle cerebral location when symptoms and signs relevant to the middle cerebral artery occur. The anterior communicating location is often difficult to predict because, weakness in both legs may occur but may be accompanied by loss of consciousness due to bilateral frontal lobe involvement. In many patients, in all locations, the bleeding may be so massive, that loss of consciousness, rapidly follows the sudden headache, and localization is not possible. Management of intracranial aneurysms will be discussed in the chapter on vascular syndromes.

CRANIAL NERVE IV - TROCHLEAR

The Trochlear nucleus *(Fig. 12-4)* is purely motor and located in the mediosuperlor part of the teg-mentum in the lower parts of the mesencepha-Ion. It extends throughout the inferior collicular levels and is nearly continuous with the lateral nuclear complex of nerve IIIl. The nucleus of nerve IV indents the medial longitudinal fasciculis (MLF). Nerve IV is a unique cranial nerve as: 1) its rootlets are on the posterior surface of the brain stem, and 2) the fibers cross as they exit the brainstem. The fibers proceed inferiorly and posteriorly in the periaqueductal gray, decussate in the anterior medullary velum, exit from the substance of the brain at the caudal end of the inferior colliculus, and then run ventrally, lateral to the nerve III rootlets. In the cavernous sinus, the nerve is lateral to the nerve III rootlets and enters the orbit through the superior orbit fissure, where it terminates in the superior oblique muscle. The major action of the superior oblique muscle occurs with the eye adducted. With the eyeball in the physiologic condition, contraction of this muscle results in rotation of the eyeball.

Clinical Disorders. Trauma is the primary cause of defects in CN IV. A nuclear lesion of nerve IV paralyzes the contralateral superior oblique muscle while a lesion in the nerve roots after decussation involves the ipsilateral muscle. A thrombosis or tumor in the cavernous sinus may also affect this nerve and cranial nerves III, V, VI.

CRANIAL NERVE V. TRIGEMINAL

The trigeminal nerve (Fig. 12-5) is the largest cranial nerve and provides sensory fibers from the face (the first branchial arch)—including the pos-terior surface of the ear, the scalp up to the vertex, and the undersurface of the lower jaw—and motor fibers to the mastication muscles. The sensory root contains the semilunar or gasserian ganglia, which is larger than the motor root. Within the brain stem are one motor and three sensory nuclei. The motor nucleus of nerve V is located at the midpontine levels medial to the trigeminal root at the lateral part of the reticular formation. The three main branches of nerve V are ophthalmic, maxillary, and mandibular. Except for the mesencephalic nuclei, the primary cell bodies for the sensory fibers are located in the trigeminal (semilunar or Gasserian) ganglion in the middle cranial fossae.

Primary Sensory Cell Bodies (Trigeminal Ganglion)

The primary sensory cell bodies of nerve V are found in the semilunar ganglion in Meckel's cave (cavum Trigeminal). They are unique as they are the only 1° sensory neurons found within the cranial cavity. They are located in the middle cranial fossa, a recess in the petrous portion of the temporal bone below the superior petrosal sinus).

Trigeminal Peripheral Branches

1. The ophthalmic branch of the trigeminal nerve (V1). This nerve originates from the medial part of the semilunar ganglion, passes into the lateral wall of the cavernous sinus, and divides into its terminal branches in the superior orbital fissure. The smallest division of CN V (V1) innervates the skin of the forehead and scalp to the vertex, the upper eyelid, the skin of the anterior and bridge of the nose, the eyeballs, cornea, and ciliary body, conjunctiva, and

iris, the mucosa in the frontal and nasal sinuses, and the cerebral tentorium.

2. The maxillary nerve springs from the middle of the semilunar ganglion in the cavernous sinus. It leaves the middle cranial fossa via the foramen rotundum and en-ters the pterygopalatine fossa where it becomes the infraorbital nerve. The maxillary division of nerve V (V2) supplies: the skin on the temples, lateral surface of the nose, the lower eyelid, the upper cheek and the upper lip, the gums and molar, premolar, and canine teeth (superior dental plexus), and the mucous membranes of the mouth, nose, and maxillary sinus. The maxillary nerve also supplies the dura in the middle cranial fossa (middle meningeal nerve).

3. The mandibular division of nerve V (V3), the largest of the three divisions is formed by the union of the sensory and motor root at the inferior border of the semilunar ganglion. The motor root exits the middle cranial fossa through the foramen ovale where it joins the sensory root. The motor portion of the trigeminal innervates the muscle of mastication and the tensor tympani and tensor veli palatini. The sensory root supplies the skin of the chin, lower jaw, and temperomandibular joint: the dura in the middle and anterior cranial fossa, the lower teeth and gums, the oral mucosa and part of the ear.

Secondary Sensory Nuclei in the brain stem.

The chief sensory nucleus of nerve V is found in the midpontine tegmentum, lateral to the root of nerve V. This nucleus is the equivalent of the gracile and cuneate nuclei. It receives cutaneous sensory information from the face and head and the nasal and oral cavities and transmits this in-formation to the thalamus via fibers that travel with or adjacent to the medial lemniscus.

The descending or spinal nucleus of nerve V is the equivalent of the substantia gelatinosa in the dorsal horn of the spinal cord. Pain and temperature fibers synapse here. External to the descending nucleus of nerve V is the descending tract of nerve V. This tract is the equivalent of Lissauer's tract of the spinal cord. Within this tract the fibers from the ophthalmic nerve descend to the third cervical level. The maxillary fibers descend to the first cervical level. The pain fibers from the mandibular and cranial nerves VII, IX, X, XI, and XII descend

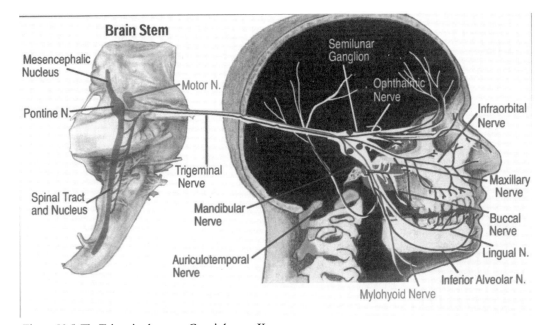

Figure 12-5. The Trigeminal nerve – Cranial nerve V

through the lower medullary lev-els. The secondary fibers are principally crossed and they ascend in the medial lemniscus to the ven-tral posterior medial nucleus in the thalamus.

The mesencephalic nucleus, located lateral to the fourth ventricle and the cerebral aqueduct in the upper pontine and midbrain levels, is proprioceptive from muscles controlled by the facial, ocular, and trigeminal nerves. This is the only primary sensory nucleus (dorsal root ganglion equivalent) in the central nervous system. The mesencephalic and motor nuclei provide a two-neuron reflex arc for the jaw jerk.

Motor Functions

The motor root originates in the motor nucleus of nerve V In the pons, and joins with the mesencephalic root, and exits from the middle cranial fossa in the foramen ovale where it then unites with the sensory root. The motor, or mandibular, nerve (V3) innervates the muscles of mastication. (Table 12-3)

Function

The combined actions of the muscles associated with the mandible are listed below in Table 12-4.

The mandibular branch of the trigeminal nerve also: innervates the tensor veil palatine and tensor tympani muscle.

1. Tensor veli palatine tenses the palate and draws it to one side, that prevents food from entering the nasal pharynx, and,

2. Tensor tympani muscle located in the inner ear pulls on the malleus that tenses the tympanic membrane and diminishes the amplitude of the vibrations caused by a loud noise.

Clinical Disorders:

Injury to the trigeminal nerve paralyzes the muscles of mastication, with the jaw deviating toward the side of the lesion. Injury may also block sensation of light touch, pain, and temperature in the face and results in the absence of the corneal and sneeze reflexes.

Localization of Motor dysfunction of the trigeminal nerve.

a. Lesions in the pons may affect only the motor nuclei and produce a paresis or paralysis of the ipsilateral muscles. If the lesion is destructive, atrophy and fasciculations of the affected muscles result. The jaw jerk is also absent.

b. Since the cortical innervation of the muscles of mastication is bilateral, unilateral supranuclear lesions may produce only a slight weakness, with a minimal increase in the jaw

TABLE 12-3. MUSCLES OF MASTICATION: INNERVATION BY V3—MANDIBULAR BRANCH OF CN5

MUSCLE	MOVEMENT OF MANDIBLE
Lateral Pterygoid	Works in unison with the other muscles - protrudes and depresses mandible. Working by itself moves the jaw laterally.
Medial Pterygoid	Works in unison with the other muscles, to elevate the mandible, assists the external pterygoid muscle in protruding mandible
Temporal	Elevates and retracts the mandible.
Masseter mandible	Elevates and slightly protrudes the
Mylohyoid	Elevates the floor of mouth and tongue.
Anterior belly of Digastric	Elevates and stabilizes the hyoid bone.

TABLE 12-4. MOVEMENTS OF THE MANDIBLE:

In mastication the jaws move up and down, forward and backward, and laterally.

ACTION OF MANDIBLE	MUSCLE GROUP
1. Elevation	Masseter, temporal, and medial pterygoid
2. Depression and opening of the mandible	- Lateral pterygoid muscle, in concert with - Suprahyoid muscles (mylohyoid, digastric, geniohyoid, infrahyoid) and, - Depressors of the hyoid (sternohyoid & mylohyoid muscles), - Along with gravity.
3. Protrusion of the jaw	- the medial and lateral pterygoid muscles, assisted by the masseter muscle
4. Retraction of the jaw	- the temporal and digas-tric muscles retract the jaw.
5. Side to side movement	- the pterygoid muscles permit side-to-side movement

jerk. A bilateral capsular or cortical lesion will produce marked paresis or even paralysis in the muscles of mastication, with an exaggerated jaw jerk.

c. Pontine lesions usually injure both the main sensory and motor nuclei, causing paralysis of the muscles of mastication and diminished tactile sensation ipsilaterally. Lesions of the lower lateral pontine tegmentum as the dorsal lateral medullary tegmentum may damage the descending tract and nucleus of V.

d. Tumors in the cerebellopontine angle or acoustic nerve may compress the trigeminal tubercle and cause an ipsilateral loss of the corneal reflex.

Localization of Sensory dysfunction of the trigeminal nerve. Trigeminal Neuralgia *(tic, douloureux)* is a disorder charac-terized by recurrent paroxysms of stabbing pains along the distribution of the involved branches. Trauma, infections, or arterial compression (superior cerebellar branch of basilar artery) may be responsible. Medication often can control the pain, but surgery is sometimes necessary for per-manent relief.

1. Clinical Signs of Trigeminal Neuralgia:

a. Etiology: "Idiopathic": with mean age of onset in the 50s

b. Trigger points that produce the pain on tactile stimulation and they are of short duration – lancinating pain, usually occurring in clusters or paroxysms. Pain may resolve spontaneously only to recur months later.

c. Location of pain -The second (maxillary division) is most frequently involved. The 3rd (mandibular division) is next in frequency. The 1st (ophthalmic division) is least common. The pain is unilateral, rarely bilateral.

d. Clinical findings. There are no sensory or motor findings on examination although after frequent repeated episodes the patient may complain of minor numbness and a more persistent aching type pain.

e. Treatment. Though many of these cases respond to medication – such as Carbamazepine, some do not and require surgical therapy. Patients with onset after age 40 often have compression of the trigeminal nerve by an anomalous loop of an artery compressing the nerve or root entry zone. Treatment in such cases that fail to respond to medications involves relocating the loop of the superior cerebellar artery or insulating the nerve from the pulsation of the vessel by and inserting a soft implant between nerve and vessel.

Other surgical procedures are designed to alter transmission through the Gasserian ganglion. Patients with an age of onset in the 20s or 30s usually have an underlying disease of brain stem, such as multiple sclerosis or arteriovenous malformation, or may have an underlying systemic disease.

e. Other causes of trigeminal neuralgia.

- Compression or invasion of nerves by nasopharyngeal tumors which have infiltrated the base of the skull through the foramen-rotundum (maxillary) or foramen – ovale – (mandibular) or other points of erosion.

-Dental pathology or maxillary sinusitis may produce pain within the appropriate divisions of the trigeminal nerve. – tingling and numbness in the mandibular or less often maxillary division may occur spontaneously or may follow dental procedures.

-*Herpes zoster* involvement of the Gasserian ganglion. Acute pain almost always in the ophthalmic division followed in 4-5 days by cutaneous lesions. Treatment with Acyclovir will decrease the period of pain. In other patients post herpetic neuralgia may develop. [Note: The H.zoster/varicella virus is often dormant in the gasserian ganglion].

The following Case is an example of Trigeminal Neuralgia

Case 12-2: This 28 year old single, black, right handed female reported approximately 21 days of sudden lancinating jabs of pain plus a duller more steady pain involving the right maxillary and mandibular distribution. She was very aware that she could trigger her pain by touching her nose, or the skin over the maxillary or mandibular area. The pain was so severe that talking; chewing or eating would all trigger episodes. As a result, she has not been eating a great deal. There was no clear-cut extension of pain into the eye, although some of the background pain did spread into the supraorbital area and back towards the interauricular line.

Two years prior to admission, she had a less severe episode in the same distribution that lasted three weeks. She denied any tearing of the eye or nasal stuffiness. She had no tingling of the face, change in hearing, facial paralysis, or diplopia.

Neurological Examination: Mental status, motor system (including gait and cerebellar system) and sensory system were intact. *Cranial nerves* were not remarkable except for findings relevant to the right fifth cranial nerve. Although touch and pain sensation over the face was normal, there were clearcut exquisite trigger points producing sharp pains on light tactile stimulation of the nose or other maxillary areas as well the mandibular skin areas as well as the lower molars.

Clinical Diagnosis: *Trigeminal neuralgia:* maxillary and mandibular divisions.

Subsequent course: Evaluation by a dentist indicated no relevant dental disease.

The following case is an example of example of a H. Zoster involving the ophthalmic division of the trigeminal nerve.

Case 12-3: This 44-year-old, right-handed physician developed over 24 hours, increasing pain in the right supraorbital and orbital areas. Edema of the eyelid developed within 36 hours. Erythematous, edematous skin lesions developed over the right supraorbital area and the anterior half of the right side of the scalp. Edema of the lid was sufficient to produce closure of the lid. Significant cervical lymphadenopathy, neck pain and low-grade fever developed. The acute pain gradually subsided over two weeks but the skin lesions persisted the skin lesions crusted and gradually cleared over four weeks. Occasional episodes of right supraorbital pain continued to occur for one year. Faded, depressed scars were still present 24 years after the acute episode.

Clinical diagnosis: Ophthalmic herpes zoster.

CRANIAL NERVE VI – ABDUCENS.

The nucleus of nerve VI *(Fig12-4)* and the fibers of nerve VII form a prominent structure in the floor of the IV ventricle of the pons, the facial colliculus *(Fig. 11-7)*. The abducens nucleus is a large purely motor nucleus surrounded by the looping fibers of nerve VII. The fibers of nerve VI leave the nucleus medial to the rootlets of nerve VII. Associated with the nucleus of nerve VI is the parabducens nucleus, or parapontine reticular formation, that integrates impulses via the medial longitudinal fasciculus. This center coordinates contractions of the lateral rectus muscle of one side with the medial rectus muscle (cranial nerve III) of the other, producing conjugate horizontal deviations of the eyes. The parapontine reticular nucleus is also referred to as the lateral gaze center.

The fibers pass inferiorly and emerge from the pons near the midline at the pontomedullary junction. The roots pierce the dura and enter the cavernous sinus (see Fig. 12-4) below and medial to nerve III and lateral to the internal carotid artery. The fibers enter the orbit via the superior orbital fissure and innervate the lateral rectus muscle, which abducts or deviates the eye laterally.

Stimulation of the frontal eye fields in the posterior part of the middle frontal gyrus produces contralateral innervation to the parapontine reticular formation. Stimulation of the occipital lobe also produces conjugate movement in the contralateral side. Movement originating from the occipital lobe is different from that in the frontal lobes as it is primarily involuntary fixation and tracking of moving objects.

Clinical Disorders: This nerve has a long intracranial course, and is thus commonly affected by: any increase in intracranial pressure (may be unilateral or bilateral), and by nasopharyngeal tumors invading the base of the skull in adults. Pontine gliomas produce similar effects in children. See Cranial Nerve Cases on CD

CRANIAL NERVE VII - FACIAL

The facial nerve *(Fig.12-6)* contains a large

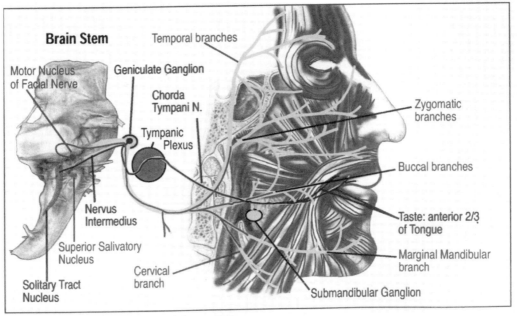

Figure 12-6. The Facial nerve – cranial nerve VII

motor nerve with a small sensory component (the intermediate nerve supplying taste, parasympathetic and somatic sensory.). In the pons there is one large mo-tor nucleus controlling the muscles of facial expression and one sensory nucleus for nerve VII, while peripherally there is one sensory ganglion the geniculate. The motor fibers supply the muscles associated with the second branchial arch (hyoid arch).

The branchial motor fibers originate from the motor nucleus of nerve VII in the lateral part of the reticular formation at the caudal pontine levels. The fibers leave the nucleus and pass medially and dorsally toward the fourth ventricle. Under the floor of the fourth ventricle, they pass over the dorsal surface of the nucleus of nerve VI and then turn anteriorly, forming the internal genu of the facial nerve. The fibers then pass laterally and anteriorly through the lateral portion of the pontine reticular formation and exit from the pons at its caudal most border between the rootlets of nerves VI and VIII in the cerebellopontine angle. The nerve enters the posterior cranial fossa and passes into the internal acoustic meatus where the motor and sensory divisions and nerve VIII lie separated from one another by only the arachnold sheaths on the nerves.

In the meatus the motor and sensory roots of nerve VII combine and enter the facial canal of the temporal bone. Then continuing until it reaches the hiatus of the facial canal, it bends around the anterior border of the vestibule of the Inner ear, forming the external genu of nerve VII. The geniculate ganglion located at the external genu, contains the primary cell bodies of the intermediate or sensory root of nerve VII that innervates the external ear.

The motor root innervates:
1. the stapedius muscle,
2. the posterior belly of the digastric muscle, and
3. the muscles of facial expression are innervated by the five branches of the facial nerve that arise in the parotid gland - temporal, zygomatic, buccal, marginal mandibular, and cervical. The mimetic muscles are: the frontal, orbicularis oculi, anterior auricular, corrugator supercilium, zygomatic, orbicularis oris, depressor Iabi, levator labi, and levator anguli oris, nasalis, mentalis, platysma, buccinator, and risorius muscles. Within the facial nucleus in pons there is a separation into subnuclei representing each of these muscle groups.

Sensory Ganglion - Geniculate. The small

sensory root originates from the geniculate sensory ganglia in the inner ear; fibers run through the facial canal into the internal acoustic meatus and then enter the pons. The fibers then pass through the lateral part of the reticular formation, and enter the tractus solitarius, which is found just lateral to the sulcus limitans. The tractus solitarius is separated from the ventricular cavity by the aqueductal gray on the floor of the fourth ventricle. The sensory root of nerve VII carries gustatory impulses from the taste buds in the ante-rior two-thirds of the tongue and cutaneous sensations from the anterior surface of the ear and the soft palate. The motor and sensory roots continue in the facial canal between the inner and middle ear and then turn and descend to the stylomascoid foramen, where they emerge from the temporal bone.

Chorda Tympani. Just before leaving the stylomastoid foramen, the chorda tympani, carrying taste fibers, separates from motor portion of nerve VII. It enters its own small canal in the petrotympanic fissure, then enters the tympanic cavity and runs on the medial surface of the tympanic membrane (from whence it gets its name), and onto the medial side of the manubrium. The nerve leaves the tympanic cavity and then emerges from the skull near the medial surface of the spine of the sphenoid and joins the lingual nerve at the medial surface of the lateral pterygoid muscle.

Taste Sensation. In the nucleus solitarius the taste sensitive cells are located as follows: 1) HCl (sour cells) are posterior, 2) sweet and salty are anterior, 3) cells responsive to bitter are diffusely located.

The secondary axons ascend in the medial lem-niscus to the parabrachial nucleus. The taste information then ascends near the central tegmental tract to the parvocellular ventral posterior medial nucleus of the thalamus. This information is then projected onto the parietal operculum. For regulation of feeding behavior, the information enters the lateral hypothala-mus and central amygdaloid nuclei.

Parasympathetic Ganglia of VII - Submandibular and Pterygopalatine.

1. Submandibular ganglia are suspended from the lingual nerve on the lateral surface of hypoglossal muscle. The *chorda tympani* contain parasympathetic preganglionic fibers from the submandiular ganglia the submaxillary and sublingual glands to the anterior 2/3 of tongue.

2. Pterygopalatine ganglia is suspended from maxillary nerve in pterygopalatine fossa and provides innervation to glands and blood vessels in the nasal cavity, palate, and paranasal air sinuses (maxillary, ethmoid, sphenoid) as well as the lacrimal glands.

Clinical Considerations:

1. Injury in the brain stem. A central injury to the facial nucleus in the pons produces ipsilateral paralysis of the facial muscles. Injury to just the nucleus or tractus solitarius, although rare, produces a loss of taste on half of the tongue because the taste fibers from nerves VII and IX are found together in the tractus solitarius.

2. Peripheral injury. Injury to the facial nerve near its origin or in the facial canal produces:

a. paralysis of the motor, secretory, and facial muscles,

b. a loss of taste in the anterior two-thirds of the tongue, and

c. abnormal secretion in the lacrimal and salivary glands.

The patient has sagging muscles in the lower half of the face and in the fold around the lips and nose and widening of the palpebral fissure. Vol-untary control of facial and platysmal muscula-ture is absent. When the patient smiles, the lower portion of the face is pulled to the unaffected side. Saliva and food tend to collect on the affected side. When the injury is distal to the ganglion, an excessive accumulation of tears occurs because the eyelids do not move but the lacrimal glands continue to secrete. Injury to the facial nerve may also interrupt the reflex arc for the corneal blink reflex, which includes the motor component to the orbicularis from nerve VII and the sensory corneal nerves of V1. The muscles eventually atrophy.

d. Injury to nerve VII at the stylomastoid

foramen produces ipsilateral paralysis of the facial muscles without affecting taste.

e Injury to the chorda tympani results in absence of taste in the anterior two-thirds of the tongue.

3. Other Peripheral Sites of Injury to VII-

a. Tumors of the parotid gland often involve the facial nerve.

b. H. zoster – involvement of the sensory geniculate ganglion, associated with the facial nerve – will result in a facial nerve paralysis. The cutaneous vesicular lesions, involve the small sensory distribution of the facial nerve in the external auditory canal and tympanic membrane. The adjacent CN VIII nerve is often involved producing vertigo, tinnitus and deafness. In some patients the dorsal root ganglion of cervical nerves 2 or 3 may also be involved.

c. Bilateral facial nerve involvement (facial diplegia) may occur in:

Guillain-Barre syndrome

Sarcoidosis – uveoparotid fever

Hemifacial Spasm: Spontaneous contractions of the facial muscles occur but often triggered by attempted movement. There is often synkinesis – an overflow – of movement so that attempts to smile result in closure of the eye, etc. Hemi-Facial Spasm relates to a previous facial nerve paralysis that has partially recovered but in which anomalous reinnervation has occurred. In the majority of cases in which the etiology is unrelated to a previous Bell's palsy, the etiology relates to a compression of the facial nerve root close to the brain stem by a blood vessel such as the anterior inferior cerebellar artery or posterior inferior cerebellar artery. With compression, demyelination will occur with subsequent ephaptic transmission between adjacent fibers. The treatment, both medical (carbamazepine) and surgical (decompression of the nerve placing a soft implant between the vessel and the nerve) is similar to the treatment of trigeminal neuralgia.

Facial myokymia is sometimes confused with hemifacial spasm. This is a fine rippling of muscles – arrhythmic involuntary fibrillations. Etiology: Pontine glioma, Multiple sclerosis – producing demyelination of the nerve within the brain stem.

Central/Cerebral Innervation of VII. The corticobulbar path-way to the lower half of the facial nucleus is a crossed unilateral Innervation, whereas the upper half of the facial nucleus receives a bilateral innervation. A unilateral lesion in the appropriate part of the precentral gyrus will thus paralyze the lower half on the opposite side, while a bilateral cortical involvement paralyzes the upper and lower facial musculature. The cerebral cortical control of the muscles of facial expression through the motor strip is especially important in communication. If these upper motor neuron fibers are destroyed other cortical regions, including the limbic system, will now cause inappropriate facial expressions (pseudobulbar palsy). Pseudobulbar palsy represents excessive uninhibited brain stem activity. External stimuli will then trigger excessive response e.g. excessive crying to a sad story or excessive laughter.

The following case is an example of one of the more common disorders of the facial nerve, Bell's palsy.

Cranial Nerve Case 12-5. 5. This 37-year old, right-handed, white, housewife was referred for emergency room consultation regarding left facial paralysis. Three days prior to evaluation, the patient had developed a dull pain above and behind the left ears. She then noted occasional muscle twitching about the left lower lip and some vague stiffness on the left side of the face. The morning of the evaluation, the patient noted she had difficulty with eye closure on the left. She noted that she was drooling from the left side of her mouth. She denied any alteration in sensation over the face and any actual deficit in hearing and denied any tinnitus. She had no crusting in the ear and no actual pain within the ear.

Neurological Examination: *Cranial Nerves:* The patient had an incomplete paralysis of the entire left side of the face. She had sufficient eye closures sitting or recumbent to cover the cornea. She was able to have minor elevation of the eyebrow in fore-head wrinkling. She had minimal elevation of the lower lip in smiling. Taste for sugar was slightly decreased but not absent on the left side of the tongue (anterior two-thirds). All other cranial nerves were Intact. Examination of the external canal and of the tym-panic membranes demonstrated no abnormality.

Motor system, reflexes, sensory system, and mental state: All intact.

Bell's Palsy related to involvement of the peripheral portion of the facial nerve. This usually transient impairment of a facial nerve is the result of swelling of the nerve within the relatively narrow bony facial canal. The etiology of this swelling is often uncertain but infections with the herpes simplex virus or the spirochetal agent producing Lyme's disease have been implicated.

More precise localization depends on whether this motor function alone is involved, as opposed to additional involvement of taste or the anterior two-thirds of the tongue, or of parasympathetic supply to the facial and salivary glands. This will be discussed above.

CRANIAL NERVE VIII - VESTIBULOCOCHLEAR NERVE

The vestibulocochlear nerve *(Fig. 12-7)* has two divisions: the vestibular and cochlear nerves, which are attached to the medulla at its border with the pons lateral to the root of nerve VII. The cochlear nerve is smaller than the vestibular nerve and is lateral to it. (See Discussion on auditory and Vestibular Pathways in Chapter 15 and on Special Sensory Systems on CD ROM).

COCHLEAR NERVE

The cochlear nerve is concerned with hearing and originates from the spiral ganglion of Corti in the petrous portion of the temporal bone. The peripheral process originates from the hair cells in the spiral ganglion.

The cochlear division of cranial nerve VIII originates from the spiral ganglion and terminates in the dorsal and ventral cochlear nuclei located on the external surface of the inferior cerebellar peduncle. The primary auditory fibers enter the cochlear nuclei and bifurcate, ending in the dorsal and ventral cochlear nuclei. Throughout the auditory system a tonotopic arrangement occurs. The secondary auditory fibers arise from the dorsal and ventral cochlear nuclei and form the three acoustic striae. The ventral acoustic stria originates from the ventral cochlear nucleus and courses through the ventral boundary of the pon-tine tegmentum, forming the trapezoid body.

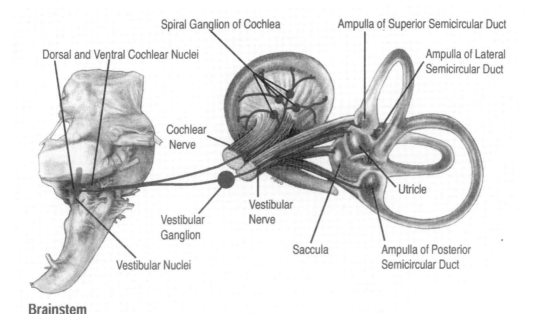

Figure 12-7. The Vestibulocochlear nerve- cranial nerve VIII

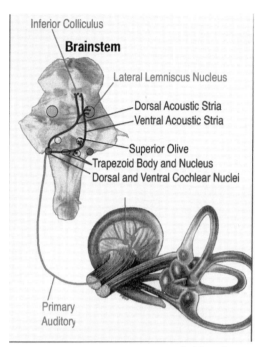

Figure 12-8 The Cochlear nerve. - a division of cranial nerve VIII

These fibers pass through the medial lemniscus, reach the superior olive, and migrate laterally to form the lateral lemniscus. The dorsal and intermediate striae arise from the dorsal cochlear nucleus and the dorsal half of the ventral cochlear nucleus. The dorsal stria crosses the midline ventral to the medial longitudinal fasciculus and joins the contralateral lateral lemniscus *(Fig. 12-8)*. The intermediate stria runs through the center of the reticular formation, crosses the midline, and enters the contralateral lateral lemniscus.

Some secondary auditory fibers end in the su-perior olivary, trapezoid and lateral lemniscus nuclei and the reticular formation. In these three nuclei tertiary axons originate which ascend, primarily crossed, to the inferior colliculus, where many of the axons terminate or send collaterals into the inferior collicular nuclei. Fibers from the Inferior colliculus and lateral lemniscus form the brachium of the inferior colliculus that ends in the medial geniculate nucleus of the thalamus. Auditory information ascends bilaterally to the medial geniculate nucleus, with the contralateral information forming the largest component. All auditory fibers synapse in the medial geniculate nucleus and are then projected into the auditory cortex, areas 41 and 42, the transverse temporal gyri of Heschl.

One of the most fascinating pathways in the central nervous system is the olivocochlear bundle of Rasmussen, or the efferent cochlear bundle. Crossed fibers originate from the areas dorsal to the accessory olivary nucleus, while uncrossed fibers originate from the superior olivary nucleus, join with the crossed pathway, exit with the inferior division of the cochlear nerve, and anastomose upon the hair cells in the cochlea. This prime example of central nervous control of peripheral receptors is most noticeable when loud sounds are quickly reduced to more moderate levels (e.g., the intensity of loud rock and roll is reduced to that of mood music). Note that similar mechanisms exist in the visual, olfactory, and gustatory systems.

Throughout the auditory system there are crossed and uncrossed fibers, so that representation is bilateral throughout.

Clinical Disorders

Auditory –Symptoms and Signs:

Deafness: Classification

Conductive: External and middle ear disease: (loss of low frequency sounds)

Sensory neural (nerve deafness – loss of high frequency sounds)

1. **Disease of Cochlea** – often bilateral, hereditary or related to loud sound exposure.

Antibiotic use – aminoglycosides, dihdrostreptomycin and Vancomycin.

If sudden and unilateral consider vascular etiology: occlusion of the internal auditory artery.

2. **Disease of Eighth Nerve:** Small tumors (acoustic neuroma) of 8th nerve within the bony canal may produce limited CNVIII symptoms. Large tumors, such as acoustic neuromas, meningiomas, etc., at the cerebellar pontine angle produce involvement of a cluster of cranial nerves (VIII, VII, ±V, ±IX, X).

Note that these "acoustic neuromas" arise from the Schwann cells of the vestibular divisions of the nerve. A more correct term is vestibular schwannoma.

Tinnitus – various forms of ringing in the ear –

Etiology – middle or inner ear or 8th nerve* Usually associated with hearing loss.

NOTE: Paroxysmal tinnitus may occur in association with various other symptoms of seizures – arising in the temporal lobe (transverse gyrus of Heschl – is the auditory projection)

Central lesions of cochlear pathway. Note that beyond the cochlear nuclei, central connections of the auditory system are both crossed and uncrossed. Therefore, a unilateral lesion is unlikely to produce a unilateral hearing defect. Massive intrinsic lesions are required to produce "brain stem deafness."

Vestibular Nerve

The vestibular nerve is concerned with equilibrium. Its primary neurons are bipolar and located in the vestibular ganglia in the internal acoustic meatus. The fibers terminate in the vestibular nuclei located in the ventricular floor of the medulla. The vestibular nerve originates from the hair cells in maculae of the scale and utricle and from the cristae of the ampullae of the three semicircular canals. It has five branches

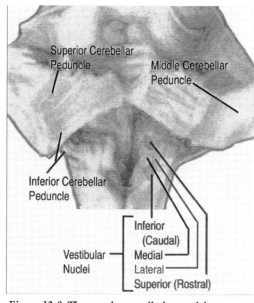

Figure 12-9. The secondary vestibular nuclei.

The *vestibular nerve* originates from the vestibular ganglia. The primary fibers terminate mostly in the four secondary vestibular nuclei in the floor of the fourth ventricle. The secondary vestibular nuclei are the inferior (descending), lateral, medial, and superior nuclei *(Fig. 12-9)*.

The *primary vestibular fibers* bifurcate into ascending and descending portions. The ascending axons end in the superior, lateral, and rostral portions of the medial nucleus, while the descending axons run in the Inferior nuclei and provide collateral's to the medial nucleus. The lateral nucleus forms the vestibulospinal tract, which ex-tends to sacral levels.

The *inferior vestibular nucleus* extends from the entrance of the vestibular nerve root, at the medullopontine junction, through the lower medullary levels. The *medial vestibular nucleus* is found medial to the inferior nucleus at the same levels. The medial and inferior vestibular nuclei form the prominent vestibular area on the floor of the fourth ventricle, lateral to the sulcus limitans. The myelinated bundles of the primary descending vestibular axons can delimit the inferior nucleus. The *lateral vestibular nucleus* is found lateral to the roots of nerve VIII in the tegmentum. The *superior vestibular* nucleus lies in the angle formed by the floor and wall of the fourth ventricle and extends through the upper pontine levels in this position. The superior cerebellar peduncle forms its dorsal border.

Primary vestibular fibers also run directly into the ipsilateral cerebellum, forming the Juxtarestiform body and projecting to the flocculus, nodulus, and uvula of the cerebellum (portions of the midline cerebellum which are old from a phylogenetic standpoint). The secondary vestibular axons originate from the secondary vestibular nuclei and project to the cerebellum, reticular formation, motor cranial nerve nuclei, and all levels of the spinal cord. Fibers from all the vestibular nuclei join the me-dial longitudinal fasciculus and are crossed and uncrossed with ascending and descending branches.

Vestibular Ascending Pathway. The vestibulothalamic pathway fibers arise from the lateral vestibular nucleus and ascend with the cervical fibers proprioceptive fibers in the medial lemniscus. They terminate in the ventral posterior nucleus and MGN. The vestibular impulses and responses have been noted in the cervical area of the postcentral gyrus and temporal lobe of the cerebrum.

Clinical Disorders: Vestibular:
Symptom:
1) Vertigo: a sensation of movement (usually rotation) of the environment or of the body and head:
 - When due to involvement of the horizontal (lateral) canals, the sense of movement is usually rotation in the horizontal plane.
 - When due to involvement of the posterior or vertical canals, the sense of movement may be in the vertical or diagonal planes with a sense of the floor rising up or receding.
- Vertigo must be distinguished from the more general symptom of dizziness – or light-headedness, which has many causes. Simple light-headedness as a symptom of pre-syncope (the sensation one is about to faint) may accompany any process which produces a general reduction in cerebral blood flow (cardiac arrhythmia, decreased cardiac output, peripheral dilatation of vessels, decreased blood pressure or hyperventilation) Hypoglycemia and various drugs also produce dizziness. Some patients cannot make the distinction and a series of operational tests may be necessary to reproduce symptoms.
 - Rotation, hyperventilation, orthostatic changes in blood pressure
Additionally, caloric testing, cardiac auscultation, EKG, and Holter monitor maybe necessary in selected cases.
2) Nausea and vomiting. Vertigo of vestibular origin is accompanied by these symptoms – because of the connections of the vestibular system to the emetic center of the medulla.
3). Nystagmus: a jerk of the eyes – slow component alternating with a fast component is associated with stimulation or disease of the vestibular system. Nystagmus is named from the fast component – and the direction of movement is specified: rotatory, horizontal or vertical.
4). Unsteadiness: Severe or mild.

Etiology – Multiple, Labyrinthine disease is the most common
 - *Meniere's disease:* Paroxysmal attacks of vertigo, ± vomiting and ataxia are associated with tinnitus and deafness. The etiology relates to endolymphatic hydrops – increase in volume and distention of the endolymphatic system of the semi circular canals.
 - *Benign positional vertigo* (nonbenign forms of positional nystagmus may occur with the brain stem disease, e.g., posterior fossa tumors). Sudden vertigo and associated symptoms of nystagmus, nausea – selectively occur on sudden tilting of head from the upright to recumbent position. This syndrome may occur as a late complication of head trauma. The proposed etiology relates to detachment of otolithic crystals that then float free in the endolymph. In particular positions of the head, this debris – intermittently blocks the flow of fluid into the cupula.
 - Antibiotic damage (streptomycin and aminoglycosides)
Vestibular Nerve-Vestibular neuropathy acute, self-limited syndrome – which follows a viral infection. Symptoms are vestibular without cochlear involvement. There is unilateral vestibular paresis.

Tumors – arising from or compressing vestibular nerve, "acoustic neuroma." Note that this tumor actually arises from the Schwann cells of the vestibular nerve.

Brain stem: Infarcts involving the vestibular nuclei will produce vertigo and/or nystagmus. Almost always other brain stem signs are present.

Cerebral cortex: Stimulation of temporal lobe in-patients with seizures of temporal lobe origin produces dizziness.

CRANIAL NERVE IX, GLOSSOPHARYNGEAL

The glossopharyngeal nerve *(Fig. 12-10)* is a mixed nerve with sensory and motor nuclei in the medulla. The functions of cranial nerve IX are often discussed with the vagus, as is

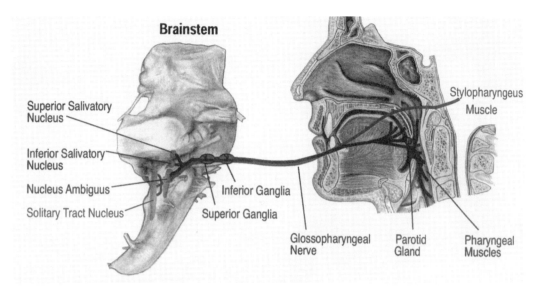

Figure 12-10. The Glossopharyngeal Nerve – Cranial Nerve IX

done here. The rootlets of nerve IX are found in the medulla in a postolivary position. The nerve trunk exits from the skull via the jugular foremen, where it lays anterolateral to the vagus nerve. In the jugular foramen are found two swellings on the nerve trunk of nerve IX, the superior and inferior ganglia.

Sensory Division. The primary sensory cell bodies are located in the superior and inferior ganglia and carry gustatory sensations from the taste buds in the poste-rior third of the tongue and cutaneous sensations from the pharynx, soft palate, auditory tube, middle ear, palatine tonsils, Eustachian tube, and carotid sinus. The general and special visceral axons enter the medulla and run in the tractus solitarius in the middle and posterior medullary level, synapsing in the nucleus solitarius. Afferent fibers are derived from baroreceptors in the wall of the carotid sinus (BP regulation) and chemoreceptors in the carotid body (ventilatory response to hypoxia).

Motor Division. The motor fibers of nerve IX originate from the most superior part of the nucleus ambiguous in the medulla. They pass posteriorly and laterally into the jugular foramen. The preganglionic parasympathetic fibers originate from the inferior salvatory nucleus in the medulla, join the rootlets of IX and run via the lesser petrosal nerve to the otic ganglion. The postganglionic fibers pass via the auriculotemporal nerve of the mandibular nerve and form the secretory pathway to the parotid gland.

Clinical Disorders.

- Lesions restricted to the nuclei or nerve roots of nerve IX are rare, but if they occur, taste is lost on the posterior third of the tongue and the gag reflex is absent ipsilaterally. There is no response when the posterior pharyngeal wall and soft palate are stimulated. The function of the parotid glands may also be impaired but can be easily evaluated by placing a highly seasoned food on the tongue and seeing if a copious flow from the duct occurs.

- Isolated lesions are uncommon. A combined involvement with vagus and accessory nerves at the jugular foramen is discussed below.

- Glossopharyngeal neuralgia is one of the few isolated syndromes: Symptoms are similar to those of trigeminal neuralgia, but localized to the posterior tongue and pharynx with swallowing, coughing or sneezing, providing the trigger. However, syncope may also occur.

CRANIAL NERVE X - VAGUS

The vagus nerve *(Fig. 12-11)* has one sensory and two motor nuclei in the medulla and

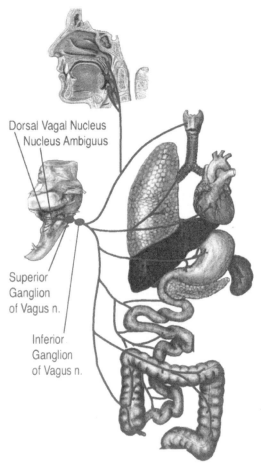

Figure 12-11. The Vagus Nerve- Cranial nerve X

also has the most extensive peripheral distribution of any cranial nerve. Its roots are located in a postolivary position in the posterolateral sulcus of the medulla, in line with the fibers of nerve IX but inferior to them. They exit from the posterior cranial fossa via the jugular foramen. The superior ganglion, the jugular, is found in the jugular foramen, while the inferior ganglion, the nodose, is found just as the nerve leaves the Jugular foremen. In the neck the vagus nerve is found in the carotid sheath close to the common carotid artery and Jugular vein.

Motor Nuclei

The dorsal motor nucleus of the vagus nerve forms the vagal trigone on the floor of the fourth ventricle at the medullary levels. It provides preganglionic parasympathetic innervation to the ganglion in the walls of the pharynx, trachea, bronchi, esophagus, stomach, small intestine and ascending portions of the large intestine, and the heart. The fibers of X end in the parasympathetic ganglia found in Auerbach's plexus (myenteric plexus). The other motor division of the vagus nerve originates from the posterior two-thirds of the nucleus ambiguous in the reticular formation of the medulla, the ventral motor nucleus, and connects to the striate muscles of the soft palate and pharynx, and the intrinsic muscles of the larynx. It is thus important in swallowing and speech. The ambiguous portion of nerve X innervates the following muscles:

- esophagus; the upper third is skeletal muscle, middle third is skeletal and smooth and the lower third smooth,
- pharynx: the superior, middle, and inferior constrictors of the pharynx
- palate: the palatoglossus, palatopharyngeus, salpingopharyngeal, and levator palatine velum of the soft palate,
- larynx: cricothyroid, the posterior cricoarytenoid, arytenoid, lateral cricoarytenoid, and thyroarytenoid.

Sensory Functions

The primary cell bodies of the sensory fibers are located in the inferior ganglion and carry gus-tatory information from taste buds in the epiglottis and pharynx, cutaneous innervation from the base of the tongue and epiglottis, and general visceral sensation from all the structures receiving motor innervation (pharynx, larynx, heart, tracheal bronchi, esophagus, small and large intestines. external ear, and dura of the sigmoid sinus). The fibers enter the medulla and synapse in the nucleus solitarius. The secondary axons subserving pain and temperature ascend in close proximity to the medial lemniscus to the ventroposterior thalamic nuclei.

The taste fibers ascend ipsilaterally to the parabrachial nuclei in the central tegmental tract and terminate in the cranial (medullary) and spinal portions. The taste fibers synapse in the parvocellular region of VPM and are then projected to the cortical taste area in the parietal operculum.

Reflexes. The vagus nerve is also important in many visceral reflexes: gag, vomiting, swallowing, coughing, sneezing, sucking, hiccuping, and yawning, and in the aortic sinus reflex. Ab-normal responses usually involve structures in addition to the vagus nerve.

Clinical Disorders. Bilateral lesions of the vagal portion of the nucleus ambiguous in the medulla are usually fatal because of its proximity to the pneumotaxic, apneustic, and medullary respira-tory centers. Unilateral lesions of the vagus nerve may or may not produce autonomic dysfunction. The heart rate, respiratory system, and gastrointestinal tracts appear to function normally, but there may be minor difficulty in swallowing. Injury to pharyngeal branches produces difficulty in swallowing.

Lesions of peripheral branches of the vagus nerve.

1. Lesions of the superior laryngeal nerve effect the motor innervation to the cricothyroid mm and the inferior constrictor of the pharynx. However, it also effects the internal laryngeal branch, the "watch dog of the larynx". This innervation of the mucosa of the laryngeal region produces coughing when foreign particles lodge in this region, thus keeping the airway open. When this reflex is not present due to injury to the nerve or an overdose of drugs, foreign objects can then enter the trachea and block the airway, which may produce aspiration pneumonia or even lead to death.

2. The recurrent laryngeal nerve supplies all of the intrinsic muscles of the larynx (except for the cricothyroid muscle) and the mucosa of the larynx below the vocal cords. Interruptions of the recurrent laryngeal nerves produce paralysis of the vocal cords, hoarseness, and dysphonia. Bilateral involvement of both recurrent laryngeal nerves produces aphonia and inspiratory stridor. Isolated peripheral lesions are more frequent and occur especially with operations on the thyroid gland. When the recurrent laryngeal is injured there is weakness in speech and anesthesia of the upper part of the larynx resulting in a weak gag reflex, paralysis of the cricothyroid muscle, and the laryngeal muscles tire easily.

The recurrent laryngeal nerve may also be injured during the following circumstances:

- Aneurysm of the thoracic aorta may involve the left recurrent laryngeal nerve.

- Apical lung (superior sulcus) tumors, pharyngeal carcinoma and metastatic involvement of cervical lymph nodes may all have effects on the vagus or recurrent laryngeal nerves.

- Diphtheria may involve the nerves.

- Myasthenia or D. botulinum toxin may affect the relevant neuromuscular junctions.

- Dissections of the neck for carotid endarterectomy or for thyroid surgery may damage the vagus or recurrent laryngeal nerves.

Central Lesions of the Vagus Nerve in the Brain Stem. Lesions in the upper medulla produce dysphagia, while lesions in the lower medulla cause dysarthria. The palate on the affected side is paralyzed, and the uvula deviates to the unaffected side (see discussion of lateral medullary syndrome in chapter 13).

The dorsal motor nucleus of the vagus nerve inhibits and depresses heart rate and constricts the coronary circulation. Paralysis of the vagus nerve produces smooth muscle contractions in the trachea, bronchi, and bronchioli with narrowing of the lumens. The dorsal motor nucleus of the vagus nerve simulates general alimentary function by causing the secretion of gastric and pancreatic juices important in peristalsis. It also stimulates the liver and spleen and inhibits the suprarenal glands. Isolated central lesions are rare. The nucleus ambiguous may be affected in syringobulbia and vascular lesions of the lateral medulla in association with other findings, as noted below, or in the poliomyelitis or the bulbar variant of amyotrophic lateral sclerosis.

Supranuclear lesions involving the extrapyramidal pathways to the dorsal motor or ambiguous nuclei usually cause difficulty in swallowing but only become clinically significant when

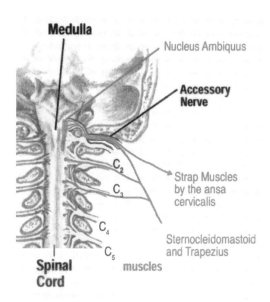

Figure 12-12. The Spinal Accessory Nerve - cranial nerve XI

they are bilateral.

CRANIAL NERVE XI - SPINAL ACCESSORY

The spinal accessory nerve *(Fig.12-12)* is a purely motor nerve that originates from both the medulla and spinal cord and is consequently divided into cranial (medullary) and spinal portions.

The *cranial root* arises from the most posterior/part of the nucleus ambiguous, exits in a postolivary position, enters the jugular forarnen, and then combines with the spinal root. This root innervates the uvula, levator veli palatini, and pharyngeal and laryngeal muscles in concert with cranial nerve X (to which it is accessory).

The *spinal portion* arises from the ventrolateral part of the ventral horn in the upper four cervical segments. The innervation to the sternocleidomastoid originates primarily from C2 while the innervation to the large trapezius muscle originates in C3 and C4. The fibers from the spinal portion of XI pass up through the foremen magnum, combines with the cranial roots, and distribute to the sternomastoid and trapezius muscles. Branches of the trigeminal also carry general sensations from these muscles. Injury to the nucleus ambiguous in the medulla causes paralysis and atrophy of the trapezius and sternomastoid muscles, resulting in weakness in head movements to the opposite side and shrugging of the shoulder.

Supranuclear lesions of the corticobulbar system produce only paresis of the muscles.

Clinical Disorders.

Isolated involvements of XI by intracranial lesions are rare, however, combined involvement at the jugular foramen is more common (See CASE 12-9 on CD). Dissections in the posterior triangle of the neck may also damage the nerve.

CRANIAL NERVE XII HYPOGLOSSAL

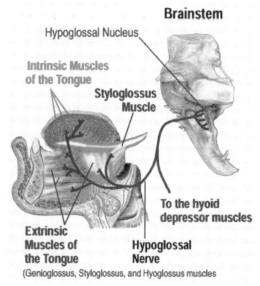

Figure 12-13. The hypoglossal nerve – cranial nerve XII.

The hypoglossal nerve *(Fig. 12-13)* is a purely mo-tor nerve that originates from the hypoglossal nucleus, which forms the hypoglossal trigone in the medullary floor of the fourth ventricle. The fibers pass anteriorly, passing inferiorly between the olive and pyramids. The fibers exit from the cranium via the hypoglossal canal and supply all the extrinsic muscles (styloglossus, hyoglossus, and genioglossus) and the intrinsic muscles (superior and inferior longitudinal, transverse, and vertical) of the tongue with the exception of the palatoglossal that is innervated by the vagus. Branches from V3 of the trigeminal nerve carry general sensations from the tongue musculature.

Clinical Disorders.

Unilateral injury to the nerve or nucleus pro-duces atrophy and paralysis in the ipsilateral tongue with the tongue protruding toward the injured side. This is due to the unopposed action of the intact genioglossus muscle pushing the tongue up and out from below.

Bilateral paralysis of the nucleus or nerve causes difficulty in eating and dysarthria.

Lesions limited to this nerve are unlikely to occur with brain stem pathology. The hypoglossal nucleus may be involved as part of a more generalized somatomotor neuron process in poliomyelitis or amyotrophic lateral sclerosis (bulbar palsy). Limited lesion may occur along the extra-medullary course of the nerve, but rarely at the hypoglossal foramen, more commonly with radical cervical dissections, or as a complication of carotid endarterectomy. Supranuclear lesions rarely produce paralysis of the tongue.

II. Effects of Extrinsic Lesions on Cranial Nerves

Cranial nerves are commonly affected by mass lesion and the syndromes produced by these extrinsic lesions are summarized in Table 12-5.

Cases demonstrating combined CRANIAL NERVE SYNDROMES are found on the CD as follows:

Cavernous Sinus Syndrome

Case 12-6: Lesion base of skull involving left cavernous sinus plus possible involvement of foramen ovale region

CEREBELLOPONTINE ANGLE SYNDROME.

Case 12-7:. Cerebellar pontine angle tumor, a vestibular Schwannoma acoustic neuroma.

Vestibular

Case 12-8: Vestibular Schwanoma

JUGULAR FORAMEN SYNDROME

Case 12-9: Jugular foramen tumor-probably neurofibroma.

III. Voluntary Control of the Cranial Nerves-Corticobulbar and Corticomesencephalic Pathways.

This system distributes to the motor nuclei of the cranial nerves (Cranial Nerves V. VII, IX, X XI, XII) and provides voluntary and involuntary con-trol of the muscles and glands.

Voluntary Control of Cranial Nerves V, VII, IX-XII. The corticobulbar pathway controls the cranial nerves in the medulla and pons, the bulb. The cranial nerves controlled in the pons are V, VI, VII, while in the medulla IX, X and XII. We have used the term corticonuclear as it reminds the student of the key role of the cranial nerve nuclei in movement. These fibers are found anterior to the corticospinal fibers in the genu of the internal capsule and medial to the corticospinal tract in the cerebral peduncle, pons, and medullary pyramids.

From the lower third of precentral gyrus, area 4 fibers supply the muscles of facial expression (motor nerve VII), and the muscles of mastica-tion (motor nerve V), and deglutination (ambiguous nuclei of nerves IX and X). From the Inferior frontal gyrus and frontal operculum, the posterior part of the par triangularis, area 44, and fibers supply the larynx.

It appears that many corticonuclear axons end on interneurons and not directly on the motor neuron of the cranial nerve. The cortical innervation of the majority of the cranial nerves is mostly bilateral with the exception for the contralateral control of the lower facial muscles; consequently unilateral lesions in the cortical bulbar system produce usually only weakness and not paralysis.

Clinical Disorder: A bilateral supranuclear lesions of the corticonuclear system produces upper motor neuron signs and also inappropriate behavioral signs, incontinence, pseudobulbar palsy.

Pseudobulbar Palsy. A bilateral corticobulbar lesion produces a pseudo bulbar syndrome – characterized by a release of bulbar functions. This is an upper motor neuron lesion with hyperactive jaw jerk, and spasticity- affecting speech, and gag reflexes. Cranial nerve components of emotions are released - emotional lability. This most likely represents the release of cortical control via the corticobulbar innervation of the muscles controlled by the cranial nerves and now the limbic control to these same nuclei takes over via descending autonomic pathways with the emotional state being undampened by the frontal lobes. ALS produces upper and lower motor paralysis.

Voluntary control of eye movements. The corticomesencephalic portion controls cranial nerves III, IV, and VI. Eye movements originate from the frontal eye fields in the caudal part of the middle frontal gyrus and the adjacent inferior frontal gyrus, area eight. In upper midbrain levels, fibers to cranial nerves III and IV leave the corticomesencephalic tract. Fibers to cranial nerve VI descend and terminate on contralateral parabducens/parapontine reticular nuclei of the pons that coordinates movements with CN III and IV through the MLF.

TABLE 12-5. SYNDROMES OF THE ADJACENT CRANIAL NERVES DUE TO EXTRINSIC MASS LESIONS

[Modified from Adams, Victor, & Ropper 1994]

SITE	CRANIAL NERVES INVOLVED	CAUSE
1. Superior orbital fissure (sphenoidal fissure)	III, IV, V, (ophthalmic), VI	Meningiomas Aneurysm
2. Optic foramen	II	Meningioma, inner 1/3 Sphenoid wing
3. Cavernous sinus Case 12-6 CD ROM	III, IV, VI, V1 ophthalmic V2 maxillary Note: with pituitary optic chasm also involved & CN findings may be bilateral.	Intracavernous carotid aneurysm Traumatic carotid cavernous fistula Thrombosis of cavernous sinus Invasive tumors of sinuses & pituitary
4. Apex of petrous bone	V, VI	Inflammation of bone (Gredenigo syndrome
5. Internal auditory canal	VII, VIII	Tumors and chronic infection
6. Cerebellar pontine angle Case 12-9	VII, VII, + V, + IX	Tumors: acoustic neuroma & meningiomas
7. Jugular foramen Case 12-7	IX, X, XI*	Tumors - e.g., neurofibroma

* Large lesions may also involve CNXII

Figure 12-14: The Twelve Cranial Nerves

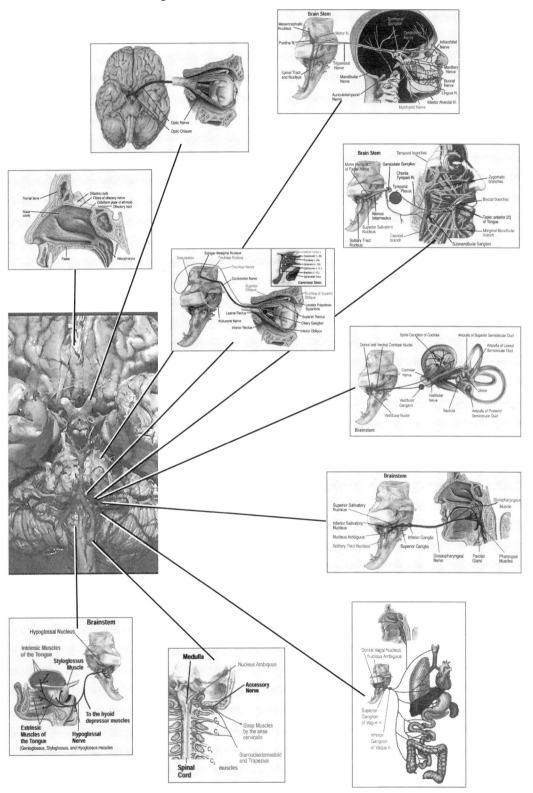

CHAPTER 13
Brain Stem: Clinical Considerations

LOCALIZATION OF THE DISEASE PROCESSES IN THE BRAIN STEM

Of all sites in the central nervous system, the brain stem provides the best example of how the anatomical pattern of involvement may allow the prediction of the actual pathologic process. The student has, of course, already encountered several examples of this phenomenon at the level of the spinal cord. For example, the clinical finding of a selective dissociated loss of pain and temperature in a cape-like distribution over the shoulders and upper extremities implies an anatomical process in a pericentral location involving the anterior commissure. One could readily deduce that an intrinsic pathological process was present and, moreover, could assume that in most cases the pathological process was that of syringomyelia.

At the level of the brain stem such examples of close correlation occur frequently. Thus, the progressive combined involvement of the ipsilateral cranial nerve VIII (auditory and vestibular), nerve VII (facial), and nerve V (trigeminal), with the manifestations of deafness, tinnitus (a buzzing or ringing sound in the ear), dizziness, peripheral facial paralysis, and unilateral facial numbness, not only indicates the location of the pathology (the cerebellar pontine angle) but also strongly suggests the actual nature of the pathology: vestibular Schwannoma (acoustic neuroma was the old term). While other pathological lesions may occur at this site, they are much less common.

There are several possible approaches to the problem of localization of disease processes in the brain stem. We have already considered in Chapter 11 a number of guidelines for localization that should be reviewed at this time. We will now consider in greater detail the particular anatomical syndromes and the various types of pathology in terms of specific extrinsic and intrinsic syndromes. Many of the extrinsic syndromes related to cranial nerve syndromes have already been considered in chapter 12 and are summarized in table 12-5. When speaking of the anatomical syndromes of the brain stem in clinical neurology, we will limit our considerations to the medulla oblongata, pons, and midbrain. The thalamus, hypothalamus, and subthalamus represent a natural rostral continuation of the midbrain and diseases involving the midbrain sometimes also involve the adjacent diencephalon. From the standpoint of vascular anatomy, the midbrain, and the diencephalon do share a vascular supply from the penetrating branches of the posterior cerebral arteries

SPECIFIC SYNDROMES AND DISEASE ENTITIES

Specific syndromes and disease entities will now be considered. Our emphasis will be on various classic syndromes, each of which implies a single level lesion. The reason for this emphasis in the teaching of clinical anatomical correlation is obvious. However, the student should realize that, in actuality, a number of the diseases involving the brain stem are not simply limited to a single level but rather affect multiple levels; that is, they are multifocal or they are system disorders. Thus, although we will consider in detail patients with limited focal infarcts, it is more frequent to have infarction at several levels of the brain stem in occlusive disease of the basilar-vertebral system (perhaps the most common disease involving the brain stem). Another rather common disease of the brain stem, multiple sclerosis, almost always follows a multifocal pattern.

DIFFERENTIATION OF EXTRINSIC VERSUS INTRINSIC DISEASES.

The same categories of disease already familiar to the student from his study of the spinal cord will be employed: extrinsic and intrinsic. At the level of the spinal cord, extrinsic diseases were much more common and were emphasized. At the level of the brain stem, the intrinsic diseases occur much more frequently. (This is also the case at the level of the cerebral hemisphere). Nevertheless, extrinsic diseases have an importance outside of their actual frequency of occurrence. As we have already noted, at the level of the spinal cord, extrinsic diseases represent progressive compression type problems such as tumors or degenerative disk disease. Early diagnosis and surgical relief of the compression will prevent the development of disability and will preserve life. All lesions compressing the brain stem are potentially life-threatening potential if they progress to the vital centers of the medulla involved in the control of respiration, blood pressure, swallowing and changes in intracranial pressure (ICP) etc. Moreover involvement of those areas essential for motor control (reticular formation of pons and medulla) or those areas in the midbrain reticular formation essential for the maintenance of consciousness will result in a severe bedridden state.

In distinguishing extrinsic and intrinsic diseases, several differential points have already been mentioned. (Table 13-1).

The development of techniques such as the CT and MRI scan has markedly increased our ability to confirm the anatomical localization of disease involving the brain stem. Magnetic resonance angiography has provided a noninvasive method of imaging the blood supply of the brain stem (See chapter 2).

EXTRINSIC DISEASES OF THE BRAIN STEM

EXTRINSIC TUMORS AND OTHER MASS LESIONS

As indicated, almost all of the problems within the extrinsic category are tumors. When we use the term "extrinsic" in the present context, we mean extrinsic to the actual substance of the brain stem. Some of the tumors to be considered are actually invasive tumors intrinsic to the cerebellum, or arising from the ependymal cells lining the 4th ventricle that are secondarily compressing or infiltrating the brain stem. Some of these tumors are actually derived from cells of the glial series,

TABLE 13-1. PATTERNS OF EXTRINSIC VS INTRINSIC DISEASE OF THE BRAIN STEM

PATTERNS OF EXTRINSIC LESIONS:	PATTERNS OF INTRINSIC LESIONS:
a. Successive unilateral involvement of multiple cranial nerves early in the course of the disease, with fiber systems involved only later in the disease.	a. Simultaneous segmental involvement of the cranial nerve nuclei and of long tracts, as in a lateral medullary infarction.
b. Early involvement of the eighth cranial nerve (as in acoustic neuroma).	b. Progressive bilateral involvement of cranial nerves and long tracts, as in infiltrating glioma.
c. A syndrome in which symptoms are limited to the midline cerebellum: loss of equilibrium and truncal ataxia.	c. Involvement of the fifth cranial nerve- descending spinal tract with selective involvement of pain sensation over the face. Involvement of the MLF, as in MS, glioma, or infarction with the occurrence of inter nuclear ophthalmoplegia and vertical nystagmus
d. Symptoms of ICP occurring early in the course of the disease - headache, nausea, and vomiting. Occasionally, as in midline cerebellar lesions, the symptoms of ICP (with or without truncal ataxia) may be the only symptom.	d. If an infiltrating tumor, the symptoms of ICP tend to occur late in the disease.

e.g., cystic astrocytomas of the cerebellum, medulloblastomas, and ependymomas. Some are locally invasive malignant tumors, e.g., nasopharyngeal carcinoma infiltrating the bone of the skull or carcinoma of the lung or breast, metastatic to the cerebellum. (In general, carcinomas only rarely spread in a metastatic manner to the brain stem per se).

(1) The Cerebellar pontine Angle Syndrome.

Vestibular schwannoma (acoustic Neuroma) (Fig. 13-1, 13-2): Tumors at the cerebellar pontine angle account for 10% of all central nervous system tumors. Most tumors encountered at the cerebellar pontine angle are

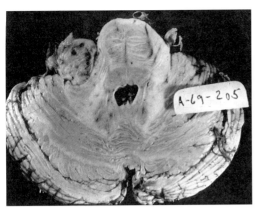

Figure 13-1. Vestibular schwannoma ("acoustic neuroma") at the cerebellar pontine angle. (Courtesy of Dr. Jose Segarra and Dr. Remedios Rosales, Boston Veterans Administration Hospital)

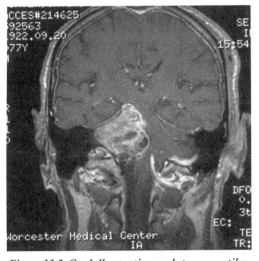

Figure 13-2. Cerebellar pontine angle tumor: vestibular schwannoma. MRI. A large mass extends from an enlarged right internal auditory canal compressing and distorting the pons and medulla. (Courtesy Drs. Milton Weiner & Alex Danylevich)

vestibular neuromas arising from the Schwann cells about cranial nerve VIII. As already noted the tumor actually arises from Schwann cells within the vestibular component of cranial nerve VIII within the internal auditory canal. This is a disease of the middle-aged or older adult. (Onset of symptoms in the teens or early 20s-should suggest the possibility of neurofibromatosis 2 in which bilateral acoustic neuromas occur (see Martuza and Eldridge 1988 and chapter 9) The early symptoms and findings relate to a progressive involvement of nerve VIII (progressive loss of hearing of a neural type and a dead labyrinth on caloric stimulation). As the tumor enlarges, symptoms referable to nerves VII and V and to the cerebellum develop. An illustrative case has been presented in Chapter 12.

Other tumors in the cerebellar pontine angle: Although less frequent, other types of pathology are found in this location: meningiomas (Fig.2-10) arising from arachnoidal cell nests) and cholesteatomas (epidermoids). A large series from the Mayo Clinic (Laird et al 1985) indicated 20 meningiomas over a six-year period compared to 160 acoustic neuromas over a similar period (Harner et al 1985). Specific neurological deficits were as frequent as in patients with acoustic neuromas. Diagnosis may be made with contrast enhanced MRI or if not available contrast enhanced CT Scan.

(2) Syndromes of the cerebellum:

a. *Midline Cerebellum (Fourth Ventricle) Syndrome).* The midline cerebellar syndrome is frequently encountered in neurology. Clinical deficit: The patient often a child presents with a disturbance of balance and ataxia of the trunk and gait but no clearly lateralized symptoms. In general, the etiology is neoplastic. The type of tumor varies as a function of age. The most common cause in an infant or child is a medulloblastoma *(Fig. 13-3).* This tumor arises from nests of external granular cells in the nodulus, an older portion of the cerebellum relating to the vestibular system. Since this portion of the cerebellum projects as a roof into the fourth ventricle, early obstruction of the ventricle is to be expected. A typical case is presented in Chapter 19 (cerebellum). Occasionally this tumor may present in adolescents and young adults. The cells may seed via the CSF, so that rarely the early symptoms may those of a lumbar radiculopathy.

In children and young adults, the most common cause is the ependymoma that arises from ependymal cells in the floor or roof of the fourth ventricle *(Fig.13-4).* The tumor mass grows into the ventricle, obstructing this cavity but also exerting upward pressure on the midline nodulus (and other aspects of the cerebel-

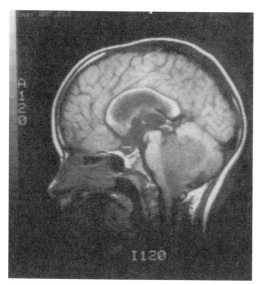

Figure 13-3. Medulloblastoma. MRI (T1). This large tumor arising from the midline cerebellum in this 4-year-old child fills the 4th ventricle producing hydrocephalus and compressing the medulla. (Courtesy Dr. Milton Weiner)

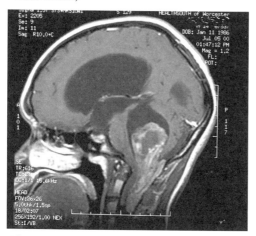

Figure 13-4. Ependymoma of the 4th ventricle. MRI (T1). This 14-year-old male had a 4-month history of headache and neck pain, with gait ataxia and nystagmus. This large mass obstructs the 4th ventricle, producing hydrocephalus, displaces the cerebellar vermis and extends out the 4th ventricular foramen then down through the foramen magnum to compress the spinal cord. (Courtesy of Drs. Gerald McGuillicuddy, Milton Weiner &Artinian).

lar vermis). At times however, the ependymoma may extensively invade the tegmentum of the medulla and pons. At times the tumor is a mixture of ependymoma and astrocytic components. Taken together, the medulloblas-toma and the ependymoma constitute the most common central nervous system tumors in the pediatric age group.

In the adolescent and young adult the low-grade cystic astrocytoma, must also be considered, although, a lateral hemisphere location is the much more common location.

In the middle-aged adult, the most common cause of the syndrome is a midline hemangioblastoma. These are tumors derived from embryonic vascular elements. In contrast to the medulloblastoma and the ependymoma, the hemangioblastoma and the cystic astrocytoma are surgically curable neoplasms. It is often found that a nodule of tumor exists within a nonneoplastic cystic cavity, allowing for removal of the tumor without sacrifice of a large amount of the cerebellum.

When the patient is 50 years old, hemangioblastomas and metastatic tumors constitute the major midline entities. Metastatic carcinomas in general, however, are more often found in a non-midline cerebellar hemisphere location.

b.*Lateral Cerebellar Hemisphere Syndrome*. Clinical Deficit: lateralized symptoms affecting the ipsilateral arm and leg have already been mentioned. The syndrome is discussed in greater detail with an illustrative case history in chapter 20 on the cerebellum. As regards the specific mass lesions, this in large part depends on the patient's age:

- In the child and young adult, the most common lesion is probably the cystic astrocytoma.

- In the middle-aged and older age groups, the most common lesions are metastatic carcinoma and the hemangioblastoma. The metastatic carcinoma may represent a solitary metastasis from a primary lesion in lung, breast, ovary, or kidney or may be only one of multiple lesions. The cerebellar lesion in any case is often surgically removed since it does represent a relatively acute life-threatening lesion. The hemangioblastoma has already been discussed. Other mass lesions of the cerebellar hemisphere in the pre antibiotic era included brain abscess (the source of infection was usually the adjacent middle ear and mastoid) and tuberculomas. . Hypertensive cerebellar hemorrhages may also present as acute mass lesions as may acute edematous cerebellar infarcts.

(3) Extrinsic Midbrain syndromes:

a. *Syndrome of compression of the pretectal area and superior colliculus (Fig.13-5, 13-6).* These are essentially three causes of this syndrome: a. tumors of the pineal (pinealomas) occurring primarily in men and boys b. hydrocephalus in infants, c. hypertensive hemorrhages of the posterior thalamus occurring most commonly in middle age or older patients. Tumors of the pineal are of several types. (1) The most frequent (50%) are germ cell tumors: germinoma, or atypical teratomas:

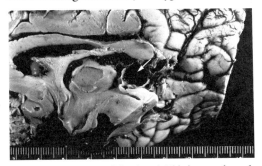

Figure 13-5. Tumor of the pineal. This is an enlarged nonmalignant cyst arising from the pineal gland (arrow). (Courtesy of Dr. John Hills, and Dr. Jose Segarra).

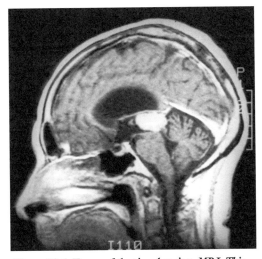

Figure 13-6. Tumor of the pineal region. MRI. This 68-year-old female presented with apraxia of gait, urinary incontinence and dementia. Symptoms resolved following a shunt procedure and radiotherapy.

arising from congenital cell nests of a mixed type and occurring at any age but most commonly in males under 10 years. Histologically these tumors are identical to congenital tumors of the testes (seminoma) or dysgerminoma of the ovary. (2) In addition to the malignant germinoma, non-invasive congenital germ cell tumors may occur and may be resected: teratomas, dermoids, and epidermoids. (3) Other types are astrocytomas and ependymomas, pineal parenchymal tumors (pineocytomas), meningiomas and arachnoid cysts.

For additional discussion see DeGirolami & Schmidek(1975), Stein (1979), and MaGieruiez et al 1983.

An illustrative case history (11-2) has been presented in the brain stem chapter on functional localization.

b. Syndromes of the tentorium

The Tentorial Notch syndrome: Herniation of the medial aspects of the temporal lobe (uncus and parahippocampal gyrus) secondary to supratentorial mass lesions is the most common syndrome found in relation to the tentorium More lateral supratentorial masses as in temporal lobe tumors produce a lateral herniation syndrome characterized initially by an ipsilateral fixed dilated pupil and then an ipsilateral CN 3 paralysis. Additional compression will shift the midbrain and compress the opposite cerebral peduncle against the rigid tentorial edge resulting in an ipsilateral hemiparesis in addition to the already present contralateral hemiparesis. As compression and displacement of the brain stem progresses, hemorrhages (Duret hemorrhages) into the substance of the midbrain and pons develop. A progressive sequence of midbrain, then pontine and eventually medullary symptoms evolve. Supratentorial masses in a more parasagittal location produce central herniation with bilateral midbrain (CN III) and subsequently pontine – medullary involvement. *(Fig.13-7, 13-8, 13-9).*

Tentorial meningiomas: These are rare. A variety of syndromes may occur: (1) midline obstructing mass with increased intracranial pressure but without definite localizing fea-

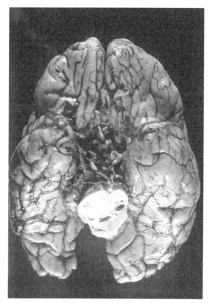

Figure 13-7. Herniation of the parahippocampal gyrus secondary to a subdural hematoma. Note residuals of Duret hemorrhages in pons (Courtesy of Pathology department St. Vincent Hospital).

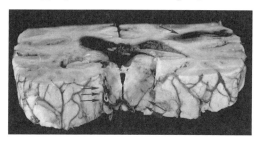

Figure 13-8. Tentorial herniation with brain stem compression. A large intracerebral hematoma and hemorrhagic infarction (single arrow) in right cerebral hemisphere has produced herniation of the medial aspect of temporal lobe (double arrow). Shift of the brain stem has occurred with compression and infarction of the left cerebral peduncle against the tentorium (triple arrow). (Courtesy of Pathology Department, Tufts New England Medical Center.)

tures, (2) mass lesion extending above the tentorium to involve the temporal and occipital lobes, (3) mass lesion extending below the tentorium, displacing or invaginating the cerebellum and compressing the brain stem and cranial nerves.

(4) Jugular foramen syndrome:

This is caused by neurofibromas or other tumors arising from the lower cranial nerves (CN IX, X and XI) as these nerves pass togeth-

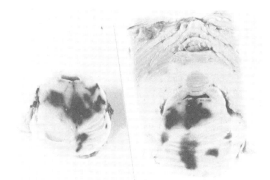

Figure 13-9. Duret hemorrhage. Multiple small hemorrhages are present in the pontine tegmentum and to a lesser extent in the basilar pons secondary to massive infarction of cerebral hemisphere, temporal lobe herniation and brain stem compression. (Courtesy of Dr. John Hills)

er through the jugular foramen. An illustrative case history is presented in Chapter 12.

(5) Foramen Magnum Syndromes.

a. *Herniation syndrome (Fig.13-10).* This is the most common process at the foramen magnum and is sometimes referred to as cerebellar pressure cone. Any mass lesion in the posterior fossa may produce herniation of the cerebellar tonsils compressing the lower end of the medulla and the upper cervical spinal cord. When gradually acquired, head tilt and stiff neck and pain in the neck may be early symptoms. When acutely acquired, a rapid and fatal arrest of respiratory and circulatory function is likely to occur. Such an acute event may be precipitated by the performance of a lumbar

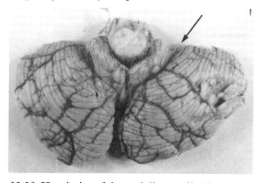

13-10. Herniation of the cerebellar tonsils. The cerebellar tonsils have been compressed downward through the foramen magnum (arrow). secondary to a teratoma of the pineal, compressing the caudal medulla and upper cervical spine. (Courtesy of Dr. John Hills and Dr. Jose Segarra)

puncture in a patient with a posterior fossa mass lesion. At times in subacute cases, this final stage of arrest of respiration and fall in blood pressure is preceded by a period of irregular respiration, slow pulse, and rising blood pressure.

Once tentorial herniation has occurred in a supratentorial mass lesion, increased mass is present in the posterior fossa. In addition, hemorrhage into the brain stem and edema of the brain stem act to increase the mass in the posterior fossa. Tonsillar herniation with medullary compression would then in this situation also be the expected terminal event. (For a more complete discussion of the clinical pathophysiology of the tentorial and tonsillar herniation syndromes, the student is referred to Plum and Posner, 1980).

b. *Meningiomas:* These are rare in this location and produce a variable picture. Their growth may be primarily in the posterior fossa with compression of the lower cranial nerves, brain stem, and cerebellum. On the other hand, the tumor may extend down through the foramen magnum, compressing the upper and middle cervical roots and then the long tracts on one or both sides of the spinal cord. In some cases, a mixture of posterior fossa and cervical cord findings will be present.

SYNDROMES DUE TO DEVELOPMENTAL AND BONY ABNORMALITIES

These syndromes primarily occur at the foramen magnum.

(1) The Arnold Chiari syndrome has been discussed and illustrated in chapter 9.

(2) Deformation of the bones at the base of the skull occurs in platybasia, a flattening of the sphenoid and occipital bones (sometimes referred to as basilar impression because the bones about the foramen magnum are pushed into the posterior fossa). There are essentially two causes for this deformation: Paget's disease, and various developmental malformations.

Paget's disease is a chronic metabolic disease of bone occurring in the adult, in which there is both excessive and disorganized bone formation and excessive bone destruction. Bony overgrowth may result in pressure on nerve roots, on cranial nerve VIII and on the facial nerve. Involvement of the vertebrae may result in spinal cord compression. Involvement of the skull is common. When the disease process affects occipital and sphenoid bones at the base of the skull, a softening, flattening, and distortion occurs. As a result, the odontoid process projects up into the posterior fossa. The posterior fossa in its normal state is a relatively small cavity. The end result is a compression of the contents of the posterior fossa -- medulla, cerebellum, and lower cranial nerves -- in addition to the upper cervical spinal cord.

Case 13-1 presented on CD ROM illustrates this problem in detail.

VASCULAR SYNDROMES AFFECTING THE BRAIN AS EXTRINSIC LESIONS:

(1) **Aneurysms:** At points of congenital weakness in the wall of an artery (usually points of bifurcation) a dilated thin-walled sac may form, termed an aneurysm. Intracranial aneurysms produce symptoms in several ways: (1) the dilated sac may act as a space-occupying lesion compressing adjacent structures such as cranial nerves, (2) rupture of the aneurysm may occur with leakage of blood into the subarachnoid space, (3) the sudden hemorrhage into this space may also dissect into the brain with the formation of a hematoma and/or rupture into the ventricular system may occur, (4) the circulation of blood in the artery beyond the point of the aneurysm may be impaired, producing ischemia and destruction of tissue supplied by the artery, (5) additionally blood in the subarachnoid space about arteries may produce vasospasm,(6) blood in this subarachnoid space and ventricles may also block the circulation of CSF resulting in hydrocephalus.

The majority (85 to 90 per cent) of aneurysms arises from what is termed the anterior circulation at essentially three locations: (1) the anterior cerebral-anterior communicating junction, (2) the middle cerebral artery bifurcation, and (3) the junction of the posterior communicating and internal carotid arteries. As already discussed in the cranial nerve chapter, Case 12-1 at this last location, an enlarging saccular aneurysm may compress cranial nerve III. The finding of a fixed, dilated pupil and ptosis of the lid in a patient with subarachnoid hemorrhage may provide a valuable clue for localization of the aneurysm at the junction of the posterior communicating –internal carotid junction. Approximately 10 to 15 per cent of saccular aneurysms are found in relation to the basilar vertebral system. Common sites include the bifurcation of the basilar artery. In addition to saccular aneurysms, fusiform dilatation of the basilar artery may occur in atherosclerosis resulting in an enlarged tortuous vessel that may compress adjacent structures.

(2) **Hypertensive cerebellar hemorrhage** (chapter20). The hematoma may secondarily compress the fourth ventricle and brain stem. The early symptoms of acute onset of headache, dizziness, vomiting, and ataxia are often followed by coma and evolving extraocular muscle findings. Similar symptoms and findings may also occur in acute large cerebellar infarcts due to the effects of edema.

INTRINSIC DISEASES OF THE BRAIN STEM

The disease entity affecting the brain stem that is most frequently encountered in clinical practice is vascular disease, predominantly of the ischemic-occlusive type. Ischemic-occlusive cerebrovascular disease usually produces its initial clinical manifestations in patients over the age of 50. The basic pathological process of atherosclerosis which underlies occlusive vascular disease (the deposition of lipids such as cholesterol in the walls of vessels) certainly may begin in an asymptomatic manner earlier in life. On the other hand, multiple sclerosis, which is the second most frequent brain stem disease entity encountered, usually produces its initial manifestations in the age group 20 to 40. Of the other intrinsic diseases affecting the brain stem, the next most frequently encountered are brain stem gliomas (occurring predominantly in the pediatric age group under 20 years and the bulbar variety of amyotrophic lateral sclerosis (occurring predominantly in the age group over 40 years).

ISCHEMIC-OCCLUSIVE DISEASE OF THE BASILAR VERTEBRAL ARTERIES

Stenosis (narrowing) or occlusion of the vertebral and basilar arteries may occur at a number of sites. In general, maximum areas of atherosclerosis disease occur at the points of origin, bifurcation, or angulation of the extracranial and intracranial portions of the large and medium size cerebral arteries. There is no one-to-one relationship of a particular site of stenosis or occlusion to a specific pattern of clinical symptoms. Thus, occlusion of a vertebral artery may occur with no resultant clinical symptoms in evidence. In other instances, such an occlusion may result in transient symptoms indicating ischemia that is a temporary decrease in blood flow (perfusion) below a critical level, within the distribution of a particular vessel or combination of vessels arising from the vertebral or basilar arteries. In still other instances, such an occlusion may result in actual tissue destruction, infarction (encephalomalacia) within the distribution of one or more vessels taking origin from the vertebral or basilar arteries; e.g., the posterior inferior cerebellar artery arising from the vertebral artery or the posterior cerebral artery from the basilar artery.

In still other instances, one may find that infarction within the distribution of a particular vessel has occurred without actual occlusion of that particular vessel or even of the parent vessel. Instead, stenosis with a decrease in blood flow has occurred.

Factors accounting for this lack of correspondence between the state of a particular vessel and the vascular status of the region supplied by that vessel include the following:

1. There are significant congenital variations with regard to the normal patency of the two vertebral arteries. One vertebral may be hypoplastic. (Fig. 2-27).

2. There are significant congenital variations in the circle of Willis often representing

persistence of an earlier fetal pattern.

3. There are significant leptomeningeal anastomoses over the surface of the cerebellum between the posterior inferior, anterior inferior and superior cerebellar arteries.

4. To a variable degree anastomoses between extracranial and intracranial arteries may occur.

5. Variations may also reflect the fact that often atherosclerosis affects more than one major artery.

6. Considerable variability may also occur because of certain general systemic and metabolic factors.

7. Certain mechanical factors may introduce a degree of variability. Thus, a patient may have stenosis of a vertebral artery without clinical symptoms when the head is in a neutral position. Hyperextension or rotation may produce symptoms. Postural effects may also occur related to a relative decrease in blood pressure in the upright position compared to the recumbent position.

Not all-occlusive disease is due to atherosclerosis. Emboli account for a significant proportion of disease involving the posterior circulation. Among the 407 well studied patients in a posterior circulation registry, (Caplan 2000) at least 40% had embolism as the single most likely diagnosis, as opposed to 32% with large vessel disease (atherosclerosis-thrombosis) and 14% with penetrating artery disease. The most common sources of embolism were cardiac (60%) and intra-arterial (35%), usually the vertebral arteries. Cardiogenic emboli were most likely to produce distal infarcts within the territory of the posterior cerebral and posterior inferior cerebellar arteries. Emboli originating from the external vertebral artery were more likely to produce proximal infarcts within the territory of the intracranial vertebral artery and the origin of the basilar artery plus or minus additional more distal infarcts. Overall of the 156 patients with embolic infarcts, 53% had distal infarcts only, (distal basilar artery-posterior cerebral and superior cerebellar), 17% had proximal infarcts only and 6% had middle territory infarcts (basilar artery and branches). The remainder were mixed including the distal field.

As regards large vessel disease, in terms of the location of sites of >50% stenosis due to atherosclerosis, among 260 patients, 44% involved the extra cranial portion of the vertebral artery, 42% the intracranial portion of the vertebral artery, 42% the basilar artery and 15% the posterior cerebral artery. Clearly some patients had stenosis or occlusion at multiple sites.

Specific vascular syndromes

Basic anatomy: We will now consider syndromes involving the territories of particular arteries. In addition to the large vessels, the vertebrals and basilar, we will deal with the major branches of these vessels. There is a general plan of brain stem vascular architecture. The basilar artery and the two vertebral arteries are placed on the basal (ventral) surface of the brain stem. One can visualize a midline or paramedian wedge-shaped area extending into the brain stem with a base on this ventral area. Penetrating vessels, termed paramedian, supply this paramedian wedge. The lateral tegmental portions of the brain stem are supplied by branches, long circumferential arteries, named for their areas of supply over the surface of the cerebellum: posterior inferior, anterior inferior, and superior cerebellar arteries. In addition, at the upper midbrain level, quadrigeminal arteries arise from the posterior cerebral arteries which function as long circumferential arteries with regard to this area of the brain stem. Variable transverse arteries, termed short circumferential, supply ventral lateral portions of the pons and midbrain arising from the long circumferential arteries of the basilar artery. Short circumflex branches also arise from the vertebral arteries to supply ventral lateral medulla. At each level of the brain stem, we will be dealing primarily with paramedian as opposed to lateral tegmental syndromes. The paramedian branches always supply the corticospinal tracts and the paramedian exiting motor cranial nerves: III, VI and XII. In the case of cranial nerve III, the paramedian vessel arises from the posterior cerebral artery.

Vertebral Arteries (Refer to Fig 2-27, 2-

28) The vertebral arteries usually arise from the subclavian artery (the left vertebral artery may arise directly from the aorta in 6 per cent of cases). The artery ascends to enter the foramen of the transverse process of the sixth cervical vertebra, and then passes upward through the foramina of the successive cervical vertebrae; then, after passing behind the superior articular process of the atlas, it enters the foramen magnum. After passing upward and medially, the artery unites with its opposite number at the lower border of the pons to form the basilar artery. During its course in the neck, the vertebral artery gives rise to radicular branches (which anastomose with the anterior and posterior spinal arteries) and to muscular branches whose anastomoses have already been considered. The intracranial branches are the posterior inferior cerebellar, the anterior spinal, the posterior spinal, and the paramedian arteries.

Syndromes associated with occlusive disease of the vertebral artery are variable:

1. Occlusion or absence of one vertebral artery, particularly at a proximal point in an otherwise intact individual, is usually well tolerated (as we have indicated, it is not unusual to find one vertebral artery hypoplastic).

2. Stenosis. a. Stenosis or Occlusion of the proximal portions of both vertebral arteries in the series of patients reported by Fisher (1970) was associated with transient symptoms: faintness, blurred vision, dizziness, loss of balance, diplopia, numbness of an arm or of the face. The symptoms are essentially those of transient insufficiency of circulation in the more distal distribution of vertebral and basilar arteries.

b. Stenosis or occlusion of the intracranial segment of the vertebral artery or arteries may result in infarction over a variable area: the distribution of the basilar artery, and/or the distribution of the paramedian branches of the vertebral and/or the distribution of the posterior inferior cerebellar artery. Actually, occlusive disease of the intracranial segment of the vertebral artery is a much more frequent cause of the lateral medullary syndrome than is limited occlusion of the posterior inferior cerebellar artery.

Medullary Syndromes

Lateral (dorsolateral) medullary syndrome (Syndrome of the posterior inferior cerebellar artery or Wallenberg's syndrome) (Fig 13-11). Of all the various classic vascular syndromes of the brain stem, this syndrome, either alone in a relatively pure form or in a modified form, is the one most frequently encountered. (It is important to emphasize that nonclassic vascular syndromes of the brain stem are more commonly encountered than the classic syndromes involving isolated vascular territories).

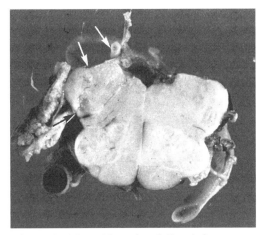

Figure 13-11 Dorsal-lateral medullary infarct (arrows) due to occlusion of the posterior inferior cerebellar artery (arrow) in a 72 year old diabetic with the correlated clinical syndrome of acute onset nystagmus on left lateral gaze, decreased pain sensation on the left side of face and right side of body, with difficulty swallowing, slurring of speech, and deviation of the uvula to the right. (Courtesy of Dr. Jose Segarra.

The brain stem territory of the posterior inferior cerebellar artery (PICA) includes the following structures (a) the restiform body (the inferior cerebellar peduncle) and the adjacent dorsal and ventral spinocerebellar pathway, (b) the descending spinal tract and nucleus of the trigeminal nerve, the lateral spinothalamic tract, (d) the descending sympathetic pathway, (e) the nucleus ambiguus and the fibers of cranial nerve X, and (f) the vestibular nuclei (primarily spinal and medial at this level). The clinical findings of a lateral medullary infarction are illustrated in the following case 13-2. It is important to note that occlusion of the posterior inferior cerebellar artery may produce infarction of the lateral medullary territory or of the PICA cerebellar territory or of both areas. Prior to CT scan and MRI, the cerebellar infarct in a patient who also had a lateral medullary syndrome would have been overlooked, since infarction of the cerebellar peduncle, as part of a lateral medullary syndrome would produce effects similar to the cerebellar infarct.

Case 13-2: This 53 year old male customer service representative on the morning of admission (01-07-2001) while seated had the sudden onset of vertigo accompanied by nausea and vomiting, incoordination of his left arm and leg and numbness of the left side of his face. On standing, he leaned and staggered to the left. Several hours later, he noted increasing difficulty in swallowing and his family described a minor dysarthria. He had a 2 pack per day cigarette smoking history for > 10 years.

Physical Examination: Elevated blood pressure of 166/90.

Neurologic Examination: *Cranial nerves*: The right pupil was 3mm in diameter; the left was 2mm in diameter. Both were reactive to light. There was partial ptosis of the left eyelid. A horizontal nystagmus was present on left lateral gaze, but extraocular movements were full. Pain and temperature sensation was decreased over the left side of the face but touch was intact. Speech was mildly dysarthric for lingual and pharyngeal consonants. Secretions pooled in the pharynx. Although the palate and uvula elevated well, the gag reflex was absent. *Motor system*: a dysmetria was present on left finger to nose test and heel to shin tests. He tended to drift to the left on sitting and was unsteady on attempting to stand, falling to the left. *Reflexes*: ankle jerks which were absent. Plantar responses were flexor. *Sensory system*: Pain and temperature sensation were decreased on the right over the right arm and leg but all other modalities were normal.

Clinical diagnosis: 1.Dorsolateral medullary infarct. 2. Decreased ankle jerks due to possible diabetic peripheral neuropathy

Laboratory Data: *Serum Studies*: Blood sugar: random elevated to 225 mg; fasting elevated to 196 mg%. Total cholesterol: elevated to 340mg% compared to normal of <200mg%, with low density lipoproteins (LDL) elevated to 261 mg%, and high density lipoproteins (HDL) 42mg%.

MRI (Fig.13-12): A significant acute or subacute infarct involved posterior inferior cerebellar artery territory of the left cerebellum and dorsolateral medulla. The MRA suggested decreased flow in the left vertebral artery.

Subsequent Course: A percutaneous endoscopic gastrostomy (PEG) tube was placed since he was unable to swallow for the first six days. By the time of transfer to a rehabilitation facility, 12 days after admission, finger to nose tests and gait had improved, he could walk with a walker. Speech and sensory examination had returned to normal. Additional discussion and analysis of this case will be found on the CD ROM.

An additional example of a lateral medullary infarct (Case 13-3) will be found on

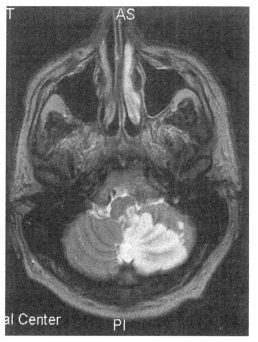

Figure 13-12: Total left posterior inferior cerebellar artery syndrome. MRI T2. Both the lateral medullary territory and the posterior inferior cerebellum are infarcted. The MRA indicated decreased flow in the vertebral artery. (Refer to Fig. 2-27C)

the CD ROM.

Paramedian Medullary Syndrome. Occlusion of the paramedian penetrating branches from the vertebral artery may result in a syndrome characterized by ipsilateral paralysis and atrophy of one-half of the tongue (hypoglossal fibers or nucleus) in addition to a contralateral hemiparesis (medullary pyramid). There may be involvement of the adjacent medial lemniscus as well, producing a contralateral deficit in position and vibratory sensation and a relative decrease in tactile sensation. This syndrome, which bears the name of Hughlings Jackson, is rarely encountered in its pure form.

Combined syndrome: lateral medullary plus paramedian ± cerebellum (Fig.13-13): An example of a patient with infarction of the paramedian vertebral territory in addition to involvement of the cerebellum and pons is presented in the MRI of *Figure 13-14.*

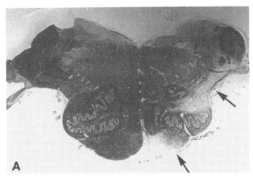

Figure 13-13. Left vertebral and posterior inferior cerebellar artery occlusions: lateral plus paramedian medullary. This 65-year-old hypertensive male expired 5 years after an acute event characterized by difficulty swallowing, "dizziness," ataxia, numbness of the left side of the face and the right side of the body plus a left-sided Horner's syndrome, a decrease in pain sensation on the left side of the face, deviation of the tongue to the left but increased deep tendon reflexes on the right side.

The Basilar Artery

Anatomy (Refer to Fig. 2-27B, 2-28): the union of the two vertebral arteries forms the basilar artery. As with the vertebral artery, it is customary to distinguish paramedian and circumferential branches. The initial long circumferential branch is the anterior inferior cerebel-

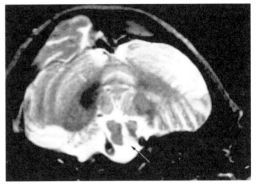

Figure 13-14. Vertebral Artery Syndrome, MRI demonstrates extensive infarction in the paramedian vertebral artery territory with lesser involvement of the cerebellum. Severe stenosis was present in vertebral artery.

lar artery which supplies the dorsolateral tegmentum of the lower one half of the pons *(Fig. 13-15A)*: including the eight nerve and the cochlear nuclei, the vestibular nuclei, the facial nerve the descending spinal tract and nucleus of the fifth nerve, the lateral spinothalamic tract, the sympathetic pathway, the middle and inferior cerebellar peduncles, and the anterior inferior portion of the cerebellum. The area of supply may also include the main sensory nucleus of the fifth nerve at a midpontine level. The internal auditory artery that supplies the inner ear may arise from this vessel or independently from the basilar artery. Whether, there is some supply as well to the nucleus ambiguous is uncertain-this area may constitute a border zone.

The next long circumferential branch is the superior cerebellar artery which supplies the tegmentum of the upper pons and caudal midbrain *(Fig.13-15B)*, including the brachium conjunctivum, the medial longitudinal fasciculus, the motor and main sensory nucleus of the trigeminal nerve, the medial lemniscus, the lateral spinothalamic tract, and the central tegmental tract, and then proceeds to supply the superior cerebellum.

Posterior Cerebral Artery. The basilar artery then terminates at the pontine-midbrain border into two posterior cerebral arteries[1]. From the top of the basilar and from each posterior cerebral artery in its proximal segment a series

BRAIN STEM: CLINICAL CONSIDERATIONS   13-13

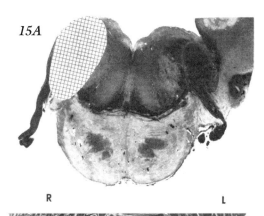

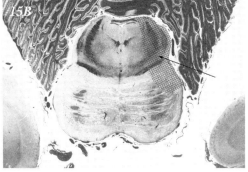

Figure 13-15 Syndromes of the basilar artery circumferential branches. A: Inferior pons. The cross hatched portion of lateral pons is the territory of the anterior inferior cerebellar artery. B: Superior pons: The cross-hatched portion of lateral pons is the territory of the superior cerebellar artery. At times, there is overlap with the paramedian tegmental territory.

of vessels arise to supply the midbrain: penetrating paramedian branches from this vessel supply the medial one-half of the cerebral peduncle, the third nerve fibers, the third nerve nucleus, and the medial one-half of the red nucleus. Other penetrating branches (thalamoperforating and thalamogeniculate) of the posterior cerebral artery and top of the basilar artery in this location supply the subthalamus, hypothalamus and much of the thalamus: pulvinar, ventral lateral, ventral posterior, medial geniculate, lateral geniculate, centromedian, and the dorsal median nuclei (see Castalgne et al 1981). Short circumferential vessels from the posterior cerebral artery supply the remainder (lateral one-half) of the cerebral peduncle and red nucleus. Somewhat longer circumferential branches also called quadrigeminal arteries supply the lateral tegmentum and tectum of the midbrain.

The main posterior cerebral artery then passes around the cerebral peduncle and midbrain close to the tentorial opening, supplying branches to the medial aspect of the temporal lobe and then dividing into the parieto-occipital and calcarine branches to supply the occipital lobe.

The syndromes of the posterior cerebral arteries relevant to the thalamus, subthalamus, and the temporal and occipital lobes are discussed in the chapters on the cerebral hemispheres and motor system[2].

SYNDROMES OF THE BASILAR ARTERY AND ITS BRANCHES

As atherosclerosis involves the basilar and vertebral arteries, there may be transient symptoms suggesting ischemia in the distribution of the various vessels originating from these major arteries. In some cases, infarction may occur within the distribution of these branches. Thus, the patient may experience episodes of vertigo or dizziness and diplopia, a bilateral blurring of vision, or a weakness or numbness of one side of the body or face, and then in a later attack, of the contralateral side of the face or body. In some cases, the transient symptoms are precipitated by extension or rotation of the head that reduces blood flow in the vertebral arteries. Sudden drop attacks without loss of consciousness are not unusual, due to ischemia of paramedian branches supplying the motor pathways: corticospinal tracts and the pontine-medullary reticular formation. Episodes of bilateral blindness may occur due to ischemia of the territory supplied by the posterior cerebral arteries.

---

[1] *Some authors refer to the short segment of posterior cerebral artery between the top of the basilar artery and the posterior communicating artery as the basilar communicating artery (see Caplan 1986).*

[2] *The student should realize that this is an artifical separation: thus 75% of patients with paramedian midbrain infarcts also have paramedian thalamic infarcts. (See Jelliger 1986).*

With total occlusion of the basilar artery, various combinations of cranial nerve findings, coma, decerebrate rigidity, and quadriplegia will be found. Before we examine the situation of total basilar artery occlusion, we will consider the various syndromes of the branch territories. In general, the partial segmental syndromes of the circumferential branches are rare. In contrast the branch syndromes of the paramedian arteries are relatively common particularity in diabetic and hypertensive patients (see Caplan 1986, Fisher and Caplan 1971). The student, moreover, will recognize that the histories presented do not really provide pure examples. There will be some signs or symptoms in each case suggesting involvement at several segmental levels of the brain stem. As with the vertebral artery we will distinguish between paramedian (penetrating branch) and lateral tegmental (circumferential branch) syndromes.

Inferior Pontine Syndromes:

Lateral Tegmentum (Syndrome of the Anterior Inferior Cerebellar Artery). Infarction of the lateral tegmentum at a caudal pontine level results in a sudden unilateral deafness in association with an ipsilateral peripheral facial weakness, ipsilateral Horner's syndrome, vertigo (due to involvement of the vestibular nucleus), and ipsilateral cerebellar symptoms. There may be involvement of the descending spinal tract and nucleus of the trigeminal nerve resulting in an ipsilateral loss of pain and temperature over the face, or the main sensory nucleus may be involved resulting in involvement of facial tactile sensation as well. If the lateral spinothalamic tract is involved, a contralateral loss of pain and temperature will be evident over the arm, leg, and trunk.

In Case 13-4 presented on the CD ROM the major area of infarction was the territory of the anterior inferior cerebellar artery involving both the brain stem and the cerebellum. However the initial area of ischemia suggested a much wider area of the territory of the basilar artery

Paramedian Lower Pontine Syndromes: Infarction within the territory supplied by the paramedian vessels of the basilar artery supplying the lower half of the pons, if unilateral results in the combination of an ipsilateral lateral rectus weakness and a contralateral hemiplegia due to involvement of the emergent fibers of nerve VI and the corticospinal tract in the basilar portion of the pons. If the area of infarction extends into the tegmentum, there would be evidence of involvement of the nucleus of nerve VI (with an ipsilateral paralysis of lateral gaze) and of the genu of nerve VII about the nucleus of nerve VI (with an ipsilateral peripheral palsy of cranial nerve VII), (syndrome of Foville). The medial longitudinal fasciculus would often be involved, with a unilateral (or bilateral) internuclear ophthalmoplegia. Involvement of the medial lemniscus would produce a contralateral hemianesthesia with deficits in position and vibratory sensation. Damage to the central tegmental tract in the pontine tegmentum may result in rhythmical contractions of the palate (palatal myoclonus). Bilateral infarcts of the pons are more frequent. A bilateral paramedian mid-pontine infarct producing a locked –in syndrome is demonstrated in *Figure 13-16*.

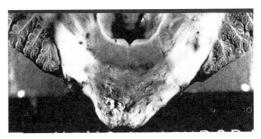

Figure 13-16. Middle-lower pontine basilar infarct with locked in syndrome. Two weeks before death, this 74-year-old female was found motionless and speechless in bed. She could, however, open and close her eyes on command. (Courtesy of Dr. John Hills).

Superior Pontine Syndromes:

Lateral tegmental superior pontine Syndrome: Territory of the Superior Cerebellar Artery *(Fig. 13-15B)*. Ischemia or infarction of the tegmentum of the rostral pons involves the medial lemniscus and the lateral spinothalamic pathways, producing a contralateral hemianesthesia and hemianalgesia. Since the secondary trigeminothalamic (quintothalamic) fibers have already crossed the midline below this level and have essentially merged with the medial lemniscus, the sensory deficit involves the contralateral side of the face as well as the contralateral arm, leg, and trunk. There is, in addition, damage to the superior cerebellar peduncle (the brachium conjunctivum). The cerebellar symptoms are ipsilateral to the lesion when the damage to the brachium conjunctivum occurs below the level of decussation. However, since the superior cerebellar artery also supplies the lateral tegmentum of the midbrain at a caudal level, the brachium conjunctivum may be involved above its decussation. If so, the cerebellar symptoms will be contralateral to the lesion. Moreover, the disturbance of movement is likely to be more variable. Thus, a coarse tremor at rest or instability of sustained posture may be seen in addition to the expected intention tremor. As at lower levels of the pons, the central tegmental tract may also be involved. The lesion (and the territory of the superior cerebellar artery) may extend into the paramedian tegmentum. If so, damage to the medial longitudinal fasciculus may occur.

Paramedian upper pons: When unilateral infarction of the corticospinal and corticobulbar tracts of the basilar pons occurs, the clinical effect is a contralateral, upper motor neuron paralysis of the face, arm, and leg *(Fig. 13-18A)*. This produces a relatively pure motor syndrome as demonstrated in the following case. Note that the more common location for a lacunar infarct producing a pure motor syndrome is the internal capsule. It is, moreover, important to realize, as demonstrated in Figures *13-16 and 13-17* that disease of the basilar artery often produces a bilateral paramedian syndrome with a bilateral infarction of the basilar pons. Basilar artery thrombosis thus occludes the paramedian branches in a bilateral manner.

Case 13-5: This 70 year old right handed white male retired salesman, on the day prior to admission while working in his garden, had the acute onset of weakness of the right arm and leg a right central facial weakness and slurring of speech. A similar episode had occurred

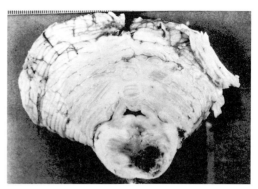

Figure 13-17. Paramedian syndrome of the upper pons Case 13-8. This 64-year old patient had transient ischemic attacks characterized by left or right-sided hemiplegia and then a thrombosis of the basilar artery. (Courtesy of Dr. John Hills and Dr. Jose Segarra.

2 months previously but had resolved over 2 weeks. There was a past history of non-insulin dependent diabetes mellitus for which the patient had been receiving an oral hypoglycemic agent. He had been taking 3 aspirin tablets per day (325mgx3).

Neurological Examination: *Cranial nerves:* flattening of the right nasolabial fold and slurring of speech. *Motor system:* pronator drift in the outstretched right arm with 4+/5 weakness in the right iliopsoas. *Reflexes:* deep tendon stretch reflexes were increased in the right arm. A Babinski sign was present on the right. *Sensory system:* intact.

Clinical diagnosis: Lacunar infarct: pure motor syndrome probable location left internal capsule with paramedian pons less likely.

Laboratory Data: The *MRI* demonstrated a well-defined left sided paramedian upper pontine infarct (Fig.13-18B). The MRA study was within normal limits.

Hospital Course: The following morning the right arm was now plegic, and the patient needed assistance with walking. By the 4th hospital day, dysarthria, and strength in right leg were improving. He was discharged to a rehabilitation facility on the 5th hospital day.

Midbrain Syndromes

Paramedian syndromes (Weber's Syndrome) (Fig. 13-19, 13-20). A unilateral infarction in the territory supplied by paramedian-penetrating branches from the proximal posterior cere-

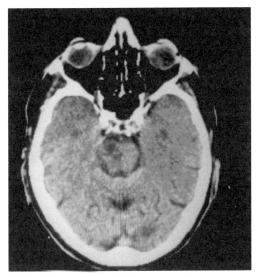

Figure 13-18A. Paramedian infarct upper pons demonstrated on CT scan.

Figure 13-19. Weber's syndrome. An area of infarction is noted in the right cerebral peduncle (arrow), so located as to involve the fibers of the right third cranial nerve. This 61-year-old patient with rheumatic heart disease, endocarditis, and auricular fibrillation had the sudden onset of paralysis of the left face, arm, and leg, plus a partial right third nerve palsy (Courtesy of Dr. John Hills, and Dr. Jose Segarra).

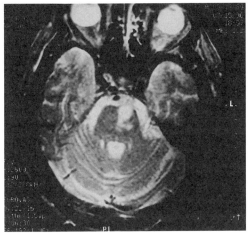

Figure 13-18B. Paramedian infarct Case 13-5. MRI T2 with contrast.

bral artery (or basilar communicating artery) involves the cerebral peduncle, the ipsilateral third nerve fibers, and, to a variable degree, the substantia nigra. The resultant clinical syndrome is that associated with the name of Weber: contralateral upper motor neuron paralysis of the face, arm, and leg, in association with an ipsilateral paralysis of third nerve function. The clinical effects of the damage to the substantia nigra may not be apparent. (Refer to Chapter19). At times, the symptoms related to the cerebral peduncles may be bilateral since the basic process of ischemia may have involved

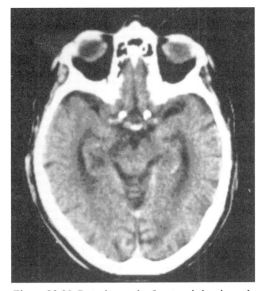

Figure 13-20: Posterior cerebral artery ischemia and infarction right paramedian midbrain and medial temporal. CT scan. This 86 year old male had a transient episode of bilateral blindness, then transient bilateral weakness and dysarthria followed by a left facial weakness, left hemiplegia, bilateral medial rectus weakness and a non responsive right pupil. (Courtesy of Dr. Joel Kaufman).

the upper basilar artery. In other cases both the penetrating (paramedian) branches and the cortical branches of the posterior cerebral artery may be involved resulting in a Weber's syndrome plus a homonymous hemianopsia (due to involvement of the calcarine artery branch supplying the occipital /visual cortex. In addition confusion may be present due to involvement of the cortical branches supplying the mesial temporal structures. As noted above, thalamic infarcts may also be found in patients with paramedian midbrain infarcts because the same process may occlude the penetrating branches to both areas.

The following case 12-6 presents an example of the more limited midbrain syndrome.

Case 12-6: This 66 year-old white male awoke two days prior to admission with weakness of both his legs. He attempted to stand but fell to the floor.

Neurological Examination: *Cranial nerves:* The right pupil was slightly larger than left; both were reactive to light. Ptosis of the right lid was present. *Motor System:* Weakness of the left upper extremity and minor weakness of both lower extremities was present, more marked on the left. The patient was unable to walk without support. When supported the patient walked with a left hemiplegic gait. *Reflexes:* A left Babinski sign was present. *Sensory System:* Intact.

Clinical diagnosis: Possible paramedian midbrain infarct: Weber's syndrome

Laboratory Data: A lumbar puncture demonstrated normal cerebrospinal fluid.

Hospital Course: Two days following admission, the patient demonstrated neurological progression with more complete third nerve palsy. During the subsequent one month improvement occurred, with the only residual a minimal weakness of the left leg and a left Babinski sign.

Case 13-7 presented on the CD ROM provides an example of a patient with more extensive infarction of the territory of the posterior cerebral artery.

Paramedian syndrome with associated tegmental involvement (Benedikt's Syndrome): Involvement of the adjacent midbrain tegmentum (red nucleus, dentatorubral thalamic fibers) will produce a contralateral tremor and movement disorder in association with an ipsilateral third nerve lesion. The involvement of the cerebral peduncle produces a contralateral hemiparesis.

Paramedian periventricular mesencephalic-diencephalic junction: This involves the periaqueductal, pretectal, posterior-medial thalamic, and subthalamic areas. Bilateral infarction of this territory supplied by the penetrating branches of the proximal posterior cerebral arteries results in a drowsy, relatively immobile state (apathetic akinetic mutism) from which the patient can be roused with strong stimulation. The lesion interrupts the ascending reticular system within the upper mesencephalon and its intralaminar diencephalic extension. Third cranial nerve findings are often present as associated findings, due to involvement of the third nerve nuclei. This syndrome has been considered in greater detail by Segarra (1970), Plum and Posner (1980) and Castaigne et al 1981 (Refer to Chapter 29 for additional discussion).

Combined Syndrome: Thrombosis Of The Basilar Artery.

Case 13-8, on the CD ROM illustrates a case of thrombosis of the basilar artery in which several midbrain and pontine syndromes previously discussed were combined.

Treatment: Prevention is of major importance. Control of hypertension and of elevations of the low density lipid component of cholesterol (LDL) and of the triglycerides should be beneficial. If dietary alterations are insufficient then the use of lipid lowering drugs (the statins) should be considered. Increase in the high density lipoprotein (HDL) produced by common sense exercise may also be beneficial. In patients with cardiac disease, (valvular disease, mural thrombi, and particularly atrial fibrillation), long term anticoagulation may prevent cerebral emboli. In addition, for cardiac arrhythmias, other measures will be considered by the cardiologist. The use of aspirin or related compounds to alter platelet aggregation may be of value in preventing additional ischemic events. In patients with symptomatic disease of the subclavian or proximal vertebral artery surgical procedures to bypass sites of occlusion may be considered. The treatment of a basilar artery thrombosis presents a major problem. Previously intravenous heparin, an anticoagulant was utilized in patients with evolving basilar artery thrombosis, but the results were often disappointing. Because of the poor prognosis of these patients, intra-arterial thrombolysis with urokinase has been utilized. In a series from the Mayo Clinic (Wijdicks et. al 1997) recanalization of the thrombosed basilar artery was achieved in 7/9 patients treated within 2-13 hours with 5 of the 7 recanalized patients recovering fully, including 2 patients who had a locked in syndrome. Failure to recanalize resulted in coma and death. A complication is the occurrence of hemorrhage in a small number of patients. An additional study of 22 patients (Cross et. al –1997) confirmed the beneficial effects of recanalization achieved by this method, in terms of survival and improvement in neurologic status. Patients with distal clots did significantly better than those with proximal or mid basilar clots.

The major vascular syndromes of the brain stem are reviewed in Table 13-2, which will be found on the CD ROM.

HEMORRHAGE

Intrapontine Hemorrhage (Fig. 13-21; 13-22). The most common location of hypertensive hemorrhages is intracerebral (putamen predominantly, thalamus and cerebral white matter somewhat less commonly). In a small percentage of cases, the pons is the primary location of the hemorrhage. A large hemorrhage in the relatively small space of the pons produces rapid effects: coma and bilateral involvement of the long tracts within seconds or minutes, death within minutes or hours. Rupture into the ventricular system is common. Although the illustrations present the typical course of such large hemorrhages, small pontine hemorrhages have been recognized in the era of CT and MRI Scan.

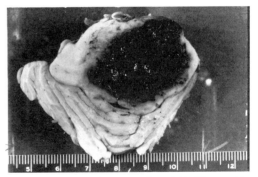

Figure 13-21. Pontine hemorrhage. This 58-year-old hypertensive patient had had a presumed brain-stem infarct 4 weeks previously then this massive pontine hemorrhage possible related to anticoagulant therapy. (Courtesy of Dr. John Hills)

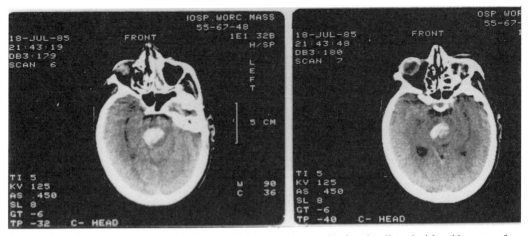

Figure 13-22. Pontine hemorrhage. This 67 year old obese and hypertensive female collapsed with sudden onset of deep coma, no purposive responses to painful stimuli. The right pupil was fixed and dilated, the left pinpoint (1mm) and minimally responsive to light. No oculocephalic or corneal responses were present. Bilateral Babinski signs were present. Respirations, which initially were spontaneous and rapid (central neurogenic hyperventilation), subsequently became irregular and shallow. She expired within 24 hours.

Arteriovenous malformations of the brain stem may produce repeated small hemorrhages and infarct with a course characterized by a variable progressive series of stroke like episodes. At times such cases have been mistaken for multiple sclerosis (Aba and Juellber, 1989, Stahl et al 1980).

Primary Intracerebellar Hemorrhages. See above and chapter 20.

SUBARACHNOID HEMORRHAGE
(see above and chapter 26)

INTRINSIC TUMORS

The substance of the brain stem is a relatively uncommon site for metastatic tumors (however see example in Chapter 11). Almost all tumors found within the substance of the brain stem are intrinsic, arising from glial elements, generally the astrocyte. Various histological grades of malignancy may be encountered. In contrast to gliomas of the cerebral hemisphere, the majority of which are of a very malignant variety (the glioblastomas), the majority of those involving the brain stem are astrocytomas with a lower grade of malignancy and with a longer course. The tumor slowly infiltrates the pons and medulla, producing a gross external enlargement of these structures *(Fig. 13-23)*. On section, the distinctions between gray and white matter are obliterated. Areas of necrosis may be found within the tumor. The essential clinical syndrome is characterized by the progressive development of bilateral long tract and bilateral cranial nerve findings. The cranial nerve findings suggest involvement over adjacent segments of the brain stem, rather than a single segmental level. This tumor occurs primarily in children, adolescents, and young adults. At the present time the diagnosis is easily confirmed by MRI scan which has replaced the earlier techniques of CT scan, pneumoencephalograms, ventriculograms and posterior fossa-myelograms.

Case 13-9: This 20 year old single female had the onset of occasional horizontal diplopia and headache. Five months later clumsiness and weakness of the left arm and leg developed accompanied by increased intracranial pressure. MRI demonstrated a mass within the pons with compression of the 4th ventricle and cerebellum *(Fig.13-24)*. Treatment with dexamethasone, (a powerful corticosteroid) ventriculoperitoneal shunt and radiotherapy resulted in shrinkage of the tumor and resolution of all neurological signs. Symptoms recurred 30 months later.

Case 13-10 presented on CD ROM records in greater detail the entire course of a patient with a pontine glioma.

DEMYELINATING DISEASES

A general discussion of demyelinating diseases particularly multiple sclerosis will be found in Chapter 9, Clinical Consideration of the Spinal Cord. In one of the illustrative cases, there were several episodes suggesting discrete lesions at various levels of the brain stem in addition to lesions involving the spinal cord.

Multiple sclerosis, the most common type of demyelinating disease, frequently involves the white matter of the brain stem and cerebellum *(Figs. 13-25, 12-26)* The classic triad of Charcot includes: nystagmus, intention tremor, and scanning speech—symptoms which are often present in the later stages of progressive cases, but which may be absent early in the disease course or in non-progressive cases.

The following example illustrates a case of multiple sclerosis with predominant involve-

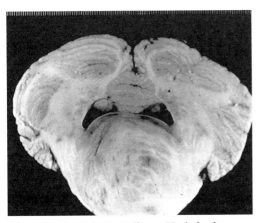

Figure 13-23. Brain stem glioma. Marked enlargement of the pons has occurred with obscuration of the usual anatomical landmarks and obstruction of the 4th ventricle. (Courtesy of Dr. John Hills)

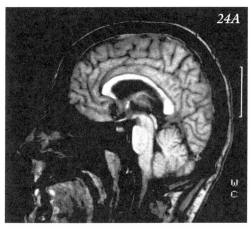

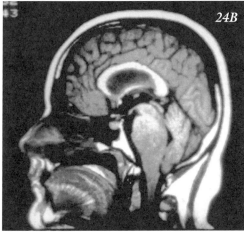

Figure 13-24. Pontine glioma. MRI scan (T1). Case 13-9: refer to text. A) Normal for comparison. B) Pontine Glioma

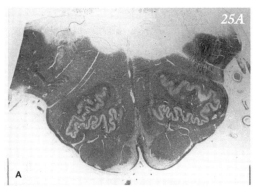

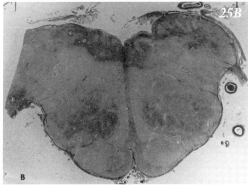

Figure 13-25A Multiple Sclerosis affecting the brain stem. Multiple irregular areas of myelin loss and gliosis are evident are evident in a 58 year old male with a long history of progressive multiple sclerosis. A. Myelin stains B. glial stain (darkly staining areas). See also spinal cord section Fig. 9-21A. Courtesy Dr. Jose Segarra).

ment of brain stem and cerebellar pathways early in the course of the disease.

Case 13-11: This 38 year old divorced right handed white female employed as a tape library supervisor had been in excellent health until, 1 year prior to admission when she developed vomiting, vertigo, numbness of her right leg and difficulty walking. All symptoms cleared over 1 month. One year later, she awoke to find that she had diplopia, divergence of both eyes, limited upgaze, blurring of vision in the left eye, and ataxia.

Neurological Examination: *Mental status:* A mild degree of indifference to her difficulties was present. *Cranial nerves:* On primary gaze, the left eye was divergent (exotropia). Neither eye could adduct past the midline. Dissociated nystagmus was present in the abducting eye on attempted lateral gaze to right or left. Vertical gaze up and down was now intact. *Motor system:* rapid alternating hand movements were somewhat slow. The gait was slightly broad based and mildly ataxic .She would fall to either side on attempted tandem gait. *Reflexes:* Deep tendon reflexes were bilaterally hyperactive. The left plantar response was extensor *Sensory System:* Intact.

Clinical diagnosis: Acute exacerbation of multiple sclerosis with predominant involvement of brain stem and cerebellar system.

Laboratory Data: *MRI (Fig 13-27)*: there were well defined foci of increased signal particularly in the T2 weighted images in the periventricular white matter, the corpus callosum, the subcortical white matter of the right parietal lobe, the left midbrain and the cerebellar peduncles. CSF: protein was high

BRAIN STEM: CLINICAL CONSIDERATIONS 13-21

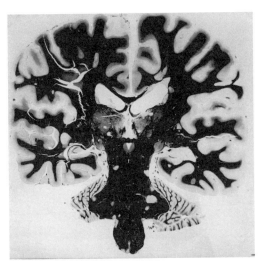

Figure 13-26. Multiple sclerosis. Multiple areas of demyelination in the white matter of brain stem and cerebral hemispheres are demonstrated in this myelin stain. (Courtesy of Dr. Harry Zimmerman, Montefiore Hospital).

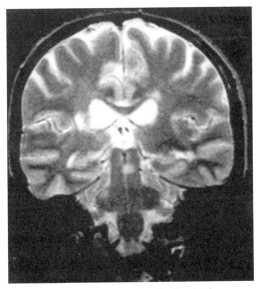

Figure 13-27. Multiple sclerosis affecting brain stem, cerebellum on clinical evaluation. MRI demonstrates brain stem, cerebellar peduncle and cerebral involvement. Case 13-11.

(60mg%), 1 oligoclonal band was present; 3 lymphocytes were present (normal).

Subsequent Course: The patient was treated with a 5-day course of high dosage intravenous methylprednisolone (1000 mg /day) followed by a tapering course of oral prednisone. Within 10 days of the intravenous therapy, the patient had significant improvement: diplopia disappeared and walking improved. When beta interferon became available, she was begun on that medication and had no significant exacerbations for the next 5 years.

Case history 13-12 presented on CD ROM concerns a patient with multiple sclerosis who began with symptoms and signs relevant to brain stem but then pursued a progressive 7 year course of increasing disability resulting in a bed ridden terminal state.

DEGENERATIVE DISEASES

1 *Focal Disease:* syringobulbia: Discussed in chapter 9 in relation to syringobulbia

2. *System Diseases.*

a. Amyotrophic lateral sclerosis: This disorder has been discussed in detail in chapter 9 on the spinal cord. Case 9-8 provides ample evidence of bulbar and corticobulbar involvement.

2. *Spinocerebellar degenerations:* Discussed in detail in chapter 20 cerebellum, and Chapter 9.

TOXIC METABOLIC DISORDERS:

1. Wernicke's encephalopathy: Discussed in detail in chapter 30 in relation to memory.

2. Central pontine myelinolysis (fig 13-28 on CD ROM)

INFECTIONS OF THE BRAIN STEM:

1. Poliomyelitis: discussed in detail in chapter 9 on the spinal cord

2. H.simplex and other infections.

CHAPTER 14
Problem Solving II: Brain Stem and Cranial Nerves

LESION DIAGRAMS: for each of the following diagrams indicate the structures involved by the cross hatched lesion(s). For each structure involved, indicate the expected clinical signs or symptoms with appropriate lateralization i.e. left or right (ipsilateral or contralateral). Where appropriate, indicate the vascular territory involved or the designation of the type of pathology and or the various names of the syndrome.

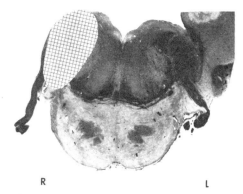

Figure 14-3.

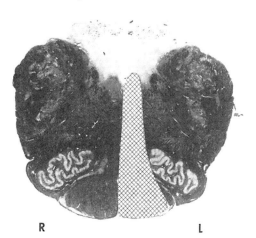

Figure 14-1.

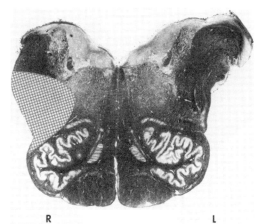

Figure 14-4.

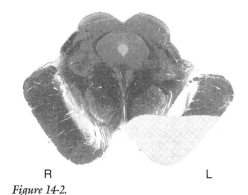

Figure 14-2.

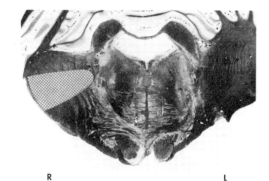

Figure 14-5.

14-2 CHAPTER 14

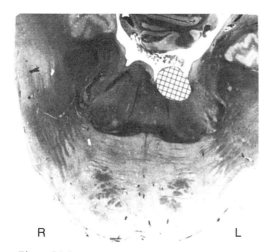

Figure 14-6.

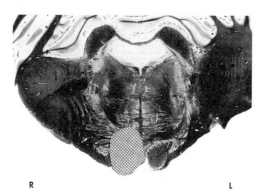

Figure 14-7.

Figure 14-8.

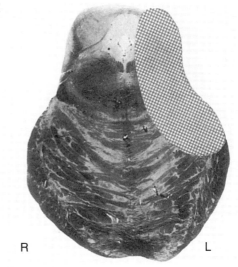

Figure 14-9.

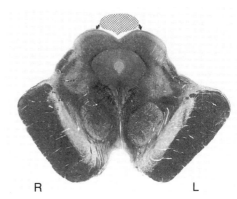

Figure 14-10.

CASE HISTORY PROBLEM SOLVING PART II - BRAIN STEM

Each of the following case histories deals with disease at the level of the brain stem or of the cranial nerves. In some cases, it will be evident that the disease process involves the spinal cord as well as the brain stem. Some of the cases deal with intrinsic disease, some with extrinsic disease. For each case indicate and diagram the location of the lesion and indicate the nature of the pathological process. Where appropriate, indicate the name(s) of the syndrome, and or the vascular territory.

Case 14-1: (Patient of Doctor John Sullivan): A 31-year-old white policeman entered the hospital with symptoms of progressive difficulty in speech, in swallowing, and with weakness of his grip in both hands. Fifteen months before admission, he first noted hoarseness, fatigability of his voice and faulty articulation. This has been slowly, steadily progressive, such that now speech was barely intelligible. Three months after onset of symptoms, he noted difficulty in swallowing both solids and liquids. He had a tendency to regurgitate liquids through his nose. Finally, 3 months before entry, the patient began to note increasing weakness of handgrip. He had no complaints referable to his legs. He denied any sensory symptoms. Six months before admission he noted the presence of muscle twitching with a diffuse distribution, but particularly in the arms and shoulders. He had lost ten pounds in weight and was easily fatigued. System review was otherwise entirely negative and general physical examination was well within normal limits.

NEUROLOGIC EXAMINATION:

Mental Status: Normal.
Cranial Nerves: Positive findings in cranial nerve function were:
 a) Stiffness and weakness of jaw muscles, with jaw clonus.
 b) Weakness of facial musculature, including eye closure, incomplete retraction of the corners of the mouth.
 c) Fasciculations were seen in the facial muscles.
 d) Gag reflex was very brisk, but palate moved weakly.
 e). Tongue would not be protruded; it was atrophic with fibrillations seen beneath the mucous membrane.
 f) Speech, as noted, was slurred, slow and strained.
Motor System:
 a) Weakness of handgrips.
 b) Atrophy was present in the thenar eminence and dorsal interosseus spaces. The proximal muscles were strong, without atrophy; leg muscles were strong and revealed no atrophy.
 c) Widespread muscle fasciculations were seen in both upper and lower extremities.

Reflexes:
 a) Deep tendon reflexes were increased symmetrically in the lower extremities.
 b) Plantar responses were extensor bilaterally (positive Babinski sign).
Sensory System: Normal.

QUESTIONS

1. Does this patient have a level lesion or a system disease?
2. Indicate the significance of the hyperactive jaw jerk and of the hyperactive gag reflex.
3. Indicate the significance of the fasciculations in the facial muscles, the atrophy of the tongue and fibrillations seen in the tongue.
4. Indicate the significance of the atrophy in the hands and the wide spread fasciculations.
5. Indicate the significance of the hyperactive deep tendon reflexes in the lower extremities and the bilateral Babinski signs
6. Where is the pathology?
7. What is the pathology?
8. Which diagnostic laboratory studies would assist in the diagnosis?
9. What would a muscle biopsy reveal?
10. Which neuroimaging studies, if any, are indicated in this case? What results are expected?
11. What is the prognosis? Discuss in terms of this patient in particular and then in terms of patients in general with this disease.

Case 14-2: [Patient of Dr. John Sullivan]

This is a 63 year old woman, who on the day before admission was suddenly seized by a sensation as though a weight had descended upon her head. She felt dizzy, staggered to a chair and called for help. It was noted that her speech was slurred and indistinct. As she looked at an object, it seemed to her to be indistinct. Since then she has had great difficulty in swallowing liquids because of a tendency to regurgitate through her nose. Solid foods did not seem to pass down. She also noted clumsiness of her right hand and of the right leg with staggering to the right side.

GENERAL EXAMINATION:

Blood pressure was 210/100 and she was obese with cardiomegaly.

NEUROLOGICAL EXAMINATION:

Mental Status: Brief mental status examination revealed her to be alert and cooperative, exactly oriented and rather apprehensive; there was no evidence of organic intellectual deficit.

Cranial Nerves:

a) examination of her fundi showed marked tortuousity of the retinal vessels; discs were normal; there are no hemorrhages or exudates

b) visual fields were full

c) Left pupil was 7 mm in size; the right 4 mm; both reacting briskly to light and accommodation; there was a partial ptosis on the right and slight enophthalmos on the right

d) There were no ocular palsies, but there were fairly well sustained quick nystagmoid jerks on gaze to either side

e) Right corneal reflex was absent; pain and temperature sensation on the right side of the face were lost, but touch sensation was intact

f) Hearing was acute bilaterally

g) Voice was somewhat hoarse; speech was slightly slurred and scanning. The right vocal cord was paralyzed; gag reflex was not elicited on the right, present on the left; uvula pulled to the left; 11th and 12th cranial nerve function were normal

Motor Examination:

a) There was instability of posture of the right arm with dyssynergia and intention tremor, which was also present in the right leg.

b) Strength of arms and legs was approximately normal

Reflexes:

a) Deep tendon reflexes in upper and lower extremities were equal.

b) Right and left plantar responses were flexor.

Sensory system:

a) There was a loss of pain and temperature sensibility throughout the left half of the body

b) Position and vibration sensation were intact

QUESTIONS:

1. Diagram the lesion using anatomical diagrams--be specific. Mark this case 1. Be certain to indicate laterality. Prepare a list of symptoms and signs in one column-anatomical structures involved in the opposite column.

2. What is the nature of the pathological process?

3. If vascular, indicate the vascular territory in terms of specific vessel.

4. Which labels or names are attached to this syndrome? Are other syndromes also present?

CASE 14-3: This 54 year old right-handed, obese white female was referred for evaluation of diplopia and ataxia. The patient had a 19-year history of diabetes mellitus initially treated with insulin and more recently with diet alone. The patient also had experienced significant pain in both lower extremities related to intermittent claudication, initially occurring on exercise but more recently occurring also at rest. Evaluation by the vascular surgery service had indicated bilateral carotid bruits.

On the night prior to admission, the patient had the acute onset of a diplopia. At the same time, she noted that she was no longer able to move her eyes upward and that in order to look up she had to turn her head back. At the same time, she developed a sense of unsteadiness. This morning she had a persistence of symptoms.

NEUROLOGICAL EXAMINATION:

Mental Status: Intact with marked anxiety

Cranial Nerves:

a) She had significant bilateral impairment of upward gaze

b) As she attempted to gaze upward, she had significant lid retraction.

c) In addition there was an indication of a weakness of right medial rectus.

d) The pupils were equal and responded to light

e) There was now no evidence of ptosis. There was no definite fatigue of the lids by repetitive movement.

Motor System:
  a) strength intact
  b) on examination of gait, patient walked on a broad base. She tended to fall to the left. She was unable to walk a tandem gait. She had a minor tremor of outstretched hands but no definite appendicular cerebellar findings.
Reflexes:
  a) deep tendon reflexes were 2+ except Achilles which were absent (probably) related to diabetes mellitus
  b) plantar responses flexor.
Sensory system:
Intact except for a decrease in vibration of toes (consistent with diabetes mellitus).

Hospital Course:
The patient had no additional progression. By day five, she had shown improvement in right medial rectus function and had improvement in her ability to look up.

QUESTIONS:

1) The bilateral lid retraction and the bilateral impairment of conjugate upward gaze in this case probably reflect involvement of _____?

2) The blood supply of this area is derived from branches of the _____ artery.

3) In this case impairment of upward gaze was due to an ischemic event, however, impairment of conjugate upward gaze is also commonly noted in relation to other pathological processes. Specify_____

CASE 14-4 [Patient of Dr. John Sullivan]: This 65-year-old man entered another hospital complaining of headaches, severe vertigo, nausea and vomiting. The diagnosis of labyrinthitis was made and the patient seemed to recover after a few days. He remained well, however, only a few days when he again had severe vertigo and vomiting. The latter was so persistent and severe that he became dehydrated. On the morning of transfer to this hospital, at about 3:00 AM, the patient noted the sudden onset of weakness and numbness of his right thigh.

GENERAL PHYSICAL EXAMINATION:
The patient had a blood pressure of 110/80. There was a cold, clammy perspiration over his entire body. No other gross physical abnormalities were discovered.

NEUROLOGICAL EXAMINATION:
Mental Status: The patient was extremely restless and confused
Cranial Nerves:
  a) He had a paralysis of left external rectus muscle
  b) There was a coarse, irregular nystagmus on gaze to either side
  c) His fundi were quite normal
  d) The pupils and pupillary reflexes were normal
  e) there was a selective diminution of pain sensation on the left side of the face. Left corneal reflex was diminished.
  f) He had a left facial paralysis of peripheral type
  g) Hearing was markedly diminished on the left
  h) Gag reflex on the left was diminished; uvula pulled to the right; patient's speech was hoarse and he had a left vocal cord paralysis; He was unable to swallow without choking
  i) His tongue protruded slightly to the right
Motor System:
  a) On examination of his limbs, there was an intention tremor on the left involving both arm and leg
  b) His arms and legs were strong and there seemed no increased resistance to passive movement
Reflexes:
  a) Deep tendon reflexes were increased on the right side throughout
  b) Both plantar responses were extensor (positive Babinski Signs)
Sensory system:
  a) There was a loss of pain and temperature sensibility throughout the entire right side
  b) Other forms of sensation could not be adequately tested because of patient's inability to cooperate

QUESTIONS:

1. Considering in isolation the left lateral rectus palsy, the left peripheral facial weakness, the decrease in hearing in the left ear, the severe vertigo and the decrease in pain sensation on the left side of the face and right side of body. Diagram this specific lesion at the proper level. Label this Case #4.

2. What specific vessel supplies the area you have outlined above?

3. Now take into account those additional findings in this case that were not included in the consideration of Question 1. Indicate vessel responsible for the entire episode (there are 2 possibilities).

CASE 14-5: This right-handed, 43-year-old white housewife had the onset of deafness in the right ear. Rapid progression of deafness was noted; some tinnitus (sensation of ringing) was also noted. Caloric testing of labyrinthine function at that time indicated no response to cold or hot water on the right side. In the following month, the onset of a minor unsteadiness of gait was noted. Approximately five months later the patient noted defects in coordination of the right hand, particularly in typing, with a progression of unsteadiness of gait. At the same time a "numbness" sensation of pins-and-needles -- "like Novocain given by a dentist" -- was noted over the entire right side of the face. Relatively continuous pain was noted extending from the right side of the neck to the sub occipital area and right post auricular area. One month later difficulty in swallowing solids and liquids was noted.

PAST HISTORY:

Episodes of vertigo 5 to 6 years prior to admission. Right frontal headache since age 33.

NEUROLOGICAL EXAMINATION:

Mental Status: Intact.
Cranial Nerves:
   a) Pain and touch sensation were decreased over all divisions of the trigeminal nerve on the right side, including the face, cornea, and the right side of the tongueb)
   b) There was minimal flattening of the right naso-labial fold and a poorer degree of eye closure on the right side than the left (orbicularis oculi).
   c) There was no perception of voice in the right ear.
   d) No vestibular response was present to ice water caloric testing in the right ear.
   e) Minimal rotatory nystagmus was present on horizontal gaze with a minor degree of vertical nystagmus on upward gaze.
   f) A minimal degree of dysarthria was apparent as regards guttural sounds.
Motor System:
   a) Strength and tone were intact.
   b) Cerebellar tests revealed a slight clumsiness in fine finger movements of the right hand.
   c) Gait was slightly ataxic when performed on a narrow base with eyes open.
Reflexes:
   a) Deep tendon reflexes were symmetrical and physiologic.
   b) Plantar responses were flexor.
Sensory syatem: All modalities were intact.

LABORATORY DATA:

Cerebrospinal fluid protein was slightly elevated to 50 mg./100 ml. Skull x-rays were negative.

QUESTIONS:

1. This patient presents a typical example of classic neurological syndrome. Locate the lesion.

2. What is the most likely pathology to be found by the neurosurgeon in this location? Which pathological processes are also possible but less likely?

3. The initial involvement of the functions of cranial nerve VIII prior to involvement of other cranial nerves should provide a clue as to the structure from which this lesion arises.

4. Does the pattern of sensory disturbance over the right side of the face indicate primary involvement of the trigeminal nerve extrinsic to the brain stem or of the descending spinal tract and associated nucleus of the trigeminal nerve within the brain stem?

5. Were long sensory and motor tracts

within the brain stem involved by this lesion?

6. Predict the clinical picture that would have occurred if the lesion had progressed.

7. During the course of surgery, aimed at total resection in these cases, the facial nerve must often be sacrificed or damaged. Based on your understanding of the anatomical considerations in these cases, indicate why this occurs. Which studies may be performed during surgery in early cases to preserve residual hearing or facial nerve function?

8. Which critical diagnostic studies would you perform?

9. When should these studies be performed?

CASE 14-6: This 57-year-old housewife one day prior to admission suddenly developed double vision and a drooping of the right eyelid. In addition, she had difficulty walking. Examination of the patient in the emergency room revealed weakness of adduction of the right eye, ptosis of the right eye and a left extensor plantar response. She was admitted to the neurology service.

PAST HISTORY:

1. Moderate hypertension had been present for many years.

2. Four years previously, the patient had begun to have sudden 30-minute episodes of right-handed weakness and dysarthria. Twenty-six months prior to admission, she had a 15-20 minute episode of bilateral blurred vision, unsteadiness of gait and tingling paresthesias of the right hand and face. Twenty-two months prior to admission, she had episode of paresthesias involving either the left leg and arm or the right arm and leg. Examination at that point demonstrated increase deep tendon reflexes on the right side. Eighteen months prior to admission, she experienced 10-minute episodes of numbness of the left hand and face accompanied by renewed numbness of the right hand. Three days prior to admission she had a brief episode of right hand weakness.

3. The patient had been receiving anticoagulant therapy (Coumadin) for phlebitis for several months prior to admission.

PHYSICAL EXAMINATION: The patient was obese and anxious with blood pressure elevated to 160/100.

NEUROLOGICAL EXAMINATION:

Mental status: the patient was alert and oriented. Recent memory was poor and delayed recall was limited to 2/4 objects. (All similar to 18 months previously)

Cranial nerves:

a) Ptosis of the right eyelid was present, but pupillary responses were intact.

b) At rest, the right eye was deviated out to the right and down. No medial or upward movement of the right eye was possible. (Note change).

c) A mild left central (supranuclear) facial weakness was now present.

Motor system: Strength was intact.

Reflexes: There were changes compared to examinations 22 and 18 months previously.

a) Deep tendon reflexes were now increased in the left lower extremity.

b) A left Babinski sign was now present with an absent left abdominal reflex.

Sensory system: All modalities were intact.

Carotid pulses: Strong bilaterally.

LABORATORY DATA:

1. Skull x rays demonstrated calcifications in the cavernous carotid arteries.

2. Electrocardiogram was normal.

3. Electroencephalogram demonstrated scattered multifocal slow waves.

4. Blood studies: Complete blood counts, serology, blood sugar and total cholesterol were normal.

5. Prothrombin time was 68% of normal.

6. The patient refused lumbar puncture and angiography.

QUESTIONS:

1. This is clearly a more complex case. However the new symptoms of diplopia, ptosis of the right eyelid plus the new findings of a paralysis of medial and upward movement of the right eye plus increased deep tendon reflexes in the left lower extremity plus a left

Babinski sign and a left central facial weakness should allow for the diagnosis and localization of at least one syndrome. What name do you assign to this syndrome? Does this syndrome involve a specific vascular territory?

2. How do you explain the earlier episodes?

a) Episodes of right hand weakness and dysarthria

b) Bilateral blurring of vision, plus unsteadiness plus tingling of the right hand and face.

c) Paresthesias left leg and arm followed by paresthesias of the right arm and leg.

3. Are there any additional questions, which you might pose for this patient?

4. If this patient had a severe headache at the time of admission, why might a lumbar puncture have been considered?

5. Why was angiography considered?

6 The patient refused both lumbar puncture and angiography. Would you have requested these studies or would you have requested other studies? If so present your plan of workup justifying each study.

7 Present your working diagnosis and final diagnosis. (There are several possible ways to tie together all of the episodes).

CASE 14-7: This 69-year-old white male was seen for outpatient neurological re-evaluation. His neurological problems began four years previously. The patient had a history of intermittent ptosis of the left lid. In addition, at times significant ptosis of both lids had been noted. Over the years, he had intermittent episodes of diplopia. He had also noted weakness of his jaw in chewing and weakness in shoulders or in his neck on exercise of these muscle groups.

NEUROLOGICAL EXAMINATION:

Mental Status: intact
Cranial Nerves:

a. The patient demonstrated a significant bilateral ptosis that was variable, it was significantly increased by exercise; it was more marked in the left eye.

b. The patient showed a significant bilateral defect in upward movement and of lateral movement of the right eye. The defect in lateral and upward movement was bilateral but was more marked on exercise.

c. The patient was able to smile for 8 seconds and then his smile began to evaporate bilaterally.

d. The patient had weakness in jaw muscles on repetitive opening and closure of jaw.

Motor System.

The patient had weakness in shoulder abduction, which developed after 10 repetitive movements. He had weakness in forward head movement which developed after 4 repetitive movements. No definite atrophy was present, no definite weakness was otherwise present in the extremities.

Reflexes: Deep tendon reflexes were intact, plantar responses were flexor.

Sensory System: intact.

QUESTIONS:

1. Indicate the diagnosis. Be specific.

2. Where is the defect located in this disease? Be specific!!

3. Discuss the underlying pathophysiology.

4. Which tests would confirm the diagnosis and produce temporary improvement?

5. Outline therapeutic approaches.

CASE 14-8: This 56-year-old white housewife was admitted for evaluation of episodes of stupor and cyanosis associated with severe laryngeal stridor (high pitched and harsh respiratory sounds) and stertorous breathing. Laryngeal stridor had been present for 20 years and had grown worse during the last 5 to 6 years. An episode of anoxia during a Cesarean section 18 years prior to admission may have been a complication of this problem. Sixteen years prior to admission numbness and weakness of the right leg had been noted. Progression had occurred in the last 2 years and episodic unsteadiness of gait had been noted. During this same period, weakness of the right hand had developed. During the one year prior to admission the patient had three hospital admissions related to episodes of coma and cyanosis. Each had followed a several-week period in which there was increased stridor and increased accumulation of tracheal-bronchial secretions with frequent periods of daytime sleepiness.

NEUROLOGICAL EXAMINATION:

Mental Status: Intact.
Cranial Nerves:

a. Pain sensation was selectively decreased over all three divisions of the right side of the face with decreased right corneal reflex.

b. Laryngoscopy revealed paralysis with atrophy of the left vocal cord. The right cord moved but was inhibited on abduction, indicating partial paresis.

c. Horizontal nystagmus was present on lateral gaze with minimal vertical nystagmus on upward gaze.

Motor System:

a. Atrophy of the right upper extremity was present including the shoulder, arm, and hand.

b. Weakness was present in the right upper extremity -- approximately 50 per cent of normal strength at shoulder, elbow, wrist, and fingers. Weakness without atrophy was present in the right lower extremity and to a lesser degree in the left lower extremity.

c. Spasticity was present on passive movement at the right knee and ankle and to a lesser degree at the left knee.

d. Gait: There was circumduction of the right leg with unsteadiness on rapid turns.

e. Cerebellar tests were negative.

Reflexes:

a. Deep tendon:

Biceps: right, 0; left, 0
Triceps: right, 0; left, 0
Radial: right, 0; left, 0
Patellar: right, 4+; left, 3+
Achilles: right, 4+; left, 3+

b. Superficial reflexes:
Plantar: extensor on the right and possibly extensor on the left
Abdominal: right, 0; left, 0

Sensory system:

a. Position, vibration, and touch were intact.

b. Pain and temperature were selectively decreased in a cape-like distribution over the shoulders.

LABORATORY DATA:

1. Cerebrospinal fluid: normal (pressure 150, cell count 0, protein 45 mg./100 ml.)

2. Skull x-rays were negative.

3. Cervical spine x-rays revealed minor non-significant degenerative changes at C5-C7.

SUBSEQUENT COURSE:

The patient was readmitted to the hospital 10 weeks later in a semi comatose and cyanotic condition. She required an emergency tracheostomy and 48 hours of respiratory assistance. Subsequent neurological examination was unchanged from that recorded earlier.

QUESTIONS:

1. Where is the lesion? Does this lesion involve a single localized segment or are several segments involved?

2. Is the pathology limited to the brain stem or to the spinal cord?

3. Indicate what structures are involved to produce:

a. laryngeal paralysis,

b. cape-like deficit in pain and temperature but sparing touch and vibration,

c. atrophy of all muscle groups in right upper extremity,

d. absence of deep tendon reflexes in both upper extremities, and

e. defective pain sensation on the right side of the face.

4. Indicate the most likely pathology and the probable prognosis.

5. What diagnostic tests should be undertaken to establish the diagnosis?

6. Why did the patient have episodes of coma and excessive daytime sleepiness? (There are several explanations)

CHAPTER 15
Diencephalon

The diencephalon appears as a large region at the upper end of the brain stem. All the ascending fiber pathways from the spinal cord and brain stem terminate upon nuclei in the diencephalon. From these nuclei the information is relayed onto the cerebral cortex.

Due to the convergence of all major ascending sensory, motor and reticular systems consciousness for general sensations are first realized at this level. In addition, the ascending motor information from the cerebellum mixes with the striatal and cortical motor fibers in the dorsal thalamus producing a motor thalamic region. Finally, the hypothalamus and limbic cortical regions have major input onto the dorsomedial nucleus of the thalamus.

I. DIENCEPHALIC BOUNDRIES

The border between the diencephalon and the midbrain is indistinct and may be arbitrarily defined by passing a line from the inferior surface of the mammillary bodies to the posterior border of the habenula. Portions of the substantia nigra and red nucleus are also seen in the posterior portion of the diencephalon.

In order to identify the majority of the nuclei in the diencephalon and basal nuclei the cerebral cortex and corpus callosum are carefully separated and removed revealing the diencephalon *(Fig 15-1)*. Then one finds the slit-like third ventricle in the midline. At the rostral end of the diencephalon we can identify the anterior tubercle (containing the anterior thalamic nuclei) and at the posterior margin the pulvinar. The major new white matter bundles at this level are the internal capsule, fornix, and stria terminalis. The basal nuclei of the cerebrum can now also be visualized *(Figure 15-2)*, and consist of caudate, putamen and globus pallidus. The globus pallidus is found lateral to the internal capsule, and the putamen is adjacent to it. The caudate nucleus is found medial to the internal capsule. In the posterior levels of the diencephalon, the caudate is relatively small, but as one proceeds anteriorly, the diencephalon becomes smaller and the caudate enlarges.

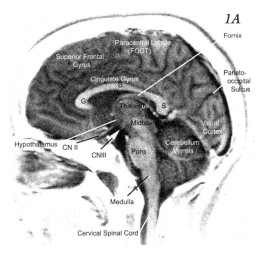

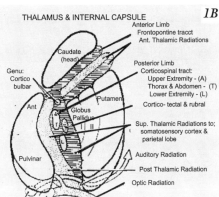

Figure 15-1. Brain in Sagital plain demonstrating. The relationship between brain stem, diencephalon and cerebrum. (MRI weighted T1.) Location of thalamus. A. MRI B. Gross Brain..

The body of the lateral ventricle along with the corpus callosum and fornix form the superior border of the diencephalon (Fig. 15-3). The posterior limb of the internal capsule and the optic tract mark its lateral boundary, and the slit-like third ventricle denotes its medial border. Inferiorly, the diencephalon is continuous with the tegmentum of the midbrain, while rostrally it ends at the lamina terminalis.

The tegmentum of the midbrain is continuous with the hypothalamus *(Fig 15-4)*. The diencephalon consists of five distinct nuclear subdivisions: thalamus, epithalamus, hypothalamus, metathalamus, and subthalamus (Figure 15-2). As can be seen from the terminology associated with the diencephalon, all structures are described by their spatial relationship to the thalamus, being above, below, or behind them. The epithalamus and hypothalamus are discussed in Chapter 16.

about a third of the way up from the floor of the third ventricle. All nuclear structures superior to the sulcus are included in the dorsal thalamus, and all nuclei below it are included in the hypothalamus *(Fig. 15-3)*.

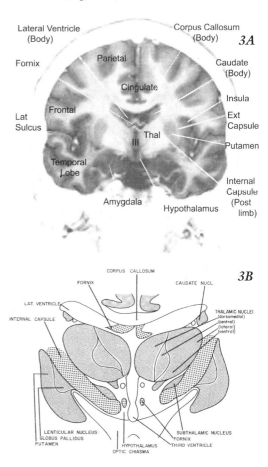

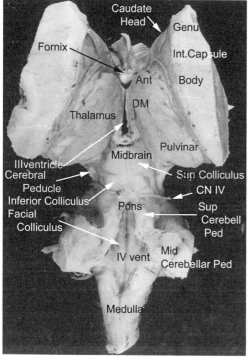

Figure 15-2. Dorsal view of gross specimen of brain stem, diencephalon, and basal ganglia. Medulla, pons, midbrain, thalamic regions and internal capsule labeled.

II. NUCLEI OF THE THALAMUS

The thalamus and metathalamus is the major relay center between the brain stem and the cerebral cortex. The thalamus is the largest portion of the diencephalon. It is ovoid in shape; the posterior limb of the internal capsule, medially by the third ventricle, bound it laterally inferiorly by the subthalamus and hypothalamus, and superiorly by the body of the lateral ventricle and corpus callosum. The boundary between the thalamus and hypothalamus (Fig 15-3) is the hypothalamic sulcus on the medial wall of the diencephalon

Figure 15-3. Mid thalamic level. A. MRI B. Schematic coronal section showing nuclei of thalamus, hypothalamus, subthalamus, basal ganglia and the white matter internal capsule.

Thalamic Borders. The external medullary lamina forms the lateral boundary of the thalamus, and separates it from the reticular nucleus of the thalamus. There is also an intrinsic bundle of white matter, the internal medullary lamina, which divides the thalamus into three major nuclear masses the anterior, medial and lateral. The anterior mass includes the anterior nuclei; the medial mass includes the dorsomedial nuclei and the midline nuclei, with the lateral mass including all of the other nuclei. Cell and fiber

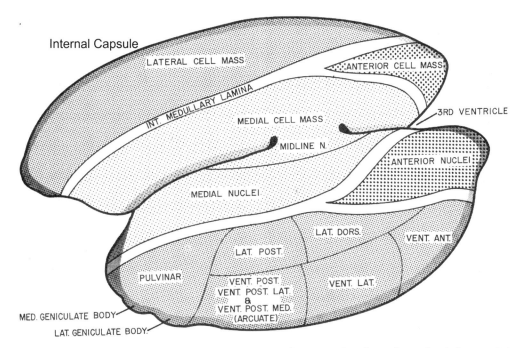

Figure 15-4. 3D Reconstruction of the thalamic nuclei with the upper portion showing the three major nuclear subdivisions and the lower portion showing individual nuclei. Modified After Carpenter Core Text, *Williams& Wilkins 1992*

stains are used to further subdivide the thalamus. With these stains the following nuclei are identifiable: anterior, medial, midline, intralaminar, lateral, posterior, reticular, and metathalamus. Each of these nuclear groupings can be further subdivided as follows Table 15-1; Nuclei in the Thalamus

III. FUNCTIONAL ORGANIZATION OF THALAMIC

A. SENSORY AND MOTOR RELAY NUCLEI-THE VENTROBASAL COMPLEX, LATERAL POSTERIOR, (FIGURE 15-4).

These nuclei *(Fig 15-4)* are part of the somatic brain as they receive direct input from the ascending sensory and motor systems, from the cerebellum, and from the optic tract.

The special sensory medial and lateral geniculate nuclei are discussed below.

The lateral nuclear mass is divided into three parts: lateral dorsal, lateral posterior, and pulvinar. It occupies the upper half of the lateral nuclear grouping throughout the thalamus. It is bounded medially by the internal medullary lamina of the thalamus and laterally by the external medullary lamina of the thalamus. In a stained preparation its ventral border with the ventral nuclear mass is distinct. The lateral nuclear mass starts near the anterior end of the thalamus and runs the length of the thalamus with its posterior portion, the pulvinar, overhanging the geniculate bodies and the midbrain.

The lateral posterior part of the lateral nuclear mass projects to the superior parietal lobule and receives input from specific thalamic nuclei. The lateral dorsal part is included above with the limbic nuclei.

The ventral basal nuclear mass consists of three distinct regions: ventral anterior, ventral lateral, and ventral posterior. The *ventral anterior* nucleus (VA) is the smallest and most rostral nucleus of this group. It can be identified by the presence of numerous myelinated bundles. The magnocellular portion of this nucleus receives fibers from the globus pallidus, via the lenticular and thalamic fasciculi, and from the intralaminar nuclei and has some projections to areas 4 and 6. Stimulation of this nucleus produces the same effects as stimulation of the intralaminar nuclei.

The ventral lateral and ventral posterior nuclei contain cell bodies that project onto the

TABLE 15-1: NUCLEI IN THE THALAMUS

A. ANTERIOR NUCLEI (Limbic)
   1. Anterior dorsal
   2. Anterior medial
   3. Anterior ventral (Anterior principalis)

B. MEDIAL NUCLEI (Limbic & Specific Associational)
   1. Nucleus medialis dorsalis
      a. Pars parvocellulares
      b. Pars magnocellularis

C. MIDLINE NUCLEI (Non specific Associational)
   1. Nucleus parataenialis
   2. Nucleus paraventricularis
   3. Nucleus reuniens
   4. Nucleus rhomboideus

D. INTRALAMINAR NUCLEI - in the internal medullary lamina (Nonspecific Associational)
   1. Nucleus centrum medianum
   2. Nucleus parafascicular
   3. Nucleus paracentralis
   4. Nucleus centralis lateralis
   5. Nucleus centralis medialis

E. LATERAL NUCLEI
   1. Pars dorsalis (Specific Associational)
      a. Nucleus lateralis posterioris
      b. Nucleus lateralis dorsalis
      c. Pulvinar
         (1) Pars inferioris
         (2) Pars lateralis
         (3) Pars medialis
   2. Pars ventralis/Ventrobasal complex (Motor & Sensory Relay).
      a. Ventralis anterior
      b. Ventralis lateralis
      c. Ventralis ventralis
      d. Ventralis posterior
         (1) Posterolateral
         (2) Posteromedial

F. RETICULAR NUCLEI (Nonspecific Associational)

G. METATHALAMIC NUCLEI (Sensory Relay)
   1. Lateral geniculate (Vision-opticothalamic)
   2. Medial geniculate (Audition-auditothalamic)

H. POSTERIOR NUCLEI (Nonspecific Associational)
   1. Limitans
   2. Suprageniculate
   3. Posterior (Nociceptive System)

motor and sensory cortex. In the thalamus the head is found medial and posterior and the leg anterior and lateral. The *ventral lateral* nucleus (VL, somatic motor thalamus) occupies the middle portion of the ventral nuclear mass. It is a specific relay nucleus in that it receives fibers from the contralateral deep cerebellar nuclei (via the superior cerebellar peduncle/ dentatorubrothalamic tract) and the ipsilateral red nucleus. This nucleus also receives fibers from the globus pallidus and from the intralaminar nuclei. This nucleus projects to area 4 and to area 6 and has a topographic projection to the motor cortex. This nucleus is important in integrative somatic motor functions because of the interplay between the cerebellum, basal ganglia, and cerebrum in this nucleus.

The *ventral posterior* nucleus occupies the posterior half of the ventral nuclear mass and consists of two parts: the ventral posterior medial nucleus and the ventral posterior lateral nucleus. Both of these nuclei receive input from the specific ascending sensory systems.

The *ventral posterior medial* nucleus (VPM, the arcuate, or semilunar, nucleus) is located lateral to the centromedian nucleus and medial to the ventral posterior lateral nucleus. The secondary fibers of the trigeminal and gustatory pathways terminate in this nucleus; thus this nucleus is concerned primarily with taste and with general sensation from the head and face. The basal region of the ventromedial nucleus receives the taste fibers and probably some of the vagal input. General sensation from the face, including pain, temperature, touch, pressure, and proprioception, terminate in the other portions of this nucleus. This nucleus projects information from the face, ears, and head, and tongue regions onto the inferior part of the postcentral gyrus, areas 3, 1, 2

The *ventral posterior lateral* (VPL) nucleus receives fibers from the posterior columns and the direct spinothalamic tracts with the bulk of the input originating in the upper extremity. The VPL nucleus projects to the body and neck regions on the postcentral gyrus (areas 3, 1, 2).

B. LIMBIC NUCLEI - THE ANTERIOR, MEDIAL, LATERAL DORSAL, MIDLINE AND INTRALAMINAR NUCLEI
(Fig 15-9).

The limbic nuclei may also be called the nuclei of the emotional brain.

The *anterior nuclei* (15-4) are at the most rostral part of the thalamus and form the prominent anterior tubercle in the floor of the lateral ventricle. The tubercle includes a large main nucleus (anteroventral) and small accessory nuclei (anterodorsal and anteromedial).

The internal medullary lamina of the thalamus surrounds the anterior nuclei. The mammillothalamic fibers form the bulk of the fibers in the internal medullary lamina. This zone is an important relay station in the limbic brain and receives input from the subiculum, presubiculum and mammillary body via the mammillothalamic tract.

The anterior nuclear complex is part of the Papez circuit: (hippocampal formation --> fornix --> mammillary body --> mammillothalamic tract --> anterior thalamic nucleus --> anterior thalamic radiation --> cingulate cortex --> cingulum --> perforant pathway -->hippocampus). This nucleus projects to the anterior cingulate gyrus (areas 23, 24, 32) and receives input from the hypothalamus, habenula, and cingulate cortex. Stimulation or ablation of this nuclear complex alters blood pressure and may have an effect on memory.

The *lateral dorsal* nucleus, like the anterior nucleus, is surrounded by the rostral portion of the internal medullary lamina of the thalamus and should be considered a caudal continuation of the anterior nuclear group. This nucleus projects to the rostral cingulate cortex and onto the parahippocampal gyrus.

The medial nuclear complex or the *dorsomedial* nucleus consists of both a large-celled and a small-celled division. This medial region is found between the internal medullary laminae and the midline nuclei lining the third ventricle *(Fig. 15-5)*. The *magnocellular* portion of the dorsomedial nucleus is interconnected with the hypothalamus, amygdala, and midline nuclei and connects to the orbital cortex; the large *parvicellular division* is interconnected with the temporal, orbitofrontal and prefrontal cortices (areas 9, 10, 11, 12; see Chapter 22)

The medial nuclear complex is an important relay station between the hypothalamus, amygdala and the prefrontal cortex and is concerned with the multimodal associations between the limbic (visceral) and somatic impulses that contribute to the emotional makeup of the individual. Destruction of this nucleus in cats results in a lower threshold for rage, so that the animal is easily aroused. In human beings, ablation of the medial nucleus or a prefrontal lobotomy has been used as a therapeutic procedure to relieve emotional distress (SEE CHAPTER 22).

C. SPECIFIC ASSOCIATIONAL- MULTIMODAL/ SOMATIC NUCLEI-THE PULVINAR NUCLEI (fig 15-4).

The Pulvinar Nuclei

The pulvinar (Fig 15-4) is the largest nucleus in the thalamus and is continuous with the posterior part of the lateral division. All the major ascending sensory pathways have some terminations in the medial and lateral nuclei. The pulvinar therefore is the major multimodal association nucleus of the thalamus integrating sensory, motor, visual, and limbic information.

The pulvinar is divided into three major nuclei: the medial pulvinar nucleus, the lateral pulvinar nucleus and the inferior pulvinar nucleus. The *medial pulvinar* nucleus projects onto the prefrontal and anterior temporal limbic cortex. The lateral pulvinar nucleus projects onto supramarginal, angular, and posterior temporal lobe. The *inferior pulvinar* nucleus is related to the visual system and projects onto the visual association cortex, areas 18 and 19. Because of the strong projections of the lateral pulvinar nucleus onto the cortex surrounding the posterior end of the lateral sulcus, lesions in the pulvinar in the dominant hemisphere produce disturbances in language.

E. SPECIAL SENSORY NUCLEI-METATHALAMUS: VISION AND AUDITION, THE LATERAL GENICULATE and MEDIAL GENICULATE Nuclei (Fig 15-4.

AUDITION (Also see Chapter 22)

The *medial geniculate* nucleus (MGN) is located on the most caudal portion of the thalamus receives the ascending auditory fibers originating from the following nuclei: cochlear, trapezoid, superior olivary, lateral lemniscal, and inferior colliculus. The auditory fibers end on the medial geniculate pars parvocellulares. Vestibular fibers end in the magnocellular portion of the medial geniculate and are projected onto postcentral gyrus. This nucleus projects to area 41 in the temporal lobe, the transverse temporal gyrus of Heschl. The ventral nucleus of the medial geniculate receives input from the inferior colliculus.

VISION (Also see Chapter 23)

The *lateral geniculate* nucleus (LGN) is a horseshoe-shaped six-layered nucleus with its hilus on its ventromedial surface. The LGN receives the optic tract and projects to areas 17 and 18. Figure 15-5 demonstrates the visual radiation onto the calcarine cortex of the occipital lobe). This nucleus contains the six neuronal layers separated by bands of myelinated axons. The four outer layers contain small- to medium-sized cells, that project onto layer iv of the striate cortex with the two innermost layers contain large cells (magnocellular) projecting onto layer I and deep portion of layer IV.

Crossed fibers of the optic tract terminate in laminae 1, 4, and 6, while uncrossed fibers end in laminae 2, 3, and 5. This nucleus also connects with the inferior pulvinar, ventral, and lateral thalamic nuclei. Some optic fibers proceed directly to the pretectal area and the superior colliculus with or without synapsing in the lateral geniculate nucleus. There they partake in the light reflex and accommodation reflex (see Chapter 21).

F. NON-SPECIFIC ASSOCIATIONAL.

The thalamic reticular nucleus forms a cap around the thalamus, medial to the internal capsule. All fibers systems leaving the thalamus and projecting onto the cerebral cortex and the fibers coming back from these same cortical areas pass through this nuclear complex with many specific systems terminating in this nucleus. This nucleus contains GABA-ergic neurons and has a modulating effect on thalamic neurons.

The *midline thalamic* nuclei (Fig. 15-4) are located in the periventricular gray above the hypothalamic sulcus. They are small and difficult to delimit in human beings, but these nuclei are intimately connected to the hypothalamus and the intralaminar nuclei, and their function must be interrelated.

The *intralaminar* nuclei (Figs. 15-3 and 15-5) are found in the internal medullary laminae of the thalamus. The most prominent intralaminar nucleus is the posteriorly placed centromedian, which is located in the midthalamic region between the medial and ventral posterior nuclei and receives many nociceptive fibers. The *parafascicular* nucleus is also easily delimited because it is at the dorsomedial edge of the habenulopeduncular tract. The intralaminar nuclei receive input from many regions throughout the central nervous system, including ascending axons from the reticular nuclei, indirect spinothalamic pathways, midbrain limbic nuclei, subthalamus, hypothalamus as well as other thalamic nuclei. The intralaminar nuclei have some direct projections onto the frontal lobe however their major influence is on the cerebral cortex by their connections to the specific thalamic nuclei.

Electrical stimulation of the intralaminar nuclei activates neurons throughout the ipsilateral cerebral hemisphere. As the stimulation continues, more and more cortical neurons fire, which is called the recruiting response, and there is a waxing and waning, the response finally peaks then decreases but may increase again. The centromedian nucleus has a strong projection to ventral anterior and ventral lateral nuclei and has

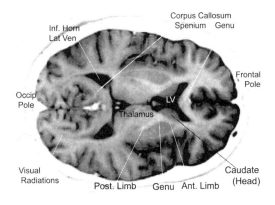

Figure 15-5. Horizontal section, demonstrating the anterior limb, genu, and posterior limb of the internal capsule. Note the putamen and globus pallidus are external to the internal capsule. (MRI weighted T2.)

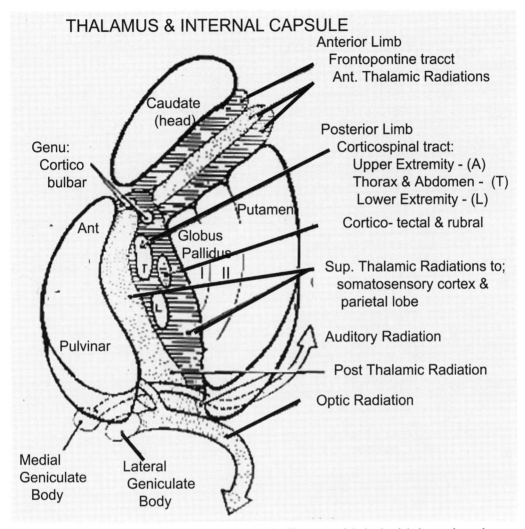

Figure 15-6. A horizontal section similar to 15-6. The major fiber tracts of the in the right internal capsule are labeled. Modified After Carpenter Core Text, Willams & Wilkins 1992.

a direct projection to the caudate and putamen which permits interrelationship with the extrapyramidal motor system (see Chapter. 17).

In Table 15-2, the principal thalamic nuclei are grouped as to the type of modality that reaches the individual nuclei. This table demonstrates that the thalamus receives input from gray matter in the spinal cord, brain stem, cerebellum and cerebrum. The type of input is identified functionally as sensory, motor, limbic or associational.

III. WHITE MATTER OF THE DIENCEPHALON

At the lateral margin of the diencephalon is the major white matter bundle that connects the diencephalon and the cerebrum, internal capsule.

Internal Capsule (Fig 15-6).

The internal capsule is a large myelinated region that marks the border between the diencephalon and telencephalon. The internal capsule, consisting of the anterior limb, genu, and posterior limb, is best visualized in a horizontal section, where it forms an obtuse angle. All fibers projecting from the thalamus onto the cerebral hemispheres must pass through this region, while all fibers leaving the cerebral cortex and going either to the diencephalon, basal nuclei, brain stem, or spinal cord must also pass through its parts. In the following diagrammatic horizon-

TABLE 15-2. FUNCTIONAL GROUPS OF THALAMIC NUCLEI

Modality	Nuclei
Sensory Relay	Nuclei: ventral posterior medial (touch and taste), ventral posterior lateral (touch & vestibular), medial geniculate (hearing), and lateral geniculate (vision). Input from ascending specific sensory modalities and project to cortical area subserving that modality.
Motor Relay	Nuclei: ventral lateral and ventral anterior. Input from the contralateral cerebellar hemispheres, superior cerebellar peduncle, medial globus pallidus, and the nigrostriatal fibers from the area compacta of the substantia nigra.
Limbic	Nuclei: dorsal medial, anterior, and lateral dorsal, Receive a well defined input from the mammillothalamic tract of the Papez circuit and project onto the prefrontal and cingulate gyrus.
Specific Associational	Nuclei: dorsal medial, lateral, and pulvinar nuclei. Forms the bulk of the thalamus and receive a multimodal input and projects onto cortical areas that also receive a multimodal intracortical associational input.
Nociceptive	Nuclei: Dorsomedial, VPL & Posterior. Receive input from ascending anterolateral system, spinothalamic, spinoreticular and spinomesencephalic
Nonspecific Associational	Nuclei: intralaminar, midline, and reticular. Receive input from diverse subcortical regions, including the reticular formation, visceral brain, and corpus striatum, project to specific thalamic nuclei or associational areas in the cortex.

tal section the three major parts of the internal capsule are identified and their constituent fiber systems discussed: the anterior limb, genu and posterior limb *(Figure 15-6)*.

The anterior limb lies between the caudate nucleus and putamen, and this portion of the capsule contains the anterior thalamic radiation and the frontopontine fibers.

The genu is found at the rostral end of the diencephalon on the lateral wall of the lateral ventricle and is the point at which the fibers are displaced laterally by the increasing bulk of the diencephalon. The corticobulbar fibers are found in the genu.

The posterior limb of the internal capsule consists of three subdivisions: thalamolentiform, sublenticular, and retrolenticular:

1. The *thalamolenticular* portion (between the thalamus and lenticular nuclei) contains the corticospinal, corticorubral, corticothalamic, thalamoparietal, and superior thalamic radiations.

2. The *sublentiform* portion (passing inferiorly and posteriorly) includes the posterior thalamic radiations that include optic radiation (Fig. 15-6), acoustic radiation, corticotectal, and temporal pontine fibers.

3. The *retrolenticular* portion (passing posteriorly into the temporal and occipital lobes) contains the posterior thalamic radiation to the occipital and temporal lobes and the parieto-occipital fasciculus.

Within the diencephalon many other myelinated fiber bundles can be identified Table 15-3.

IV. RELATIONSHIP BETWEEN THE THALAMUS AND THE CEREBRAL CORTEX. FIG 15-7.

The thalamus has several major roles;

1) To receive input from the brain stem, cerebellum and corpus striatum,

2) To project this information, after some processing, onto the cerebrum, and

3) To receive reciprocal projections from the same cerebral areas. The relationship between many thalamic nuclei and specific cortical areas is so intimate that when a cortical region is destroyed the neurons in a particular thalamic nucleus atrophy (Walker 1938). This degeneration is a consequence of the thalamic projection and its reciprocal from the cortex, being restricted to only a single specific cortical region and is seen in portions of the following nuclei: anterior, ventral lateral, ventral posterior, lateral geniculate, medial geniculate, and magnocellular portion of the medial dorsal. and pulvinar. The intralaminar, midline, posterior, portions of the pulvinar, and medial nuclei are unaffected by cortical lesions.

DIENCEPHALON 15-9

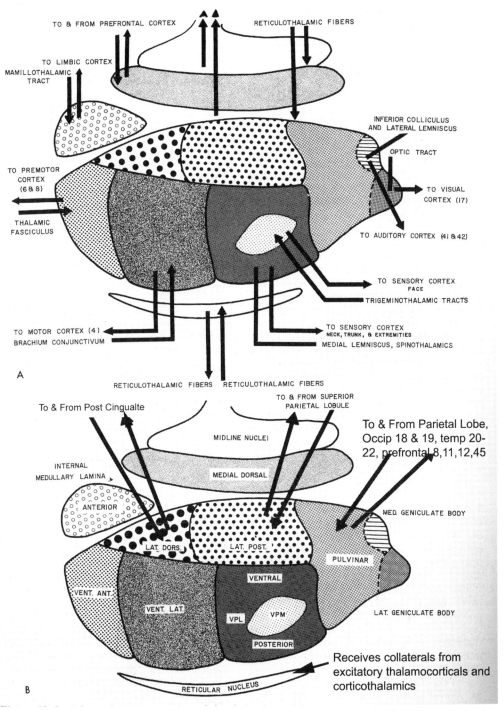

Figure 15-7; Representation of the thalamic nuclei: A- showing connections of ventrobasal, midline, anterior, lateral nuclei and geniculate nuclei. B- demonstrating connections of lateral dorsal, lateral posterior and pulvinar nuclei.. (After Truex, R.C., and Carpenter, M.B.: Human Neuroanatomy. Baltimore, Williams and Wilkins, 1970)

TABLE 15-3: MAJOR FIBER BUNDLES ASSOCIATED WITH THE DIENCEPHALON:

Pathway	Connections
External medullary lamina	Includes fibers from medial lemniscus, superior cerebellar peduncle, spinothalamics, and pathways between cortex and thalamus
Fornix	Subiculum and hippocampus to septum & hypothalamus (Part of Papez Circuit)
Habenulo-peduncular	Habenula to interpeduncular nucleus and tegmentum of midbrain
Internal capsule includes: Anterior limb, genu, posterior limb	Thalamus and cerebral cortex, and cerebral cortex to brain stem and spinal cord
Internal medullary lamina	Contains mammillothalamic, ascending reticular and spinothalamic fibers.
Mammillo-thalamic	Mammillary bodies with anterior thalamic nuclei (Part of Papez Circuit)
Mammillary peduncle	Mammillary nuclei with tegmentum of midbrain

Thalamic input onto the cortical layers.

The most important input the cerebral cortex receives is from the thalamus. The densest input from the specific relay projectional nuclei (e.g. VPL, VPM) is to layer 4, but as indicated above, pyramidal cells in layers 3 and 5 many also receive direct or indirect inputs. There is a modality specific columnar arrangement. With many nuclei having an extensive input onto the cortex there is always a reciprocal cortical input. The nonspecific nuclei (midline & intralaminar) have a more diffuse input and project onto layer I.

Thalamic Radiations and the Internal capsule.

The fibers that reciprocally interconnect the cortex and the thalamus form the thalamic radiations *(Fig 15-6)*. The thalamic radiations are grouped in the internal capsule into 4 peduncles: anterior, superior, posterior and inferior.

Anterior radiations- connects frontal lobe and cingulate cortex with median and anterior thalamic nuclei. This is an extensive projection.

Superior radiations- connects motor sensory strip and adjacent frontal and parietal lobes with fibers from the ventrobasal nuclei. This is also a large projection.

Posterior radiation- connects occipital and posterior parietal areas with pulvinar, and lateral geniculate nucleus (geniculocalcarine radiations - optic radiations). This is an extensive radiation due to the optic radiations.

Inferior radiations- connections from medial geniculate (auditory radiations) and pulvinar with temporal lobe. This is the smallest group of fibers between the thalamus and cortex.

In *Figure 15-7* the afferents and efferents of the individual thalamic nuclei are presented.

In *figure 15-8* we have shown the cortical projections of the same thalamic nuclei.

In *figures 15-9, 15-10, 15-11,* and *15-12* we discuss the major afferent pathways to the thalamus and metathalamus.

In the table that follows, Table 15-4, we have summarized the cortical projections of the major thalamic nuclei and also included the functions of these regions.

Other Possible Inputs to Thalamus. In addition to the major pathways discussed above there are other subcortical sources of input onto the cerebral cortex that pass through the diencephalon, and they may well have modulating effects on the thalamus (see discussion in chapter 17) these include:

1. *Noradrenergic* (norepinephrine) pathway from the locus ceruleus of the midbrain projecting in primates predominantly to layer 6 of the motor and somatosensory cortex and related frontal and parietal association cortex.

2. *Serotoninergic* pathway from the raphe nuclei in the pons and medulla and the pontine reticular formation.

3. *Dopaminergic* pathways from ventral tegmental – rostral mesencephalic nuclear groups. The strongest projection is to the prefrontal cortex, and limbic system.

4. *Cholinergic* pathways from the basal forebrain nucleus of Meynert project widely to cerebral cortex.

5. *GABA-ergic* pathways from basal forebrain, ventral tegmental area and zona incerta to sensory and motor cortex.

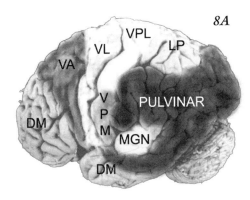

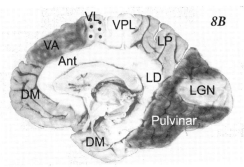

Figure 15-8. Thalamic Projections onto Cerebral Cortex. A. Projections onto Lateral surface of a cerebral hemisphere (B), Projections onto medial surface of cerebral hemisphere

Clinical Considerations.

The thalamus is the final processing station for systems that project to the cerebral cortex; thus it serves an important integrative function. Many sensations are first crudely appreciated at thalamic levels, including pain, touch, taste, and vibration. The discriminative processes associated with these sensations, as well as tactile discrimination, vision, audition, and taste, are elevated to consciousness in the cerebral hemisphere. Glutamate is the principal transmitter in the thalamus.

Thalamic syndrome (of Dejerine). In the human a large lesion in the thalamus results in the thalamic syndrome, causing diminished sensation on the contralateral half of the head and body. (Complete anesthesia results from injury to the ventral posterior lateral nucleus.) Some pain and temperature sensation from the contralateral side of the body may be retained.

The internal capsule is usually involved, producing an upper-motor-neuron lesion of the contralateral limb. The thresholds for pain, temperature, touch, and pressure are usually elevated contralaterally. A mild sensory stimulus now produces exaggerated sensory responses on the affected side and may even cause intractable pain. A change in emotional response may also be noted.

Lesions restricted to the medial thalamic nucleus produce memory and personality disturbances; lesions in the lateral nuclei produce sensory deficits (see Chapter 20).

SUBTHALAMUS (FIG. 15-3)

The subthalamus region is included in the diencephalon, but functionally it is part of the basal ganglia. The subthalamic nucleus, zona incerta, and prerubral field together form the subthalamus. This region is medial to the posterior limb of the internal capsule, lateral to the hypothalamus, and below the thalamus. The ansa lenticularis, subthalamic fasciculus, lenticular fasciculus, and thalamic fasciculus provide input

TABLE 15-4: CORTICAL PROJECTION OF MAJOR THALAMIC NUCLEI

NUCLEUS	CORTICAL PROJECTION	FUNCTION
Anterior & Lateral Dorsal	Cingulate and parahippocampal gyrus	Limbic
Medial Dorsal	Prefrontal & rostral temporal	Limbic
Lateral Geniculate	Calcarine Cortex	Vision
Lateral Posterior	Superior parietal lobule	Sensory Associational
Medial Geniculate	Transverse Temporal of Superior Temporal	Hearing
Pulvinar	Inferior parietal lobule, posterior temporal lobe, and occipital association cortex	Multimodal Associational
Ventral Anterior	Premotor	Motor Associational
Ventral Lateral	Precentral	Motor Cortex
Ventral Posterior	Postcentral	Sensory Cortex

from the globus pallidus to the subthalamus. The zona incerta lies between the thalamus and lenticular fasciculus. The adjacent substantia nigra also projects onto the subthalamus.

The medial portion of the globus pallidus provides the principal input to the subthalamus via the subthalamic fasciculus, which runs through the internal capsule. The motor and premotor cortex also project to this region. The subthalamus projects back to both segments of the globus pallidus, and substantia nigra via the subthalamic fasciculus, and projects to the thalamus and contralateral subthalamus. The glutaminergic cells in the subthalamus form the main excitatory projection from the basal ganglia.

In the human, a lesion in the subthalamus, usually vascular in origin, produces hemiballismus: purposeless, involuntary, violent, flinging movements of the contralateral extremity. These movements persist during wakefulness, but disappear during sleep. This lesion represents underactivity in the indirect pathway in contrast to over activity that produces Parkinsonism. (See Chapter 18).

THE MAJOR SENSORY PATHWAYS OF THE CENTRAL NERVOUS SYSTEM

This discussion will focus on the sensory systems that form the bulk of the afferents to the thalamus. The voluntary motor pathways are discussed in the end of this chapter while the other motor pathways are discussed in Chapters 11, 12, and 17.

Basic Principal of Sensory System: All sensory systems have three neurons:

1st neuron in periphery the dorsal root ganglion or ganglion of cranial nerves V, Vii, IX, or X,

2nd neuron within CNS spinal cord or brain stem; its axon u crosses to the contralateral side, and

3rd order neuron is the final neuron in the sensory systems and its axons reach the cerebral cortex.

Medial Lemniscus--General Somatic Tactile Sensation *(Fig 15-9).*

The ascending sensory fibers from the body for touch (posterior columns), from the face for touch and pain (trigeminothalamic), and from the viscera for general and special sensation ascend in the medial lemniscus to the dorsal thalamus.

Posterior Columns (Fasciculus Gracilis and Cuneatus).

The posterior columns--the fasciculus gracilis, and the fasciculus cuneatus (Fig. 15-9)--conduct proprioception (position sense), vibration sensation, tactile discrimination, object recognition, deep touch (pressure) awareness, and two-point discrimination from the neck, thorax, abdomen, pelvis, and extremities. The sensory receptors for the system are the Golgi tendon organs, muscle spindles, proprioceptors, tactile discs, and Pacinian corpuscles (deep touch, or pressure). The primary cell body is located in the dorsal root ganglion.

The well-myelinated fibers of this system enter the spinal cord as the medial division of the dorsal root and bifurcate into ascending and descending portions, which enter the dorsal column.

The fasciculus gracilis contains fibers from the sacral, lumbar, and lower thoracic levels, while the fasciculus cuneatus contains fibers from the upper thoracic and cervical levels. Fibers from the sacral levels are the first to enter and lie most medial, followed by lumbar, thoracic, and finally, cervical fibers. The primary axons ascend in the dorsal columns of the spinal cord to the secondary cell body of this system located in the nucleus gracilis and the nucleus cuneatus in the lower medullary levels.

From the secondary cell bodies the fibers cross the midline, join the medial lemniscus, and ascend to the ventral posterior lateral nucleus in the thalamus. From this thalamic nucleus these fibers are projected to the postcentral gyrus (areas 1, 2, 3).

Fibers also descend in the dorsal columns, but their functional significance is unknown.

The fibers responsible for proprioception cross in the medial lemniscus. The fibers for vibration sensation and tactile discrimination ascend bilaterally in the medial lemniscus to the ventral posterior lateral nucleus. Consequently, a unilateral lesion can abolish proprioception, but tactile discrimination and vibration sensation will not be entirely lost.

Clinical Lesions.

Injury to the posterior column appears not to affect pressure sense, but vibration sense, two-point discrimination, and tactile discrimination are diminished or abolished, depending on the extent of the lesion. Interruption of the medial fibers (cervical-hand region) affect the ability to recognize differences in the shape and weight of objects placed in the hand is impaired. Since the extremities are more sensitive to these modalities than any other body regions, position sense is impaired more severely in the extremities than elsewhere, and the person has trouble identifying small passive movements of the limbs. Consequently, performance of voluntary acts is impaired and movements are clumsy (sensory ataxia). Lesions in lateral fibers of the posterior column, gracilis, may be devastating due to the interruption of one of the most important sensory mechanisms the ability to detect the sole of the foot. This is a major handicap in walking in dim lighted or a dark room, in driving a car etc.

The discussion of the analysis of tactile information in the cerebral cortex is found in Chapter 21 with several cases also illustrating lesions in this region of the brain.

Tracts Originating from Secondary Trigeminal Nuclei in the Brain Stem (Fig. 15-10). Above all, remember that the primary cells of the trigeminal nerve are located in the trigeminal ganglion in Meckel's cave in the middle cranial fossa.

Proprioception from the Head.

Mesencephalic Nucleus. These unique primary cell bodies are located not only in the trigeminal ganglion but also within the pons and midbrain along the ascending trigeminal rootlets, where they form the mesencephalic nucleus of nerve V. The primary axons project to the motor nucleus of nerve V and the reticular formation. Axons are also projected to the cerebellum and inferior olive.

The origin of the secondary neuron is unclear (but probably the descending nucleus of nerve V). The secondary axons ascend near the medial lemniscus to the ventral posterior medial nucleus of the thalamus.

Pain and Temperature from the Head.

Descending Nucleus of Nerve V. The cell bodies of primary neurons are located in the trigeminal ganglion, and the primary axons are numerous. These fibers enter the pons, lie on the external surface of the descending tract of nerve V, and then terminate in the descending nucleus of nerve V. The ophthalmic fibers descend to C3, the maxillary to C1, and the mandibular to

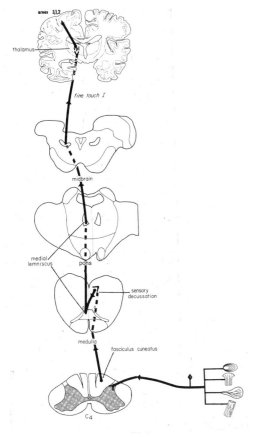

Figure 15-9 Posterior columns, the funiculus gracilis and the funiculus cuneatus. This fiber bundle originates from tactile and proprioceptive receptors. The primary and secondary cell bodies are on the ipsilateral side of the spinal cord. The secondary neurons are found in the medulla. The secondary axons cross the midline and ascend in the contralateral medial lemniscus to the ventral posterior lateral nucleus (VPL) in the thalamus. In the spinal cord the posterior columns are uncrossed and divided into the medial gracile fasciculus (lower extremity) and the lateral cuneate fasciculus (upper extremity). Fibers from the upper extremity form 50 % of the posterior columns with the lower extremity 25% and the remainder from the thorax and abdomen. These fibers cross in the sensory decussation and form the bulk of the medial lemniscus along with trigeminal and other fibers.

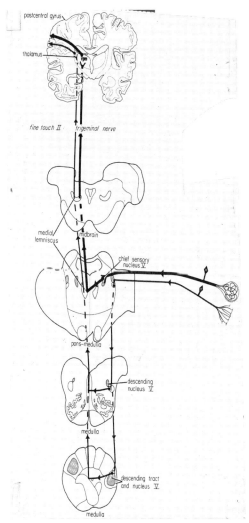

Figure 15-10 Sensory Portion of Trigeminal System. Touch and Pain from the head. This is the largest cranial nerve in the brain stem and has three divisions: Ophthalmic (V1), Maxillary(V2) and Mandibular (V3). Each of these divisions brings in sensation from pain, temperature, touch and pressure receptors in the skin, muscles and sinuses they innervate. The tactile fibers synapse with the chief sensory nucleus in the pons. The Proprioceptive fibers (not shown) enter and ascend through pontine and midbrain levels and synapse on the mesencephalic nucleus of nerve V which then synapses on the motor nucleus of V in the pons for the "jaw jerk". The secondary axons ascend in the medial lemniscus to the ventral posterior medial nucleus (VPM) in the thalamus. The pain fibers cross while the touch fibers run bilateral up to the thalamus, and end in VPM and are then projected onto the lower third of the postcentral gyrus.

the lower medulla. The secondary axons leave the descending nucleus of nerve V and ascend bilaterally in the medial lemniscus to the ventral posterior medial nucleus of the thalamus. The predominant component is crossed.

Tactile Discrimination, Vibration Sensation, and Pressure Sense from the Head,

Chief Nucleus of V. The cell bodies of primary neurons are located in the trigeminal ganglion. The primary axons enter the pons and end on the main sensory nucleus of nerve V, from which the secondary axons ascend contralaterally in the medial lemniscus or ipsilaterally in the dorsal trigeminothalamic tract to the ventral posterior medial nucleus, primarily on the contralateral side of the brain stem.

Lesions in the sensory nuclei cause an absence of sensation on the same side of the head, while lesions of the tracts produce diminished sensation.

Anterolateral Pathways- Spinothalamic and Trigeminothalamic Tract--Pain Pathways *(Fig. 15-10)*

Pain from the Body. Fig 15-11.

Anterolateral Pathway/Spinothalamic. This tract conducts pain and temperature sensation from the neck, thorax, abdomen, pelvis, and extremities. The receptors for pain are the naked free nerve endings, and the corpuscles of Ruffini and of Krause detect warmth and cold. The primary cell bodies are located in the dorsal root ganglion; the axons enter the spinal cord and ascend or descend one segment ipsilaterally before ending on neurons in the dorsal horn lamina 1 to 3. The axons of the secondary cells cross in the anterior commissure of the spinal cord and ascend in the ventrolateral part of the lateral funiculus in the spinal cord and medulla. In the pons they are found at the lateral margin of the medial lemniscus and continue up to the ventral posterior lateral nucleus and the intralaminar nuclei of the dorsal thalamus. The axons in this tract convey information from the contralateral side.

The pain fibers in this tract are arranged segmentally with the most lateroposterior fibers representing the lowest part of the sacral levels of the body, and the more medioanterior fibers representing the upper extremities and neck. The temperature fibers have this same arrangement but are internal to the pain fibers. For surgical relief of pain this tract is located by identifying the denticulate ligament and then sectioning about 1 mm. below the ligament (Fig7-29).

The sharp pain noted with most injuries is associated with the heavily myelinated direct, or spinothalamic, pathway to the ventral posterior lateral nucleus. In addition to the stabbing pain, dull throbbing pain is usually noted; it is carried up more slowly by a multisynaptic pathway, probably via the reticular system, and is called the spinoreticulothalamic pathway.

A unilateral lesion of the spinothalamic tract in the spinal cord produces an almost complete absence of pain (analgesia) and temperature sensation (thermoanesthesia) on the contralateral side. At the upper pontine levels, the lateral spinothalamic tract is usually closely associated with the medial lemniscus. Consequently, a unilateral lesion at these levels diminishes pain and temperature sensation from the opposite side of the body--as well as touch, vibration sensation, and proprioception--.

Anterior Spinothalamic Tract--Light Touch

Light touch awareness from the neck, thorax, pelvis, and extremities is carried up the anterior spinothalamic tract *(Fig 15-11)*. The primary neuron is located in the dorsal root ganglion; the secondary neuron is located in the dorsal horn of the same side. The secondary fibers ascend contralaterally in the lateral funiculus, joining the medial lemniscus at the upper pontine levels. Stroking a hairless area of the skin with cotton evokes light touch. Lesions involving this tract produce no definite clinical deficiencies, probably because somewhat similar sensations are also carried in the uncrossed dorsal columns (fine touch, pressure).

Pain and Temperature Sensation from the Head. This information is carried by the descending nucleus and root of nerve V. The primary cell bodies are found in the trigeminal ganglion located in the middle cranial fossa, and the secondary cell bodies originate in the descending nucleus of nerve V. The descending nucleus is very large and extends from pons into medulla and upper cervical spinal cord. This nucleus is divided into three regions pars – oralis

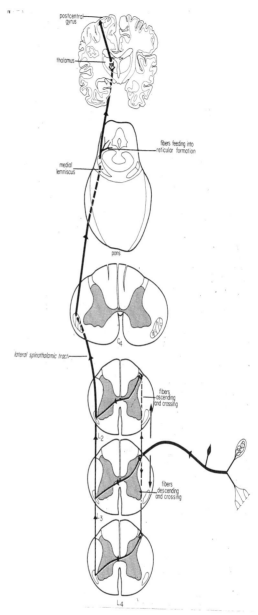

Figure 15-11 Anterolateral pathway-lateral spinothalamic fibers. The Pain Pathway. Originate from pain and temperature receptors in the skin. Primary cell bodies are dorsal root ganglia of the spinal cord. The axons synapse in the dorsal horn, cross within 3 spinal cord segments and form the spinothalamic fibers. These fibers pass into the thalamus. The direct pathway, for localizing the pain on the body, fibers end in VPL which projects onto postcentral gyrus. How one responds to pain comes from the indirect pain pathway that ascends to the dorsomedial thalamic nucleus and is then projected onto the prefrontal gyrus.

(in pons and upper medulla), pars – interpolar (in medulla), and pars – caudalis (in cervical cord). These fibers enter the spinal trigeminal nucleus external to the descending nucleus on the lateral surface of the pons and are somatotopically organized with V3 dorsal, V1 ventral, and V2 intermediate. Fibers of cranial nerves VII, IX and X are dorsal to V3. These fibers descend and end in the spinal trigeminal nucleus with V1 ending in pars caudalis in cervical levels, V2 to interpolar nucleus in medulla, and V3 to pars oralis in upper medulla and pons. The secondary axons ascend contralaterally in the medial lemniscus to the ventral posterior medial nucleus of the thalamus or to the intralaminar nuclei.

Visceral Referred Pain.

Referred pain is caused by painful impulses originating in the viscera that are transmitted by the sympathetic nerves and referred to the periphery of the body. The organ in which the pain is felt belongs to the same sensory dermatome as the visceral structure where it originates as follows: lungs, C8 to T8; heart, C8 to T8; bladder, T1 to T9; testes, T10 to T12; kidneys, T11 to L1, and rectum, S2 to S4.

Unconscious proprioception from the lower extremity, Thorax, Abdomen and Pelvis

This information is carried up to the cerebellum by the anterior and posterior spinocerebellar tracts. The receptors for both tracts are the neuromuscular spindles and the Golgi tendon organs.

Unconscious Proprioception from the Lower Extremity, Thorax, Abdomen, and Pelvis. This information is carried up to the cerebellum by the anterior and posterior spinocerebellar tracts.

Unconscious Proprioception from the Upper Extremity and Neck and Head.

Cuneocerebellar Pathway. The primary neurons are found in the dorsal root ganglion at cervical levels and in the trigeminal proprioceptive fibers. Their axons ascend in the dorsal columns and end in the external cuneate nuclei (external to the cuneate nucleus) in the lower medullary levels. The secondary axons ascend bilaterally with the posterior spinocerebellar tract and enter the cerebellum primarily uncrossed.

Cerebellar Tracts--Unconscious Proprioception See chapter 21.

The cerebellar cortex must have proprioceptive information to insure proper muscle tone and coordination. Fibers carrying this information run to the cerebellum in the following pathways: posterior spinocerebellar, anterior spinocerebellar, cuneocerebellar.

Posterior Spinocerebellar Tract (Fig. 15-12). The primary neuron is located in

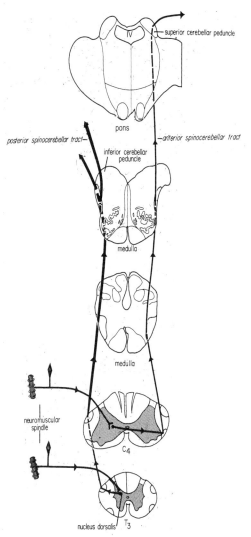

Figure 15-12. Anterior and posterior Spinocerebellar Pathways. -Unconscious proprioception to the cerebellum. These tracts carry information from Proprioceptive receptors in the body into the cerebellum with the anterior pathway crossed and the posterior uncrossed.

the dorsal root ganglion; the secondary neuron is located in Clarke's column in the dorsal horn at levels C8 to L3 of the cord. The secondary axons ascend uncrossed in the lateral funiculus, enter the cerebellum via the inferior cerebellar peduncle, and terminate in the ipsilateral vermis (midline portion of the cerebellum).

Anterior Spinocerebellar Tract (Fig. 15-12). The primary neuron is in the dorsal root ganglion, while the secondary neuron is in cells at the margin of the dorsal horn that are scattered throughout the spinal cord, especially in the lumbar levels. The secondary axon crosses the midline and ascends anterior to the posterior spinocerebellar tract in the lateral funiculus and medulla; it enters the cerebellum via the superior cerebellar peduncle and terminates in the same parts of the cerebellar vermis as the posterior spinocerebellar tract except it is located on the contralateral side.

The receptors for both tracts are the neuromuscular spindles and Golgi tendon organs.

Cerebellar Peduncles (see Chapter 20). The cerebellum is interconnected to the brain stem by three major fiber bundles, or brachium: the superior cerebellar, middle cerebellar, and inferior cerebellar peduncles.

The axons in the *superior cerebellar peduncle* (brachium conjunctivum) are primarily efferent and ascending axons originating in the deep cerebellar nuclei - dentate, emboliform,

These fibers cross in the tegmentum of the lower midbrain (decussation of the superior cerebellar peduncle) with some fibers ending on the red nucleus and oculomotor complex. However, the majority of the fibers in the superior cerebellar peduncle distribute to the motor nuclei of the thalamus - ventral lateral and ventral anterior nuclei. This cerebellar projection to the motor cortex permits the well-controlled volitional acts of the corticospinal and corticobulbar systems. The ventral spinocerebellar fibers enter the cerebellum through the superior cerebellar peduncle.

The *middle cerebellar peduncle* (brachium pontis) is the largest cerebellar peduncle and is purely afferent (Fig. 15-12). The largest group of descending fibers from the cerebral hemispheres is directed upon the pontine gray. These fibers form the bulk of the cerebral peduncle and descend from all cortical lobes and the cingulate and insular gyrus, with the largest group of fibers coming from the somatosensory strip and premotor cortex. These corticopontine fibers synapse upon the pontine gray and then pass into the contralateral cerebellar hemisphere. Lesions in these fibers systems do not cause any clinical deficit although these fibers must be important in all fine movements.

The corticopontine fibers end on the pontine nuclei. The axons from the pontine nuclei are crossed and constitute part of the mossy-fiber input to the granule cells.

The *inferior cerebellar peduncle* (restiform body) connects the brain stem and spinal cord with the cerebellum (Fig. 15-12). The climbing fibers from the inferior olive enter this peduncle and extend to the Purkinje cells in the deep cerebellar nuclei. The fibers from the vestibular nuclei, reticular formation, and trigeminal nuclei, as well as the dorsal spinocerebellar fibers, enter through this peduncle. Many of these fibers contribute to the mossy-fiber input to the cerebellum. The aminergic fibers from the raphe and locus ceruleus also enter the cerebellum through the inferior cerebellar peduncle. There is a major cerebellar projection onto the reticular formation of the pons and medulla and onto the vestibular nuclei from the fastigial nucleus, as well as a strong cerebellar input onto the vestibular nuclei from the flocculonodular lobe.

SPECIAL SENSORY
CRANIAL NERVE VIII

Visual Pathway. Discussed in Chapter 21 on Vision.

The inner ear that contains the receptors for the auditory and vestibular systems and their central projections are discussed in the Special Sensory Section on the CD that also includes the Chapter on Vision.

CN VIII - Auditory Pathway. The auditory nerve, Cranial Nerve VIII, terminates in the dorsal and ventral cochlear nuclei in the medulla. Each sound has a specific location on the cochlea. The termination is in a tonotopic pattern in each nucleus with the low frequency fibers terminating ventrally and the high frequency fibers dorsally creating a series of frequency sensitive layers. The frequency of sounds is from 100 - 15,000Hz.

The dorsal cochlear nucleus projects contralaterally through the lateral lemniscus to the inferior colliculus while the ventral cochlear nucleus projects bilaterally to the medial superior olivary nucleus and ipsilateral lateral superior olivary nucleus. Frequencies above 2000 Hz are recorded in the lateral superior olivary nucleus. The lateral superior olive projects to the contralateral inferior colliculus while the medial superior olivary nucleus projects ipsilaterally to inferior colliculus, medial geniculate and auditory cortex.

Cortical Projection. The laminated ventral medial geniculate projects onto the primary auditory cortex area 41 while the dorsal medial geniculate projects onto auditory association areas 42.

CN VIII - Vestibular Pathway (These fibers should be considered part of the proprioceptive system).

The vestibular fibers originated from two ganglia the superior and inferior ganglia of Scarpa. There are 4 major vestibular nuclei: superior (SVN), lateral (LVN), medial (MVN) and inferior (IVN). Trunk and limb somatosensory pathways send information to LVN, MVN and IVN. Visual information also indirectly reaches these same nuclei. The semicircular input reaches SVN, MVN, and LVN. The utricle projects mainly to LVN, the saccule to IVN. The fastigial nucleus of the cerebellum is excitatory to the vestibular nuclei while the cerebellar cortex is inhibitory. The vestibular nuclei project to deep layers of the superior colliculus adjacent to projections from cervical proprioceptive endings.

Cortical Projection. From the LVN there is a small contralateral input to ventroposterior and medial magnocellular of MGN. Ventroposterior projects to SI arm region of 3a, while the MGN projects to area 5 and the neck region of area 2. Stimulation of either of these cortical regions produces vestibular sensations. The cortical neurons respond to both vestibular and proprioceptive stimuli from neck trunk and proximal joints and the response are excitatory for one rotation in one direction and inhibitory in the other.

Taste and General Sensation from the Viscera

General sensation from the gastrointestinal system (cranial nerve X) and from the taste buds on the tongue (cranial nerves, VII, IX, and X) enters the brain stem to terminate on the nucleus solitarius in the medulla. Fibers ascend primarily uncrossed either in the central tegmental pathway or in the medial lemniscus to reach basomedial portion of ventral posterior medial (VPM) nucleus in the dorsal thalamus. Many of the gustatory fibers terminate in the ipsilateral parabrachial nucleus before reaching the ventrobasal portion of VPM nucleus. The cortical region for taste is in the parietal lobe in the upper banks of the lateral sulcus.

MAJOR VOLUNTARY MOTOR PATHWAYS

In the older literature the motor pathways were divided into pyramidal (corticospinal and corticonuclear) and extrapyramidal (rubrospinal, tectospinal, and so on). These terms are no longer used, and each pathway is now usually described in terms of its function.

Basic Principal of Motor System. The motor system consists of two portions: an upper and lower motor neuron. The portion originating in the gray matter of the motor sensory strip of the cerebral cortex is called upper motor neurons. The pathway from the upper motor neurons, corticospinal or corticonuclear, descends and crosses over to innervate the lower motor neurons in the brain stem or spinal cord.

The axons of the upper motor neurons either synapse on interneurons or end directly on the lower motor neurons. The axons of the lower motor neurons leave the central nervous system and form the motor division of the peripheral nervous system.

CORTICOSPINAL TRACTS-- VOLUNTARY CONTROL OF THE LIMBS, THORAX, AND ABDOMEN. *(Fig. 15-13)*

This tract innervates the motor neurons that control the skeletal muscles in the neck, thorax, abdomen, pelvis, and the extremities. It is essential for accurate voluntary movements and is the only direct tract from the cortex to the spinal motor neurons.

fibers pass through the genu of the internal capsule into the middle third of the cerebral peduncles and enter the pons, where they are broken into many fascicles and are covered by pontine gray and white matter. In the medulla, the fibers are again united and are found on the anterior medial surface of the brain stem in the medullary pyramids.

About 75 to 90% of the corticospinal tract fibers cross at the medullospinal junction and are thereafter found in the lateral funiculus of the spinal cord. (The majority of fibers from the dominant hemisphere cross at the medullospinal junction.) Some corticospinal fibers remain uncrossed in the anterior funiculus but will slowly cross at cervical levels. About 50% of the corticospinal fibers end at the cervical levels--about 30% go to the lumbosacral levels, and the remainder to the thoracic levels.

Most of the corticospinal fibers end on internuncial neurons in laminae 7 and 8 of the spinal cord, but in regions in the spinal cord where the digits are represented (in the cervical and lumbosacral enlargements) the corticospinal fibers sometimes end directly on the motor horn cells. The internuncial neurons are located in lamina 7 of the spinal cord. Lesions in the corticospinal tract produce contralateral upper motor neuron symptoms, while lesions in the spinal motor neurons or rootlets cause lower motor neuron symptoms.

CORTICONUCLEAR SYSTEM-- VOLUNTARY CONTROL OF THE MUSCLES CONTROLLED BY CRANIAL NERVES V, VII, AND IX TO XII *(Figs. 15-14)*

This system distributes to the motor nuclei of the cranial nerves V, VII, IX, X, XI, and XII and provides voluntary and involuntary control of the muscles and glands innervated by these nerves. In the older literature this pathway was called corticobulbar because the medulla and pons containing these nuclei are collectively called the bulb. We prefer the term corticonuclear because it makes the student realize the key role of the cranial nerve nuclei in movement. These fibers are found anterior to the corticospinal fibers in the genu of the internal capsule and medial to the corticospinal tract in the cere-

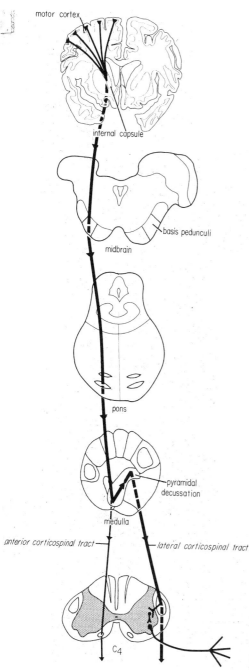

Figure 15-13. Corticospinal Pathway. Voluntary Movement of the Muscles in the upper and lower extremities, thorax and abdomen.

This system originates from the pyramidal cells and giant cells of Betz in the upper two-thirds of the precentral gyrus, area 4, and to a lesser degree from area 6 and parts of the frontal, parietal, temporal, and cingulate cortices. The

bral peduncle, pons, and medullary pyramids.

The fibers supplying the cranial nerves have the following cortical origin:

1. Muscles of facial expression (motor nerve VII), mastication (motor nerve V), and deglutition (ambiguous nuclei of nerves IX and X) originate from pyramidal cells in the inferior part of the precentral gyrus, area 4.

2. Muscles in the larynx are controlled from the inferior frontal gyrus and from the frontal operculum (the posterior part of the pars triangularis, area 44). It appears that many corticonuclear axons end on interneurons and not directly on the motor neuron of the cranial nerve.

3. The cortical innervation of the cranial nerves is bilateral, with the exception of the lower facial muscles, which are innervated by the contralateral cortex. Consequently unilateral lesions in the corticonuclear system produce only weakness and not paralysis. Paralysis results only from bilateral lesions in the corticonuclear system.

4. Lower motor neuron lesions of motor cranial nerves. A lesion of the motor nucleus or rootlet produces a lower motor neuron syndrome with paralysis, muscle atrophy and decreased reflexes.

Supranuclear lesions produce upper motor neuron symptoms. Associated with bilateral lesions of the corticonuclear system is pseudobulbar palsy with bilateral UMN signs and inappropriate behavioral. This probably represents the release of cortical control via the corticonuclear innervation of the muscles controlled by the cranial nerves. Limbic control to these nuclei takes over, via descending autonomic pathways, and the emotional state is undamped by the frontal lobes

CORTICOMESENCEPHALIC SYSTEM-- VOLUNTARY CONTROL OF MUSCLES ASSOCIATED WITH EYE MOVEMENTS--CRANIAL NERVES III, IV, AND VI.

The fibers controlling eye movements originate from the frontal eye fields in the caudal part of the middle frontal gyrus and the adjacent inferior frontal gyrus (area 8). At upper midbrain levels, fibers to cranial nerves III and IV leave the cerebral peduncles and take a descending path through the tegmentum, usually in the medial lemniscus (corticomesencephalic tract). Fibers to cranial nerve VI also descend in this way to the parabducens and parapontine reticular nuclei of

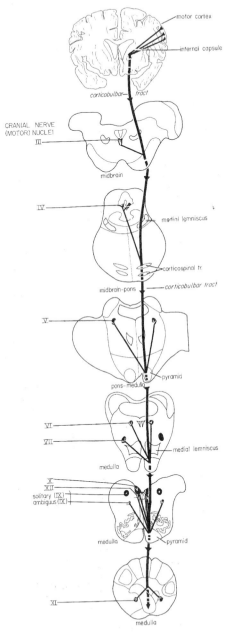

Figure 15-14. Corticobulbar/Corticonuclear Pathway. Voluntary Movements of the muscles associated with Cranial Nerves V, VII, and IX, X, XI, XII. Control of the cranial nerves that move the muscles (III, IV & VI) of the eyes is through the Corticomesencephalic pathway

the pons, which coordinate movements under the control of cranial nerves III and VI ascending through the medial longitudinal fasciculus.

The table that follows, Table 15-5, lists the Major Sensory and Motor Pathways in the human central nervous system

TABLE 15-5. MAJOR SUBCORTICAL TRACTS WITHIN THE BRAIN STEM AND DIENCEPHALON

Pathway	Function
Sensory Pathways	Functions
Spinothalamics and Trigeminothalamics	Pain and temperature from body, head and neck
Solitary tract	Taste and general sensation from the viscera
Cerebellar Peduncles: Superior (Cerebellum-Cerebrum) Middle (Cerebrum-Cerebellum) Inferior (Cerebellum with brain stem and spinal cord & CN VIII)	Unconscious sensory and motor
Optic Pathway - from retina via lateral geniculate to calcarine cortex (Chapter 21).	Vision & light reflexes
Auditory Pathway – cochlear nuclei via lateral lemniscus to inferior colliculus, brachium of inferior colliculus to medial geniculate and to auditory cortex, (Chapter 11 & Inner Ear- Special Senses on CD).	Auditory
Vestibular Pathway (Part of Proprioceptive Pathway)- lateral vestibular nucleus to Ventral posterior and MGN then to postcentral gyrus. (Chapter 11 & inner ear –Special Senses on CD)	Proprioceptive
Motor Pathways	Functions
Corticospinal.	Voluntary control of muscles in limbs thorax and abdomen
Corticonuclear	Voluntary control of: Facial expression by innervation of cranial nerves V, VII, IX-XII. Larynx by innervation of cranial nerves
Corticomesencephalic	Voluntary control of eye movements by innervation of cranial nerves III, IV and VI.
Rubrospinal, rubronuclear, tectospinal & tectonuclear	Backup to voluntary motor system
Reticular	Limbic (emotional)
Medial Longitudinal Fasciculus	Coordinate eye movements in brain stem via cranial nerves III, IV & VI
Monamine Containing: Norepinephrine, dopaminergic, serotonergic	Set tone in motor system and effect limbic system.
Descending sympathetics & parasympathetics	Autonomic functions

CHAPTER 16
Hypothalamus, Neuroendocrine System, and Autonomic Nervous System

Case History 16-1. A 25-year old white male auto body mechanic in July, developed numbness in the median nerve distribution of the left hand with pain at the left wrist, shooting into the index, middle and ring fi8ngers. And aching pain in finger joints. To a lesser degree, similar symptoms had developed in the right hand. The clinical impression of bilateral carpal tunnel syndrome was confirmed by nerve conduction studies. In addition the patient presented many of the facial features of acromegaly; coarse lips, large jaw and nose. None of his family had similar facial features and according to his mother, these features had emerged about age 12.

Growth hormone levels were consistently elevated to 17-20 ng/ml (normal is less that 5 ng/ml fasting and recumbent). A glucose load produced no suppression. Prolactin level was moderately increased to 37 ng/ml (normal in men if less that 1-20 ng/ml). MRI demonstrated an enlarged pituitary gland with a hypodense lesion, consistent with an adenoma *(Fig 16-1)*. A partial agenesis of the corpus callosum (splenium and posterior one-third of the body) was also present and may have related to long-standing learning disabilities. A trans-sphenoid excision of the pituitary adenoma was performed. Except for transient diabetes insipidus, he did well and all symptoms disappeared.

Comments.

The patient presented with a chief complains relevant to compression of the median nerve at the carpal tunnel. Usually the carpal tunnel syndrome occurs in one hand or the other and may reflect local swelling in the tendons and soft tissues. This is at times a result of occupation-related repetitive motion at the wrist. In occasional case, an underlying systematic conditions is present, these include; rheumatoid arthritis, pregnancy, myxedema of the hypothyroid state and acromegaly associated with excessive secretion of growth hormone by pituitary adenomas.

In this case, the neurologist noted the facial feature characteristic of acromegaly. In retrospect, the patient's mother was able to date these changes to his teen years. Excessive production of growth hormone, prior to closure of the epiphyseal plates in long bones, result in excessive growth or gigantism. While excessive e production of growth hormone after epiphyseal closure

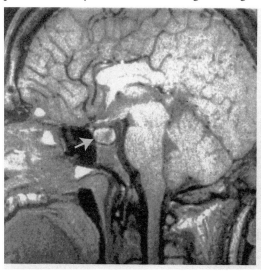 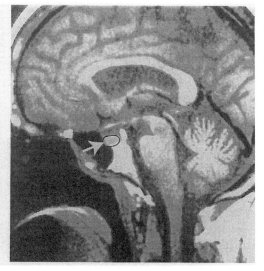

Figure 16-1. Adenomas. A) MRI, T1-weighted midline sagital sections; B) Control MRI, T1, weighted

results in enlargement of facial bones, jaw, hand and feet. The pain in the finger joints may have reflected an early stage of enlargement of the fingers.

In the present case, the adenoma had enlarged the pituitary gland, but the tumor had not expanded out of the sella turcica. The pituitary gland sits in the sella turcica, which is part of the sphenoid bone. Suprasellar, or upward extension of the tumor may result in compression of the optic chiasm. A characteristic visual field defect occurs; a bitemporal hemianopsia (loss of the outer halves of both visual fields). Lateral extension of the tumor into the cavernous sinus may compress the III, IV and VI cranial nerves, resulting in diplopia (double vision). Headaches may reflect unusual pressure on the diaphragm sella. Acute headache may reflect sudden increase in mass within the tumor due to infarctions of hemorrhage (so called "pituitary apoplexy". Pain in the eye of face may reflect compression of the maxillary or ophthalmic division of the trigeminal nerve with the cavernous sinus.

Careful removal of the tumor, in this case utilizing the trans-sphenoidal approach and the operating microscope, allowed preservation of residual pituitary function. Transient diabetes insipidus often is present reflecting temporary damage to the posterior pituitary or pituitary stalk.

HYPOTHALAMUS

The hypothalamus is a very small nuclear mass found at the base of the third ventricle in the diencephalon. Its functional potency is related to its control of the hypophysis and its descending autonomic fibers, which affect preganglionic parasympathetic and sympathetic nuclei in the brain stem and the thoracic, lumbar and sacral spinal cord. In addition, this diencephalic region with its important tracts form part of the Papez circuit and must be included in any discussion of the Limbic (or emotional) brain in Chapter 22.

The hypothalamus bilaterally forms the medial inferior part of the floor of the diencephalon and is located on the wall of the third ventricle, extending from the anterior margin of the optic chiasm to the posterior margin of the mammillary bodies. It is bounded dorsally by the dorsal thalamus and laterally by the subthalamus and internal capsule. It is connected to the hypophysis by the hypothalamic-hypophyseal tract and the hypothalamic-hypophyseal portal system. The hypothalamus consists of more than a dozen nuclei with their associated fiber systems.

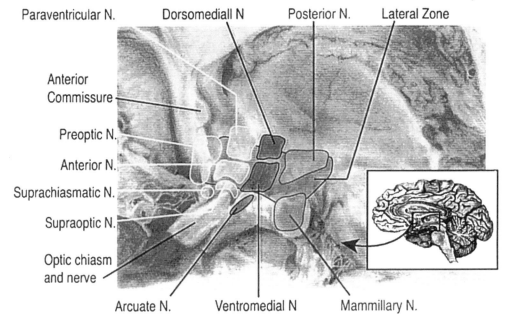

Fig 16-2. The zones and nuclei of the hypothalamus. The hypophysis and the nuclei of the hypothalamus.

Although the hypothalamus weighs only 4 gm, this small area controls our internal homeostasis and influences emotional and behavior patterns.

Hypothalamic Nuclei

Anatomically the hypothalamus is divided into a medial and a lateral zone *(Fig. 16-2)*.

The medial zone consists of three regions:

1. The anterior zone, including the preoptic, supraoptic, pariventricular, anterior, and suprachiasmatic nuclei (Fig. 16-2).

2. The middle zone, including the dorsomedial, ventromedial, lateral, and tuberal nuclei *(Fig. 16-3)*.

3. The posterior zone *(Fig. 16-4)*, including the posterior hypothalamic nuclei and mammillary nuclei (both large- and small-celled).

The lateral zone includes the lateral nucleus, the lateral tuberal nucleus, and the medial forebrain bundle. In addition, Periventricular nuclei are adjacent to the third ventricle at all levels of the medial region.

Anterior Group *(Fig 16-3)*. The preoptic, suprachiasmatic, supraoptic, and paraventricular

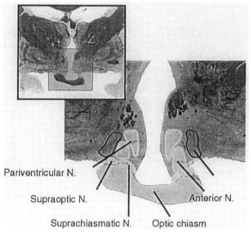

Figure 16-3. Myelin stain of diencephalon at the level of the anterior hypothalamus and optic chiasm showing corpus callosum, precommissural anterior commissure, and corpus striatum.

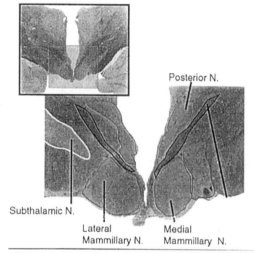

Figure 16-5. Myelin and cell stain (Kluver-Berrera) of diencephalon at posterior hypothalamus. (Corpus callosum removed.)

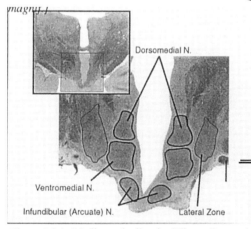

Figure 16-4. Myelin and cell stain (Kluver-Berrera) of diencephalon at mid-hypothalamus, demonstrating tuberal nuclei and mammillary bodies. (Inset magnifications)

nuclei are found in this region. The supraoptic and paraventricular nuclei contain secretory granules (containing oxytocin and vasopressin) that leave these nuclei and pass down the hypothalamic-hypophyseal pathway, which terminates at the capillaries in the posterior lobe of the pituitary, or neurohypophysis.

The supraoptic nucleus functions as an osmoreceptor; it releases vasopressin in response to an increase in osmotic pressure in the capillaries. Vasopressin or ADH (antidiuretic hormone) that increases water resorption in the kidneys.

The suprachiasmatic nucleus, which receives direct input from optic fibers in the chiasm, is important in the endocrine response to light levels and may well be the location of the biological clock.

Middle Group *(Fig 16-4)*. This region includes the dorsomedial, ventromedial arcuate (infundibular), lateral, and lateral tuberal nuclei. The ventromedial nucleus has extensive projections onto the diencephalon, brain stem, and spinal cord. Lesions here affect feeding behavior (e.g., obesity or anorexia). Arising from the tuberal nuclei, the tuberoinfundibular pathway terminates on the hypophyseal portal veins, permitting releasing factors produced in the hypothalamus to reach the anterior lobe of the pituitary.

Posterior Group (fig 16-5) consists of the mammillary body (medial and lateral nuclei) and posterior hypothalamic nuclei. A major part of the fornix projects onto this region with the mammillary nuclei then projecting via the mammillothalamic tract onto the anterior thalamic nuclei. The mammillary nuclei also project onto the brain stem tegmentum through the mam-

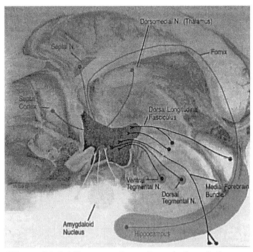

Figure 16-6 Afferent pathways into the hypothalamus.

millary peduncle. Lesions in these nuclei are seen in Korsakoff's psychosis with accompanying retrograde amnesia (See Chapter 30).

Table 16-1 Review functional localization in the Hypothalamic Nuclei.

Afferent Pathways

The hypothalamus is so situated that it functions as an important integrating center between the brain stem, reticular system and the limbic forebrain structures.

The major subcortical (afferent) inputs into the hypothalamus include *(FIG 16-6:*
FIG 16-6

1. Reticular input from the ascending multisynaptic pathways in the mesencephalic reticular nuclei via the medial forebrain bundle, dorsal longitudinal fasciculus, and mammillary peduncle; 2.The globus pallidus through lenticular fasciculus and ansa lenticularis (see chapter 18).

3. Thalamic input from the dorsal medial nucleus and intralaminar nuclei via direct connections onto the hypothalamus.

4. Fibers from the optic nerve, retinohypothalamic fibers, which end in the suprachiasmat-

TABLE 16-1. REVIEW OF FUNCTIONAL LOCALIZATION IN HYPOTHALAMIC NUCLEI:

Hypothalamic Nuclei	Function
Supraoptic & Paraventricular nucleus	Osmoregulation--release of vasopressin or oxytocin. Lesion here produces diabetes insipidus
Ventromedial nucleus and lateral hypothalamus	Feeding centers. Lesion in ventromedial produces overeating; lesion in lateral hypothalamus produces anorexia.
Anterolateral region and posterior hypothalamus	Stimulation of anterolateral region excites parasympathetic nervous system and inhibits sympathetic nervous system (posterior region) and produces slow wave sleep. Stimulation of posterior hypothalamus excites sympathetic and inhibits parasympathetic and is important in arousal and waking from sleep. Lesions in anterolateral produce sympathetic response while lesion in posterior produces parasympathetic response.

ic nucleus (see chapter 20).

5. The major cortical input into the hypothalamus comes from the temporal lobe - hippocampal formation (fornix), amygdala (stria terminalis), from cingulate gyrus (cingulohypothalamic), and frontal associational cortex (corticohypothalamic fibers). Thus we see that the hypothalamus is so situated that it serves as an important integrating center between the brain stem reticular system and the limbic forebrain structures.

Efferent Pathways (Figs. 16-7)

Fiber tracts and the neurosecretory system mediate the output of the hypothalamus. The efferent pathways from the hypothalamus are:

1. *Mammillothalamic tract,* which interconnects the medial mammillary nucleus with the anterior thalamic nucleus and, ultimately, with the cingulate cortex.

2. *Periventricular system,* which arises in the supraoptic, tuberal, and posterior nuclei in the hypothalamus to supply the medial thalamic nucleus and joins the medial forebrain bundle and dorsal longitudinal fasciculus.

3. *Mammillotegmental tract,* which connects the medial mammillary nuclei with the midbrain reticular nuclei and autonomic nuclei in the brain stem.

4. *Medial forebrain bundle and dorsal longitudinal fasciculus,* which receive fibers from the lateral hypothalamic nucleus and paraventricular nuclei and connect to the midbrain tegmentum and autonomic neurons in the brain stem associated with cranial nerves VII, IX, and X and the spinal cord.

5. *Hypothalamic-hypophyseal tracts, or neurosecretory system* (Fig. 16-8), which connects the supraoptic, paraventricular, and magnocellular nuclei via the hypophyseal stalk with the neural hypophysis.

6. *Hypothalamic-hypophyseal portal system,* which permits releasing factors made in the hypothalamus to reach the adenohypophysis (see below).

Functional Stability.

The hypothalamus through its relation with the autonomic nervous system and hypophysis, maintains a more-or-less stable, controlled, internal environment regardless of the external environment. It affects body temperature, water balance, neurosecretion, food intake, sleep and the parasympathetic and sympathetic nervous systems.

NEUROENDOCRINE SYSTEM, THE HYPOTHALAMUS AND ITS RELATION TO THE HYPOPHYSIS

In addition to the direct control of the autonomic nervous system by central pathways, the hypothalamus has a powerful influence on other homeostatic mechanisms of the human by its

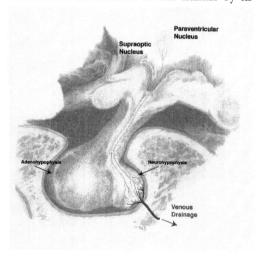

Figure 16-8. The hypothalamic-hypophyseal tract (neurosecretory system). This tract provides a direct connection between the hypothalamus and the neurohypophysis. The neurosecretory granules pass down the axons of this tract and are stored in the neural lobe until they are released into the bloodstream.

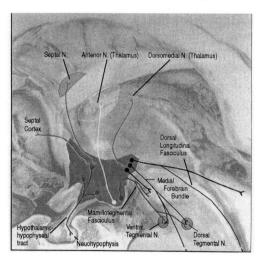

Figure 16-7. Hypothalamic Efferent Pathways

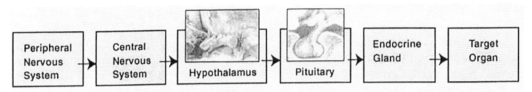

Figure 16-9. Information flow between the brain and endocrine system.

direct control of the hypophysis. The discussion that follows is an overview of this information flow that is the basis of the neuroendocrine system *(Fig 16-9)*

Hypophysis Cerebri

The hypophyseal gland or pituitary gland consists of two major divisions the anterior lobe or adenohypophysis and the posterior lobe, the neurohypophysis. There is also a smaller pars intermedia associated with the anterior lobe. The adenohypophysis is the larger component and contains cords of different cell types closely associated with capillaries of the hypothalamic-hypophyseal portal system. The neurohypophysis consists of the infundibular stalk and the bulbous neurohypophysis and includes neuroglia like pituicytes, blood vessels and some finely myelinated axons. The hypophysis is directly connected to the hypothalamus by the hypophyseal stalk (Figs. 16-5 and 16-7) and indirectly connects to the hypothalamus by the hypothalamic hypophyseal venous portal system. The hypophysiotrophic area of the hypothalamus is necessary for hormone production in the adenohypophysis. The hypophysiotropic area includes parts of the ventromedial and other hypothalamic nuclei. These peptidergic neurons are termed parvicellular (composed of small cells). The cells forming the neurohypophyseal tract are magnocellular neurons.

Hypothalamic-Hypophyseal Portal System.

The *hypothalamus and adenohypophysis* are connected by the hypothalamic- hypophyseal venous portal system *(Fig. 16-10)*. Releasing factors travel from the hypothalamus to the adenohypophysis by this means.

The *hypophysis* receives its blood supply from two pairs of arteries: the superior hypophyseal arteries and the inferior hypophyseal arteries. The superior hypophyseal arteries leave the internal carotid and posterior communicating arteries

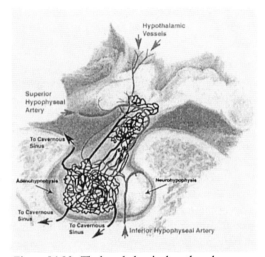

Figure 16-10. The hypothalamic- hypophyseal venous portal system. These venous channels provide vascular continuity between the hypothalamus and the adenohypophysis with releasing factors from the hypophysiotrophic area of the hypothalamus drain into the veins of the hypothalamus, which connect to the capillary bed in the infundibular stalk and median eminence. The factors are then carried to the adenohypophysis.

and supply the median eminence, infundibular stalk, and adenohypophysis. In the infundibular stalk and median eminence the arteries terminate in a capillary network, which empties into veins that run down the infundibular stalk. In the anterior lobe these veins form another capillary network, or sinusoid.

The *inferior hypophyseal arteries* leave the internal carotid and distribute to the posterior lobe, where they form a capillary network that flows into the sinus in the adenohypophysis. The hypophyseal veins finally drain into the cavernous sinus. The releasing factors enter the venous drainage of the hypothalamus, which drains into the median eminence and hypophyseal stalk. These factors are then carried via the hypophyseal portal system to the adenohypophysis.

The *capillary loop*s in the infundibular stalk and neural hypophysis drain in close proximity to the neurosecretory axons of the hypothalamic-hypophyseal tract.

Hypophysiotrophic Area.

The hypophysiotrophic area is a paramedian zone that extends rostrally at the level of the optic chiasm dorsally from the ventral surface to the paraventricular nuclei. Caudally this area narrows down to include only the ventral surface of the hypothalamus at the level of the mammillary recess of the third ventricle. The hypophysiotrophic center includes neurons in the following hypothalamic nuclei: paraventricular, anterior, arcuate, ventromedial, and tuberal.

Releasing and inhibitory factors, listed in Table 16-2, are produced in the parvicellular hypophysiotrophic area and stimulate the adenohypophysis to produce and release hormones into the circulation.

TABLE 16-2. RELEASING FACTORS AND HORMONES PRODUCED IN THE HYPOTHALAMUS

Releasing Factors Produced in parvicellular region of hypothalamus:
Corticotrophin
Growth hormone (somatotropin hormone)
Growth hormone<->inhibiting factor
Luteinizing hormone
Melanocyte-stimulating hormone
Melanocyte-stimulating hormone<->inhibiting factor
Prolactin-inhibitory factor
Thyrotropin
Hormones produced or stored in neurohypophysis:
Vasopressin from paraventricular nucleus
Oxytocin from supraoptic nucleus

Hormones Produced by Hypothalamus.

Two hormones, vasopressin and oxytocin, are manufactured in the hypothalamus, carried down the hypothalamic-hypophyseal tract, and stored in the neurohypophysis before being released into the bloodstream.

Vasopressin is an antidiuretic hormone (ADH) formed primarily in the supraoptic nucleus and to a lessor degree in the paraventricular nucleus and carried down the hypothalamic-hypophyseal tract. This substance acts on the distal convoluted tubules and collecting ducts in the kidney causing resorption of water.

Oxytocin is formed primarily in the paraventricular nucleus and to a lesser degree in the supraoptic nucleus and is carried down the hypothalamic-hypophyseal tract. This substance stimulates the mammary gland, causing it to contract and expel milk concomitant with suction on the nipple. During delivery, oxytocin also contracts the smooth muscle in the uterus and dilates the uterus, cervix, and vagina.

Hormones Produced in Adenohypophysis.

Releasing factors from the hypothalamus traverse the hypothalamic-hypophyseal portal system and trigger the formation and eventual release of the glandotropic or tropic hormones from the adenohypophysis listed in Table 16-3.

TABLE 16-3. HORMONES PRODUCED IN ADENOHYPOHYSIS

RELEASING HORMONES:
-Adrenocorticotropic (ACTH)
-lactogenic (prolactin);
-gonadotrophic:
Luteinizing (LH) in females &
Interstitial cell stimulating in males (ICSH)
-Follicle-stimulating (FSH) in females
-Melanocyte-stimulating (MSH);
-Somatotropin-somatotropin (STH);
-Thyrotropic hormone (TSH);
INHIBITORY HORMONES:
-lactogenic(dopamine)
-Melanocyte
-somatotopic

Hormones Produced in Adenohypophysis

Releasing factors from the hypothalamus traverse the hypothalamic-hypophyseal portal system and trigger the formation and eventual release of the glandotropic or tropic hormone (listed in table 16-2 from the adenohypophysis. All of the tropic (switch on) hormones produced in the anterior lobe are at the same time trophic (growth promoting or maintaining) hormones, in whose absence their target organs atrophy Fig 16-11).

Hormones produced in the adenohypophysis:

1. *Adrenocorticotrophic Hormone (ACTH).*

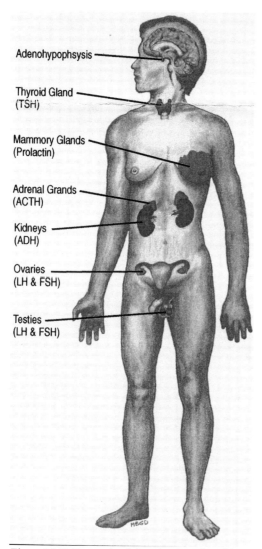

Figure 16-11 Target glands for adenohypophyseal hormones.

This hormone is produced by basophilic cells and stimulates the formation of adrenal glucocorticoids (hydrocortisone) and, to a lesser degree, mineral corticoids (aldosterone). In conditions of stress (cold, heat, pain, fright, flight, inflammation, or infection), ACTH is produced. Cushing's disease is caused by excessive secretion of ACTH while Addison's disease results from ACTH insufficiency.

2. *Lactogenic Hormone (Prolactin)*. This substance in combination with the growth hormone promotes development of the mammary glands and lactation.

3. *Gonadotropin (luteinizing hormone* [LH] in females; or interstitial cell<->stimulating hormone [ICSH] in males). In females gonadotropin is necessary for ovulation. It causes luteinization of follicles after they are ripened by FSH (follicle-stimulating hormone). In males gonadotropin activates the interstitial cells of the testis to produce testosterone that is essential for the development and maturation of the sperm.

4. *Follicle-Stimulating Hormone* (FSH). This substance stimulates growth of the ovarian follicle in the female and stimulates spermatogenesis in males.

5. *Growth Hormone* (somatotropin, or STH). This hormone is important for growth after birth. It specifically affects the growth of the epiphyseal cartilage but is also important in the metabolism of fat, protein, and carbohydrates. Cessation of growth follows hypophysectomy. The inhibitory form of this hormone controls the release of the hormone. Gigantism results from an excessive somatotropin before the epiphyseal plate's close. The following case 16-2 is different from Case 16-1 as the pituitary tumor in this instance produces defects in vision and an increase in the growth hormone after the epiphyseal plates have closed with resultant acromegalic (enlarged extremities) gigantism and also effects the libido.

6. *Thyrotrophic Hormone (TSH)*. TSH stimulates the thyroid gland to produce triiodothyronine (T3) and thyroxine (T4), which regulate the rate of breakdown of fats and carbohydrates. When circulating levels of T3 and T4 rise, thyrotropic-releasing hormone (TRH) release from the adenohypophysis is inhibited by a drop in TRH production in the hypothalamic neurons until the preferred levels returns

Case History 16-2. Patient of Dr. Elliott Marcus.

This 51-year-old married lumber salesman was admitted for evaluation of a progressive disturbance of vision. Approximately 8 years previously, the patient noted blurring of print on the labels of paint cans. Glasses were obtained and produced improvement, but within a short time, symptoms recurred. During the succeeding years, the glasses had to be changed eight times. For 6 to 7 months, the patient had noted diplopia, especially on right lateral gaze when driving at night. For 3 to 4 months, a left frontoparietal and left orbital headache had been present. One month prior to admission, progression of visual symptoms occurred. The patient had a marked decrease in visual acuity and was unable to find objects on shelves. He reported also that he was unable to drive his automobile because he was unable to see the sides of the road. The patient consulted an ophthal-mologist who noted visual field defects and referred the patient for neurological evaluation.

History:

The patient denied any significant change in physical appearance, appetite, or fluid intake. The patient's sexual interests had declined. He had no sexual intercourse for the last ten years.

General Physical Examination:

1. The patient was obese with large puffy hands and feet, a prominent jaw, and prominent frontal skull area.

2. Pallor of skin and mucous membranes was noted.

3. Axillary hair was absent; pubic hair was normal.

4. Testes were of normal size but were described as soft.

Neurological Examination:

1. *Cranial Nerves:*

a. *Visual acuity wa*s reduced bilaterally, however, the patient could read 1 cm print at 14 inches.

b. On visual field examination a complete bitemporal defect was present, cutting the central vision (Fig. 23-11).

c. Examination of the fundi revealed pallor of the left optic disk, suggesting optic atrophy.

d. On testing extraocular movements, a diplopia was present in a pattern that suggested bilateral medial rectus involvement (that is, bilateral third-nerve damage).

Laboratory Data:

The following abnormal levels were:

1. Thyroid studies indicated borderline low function with a protein bound iodine of 4.0 µg/dl. (Normal =

8 µg/dl) and radioactive iodine of 23.3% in 24 hours (normal = 20 to 50%).

2. Adrenal-function studies indicated subnormal values. Urinary 17-hydroxysteroids were 2.1-to 2.3 mg/24 hr. (normal = 3 to 10 mg/24 hr).

3. Urinary 17-ketosteroids were 4.0 mg/24 hr normal = 10-25 mg/24 hr).

4. Despite these low urinary steroid levels, no FSH was detec-table in the urine.

5. Skull x-ray revealed marked enlargement of the sella turcica (length 3 cm, depth 2 cm) without calcification.

6. Electroencephalogram was abnormal with bilateral frontal, slow (5 cps) waves, suggesting a bilateral frontal dysfunction.

7. Bilateral carotid arteriograms demonstrated a marked elevation of the distal segment of the internal carotid arteries, with elevation of the origins of both the anterior and middle cerebral arteries; consis-tent with a large tumor extending superiorly out of the pituitary fossa.

8. Pneumoencephalogram large lesion extending out of the pituitary fossa was again demonstrated.

9. Cerebrospinal-fluid protein was elevated to 85 mg/dl.

Hospital Course

Replacement therapy with cortisone acetate and thyroid was begun. Doctor Bertram Selverstone performed a right frontal craniotomy.

Clinical Diagnosis –Necrotic chromophobe adenoma.

Comment

Although the patient did not emphasize an acute history of a bitemporal hemianopsia, he did mention that his ability to drive had been limited not only by the decrease in visual acuity but also by his inability to see the road at the sides of the car, that is, the peripheral temporal fields.

The patient presented with many of the stigmata of acromegaly but claimed that he had large hands, feet, jaw, etc., for all of his adult life.

Changes in libido had apparently been present for at least 16 years. Axillary hair had never been present. One may only speculate that perhaps a mixed tumor of the pituitary had been present almost all of the patient's life, only becoming symptomatic once extrasellar (that is, suprasellar) extension had compressed the optic nerves and chiasm.

Today, blood levels of growth hormone and

prolactin would be obtained and, in all likelihood, would have been found to be elevated long before the visual symptoms appeared. Today such symptoms would have prompted neuroimaging studies such as MRI or CT with the diagnosis probably being established much earlier.

Tables 16-2 and 16-3 review the functions of the anterior and posterior lobes of the pituitary.

Hypothalamus and the Autonomic Nervous System

An individual performing daily functions usually has a balance between the parasympathetic and sympathetic nervous system. Stimulation of the anterior hypothalamus (anterolateral region) excites the parasympathetic nervous system and inhibits the sympathetic nervous system. The heart beat and blood pressure decrease (the vagal response), the visceral vessels dilate, peristalsis and secretion of digestive juices increase, the pupils constrict, and salivation increases.

Stimulation of the posterior hypothalamus (posteromedial region) excites the sympathetic nervous system and inhibits the parasympathetic nervous system. The heartbeat and blood pressure increase, the visceral vessels constrict, peristalsis and secretion of gastric juices decrease, pupils dilate, and sweating and piloerection occur.

A lesion in the anterior or parasympathetic zone produces a sympathetic response; destruction of the posterior or sympathetic zone may produce a parasympathetic response. The effects on blood vessels are produced by stimulation of the smooth muscles in the tunica media, and the effects on the sweat glands result from the stimulation of the glands through the cranial or peripheral nerves.

Functional Localization

The hypothalamus, through its relation with the autonomic nervous system and hypophysis, maintains a more-or-less stable, controlled, internal environment regardless of the external environment. It affects body temperature, water balance, neurosecretion, food intake, sleep, and the parasympathetic and sympathetic nervous system.

Food Intake. Bilateral lesions restricted to the ventromedial hypothalamic nuclei produce hyperphagia, resulting in obesity. Stimulation of this center inhibits eating. Experiments have shown that animals with ventromedial lesions are not motivated to eat more; they just do not know when to stop eating. For this reason it is thought that the satiety center is located here.

Bilateral lesions in the lateral hypothalamic region produce anorexia; the animal refuses to eat and eventually dies. However, if this region is stimulated but not destroyed, the animal overeats. Appetite, the stimulus for eating, is dependent on many factors, including glucose concentration in the blood, stomach distension, smell, sight, taste, and mood. Lesions in other portions of the limbic system may also affect appetite.

Although obesity in humans usually does not have a neurological cause, in rare cases some overweight individuals have had a ventromedial nuclei lesion.

Sleep Cycle. The hypothalamus helps set our level of alertness with the assistance of input from the reticular system and cerebral cortex (see Chapter 11 & 30). The medial preoptic region can initiate slow wave sleep. The lateral hypothalamus receives input from the ascending reticular system that produces arousal. The posterior portion, including the mammillary body, is important in controlling the normal sleep cycle. Lesions in this region may produce an inversion in the sleep cycle, or hypersomnia. Lesions in the anterior hypothalamus may produce insomnia.

The amount of light is monitored in the hypothalamus by input from the optic nerve directly to the suprachiasmatic nucleus. The sympathetic innervation to the pineal monitors the dark levels, and melatonin release from the pineal is part of the sleep mechanism.

Body Temperature. Maintenance of a constant body temperature is vital to a warm-blooded animal. The hypothalamus acts as a thermostat, establishing a balance between vasodilation, vasoconstriction, sweating, and shivering. It does this by means of two opposing thermoregulatory centers: one regulates heat loss, and the other regulates heat production. Lesions in either of these centers interfere with the regulation of body temperature.

Control Center for Heat Loss. A destructive lesion in the anterior hypothalamus at the level of the optic chiasm or in the tuberal region renders an animal incapable of controlling its body temperature in a warm environment; a tumor in the third ventricle may produce a similar result. The human can no longer control the heat-loss mechanism, which consists of cutaneous vasodilation and sweating. Without this mechanism the body temperature rises (hyperthermia), and the patient becomes comatose and may even die.

Control Center for Heat Production and Conservation. Lesions in the posterior hypothalamus above and lateral to the mammillary bodies disrupt an animal's ability to maintain normal body temperature in either a warm or cold environment. Normally, exposure to cold produces vasoconstriction, shivering, piloerection, and secretion of epinephrine in mammals, all of which tend to increase body temperature. Following a lesion in the posterior hypothalamus, the body temperature tends to match that of the environment, whether warmer or colder, relative to 37°C. (poikilothermy). Poikilothermy probably reflects damage to the fibers descending from the heat-loss center as well as direct injury to the heat production and conservation center.

The thermoreceptors are located in the skin (end bulbs of Krause and end bulbs of Golgi-Mazzoni; see Fig. 3-22) or in the hypothalamus itself, which monitors the temperature of the cranial blood. Some receptors respond to cold and others to warmth. Cold receptors excite the caudal heat production and conservation center and inhibit the rostral heat-loss center. Warm receptors reverse this pattern.

Water Balance and Neurosecretion (Fig.16-10). Water balance is regulated by the hypothalamic-hypophyseal tract, which originates in the supraoptic and paraventricular nuclei and runs through the median eminence and hypophyseal stalk to terminate in close proximity to blood vessels in the neural hypophysis. A lesion in the region of the supraoptic nucleus or in the hypothalamic-hypophyseal system causes diabetes insipidus with polyuria (excessive formation of urine). The patient drinks copious amounts of water (polydipsia) and may excrete 20 liters of urine a day. The symptoms of diabetes insipidus may be relieved by the administration of vasopressin (Pitressin), an antidiuretic hormone (ADH), or extract from the hypophysis.

The hypothalamic-hypophyseal tract controls the formation and release of ADH, which causes water to be reabsorbed from the distal convoluted tubules and collecting duct of the kidneys into the bloodstream, thereby limiting the amount of water lost in the urine. Ingestion of large amounts of water causes a drop in the ADH level reduces tubular reabsorption with resultant diuresis. Water deprivation increases the ADH level and causes increased reabsorption and concentrated urine.

The cells in the supraoptic nuclei form much of the ADH or its precursors. ADH is formed in the cell body, travels down the axons of the hypothalamic-hypophyseal tract, and is stored in the posterior lobe of the hypophysis as neurosecretory granules. The osmoreceptors in the aortic and carotid body and in the hypothalamus itself detect the concentration of water in the blood. This information is passed on to the hypothalamic nuclei, which then control the release of ADH into the circulation.

Water intake is also controlled by hypothalamic neurons, with the area dorsal to the paraventricular nuclei being one of the water intake centers.

HYPOTHALAMUS AND EMOTIONS. As noted in Case 16-2 above and in Chapter 21, the hypothalamus sets the foundation for emotional responses and controls visceral functions. It exerts these powerful influences through the multisynaptic descending autonomic pathways in the lateral margins of the reticular formation and by controlling the release of the catecholamine--epinephrine (adrenaline) and norepinephrine (noradrenalin)--from the adrenal medulla.

Most postganglionic sympathetic fibers release a substance that is 80% epinephrine and 20% norepinephrine, which acts on alpha- and beta-adrenergic receptors in the organs; these are called adrenergic fibers. Small doses of acetylcholine injected into the bloodstream produce the same effect as stimulation of the cholinergic endings of craniosacral column.

Norepinephrine causes constriction of the peripheral vessels, with resultant increased blood flow in the large vessels. The blood pressure and pulse rate rise and blood flow to the coronary muscles increases. Epinephrine increases the heart rate and the force, amplitude, and frequency of heart contractions. It also dilates the pupils and the urinary and rectal sphincters, causing voiding and defecation. It inhibits the motility of the gut and causes the bronchial musculature to relax, thereby dilating the bronchial passages.

Epinephrine increases oxygen metabolism and the basic metabolic rate, causes hyperglycemia by increasing phosphorylase activity, which accelerates glycogenesis in the muscles and liver. And it increases the rate of synthesis of lactic acid to glycogen, thereby prolonging the contraction of the skeletal muscle.

HYPOTHALAMUS AND LIGHT LEVELS

Light to produces dramatic effects that are mediated by the hypothalamus via the retinohypothalamic tract to the suprachiasmatic nucleus. In many animals abnormal variations in light levels alter the sexual cycle and prevent estrus. Light levels also affect the formation of melanin.

Pineal Body (Fig. 1-17). The pineal, or epiphysis cerebri, is a small pinecone<->shaped gland found in the roof of the third ventricle. This richly vascularized gland contains glial cells and pinealocytes and is innervated by the sympathetic nervous system from the superior cervical sympathetic ganglion with synapses on the pinealocytes.

Pinealocytes contain serotonin, melatonin, and norepinephrine. The pineal makes melatonin from serotonin; the enzyme levels necessary for this transformation are controlled by the suprachaismatic nucleus. This nucleus sends fibers via the descending sympathetic pathway into the medial forebrain bundle, and via the descending spinal fibers to the intermediolateral cell column of the upper thoracic cord and then on to the superior cervical ganglion. The pinealocytes are related to the photoreceptor cells seen in this region in fish and amphibians. Pinocyte secretion of serotonin and melatonin is a cyclic response to stimulation from the optic system. These 24-hour circadian rhythmns may well be the biological clock.

AUTONOMIC NERVOUS SYSTEM (Fig. 16-12)

Langley's description of the autonomic nervous system includes three parts: the sympathetic, parasympathetic and enteric nervous systems.

The enteric nervous system is recognized as a separate portion of the autonomic nervous system because peristalsis and other spontaneous movements persist after it is isolation from all nervous input (Gershon 1981).

Much of the effectiveness of the hypothalamus is due to its innervation of the nuclei in the central nervous system that form the autonomic nervous system. In this section we identify these neurons in the central and peripheral nervous systems and list the responses produced by the autonomic nervous system in each organ. Note that normally an antagonistic balance exists between the functions of the parasympathetic and sympathetic systems. However, this equilibrium may be modified by environmental stresses or disease processes.

The somatic nervous system has its own receptors and effectors, which provide cutaneous and motor innervation to the skin and muscles in the head, neck, and body. The visceral, or autonomic, nervous system also has its own receptors and effectors, which provide dual motor and sensory innervation to the glands, smooth muscles, cardiac muscles, viscera, and blood vessels. These axons run part of the way with somatic fibers, but they also have separate pathways.

In the somatic nervous system the efferent neuron is the ventral horn cell or motor cranial-nerve nucleus. Its axon leaves the central nervous system to innervate a skeletal muscle.

In the visceral nervous system two efferent neurons are found: the preganglionic neuron and the postganglionic neuron.

The preganglionic neuron is found in the intermediate horn at the thoracic and lumbar levels of the spinal cord. The axon of this cell leaves the central nervous system via the ventral root and ends in an autonomic ganglion, which parallels the spinal cord or is found near the target organ

The preganglionic cell bodies are found in the CNS at the following sites:

Parasympathetic (Craniosacral):

1. Cranial parasympathetic (cranial nerves III, VII, IX, and X).

2. Sacral parasympathetic (sacral segments 2

to 4).

Sympathetic (Thoracolumbar):

1. All thoracic segments and lumbar segments 1 and 2.

The postganglionic neuron, or the motor ganglion cell, is located outside the central nervous system and ends in the appropriate gland, smooth muscle, or cardiac muscle.

The nerve fibers are further classified as being adrenergic or cholinergic, depending on which nerve transmitter is involved--epinephrine or acetylcholine. The lumbar sympathetic preganglionic axons to the adrenal gland activate chromaffin cells (modified nerve cells) in the adrenal medulla, which release epinephrine (80%) and norepinephrine (20%) into the circulation and act on alpha and beta-adrenergic receptors in the hollow organs, glands, and skin. Injections of epinephrine into the blood stream produce an effect similar to the stimulation of the thoracolumbar or sympathetic nervous system. Small doses of acetylcholine injected into the bloodstream produce the same effect as stimulation of the craniosacral column. Most postganglionic parasympathetic fibers release acetylcholine at their terminals.

The axons ending on sweat glands, though receiving innervation from the thoracolumbar system, are cholinergic. Pilomotor axons, which also receive thoracolumbar innervation, are adrenergic.

The two systems provide dual innervation to glands and viscera and act to produce optimal balances of internal conditions under any environmental condition.

ENTERIC NERVOUS SYSTEM

In the alimentary canal (esophagus, stomach, intestines, and rectum) there are about 100 million neurons (Gershon, 1981). One network of neurons, Meissner's (submucosal) plexus, lies in the connective tissue between the longitudinal muscle coat and the circular musculature coat in the alimentary canal. Another, Auerbach's (myenteric) plexus, lies in the connective tissue between the circular muscle coat and the muscularis mucosa. Each plexus consists of multipolar, bipolar, and unipolar neurons. The enteric nervous system includes glia that appears similar to the astrocytes in the central nervous system. The astrocytes enwrap the neurons and their processes and also form a barrier between the blood and the enteric nervous system. There are extensive connections within the enteric nervous system and projections onto the pancreas, gallbladder, and elsewhere. The unipolar and bipolar neurons appear to function as sensory neurons with the multipolar cells forming many of the connections.

The noradrenergic sympathetic postganglionic fibers project onto the submucosal and myenteric plexus, where they are inhibitory to the cholinergic neurons. (The cholinergic vagal input from the dorsal motor nucleus of the vagus to the submucosa and mucous membrane is excitatory.)

PARASYMPATHETIC SYSTEM (FIG. 16-12)

CRANIAL NERVES

Cranial Nerve III. The fibers originate in the Edinger-Westphal nucleus in the midbrain and innervate the constrictors of the pupil via the ciliary ganglion.

Cranial Nerve VII. The axons originate in the superior salivatory nuclei of the pons and innervate glands in the nasal cavity and palate, and in the lacrimal (pterygopalatine), submandibular, and sublingual (submandibular) glands.

Cranial Nerve IX. The fibers originate in the inferior salivatory nucleus in the medulla. The axons exit with nerve IX and end in the otic ganglion, which innervates the parotid gland.

Cranial Nerve X. The dorsal motor nucleus of the vagus nerve provides preganglionic parasympathetic innervation to the pharynx, larynx, esophagus, lungs, heart, stomach, and intestines up to the transverse colon. The postganglionic fibers originate from ganglia in the respective organs and in the nodes of the submucosal plexus of Meissner and the myenteric plexus of Auerbach in the enteric nervous system associated with the alimentary tract.

Sacral Plexus

The parasympathetic sacral plexus originates from sacral segments S2 and S4 and supplies the genitalia, sigmoid colon, rectum, bladder, and ureter. The ganglia are located in the organ and are named for the organ.

Sympathetic System (Fig. 16-12)

The preganglionic fibers originate in the intermediolateral column of gray matter in the spinal cord from T1 to L2; they exit in the white rami and enter the paravertebral sympathetic ganglionic chain. The student is reminded that cranial nerves are not part of the sympathetic nervous system. The ganglia form a continuous paravertebral trunk in the cervical, thoracic, lumbar, and sacral segments.

The postganglionic fibers originate in the ganglia and exit via the gray communicating rami. The postganglionic fibers reach their destinations via plexuses enwrapping the arteries or via cranial and spinal nerves.

Cervical Sympathetic Ganglia

The cervical ganglia extend posterior to the great vessels from the level of the subclavian artery to the base of the skull. There are usually three ganglia in the cervical region: superior, middle, and inferior.

The superior cervical ganglion is flat and the largest, (25 to 35 mm). The internal carotid nerves leave the superior border of the superior cervical ganglion and pass into the cranial cavity with the carotid artery in the carotid canal. These fibers form the plexus around the carotid artery and in the cavernous sinus. This ganglion provides postganglionic innervation to the eye, lacrimal gland, parotid gland (otic ganglia), submaxillary and sublingual glands (sublingual ganglia), and some branches to the heart.

The middle cervical ganglion is small and variable and is found at the level of the cricoid cartilage. It provides postganglionic fibers to the heart.

The inferior cervical ganglion is usually fused with the first thoracic ganglion to form the stellate ganglion. It is also small and irregular in shape and lies posterior to the vertebral artery. The postganglionic fibers innervate the heart.

Thoracic Sympathetic Ganglia

The thoracic portion of the sympathetic trunk extends anterior to the heads of the first ten ribs and then passes along the sides of the lower two thoracic vertebrae. The number of thoracic ganglia varies from 10 to 12.

The ganglia of T1 to T4 provide postganglionic fibers to the heart and lungs. The postganglionic fibers from T5 to T12 form the splanchnic nerves. The greater splanchnic nerve is formed by preganglionic rootlets from T5 to T10; the axons pass through the thoracic ganglia and end in the celiac ganglion. From there postganglionic axons run above the aorta, esophagus, thoracic duct, and azygos vein to reach the stomach, small intestine, and adrenal glands. The lesser and least splanchnic nerves arise from spinal cord segments T11 and T12 and run through the paravertebral ganglion to the celiac and superior mesenteric ganglia. From there the postganglionic axons run to the small intestine, the ascending portion of the colon, and the kidneys.

Lumbar Sympathetic Ganglia

The lumbar sympathetic trunk consists of small ganglia that lie along the anterior border of the psoas muscle. The preganglionic fibers originate from the intermediate column of spinal cord, segments L1 and L2, pass through the lumbar ganglia, and end in the inferior mesenteric ganglion, where they innervate the descending colon, rectum, urinary bladder, ureter, and external genitalia.

Autonomic Dual Innervation of Specific Structures (Fig. 16-12)

EYE

Parasympathetic. The preganglionic cell bodies are found in the Edinger-Westphal nucleus associated with cranial nerve III. These fibers run on the outside of the third nerve and end in the ciliary ganglion. The postganglionic fibers run in the short ciliary nerve, providing innervation to the ciliary muscle and the constrictor of the iris, which permit accommodation for near vision.

Sympathetic. Preganglionic axons originate from the intermediolateral nucleus at cord levels C8 to T2. The fibers ascend in the sympathetic chain to the postganglionic neurons in the superior cervical ganglion. The postganglionic axons join the carotid plexus and then the ophthalmic branch of nerve V to innervate the dilator muscle of the iris and the levator palpebrae and radial ciliary muscles, producing eye opening (lifting of the lid) and dilation of the pupil for distant vision.

LACRIMAL GLANDS

Parasympathetic. Preganglionic fibers originate in the superior salivatory nucleus of nerve VII and run to the pterygopalatine ganglion. The postganglionic fibers join the maxillary division of nerve V, resulting in vasodilation and secretion.

Salivary Glands

Parasympathetic. The preganglionic axons to the submaxillary and sublingual glands originate in the superior salivary nucleus of nerve VII and enter the submandiublar gland from which the postganglion axons enter the salivary glands. The preganglionic axons to the parotid originate in the inferior salivatory nucleus of nerve IX (medulla) and leave the central nervous system cns and enter the otic ganglion. The resulting action is vasodilation and secretion.

Sympathetic. These fibers originate from the intermediolateral column at the upper thoracic level, exit from the central nervous system, run up to the superior cervical ganglion, and then proceed to the glands. Vasoconstriction and reduced secretion result.

HEART

Parasympathetic. The preganglionic fibers originate in the dorsal nucleus of nerve X, exit from the central nervous system, and run in the vagus nerve to the postganglionic ganglia in the cardiac plexus and atria. These axons reach coronary vessels and atrial musculature, causing cardiac deceleration.

Sympathetic. The preganglionic fibers originate in the T1 to T4 segments, exit from the central nervous system, and reach the postganglionic neurons in the upper thoracic ganglion and all three cervical ganglia. The fibers form the cardiac nerves and enter the cardiac plexus. The resulting effect is cardiac acceleration.

LUNGS

Parasympathetic. The preganglionic fibers originate in the dorsal motor nucleus of nerve X and run in the vagus to the pulmonary plexus in the bronchi and blood vessels. They cause constriction of the bronchi and decreased respiration.

Sympathetic. The preganglionic fibers originate in T1 to T4. The postganglionic fibers originate in the upper thoracic and lowest cervical ganglia. The result in this case is dilation of the bronchi and increased respiration.

ABDOMINAL VISCERA

Parasympathetic. These fibers originate from the dorsal motor nucleus of nerve X and S2 to S4, and end in the intrinsic plexus in the organ. They stimulate peristalsis and gastrointestinal secretions.

Sympathetic. The sympathetic fibers originate from the lower thoracic and lumbar segments, pass through the paravertebral ganglia and form the splanchnic nerves, which end in the celiac and superior and inferior mesenteric ganglia. Postganglionic fibers from these ganglia end on smooth muscle or glands. They inhibit peristalsis and gastrointestinal secretion; stimulate secretion from the adrenal medulla and cause vasoconstriction of the visceral vessels (see discussion of enteric nervous system above).

PELVIS

Parasympathetic. These fibers originate from segments S2 to S4 and end in ganglia in the organs. Postganglionic fibers supply the uterus, vagina, testes, erectile tissue (penis and clitoris), sigmoid colon, rectum, and bladder. They cause contraction and emptying of the bladder and erection of the clitoris or penis.

Sympathetic. These fibers originate from the lowest thoracic segments and segments L1 to L3 and run in the hypogastric nerves to the target viscera, where the ganglia are found. They cause vasoconstriction and ejaculation of semen, and inhibit peristalsis in the sigmoid colon and rectum.

CUTANEOUS AND DEEP VESSELS, GLANDS, AND HAIR

Blood vessels receive only sympathetic postganglionic innervation from the appropriate paravertebral ganglion. Activation of this system produces vasoconstriction (cooling of the skin), sweating, and piloerection.

The innervation of the peripheral blood vessels and glands is summarized in Table 16-1.

Disorders of the Autonomic Nervous System

Disorders of the autonomic nervous system can affect any of the organs innervated by the autonomic nervous system. Only a few will be discussed here. The student is referred to any standard neurology text for further details.

Eye. A lesion in the nucleus or nerve root of cranial nerve III injuring the fibers from the Edinger-Westphal nucleus produces pupillary dilation (due to only sympathetic input). Injury to the descending sympathetic fibers produces an ipsilateral Horner's syndrome (pupillary constriction and drooping of the eyelid). The Argyll-Robertson pupil does not react to light but will accommodate and is commonly seen in neurosyphilis in the central nervous system. (See discussion of tabes dorsalis in Chapter 9.)

Blood Vessels. Reynard's disease (vasomotor disease), most common in young women, causes the patient to be abnormally sensitive to cold. On exposure to cold the extremities become cyanotic and then red and very painful. Sympathetic innervation does not seem to dilate the vessels. Vasodilators help, but resection of the preganglionic sympathetic innervation to the limb usually brings permanent relief.

Heart and Gastrointestinal Tract

These organs respond to any change in our emotional state. Consequently, psychosomatic diseases, e.g., hypertension and ulcers, affect them.

BLADDER

The bladder is a reflex organ and contracts in response to stretch. The afferents are carried into the sacral spinal cord; efferent are supplied by the parasympathetic (S2<->S4) and sympathetic (L1<->L2) ganglia. The parasympathetic axons initiate the reflex contraction of the bladder during urination (micturition)--a voluntary action initiated through the nerve to the external sphincter. Sympathetic axons do not participate in micturition but control blood vessels and closure of the internal sphincter during sexual intercourse.

Normally, the bladder fills and expands, stretching the receptors in the bladder walls. The pressure buildup is carried into the central nervous system by the sensory nerves. Following the voluntary contraction of the external sphincter, the bladder contracts reflexively; the internal and external sphincters open and the muscles in the pelvic floor relax (control from cerebral cortex). The urine passing through the urethra continues the stimulation until the bladder is empty. The external sphincter then closes. We can also initiate closing voluntarily with the assistance of the abdominal muscles.

Interruption of the cortical pathways (bilateral involvement of the cerebral hemispheres) produces the uninhibited bladder, in which sensation is normal, but the patient no longer has control over voiding. Thus, as the bladder fills and becomes distended, it voids suddenly without cortical control.

Destruction of the parasympathetic supply by injury to the motor roots produces the autonomic bladder, in which there is no sensation and no longer any reflex or voluntary control. The need to void arises from abdominal discomfort produced by increasing intra-abdominal pressure.

Injury to the sensory roots or ganglia or to the dorsal column produces the sensory paralytic bladder, in which there is no sensation and no reflex ability to void. The bladder visibly enlarges and then dribbles. Voiding is also difficult without sensory feedback.

Isolation of the bladder from the upper motor neurons either by transection of the spinal cord above the sacral level or by extensive brain disease produces the reflex bladder. In this condition the descending autonomic fibers are interrupted as well as the corticospinal fibers, but the spinal reflex circuit remains intact. Micturition is involuntary and results from just the stretch reflex alone. Voiding is sudden and uncontrollable.

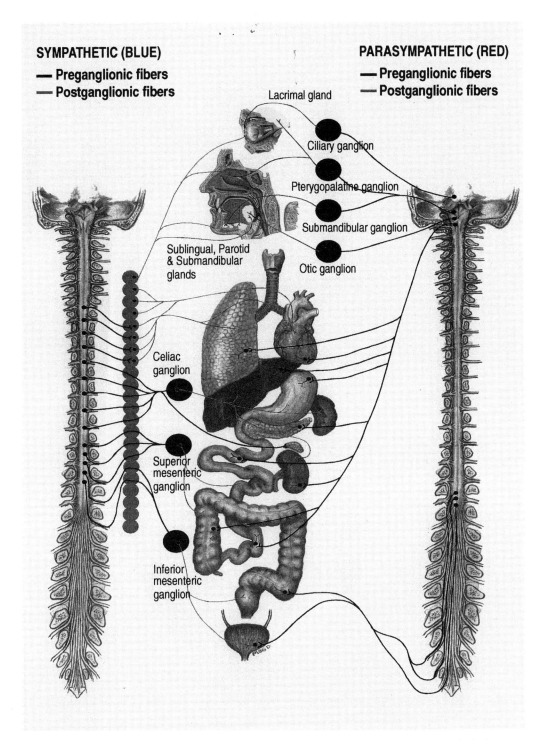

Figure 16-12. Autonomic Nervous System: Left, Sympathetic (thoracolumbar); right, parasympathetic (craniosacral).

CHAPTER 17
Cerebral Cortex: Cytoarchitecture, Physiology and Overview of Functional Localization

PART I. ANATOMICAL CONSIDERATIONS

A firm knowledge of the structure and function of the cerebral cortex and the relationship of the cerebral cortex to the subcortical centers is of crucial importance for all who wish to understand the behavior and accomplishments of man as opposed to other animal forms. The use of the thumb and fingers for fine manipulations with tools, the use of linguistic and mathematical symbols for communication in the auditory and visual spheres, and the capacity for postponement of gratification all reflect the evolution of the cerebral cortex. This development of the cerebral cortex has led to a laminar arrangement of 16 billion nerve cells with an almost infinite number of synapses. Beaulieu and Colonnier have estimated that in a cubic millimeter of neocortex there are 2.78 times 10\8 synapses 84% of which are type I (excitatory) and 16% type II (inhibitory).

The almost infinite number and variety of circuits present not only has provided the anatomical substrate for the recording of an infinite number of past experiences but has also allowed for a plasticity of future function projected both in time and in space.

The attempt to relate functional differences to the differences in structure of the various areas of the cerebral cortex was a scientific outgrowth of the earlier philosophical arguments concerning the relationship of mind and body. We may indicate at the onset that to a certain degree cytoarchitectural differences do reflect functional differences. This is never, however, a one-to-one relationship. Moreover, at times, architectural differences are not sharp; strict borders may not be present.

The various areas of cerebral cortex differ as regards several parameters:

1. Thickness, evident on simple visual inspection (average thickness is 2.5 mm., motor cortex: 4.5 mm., visual cortex 1.45 to 2.0 mm.). In addition to differences in overall thickness, some areas differ in the thickness of the various layers that constitute the laminar pattern.

2. Relative density of the various cell types (evident in Nissl stains for cells or Golgi silver stains) to be discussed below e.g.: pyramidal, vs stellate, It must also be noted that within a given area of cortex, the various layers of cortex differ as regards the relative distribution of these various types of cells *(Fig.17-1)*

3. Density of horizontal fiber plexuses (stripes) (evident on myelin and silver stains) and density of axodendritic and other synapses (evident on Golgi stain and Fink-Heimer stains).

4. Degree of myelination of intracortical fiber systems (evident on myelin stain). To some extent the various myelinated bands may be seen with the naked eye in freshly cut sections of the cerebral cortex. For example, Gennari in 1782 and Vicq d'Azyr in 1786 independently had noted a white line in the cortex near the calcarine fissure. The various features to be noted in the several types of stains are demonstrated in *Figure 17-1*.

CYTOLOGY

From the standpoint of cytology the neurons within the cerebral cortex may be classified into two major categories pyramidal and stellate *(Fig.17-2)*. The axons of pyramidal cells (Type I cells/neurons with long axons) form the majority of association, callosal and subcortical projections while the stellate cells (type II/ neurons with short axons) form the local circuits (Refer to chapter 3 for additional discussion).

Neurons may also be classified on the basis of whether the dendrites have spines or an

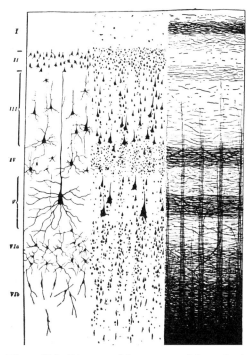

Figure 17-1. Diagram of the structure of the cerebral cortex. The results obtained with (a) a Golgi stain, (b) a Nissl stain or other cellular stain, and (c) a myelin stain are contrasted. I = molecular layer; II = external granular layer; III = external pyramidal layer; IV = internal granular layer; V = large or giant pyramidal layer (ganglionic layer); VI = fusiform layer. The following features should be noted in the myelin stain: 1a = molecular layer 3a1 = band of Kaes Bechterew; 4 = outer band of Baillarger; 5b = inner band of Baillarger. (After Brodmann, from Ranson, S. W., and Clark, S. L.: The Anatomy of the Nervous System, Philadelphia, W. B. Saunders, 1959, p. 350).

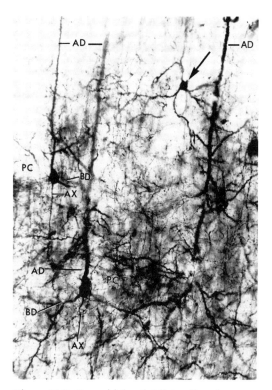

Figure 17-2. Pyramidal and stellate cells as demonstrated in Golgi stain of cat cerebral cortex. PC = pyramidal cells; arrow = stellate cell; AD = apical dendrite of pyramidal cell; BD = basal dendrite of pyramidal cells; AX = axons of pyramidal cell.

absence of spines.

From the standpoint of electrophysiology the neurons may be classified as fast spiking, regular spiking or bursting (repetitive discharge).

I. Neurons whose dendrites have spines (spiny neurons) account for the majority of the neurons in cerebral cortex. These are excitatory and are of two types:

a. Pyramidal cells: constitute 2/3 of all neurons in the neocortex. These cells are defined by a prominent apical dendrite extending through all layers of the cortex to layer 1. Pyramidal neurons have cell bodies of variable height--small 1u to 12 u; medium 20 to 25 u; large 45 to 50 u, and giant 70 to 100 u. The giant pyramidal cells are characteristic of the motor cortex. The upper end of the pyramidal cell continues on toward the surface as the apical dendrite. Numerous spines are found on the dendrite providing sites for synaptic contact. The pyramidal neurons have the largest dendritic trees and also contain axons that form associational, callosal and projectional connections. In general, the larger the cell body, the larger the apical dendrite and the wider the spread in terms of ramifications of the terminal horizontal branches. The wider the spread of the apical dendrite, the greater the number of possible axodendritic synapses. The larger the cell body, the greater the number of possible axosomatic synapses. In addition to the apical dendrite, shorter basal dendrites arise from the base of the cell body and arborize in the vicinity of the cell body. The axon emerges from the base of the cell body and descends toward the deeper white matter. In general the axons of small- and medium-sized pyramidal cells terminate as association fascicles within the cortex. The axons of large and giant pyramidal cells enter the deeper white matter as (a) association fibers to other cortical areas, (b) commissural fibers to the contralateral hemisphere, or (c) projection (efferent) fibers to subcortical, brain stem, and spinal cord areas. In addition, recurrent collateral association fibers may branch off from these axons within the cortex. In addition to functioning as the main efferent outflow of the cerebral cortex to subcortical areas (thalamus, basal ganglia, brain stem and spinal cord), and to other cerebral cortical areas (ipsilateral, and contralateral hemispheres), the pyramidal cells also provide a massive collateral net work to other neurons with in the local area. Each pyramidal cell also receives multiple inputs on to its dendritic spines. The Betz cells, the giant pyramidal neuron of the motor cortex have at least 10,000 spines each with a type I excitatory synapse.

b. Spiny stellate neurons: the prominent apical dendrite is not present. Instead dendrites of almost equal length radiate out from the cell body (soma). These neurons are found primarily in layer 4 of sensory cortex. Their axons project locally within the area and not to distant sites.

Physiology: Most spiny neurons/pyramidal cells utilize glutamate as a synaptic transmitter and demonstrate "regular" spiking that is an adapting pattern of discharge in response to a constant current injection. These cells generally demonstrate a significant after hyperpolarization (AHP). Some of the pyramidal cells located in deeper layers, primarily layer 5 may demonstrate a bursting pattern at low threshold. These neurons respond to depolarization by generating repetitive spikes, usually 3 in sequence. The regular firing pyramidal cells are characterized by thin apical dendrites that do not branch extensively in layer 1. In contrast, the burst firing pyramidal cells have thick apical dendrites and extensive branching in layer 1. As we will discuss below, this may have considerable significance for the difference in excitability of the various cortical areas for example motor cortex compared to visual cortex.

II. Smooth neurons without spines formerly referred to as smooth stellate cells. Multiple types have been described. The most common

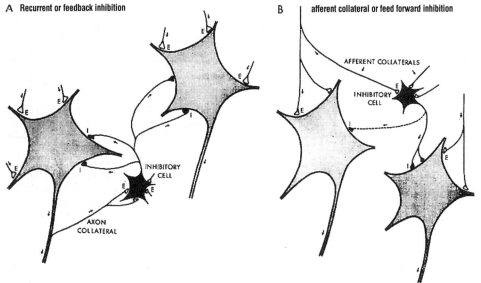

Figure 17-3 Diagram indicating possible function of stellate cells as inhibitory interneurons. Two mechanisms are demonstrated. A) Recurrent or feedback inhibition and B) afferent collateral or feed forward inhibition. Modified from Eccles, J.C. in Jasper, H.et al Basic Mechanisms of the Epilepsies Boston, Little Brown p.249, 1969.

is the basket cell that has axons that form baskets around the soma of the pyramidal cells. The Retzius-Cajal neurons are horizontally oriented cells found in layer 1. The Martinotti cell, the double bouquet cell of Cajal and the chandelier cell have a vertical orientation.

Physiology: The smooth cells are generally fast spiking and make type II inhibitory connections utilizing gamma aminobutyric acid (GABA) as the transmitter. In contrast to pyramidal neurons, the action potentials of these fast-spiking cells are brief; the repolarization phase is rapid and followed by a significant under shoot. This has been correlated with very large, fast repolarizing potassium current.

The inter-relationship of neurons within the neocortex is demonstrated in *Figure 17-3*.

BASIC DESIGN AND FUNCTIONAL ORGANIZATION OF CEREBRAL CORTEX:

The cerebral cortex is characterized by two essential design principles: 1) the neurons have a horizontal laminar arrangement, 2) connections occur in a vertical columnar manner. These vertical columns extend from pial surface to white matter. A specific region, a specific stimulus, or a specific stimulus orientation activates each cylindrical column. Activation of a specific column also activates inhibition in adjacent columns, the concept of inhibitory surround. (See Angevine and Smith 1982). In the primary sensory each column is concerned with a particular modality of sensation for a particular representation of the body or specific sector of a field of stimulation. This columnar organization first was described by Mountcastle for somatosensory cortex and has been subsequently found in other sensory and associational cortex. As discussed by Rakic (1988) this columnar or radial organization maybe a reflection of the ontogenetic ingrowth of neurons from the ependymal surface via glial guide to the cortical surface to be discussed in chapter 4. A similar columnar organization of the callosal system has also been demonstrated (See Jones -1985).

FUNDAMENTAL TYPES OF CEREBRAL CORTEX:

Brodmann in 1903 distinguished 2 fundamental types of cerebral cortex: Homogenetic or Heterogenetic.

1. *Homogenetic.* This type as already illustrated in Figure 17-1 has a 6-layer pattern at some time during development recognizable by the end of the 3rd fetal month. This type of cortex is also called neocortex or isocortex, or neopallium, or supra limbic.

2. *Heterogenetic:* This type of cortex does not have 6 layers at any time during development or during adult life. This type of cortex is also called allocortex and has 3 layers. There is a transitional zone between allocortex and the neocortex: the mesocortex that corresponds to the cingulate gyrus, the parahippocampal gyrus, the piriform area and the anterior perforated substance (*Fig.17-4*).

SUMMARY OF THE FUNDAMENTAL 6 LAYERED SCHEME OF NEOCORTEX (refer to Figure 17-1):

Layer I: The molecular or plexiform layers. This layer primarily consists of dendrites and axons from other cortical area, deeper layers and nonspecific thalamic input. A tangential plexus of fibers composed of the apical dendrites, ascending axons and axon collaterals provides a dense collection of axodendritic synapses.

Layer II: The external granular layer. This layer is a relatively densely packed layer of small stellate granule cells and small pyramidal cells whose apical dendrites terminate in the molecular layer and whose axons are sent to lower cortical layers.

Layer III: The external pyramidal layer. This layer is composed of medium and large pyramidal cells whose apical dendrites extend to layer I. The axons of these cells may function as association, commissural or intra cortical association fibers.

Layer IV: The internal granular layer. This layer is densely packed with granular/stellate

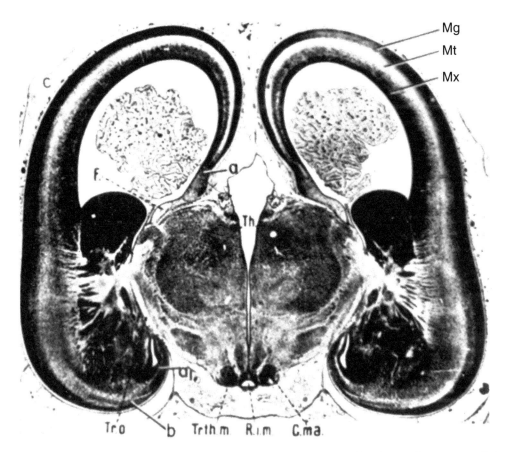

Figure 17-4. Frontal section through the diencephalic part of the prosencephalon of an 87 mm. (crown to rump), 3-1/2-month embryo. a and al = anlage of the hippocampus (heterogenetic or rhinal allocortex); b = anlage of the parahippocampal cortex (limbic mesocortex); c = supra limbic isocortex (or homogenetic). Mg = marginal layer; Mt = mantle layer; Mx = matrix layer. From Yakovlev, P.I.: Res. Publ. Assn. nerv. ment. Dis., 39:3-46, 1962 (Wiliams and Wilkins).

cells. A dense a horizontal plexus of myelinated fibers, the external band of Baillarger is also present composed of the branches of the specific thalamo-cortical projection system. These fibers synapse with the stellate cells of this layer or the basilar dendrites of the pyramidal cells of layer III or with the apical dendrites of the pyramidal cells of layers V and VI.

Layer V: Internal or large and giant pyramidal cell layer also called the ganglionic layer This layer serves as the major source of outflow fibers particularly to motor areas of the basal ganglia, brain stem, and spinal cord and possibly to the projection nuclei of the thalamus. Although cortico- cortical out put arises primarily from the more superficial pyramidal cells of layer 3, layers 5 and 6 also provide such an output. Note that there is some overlap of the functions of pyramidal cell of layers 5 and 6. Collaterals of the axons also function as intra cortical association fibers. In the deeper portion of this layer a dense horizontal plexus of fibers is present the internal band of Baillarger.

Layer VI: The fusiform or spindle cell multiform layer. This layer is composed of a mixture of spindle shaped cells, pyramidal and stellate cells. Dendrites ascend to various cortical levels. The axons enter the white matter as short association fibers or ascend to other cortical layers. However the corticothalamic outflow usually occurs from pyramidal cells in this layer.

CLASSIFICATION OF THE VARIOUS TYPES OF NEOCORTEX:

It is evident that in the adult not all areas of neocortex have the same appearance. Brodmann termed those areas of homogenetic neocortex that demonstrated the typical 6-layered pattern seen in Figure 17-1 as homotypical.

At one extreme is the agranular motor cortex. This has many giant pyramidal cells present in layer 5 but with a virtual absence of an internal granular layer 4. At the opposite extreme is the granular primary sensory projection cortex. Layers 2, 3, and 4 appear as an almost continuous granular layer. Layers of 5 and 6 are thin and few large pyramidal cells are present.

Von Economo divided the cerebral cortex into 5 basic categories (Fig.17-5):

Type 1 is the agranular motor cortex.

Type 2 is referred to as homotypical or frontal granular cortex and is found in the superior frontal gyrus of the prefrontal area.

Type 3 is the midpoint and is best seen in the inferior parietal lobule.

Type 4 is referred to as polar and is represented by the non striate visual association cortex of area 18 close to the occipital pole

Type 5 is the granular striate occipital visu-

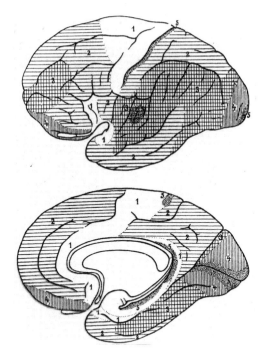

Figure 17-6. Distribution of the five cortical types of Figure 17-5 over the lateral and medial surfaces of the cerebral surface. The Sylvian fissure has been laid open. (From Von Economo, C.: The Cytoarchitectonics of the Human Cerebral Cortex. London, Oxford University Press, 1929, p. 18).

al projection cortex. Because the large number of granule cells in this type of cortex resembles- - a cloud of dust, - the Greek word konios is employed –thus the term koniocortex.

Figure 17-6 demonstrates the distribution of these various cortical types. Von Economo and Koskinis (1925) then developed a classification that dealt in a logical manner with the variations of cytoarchitecture in terms of definable histologic gradations using a letter terminology. Thus motor cortex received the designation FA, premotor cortex FB, and the frontal eye field, FC etc. In this textbook we will follow the classification of Brodmann that has been most commonly employed *(Fig.17-7)**. Numbers were initially assigned as histologically distinct areas were identified in successive horizontal slices moving in anterior and posterior directions from the central sulcus. The crest of the postcentral gyrus would then

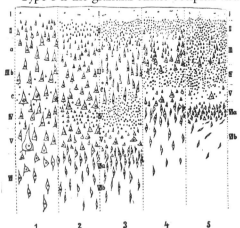

Figure 17-5. The five fundamental types of cortical structure: Type 1 = agranular motor cortex. Type 2 = frontal homotypical (frontal granular). Type 3 = parietal homotypical. Type 4 = polar, e.g., area 18. Type 5 = granular koniocortex. (From Von Economo, C.: The Cytoarchitectonics of the Human Cerebral Cortex. London, Oxford University Press, 1929, p. 16).

---

* For a comparison of the 2 systems in the monkey see fig. 17-8.

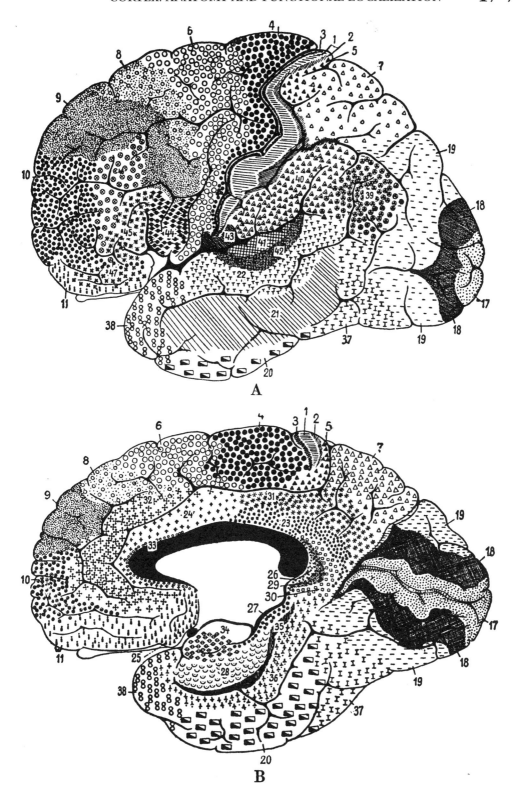

Figure 17-7. Cytoarchitectural areas of cerebral cortex as designated by Brodmann. (From Ranson and Clark: The Anatomy of the Nervous System. Philadelphia, W. B. Saunders, 1959, p. 356).

appear in the 1st horizontal section and would be assigned the number 1. The primary motor cortex is assigned 4, the premotor cortex 6 and the frontal eye field 8. Thus there is no logical reason why the numbers 6 and 8 are assigned to areas of the frontal lobe and 5 and 7 to areas in the parietal lobes. Nevertheless, some of the numbers are used with sufficient frequency in everyday neurological language that the student should commit these to memory. Before we review these areas with their correlated function, it is necessary to consider the methods employed in the study of functional localization.

PART II. METHODS FOR THE STUDY OF FUNCTIONAL LOCALIZATION IN CEREBRAL CORTEX

A. HOW DO WE STUDY FUNCTION?

In considering functional localization, and the correlation with cytoarchitecture, it has been customary to consider the effects of two general categories: stimulation and ablation.

Stimulation: Various disease processes in man involving the cerebral cortex may produce a local area of excessive discharge, in a sense, a local area of stimulation resulting in focal seizures

Actually our first understanding of cortical function was derived from the systematic study of such cases of focal discharge *(termed focal seizures, or focal convulsions, or focal epilepsy, or epileptiform seizures)* by Hughlings Jackson in 1863 and 1870. Jackson was able to predict that an area existed in the cerebral cortex that governed isolated movements of the contralateral extremities. *Such focal seizures are also referred to as partial seizures in the International Classification (see Commission, 1981, 1989).* These partial seizures are subdivided into *simple partial* (simple motor or sensory or psychic symptoms) as contrasted to *complex partial* (complex motor sequences of automatic behavior and/or alterations in mental function as regards perception, memory and affect) accompanied by impairment of consciousness, confusion and amnesia.

Partial seizures or partial epilepsy must be distinguished from those seizures which are nonfocal that is bilateral or generalized in their onset *(generalized or idiopathic epilepsy in the International Classifications)*.

Other methods of stimulation:

Electrical stimulation, of the cerebral cortex may be carried out, in the human patient undergoing operative procedures on the cerebral cortex to identify specific cortical areas or may be employed on animal preparations in the experimental laboratory. Transcranial electrical and magnetic stimulation of cerebral cortex has been developed as a clinical technique.

Evoked potentials: stimulation of afferent pathways may be utilized to identify short latency responses evoked from specific cortical areas. When stimulating the lower extremity (or sciatic nerve) response is limited to that specific area of the contralateral postcentral gyrus devoted to the representation of the lower extremity. Stimulation of the upper extremity will elicit a similar response from a different portion of the contralateral postcentral gyrus: that devoted to representation of the upper extremity. Eventually it is possible to prepare a map of the somatosensory projection cortex of the postcentral gyrus. It is possible to refine this technique at the response end by using microelectrode recordings from within or just outside neurons at particular depths of the cortex. It is also possible to define more carefully the parameters of stimulation as Hubel and Wiesel have done in their analysis of responses in the visual cortex to specific types of visual stimuli (see Chapter 23). The evoked potential method has been subsequently developed as a practical noninvasive method to study conduction in the visual, auditory and somatosensory systems.

By stimulating particular thalamic nuclei, one may map out specific thalamocortical relationships.

A modification of the technique is that of antidromic stimulation of an efferent pathway, such as the pyramidal tract, to map out the cortical nerve cells of origin of fibers in this tract.

Chemical agents locally applied to the pial surface of cerebral cortex may be employed in the laboratory preparation to produce an acute or chronic focus of discharge. The most common agent employed for the acute focus penicillin. Chronic foci are produced by cobalt, or alumina cream.

Interpretation of results: The results obtained by the method of stimulation are subject to several possible interpretations from the standpoint of functional localization:

1. The effect observed may be clearly related to the actual function of the area stimulated, e.g., stimulation of the motor cortex results in discrete movements.

2. The effect observed may be in part related to the spread of discharge to other cortical areas via corpus callosum to the opposite hemisphere or association pathways, within the same hemisphere.

The effects observed may in part reflect the spread of discharge to various subcortical and brain stem centers. The generalized convulsive seizure that may develop secondary to the focal seizure represents such a spread.

At times, the secondary spread to other cortical and subcortical areas may be so rapid that the focal origin of the seizure discharge may be difficult to ascertain and it may be difficult on clinical grounds to separate such seizures from non-focal (primary, generalized) epilepsy.

3. The effects observed may indicate that the remainder of the central nervous system is functioning without the participation of the area stimulated. Thus, if all the neurons of a given cortical area are involved in a seizure discharge, they may be unable to participate in their usual activities, be they of an afferent, integrative, or efferent nature. Thus, the effects of seizure discharges involving prefrontal or hippocampal areas may be in some aspects quite similar to the effect of ablation of these areas as regards intellectual functions, personality, and memory or learning. This concept also implies, in a sense, to release of function in other areas (cortical and subcortical). Therefore, certain of the emotional responses of the patient with seizure discharges involving the prefrontal or temporal lobe areas are also similar to the effects obtained on ablation of these areas and suggest release of other centers.

4. In the interpretation of the relationship of location of the pathology to the seizure manifestation, the threshold for discharge of the underlying cortex must be considered. Thus, in the large series of parasagittal meningiomas reported by Cushing and Eisenhardt (1938), seizure invariably accompanied those overlying the Rolandic cortex ("middle third parasagittal") at a relatively early point in the clinical history, whereas, anterior or posterior one-third parasagittal lesions, had seizure manifestation at a much later point corresponding to a larger size of tumor. Motor cortex has a much lower threshold for seizure discharge than other areas of cerebral cortex.

Ablation: This method involves destruction of a specific area by operative resection, coagulation, or freezing to the point of tissue death. In man, the method of clinical pathological correlation has produced considerable information. The destructive lesions have been the result of a local area of tissue destruction due to blood vessel occlusion, hemorrhage, solitary metastatic tumor, or intrinsic tumor. In the laboratory transient focal depression may be produced by focal cooling or by application of KCl.

Interpretation of results:

1. The observed effects may indicate deficits directly related to a failure of a positive function normally sub-served by the area destroyed. For example, after destruction of parietal somatosensory projection areas, deficits in stereognosis and position sense occur.

2. The effects may reflect the release of centers in other cortical areas or at other levels of the neural axis. This presumes that the area ablated usually inhibits certain functions of these other centers. The end result then is an enhancement of certain functions. Spasticity several days' weeks or months after ablation of the motor cortex is an example.

3. A temporary loss of function of a lower center may follow acute ablation of higher centers. For example, following acute unilateral ablation of all motor cortex in man, there will be observed a temporary loss or depression of deep tendon reflexes and of other stretch reflexes in the contralateral extremities. There will be initially a flaccid paralysis. This state (referred to as the "diaschisis of Von Monakow") is analogous to spinal shock or pyramidal shock. As recovery from this state occurs, deep tendon reflexes return and become progressively more active. The flaccid paralysis is replaced by a spastic paresis. Trans hemispheric effects also occur as discussed by Andres (1991).

B. HOW DO WE CONFIRM THE LOCATION OF THE PATHOLOGY?

Initially, data as to the actual location of the disease process and its nature (old scar from traumatic injury, old area of tissue damage due to ischemia, infection, local compression, or invasive brain tumor) were dependent on final examination of the brain at autopsy. One may then speak of the method of clinical pathological correlation. With the development of neurosurgery it became possible to define and to confirm during life the anatomical boundaries of some disease process, often allowing for treatment or cure of the basic disease process.

The specialized techniques of electroencephalography, arteriography, pneumoencephalography, and radioactive brain scan were developed during the 1930s, 1940s, 1950s, to provide some information about localization prior to surgery.

The subsequent development of the more precise noninvasive imaging techniques of computerized axial tomography, (CT scan), magnetic resonance imaging (MRI scan) and positron emission tomography (PET scan) during the 1970s and 1980s now provide the closest correlation of clinical findings and anatomical location of the lesion in the living patient, supplanting the earlier radiological techniques of arteriography, pneumoencephalography and radioactive brain scans.

The techniques of CT scan, MRI, PET and SPECT have been discussed previously (Chapter 2).

Concept of representation and re representation of function: In considering functional localization, one must keep in mind that a given function may be represented and re represented at the various levels of the neural axis.

Ontogenetic factor. In considering the effects of ablation, one must also consider the ontogenetic factor. Bilateral ablation of the prefrontal cortex in the adult monkey produces impairment in the capacity to delay responses. Bilateral prefrontal ablation at one week of age in the infant monkey does not have this effect. Such a monkey when tested at 4 months, the usual age when delay response appears, is able to manifest this capacity. Similar observations may be made with regard to somatosensory and visual pattern discrimination in the infant cat as compared to adult cats following the appropriate resection of the somatosensory or visual projection cortex. Similar observations may be made in man. Thus, extensive destruction in most adults of the speech areas in the left hemisphere will result in a marked and lasting deficit in linguistic expression (so-called Broca's expressive aphasia). Destruction of this same area in the infant or young child does not produce an enduring defect.

C. CORRELATION OF NEOCORTICAL CYTOARCHITECTURE AND FUNCTION

General observation: Those neocortical areas with specialized cytoarchitecture deviating from the classical 6-layered cortex have very specialized function for example primary motor or sensory cortex. In contrast the classical 6-layered pattern tends to occur in association areas such as prefrontal, which have more complex and not such limited highly specialized functions.

The more commonly used Brodmann numbers are listed in the following outline with a brief note as to function. The corresponding terminology of Von Economo and Koskinis is

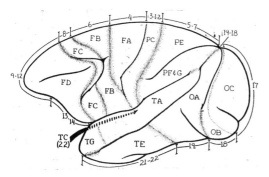

Figure 17-8. Reference diagrammatic cytoarchitectural map of monkey brain relating number system terminology of Brodmann to the letter system terminology of Von Bonin and Bailey (after Von Economo and Koskinis). The borders are only approximate. (From Ruch, T.C., and Patton, H.D.: Physiology and Biophysics. 19th Edition, Philadelphia, W.B.Saunders, 1965, p.156.)

indicated in parentheses. The list is not all inclusive and does not include the areas of mesial temporal lobe. Since many studies have been performed in the monkey a comparison guide for cytoarchitecture is provided in *Figure 17-8*.

Figure 17-9 demonstrates the lobes, gyri and major sulci visible on the lateral surface of the cerebral hemisphere. Figure 17-10 demonstrates the lobes, gyri and major sulci visible on the medial surface of the cerebral hemisphere. The orbital surface is illustrated in Figure 22-1.

FRONTAL LOBE

The following cortical areas according to

TABLE 17-1. MAJOR GYRI IN THE FRONTAL LOBE.

Gyrus	Major Function
Precentral	Origin of volitional motor pathways, the corticospinal and corticonuclear pathways (upper motor neurons)
Superior frontal	Premotor and prefrontal regions
Middle frontal	Prefrontal and frontal eye fields
Inferior frontal	Broca's motor and speech area in the dominant hemisphere
Orbital	Limbic
Gyrus rectus	Limbic

Brodmann *(Fig 17-7)* are noted in the frontal lobe: Primary motor area - 4, Premotor - 6 & 8, inferior frontal (Broca's) 44, 45, 47; Frontal eye fields area 8, Prefrontal/Orbital - 46, 9, 10, 11, 12, 13, 14. It is customary to divide the frontal lobe into those more posterior areas devoted to motor functions and those more anterior areas devoted to non-motor functions.

MOTOR AREAS:

Area 4 (FA) corresponds in general to the precentral gyrus and functions as the primary motor cortex. It is continued onto the medial surface as the paracentral lobule (*Fig.17-10, 17-11*).

Stimulation: discrete repetitive focal movements are produced based on the area stimulated for example repetitive jerks of the thumb or of the foot. A march of focal movements may occur, for example starting in the thumb, then spreading to hand, then to arm, then to face and leg etc.

Ablation: an upper motor neuron type weakness occurs with a contralateral monoplegia or hemiparesis.

Area 6 (FB) located anterior to area 4 and thus referred to as premotor cortex, functions as motor association or elaboration area (Fig. 17-8, 18-6). This area also functions to inhibit certain functions of the primary motor cortex. The supplementary motor cortex represents the continuation of this area onto the medial aspect of the hemisphere (*Fig.17-11, 18-12*).

Stimulation: patterns of movement occur, for example tonic rotation of head eyes and trunk to the opposite side associated with tonic abduction of the arm at the shoulder and flexion of the arm at elbow. Stimulation of the supplementary motor cortex produces similar complex patterns of tonic movement .In addition there are bilateral movements of the trunk and lower extremities.

Ablation: Transient release of "primitive" automatic reflexes mediated by the primary motor cortex such as instinctive grasp, and suck occurs. A more persistent release of these instinctive reflexes follows the combined abla-

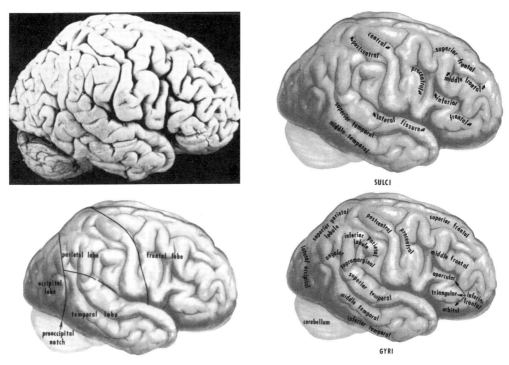

Figure 17-9. Lobes, gyri and major sulci. Lateral surface.

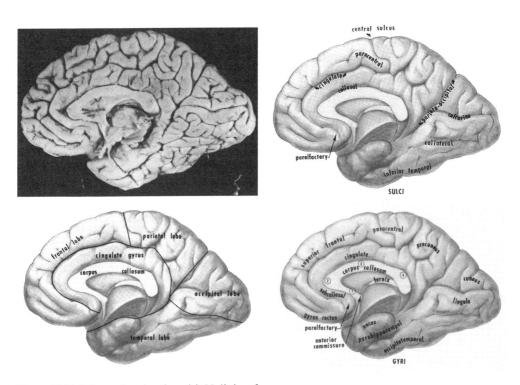

Figure 17-10. Lobes, gyri and major sulci. Medial surface

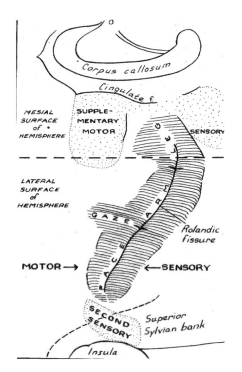

Figure 17-11. Map of somatic motor and sensory areas. Note that the primary motor and sensory areas extend into the depth of the rolandic fissure and extend onto the medial surface of the hemisphere. Note that the premotor cortex, area 6 also extends onto the medial surface of the hemisphere as the supplementary motor cortex. Note the location of the adversive gaze field in relation to motor cortex. (From Penfield, W., and Jasper, H.: Epilepsy and the Functional Anatomy of the Human Brain. Boston, Little, Brown and Company, 1954, p.103.)

tion of areas 6, 8 and supplementary motor cortex. Such lesions also produce a significant gait apraxia –an impairment of the ability to walk without actual weakness or sensory deficit. Other deficits in the initiation of patterns of movement in relationship to somatosensory or visual stimuli may also occur.

Area 8 (FC) (Fig.18-18) located anterior to area 6 in the middle frontal gyrus in the human is often grouped with area 6 as premotor cortex. This area functions in relation to conscious eye turning and is often referred to as the frontal eye field or frontal center for adversive or contraversive eye movement. This area is involved when the subject responds to the command "turn your eyes to the left (or right)" The effects are mediated by descending crossed connections to the pontine center for lateral gaze

Stimulation: (Fig.18-18) a conscious repetitive conjugate eye movement to the opposite field occurs. More specific effects can also be achieved by discrete stimulation within this area for example eyelid opening, repetitive eyelid movements' pupillary dilatation. (Fig.18-19). Bilateral stimulation will result in conjugate upward eye movements.

Ablation: transient paralysis of voluntary conjugate gaze to the contralateral visual field occurs. In general, in man, this paralysis of voluntary gaze does not occur in isolation but is associated with those processes such as infarction which have also produced a severe hemiparesis. The patient then lies in bed with the contralateral limbs in a hemiplegic posture and with the head and eyes deviated toward the intact arm and leg. This deviation most likely reflects the unbalanced effect of the adversive eye center of the opposite intact hemisphere. The effect is usually transient clearing in a matter of days or weeks. A transient neglect syndrome primarily involving the contralateral visual field occurs although contralateral tactile and auditory stimuli may also be neglected. Bilateral ablations within this area in the monkey, resulted in animals that had a bilateral neglect of the environment, remained apathetic and continued to have a "wooden expression". Similar findings may occur in the human patient with degenerative disorders involving these frontal areas.

Areas 44 and 45 (FCB) These areas correspond in general to the inferior frontal gyrus triangular and opercular portions. In the dominant (usually left) hemisphere this constitutes Broca's motor speech center or the anterior speech center (Fig.24-1).

Stimulation (Fig. 24-2): arrest of speech occurs. Occasionally, simple vocalization occurs. A similar arrest of speech may also occur on stimulation of the other speech centers –posterior and superior.

Ablation: a non-fluent aphasia occurs. The patient is mute. With limited lesions, considerable language may return. While isolated lesions (for example embolic infarcts) involving this area may occur, more often this type of language defect is associated with a contralateral hemiparesis.

PREFRONTAL –NONMOTOR AREAS

Areas 9, 10, 11 and 12 (FD), the prefrontal areas, are concerned with emotional control and other aspects of cognitive function (Fig. 18-21). Areas 13,14 (FE, FF, FG, FH) are also often included in this group.

Stimulation: Complex partial seizures with alteration of personality, emotion, and behavior occur often accompanied by tonic motor components. Note that 25% of complex partial seizures originate in frontal lobe structures but 75% originate in temporal lobe.

Ablation: An alteration in personality, affect and control of emotion may occur and/or there may be an alteration in cognitive/executive function.

PARIETAL LOBE

Areas 3, 1, 2 (PA, PB, PC) These areas correspond to the *postcentral gyrus* and function as the somatosensory projection areas (continue into paracentral lobule).

Stimulation: Episodes of localized tingling paresthesias occur which may spread as with focal motor seizures.

Ablation: Deficits in cortical sensory (discriminative sensory) modalities will occur, for example stereognosis, position sense, graphesthesia, and tactile localization.

Areas 5, 7 plus in human 39, 40 (PD, PE, PF, PG): The Parietal Lobules About half way up the postcentral sulcus, the *intraparietal sulcus* extend posteriorly, dividing the parietal lobe into superior and inferior parietal lobules. The *inferior parietal lobule* consists of the supramarginal and angular gyri. The *supramarginal gyrus* surrounds the posterior ascending limb of the lateral sulcus. The *angular gyrus* is behind the supramarginal gyrus and surrounds the posterior end of the superior temporal sulcus *(Table 17-2)*.

Areas 39, 40 (PG, PF) correspond to the

TABLE 17-2. MAJOR GYRI IN THE PARIETAL LOBE.

Gyrus	Major Functions
Postcentral	Sensory cortex
Superior parietal lobule	Sensory associational
Inferior parietal lobule: --Pars supramarginal --Pars angular	Sensory language areas in the dominant hemisphere
Precuneus	Sensory associational

inferior parietal lobule (the angular and supramarginal gyri). These areas in the dominant hemisphere function in relation to reading and writing as higher integrative areas for language. This area is part of the posterior speech area. In the nondominant hemisphere these areas relate to our concepts of visual space.

Stimulation: In the dominant hemisphere, arrest of speech occurs.

Ablation: In the dominant hemisphere, defects in reading, writing and calculations occur. These deficits along with difficulty in finger identification and left/right confusion constitute aspects of the Gerstmann syndrome. A variety of fluent aphasia occurs. In the non-dominant hemisphere, a denial or neglect syndrome occurs.

TEMPORAL LOBE:

The temporal lobe is a complex structure that includes neocortex, allocortex, mesocortex and a subcortical nucleus (the amygdala). At its posterior borders it merges into the parietal and occipital lobes.

Major areas in the Temporal Lobe from figure 17-7:

Auditory (Transverse Temporal) 41; Auditory associational 42 and 22; middle temporal 21; inferior temporal 20; posterior temporal 37; entorhinal 27, 28, 35.

The temporal lobe lies below the lateral sulcus and on its lateral surface has a sequence of three anterior to posterior arranged gyri: the superior, middle, and inferior temporal gyri.

The inferior temporal gyrus extends onto the ventral surface of the cerebrum. The supe-

TABLE 17-3 GYRI IN THE TEMPORAL LOBE

Gyrus	Major Functions
- Hippocampal formation - Amygdala	- Limbic and memory - Limbic
Parahippocampal	Limbic
Occipitotemporal	Limbic
Inferior temporal	Limbic
Middle temporal	Limbic and facial recognition
- Superior temporal - Transverse temporal gyri	- Sensory language area in the dominant hemisphere; - Auditory area

rior temporal gyrus forms the temporal operculum. Near its posterior end two gyri are seen running into the lateral fissure (Table 17-3). These are the transverse temporal gyri, consisting of the primary receptive auditory cortex, area 41.

One of the first conclusive bits of evidence on the significance of anatomic asymmetry in the cerebrum was noted by Geschwind and Levitsky (1968) in the temporal lobe. They made a horizontal section through the lateral sulcus and removed the overlying parietal and occipital cortex. They called this exposed region of the superior temporal gyrus, the *planum temporale*. They noted a larger planum temporale in the left temporal area than on the right side in right handed humans. In left-handed individuals the area was the same on both sides. Their observations on the anatomical basis of cerebral dominance have since been supported by radiological and other anatomical studies.

Areas 41, 42 (TC, TB) the transverse gyri of Heschl function as primary auditory projection areas.

Stimulation: Episodic tinnitus (a ringing sensation) occurs. Usually such seizures are not limited to this symptom in isolation, since other aspects of temporal lobe are involved to produce the more complex phenomena of the temporal lobe seizure.

Ablation: limited unilateral lesions may produce a disturbance in the ability to localize sounds. Bilateral lesions of a limited nature would be rare but could produce "cortical deafness"

Area 22 (TA) corresponds to the superior temporal gyrus and surrounds areas 41-42. This is an auditory higher association center. In the dominant hemisphere, the posterior half of this area represents an auditory association area concerned with the reception and interpretation of spoken language. This is one component of the posterior speech area. The area is often referred to as Wernicke's receptive aphasia center. In the nondominant hemisphere this area is more concerned with visual aspects of space.

Stimulation: Stimulation of the posterior speech (Wernicke's) area (Fig.24-2) produces an arrest of speech. Seizures originating in the superior temporal gyrus are also characterized by "experiential "phenomena: distortions of auditory and visual perception alterations in the sense of time and in well formed visual and auditory hallucinations (psychical seizures). As will be discussed in chapter 22,complex partial seizures that may begin with these phenomena but subsequently proceed to impairment of awareness; amnesia and automatisms (unconscious stereotyped patterns of movement) reflect involvement of or spread to the mesial temporal areas of hippocampus and amygdala. Fear that may also accompany these seizures reflects involvement of the amygdala. Olfactory hallucinations that may also accompany these seizures reflects involvement of the mesial temporal structure, the uncus.

Ablation: Damage to Wernicke's area a part of the posterior speech area produces a deficit in the comprehension of speech, a type of fluent aphasia.

Additional information about more extensive unilateral or bilateral damage to the temporal lobe will be considered in chapters 22 and 30. Suffice it to say, that bilateral damage to the hippocampus will produce severe deficits in the ability to record new memories. Damage to the amygdala will alter emotional control

TABLE 17-4 MAJOR GYRI IN OCCIPITAL LOBE.

Gyrus	Functions
Lateral occipital	Visual associational
Lingula	Superior visual field
Cuneus	Inferior visual field

OCCIPITAL LOBE

The lateral occipital gyri consist of visual associative cortex. The visual receptive cortex (calcarine-cortex) is found on the medial surface of the hemisphere (Table 17-4).

Areas in the Occipital Lobe: visual (calcarine) 17; visual association 18 & 19.

Area 17 (OC) corresponds to the striate cortex bordering the calcarine fissure and functions as the primary visual projection area. In modern physiological terms the designation V1 is utilized.

Stimulation: Seizures originating in this area produce simple unformed visual hallucinations such as flashing lights, stars or jagged lines. The phenomena may be localized to the contralateral visual field or at times to the contralateral eye.

Ablation: Homonymous visual deficits are produced for example, a homonymous hemianopsia or quadrantanopsia.

Areas 18, 19 (OB, OA) form surrounding stripes around area 17. These areas function as visual association areas of varying complexity. The terms V2-V4 are utilized. These areas are also involved in the fixation and following of objects in the contralateral visual field. Note that visual areas V4 and V5 extend into the adjacent temporal-parietal cortex.

Stimulation: Some of the effects are similar to stimulation of area 17. In addition conjugate deviation of the eyes to the contralateral field will occur.

Ablation: Selective lesions may produce deficits in some of the more complex visual activities. In addition defects in visual fixation and following may occur.

III. DEVELOPMENTAL ASPECTS OF NEOCORTEX:

The topics of prenatal development, postnatal development and abnormalities of cortical neuronal migration are discussed in chapter 4 and on this chapter's CD ROM.

IV. SUBCORTICAL WHITE MATTER AFFERENTS AND EFFERENTS

Essentially three types of fiber systems occupy the subcortical white matter: (1) projection fibers, (2) commissural fibers, and (3) association fibers.

Projection fibers consist of the corticopetal afferent fibers, such as the thalamocortical radiations, and the corticofugal efferent fibers such as the corticospinal, corticoreticular, and corticorubral tracts *(Fig. 17-12)*.

The major commissural systems are the corpus callosum, anterior commissure, and hippocampal commissure. The corpus callosum is the major commissure for the neocortical areas (except the middle and inferior temporal gyri) *(Fig. 17-13)* and the rostral portions

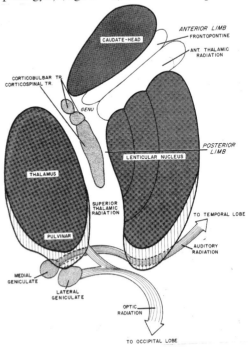

Figure 17-12. Projection fiber systems passing through the internal capsule. The superior thalamic radiation includes fibers projecting from ventral thalamus to sensory and motor cortex.

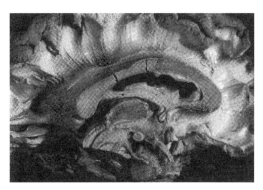

Figure 17-13. Radiation of fibers from corpus callosum as dissected in human brain.

of the superior and inferior temporal gyrus (see Pandya and Posene 1985- for a more detailed discussion). In general, homologous areas of the two hemispheres are interconnected. However, there are significant regional variations between homologous areas as regards fiber density *(Fig. 17-14)*. Thus, there is a high density of callosal fibers connecting the premotor areas of the two cerebral hemispheres. On

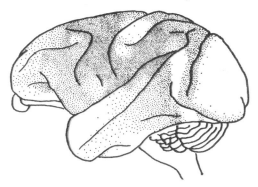

Figure 17-14. Pattern of distribution of commissural fibers over the cortical surface of the left hemisphere of the monkey (Macaca mulatta) as studied following section of major commissures. (From Myers, R. E.: In Ettlinger, E. G. (ed.): Functions of the Corpus Callosum. Boston, Little, Brown & Company, 1965, p. 142).

the other hand, a primary visual projection area - area 17 - has almost no direct callosal connection to the contralateral hemisphere. Area 17 transmits to the adjacent area 18 that has connection to contralateral area 18[1].

Cortical areas also differ as regards the spread of fibers to asymmetrical as well as to symmetrical points in the contralateral hemisphere. Thus area 6 has widespread connections in the contralateral hemisphere not only to area 6 but also to areas, 4, 5, 7, and 39, whereas area 4 has discrete contralateral connection only to the homotypic points.

The anterior commissure interconnects the rostral portions of the superior, middle and inferior temporal gyri, inferior-posterior orbital gyri and paleocortex (parahippocampal gyrus). The hippocampal commissure[2] interconnects the hippocampal formation and dentate gyri (archicortex) and the surrounding presubiculum, entorhinal and adjacent inferior temporal gyrus. These fibers are conveyed via the fimbria of the fornix and cross at the point beneath the splenium of the corpus callosum, where the posterior pillars of the fornix converge.

Two types of subcortical association fibers may be distinguished (excluding from this discussion, the intracortical association fibers): (a) short subcortical U or arcuate fibers which interconnect adjacent gyri and (b) long fiber bundles which reciprocally interconnect distant cortical areas. The following long fiber bun-

---

[1]*The organization of callosal projections is more complex than the picture provided by the early degeneration maps. Thus within somatosensory cortex callosal connections are most dense in areas 2, less dense in area 1 and least dense in areas 3B. Within each of those areas differences also occur. In area 3B, relatively dense connections are present for trunk and head but relatively sparse connections for hand and foot. In contrast within area 2 relatively dense connections are present for all parts of the body. Significant changes occur during development in part related to the overlap with thalamocortical afferent input. For a more complete discussion see Killackey 1985, Killackey and Chalupa 1986, Dehay et al 1988. The end result is in part a complementary relationship to the distribution of thalamocortical fibers.*

[2]*In the monkey two hippocampal commissures have been identified-based on relationship to the fornix. (For a more complete discussion of the topography of commissural fibers see Pandya and Rosene, 1985).*

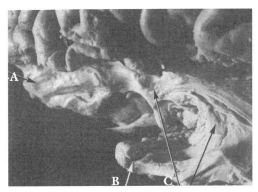

Figure 17-15. Dissection of long fiber systems; the uncinate fasciculus (C) passing from orbital frontal (A) to anterior temporal areas (B).

dles may be distinguished on blunt dissection of the cerebral hemisphere:

1. The *uncinate fasciculus* interconnects the orbital and medial prefrontal areas and the anterior temporal area *(Fig. 17-15)*.
2. The *superior longitudinal fasciculus* interconnects the superior and lateral frontal, parietal and temporal, and occipital areas (Fig. 24-5). The extension of this fiber system into the temporal area, passing through the subcortical white matter of supramarginal and angular gyri, is often distinguished as the arcuate fasciculus. This fasciculus has considerable importance because of its role in connecting the receptive language centers of the temporal lobe with the expressive motor speech centers of the inferior frontal gyrus. Such a connection must be made if a sentence that has been heard is to be repeated.
3. The *cingulum* passes within the subcortical white matter of the cingulate gyrus; it interconnects the subcallosal, medial-frontal, and orbital-frontal with the temporal lobe, occipital areas, and cingulate cortex.
4. The *inferior longitudinal fasciculus* interconnects the occipital and inferior temporal areas.
5. The *inferior frontal occipital fasciculus* interconnects the frontal and occipital areas. It is often difficult to clearly differentiate this fiber system from the uncinate fasciculus.

MAJOR AFFERENT INPUTS AND EFFERENT PROJECTIONS OF NEOCORTEX:

AFFERENT INPUTS:

THALAMUS:

The thalamus is the major source. The transmitter is glutamate. With the exception of the olfactory system, all sensory information passes through the thalamus`.

In many classifications, the thalamus has been divided into three major divisions: 1) the dorsal thalamus, 2) the ventral thalamus, and 3) the epithalamus.

The **dorsal thalamus** is by far the largest component and at times the term thalamus has been applied only to the dorsal thalamus. This structure is composed of a series of relay nuclei that connect in a reciprocal manner with the cerebral cortex and the striatum (Refer to chapter 15) these nuclei can be subdivided into specific projection (or relay) nuclei and nonspecific or diffuse projection (or relay) nuclei.

The *specific nuclei* have specific afferent inputs and project in a relatively precise topographic manner to specific areas of cerebral cortex. Experience, or functional disuse may modify the precise cortical areas devoted to the topographic representation of a given sensory area. Examples of specific nuclei and their cortical projections have been provided in chapter 15.

The densest input from the specific relay projection nuclei is to layer 4, but as indicated above, pyramidal cells in layers 3 and 5 many also receive direct or indirect inputs. There is a modality specific columnar arrangement.

The *nonspecific nuclei* have a more diffuse input and a more diffuse but not necessarily a generalized cortical interaction. Examples include the following: midline nuclei (massa intermedia), central-medial, intralaminar nuclei, and the medial dorsal (which is diffuse to frontal areas).

The nonspecific nuclei project mainly to layer 1.

The **ventral thalamus** is a thin shell nucleus that is external to the external medullary lamina. It does not project directly to cerebral cortex, but receives innervation from cerebral cortex. The subnuclei grouped as the thalamic reticular nucleus have specific reciprocal relationships with projection (relay) nuclei in the dorsal thalamus. As we will see below in discussing the basis of the electroencephalogram these nuclei have a significant effect on thalamocortical interactions. These nuclei in the rodent (but not in the primate) have a high concentration of GABA-ergic neurons. The drive on the relay nuclei is therefore inhibitory.

The **epithalamus** includes the pineal body, the stria medullaris and the habenular trigone. These structures relate primarily to the hypothalamus and are not relevant to a discussion of thalamocortical interactions.

NON-THALAMIC SOURCES OF INPUT:

There are multiple other subcortical sources of input including the following.

1. **noradrenergic (norepinephrine)** pathway from the locus ceruleus of the midbrain projecting in primates predominantly to layer 6 of the motor and somatosensory cortex and related frontal and parietal association cortex. This pathway functions in relationship to arousal responses induced by sensory stimuli and activity in this system is altered during the rapid eye movement stage of sleep
2. **serotoninergic pathway** from the raphe nuclei in the pons and medulla and the pontine reticular formation. Although these neurons project to all cortical areas, the predominant projection in the primate is to layer 4 of area 17, the primary visual cortex. This system may be involved in the onset of the slow wave stage of sleep. During rapid eye movement sleep there is decreased activity in this system. Damage to the system produces insomnia.
3. **dopaminergic pathways** from ventral tegmental – rostral mesencephalic nuclear groups. The strongest projection is to the prefrontal cortex, and limbic system. The fibers extend to all cortical layers except for IV. The projection is primarily inhibitory. This system is probably involved when psychotic behavior occurs in schizophrenia.
4. **cholinergic pathways** from the basal fore brain nucleus of Meynert project widely to cerebral cortex. In addition cholinergic neurons in the septum project to the hippocampus. In addition, cholinergic neurons in the brain stem tegmentum project to the thalamus and will be discussed in the chapter on sleep (chapter 29). Electrical stimulation of brain stem reticular formation (resulting in desynchronization /arousal of the EEG) results in an increase in rate of liberation of acetylcholine from the surface of cerebral cortex. This effect may be mediated through the reticular formation projection to the basal fore- brain.
5. **GABAergic pathways** from basal forebrain, ventral tegmental area and zona incerta to sensory and motor cortex.
6. **claustrum** projects to all sensory, limbic, and motor areas with an excitatory input.

EFFERENT PROJECTIONS:

Projections occur in a reciprocal manner with many subcortical areas (thalamus, basal ganglia, brain stem and spinal cord). However these projections are estimated to account for only 0.1 to 1% of all fibers in the white matter. Instead most fibers are involved in intra- hemispheric and interhemispheric connections, pyramidal neurons to pyramidal neurons and to interneurons.

Part V: Neurophysiology Of The Cerebral Cortex: Correlates Of Cortical Cytoarchitecture And The Basis Of The Electro-Encephalogram will be found on the CD ROM chapter 17

Part VI: Clinical And Physiological Correlates Of Cytoarchitectural Variation will be found on the CD ROM chapter 17

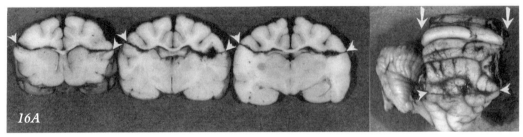

PART V: NEUROPHYSIOLOGY OF THE CEREBRAL CORTEX: CORRELATES OF CORTICAL CYTOARCHITECTURE AND THE BASIS OF THE ELECTRO-ENCEPHALOGRAM:

The cerebral cortex of man and of other mammals is characterized by continuous rhythmic sinusoidal electrical activity of variable frequency. This electrical activity may be recorded through the scalp of man in the form of the electroencephalogram (EEG) or directly from the cerebral cortex during surgery in the form of the electrocorticogram (ECG). The term EEG will be used during the subsequent discussion to refer to the activity found in either the EEG or the ECG. The electroencephalogram provides a physiologic correlate of the various states of consciousness.

The normal electroencephalogram (EEG) in the awake adult who is recorded while resting with eyes lightly closed is characterized by a dominant activity in the posterior (parietal-occipital) recording areas of continuous 8-13 Hz sinusoidal waves of 25-75uv (Fig.2-18). Amplitudes recorded directly from the cortex are higher. This pattern is referred to as the alpha rhythm since this was the first pattern discovered. The second described pattern the beta rhythm is found predominantly in the frontal recording areas and consists of low amplitude 14-30Hz activities. Many sedative medications will produce increased amounts of generalized beta activity and eventually generalized slow wave activity. The frequency of the basic background activity is also influenced by the age and level of consciousness of the patient. Any pattern slower than alpha is referred to as composed of slow waves (theta=4-7Hz; delta =0.5-3Hz) as illustrated in figures 2-24, 2-25. Many

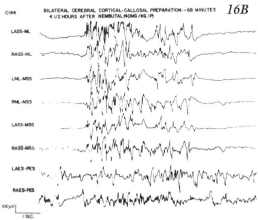

Figure 17-16. Isolated cerebral cortex. A) The anatomical bilateral cortical callosal preparation. B) The electrical activity of the preparation. The upper 6 channels are bipolar recordings from the isolated areas. The lower 2 channels are recorded from the non-isolated cortex. From Marcus, E. M. In Reeves, A. Epilepsy and the Corpus Callosum, New York Plenum Press, p.158. 1985.

sedative medications will produce increased amounts of generalized beta activity and eventually generalized slow wave activity. (Fig. 2-21). The frequency of the basic background activity is also influenced by the age and level of consciousness of the patient. (Fig. 2-19, 2-20) In general the activity of the two cerebral hemispheres is relatively symmetrical and synchronous.

The electrical activity recorded is not a result of axon potentials and does not represent the activity of individual neurons and synapses. Instead EEG rhythms represent the summated electrical activity of a large number of synapses located in the more superficial layers of the cerebral cortex. Seizure discharges do not originate in single neurons but instead are generated in a group of neurons that manifest increased excitability and synchronization (the epileptic focus). Never the less certain mutations involving the neuronal membrane and the synaptic receptor provide the basis for several types of genetically determined seizure disorders (see chapter 29). The difference between axonal and synaptic activity has been discussed in chapter 5.

THE ACTIVITY OF ISOLATED CEREBRAL CORTEX:

In considering the basis of the electroencephalogram, it is of value to first consider the activity of large blocks of cerebral cortex isolated from all subcortical interactions, but retaining pial blood supply. The basic activity of such preparations in the cat, the monkey and in the human is characterized by bursts of electrical activity composed of mixed frequency sharp waves, spikes, slow waves and fast activity alternating with periods of relative electrical silence (Ingvar and Echlin). In the chronic studies of Kellaway et al (1966) the bursts of activity were correlated in large part with unit discharges in the depths of cortex at the level of the large pyramidal cell bodies. In a related preparation large homologous blocks of cerebral cortex were isolated in each hemisphere along with the interconnecting corpus callosum *(Fig.17-16A)*, (Swank (1949, Marcus&Watson, 1966). The basic pattern is composed of bilateral relatively synchronous and symmetrical bursts of activity with intervening periods of relative electrical silence *(Fig.17-16B)*. Similar burst suppression type activity is obtained in preparations in which all thalamic, hypothalamic and rostral mesencephalic structures have been ablated. The burst suppression pattern may be seen in several clinical circumstances: 1) very deep anesthesia, 2) diffuse encephalopathies as in anoxic encephalopathy or encephalitis, 3) following intractable status epilepticus or 4) in the premature brain.

THE ACTIVITY OF ISOLATED THALAMUS

In contrast in the studies of Kellaway et al (1966), the electrical activity of isolated thalamus was characterized by continuous sinusoidal waves of 8Hz. The more recent studies of Steriade and Llinas (1988) have demonstrated that all thalamic reticular neurons can spontaneously generate rhythmic discharges at a rate of approximately 10 Hz. Groups of such cells even when deafferentate can generate synchronized oscillations. These neurons have a strong inhibitory input to the thalamic relay (projection) neurons and thus could entrain these neurons in a similar oscillation. This could then be reflected in a similar entrainment of cortical neurons and synapses in a 10 Hz oscillation. (See below).

In brief then the continuous rhythmic activity seen in the electroencephalogram must indicate the effects of thalamic neurons on the activity of cortical neurons and synapses. As we will see below, brain stem structures are also of importance in modifying cortical activity.

MODIFICATION OF CORTICAL ACTIVITY BY STIMULATION OF VARIOUS STRUCTURES: EVOKED POTENTIALS.

The background cortical electrical activity may be modified by stimulation at various levels. These evoked responses include the superficial cortical response, the direct response, the transcallosal response, the primary evoked response, the recruiting response and the generalized arousal response.

Primary evoked response. Stimulation of the sciatic nerve or of other specific sensory pathways produces a similar surface positive wave followed by a surface negative wave. *(Fig.17-17)* A similar response of much shorter latency (1-5msec) is produced by stimulation of the specific thalamic nucleus involved in the primary evoked response.

Augmenting response: With repetitive stimulation of the same nucleus at 6-12 Hz the amplitude and latency of each successive response increases *(Fig. 17-17)*.

Recruiting response: In contrast, repetitive 6 to 12 Hz stimulation of the nonspecific thalamic nuclei such as the intralaminar, the medial or midline groups or the thalamic reticular

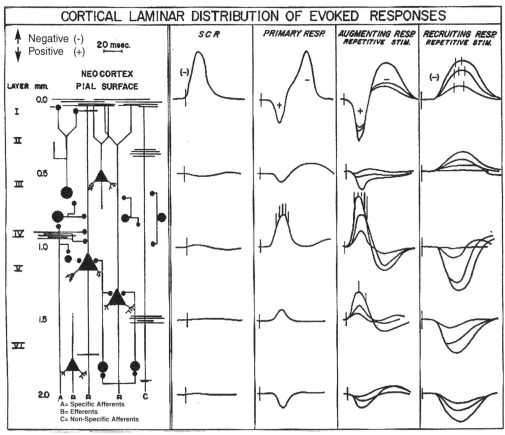

Figure 17-17. Laminar analysis of the various cortical evoked responses in terms of the type and magnitude of response recorded with an external microelectrode at various depths. Specific afferent fibers terminate in layer IV and are shown making synaptic contact with both granule cell interneurons and adjacent pyramidal cells. Nonspecific afferent fibers are shown terminating in many layers including layer I. The SCR (superficial cortical response) is limited to the superficial layers of cerebral cortex. The initial surface positive component of the primary evoked response (specific stimulus, recording from specific cortical projection) originates in the deeper cortical layers (the vertical lines represent cell discharges). The subsequent surface negative wave originates in superficial cortical layers. The transcallosal response would have a similar form and analysis. Specific thalamic nucleus stimulation results in the augmenting response. The origin of the recruiting response to stimulation of nonspecific thalamic nuclei is complex (see text). (Reproduced with modifications from an unpublished diagram by Purpura)

nucleus) produces a predominantly surface negative response of greater latency (15-60 msec) termed the recruiting response *(Fig.17-17, 17-18)*. The response waxes and wanes and is similar to the 8 to 12 Hz spindle response that characterizes light anesthesia in the cat. In the human, a similar spindle occurs at 14 - 15 Hz during stage 2 sleep. Although this system is referred to as the diffuse thalamocortical system, in the monkey, the response is obtained predominantly from the frontal and parietal association cortex. The recruiting response represents summated EPSPs (excitatory post synaptic potentials) generated in the more superficial layers of cerebral cortex and the waves are therefore surface negative. At times a small initial surface positivity may be present. Under particular conditions of anesthesia with stimulation of these same thalamic nuclei, a longer duration (100-200 msec) surface negative wave may follow the initial wave of the recruiting response. This is however correlated with a prolonged hyperpolarization in the deeper cortical layers. In the cat preparation

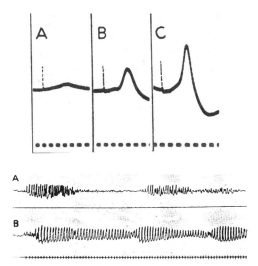

Figure 17-18. The recruiting response. Cortical responses in anterior sigmoid gyrus of cat to successive stimulation of intralaminar nuclei of thalamus (A, B, C) at 8 Hz characterized by long latency and a progressive increase in amplitude of the surface negative response. The time marks indicate 10 msec. (From Morison, R.S., and Dempsey, E.W.: Amer. J. Physiol., 135:288, 1942) II. The similarity of the waxing and waning amplitude of the recruiting response to the spontaneous 8-12 cps spindles, which characterize sedation or anesthesia with Nembutal or other barbiturates, is also demonstrated. Recordings from middle suprasylvian gyrus of cat. A) Spontaneous spindle bursts. B) Recruiting response to stimulation (stimulus marker) or intralaminar areas of thalamus. From Dempsey, E. W., and Morison, R. S.: Amer. J. Physiol., 135:297, 1942 (Amer. Physiol. Soc.)

this combination of the initial recruiting wave (the spike) and the subsequent slow wave provides a model of the thalamic induced spike wave complex and has been correlated with behavioral changes (arrest of activity, myoclonus) similar to the absence seizures of idiopathic epilepsy (see chapter 29).

Arousal or activation or desynchronization response: Stimulation of the sciatic nerve or other sensory pathways or eye opening or intense mental activity will also have a significant generalized effect on the background activity of the EEG. The alpha rhythm is replaced by a low-voltage fast activity (Fig. 2-18C). This response can be reproduced by high frequency (50-300Hz) stimulation of the brain stem ascending reticular activating system. The response occurs with a latency of approximately 40 milliseconds and outlasts the period of stimulation. The long latency suggests a pathway involving several synapses as opposed to the classic short latency lemniscal sensory pathway. This system has diffuse effects on the cerebral cortex via 2 systems: the nonspecific thalamic nuclei and the pathway from basal diencephalon to the basal forebrain. The intralaminar nuclei of the thalamus represent an upward continuation of the ascending reticular formation. Stimulation of these thalamic nuclei, e.g., centrum medianum, at high frequency (300 Hz) produces the same effects on cortical electrical activity as stimulation of ascending reticular formation at a brain stem level Thus sciatic stimulation produces not only the short latency, localized, primary evoked response but also the long latency, generalized, arousal response. This reflects the fact that sensory data enters the reticular formation via collaterals from the specific sensory pathways, particularly the spinothalamic system. Moreover, multiple sensory modalities may synapse on the same neuron in the reticular formation. The system then is in a sense nonspecific as to the modality identity of the incoming stimulus.

The role of the reticular formation and

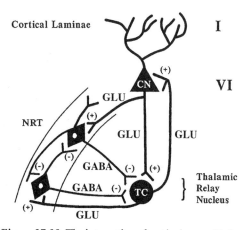

Figure 17-19. The interaction of cortical pyramidal cell (CN), thalamic relay nucleus (TC) and thalamic reticular nucleus (NRT), GLU=Glutamate, GABA= Gamma amino butyric acid. Modified from Snead, O.C. III, In Malafossa, A. Idiopathic Generalized Epilepsies: Clinical, Experimental and Genetic Aspects. London John Libbey

other brain stem areas in sleep, consciousness, and attention will be considered in greater detail in chapter 29.

THE INTERACTION OF CORTICAL NEURONS AND THALAMIC RELAY NUCLEI:

The interaction is complex. Thalamic relay nuclei function in two possible states: (1) a transmission mode when the neuron is near firing threshold in which the discharge reflects the sensory input. (2) A burst mode when the neuron is hyperpolarized by inhibitory input. The thalamic relay neurons have a special voltage gated calcium channel (transient or T type) that is inactive when the membrane potential is near threshold. However when the cell is hyperpolarized, incoming excitatory synaptic potentials trigger transient opening of this calcium channel. The resulting calcium current then brings the neuron potential above threshold. The neuron then fires a burst of action potential until the calcium that has entered the cell activates potassium current that again produces a hyperpolarized state. During burst firing the thalamic relay neurons cannot transmit sensory information to the cerebral cortex. The source of the hyper polarization of the thalamic relay neurons is the reticular thalamic nucleus that is composed of GABA-ergic inhibitory neurons. In turn the neurons of the reticular nucleus receiving collateral input from cortical thalamic and thalamocortical relay neurons (Fig.17-19). The thalamic reticular nucleus also has the property of the burst mode of discharge when hyperpolarized. The burst firing of the thalamic relay cells is reflected in rhythmic waves of excitatory postsynaptic potentials in the dendrites of cortical neurons and is expressed in the surface EEG as rhythmic slow waves. This pattern of slow wave activity may be seen in sleep or in diffuse encephalopathies

During states of wakefulness, the thalamus remains in the transmission mode because of cholinergic input from rostral pons to the thalamic relay and thalamic reticular neurons. There are also cholinergic inputs to the thalamic reticular nucleus from the basal forebrain.

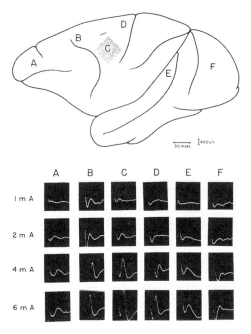

Figure 17-20. Variations in thresholds and amplitude of superficial cortical response (local cortical response) as a function of locus of stimulation in the monkey. Note the low threshold and relatively higher amplitude of the premotor and motor cortex responses (B, C, D) compared to those obtained in area 17 (F) posterior temporal (E) and prefrontal areas (A). The stimulation intensities were the same in all areas (0.10 msec. single shocks). From Eidelberg, E., Konigsmark, B., and French, J.D.: Electroenceph. clin. Neurophysiol. 11:123, 1959 (Elsevier).

PART VI: CLINICAL AND PHYSIOLOGICAL CORRELATES OF CYTOARCHITECTURAL VARIATION

1) The capacity for focal seizure discharge:

Anatomical Observation: In the monkey as in man the motor cortex (area 4) of the precentral gyrus is characterized by the presence of giant and large pyramidal cells. These giant and large pyramidal cells are characterized by a long apical dendrite. Associated with these pyramidal cells, there is noted in the Golgi stain, the presence of a massive tangential plexus in the molecular layer with the presence of many axodendritic synapses. In contrast, the striate occipital cortex (area 17) has a dense external line of Baillarger in layer IV but only a thin tangential plexus in the molecular layer.

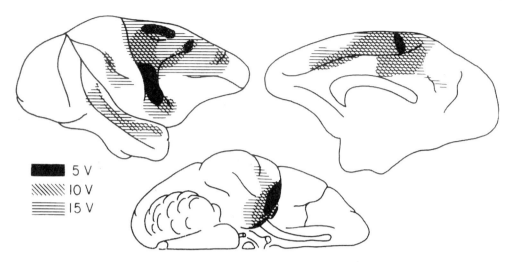

Figure 17-21. Regional Differences in Seizure Susceptibility in Monkey Cortex. The thresholds for electrical seizure after discharge in response to direct stimulation of cerebral cortex are shown. Note the low threshold of motor cortex of sectors in premotor cortex and of anterior medial temporal area as contrasted to striate occipital, (area 17) prefrontal posterior temporal, and parietal areas. Compare to Figure 17-30, note the similarity as regards pattern of regional variation. From French, J.D., Gernandt, B.E., and Livingston, R. B.: Arch. Neurol. Psychiat., 75:270, 1956 [American Medical Association (AMA)].

Giant pyramidal cells are absent and large pyramidal cells are sparse.

Physiological Observations: The superficial cortical response represents the summated postsynaptic potentials generated at superficial axodendritic synapses. In the studies of Eitelberg et al. (1959) this response could be easily obtained at low threshold in the motor cortex. On the other hand, only a feeble response was produced in the striate occipital cortex *(Fig. 17-20).*

French et al. (1956) studied the threshold for generation of propagated seizure after discharge following electrical stimulation of monkey cerebral cortex. Low thresholds were found in the motor cortex with the ready development of generalized seizure discharge. On the other hand, high thresholds were present in the striate occipital cortex *(Fig. 17-21).*

Clinical Correlation: The capacity for a clinical seizure is directly related to the capacity for generation of repetitive discharge and of after-discharge. Thus whether a tumor involving the cerebral cortex announces its location with a seizure will be related to its location. There is, then, an extremely high incidence of focal or generalized seizures in patients with parasagittal meningiomas overlying the motor cortex. On the other hand, tumors in the prefrontal, posterior-temporal parietal, or occipital areas may grow to a large size without producing focal seizures. When focal seizures occur they may indicate a compromise of low threshold motor cortex with discharge originating at that area rather than at the primary site of involvement.

2) The capacity for contralateral spread of discharge as related to callosal fiber system density:

Anatomical Observation: Using the Nauta method to study degeneration of axons, Ebner and Myers (Myers, 1965) have evaluated the density of callosal projection approximately 10 days after section of the major interhemispheric commissures. A dense callosal projection was found in the precentral, premotor and inferior parietal areas. Only a sparse callosal projection was present in striate occipital and superior temporal areas *(Fig. 17-14).*

Physiological Observations: The original studies of Curtis (1940) on the transcallosal response in the monkey indicated significant regional differences in the capacity for generation of this response. The response was easily obtained on stimulation of the premotor, precentral, and parietal areas but could not be

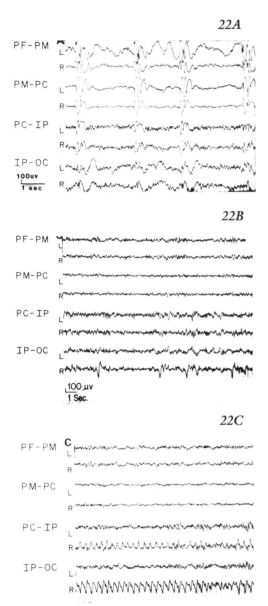

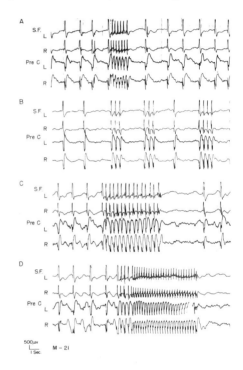

found in studies of McCulloch and his associates (1944) employing a local area of seizure discharge (strychnine neuronography) rather than direct electrical stimulation. Both series of studies also indicated that significant regional differences existed as regards the diffuseness of callosal projection. Thus, certain cortical areas (e.g., motor cortex) had projection only to the symmetrical area of the contralateral hemisphere. Other areas, such as the premotor area

Figure 17-22. Capacity for spread of an experimental focal discharge. A) Unilateral focus in the left premotor area 6 of the monkey results in a wide intra and inter hemispheric spread. B) Production of the same unilateral focus in area 17 of the right occipital area results in a restricted unilateral discharge which even when prolonged as in C) fails to spread. PF=prefrontal, PM= premotor, PC= precentral, IP= intraparietal, OC=Occipital.

obtained on stimulation of the striate occipital cortex. In general, similar differences were

Figure 17-23. Capacity for bilateral synchrony in premotor area of monkey cerebral cortex. Experimental design of bilateral symmetrical discharging premotor foci (1% conjugated estrogen). Recordings from superior frontal (S.F.) and precentral gyri (Pre C). Note the repetitive bilaterally synchronous discharges of spikes and spike slow wave complexes. Compare to Figure 17-24. From Marcus, E. M., and Watson, C. W.: Arch. Neurol., 19:102, 1968 (AMA). See also chapter 29.

(area 6) had widespread projection to many points in the contralateral hemisphere (areas 6, 4, 1, 5, and 39 as regards area 6).

Clinical correlation: Experimental or clinical seizures originating from the premotor cortex will have widespread bilateral effects where-

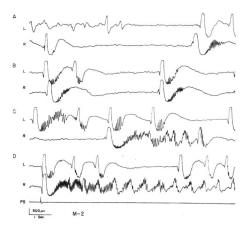

Figure 17-24. Lack of capacity for bilateral synchrony in area 17 of monkey cerebral cortex. Experimental design of bilateral symmetrical discharging foci as in figure 17-23. Recordings from left and right occipital area. Note independence of discharge particularly when prolonged (A and C). Occasional bilateral discharges relatively synchronous at onset could occur (B), particularly discharges when triggered as in D) by an extrinsic source, e.g., photic stimulation (PS). From Marcus, E. M., and Watson, C. W.: Arch. Neurol., 19:107, 1968 (AMA).

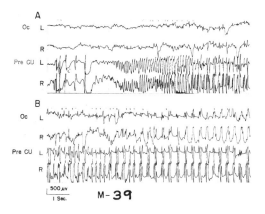

Figure 17-25. Regional variations in capacity for discharge and bilateral synchrony following administration of a threshold dosage of an intravenous convulsant agent (15mg/kg) in a monkey. Discharges begin in a bilateral synchronous manner in the precentral areas (Pre C U). Only later do discharges begin in the occipital area (Oc) and these are not synchronous or symmetrical. From Marcus, E. M. in Reeves, A. G. Epilepsy and the Corpus Callosum. New York Plenum Press p.169. 1985.

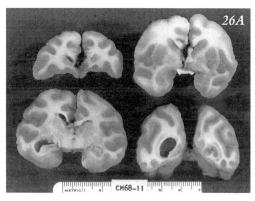

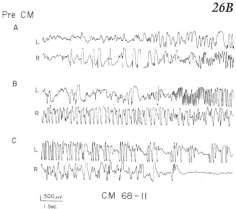

Figure 17-26: Intravenous pentylenetetrazol at threshold dosage 6 days after complete section of major commissures in the monkey. Bipolar recordings from upper –middle precentral gyrus. All bilateral synchrony is lost. A) Anatomy. B) EEG recording. From Marcus, E.M. in Reeves, A.G. Epilepsy and the Corpus Callosum New York Plenum Press. P183.

as seizures originating from the striate occipital cortex will have limited focal effects. The wide spread intra- and interhemispheric spread of discharge from a unilateral experimental focus in area 6 is demonstrated in Figure *17-22A.* Compare this to the limited spread of discharge from a similar focus in areas 17 of occipital cortex *(Figure 17-22 BC)*.

3) **The capacity for bilateral synchronous discharges:** *Anatomical and Physiological observations: see above*

Experimental epilepsy correlation:

A) Bilateral foci of epileptic discharge in symmetrical cortical areas of the two hemispheres: an interaction soon occurs, resulting in the synchronization of the seizure discharges in the two hemispheres. However in the monkey there are significant regional variations in the capacity for this interaction. In the premotor and precentral areas a close and well-developed interhemispheric synchrony is evident *(Fig. 17-23)*; in striate occipital and superior temporal areas, bilateral synchrony is poorly sustained *(Fig.17-24)*. Section of major commissures markedly disrupts the synchrony of discharge. Refer to Marcus (1985).

B) Intravenous injection of a threshold amount of pentylenetetrazol (Metrazol) provides an example of a diffuse toxic disease (Marcus- 1985): Following injection in the monkey, bilateral synchronous discharges developed first in the premotor and precentral areas. Discharges in the temporal and striate occipital areas develop later and are independent and multifocal *(Fig 17-25)*. Synchrony is lost following section of the corpus callosum (Fig 17-26), but is maintained in the cortical callosal preparation *(Fig. 17-27)*.

The role of the corpus callosum in other generalized models of epilepsy is discussed in chapter 28.

Clinical Correlation: Seizure discharges, as recorded in the human electroencephalogram, may be focal, multifocal, or generalized with bilateral symmetry and synchrony of discharge. When the latter category of bilateral discharges is examined it is evident that the bilateral synchrony is usually best developed in the frontal and central and parietal parasagittal recording areas. On the other hand, bilateral discharges in the temporal and occipital areas often appear to be independent, that is, multifocal. Section of the corpus callosum has been employed to limit spread of seizure discharge or to prevent the interaction of multiple foci of epileptic seizure discharge when seizures could not otherwise be controlled (See Reeves, 1985).

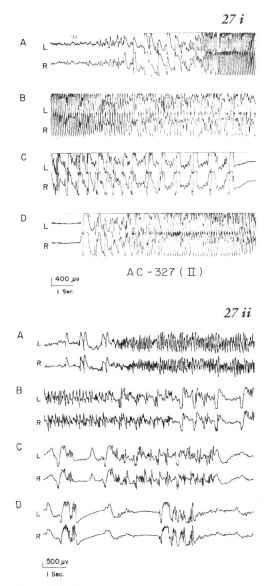

Figure 17-27. Intravenous pentylenetetrazol at threshold dosage (20 mg/Kg) I) Intact cat II) Bilateral cortical callosal isolation. The apparent amplitude differences between the intact and the isolate reflect differences in the degree of amplification.

CHAPTER 18
Motor System and Movement I: Reflex Activity, Central Pattern Generators and Cerebral Cortical Motor Functions

INTRODUCTION

In studying the motor system, we will consider reflex activity, central generators of patterns of movement, voluntary movement and learned movements .We will also consider two inter-related aspects: posture and movement. Under posture we will be studying static or tonic reactions. Under movement we will be studying short duration phasic reactions. We should keep in mind, as Sherrington has indicated, that the reflexes involved in posture and movement are the same. There are no reflexes exclusively for the maintenance of a correct posture, as opposed to those reflexes involved in a movement, We should note that with more detailed microelectrode studies we may find that some neurons in the cerebral cortex or spinal cord are predominantly involved in phasic activities and other predominantly in tonic activities. In the following discussion, we will examine motor function at the level of the spinal cord, the brain stem and cerebral cortex .It is important to realize that motor functions are represented at successively higher physiological and anatomical levels of the neural axis. As we go higher in the neural axis, we are utilizing and modifying mechanisms that have been integrated at a lower level of the neural axis a concept first expressed in the modern era by Jackson. Thus pattern generators make use of the motor mechanisms involved in reflexes without the necessity of afferent input. In turn voluntary and learned movements incorporate or impose a higher level of a more complex cortical control of these reflex and central pattern mechanisms.

In subsequent chapters the role of basal ganglia and the cerebellum in modulating movement will be considered.

REFLEX ACTIVITY.

The terms utilized in defining a reflex are presented in Table 18-1

CONCEPT OF CENTRAL PATTERN GENERATORS

Specific neural circuits or neuronal clusters or centers may be responsible for the generation of simple and complex patterns of movement in the absence of afferent input. In actuality, such central patterns are modified or modulated by afferent input, utilize many of the reflex components to be discussed below and are controlled and/or modified by descending control motor systems from higher centers. Such pattern generators have been described at the level of the spinal cord and the brain stem (mesencephalic locomotor system) for more complex behavior related to locomotion. The latter system is also involved in the motor components of emotional expression, as well as chewing, licking and sucking behavior. For the accurate performance of complex patterned movements, involving the limbs, afferent input and reflex activity, is essential. In the absence of such afferent input, the movements while repetitive are not as well coordinated. (See DeLong 1971, Harris-Warrick and Johnson 1989).

1. *The spinal pattern generator.* This is probably composed of bilateral clusters of interneurons in the intermediate gray matter at the base of the posterior horn. These interneurons are the same interneurons involved in the flexion reflex to be described below. In the spinal cord preparation to be discussed below the effects of dopaminergic and adrenergic agents in triggering behavior can be demonstrated.

2. *The mesencephalic locomotion pattern generator:* this center is located in the periventricular tegmentum at the junction of midbrain and pons. Axons from this region descend to the medullary medial reticular formation.

TABLE 18-1: DEFINITION OF A REFLEX

TERM	EXAMPLES
Adequate stimulus	Nociceptive, proprioceptive, tactile, visual
Synapses involved	Monosynaptic in the stretch reflex. Polysynaptic in the flexion reflex.
Segments involved	Segmental-stretch reflex, Intersegmental-inter limb reflexes. Suprasegmental- brain stem–tonic neck/labyrinthine. Cerebral cortex, placing and long loop reflexes.
Movement or response	Flexion, extension, righting, standing, walking, grasp, avoidance, placing
Aim or purpose	Avoidance of pain, escape from predator, acquisition of food

From this region axons descend as the medial reticulospinal tract in the ventrolateral funiculus to the spinal locomotor system of the lumbar spinal cord. In the decerebrate preparation to be discussed below, glutaminergic effects can be demonstrated. Glutamate receptor antagonists will prevent the locomotion effects that occur on stimulation of the mesencephalic center.

3. *Other motor pattern centers:* the premotor cortex for visually guided movements, the subthalamic nucleus, the pontine reticular formation and the cerebellum will be discussed below.

EFFECTS OF SPINAL AND BRAINSTEM LESIONS ON THE MOTOR SYSTEM

TRANSECTION OF THE SPINAL CORD IN THE HUMAN:

In humans transection of the spinal cord reflects (1) the effects of spinal shock and (2) the prepotency of flexion reflexes as reflex recovery occurs.

Transection of the spinal cord in the human is reviewed in Table 18-2 and in chapter 9

TRANSECTION OF THE BRAIN STEM- THE DECEREBRATE PREPARATION

The decerebrate preparation, involves a transection of the brain stem a level between the vestibular nuclei and the red nucleus usually at the intercollicular midbrain level.

In man the decerebrate state may reflect several pathological processes such as basilar artery thrombosis with brain stem infarction, temporal lobe herniation with midbrain compression, or massive destruction of both cerebral hemispheres.

Decerebrate Rigidity - A state described as decerebrate rigidity develops almost immediately. Decerebrate rigidity may be defined as the exaggerated posture of extension of the antigravity muscles due to the enhancement of proprioceptive stretch reflexes. In four-legged animals, such as the cat and dog, the enhancement of these stretch reflexes results in extensor posture in all four limbs with extension of the tail and arching of the back and neck. This posture is also referred to as opisthotonos *(Fig. 18-2)*.

Sherrington has described the posture of decerebrate rigidity as "an exaggerated caricature of standing". This is most intense in those muscles that normally counteract the effect of gravity. In an animal such as the sloth that normally hangs upside down, it is the flexor posture that is exaggerated.

It is important to realize that with transection of the brain stem, the decerebrate rigidity develops almost immediately. Spinal shock does not result. It is important to note that though the extensor tone is sufficient to allow the animal to stand, the animal is pillar--like. The pillar-like limb seen in this extensor posture is indicative of the positive supporting reaction already seen in the "spinal" cat and dog. This is the tonic form of the extensor thrust reaction previously noted. The animal lacks righting reflexes and has no reactions to sudden displacement.

Modification of Decerebrate Rigidity - The posture of decerebrate rigidity may be modified by several influences: tonic neck reflexes, tonic labyrinthine reflexes noxious stimuli, and time .

Tonic Neck Reflexes (Fig.18-3). Tonic neck reflexes are studied best after destruction of the

TABLE 18-2: COMPLETE TRANSECTION OF THE SPINAL CORD IN MAN

ONSET	EVENT	MANIFESTATIONS
Immediate	Spinal shock	Below lesion: flaccid paralysis of limbs, no DTR's, Atonic bladder and bowel (overflow incontinence), Loss of autonomic control: hypotensive if erect[1]
	Sensory loss	Absence of all sensation below lesion No spontaneous
	If section C1-C4	respiration (phrenic outflow -C4)
1-6 weeks	Recovery of flexion reflexes Note that flexion reflexes usually remain prepotent	Sign of Babinski, initially dorsiflexion of great toe and fanning of small toes, later flexor withdrawal of foot, leg and thigh
3-4 weeks	Onset of detrusor & bowel function	Reflex voiding and defecation begins to occur
	Minimal return of DTR's	Achilles and patellar DTR's now present
2-3 months	Mass reflex	Flexion reflexes now exaggerated without local sign with flexor spasms, bladder emptying and profuse sweating, erection & ejaculation
	Autonomic reflexes	Sweating below level of lesion returns
6 +months	Local sign develops	Mass reflex less prominent, flexion reflexes more specific for point of stimulus (may recur with fever)
	Hyperactive extensor reflexes	Deep tendon reflexes become hyperactive, spasticity develops. Crossed extension (Fig 18-1) and alternate stepping may occur. Occasionally positive support reaction is sufficient for spinal standing

[1] Other aspects of autonomic control are also affected including sympathetic control of blood pressure - noted particularly with hypotension occurring in the upright position (orthostatic) and parasympathetic control of intestinal -peristalsis (ileus occurs).

labyrinth or after bilateral section of nerve VIII. These procedures eliminate various labyrinthine influences. The afferents for the tonic neck reflexes are conveyed via the upper cervical dorsal roots from joint receptors - for example, those located in the atlanto-occipital joint. The tonic neck reflexes produce several types of modification of the decerebrate posture. Rotation of the head produces a fencer's posture with extension of the forelimb on the side to which chin, nose, and eyes have rotated with flexion of the forelimb on the contralateral side. Correlated inter limb adjustments may occur in the lower extremities. Extension of the head at the neck produces extension of the forelimbs and flexion of the hindlimbs. One may imagine the posture of a cat that is attempting to look at objects upon a table. Flexion of the head onto the chest produces flexion of the forelimbs and extension of the hindlimbs. One may imagine the posture of a cat that is attempting to look under a shelf or door.

Tonic Labyrinthine Reflexes (Fig. 18-4). These tonic labyrinthine reflexes are studied best after the head has been immobilized in a plaster cast or after section of the upper cervical dorsal roots. The afferents are conveyed from the otoliths via the vestibular nerves to the vestibular nuclei.

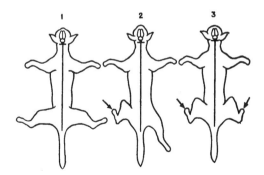

Figure 18-1. High spinal cat: flexion and crossed extension reflexes. (1) The initial posture of the animal. (2) Nociceptive stimulation of the left hind foot produces flexion of the left hind limb and crossed extension of the right hind limb. (3) Nociceptive stimulation of both right and left hind limbs leads to flexion of both hind limbs. Flexion reflex then remains prepotent. (From Sherrington, C.: The Integrative Action of the Nervous System. New Haven, Yale University Press, 1947, p.226).

Figure 18-2 The posture of the decerebrate cat when suspended. The hyperextended, rigid posture of neck, back, limbs, and tail is to be noted (opisthotonos). From Pollack, L.J., and Davis, L.: J.Comp.Neurol., 50:384,1930 (Wiley).

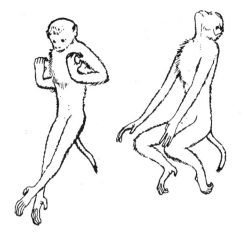

Noxious stimulus to a limb produces the classic inter limb reflex figure *(18-4)*.

Time: With time (7-14 days), rigidity decreases and righting reflexes begin to emerge.

The anatomical basis of decerebrate rigidity: Refer to chapter 11.

The Midbrain Preparation (transected above the superior colliculus with red nucleus and portions of subthalamic nucleus and posterior hypothalamus intact). In this preparation, propriocentive reflexes are modified by contactual stimuli.

In the dog or cat midbrain preparation, a sequence of righting reflexes occurs. Righting responses in the animal with intact central nervous system are demonstarted in. *Fig. 18-6, 18-7.* Kinetic reactions such as standing and walking may also occur. In contrast, the primate midbrain preparation is unable to stand. There are some fragments of and of a traction grasp

Role of hypothalamus: The motor responses for rage are represented at a brain stem level. Stimulation of the midbrain of the decerebrate preparation will produce all of the motor components of anger. The hypothalamus however is essential for occurrenses of the sham rage of the decorticate animal. (Bard).

THE DECORTICATE PREPARATION

The cat and dog preparations are similar to the midbrain animal. In the primate, the tonic grasp reflex may be present. The decorticate

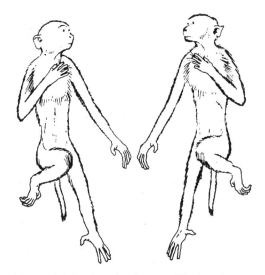

Figure 18-3. Tonic neck reflexes modify decerebrate rigidity (monkey preparation). Upper figures: flexion of head results in flexion of upper extremities; extension of head produces extension of upper extremities. Lower figures - rotation of head produces extension of upper limb on side to which face is turned and flexion of contralateral upper limb. Correlated adjustments occur in the lower extremities. (From Twitchell, T.E.: J.Amer.Phys.Ther.Ass., 45:413,1965).

human or primate manifests a considerable degree of spasticity often referred to as *decorticate rigidity*. The posture of the decorticate primate or human involves extension of the lower extremities and flexion of the upper extremities *(Fig.18-8)*. In a sense this is a double hemiplegic posture.

MOTOR SYSTEM AND MOVEMENT I

Figure 18-4. Tonic labyrinthine reflexes modify decerebrate rigidity: Upper Figure: With the animal held in the supine position and the head facing upwards, the limbs extend. Lower Figure - With the animal held in the prone position (face downwards) the limbs flex. The tonic neck influences must be eliminated by immobilization of the neck or by section of the upper cervical dorsal roots. Note that in man with neurological disease, the opposite pattern may occur so that a flexion posture predominates when the patient is supine and an extensor posture when prone. (From Twitchell, T.E.: J.Amer.Phys.Ther.Ass. 45:414, 1965).

Figure 18-5. Reflex figures in the decerebrate cat preparation. Effects of nociceptive stimulation. a. Prior to stimulation. b. Change produced by stimulation of left fore foot. c. Change produced by stimulation of left hind foot. (From Sherrington, C.S.: The Integrative Action of the Nervous System. New Haven, Yale University Press, 1947, p.167).

REACTIONS DEPENDENT ON CEREBRAL CORTEX

The cerebral cortex may be considered as providing for a complex reaction to the external environment. The cerebral cortex analyzes afferent information from many sources and utilizes reactions which have been integrated at lower levels of the neural axis. Thus, the presence of the cerebral cortex allows for the accurate projection of the limb in space and for the interaction of various reflex activities with visual and tactile stimuli.

The following reflexes are associated with the cerebral cortex:

1. The *optical righting reflex. (Fig. 18-7)*. This results in the righting of the head in relation to a visual stimulus. This response continues to occur after elimination of the labyrinths and dorsal roots from the upper cervical neck proprioceptors. The response can still be demonstrated after bilateral ablation of the motor and pre-motor areas in the primate. We

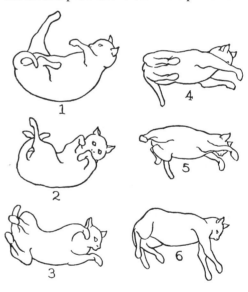

Figure 18-6. Righting reflexes in intact cat. Diagram showing series of maneuvers which a cat executes in order to turn itself in the air: (1) free fall; (2) head turns, forelimbs drawn in, hindlimbs extended; (3), (4), (5), continued turning with gradual extension of forelimbs while hindlimbs are drawn closer to axis of rotation; (6) turning completed. (Redrawn from Marey, C.R.: Acad.Sci.Paris, 1894, 119:714-717: From Fulton, J.F.: A Textbook of Physiology, 17th Edition. Phila-delphia, W.B. Saunders, 1955, p.221).

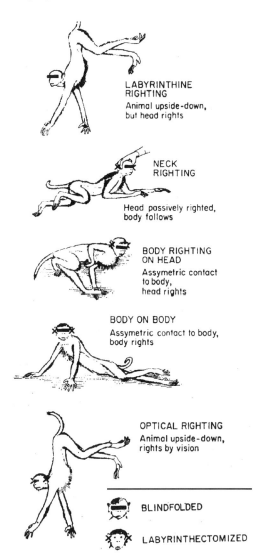

Figure 18-7. Righting reflexes in the monkey. With the exception of the optical righting reflex (which depends on cerebral cortex), all of the righting reflexes demonstrated above are within the capacity of the midbrain preparation. (From Twitchell, T.E.: J.Amer.Phys.Ther.Ass. 45:415, 1965).

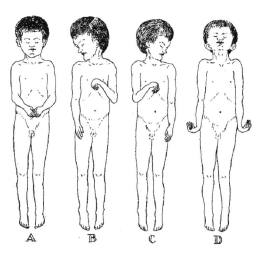

Figure 18-8 Decorticate rigidity in the human. The differences in man between (A) decorticate rigidity in neutral position (both arms are flexed) and (D) true decerebrate rigidity (legs are extended arms are rigidly extended and pronated.). (B) and (C), alterations produced in the decorticate human by head turning due to tonic neck reflexes. (From Fulton, J.F., A Textbook of Physiology, Philadelphia, W.B. Saunders, 1955, p.217).

should, of course, note that righting of the head when the head has been turned on to the side, is dependent not only on visual cues but also on labyrinthine cues and asymmetrical body contact stimuli.

2. *Placing Reactions.* A number of placing reactions are also noted in the intact animal. When the animal approaches a visible edge, the forelimb is advanced to be precisely placed on the surface. This is termed the visual placing reaction. On the other hand, if the animal is blindfolded and the dorsum of the foot or hand touches the edge of a surface, appropriate adjustments are then made to place the plantar or palmar surface of the extremity on the surface. This is termed the tactile placing reaction. Also noted in the intact preparation is a hopping response. If the animal's body is displaced to one side, the animal abducts the leg to that side to regain stability.

3. The *grasp reflex*, as defined by Seyffarth and Denny-Brown, consists of a stereotyped progressive forced closure of the subjects hand on the slowly moving stimulus of the examiner's fingers across the palm. These authors considered the tactile pressure to be the appropriate stimulus. This reflex had emerged following frontal lobe ablation in the monkey (Rushworth and Denny-Brown). Local anesthesia infiltration of the palm abolished the reflex.

4. *Instinctive Grasp.* A more complex and variable response was also described by Seyffarth and Denny-Brown: the instinctive

tactile grasp reaction *(Fig. 18-9)*[2]. This response is also triggered by a tactile (or contactual) stimulus to the hand. The hand orients so as to grasp the object. Often the hand follows a moving object making moment-to-moment adjustments so as to grasp onto the object. This instinctive tactile grasping reaction allows for the exploration of space with the hand. The grasp reflex and instinctive tactile grasping reaction are of course, seen normally when the intact individual attempts to grasp onto an object or to reach for an object. It is not surprising that intact motor and parietal cortex is required. The movements involve distal finger and thumb movement and require precise feedback information.

At times, however, the reaction is seen to be released in abnormal form and degree and to occur in a context where voluntary grasping is inappropriate. Patients with release of the tactile grasp reaction usually have damage to the premotor cortex (areas 6 and 8) and the adjacent supplementary motor cortex and cingulate area on the medial surface of the hemisphere. The instinctive tactile grasp reaction, then, although requiring an intact motor and parietal cortex is released in an abnormal form by lesions of the premotor, supplementary motor, and cingulate areas. This is a form of transcortical release. The same lesion, of course, also acts to release various subcortical areas[3]

The instinctive tactile grasp reaction is to be compared to the instinctive tactile avoiding reaction (Fig. 18-10). The stimulus for this reaction is a distally moving tactile stimulus to the outer (ulnar) border of the hand or to the palmar

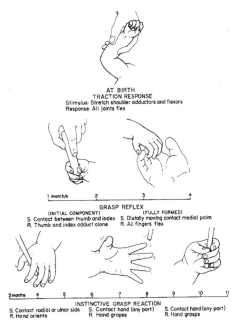

Figure 18-9. Evolution of the automatic grasping responses of infants. From Twitchell, T.E., Neuropsychologia, 3: 251, 1965 (Elsevier).

surface of the hand or to the dorsum of the hand. This stimulus leads to a non-tonic orientation of the hand away from the stimulus. This is a more precise form of the tonic avoiding response noted in a decorticate preparation. The instinctive tactile avoiding response does not require the pyramidal system. The integrity of the cingulate area, area 8, and the supple-

---

[2]*Seyffarth and Denny-Brown (1948) also described a group of behavioral responses to the same stimulus, which are more variable and appear to be more subject to factors of attention, distraction, etc.: These were grouped together as the* **instinctive grasp reaction:** *(a) the closing reaction, (b) the final grip, (c) the trap reaction, (d) the magnet reaction (the hand moves to follow the object), (e) instinctive groping (if the stimulus moves without capture, the hand pursues).*

[3]*A recent clinical review of the grasp reflex, in which CT scans were used to locate lesions (DeRenzi and Barbieri), indicates the strongest association with damage to medial frontal areas: predominantly, anterior cingulate gyrus and supplementary motor cortex and a lesser correlation with damage to the lateral premotor areas. Usually, the release was bilateral, unless the damage to adjacent motor cortex had produced a severe paresis of the contralateral hand. In such cases, only the ipsilateral hand could be tested. Considerable research data reviewed by DeRenzi and Barbieri suggests that the anterior cingulate area not only interconnects with the supplementary motor cortex but also has a similar pattern of connection to basal ganglia, cortical areas and spinal cord.*

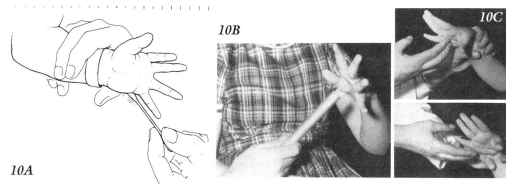

Figure 18-10. Avoiding responses. A. Elicitation of the avoiding response by contact stimulus to the ulnar border of hand as a normal response in infant of 4 to 6 weeks. From Twitchell, T.E., Neuropsychologia, 3:250, 1965 (Elsevier). B. Avoiding response as a pathological phenomenon. Over-extension and abduction of fingers as hand approaches object. (child with infantile spastic hemiparesis.) C. Avoiding response as a pathological phenomenon in another patient with infantile spastic hemiparesis. Reaction of hand to contact stimulation of the palm. From Twitchell, T.E.: Neurology, 8:17, 1958 (Lippincott/Williams and Wilkins).

mentary motor cortex are required. The reaction is released in abnormal degree by damage to the parietal lobe (and adjacent motor cortex). The instinctive grasp and avoidance responses normally are in equilibrium from an anatomical and physiological standpoint.

5. *Other Reflexes Associated with the Cerebral Cortex.* In addition to these tactile instinctive grasp and instinctive avoiding responses, various visual triggered responses of a similar nature may be seen. One may speak of a visual instinctive grasp reaction. This requires the integrity of the posterior parietal areas, the visual mechanisms, and the motor cortex. As discussed below the integrity of the lateral premotor cortex is undoubtedly required for visually triggered reaching and grasping. Damage to the temporal lobe releases this reaction in an abnormal form. The response does occur, of course, in a normal form as one normally reaches for an object, using visual stimuli. This type of phenomenon occurs in the Kluver-Bucy syndrome, which follows the production of bilateral lesions of the temporal lobe in the monkey. The release of various visual automatisms occurs. The animals tend to pick up all objects that appear in their field of vision and to bring these objects to their mouth. (See limbic system chapter for additional discussion). Magnani et al, studied the anatomical correlation of release of visually triggered grasping and groping response in man. In contrast to the experimental studies, the CT scan evidence suggested a medial frontal location of lesion responsible for such release.

Another variety of visually cued reaction is the visual instinctive avoiding reaction. This requires the integrity of the temporal lobe and visual mechanisms. The response apparently is independent of the pyramidal system. Again, in the intact individual, equilibrium of grasp and avoidance is to be noted.

While our discussion has focused on the upper extremity, similar grasp and avoiding reactions may also be noted in the lower extremities. Lesions of premotor cortex also affect gait: producing an apraxia in walking termed a frontal lobe gait apraxia. This will be discussed later in this chapter.

POSTNATAL DEVELOPMENT OF MOTOR REFLEXES

Some reflexes are present at birth and disappear with time: the startle reflex, sucking, rooting and the extensor plantar response. A sequence of reflex activities is noted in the developing human infant as regards the use of hands and fingers .In addition a sequence of changes occurs as regards sitting, standing and walking as summarized in table 18-2.

OVERVIEW OF THE RELATIONSHIP OF PRIMARY MOTOR, PREMOTOR AND PREFRONTAL CORTEX.

In order to understand voluntary and

learned behavior, it is necessary to explore the relationship of primary motor, premotor (including supplementary motor) and the prefrontal areas *(Fig.18-11)*. Voluntary and learned behaviors also reflect the motivational role of limbic structures to be considered in subsequent chapters.

The large and giant pyramidal cells in the *primary motor cortex* discharge producing movements in specific directions (flexion, extension, abduction, adduction) resulting from the contraction of muscles at a specific joint. Among other sources of input, the primary motor cortex receives information from somatosensory cortex of the postcentral gyrus and the ventral lateral nucleus of the thalamus. This input may allow the response of motor cortex to peripheral stimuli in what are termed long loop reflexes.

In contrast, *the premotor cortex* is concerned with patterns of movements that are motor programs Neurons in the premotor cortex discharge during the preparation phase prior to movement. The premotor cortex projects in parallel pathways to motor cortex, to basal ganglia and to spinal cord. The premotor cortex receives projections among other sources from the unimodal sensory association cortex (somatosensory auditory, and visual) as well as the primary somatosensory cortex and the ven-

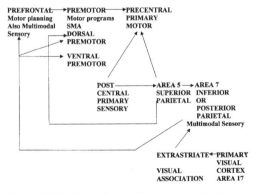

Figure 18-11. Interrelation of motor, premotor, prefrontal and posterior parietal multimodal sensory areas. The input of the somatosensory and visual systems to the multimodal posterior parietal area are presented. The input of the auditory system and the relationship of the multimodal sensory system to the limbic system are not demonstrated. In a general sense the basis for reaching and grasp are demonstrated.

TABLE 18-2: DEVELOPMENTAL CHANGES IN MOTOR RESPONSES[4]

RESPONSE	POSNATAL TIME
Sucking and rooting reflexes	Present at birth.
The Moro (startle) reflex	Present at birth and persisting until 2 months
Tonic neck reflexes	May be present at birth, fragments at 1-2 months
Traction response grasp elicited by traction at shoulder	May be present at birth
Grasp reflex: all fingers flex briefly as unit	Birth - 8 weeks
Instinctive grasp reaction	Begins at 4 - 5 months.
Projected grasp (pincer). Then thumb-finger opposition. Subsequently this is voluntary	Begins at 6 to 8 months
Plantar grasp response	Present at birth disappears at 9-12 months
Extensor plantar response on lateral plantar stimulation (due to immature corticospinal pathway)	Present at birth, disappears at 9-12 months
Plantar flexion on lateral plantar stimulus (dorsiflexion)	Present from 9-12 months and normal through life
Sitting	Infants usually sit unsupported at 6 to 7 months,
Standing with support / Pulls self to standing position holding onto object	-At 8 to 9 months / -At 10-12 months
Walking	Begins to walk with support at 10 to 12 months. Begins to walk independently at 9 to 16 months

[4]We have not considered here the prenatal motor development of the human fetus. Wolff and Ferber (1979) review many of the earlier studies (which demonstrated considerable capacity for swimming movements) and discuss the use of high-resolution ultrasound to study in a non-invasive manner movement's intra-utero. Such studies allow more precise establishment of actual fetal maturation. The suck reflex is actually present in the prenatal state.

tral lateral nucleus of the thalamus.

The *prefrontal areas* have functions that relate to the limbic system and the motor system. While the orbital and mesial regions of the prefrontal cortex are related to the limbic system and will be considered in a later chapter, the dorsal region of the prefrontal cortex is concerned with the executive functions involved in the planning, regulation and sequencing of behavior and movement over time. In this regard movement is simply the expression of behavior. This same area of prefrontal cortex functions as the anterior multimodal association cortex and receives projection from the posterior multimodal association cortex located at the intersection of the major sensory association areas. In turn this dorsal prefrontal area projects to the premotor cortex.

PRIMARY MOTOR CORTEX (Area 4)

Area 4 is distinguished by the presence of giant pyramidal cells in layer V. Occasionally giant pyramidal cells may also be found more anteriorly in area 6. In humans, much of area 4, particularly its lower half, is found on the anterior wall of the central sulcus. More superiorly, area 4 appears on the lateral precentral gyrus and continues onto the medial aspect of the hemisphere in the anterior half of the paracentral lobule.

It is important to note that the pyramidal tract is not named because of its cells of origin but rather on the basis of its passage through the medullary pyramid. Of the approximately one million fibers in the pyramidal tract, 31 per cent arise from area 4; 29 per cent arise from the premotor cortex of area 6; and 40 per cent arise from areas 3, 1, 2, 5, and 7 of the parietal lobe. Fibers with a large diameter (9 to 22 u) number about 30,000 and correspond roughly to the number of giant pyramidal cells. Most of these large fibers originate from area 4. It is also important to note that the same cortical areas that give rise to the pyramidal tract also give rise to other descending motor pathways: corticoreticular, and corticorubral, (as well as corticostriate and corticothalamic). From a phylogenetic standpoint this is a logical arrangement since this relatively new control center, the motor cortex, has available a relatively new fast conducting pathway to the anterior horn cells. At the same time the motor cortex also has available pathways to certain older brain stem centers, which formerly exercised primary control, e.g., the reticular formation and the red nucleus. Note that 70-90% of the fibers passing through the medullary pyramid will decussate and continue as the lateral corticospinal tract. These fibers originate primarily from those portions of area 4 representing the distal extremities and synapse directly on alpha motor neurons in the lateral sector of the ventral horn.

In contrast the remaining uncrossed fibers arise from area 6 and the portions of area 4 representing the trunk and neck areas. These uncrossed fibers as they descend through the brain stem, give off collaterals to the bilateral medial brain stem areas of the reticular formation. These uncrossed fibers continue as the anterior corticospinal tract and terminate bilaterally on interneurons, which in turn will synapse on neurons in the ventral medial cell column of motor neurons.

The fibers originating in the parietal cortex synapse in relationship to interneurons that in turn synapse in relationship to neurons of the dorsal horn.

The premotor sector of area 6 projects primarily to the medial pontomedullary areas. The spinal projections of this area consist of the ventromedial descending reticulospinal system and the vestibulospinal tracts.

The red nucleus receives inputs from the rostral (proximal) arm and leg sectors of area 4 and the adjacent premotor cortex and projects as the rubral-spinal system.

Area 4: Stimulation. Electrical stimulation of the motor cortex at threshold level produces isolated movements of the contralateral portions of the body. The movements may be described as a twitch or small jerk. With increased strength or increased frequency of stimulation, repetitive jerks occur; the movements involve a larger part of the contralateral side of the body and an orderly sequence of spread may occur. The repetitive jerks are similar to the movements of a focal motor seizure. The orderly sequence of spread is duplicated in the "Jacksonian March" of a focal motor seizure.

In the conscious patient, these movements,

which can be evoked by stimulation of the motor cortex during surgery, are not clearly the same as consciously willed voluntary movements. According to Penfield, the patient recognizes that these are not voluntary movements of his own and will often attribute them to the actions of the neurosurgeon.

There was considerable controversy as to whether muscles or movements were represented in the motor cortex. Single-unit recording studies have suggested a close topographic grouping of all the neurons for a given muscle. Occasionally stimulation will evoke the movement of a single muscle, usually a distal finger muscle, such as an interosseous. In general, however, stimulation produces contraction of a group of muscles concerned with a specific movement. Voluntary actions also involve the integrated contraction of several muscles rather than the isolated contraction of a single muscle. It should then be apparent that the motor cortex "thinks" in terms of movements. This is, in a sense, inherent in the intracortical connections between the neurons representing the various muscles related to a given movement.

Although there is considerable overlap and plasticity in the representation of movements on the surface of the motor cortex, a certain sequence is apparent *(Fig.18-12)*. In the classical homunculus, the representation of the pharyngeal and tongue movements is located close to the sylvian fissure. Unilateral stimulation of these areas results in bilateral movements. Next in upward sequence are movements of the lips, face, brow, and neck. In the middle third of the precentral gyrus, a large area is devoted to the representation of the contralateral thumb, fingers, and hand. In the upper third, the sequence of the contralateral wrist, elbow, shoulder, trunk, and hip is found. The remainder of the lower extremity in the sequence (knee, ankle, and toes) is represented on the medial surface of the hemisphere in the paracentral lobule. The representation of the anal and vesicle sphincters has been located lowest on the paracentral lobule. It is important to note how large the areas devoted to the thumb, fingers, and face (particularly the lips and tongue) are in the human, reflecting the

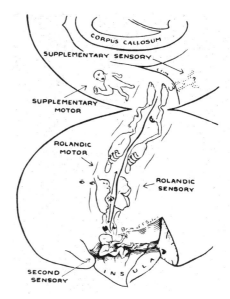

Figure 18-12. Somatic figurines. The primary motor and sensory representation and the effects produced by supplementary motor and second sensory stimulation are demonstrated. (From Penfield, W. and Jasper, H. Epilepsy and the Functional Anatomy of the Human Brain. Boston, Little Brown and Company 1954,p105).

evolution in use of these structures as regards speech, writing, tool making and manipulation.

It is also important to note that not all areas of the motor cortex have the same threshold for discharge. As studied in the baboon (Phillips, 1966) the thumb and index finger are the areas of lowest threshold. Slightly stronger shocks are required to produce a movement on stimulation of the great toe area; at slightly greater intensities of shock, movements of the face begin to occur on stimulation of the appropriate area. These findings may relate in part to the fact that those pyramidal fibers making direct connection to anterior horn cells are primarily those mediating cortical control of the distal finger and thumb muscles.

It must also be noted that at times on stimulation of the postcentral gyrus discrete movements will be produced similar to those elicited from the motor cortex. There is some evidence that such discrete movements are produced at a higher threshold than in the precentral cortex and that the ability to obtain such discrete movements is lost after ablation of the primary motor cortex. However, not all of this neurosurgical data on stimulation of postcentral gyrus in the human is consistent.

Modern concepts of the plasticity of the primary motor cortex: More recent studies reviewed by Sanes and Donoghue (2000) indicate a considerable dynamic plasticity in the primate primary motor cortex. Thus although modules for the head region, upper extremity and lower extremity can be distinguished, there is within each module, a considerable overlap and plasticity with less of the specific localization for each movement suggested by the earlier maps. Instead a mosaic is present. The area devoted to specific movement representations may be modified by trauma to peripheral nerves, central nervous system damage, and most importantly by motor skill learning and cognitive motor actions.

Area 4: Ablation. The effects of ablation of the motor cortex must be distinguished from the more selective effects of limited section of the pyramidal tract in the cerebral peduncle or medullary pyramid. Damage to or ablation of the motor cortex affects not only the pyramidal tracts, but also the corticorubral, corticostriate, corticothalamic and corticoreticular pathways.

Studies of the recovery of function following motor cortex lesions in the monkey) are most useful in illustrating many aspects of the integration of reflex activity at successive levels of the neural axis and are reviewed on the CD ROM in table 18-3.

Studies of recovery of function in the human: The sequence of events in man following partial vascular lesions (infarcts) of the precentral cortex or of the motor pathways in the internal capsule has been studied by Twitchell (1951).

1. *Flaccid Paralysis.* Particularly with a large lesion there is a transient period of flaccid paralysis of the contralateral arm and leg and the lower half of the face, with a transient depression of stretch reflexes. The duration of this flaccid state (reflecting "cortical shock") is variable; it may be quite fleeting depending on the extent of the lesion (which is usually less than total). The sign of Babinski is present with an extensor plantar response on tactile or painful stimulation of the lateral border of the foot.

2. *Return of Deep Tendon Reflexes/Stretch Reflexes.* After a variable period of time, the deep tendon reflexes in the affected limbs return and become hyperactive. There is increased resistance on passive motion at the joints (when there are severe lesions even this stage may not be reached). This does not affect all movements equally but tends to have its greatest effect on the flexors of the upper extremity. *A hemiplegic posture and gait is noted:* the upper extremity is flexed at the elbow, wrist, and fingers and the leg is extended at the hip, knee, and ankle and externally rotated at the hip with circumduction in walking. Walking is possible with the hemiplegic limb as movement of the proximal limb begins to return. In general, recovery of function in the lower extremity is usually greater than in the upper extremity. Following the return of stretch reflexes, variation in the relative degree of spasticity in the flexor and extensor muscles in relation to tonic neck reflexes may be noted. With the return of tonic neck reflexes one will begin to note proximal limb movements.

3. *Return of Traction Response-* Abduction of the arm and shoulder will result in stretch of the abductors. There will then be an adduction of the arm at the shoulder. In addition, synergistic flexion of the arm at the elbow, wrist, and fingers will occur, and objects may be grasped in this manner. This flexion synergy, however, results in a whole hand grasp; all of the fingers flex together.

4. *Return of Selective Ability to Grasp.* This stage of recovery involves a more selective grasp of the entire hand albeit a tonic grasp. There is some dissociated synergy so that the hand alone tends to grasp. Finger and thumb opposition begin to return.

5. *Return of the instinctive tactile grasp reaction with the capacity for projected movement.* This occurs with incomplete lesions. If a massive lesion is present, the instinctive grasp reflex does not return.

6. *Return of distal hand movement.* Eventually, depending on the extent of the lesion, some return of distal hand movement

may occur but such recovery is usually less than that which occurs for movements at more proximal joints, e.g., at the shoulder and hip.

From a physiological and anatomical standpoint, the logic of this pattern of recovery should be evident. One sees in this sequence a recovery of spinal reflex activities, then of brain stem reflex activities, as noted in the decerebrate preparation, then those noted in the decorticate preparation, and finally those reflexes integrated at a cortical level. (This analysis of recovery of function is to some extent an oversimplification: A small hemispheric lesion of the internal capsule in the adult may produce as great a deficit as a hemispherectomy performed on a hemisphere damaged extensively early in life; Freund 1991).

More recent studies by Freund (1987) of lesions limited to the precentral gyrus in the human indicate that the essential deficit consists of deficits in force generation and execution of independent finger movements. After partial recovery, force is regained and finger movements, although clumsy and slow, are purposive and appropriate to the object. It should be noted that in man, following massive lesions of the motor cortex, a long delay may occur before spasticity develops and deep tendon reflexes return. Such a long delay in the reappearance of stretch reflexes usually signifies a poor prognosis for complete recovery.[5]

The recent study of Warabi et al, 1990 utilized the CT scan to predict recovery of voluntary movement based on the degree of shrinkage of the cerebral peduncle. When the peduncle had shrunk to less than 60 percent of normal, recovery of distal limb function was incomplete. When cerebral peduncle was greater than 60 percent of normal, recovery of distal movement occurred.

Additional discussion regarding the possible reorganization of sensory and motor function at a cortical level that follows damage to these systems is provided in the review of Kaas, 1991 and the study of Jacobs and Donoghue (1991). Damage may unmask a considerable degree of latent cortical plasticity as discussed above. The intrinsic horizontal fibers within the primary motor cortex interconnect large areas and could provide the substrate for this reorganization. Some of the changes occur within hours suggesting some degree of prewiring. In the studies of Nudo et al (1996), experimental destruction of part of the hand area in the monkey resulted in loss of discrete finger movements. The monkey relied on proximal arm movements and the area of representation of the shoulder and elbow expanded into the otherwise undamaged hand and forearm areas. However if the lesioned animals were trained to use the hand, the remaining undamaged area of cortex representing the hand and digits expanded into the area of representation of elbow and shoulder. The trained animals regained the ability to use discrete finger movements in retrieving small objects in 3-4 weeks. The untrained animals did not achieve this level of recovery. These studies certainly indicate the potential role of rehabilitation (occupational and physical therapy) in helping patients recover from the effects of cerebrovascular accidents, trauma and other pathology involving the motor cortex. PET scan studies indicate increased functional activity in other areas of the same and contralateral hemisphere in patients who have recovered from a hemiparesis due to an infarct involving the internal capsule. This was noted in the posterior cingulate area on the side of the lesion and the premotor and caudate nucleus of the opposite intact hemisphere. During movement there was increased bilateral activation of anterior insular cortex, ventral premotor, anterior cingulate, and inferior parietal areas (39 and 40), as well as premotor cortex of the intact hemisphere (Weiller et al, 1992).

The following case history illustrates the effects of focal disease involving the motor cortex.

Case 18-1: Three months prior to admis-

---

[5]*The studies of Patano et al (1995) utilizing CT/MRI and SPECT scan have demonstrated that patients with a prolonged flaccid hemiparesis as opposed to a spastic hemiparesis had a greater structural involvement of the lentiform nucleus and lower perfusion in the lentiform nucleus, thalamus and cerebellum as well as cortical motor areas.*

sion, this 29-year-old, right-handed, white female manager developed an intermittent "pulling sensation" over the anterior aspect of the left thigh, lasting a few minutes which recurred about twice a week. On the evening of admission she again had the onset of this same sensation, then had numbness followed by "jerky movements" in her left foot which progressed to involve the whole left lower extremity. She was then observed to have tonic extension then clonus of all four limbs with loss of consciousness for a minute or less. She still appeared dazed and slightly confused for the next 5 minutes, and was subsequently amnestic for the secondarily generalized seizure

Physical Examination: Normal except for bite marks on the left lateral aspect of the tongue.

Neurologic Examination: Significant only for a slight asymmetry of deep tendon reflexes, left > right.

Clinical Diagnosis: Focal seizures originating right parasagittal motor cortex (foot area), with question of parasagittal meningioma.

Laboratory Data: *Electroencephalogram* (Fig.29-1B) demonstrated rare focal spikes, right Rolandic parasagittal consistent with the focal source of the secondarily generalized seizure.

CT scan (Fig.18-13) demonstrated an enhancing lesion in the right parasagittal region just anterior to the central sulcus.

Arteriograms (Fig.18-14) were consistent with a right parasagittal meningioma.

Subsequent Course: The patient did well after removal of a well-circumscribed right parasagittal meningioma by Doctor Bernard Stone. The patient received diphenylhydantoin (Phenytoin) to prevent additional seizures.

SELECTIVE LESIONS OF THE PYRAMIDAL TRACT

We have previously indicated that the student should consider as the primary manifestations of an upper motor neuron lesion a spastic paralysis with increased deep tendon reflexes and the sign of Babinski. A more detailed analysis indicates that the effects of selective pyramidal tract damage are not quite the same

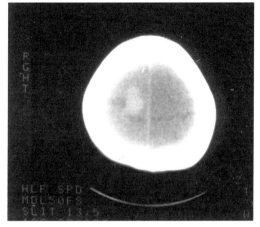

Fig. 18-13. Parasagittal meningioma. Case 18-1. CT scan demonstrates an enhancing lesion in the most rostral sections of the CT scan, just anterior to the right rolandic sulcus. See text for details.

or equivalent to those, which follow, in a general sense, upper motor neuron lesions. We should note that the ablation of the motor cortex destroys not only the origin of the pyramidal tract (the corticospinal tract) but also the origin of the cortico-reticular, the corticorubral, the corticothalamic, and the corticostriatal fibers. The effects of such a cortical ablation are to release these various subcortical and brain stem centers from cortical control and also to produce the effects of damage to the corticospinal tract. We should note that section of the pyramidal tract in the medullary pyramid or in the cerebral peduncle does not have the same anatomical and physiological effects as ablation of the motor cortex, which has been previously summarized

Furthermore selective section of the pyramidal tract or cerebral peduncle may yield different effects than section of the lateral column in the spinal cord. The descending rubrospinal and the lateral reticulospinal tract (from the bulbar inhibitory center of the medulla) are closely related to the corticospinal tract in the lateral column. In addition the medial reticulospinal tract from the pontine extensor facilitatory center descends in the anterior column and thus would be left undamaged by a lesion of the lateral column.

The studies of Bucy suggest that both man and the monkey were capable of useful move-

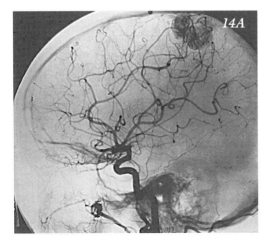
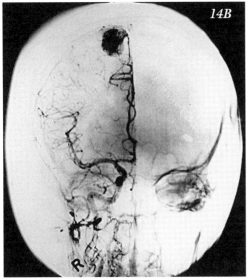
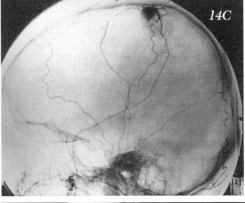
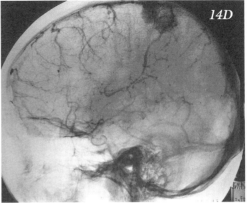

Fig. 18-14. Parasagittal meningioma. Case 18-1. Arteriograms demonstrate the characteristic features of a circumscribed vascular tumor in the parasagittal rolandic area, supplied by the right anterior cerebral artery and meningeal branches of the external carotid artery. A. Arterial phase lateral common carotid injection. B. Arterial phase AP view common carotid injection. C. Selective injection of the external carotid artery. D. Common carotid injection: venous phase - tumor blush with large draining vein. See text for details.

ments after section of the pyramidal tract in the cerebral peduncle. Thus, locomotor abilities returned. The monkey could use his hand in feeding and in manipulating small objects. With bilateral section, however, the hand was used as a claw with some deficiency in independent finger movements. It should be stressed that spasticity and increased deep tendon reflexes were not noted. The sign of Babinski was found in the human patient. In the human patient, the retention of discrete finger movements was remarkable. Somewhat different results were obtained by Denny-Brown (1966) following section of the pyramidal tract in the medul-lary pyramid. There was a moderate increase of deep tendon reflexes and a mild spasticity. Schwartzman (1978) found no spasticity, 3 years after section of the medullary pyramid in the monkey (A recent study by Jagiella and Sung (1989) does suggest the eventual development of spasticity in a patient with the bilateral ischemic infarction of the medullary pyramids).

What was clearly defective after section in all of these studies was the discrete fractionated distal finger movements required for the instinctive tactile grasp reaction that normally would allow for the precise orientation of fingers to a moving three-dimensional stimulus. The pyramidal tract fibers mediating cortical control of these distal muscles make direct connection to the anterior horn cell. These direct fibers of the corticospinal system elicit predominantly flexor facilitation and extensor inhibition. In contrast, the indirect corticospinal fibers (those which do not synapse directly on anterior horns cells) synapse predominantly in relation to interneurons, which make synaptic contact with anterior horn cells innervating more proximal muscles. Additional discussion will be found in Kuypers 1987.

We have not considered in our discussion the possible role of the pyramidal tract in modifying transmission of sensory information at the spinal cord (posterior horn) and brain stem levels (nucleus gracilis and nucleus cuneatus). The various studies, which have suggested such a role, are considered in the review of Bizzi and Evarts (1971).

CORTICORUBRAL SPINAL SYSTEM

In the monkey and cat, if the pyramidal tracts are intact, destruction of the rubral spinal system produces no definite motor deficit. However, in monkeys and cats that have already had bilateral section of the pyramidal tract (a section which has produced very little qualitative change in limb movement) a complete loss of distal limb movement may occur after additional bilateral destruction of the rubrospinal tract. In the intact monkey, complex coordinated movements may be evoked by stimulation of the red nucleus. Flexor excitation predominates with extensor inhibition. The studies of Lawrence and Kuypers (1968) suggested that the rubrospinal tract is mainly concerned with steering movements of the extremities, particularly distal segments. On the other hand, the ventromedial brain stem reticulospinal tract is mainly concerned with steering movements and posture of head or body and synergistic movements of the extremities (Kuypers, 1987).

PREMOTOR CORTEX (Areas 6 and 8):

Overview of the functions of the premotor and supplementary motor cortex *(Fig. 18-15)*.

Penfield grouped area 6, area 8, and the supplementary motor cortex (the continuation of areas 6 and 8 onto the medial surface of the hemisphere) on clinical grounds as intermediate frontal areas. These areas may also be grouped from the functional standpoint as concerned with the preparation for movement and the temporal sequencing of movements (Halsband 1993, Halsband and Freund 1990, Passingham 1987, Rizzolatti 1987, Rome 1986, Shibasaki et al 1993, Wise 1988, Wiesendanger 1987,Wiesendanger and Wise 1992)

These areas receive a polysensory input from sensory association areas. These are widespread connections of these areas both within and between the hemispheres (see Pandya and Kuypers, 1969, Pandya and Vignolo 1971, Ward et al 1946).

There is considerable rationale for grouping these areas from the standpoint of the results of stimulation studies. Moreover, there is also considerable justification for such an approach from the cytoarchitectural standpoint. Histologically, area 6 is similar to area 4 but lacks, in general, giant pyramidal cells. Area 8 has some similarity to area 6 but is also transitional to the granular prefrontal areas. There are considerable variations when the several cytoarchitectural maps are compared as to

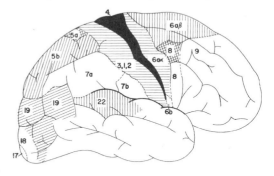

Figure 18-15. The motor cortical fields of man as determined by Foerster employing cortical stimulation. Stimulation of the area indicated in black produced discrete movements at low threshold and was designated pyramidal. Lines or cross-hatching indicate other areas producing movements and were designated in a broad sense as extrapyramidal. These movements were usually more complex synergistic and adversive movements and included the adversive responses obtained by stimulation of area 19. Each of the areas extending to the midline is continued on to the medial surface. Modified from Foerster, O.: Brain, 59:137, 1936 (Oxford Univ. Press).

the exact boundaries of area 8 in both man and monkey. Areas 44 and 45 in the inferior frontal convolution (Broca's motor speech or expressive aphasia centers) should also be included within this motor association or intermediate frontal grouping, from both the functional and cytoarchitectural standpoints. For simplicity of presentation, these areas will be considered later in relation to language function.

Note that clinical lesions, such as vascular infarcts and tumors, usually do not respect cytoarchitectural borders. Thus, vascular lesions that involve the lateral premotor aspect of area 6, usually also involve the primary motor cortex. Tumors such as parasagittal meningiomas may involve both the lateral premotor and the supplementary motor cortex or may involve the prefrontal and the several premotor areas.

With increasing research, there has come recognition that there are differential functions for the various sectors of these premotor areas, (Wise etal, 1997, Krakauer &Ghez, 2000).

An initial distinction may be made between the supplementary motor cortex (SMA) and the lateral premotor areas (PMA).

1. *The Supplemental Motor Area.* SMA is concerned with the sequencing of movements that are initiated internally without triggering from external sensory stimuli. SMA is concerned with the performance of learned sequences of movement. The area just anterior to the SMA, the pre SMA is involved along with other prefrontal areas such as area 46 in the actual process of learning the sequences (training of a motor skill). This same pre SMA area is also involved when the same complex movement sequence is imagined. Once a sequence has been fully and usually over learned, control may be centered in primary motor cortex.

The Readiness Potential (Bereitschafts Potential): In man, surface electrodes placed on scalp are able to record 1-2 seconds prior to voluntary movements, a significant surface negative potential with maximum amplitude over the midline of the frontal area corresponding primarily to the supplementary cortex. The process of initiation of more complex actions enhances this potential. In contrast, when the same movements were triggered by external stimuli, there was no such readiness potential. These non-invasive studies in man (as well as blood flow studies) confirm much earlier experimental data derived in studies of the monkey (Papa et al 1991, Deeke, 1987, Wiesendanger 1987).

2. *The Lateral Premotor Area.* In contrast to the SMA, the lateral PMA is concerned with the sequencing of movements that are triggered by external stimuli. In this regard, there is a close relationship of the lateral PMA to the sensory association cortex of the superior parietal cortex (area 5) and the posterior parietal multimodal sensory areas, which are directly connected, with the extra striate visual cortex. This relationship is complex as demonstrated in reaching for and grasping precisely an object presented at a particular point in visual space.

The lateral PMA has been divided into two major functional areas: the dorsal and the ventral.

a. *The dorsal lateral PMA* neurons are involved in reaching for that object. This area is also involved in learning to associate specific stimuli particularly visual with the motor response of reaching for that stimulus. As opposed to the learning of a repetitive motor act (pressing a key) involving the supplementary motor cortex, this is a different type of learning referred to as associative learning. Multiple bits of information must be integrated for this reaching activity, visual position of object, position of gaze, position of arm etc and this involves information received from various posterior parietal areas adjacent to the extrastriate cortex discussed in greater detail by Wise et al (1997.

b. *The ventral lateral PMA* is concerned with grasping. This area also must receive information from the extrastriate cortex regarding size and shape of the object as well as other multimodal and sensory association data so that an effective grasp of the object may occur.

Essentially stimulation of PMA and SMA results in a pattern of movement characterized by tonic abduction of the contralateral arm at shoulder with flexion at elbow plus deviation of head and eyes to the contralateral field.. In addition with SMA stimulation bilateral lower extremity movements may occur. *(Fig. 21-12)* Ablation of these areas will result in a release of grasp reflex and an apraxia in limb movements. Bilateral lesions of the SMA may result in an akinetic mute state.

Case 18-2 provides an example of a low-grade glioma involving the premotor and supplementary motor cortex from the standpoint of focal seizure activity. The motor effects of frontal lesions are presented in case histories 22-2 and 22-3, which follows the discussion of the prefrontal areas.

Case 18-2: This 57-year-old right-handed single nun and occupational therapist, at age 47 had the onset of observed generalized convulsive seizures, which occurred without warning with sudden loss of consciousness. Shortly after the onset of the grand mal seizures, the patient began to have focal seizures, which increase in frequency to one per week. She described these as beginning with the abduction and raising of the right arm at the side, over her head, then a jerking of the right arm, and then she would fall to the ground. She would remain fully conscious but she would be unable to talk. She would have an awareness that something strange was occurring. She had a sinking sensation. She denied any impairment of memory after the episode. However, she would be aware that her right arm and leg were tingling and weak for perhaps 5 minutes afterwards and she would be unable to speak for 10-30 minutes afterwards. These focal seizures occurred once a month, but for the last 2 years, they had been occurring once a week. The patient indicated that over the last 10 years she has had problems with her memory.

She was reported to have had three "negative" CT scans over the years, and yearly EEG's, which did not reveal focal features or electrical seizure discharges.

Neurological examination: *Mental Status:* delayed recall without assistance, was 0 out of 5 and with assistance 2 out of 5. *Motor System:* intact except for a decreased swing of the right arm in walking. *Reflexes:* deep tendon reflexes were increased on the right at biceps and patellar. Plantar response was extensor on the right and equivocally extensor on the left.

Clinical diagnosis: Seizures of focal origin left premotor/supplementary motor cortex possibly secondary to low grade glioma. A meningioma was less likely due to the reports of "normal CT scans."

Laboratory data: *CT scan (Fig.18-16)* demonstrated an area of calcification with surrounding edema in the left premotor area.

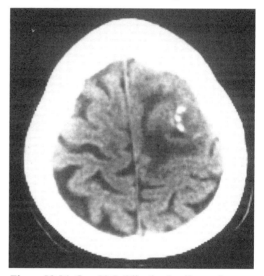

Figure 18-16. Case 18-2. Oligodendroglioma of premotor and supplementary motor cortex. CT scan. The non-contrast CT scan demonstrated an area of calcification with surrounding edema, involving the premotor cortex. See text.

MRI scan (Fig.18-17), demonstrated extensive involvement of the premotor and supplementary areas by tumor and edema.

Subsequent course: Doctor Bernard Stone performed subtotal resection of this tumor. Histological examination of the tissue confirmed the preoperative impression of an infiltrating glioma - predominantly oligodendroglioma in type. The patient received postoperative radiotherapy, 5000 rads to the left hemisphere and an additional 1000 rads to area of tumor. When last seen in follow-up 4 years after surgery, the patient was still experiencing infrequent (1-2 per month) short focal seizures of the same type involving focal movements of right arm, plus speech arrest. The neurological

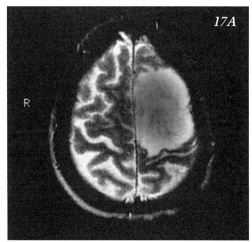

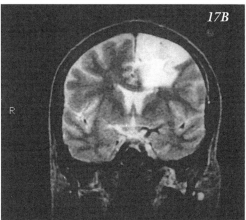

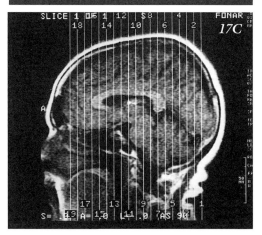

Fig. 18-17 Case History 18-2. Oligodendroglioma of premotor and supplementary motor cortex, MRI non-enhanced A) Horizontal T 2 weighted study demonstrates the wide extent of tumor and edema involving the premotor and frontal cortex. B) Coronal section T2 weighted demonstrates the involvement of the supplementary motor cortex. C. Reference diagram for coronal section (slice 13). See text.

examination demonstrated some decrease in spontaneous speech some blunting of affect. Bilateral extensor plantar responses were present. She had a moderate frontal type apraxia of gait and required a cane to avoid falls.

CORTICAL CONTROL OF EYE MOVEMENT: FRONTAL (AREA 8) AND PARIETO-OCCIPITAL EYE FIELDS.

Overview of eye movements: In chapters 11 and 12 –the basic cranial nerve and brain stem mechanisms for eye movement were considered. In this section, the mechanisms for control from other centers will be considered. This is a complex subject and has been presented in great detail in the studies of Goldberg (2000). More specific information regarding the frontal eye field will be found in Schall & Thompson (1999).

Vision is sharpest at the fovea, the center of the macular area of the retina. There are various mechanisms for moving the eyes so that points of interest are centered on the fovea of the two eyes. These may be listed as follows: saccadic eye movements, smooth eye movements, vergence eye movements, fixation system, vestibulo-ocular responses, and opticokinetic responses.

Definitions and anatomical localization:

1. *Saccadic eye movements: Rapid movements of gaze that shift the fovea rapidly to a point of interest are termed saccades.* These movements may be triggered by attending to a visual point of interest, by memories or by commands. The brain stem substrate for all horizontal saccades is the lateral gaze center of the paramedian pontine reticular formation. Several types of neurons are present in this center, Neurons referred to as medium lead burst cells provide direct excitation to motor neurons in the ipsilateral abducens nucleus and to the interneurons in that nucleus which give rise to the medial longitudinal fasciculus. Long lead burst cells drive the medium lead burst cells and receive excitatory input from the higher centers to be discussed below, The medium lead burst neurons also excite inhibitory burst neurons which in turn suppress the discharge

of contralateral abducens neurons. In addition, there are neurons located in the dorsal raphe nucleus of the pontomedullary area, which cease firing during saccades (omni pause neurons). These neurons if stimulated during a saccade will stop the saccadic movement. Additional neurons provide tonic discharges relevant to eye position and maintain the eyes in a position set by the saccade: the flocculus of the old vestibular portion of the cerebellum, the vestibular nucleus and the nucleus prepositus hypoglossi.

For vertical saccades, burst and tonic neurons are found in the rostral interstitial nucleus of the MLF in the mesencephalic reticular formation.

Central control of saccades: A complex system with redundant features is present. Essentially, two major cortical areas are involved: the contralateral frontal eye field (area 8) and the contralateral lateral intraparietal area of the posterior parietal area at the border of the extrastriate cortex. Area 8 provides the motor input for the response to the command e.g. "look to the left or right field". The frontal eye field has major interconnections in a topographic manner with areas that participate in the streams of information that flow out of the extrastriate visual cortex. These two streams, the dorsal and ventral will be discussed in the visual system chapter. The posterior parietal area in contrast is involved with visual attention, with representations of the location of objects of interest. Beyond a simple representation, there is a transformation of this information by parietal neurons into motor coordinates (Colby & Goldberg-1999). There is a significant interconnection with area 8 with an exchange of information. Both areas project to the intermediate and deep cell layers of the superior colliculus. The superior colliculus in turn projects to the mesencephalic and pontine reticular formation. The same areas of the superior colliculus also receive inputs from the prestriate areas of the posterior temporal gyri. The superficial layers of the superior colliculi receive input from both the striate cortex and the retina. The frontal eye field (area 8) projects not only to the superior colliculus but also directly to the pontine and mesencephalic reticular formation. Lesions of either area 8 or of the superior colliculus would produce a defect in saccades to the contralateral field and a transient visual neglect. The defect however would be temporary, since the alternative system would be intact. Lesions of the parietal cortex tend to produce longer lasting visual neglect syndromes. The interaction of area 8 and the superior colliculus is even more complex. Area 8 projects to the caudate nucleus. The caudate nucleus inhibits the S. nigra pars reticularis, which in turn acts to inhibit the superior colliculus.

Other cortical areas are also involved in saccades: the SMA eye field and the dorsolateral frontal cortex. The former area has neurons that direct movement to part of a target. The latter contains neurons that discharge when the saccade is made to a remembered target. Additional discussion of the role of the frontal eye fields will be found in Pierrot-Deseilligny et al, 1995.

2. *Smooth pursuit in contrast to saccades keeps the image of a moving target on the fovea.* Eye movement velocity must match the velocity of the target. The medial vestibular nucleus, the nucleus prepositus hypoglossi and the paramedian pontine reticular formation receive information from the vermis and flocculus of the cerebellum and project to the abducens nucleus and the oculomotor nucleus. Morrow & Sharpe, 1995 discuss the deficits in smooth pursuit after unilateral frontal lesions.

The cortical inputs for smooth pursuit arise from area 8:the frontal eye field. In the monkey, stimulation of area 8 produces ipsilateral pursuit. Area 8 receives input from the posterior sectors of the superior and middle temporal gyri. This area in turn receives input from the visual cortex. In man the posterior parietal area may serve this same function. Both the frontal and temporal (or posterior parietal) areas project to the dorsolateral pontine areas.

3. *The fixation system holds the eyes still during intentional gaze on an object.* During fixation, saccades must be suppressed. The most rostral segment of the superior colliculus has a representation of the fovea and neurons here

are active during fixation. These neurons act to inhibit neurons in the more caudal superior colliculus. In addition, these neurons excite the omni pause neurons in the lateral gaze center, which as discussed above inhibit saccades.

4. *Vergence movements (convergence and divergence) and accommodation* are controlled by neurons within the third nerve complex of the midbrain.

5. *Vestibulo-ocular movements* hold images stable on the retina during brief head movements. In put occurs from the vestibular system.

6. *Opticokinetic movements hold images stable during sustained head rotation.* Visual input is required. In rabbits, retinal neurons project to the nucleus of the optic tract in the pretectum which then projects to the medial vestibular nucleus. These neurons can distinguish between visual and vestibular stimuli. In primates, there is in addition, a "cortical system" that includes the magnocellular layers of the lateral geniculate, striate cortex, and the posterior portions of the middle and superior temporal gyri. Compared to the non-cortical system, which responds to stimuli moving slowly in a temporal to nasal direction, this "cortical system" can respond to stimuli moving with higher velocities in a nasal to temporal direction. This evolution in the system may reflect the changes, which have occurred in the placement of the orbits in the skull when the primates are compared to the rabbit.

Opticokinetic nystagmus occurs when a pattern of vertical black and white lines is moved slowly in front of the eyes. A similar nystagmus occurs as one sits in a moving train looking out the window at the series of telephone poles flashing by. A lesion in the cortical system produces defective opticokinetic nystagmus to visual stimuli moving towards the side of the lesion.

Area 8: Stimulation: there is disagreement regarding the exact boundaries of area 8 in the human. Some investigators (Vogts, Foerster, Penfield) *(Fig.17 -10, 18-10, 21-15,)* have limited area 8 to those portions of middle frontal and inferior frontal gyri just anterior to the precentral gyrus. These authors have then designated all of the superior frontal gyrus between the prefrontal areas and the motor cortex as area 6. It is clear that conscious, adversive eye and head movements to the contra-lateral field occur most frequently in relation to stimulation of that portion of the middle frontal gyrus just anterior to the precentral gyrus. In actuality in the studies of Penfield and Jasper 1954 and confirmed more recently by Luders et al (1988), the eye field for gaze actually extended onto the motor strip between the face and hand areas as demonstrated most particularly in *Fig. 18-10*. Adversive head and eye movements may also occur, of course, as part of a more complex movement on stimulation of area 6 or of the supplementary motor cortex. Moreover, adversive head and eye movements, as a seizure phenomenon following loss of consciousness, may occur with foci of seizure discharge more anteriorly placed in the prefrontal areas (compare *Fig. 18-18a and 18-18b)* but these have less localizing significance. Quesney et al (1990) concluded that seizures arising in parasagittal frontal area are more likely to involved conscious head turning; seizures originating in anterolateral dorsal-frontal convexity - unconscious head turning. In addition, stimulation in areas 18 and 19 will also produce eye turning. Thus, the clinical neurologist must in each case weigh the localizing significance of seizures characterized by head and eye turning. Such seizure components do not always have the more specific localizing significance of the focal motor seizure or of the olfactory hallucination (Refer also to the paper of Ochs et al, 1984, which discusses the variability of localization in terms of ipsilateral versus contralateral deviation of the eyes)

In addition to head and eye turning, opening and closing of the eyelids may occur on area 8 stimulation. At times, pupillary dilatation and eye movements, other than conjugate aversion, may occur, e.g., upward deviation of the eyes *(Fig.18-19)*.

Area 8: Ablation. Ablation of area 8, or (in man) infarction of this area following occlusion of the blood supply, results in a transient paralysis of voluntary conjugate gaze to the contralateral visual field. In general, in man, this

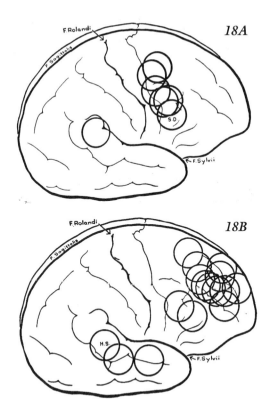

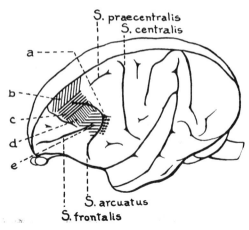

Figure 18-19. Functional subdivision of the frontal eye fields in the monkey under relatively light anesthesia: a) closure of eyes; b) pupillary dilatation; c) eyelid opening (awakening reaction"); d) conjugate deviation to the opposite side; and e) nystagmus to opposite side. (From Smith, W.K., (in) Bucy, P.: The Precentral Motor Cortex, Urbana, University of Illinois Press, 1944.) In the studies of Crosby, E.C., et al (J.Neurol. 97:357-383, 1952) oblique and vertical movements were also observed.

Figure 18-18 A. Simple (conscious) adversive seizures at onset. The epileptogenic lesion localizes to the middle frontal gyrus anterior to motor cortex. (From Penfield, W., and Kristiansen, K.: Epileptic Seizure Patterns, Springfield, Ill., Charles C. Thomas, 1951, p.30.) B. Unconscious Adversive Seizures at onset. The epileptogenic focus in 16 patients whose seizures began with unconscious aversion of head and eyes localizes predominantly to the anterior portion of the frontal lobe (From Penfield, W., and Kristiansen, K.: Epileptic Seizure Patterns. Springfield, Ill., Charles C. Thomas, 1951, p.22.)

paralysis of voluntary gaze usually does not occur in isolation but is associated with a sufficient degree of infarction to have produced a severe hemiparesis. The patient then lies in bed with the contralateral limbs in a hemiplegic posture and with the head and eyes deviated towards the intact arm and leg. This deviation may perhaps reflect the unbalanced effect of the adversive eye center of the intact hemisphere. The effect is usually transient, clearing in a matter of days or weeks. These patients at times also appear to neglect objects introduced into the visual field contralateral to the lesion. The patients also neglect contralateral tactile and auditory stimuli. In the past, such a unilateral neglect has been attributed to associated involvement of the inferior parietal lobule. The studies of Kennard (1938) and of Welch and Stuteville (1958), however, suggest that these unilateral neglect syndromes may also occur as transient phenomena following ablation of area 8 in monkeys (area within the superior limb of the arcuate sulcus). These animals also had conjugate deviation of head and eyes to the side of the lesion and forced circling. Bilateral ablations within this area resulted in animals that, in a sense, had a bilateral neglect of the environment, remained apathetic, and neglected visual and auditory and tactile stimuli, although they could follow moving objects. Although some recovery occurred over a period of weeks to months, the animals continued to have a "wooden expression" and a fixed gaze. Some of these findings are seen in the human following bilateral frontal lobe damage. In addition, however, a similar state may occur in Parkinson's disease or in brain stem lesions producing akinetic mutism, which will be discussed later. More recent discussions of the "neglect" syndrome will be found in Watson et al 1978 and Heilman et al 1983. Note also that bilateral discharge of this area may produce a similar state of "neglect of the environment" and arrest of movement (Marcus et al 1970).

SUPPRESSOR AREAS FOR MOTOR ACTIVITY (Negative Motor Response)

Dusser de Barenne and McCulloch found in the monkey that stimulation of that portion of area 4, adjacent to area 6 (designated area 4S), resulted in a suppression of the motor response elicited by stimulation of area 4. In addition, thresholds for obtaining responses from area 4 were raised, and after discharge was aborted. Stretch reflexes and muscle tone were decreased. It was soon discovered that these effects could be obtained by stimulation of a number of other cortical areas: area 8 (8S), 2S and 19S. On the medial aspect of the hemisphere, similar effects could be obtained from the anterior sector of the cingulate gyrus: areas 24S and 32. These "suppressor" areas cannot be distinguished on cytoarchitectural grounds from adjacent cortical areas. Similar effects have been elicited from stimulation of the caudate nucleus and of the bulbar medullary inhibitory center. There has been some evidence that the cortical motor suppressor effects are mediated via cortical reticular and corticostriate connections. In subsequent studies, Fangel and Kaada, 1960 indicated that arrest of movement could occur from stimulation of a number of distinct neocortical and rhinencephalic points. Note must of course be made that a nonspecific interference with voluntary movement may occur in relation to stimulation of the motor cortex. Thus, the patient who is experiencing a focal motor seizure involving the hand cannot use that hand in the performance of voluntary skilled movements. Similarly, arrest of speech is frequently produced by stimulation of the various speech areas of the dominant hemisphere, whereas vocalization is only rarely produced.

Recent studies by Luders et al (1988) with precise stimulation of the cerebral cortex of man in preparation for epilepsy surgery have revisited the concept of suppressor areas - renaming the phenomena as "Negative Motor Response" - an inability to perform or sustain a voluntary movement at a stimulation intensity that did not produce any symptoms or signs. The areas of cortex included (a) the inferior frontal gyrus, immediately anterior to primary motor area for face; (b) less often-premotor cortex just anterior to the frontal eye field or the hand area; (c) The supplementary motor cortex. The authors speculate that these effects are produced because the stimulation interferes with the preparation for movement.

PREFRONTAL CORTEX (Areas 9, 10, 11, 12, 46, 13, 14). Additional discussion will be found in chapter 22

The term prefrontal refers to those portions of the frontal lobe anterior to the agranular motor and premotor areas (areas 9,10,11,12). Three surfaces or divisions are usually delineated: 1) lateral - dorsal convexity, 2) medial and 3) orbital or ventral .All three are concerned with executive function and relate to the mediodorsal thalamic nuclei. In general the lateral-dorsal relates to motor association executive function, the orbital –ventral and the medial to the limbic –emotional control executive system. Recent studies have questioned this simplistic approach. Thus certain functions previously considered to be associated with dorsolateral location have been now localized to a medial location. (Stuss et al 2000). In addition, there is evidence that dorsal lateral lesions also alter affect and motivation. Moreover as the clinical examples of chapter 22 will demonstrate, most pathological processes involve the functions associated with more than one division either directly or through pressure effects.

Most studies of stimulation or of ablation lesions involving the prefrontal areas have included, within the general meaning of the term prefrontal, areas beyond areas 9, 10, 11, and 12. Posterior orbital cortex (area 13) and the posteromedial orbital cortex (area 14) have been added to the orbital frontal group (Fig. 18-20). These areas all share a common relationship to the medial dorsal nucleus of the thalamus. Areas 46 and 47, located in man on the lateral surface of the hemisphere, are also included. The anterior cingulate gyrus (area 24) is also included in the prefrontal area, although it may be considered part of the limbic system and relates to the anterior thalamic nuclei.

Overview of the role of the prefrontal area in motor and cognitive function: *Our concern here is primarily with the dorsal prefrontal subdivision.* The ventral/orbital and medial subdivisions also will be considered here but also in the limbic system chapter. We are concerned with the sequencing of behavior, the planning of motor function and with a very short type of memory referred to as working memory. *Working memory* provides for the temporary storage of information that is needed for ongoing behavior. This may be verbal, visual or central executive. The concept of working memory is best demonstrated in the delayed response test in the monkey. Bilateral removal of prefrontal areas in monkeys and chimpanzees produced significant effects on two separate aspects of behavior: (1) delayed response and delayed alternation response, (b) emotional responses, particularly aggression (Jacobsen 1935, Fulton and Jacobsen 1935, Jacobsen and Nissen 1937). In the delayed response test, the hungry test subject observes food placed under one of two containers. A solid opaque screen is then interposed and after a delay of at least 5 seconds, the screen is raised - allowing the animal to select the appropriate container. Normal primates rapidly learn this task and a more complex task of delayed spatial alternation in which the choices must be alternated between the left and right containers. Both tasks, however, are difficult for the lesioned animal to perform if delays are imposed. Subsequent studies (see Butters and Pandya 1969) have demonstrated that a relatively small lesion around the middle one-third of the principal sulcus is sufficient to produce the deficit. Although Jacobsen initially attributed the disorder to a general deficit in short term memory, subsequent studies have indicated that this is not necessarily the case, since such animals can learn discriminations involving the selection of a specific object from among several presented. When appropriate tests are selected, similar deficits can be demonstrated in man. A more detailed discussion of this aspect of prefrontal disorder is provided by Milner and Teuber 1968 and Stuss and Benson (1986). In the human, the digit span test utilizes working memory.

Additional studies have defined three sectors of the dorsal subdivision, which are concerned with different aspects of working memory and motor planning. These areas may be specified based on their relationship to the principal sulcus in the monkey dorsolateral prefrontal cortex.(Fig. 18-20)

The *area surrounding the principal sulcus* contains neurons that begin to fire as the initial visual cue of food is presented in a particular position in the contralateral visual field, and continue to fire during the delay period. Lesions in the area of the principal sulcus will interfere with delayed response.

Neurons in the *cortex ventral to the principal sulcus* code information about what the object is in terms of shape and color. This area receives information via a ventral pathway from the visual cortex involving the inferior temporal lobe.

The *cortex dorsal to the principal sulcus* codes information regarding the location of the object. This area receives visual information via the posterior parietal area adjacent to the extrastriate cortex as discussed above.

There are other neurons that respond to both types of stimulus information.

In the human, aspects of prefrontal function are tested with a variety of tests. These include more complex forms of the delayed response and delayed alternation tests. The Luria motor sequences can be easily utilized in the office: the patient must reproduce a demonstrated sequence of hand movements (strike the thigh with the open hand then with the ulnar aspect of the open hand and then with the ulnar aspect of the closed fist). In the Wisconsin Card Sorting test, the patient must sort 60 cards based on color, or symbol or number of symbols. By changing the rules once the patient has developed a pattern set, it is possible to also determine how readily set can be changed. Failure to change set is termed

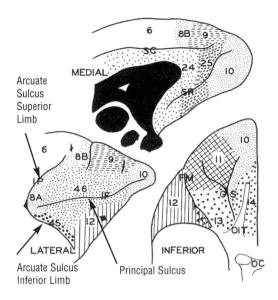

Fig. 18-20 Prefrontal areas of the Monkey (Macaca Mulatta). Note location of orbital areas 13 and 14, anterior cingulate area 28, in addition to motor prefrontal areas 9, 10, 11, 12. Modified from Walker, E.A.: J.Comp.Neurol. 73:81, 1940 (Wiley).

rigidity.

Fletcher & Henson (2001) have summarized PET scan and functional MRI studies, which suggest a role for lateral prefrontal areas both in working memory and long-term memory. Activation of the ventrolateral region (inferior to the inferior frontal sulcus) has been correlated with the updating and maintenance of information. Activation of the dorsolateral region (dorsal to the inferior frontal sulcus) has been correlated with the selection /manipulation /monitoring of that information. Activation of the anterior prefrontal cortex (anterior to a line drawn vertically from the anterior edge of the inferior frontal sulcus) has been correlated with the selection of processes/sub goals.

DISORDERS OF MOTOR DEVELOPMENT.

Non-progressive disorders of motor function or control recognized at birth or shortly after birth are referred to as cerebral palsy. Little, in 1862, first described "spastic rigidity". He was describing the most common form of cerebral palsy: spastic diplegia: severe bilateral spastic involvement of the lower extremities or a double hemiplegia. Other types of spastic cerebral palsy include: a. quadriplegia, and b. infantile hemiplegia.

In addition to the spastic varieties of cerebral palsy, other types have been identified: a. double athetosis, b. ataxic c. dystonic.

As reviewed by Paneth (1986), cerebral palsy affects at least 1 of 500 school age children. More severely affected patients may die in infancy. Severely affected children may have subnormal intelligence (50%) and seizure disorders (25%).

Little related the motor deficits to the circumstances of birth: premature delivery, breech presentations, or prolonged labor. Often respiration was impaired at birth. Convulsions often occurred in the neonatal period. Little's hypothesis regarding causation has had serious medical and legal implications for the practice of obstetrics.

Sigmund Freud (1897) prior to his investigations into hysteria and the psychoneuroses undertook detailed clinical and neuropathological studies of these developmental motor syndromes. Freud unified the various disorders into a single syndrome, and proposed the classification, which is presented above. He also proposed that the abnormalities of the birth process - noted by little, were in actuality the consequence and not the cause of the perinatal pathology.

The National Collaborative Perinatal Project followed 54,000 pregnancies and the subsequent offspring to age 7. The conclusions of that study (Nelson & Ellenberg 1986) tend to support Freud's hypothesis. Major prenatal malformations such as microcephaly and prenatal rubella infection, were more common in the 189 children with cerebral palsy. Maternal mental retardation, and low birth weight below 2000 grams were also significant predictors of cerebral palsy as was breech presentation but not breech delivery. In contrast, birth asphyxia did not add predictive power in the analysis.

GAIT DISORDERS OF THE ELDERLY.

The older patient is frequently at risk for

progressive disability in two areas of neurological function: (a) motor function: primarily manifested by disturbance of stance and gait, (b) higher cortical functions of cognition and memory manifested by dementia. (The latter problem will be considered in Chapter 30.) These major problems of geriatrics - both alone or in combination directly and indirectly, account for many of the prolonged admissions to acute care hospitals; and constitute the major reason for eventual placement of the elderly patient in chronic care facilities such as nursing homes. The cost impact on the health care system is massive because of the increasing age of the population and because of the frequency of the disorders.

Sudarsky (1990) discusses the several studies that indicate (a) 15% of patients over the age of 60 have some abnormality of gait; (b) 200,000 hip fractures occur annually in the United States - most due to falls by older individuals. (c) Accidental death is the sixth leading cause of death among the elderly - with the majority of these deaths the results of falls; (d) 40-50% of patients in nursing homes have difficulty with walking; (e) surveys of older patients (of average age 78) indicate a considerable fear of falling and insecurity of gait resulting in a self-imposed limitation of activities in 50%. [Tinettei and Speechley (1989) estimate the annual incidence of falls in the community: 25% at age 70 and 35% after age 75. For institutionalized patients fall incidence is 50%.];

Not all falls and not all gait disorders are related solely to neurological disorders. Joint and skeletal disorders may impair gait and result in falls. Age-related changes in vision, hearing, and vestibular function may all impair gait and postural stability.

In terms of the nervous system, changes in the peripheral nervous system - those related to aging and to such disorders as diabetes mellitus, and nutrition may all result in impaired proprioception with resultant changes in balance and gait. The deficits are compounded by concurrent changes in vision, hearing and vestibular dysfunction, (see also Tinetti et al 1988).

It is the frequent central nervous system changes resulting in falls, which concern us here. Sixteen % of these patients have a cervical myelopathy, 10% Parkinson's disease, 20% a frontal gait disorder due to normal pressure hydrocephalus or to multiple strokes or to involvement of white matter (Leukoaraiosis - or Binswanger's disease, Thompson and Marsden, 1987). Cerebellar degeneration occurs in 8%, sensory imbalance in 18% and toxic metabolic in 6%.

An example of a treatable cause: normal pressure hydro-cephalus (NPH) will be found in case history 18-3 and Figures 18-21, 18-22 presented on CD ROM. These patients present with a triad of gait impairment (apraxia or ataxia), urinary incontinence, and dementia. The gait impairment usually predates and is more severe that than the disorder of memory. In normal pressure hydrocephalus, a block is present in the subarachnoid space high over the convexity so that absorption of cerebrospinal fluid (CSF) fails to occur. In such cases, there is no obstruction to CSF flow within the ventricular system or within the CSF cisterns at the base of the brain. Such cases may be secondary (previous head traumatic or subarachnoid hemorrhage with the late effects of blood within the subarachnoid space) or primary, no prior history. The designation of "normal pressure" refers to the measurement of CSF pressure at lumbar puncture in those cases. Classically, removal of 20-30 cc of CSF produces a remarkable improvement. Not all cases of hydrocephalus in the elderly are of the "normal pressure" communicating variety. In some cases, also presenting with a similar gait disorder, there is a long-standing obstructive hydrocephalus, secondary to stenosis of the aqueduct of Sylvius, which was well compensated earlier in life, but has become symptomatic late in life. Rarely, a pineal region tumor or midline cerebellar tumor may produce obstructive hydrocephalus in the elderly. For additional discussion of NPH refer to Adams et al 1965, Fisher 1982.

CHAPTER 19
Motor Systems II: Basal Ganglia and Movement Disorders

I. ANATOMICAL BACKGROUND
(Fig. 19-1, 19-2)

The term "basal ganglia" originally included the deep telencephalic nuclei: the caudate, putamen, globus pallidus, the claustrum, and nucleus accumbens. The globus pallidus and putamen are lens shaped and are called lenticular nuclei. Collectively the putamen and caudate are called the corpus striatum. Additional structures now included within this group are the substantia nigra, the subthalamic nuclei, the ventral tegmental area and the ventral pallidum. The caudate and putamen have the same structure and are continuous anteriorly. The globus pallidus has two sectors: a medial or inner and a lateral or outer. The substantia nigra has two components: a ventral pars reticularis which is identical in structure and function to the medial sector of the globus pallidus and a dorsal darkly staining component the pars compacta which contains large dopamine

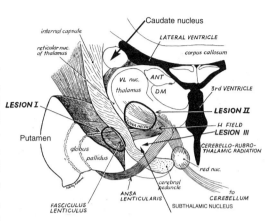

Figure 19-2 Major connections of the basal ganglia with sites for surgical lesions or implantation of stimulators in Parkinson's disease. Lesion I-globus pallidus. Lesion II-Ventral lateral nucleus of the thalamus (the sector of VL involvement is termed VIM= nuc. ventral inferior medial). Lesion III- subthalamic nucleus, (Modified from Lin, F.H., Okumura, S., and Cooper, I.S.: Electroenceph. Clin. Neurophysiol. 13:633, 1961)

and melanin containing neurons.

The nuclei of the basal ganglia may be categorized as (1) input nuclei (caudate, putamen and accumbens), (2) intrinsic nuclei (lateral segment of the globus pallidus, subthalamic nucleus, pars compacta of the substantia nigra and the ventral tegmental area) and (3) output nuclei (medial segment of the globus pallidus, pars reticularis of the substantia nigra and the ventral pallium. The consideration of the chemical and pharmacological anatomy of transmitters and circuits within the system will provide an understanding of function and dysfunction within this system.

It should be noted that the nuclei of the basal ganglia, the circuits involving the basal ganglia, the cortical areas projecting to the basal ganglia, the cerebellar nuclei relating to the basal ganglia, and the reticular formation (which has connections with both the cortex and the basal ganglia) were once grouped

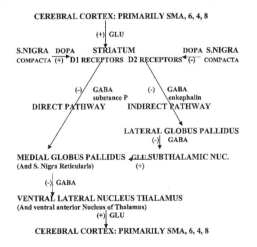

Figure 19-1. Diagram of the connections of the basal ganglia with transmitters and major transmitter action (+) excitatory or (-) inhibitory. GABA = gamma aminobutyric acid (-); GLU = Glutamate (+); Dopamine = Dopa (+). The action of acetylcholine within the striatum has been omitted. See text for details.

together as an "extrapyramidal system". The term "extrapyramidal disorders" - was used to refer to the effects of lesions within the basal ganglia system. However, the term "extrapyramidal" was also utilized to refer to the descending pathways: corticorubral spinal and cortical-reticulospinal that were alternative descending systems to the pyramidal system. The term extrapyramidal then becomes very non-specific and confusing and had best not be utilized. It is important to note, moreover that the cortical areas giving rise to this extrapyramidal system are also, in part, the same areas giving rise to the pyramidal system.

Additional discussions of the gross anatomy will be found in chapter 1, and in Noback, et al (1991); Alexander, et al (1986); Young and Penney (1988).

The connections and transmitters within this system are presented in the diagram of Figure 19-1 and are summarized below.

The essential pathways are the following: cerebral cortex (+) to striatum (caudate / putamen) (-) to globus Pallidus (-) to ventrolateral thalamus (+) to cerebral cortex. Additional circuits involve subthalamus and s. nigra.

Adding greater detail to the outline indicates the following sequence:

1. The major input into the basal ganglia is from the cerebral cortex to the striatum (caudate nucleus/putamen/accumbens). This input is excitatory - utilizing the transmitter glutamate and arises primarily from the supplementary motor premotor and motor cortices (areas 8, 6, and 4). However, the projection from the neocortex occurs from many other areas. In general, frontal and parietal, lateral temporal and occipital areas project to the caudate nucleus, supplementary motor, premotor and primary motor cortex and primary somatosensory cortex project to the putamen. The hippocampal areas, medial and lateral temporal project to the nucleus accumbens. Alexander, Delong and Strick (1986) have distinguished the following segregated basal ganglia - thalamocortical circuits based on the origin of the initial cortical input: 1. Motor 2. Oculomotor 3. Prefrontal 4. Orbital frontal 5. Cingulate-limbic. These are outlined in Table 19-2

2. There are local circuits within the striatum, which involve the excitatory transmitter acetylcholine.

3. The striatum also receives a major dopaminergic input from the substantia nigra compacta. Based on the type of receptor, this input may be either excitatory (D1) or inhibitory (D2). Dopamine receptors have been classified based on the dopamine mediated effects on the enzyme adenylate cyclase (which is involved in the ATP to cyclic AMP transformation). D2 receptors are involved in the inhibition of the enzyme; D1 receptors in stimulation of the enzyme. Note that 5 dopamine receptors have now been identified The complex and still evolving topic has been reviewed by Cooper et al, 1991, Guttman, 1992, Gerfen and Engber, 1992.

4. Because of this differential input two pathways have been identified: the direct and the indirect

5. The major direct outflow pathway arises

TABLE 19-1 ANATOMICAL DIVISIONS OF CORPUS STRIATUM

Basal Ganglia – Telencephalic Nuclei	Function
NEOSTRIATUM	
Dorsal-Caudate /putamen	Input nuclei-neocortical and s. Nigra .Nigra compacta
Ventral-Nucleus accumbens	Input nucleus limbic system
PALEOSTRIATUM (Globus pallidus)	
-Dorsal portion:	-Intrinsic nucleus
-Lateral segment	-Output nucleus
-Medial segment	-Output nucleus
Ventral portion Ventral pallidum	

Basal Ganglia – Associated Nuclei	Function
Subthalamic nucleus of diencephalon	Intrinsic nucleus
Substantia Nigra of Midbrain	
- dorsal -pars compacta (dopaminergic)	-Intrinsic nucleus
- ventral- pars reticularis	-Output nucleus

TABLE 19-2 CEREBRAL CORTEX TO BASAL GANGLIA TO THALAMUS TO CEREBRAL CORTEX CIRCUITS MODIFIED FROM ALEXANDER ET AL,)

LOOP SEQUENCE	MOTOR	OCULOMOTOR	PREFRONTAL	ORBITAL	LIMBIC
1A. Major cortical input	Area 6/ SMA	Frontal eye field/ Area 8	Dorsolateral Prefrontal	Lateral orbital frontal	Anterior cingulate (area 24)
1B. Additional Cortical inputs	Areas 4, 6/PMA, 1,2,3, (post central)	Dorsolateral prefrontal, Posterior parietal	Posterior parietal, arcuate/frontal, PMA	Limbic / Anterior cingulate See limbic	Hippocampus, parahippocampal amygdala entorhinal, temporal neocortical
2. Striatum	Putamen	Caudate (body)	Caudate (head)	Caudate(head)	N. accumbans
3. Pallidum/ S. Nigra reticularis	Globus Pallidus/ S.Nigra/ reticularis	Globus pallidus (caudal) dorsomedial), S. Nigra reticularis	Globus pallidus (lat.dorsalmedial). / S. Nigra reticularis	Globus Pallidus (dorsomedial)/ S. Nigra reticularis	Ventral pallidum, , globus pallidus, (rostral lat.) S.N.reticularis
4. Thalamic nuclei	Ventro-Lateral (pars oralis/ medialis)	Ventral anterior (magnocellularis) Medial dorsal (parvocellularis)	Medial dorsal (parvocellularis), Ventral anterior (parvo Cellularis)	Medial dorsal (magnocellularis) Ventral Anterior (parvo Cellularis)	Medial dorsal (postmedial)
5. Projection to Cortical Area	6/SMA	Frontal eye field/ 8	Dorsolateral prefrontal	Lateral orbital	Anterior Cingulate

from those striatal neurons receiving excitatory dopaminergic input (D1) and passes to the medial globus pallidus - mediated by the inhibitory transmitter gamma aminobutyric acid (GABA). Substance P is also involved as a transmitter or modulator in this pathway. The major outflow is to the inner (medial) segment of the globus pallidus. There is a lesser outflow of fibers from the striatum to the substantia nigra reticularis, which has the same microscopic structure as the globus pallidus. The outflow of the neurons of these structures is to the ventral lateral and ventral anterior thalamic nuclei, mediated again by the inhibitory transmitter GABA This outflow passes via two fiber systems: the fasciculus lenticularis and the ansa lenticularis. The fasciculus lenticularis penetrates through the posterior limb of the internal capsule; the ansa lenticularis loops around the undersurface of the most inferior part of the internal capsule and then passes upward to join the fasciculus lenticularis. These two fiber systems join at a point known as the Field H2 of Forel, then curve back towards the thalamus in the Field Hl of Forel where they are known as the thalamic fasciculus. This fasciculus then enters the ventrolateral and ventral ant-erior nuclei of the thalamus. The outflow of these thalamic nuclei is excitatory to the motor areas of the cerebral cortex predominantly to the premotor and supplementary motor areas.

6. The indirect outflow pathway arises from those striatal neurons receiving inhibitory dopaminergic input (D2) and passes to the lateral (or external) sector of the globus pallidus mediated by the inhibitory transmitter GABA. Enkephalin is also involved as a transmitter or modulator in this pathway. The neurons of the lateral globus pallidus give rise to inhibitory fibers again utilizing GABA which connect to the subthalamic nucleus. This nucleus gives rise to excitatory fibers utilizing glutamate which provide additional input to medial (inner) sector of the globus pallidus and the substantia

nigra reticularis. As in the direct pathway the subsequent connection of these structures is inhibitory to the ventral lateral and ventral anterior nuclei of the thalamus. The subsequent step in the sequence as in the direct pathway is excitatory to the motor areas of the cerebral cortex utilizing the transmitter glutamate. In the case of both the direct and the indirect pathways a loop has been completed: cerebral cortex to striatum to globus pallidus to ventrolateral and ventral anterior thalamus back to cortical motor areas

7. In terms of final effect of this system on the thalamus, if one starts the analysis at the striatum, the end result of the direct system on the thalamus is excitatory. Inhibition of inhibitory drive is referred to as disinhibition In contrast following through the sequences of the indirect system; one finds that the end result is inhibitory

If one begins at the substantia nigra compacta and follows through the direct and indirect circuits, it is evident that both have the same end result as regards the final effect at the thalamic and the cortical level.

8. It is then possible to analyze the eventual thalamic and cortical effects of a decrease in dopaminergic input at the level of the striatum: the thalamus and cortex will receive less excitation. The results are the same whether the direct or indirect pathway is considered. Movement then will be decreased or slowed down. The terms akinesia or bradykinesia are employed.

On the other hand, if there is increased dopaminergic activity, the end results at the thalamic or cortical level will be an increase in motor activity irrespective of whether the direct or indirect pathway are considered. The terms hyperkinesia or dyskinesia are employed

The ultimate effect of dopamine then is to facilitate movement. (Cote &Crutcher, 1991 and Bergman, et al, 1990).

The striatum also has GABA mediated inhibitory outflow to S.nigra compacta. Specific sites in the striatum project to the S.nigra compacta (and receive afferent input from the S.nigra compacta)[1].

Now let us consider the more complex nature of the microanatomy of the striatum.

1. From a histochemical and histological standpoint, distinct areas may be noted, referred to as patches within the larger background matrix. The patch areas have a high density of neurons; matrix areas are less dense. Patch zones contain little acetylcholine esterase; matrix areas are rich in acetylcholine esterase. Patch areas are associated with bundles of dopamine fibers early in development. Later in development there are less dense dopaminergic inputs to the matrix. (Graybiel & Ragsdale, 1978). Neurons within the striatum may be additionally categorized on the basis of size and morphology of the dendrite and size as (a) medium-sized, spiny neurons (b) large or small aspiny neurons. The different properties of these regions and neuron types are presented in table 19-3.

There is considerable evidence that the dopamine-containing terminals and their related synapses and the cholinergic synapses in the striatum have essentially antagonistic actions. Normally, however, they remain in a particular balance. We will discuss the problem of an imbalance when we come to discuss Parkinson's disease (Duvoisin, 1967; Hornykiewicz, 1970). As discussed above both tend to be found in specific sites in striatum (striasomes). These sites exist in a larger matrix of the striatum.

Several additional loops and circuits must now be considered:

1. In addition to its major input from the cerebral cortex, the caudate-putamen also receives a lesser afferent input from the nucleus centrum medianum (C.M.) (excitatory transmitter presumed to be glutamate or aspartate). It should be noted that the C.M. represents a major interlaminar nucleus of the thalamus and is the main thalamic extension of the

---

[1] *A more detailed examination of the striatum in terms of microscopic neuroanatomy and neuropharmacology suggests a much more complex story (Wexler et al, 1991, Albin et al, 1989, Martin and Gusella 1986).*

reticular formation. The centrum medianum, however, also receives fibers from the globus pallidus (inhibitory transmitter: GABA). This again completes an additional loop; relating the basal ganglia to the reticular formation.

2. It should also be noted that some fibers leave the ansa lenticularis and, rather than passing into the thalamus, descend to the pedunculopontine nucleus within the tegmentum of the mesencephalon. This nucleus then projects back to the globus pallidus.

3. There is also evidence for a direct excitatory input from motor and premotor areas to the subthalamic nucleus providing an additional circuit for modulating motor activity.

Overview of the dopaminergic systems: In addition to the dopaminergic nigral- striatal pathways, which are of major concern in our discussion of motor function, there are other major dopaminergic pathways. These other pathways, however, are of importance when one begins to utilize L-Dopa (the precursor of Dopamine) in therapy of Parkinson's disease. In addition, these pathways are of importance when agents (such as neuroleptics employed in the treatment of psychoses or other psychiatric syndromes) are used to block the dopamine receptor or to prevent the storage of dopamine In addition, the overall disease state in Parkinson's disease may also involve these pathways.

(a) The mesolimbic system - originating in the ventral tegmental area - medial and superior to the S. nigra and projecting to the nucleus accumbens (ventral striatum); the stria terminalis septal nuclei; the amygdala, hippocampus; mesal frontal, anterior cingulate, and entorhinal cortex. A major role in emotion, memory and perception is postulated for this system. Hallucinations are a side effect of a high dose of L-Dopa in the Parkinsonian patient.

(b) The mesocortical system- originating in the ventral tegmental area and projecting to neocortex particularly dorsal prefrontal cortex with a role in motivation, attention, & organization of behavior.

TABLE 19-3: THE COMPLEX NATURE OF THE MICROANATOMY OF THE STRIATUM

MATRIX REGIONS	PATCH REGIONS
1. Matrix areas form large background regions with a low density of neurons	1. The patch areas are smaller and have a high density of neurons.
2. Receive afferent input from superficial cortical layers and multiple sites (motor, sensory, motor, premotor, frontal, parietal and occipital) and Intralaminar thalamic sites Young and Penney 1988).	2. Patch neurons receive input from deeper cortical layers in prefrontal and limbic cortex.
3. Matrix areas are rich in acetylcholine esterase	3. Patch zones contain little acetylcholine esterase. Patch neurons contain GABA and substance P and project to the S. nigra compacta. Patch neurons receive a dense dopaminergic input from the substantia nigra compacta.
SPINY NEURONS	ASPINY NEURONS
Spiny neurons in general project outside the striatum and account for 90% of striatal neurons.	Aspiny neurons in general are interneurons, form 10% of the neurons and are intrinsic to the striatum.
Spiny neurons contain a transmitter and a neuropeptide modulator Types of Spiny Neurons: **a.** Spiny matrix neurons containing GABA and the neuropeptide substance P and projecting to medial globus pallidus and substantia nigra pars reticularis. **b.** Spiny, matrix neurons, containing GABA and another neuropeptide: dynorphin projecting to the medial globus pallidus and substantia nigra pars reticularis. **c.** Spiny Matrix neurons containing GABA and the neuropeptide enkephalin and projecting to the lateral globus pallidus. **d.** Patch neurons containing GABA and substance P and projecting to the S. nigra compacta.	Large aspiny interneurons within the matrix contain the transmitter acetylcholine Small aspiny interneurons within the matrix contain the neuropeptides Somatostatin and Y.

(c) Hypothalamic (arcuate nucleus) to infundibular stalk of pituitary involved in hypothalamic inhibition of prolactin release.

Overlap with the cerebellar system. It is also important to remember that the major outflow from the cerebellum (via the superior cerebellar peduncle) terminates in relation to the same ventrolateral and ventral anterior nuclei of the thalamus[2]. This thalamic nucleus receives a projection from both the cerebellum (predominantly the dentate nucleus) and the globus pallidus. However, the information from the cerebellum is projected in a more limited manner primarily via ventrolateral to the motor areas of cerebral cortex, predominantly the pre motor and motor cortices. The basal ganglia differ from the cerebellum in several important respects. In contrast to the more limited relationship of cerebellum to ventral lateral nucleus of thalamus (projection to motor areas of cortex - and projections from motor areas of cortex via pontine nuclei to the cerebellum), the basal ganglia have widespread inputs from the cerebral cortex and via several thalamic nuclei back to multiple cortical areas. These cortical basal ganglia relationships are not diffuse or overlapping but instead, are organized in parallel systems.

It should be very clear then that patients with disease of the basal ganglia may have significant manifestations not only of motor function but also of cognitive and emotional function (Cooper, et al [1991] and Starkstein, et al (1989) in relationship to Parkinson's Disease and Laplane, et al [1989] with regard to obsessive compulsive behavior change.

There is considerable overlap with frontal lobe, premotor and supplementary motor dysfunction as one might predict, based on the anatomical connections discussed above. An understanding of the basic anatomical interrelationships will aid in an understanding of the effects of the various medical and surgical modalities of treatment in Parkinson's disease. Thus, when dysfunction develops in the circuit, lesions at specific critical points, such as ventrolateral thalamic entry area or the subthalamic nucleus may allow restoration of considerable function. The usual location of surgical lesions for the relief of the tremor and rigidity of Parkinson's disease are shown in *Fig. 19-2*. It should be evident that the lesion in the ventral lateral nucleus of the thalamus would also be extremely effective in eliminating tremor originating in the cerebellum since such a lesion would prevent a disordered cerebellum from acting on the circuits passing through this nucleus and this disordered cerebellum would no longer influence motor activities originating at a cortical level.

We must consider the circuits through the basal ganglia modulators of cortical function, particularly as regards movement. For voluntary movement, these circuits clearly provide complex modulation acting on the supplementary motor cortex. This modulation represents a focused equilibrium of opposing influences. When one of these influences acts to a disproportionate degree, dysfunction in the main circuit from the cortical motor cortex to the anterior horn cells and to the reticular formation will result.

CLINICAL SYMPTOMS AND SIGNS OF DYSFUNCTION

General - overview Based on the anatomical circuits considered above a variety of symptoms and signs may occur.

(1) A lack of movement, an inability to initiate movement or a slowness of movement because of excessive inhibition (a lack of excitation) so that the excitatory effect of ventral lateral nucleus - of thalamus on supplementary motor neurons fails to occur due to a decrease in dopamine. This lack of movement and slowness of movement is defined as akinesia and bradykinesia as discussed above.

---

[2] *It should be noted that the dentate nucleus of the cerebellum also has some outflow (via the superior cerebellar peduncle) to the red nucleus. After synapsing at the red nucleus, fibers are also conveyed to the ventrolateral nucleus of the thalamus. The major outflow, however, apparently is directly to these thalamic nuclei from the dentate and, to a lesser degree, from the emboliform nucleus of the cerebellum.*

(2) Elimination of inhibition at a critical point in the system may result in a release of disordered movement. The disordered movements are defined as dyskinesia and tremor.

(3) The relationship of muscle tone in antagonists and agonists may be altered so that excessive tone is present, defined as rigidity. Tremor superimpose on this rigidity produces cogwheel rigidity.

(4) A specific extreme of posture may be maintained defined as dystonia.

(5) Problems in the sequencing of movements as in walking or of posture as in standing may occur.

(6) Disorganization of those higher level-righting reflexes discussed in the previous Chapter 18 may occur.

Not all of these symptoms and signs occur in a given disease or case. Some of these symptoms and signs occur with disease at other sites. As already discussed, akinesia or bradykinesia is frequently seen in patients with disease of frontal - premotor and supplementary motor areas. Refer to Delwaide&Gance (1988) for additional discussion of correlation of symptoms and signs with anatomy/physiology).

It should also be very clear then that patients with disease of the basal ganglia might have significant manifestations not only of motor function but also of cognitive and emotional function. (see Cooper, et al, 1991 and Starkstein, et al, 1989 in relationship to Parkinson's Disease and Laplane, et al (1989) with regard to obsessive compulsive behavior change.

There is considerable overlap with frontal lobe, premotor and supplementary motor dysfunction as one might predict, based on the anatomical connections discussed above. An understanding of the basic anatomical interrelationships will aid in an understanding of the effects of the various medical and surgical modalities of treatment in Parkinson's disease. Thus, when dysfunction develops in the circuit, lesions at specific critical points, such as ventrolateral thalamic entry area or the subthalamic nucleus may allow restoration of considerable function. It should be evident that this lesion in the ventral lateral nucleus of the thalamus would also be extremely effective in eliminating tremor originating in the cerebellum since such a lesion would prevent a disordered cerebellum from acting on the circuits passing through this nucleus and this disordered cerebellum would no longer influence motor activities originating at a cortical level.

We must consider the circuits through the basal ganglia modulators of cortical function, particularly as regards movement. For voluntary movement, these circuits clearly provide complex modulation acting on the supplementary motor cortex. This modulation represents a focused equilibrium of opposing influences. When one of these influences acts to a disproportionate degree, dysfunction in the main circuit from the cortical motor cortex to the anterior horn cells and to the reticular formation will result. The following table lists clinical symptoms base based on the anatomical consideration of multiple sequential circuits involving inhibitory synapses, certain symptoms could be predicted.

SPECIFIC SYNDROMES
PARKINSON'S DISEASE AND THE PARKINSONIAN SYNDROME (Recent reviews are provided by Dunnett & Bjorklund, 1999, Lang&Lozano, 1998, Olanow &Tanner, 1999, Quinn, 1995, Riley&Lang, 2000)

The most common disease involving the basal ganglia is Parkinson's disease, first described by James Parkinson in 1817[3]. Parkinson's disease after Alzheimer's disease is the most common degenerative disorder of the central nervous system affecting approximately 1 million individuals in the United States. The basic pathology is the progressive loss of dopamine producing neurons in the pars compacta of the substantia nigra. A certain mixture

---
[3] *As we will discuss below, other disease entities may overlap with many of the same symptoms and signs and it is appropriate to refer to the larger group as the Parkinsonian Syndrome. The term "Parkinsonism" has also been employed for this larger group.*

of symptoms then develops each of which may vary in severity. The essential (cardinal) signs and symptoms consist of: tremor, rigidity, akinesia and defects in postural and righting reflexes. Charcot essentially described this total picture and he named the disease after Parkinson (Goetz, 1986). The cardinal signs and symptoms are summarized in table 19-4.

Some cases, throughout their course may continue to manifest primarily tremor as the major symptom and such cases tend to have a more favorable prognosis. Several types of Parkinson's disease and syndromes may be specified. The most common variety is the idiopathic or primary Parkinson's disease, a degenerative disease arising insidiously in the patient of age 40, 50 or 60. In a series of 1644 patients with Parkinsonism, evaluated by Jankovic (1989) at a movement disorder clinic, 82% were found to have idiopathic Parkinson's disease. At times, there is a family history of the same disease.

1. Pathology of Parkinsonian Lesions. The basic pathology involves a progressive loss of the pigmented neuromelanin containing neurons of the pars compacta of the substantia nigra *(Fig.19-3)*. (The pigmented neurons of the locus ceruleus are also affected.)[4] (The student will recall that both melanin and dopamine are steps in the metabolic pathway involving phenylalanine.) A long pre-clinical period estimated at 5 years occurs in this progressive disease. Various estimates suggest that when symptoms first occur, 70-80% of striatal

TABLE 19-4: CLINCAL SIGNS IN PARKINSON'S DISEASE

Clinical Signs	Clinical Appearance
Rhythmic tremor at rest: Alternating type, initially fine, 4-5 Hz, and then, as the disease progresses, often Coarse. The tremor disappears on movement, but may re-emerge when a posture is maintained	The tremor affects not only the extremities, but also the eyelids, the tongue and voice. The alternate movements of thumb against opposing index finger are often referred to as "pill rolling".
Clinical rigidity which must be differentiated from decerebrate rigidity which is primarily spasticity with the jack-knife quality of spasticity (a sudden resistance is encountered, which, as additional force is applied, suddenly gives way).	Comparable to the resistance of a lead pipe bending under force in which the resistance is relatively constant throughout the range of motion. As with spasticity, rigidity may reflect increased activity of the gamma system.
Akinesia- lack of spontaneous movement and difficulty in initiating movement. A corollary term is bradykinesia which refers to a slowness of spontaneous movement.	Seen in the fixed facies of the Parkinsonian patient, lack of spontaneous blinking and movement, loss of associated movements (the swing of the arms in walking). Correlated as discussed above with the reduction in dopaminergic input.
Defects in posture righting reflexes are best noted as the patient stands, walks and attempts to turn.	Normally, as an individual turns, the eyes move initially in the direction of the turn, his head moves, then shoulders and arms, and the body then follows. The patient with Parkinson's disease, however, turns "en bloc", In walking at times, develops forward propulsion, seems to tilt forward and develop increasing speed – producing loss of balance with falls and injuries, and instability when turning.

---

[4] *Whether the Dopamine-nigral-striatal-(mesostriatal) system involvement is the entire story in Parkinson's disease is unclear. The mesolimbic cortical Dopaminergic system is also affected (50-60% decrease). Noradrenergic and serotoninergic and cholinergic systems are also affected; e.g., 50-60% decrease in nor-adrenalin concentrations in cerebral cortex, locus ceruleus, limbic system, hypothalamus and cerebellum; 58% loss of serotonin in dorsal raphe nucleus. There is also cholinergic nerve cell loss of 32-87% in nucleus basalis of Meynert. In general, none of these systems show the degree of degeneration found in the nigral-striatal-dopaminergic system (Estimated as 80-90% depletion of striatal dopamine and 60-70% degeneration of S. Nigral neurons when symptoms first emerge). For extensive reviews and discussions, see Agid et al, 1987, 1989.*

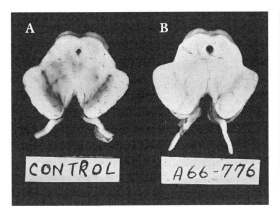

FIGURE 19-3. Parkinson's Disease. The substantia nigra in Parkinson's disease. A, Normal substantia nigra. B, Similar region in case of idiopathic Parkinson's disease. A marked loss of pigmentation is evident. (Courtesy Dr. Thomas Smith. Neuropathology University of Massachusetts)

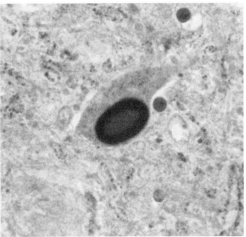

Figure 19-5. Lewy body. Slide stained for alpha synuclein.100X. (Courtesy of Dr. Thomas Smith).

dopamine is depleted. In terms of neurons in the pars compacta of the substantia nigra 50% (compared to age matched controls) to 70% (compared to young individuals) are already lost. Normally the rate of cell loss in the nigra during normal aging is 5% per decade. In patients with Parkinson's disease the rate of cell death is subsequently 45% per decade.

On histologic examination (Fig. 19-4), there is a characteristic, specific finding of Lewy bodies: round, eosinophilic bodies surrounded by a clear halo within the cytoplasm of surviv-

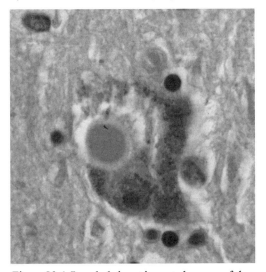

Figure 19-4. Lewy body in a pigmented neuron of the substantia Nigra. H&E stain, original at 125X. (Courtesy of DR. Thomas Smith).

ing neurons of the substantia nigra. In many cases, these Lewy bodies are found in other neurons, e.g., cerebral cortex. Immunocytochemistry has demonstrated that a small protein a-synuclein is a major component of the Lewy body (Fig.19-5). Under normal circumstances this protein is located in presynaptic nerve terminals. It is therefore likely that the Lewy body represents degenerated and aggregated neurofilaments. The Lewy body may also be found diffusely in cerebral cortex in Lewy body dementia.

2. Etiology of Parkinson Disease. Why do the neurons degenerate? The majority of cases with L-DOPA responsive Parkinson's disease do not have a clear-cut genetic basis. However mutations in the alpha synuclein gene have been found in several families of Italian or Greek origin with early onset autosomal dominant Parkinson's disease mapping to locus 4Q21-25 (Goedert et al, 1998 and Duvoisin, 1998). Other families with autosomal recessive juvenile onset disease have implicated the parkin gene at locus 6q25.2-27.Other gene loci are discussed by Dunnett & Bjorklund, 1999.

The major etiologic factors have involved a genetic predisposition and possible toxic exposures.

Neuroleptic agents: The use of neuroleptic agents may unmask an underlying genetic predisposition to Parkinson's disease. A significant percentage (15-60% depending on age), of patients who receive neuroleptic tranquilizers also will develop symptoms of Parkinsonism. These agents include pheno thiazides such as chlorpromazine, and haloperidol and reserpine. Reserpine interferes with the storage of dopamine within the nerve cells and, thus, depletes the brain of this neurotransmitter. Pheno thiazides and related compounds, on the other hand, block the post-synaptic receptors at dopaminergic synapses. In general, of the symptoms produced by these agents, the akinesia, rigidity and tremor will disappear when the agent has been cleared from the body. However, in some cases, the Parkinsonian features appear to remain as a permanent deficit. In such patients, one may find evidence in the family history of other cases of Parkinson's disease (Schmidt and Jarcho, 1966). As discussed by Koller (1992), there is then a clear indication that neuroleptic induced Parkinsonism represents early or latent Parkinson's disease made evident by anti-dopaminergic medications.

Toxic Agents. The major impetus for the study of possible toxic exposure has come from the study of cases of Parkinson's disease developing in miners exposed to manganese and more recently in cases exposed to the agent MPTP.

Manganese Poisoning. A relatively rare cause of a Parkinsonian syndrome is manganese poisoning. Manganese apparently accumulates in the melanin containing neurons of the substantia nigra and interferes with the enzyme systems involved in the production of dopamine (Mena et al, 1970),

MPTP Toxicity. Another form of Parkinson's disease induced by the drug MPTP, has provided major insights in the possible etiology of the degeneration of the dopamine containing neurons in the substantia nigra. (MPTP is the abbreviated term for 1-Methyl- 4 phenyl-1, 2,3,6, tetrahydropyridine). This agent is a by-product of the chemical process for the production of a reverse ester of meperidine (demerol): 1-methyl-4 propi-onoxypiperidine (MPPP). Meperidine is a controlled narcotic; MPPP is not controlled, has narcotic action, and can be produced in clandestine labora-tories. MPTP had been recognized shortly after its legitimate synthesis in 1947 to produce a severe rigid - akinetic state in monkeys. However, it was not until the summer of 1982, when MPPP was produced on a large scale in Northern California, for illicit mass distribution, that significant numbers of young drug abusers began appearing at emergency rooms with an acute Parkinsonian syndrome. The investigation of these cases has been presented in a series of papers by (Langston et al, 1983; 1992; and Ballard et al, 1985). Essentially, the compound MPTP has selective toxicity on the dopaminergic neurons of the pars compacta of the s. Nigra, not only in human but also in subhuman primates, such as the Rhesus monkey, Macaca mulatta, and the squirrel monkey. Both in the human and the monkey, MPTP reproduces the clinical disease, the pathological findings and the effects of pharmacological therapy of Parkinson's disease to a relatively complete degree, including in older animals, the presence of Lewy bodies. With such an animal model, experimental medical and surgical therapy can be studied with a clear-cut correlation with the human disease (for example, see Aziz et al, 1991; Langston, et al, 1984,Kopin 1988). Moreover, since there were many patients exposed to the agent but not all exposed patients and not all acute cases developed the chronic Parkinsonian state, prospective studies can be performed as to the natural history and factors influencing the onset of the disease (Stone, 1992). In part because of MPTP, increased attention has been focused on other environmental exposures that may be involved in producing neurotoxic effects culminating in Parkinson's disease (Tanner&Langston, 1990,Olanow &Tatton, 1999).

Other pathological processes These disorders also will produce some of the symptoms of Parkinson's disease without the presence of Lewy bodies (Gibbs, 1988).

1. Bilateral necrosis of the globus pallidus

may produce akinesia. Such a necrosis could occur in carbon monoxide poisoning, or anoxia, or in relation to infarction. Rigidity in flexion may also follow bilateral necrosis of the putamen or the globus pallidus.

2. A tremor at rest may occur with destructive lesions involving the ventro-medial tegmentum of the midbrain. Such lesions, apparently, interrupt the pathway from the substantia nigra to the striatum. A degeneration of neurons in the substantia nigra and a decrease in dopamine in the corpus striatum occurs.

3. "Arteriosclerotic or vascular (multi infarct)" variety in which bilateral infarcts occur in the putamen Usually this is the accompaniment of a lacunar state in which multiple small infarcts occur in the basal ganglia and internal capsule. The resultant syndrome of rigidity and akinesia affects the lower half of the body and thus can be clearly distinguished from the classical idiopathic l-dopa responsive Parkinson's disease.

4. Von Economo's Encephalitis. A certain proportion of cases may be termed post-encephalitic. These cases had a clear onset in relation to the epidemic of Von Economo's encephalitis lethargica in the years 1916-1926. Approximately 50% of the survivors of this devastating disease developed a Parkinson's syndrome due to severe loss of neurons in the substantia nigra. In such cases, Lewy bodies are not found but neurofibrillary tangles are present in surviving cells (see Alzheimer's disease in Chapter 30 for a discussion of neurofibrillary tangles). In addition to the aforementioned Parkinsonian symptoms and signs, these patients presented other evidence of involvement of the central nervous system, including oculogyric crises (tonic vertical gaze), chorea, dystonia and sleep disorders. The symptoms and signs indicated the involvement not only of the substantia nigra, but also of multiple other sites including the midbrain, hypothalamus and hippocampus (see Gibb, 1988). The essential distribution of the pathology was in the periventricular areas about the aqueduct and third ventricle A dramatic example of post encephalitic Parkinsonism and its treatment is presented in the movie "Awakenings", based on the book by Oliver Sacks.

Management (Refer to Lang & Lozano, 1998 and Riley &Lang, 2000).

When one considers that the basic pathology in Parkinson's disease represents a marked loss of Dopamine containing neurons in the S.Nigra and a marked decrease in Dopamine concen-tration in the striatum whereas, other transmitters and neurons are affected to a much lesser degree (e.g., cholinergic system within corpus striatum now has a relatively unopposed action) then the possible types of treatment are very evident.

1. *Early medical treatment:* For many years, the standard treatment involved the use of anticholinergic compounds such as belladonna and atropine or of synthetic analogues. Trihexyphenidyl (Artane) and benztropine (Cogentin) are examples of such agents. The anticholinergic compounds affect primarily the tremor and to some extent the rigidity of Parkinson's disease and in many cases may produce limited improvement.

2. *Early approaches to surgical intervention.* Various surgical procedures were attempted in the 1940s and 1950s, involving limited ablations of area 4 or area 6 or section of the cerebral peduncle. In 1952, while attempting to section the cerebral peduncle, Cooper accidentally occluded, the anterior choroidal artery, and the Parkinsonian patient showed a significant improvement in contralateral symptoms, presumably as a result of infarction of the inner section of the globus pallidus. Cooper soon came to employ stereotaxic techniques for the direct production of lesions in the globus pallidus (Fig. 19-2 lesion 1). The procedure produced a significant reduction in contralateral tremor and to a lesser extent, a decrease in the contralateral rigidity. Cooper was able to demonstrate later that lesions in the ventrolateral thalamus were much more effective (Fig.19-2, lesion 2). Initially, the stereotaxic lesions were produced by the injection of alcohol or coagulation. In later procedures, a freezing cannula was employed.

3. *Transmitter replacement:* With the increasing knowledge of the neurotransmitters involved in the substantia nigra and corpus striatum and with reports of deficits of dopamine in these structures in patients with Parkinson's disease, the use of replacement therapy with L-Dopa (dihydroxyphenylalanine) subsequently developed (Cotzias et al, 1967, 1969). As we have indicated, the anticholinergic drugs had a limited effect primarily on the tremor and rigidity; surgery affected primarily the tremor; the use of L-Dopa, which was more a physiological technique, improved the severe akinesia in addition to decreasing the tremor and rigidity. The use of L-Dopa required large amounts of medication since much of the administered dosage was decarboxylated to dopamine at peripheral systemic sites and never crossed the blood brain barrier to be converted to dopamine (note that L-Dopa-crosses the blood brain barrier; dopamine does not). There were significant side effects due to the peripheral actions of the drug. By combining L-Dopa with a peripheral Dopa decarboxylase inhibitor, Carbidopa (in the form of Sinemet) a much lower dosage of L-Dopa was required with fewer gastrointestinal and blood pressure side effects. Central side effects continued to occur, e.g., peak dose related dyskinesias such as chorea and myoclonus. The short duration of action also continued to provide problems in terms of wearing off of dose. Other sudden on-off effects with sudden arrests of movement that were not entirely related to time of dose also occurred as the disease progressed. Individual titration of dosage, specific to the particular patient is often required. The basic problem is that Parkinson's disease is progressive. Once the neurons in the s. Nigra have all died off or have become dysfunctional, there is no capability to convert L dopa to dopamine.

4. *Other therapeutic maneuvers* have been developed for such patients. Various dopamine agonists have been developed to bypass the problem of conversion of L-dopa to dopamine. Additional surgical procedures involving the implantation of stimulators in deep brain structures such as the globus pallidus, thalamus and subthalamic nucleus have also been developed. Transplantation of fetal mesencephalic cells into the striatum has benefit in younger patients but may increase dyskinesias. The problems of management are reviewed in the following patient who was followed for 19 years.

Case 19-1: This 48-year-old married left-handed white male schoolteacher and administrator, 3 months prior to evaluation in 1982 developed a sense of fatigue, stiffness and lack of control in the left arm. After a period of prolonged writing, he would have to make a conscious effort to continue writing Approximately one month prior to the evaluation; he first noted a slowing down in movements of the left leg. "I have to remember to lift it."

Neurological examination: *Cranial Nerves:* He had a tremor of the closed eyelids and a positive glabellar sign (he was unable to suppress eyelid blinking on tap of forehead). *Motor System:* In walking there was a tendency to turn en bloc and a decreased swing of the left arm. There was a slight increase in resistance to passive motion at the left elbow, wrist and knee (rigidity without definite cogwheel component). Although the patient was left-handed, there was a slowness of alternating finger movements of the left hand. No tremor was present. Handwriting was intact with no evidence of micrographia.

Clinical diagnosis: Early Parkinson's disease, predominantly unilateral.

Laboratory data: All studies were normal including CT Scan of the brain, and thyroid studies.

Subsequent course: The patient did not

---

[6] *This is a combination of 10 mg. of carbidopa (Dopa decarboxylase inhibitor) and 100 mg. of L-Dopa. Higher dosage combinations are a) 25 mg. of carbidopa with 100 mg L-Dopa, and b) 25 mg. carbidopa with 250 mg. L-Dopa. A sustained release form is also available (50 mg. of carbidopa with 200 mg. of L-Dopa), and is marketed as Sinemet CR.*

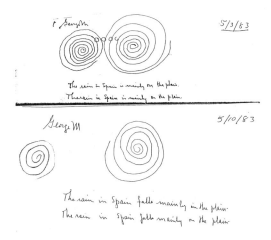

Figure 19-6. Parkinson's disease. Case 19-1. See text. Handwriting sample: on 05/03/83 before therapy with L-dopa/Carbidopa and 05/10/83-one week after starting therapy. In each example, the larger circles have been drawn by the examiner.

wish to begin any treatment at that point in time. Re-evaluation at 3 months indicated progression. One year after onset of symptoms there was significant micrographia *(Fig. 19-6A)* and cogwheel rigidity at left shoulder, elbow and wrist. Therapy with Sinemet (10-100 mg, 3 times per day.)[6] was begun. Re-evaluation one week later demonstrated a marked improvement in handwriting *(Fig.19-6B)* and decreased cogwheel rigidity.

As the dose of Sinemet was increased to (10/100) mg 5x/day over the next week, occasional choreiform movements of the fingers of the left hand and occasional backward flinging (hemi ballistic) movements of the left arm occurred at 60-90minutes after later doses of Sinemet. As the dose reached 10-100 mg. 6x/day, the patient reported that his handwriting now was back to normal. Examination confirmed the significant improvement in all findings. The patient did experience 90 minutes after a Sinemet dose - a tremor of the left shoulder and an exaggerated grasp of the left foot toes on lifting the leg (possible form of dyskinetic or dystonic reaction). He was aware that some symptoms of his underlying disease would emerge 3-4 hours after a dose of Sinemet. Five years after onset of symptoms, despite a higher dosage of Sinemet of 25/100 mg. tablets, 7 or 8 x/day, more resting pill-rolling tremor of the left hand had emerged. (In actuality 75-100mg of carbidopa is probably sufficient to saturate the peripheral dopa decarboxylase system). Moreover, choreiform -movements of the left arm occurred 60 to 90 minutes after each dose of Sinemet. The choreiform movements disappeared when the dosage of Sinemet was reduced to 25/100mg 6/day. The increased rigidity and tremor responded to the addition of a DOPA agonist Bromocriptine (Parlodel). Over the subsequent years other agents were employed but the disease continued to progress with all of the problems in management discussed above. Eventually, in 1997 he underwent mesencephalic fetal cell transplantation. Evaluation 3 years after the transplant demonstrated a little improvement but a marked increase in the dyskinesias. Some improvement in dyskinesias occurred after dopaminergic medications again were reduced.

DIFFERENTIAL DIAGNOSIS OF PARKINSON'S DISEASE:

1. *In considering the akinesia and bradykinesia of Parkinson's disease, the major differential diagnosis in the older population are the syndromes of gait apraxia related to frontal lobe,* as discussed in Chapter 18. These may be outlined as follows:

   a. Normal pressure hydrocephalus

   b. Other types of hydrocephalus: late decompensation of aqueductal stenosis

   c. Multiple lacunar infarcts (the lacunar state)

   d. Leukoariosis - or Binswanger's Disease involving the periventricular white matter and seen in elderly hypertensive patients

   e. Alzheimer's disease: Note, however that Mayeux et al, 1985, found that 34% of patients with Alzheimer's disease had some extrapyramidal signs. To confound the picture, 44% of patients with Parkinson's disease have dementia, with 29% found to have Alzheimer's disease and 10 % Lewy body dementia.

   f. Frontal lobe tumors, e.g., subfrontal or parasagittal meningiomas.

   g. Myxedema with hypothyroid state.

h. General paresis of neurosyphilis, now uncommon but refer to case 30-5.

In general, frontal lobe syndromes do not have a prominent tremor (unless a coincident essential or senile tremor is present). Rigidity may be present but this variable rigidity is of the frontal lobe type referred to as gegenhalten. A cogwheel component (to this frontal lobe rigidity) is usually not present unless there is a superimposed tremor of other cause. The rigidity and akinesia predominantly affect the lower extremities.

2. *In considering the tremor of Parkinson's disease - the major differential diagnosis is essential (or senile or familial tremor.)* In general, this is a tremor of the outstretched hand with some increase as termination of movement or when a sustained posture is maintained. The tremor is not present at rest and there is no pill rolling quality. Rigidity is not present, and gait is normal with a good swing of the arm. The tremor is usually long-standing without major disability. Note that there is a small group of Parkinson's disease patients with favorable outcome who have a resting tremor as a predominant symptom over many years. They will, however, have to a minor degree, on careful examination, many of the other cardinal symptoms of Parkinson's disease.

3. *Secondary Parkinsonism: Neuroleptic, toxic or vascular induced.* In the series of Jankovic (1989), and Siemer and Reddy (1991), these cases accounted for 10% of all cases of Parkinsonism.

4. *Parkinsonism plus syndromes.* (Table 19-5) In the same series these cases accounted for 10% of all cases of Parkinsonism. These latter patients all share the following characteristics:

a. A progressive disorder is present.

b. From a clinical and pathological standpoint, the pathology of primary neuronal degeneration is much more widespread than in idiopathic Parkinson's disease. The term multisystem atrophy is often applied to some of these cases.[7]

c. Lewy bodies are not found.

d. In general, the patients are all poorly responsive to Levodopa. (Since the pathology extends far beyond the Dopaminergic producing neurons of the S. Nigra.)

e. Some of the patients have a familial disorder.

The more frequent types may be briefly noted (excellent, more detailed discussions are found in Jankovic, 1989, Riley &Lang, 2000).

5. Other neurological syndromes with rigidity and akinesia:

a) Early onset Huntington's Disease and Wilson's disease (see below).

b) Hallervorden Spatz Disease: A rare familial (autosomal recessive disorder with linkage to chromosome 20P) in which prominent iron deposits and neuronal degeneration in globus pallidus and substantia nigra are associated with childhood onset of Parkinsonian syndrome, dementia, spasticity, dystonia and chorea.

c) Parkinsonism - dementia - ALS complex found in the Chamorros of Guam (and several other Pacific groups)- a probable toxic disease involving an excitatory amino-acid (beta N methyl-amino L - alanine) - derived from the cycad seed once used in the production of floor.

d) Creutzfeldt -Jakob Disease: See discussion of dementia (chapter 30)

CHOREA, HEMICHOREA AND HEMIBALLISMUS:

Chorea[8] may be defined (Committee 1981) as "excessive spontaneous movements irregularly repeated randomly distributed and abrupt in character". These movements,

---

[7] *Many patients with Parkinson's Disease have some symptoms, usually minor, that suggest some degree of involvement beyond the S.Nigra and beyond the Dopaminergic nigral - striatal system, e.g., orthostatic hypotension which is present in the untreated patient. Idiopathic Parkinson's patients also may have some minor degree of upward gaze impairment.*

[8] *At times the adjective choreiform is used to describe the movement rather than employing the noun: chorea*

TABLE 19-5 PARKINSONISM PLUS SYNDROMES

Name	Abbreviation	Pathology-type	Location	Clinical syndrome
1. Progressive supranuclear palsy (the most common)	PSP	Neuron loss and abnormal tau protein in neurofibrillary tangles in neurons and glia Cortex and subcortex	Midbrain tegmentum and ectum, substantia nigra, globus pallidus, subthalamic nuc. basal nuc., cortex	Parkinsonism plus supranuclear impairment of gaze particularly in vertical plane Imaging=atrophy of midbrain
2. Corticobasal degeneration	CBD	Ballooned and achromatic neurons plus abnormal neurofibrillary tangles with tau protein. Linked to chromosome 17 in familial cases	Asymmetrical involvement of frontal –parietal cortex (superior frontal), S. nigra, caudate, thalamus	Akinetic-rigid Parkinsonism plus, dementia plus lateralized apraxia, useless hand, and lateralized myoclonus Overlap with Pick's, Frontal dementia and PSP. Imaging= focal atrophy of frontal- parietal cortex.
3. Multiple system atrophy Incorporates 3 previously separate disorders a. striatal nigral degeneration b. olivopontocerebellar atrophy c. Shy-Drager Syndrome	MSA	Neuronal loss, gliosis, and inclusions in oligodendroglia neuronal cytoplasm, nuclei and glia containing filaments derived from cytoskeletal proteins: tau, ubiquitin and alpha synnuclein	S.nigra, striatum, olivopontocerebellar pathways plus intermediolateral cell column of spinal cord	Akinetic –rigid Parkinsonism (predominates in 80%) plus cerebellar ataxia (predominates in 20%) plus dysautonomia (with neurogenic orthostatic hypotension) Imaging = atrophy of striatum or pons/cerebellum depending on predominant symptoms

although involuntary, consist of fragments or sequences of normal, coordinated movements. Face, mouth, head, proximal or distal limbs may be involved.

Two types of processes must be distinguished – (1) lateralized hemichorea and (2) generalized chorea. The lateralized hemichorea merges into a remarkable disorder hemiballismus characterized by sudden flinging or ballistic movements at the proximal joints such as. throwing or sudden swinging movements at the shoulder joint. Patients may start with hemichorea and evolve into hemiballismus. In discussing this topic, we will deal first with the focal problem and then with the generalized type. We will see in both varieties abnormalities of the anatomic sequences discussed earlier in this chapter.

HEMICHOREA AND HEMIBALLISM:

Initial clinical and experimental studies, correlated these disorders with lesions of the contralateral subthalamic nucleus (see Carpenter, et al, 1950; Whittier and Meller, 1947; Whittier, et al, 1949). It was recognized that the underlying lesion could be a small infarct due to posterior cerebral penetrating branch involvement, often on a lacunar basis, in a patient with hypertension or diabetes mellitus. At times, the adjacent thalamic areas were involved, as well, e.g., the ventral posterior lateral or ventral lateral nucleus. At times, the responsible lesion was a small hemorrhage in the subthalamic nucleus *(Fig.19-10)* and, rarely, a small metastatic tumor.

As regards the explanation for such contralateral excessive movement, hemichorea or hemiballismus, one needs only to consult Figure 19-1. Decreasing the excitatory drive from the subthalamic nucleus actingoin the medial globus pallidus and the substantia nigra reticularis would decrease the inhibitory drive from those structures acting in the thalamus.

Therefore the thalamic/premotor-SMA circuit would be more active and more movement would result.

There were other clinical studies that did suggest other possible localization in the striatum (Cooper, 1969; Goldblatt, et al, 1974; Martin, 1957; Meyers, 1947; Schwarz and Barrows, 1949). In experimental studies Crossman, et al, 1988 demonstrated that local injection of a GABA antagonist into the border of the lateral segment of the globus pallidus and the adjacent medial segment of the putamen would produce hemichorea in the monkey. Such lesions interrupt the inhibitory afferent input to the lateral globus pallidus from the putamen. The lateral globus pallidus would then have greater inhibitory drive on the subthalamic nucleus. The overall effect then would be equivalent to producing a lesion of the subthalamic nucleus; hemichorea or hemiballismus would result. (See Mitchell et al, 1989).

The responsible lesion in these cases would be in the territory of the lenticulo striate penetrating branches of the middle cerebral artery.

Another factor to be considered in chorea and other induced dyskinesias is the possible unopposed action of the nigral striatal Dopaminergic circuits as discussed earlier in this chapter. L-Dopa will exacerbate choreiform disorders[9]. Agents that block dopamine receptors, e.g., the butyrophenones (such as haloperidol) are often very effective in alleviating hemichorea, hemiballismus and chorea (see Crossman, 1990; Klawans, et al, Albin Young and Penney, 1989) for additional discussion.

Hemichorea: Case history 19-2 presented below provides an example of hemichorea with MRI correlation of lesion location in the striatum (caudate/putamen) following a hemorrhage into those structures.

Case 19-2: This 88-year-old, right-handed, widowed white female with a past medical history of profound hearing impairment, adult onset diabetes, euthyroid goiter, coronary artery disease, and hypertension, 5 days prior to admission, developed the insidious onset of involuntary rotary movement in the left foot which increased in intensity and progressed to involve the left shoulder. The movements were constantly present, even during sleep and with progression; her ability to ambulate with a walker was impaired. She also had a dull headache greater on the left side than on the right during the last few days.

Neurological examination: *Cranial Nerves:* Intact except: for left eye blindness (cataracts), and gross impairment of hearing such that shouting was necessary to communicate. *Motor System:* There was minimal weakness in the left upper extremity (5-/5). There was a persistent hemichorea: involuntary twisting movements of her left foot and upward and rotary movements of her left shoulder. The movement was exacerbated by motor-- tasks but did dissipate with sleep. Gait was slightly unsteady due to hesitant placement of the left foot. She did quite well with minimal assistance. *Reflexes:* Deep tendon reflexes were symmetrical, and physiologic except for decreased Achilles secondary to diabetes. the left plantar response was extensor. *Sensation:* Symmetrically diminished vibration sense in toes..

Clinical diagnosis: hemichorea

Laboratory data: *MRI scan* 3 days after admission, revealed mild cortical atrophy and a hyperdense area the right basal ganglia (particularly the head of caudate and putamen) and external capsule consistent with subacute hemorrhage *(Fig.19-7)*.

Hospital course: The patient was begun on haloperidol 0.5 mg by mouth every morning and advanced to 1 mg., by mouth three times per day with improvement in the hemichorea, particularly in the upper extremity over the next two weeks

Hemiballism: As we have previously indi-

---

[9] *L-Dopa induced dyskinesias are not observed in intact monkeys and humans who have normal dopaminergic and normal basal ganglia function. Parkinsonian patients, monkeys with MPTP lesions and patients with a predisposition to Huntington disease will develop dyskinesias after receiving L-Dopa (see discussion below and Luquin et al, 1992).*

cated, the movements of hemi-chorea may evolve or merge into more violent, uncoordinated, rotatory, flinging movements at the shoulder joint and other proximal joints; termed hemi ballistic. The responsible lesion was located in the subthalamic nucleus *(Fig. 19-8)* or in the putamen or at the border of lateral pallidum - putamen, as discussed above.

The following case history illustrates this movement disorder in a patient seen prior to the era of CT and MRI scans in which the localization is relatively clear.

Case 19-3: This 79-year-old, right-handed, white housewife had the abrupt onset in the early morning hours, 2 days prior to admission, of almost constant flinging movements of the right arm over which she had no control. At the same time, she noted that her right arm felt numb and heavy. Over the next 2 days, the movements decreased markedly and she regained more control of the arm.

Past history: The patient had been a known diabetic for 21 years, receiving insulin - most recently, lente insulin, 25 units each morning.

Neurological examination: *Motor system:* Minimal weakness was present in the right upper extremity at the elbow, wrist and fingers. Alternating movements and finger-to-nose testing in the right hand were markedly impaired. Occasional involuntary flinging movements occurred at the right shoulder. *Sensory System:* Pain, touch, vibration and position sense were all absent in the right upper extremity to the elbow and decreased in this extremity above the elbow.

Clinical diagnosis: hemiballismus due to a lesion of penetrating branches of the posterior cerebral artery, which involved the subthalamic nucleus and ventral posterior lateral nucleus of thalamus.

Laboratory data: Chest and skull X-rays EEG, and cerebrospinal fluid studies were all normal.

Subsequent course: The flinging movements of the right arm subsided spontaneously, shortly after admission. Position sense returned and pain sensation showed a mild improvement. The ataxia on finger-to-nose testing disappeared.

GENERALIZED CHOREA: Bilateral generalized chorea may occur under the following circumstances:

(1). *Acute onset with an immunological basis* (a) one aspect of rheumatic fever (plus or minus carditis and arthritis). (b) In relationship to various autoimmune disease such as lupus erythematosus.

(2). *As an acute complication of L-dopa therapy* in patients with compromise of the dopaminergic system or with disease of the basal ganglia - as discussed above.

(3). *As an acute or chronic complication of various metabolic disorders:* hepatic, renal, hypocalcemia, etc. rarely occurring in pregnancy. In acute hepatic encephalopathy, a number of toxic substances that would normally be removed from the portal venous system on passage through the liver are not removed. Instead, in patients with liver disease, portal hypertension may develop, portal - systemic shunting develops and toxic substances such as ammonia may accumulate in blood and brain. In addition to depression of consciousness (hepatic coma), a hepatic "flap" or "asterixis" may develop. With the arms outstretched, extended at the elbows, and the hands extended at wrist, an irregular flexion extension movement will develop at the wrists and at the metacarpal - phalangeal joints. A similar "asterixis" may also be noted in other metabolic diseases such as uremia and pulmonary disease with CO_2 retention. In all of these disorders: ataxia, dysarthria and action tremor of hands may also occur. With uremia and hypocalcemia, myoclonus (sudden jerks of the extremities) or generalized convulsions may occur.

With repeated episodes or prolonged periods of hepatic encephalopathy, more profound and lasting changes in neurological function occur producing a syndrome characterized by chorea and athetosis, ataxia, dysarthria, tremor and dementia. Neuro pathological examination of the brain demonstrates cavities within the basal ganglia, cerebellum, and cerebral cortex. Swollen astrocytes are found, particularly

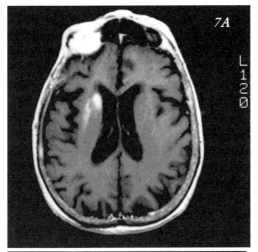

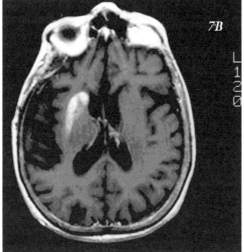

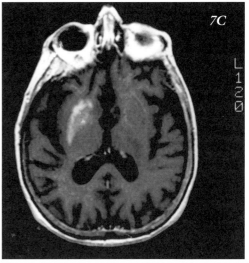

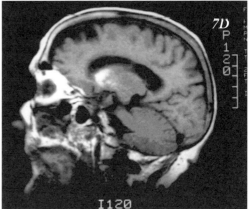

Figure 19-7. Hemichorea. Case 19-2 See Text. MRI demonstrates a hyperdense area of the head of the right caudate, adjacent putamen and external capsule - consistent with a hemorrhage. A, B, C - sequential horizontal sections at 5 mm. intervals - T1; weighted D - Sagittal section 12.5 mm to right of midline.

within basal ganglia. This chronic irreversible syndrome was initially described as the acquired non-Wilsonian type of hepatocerebral degeneration. (Victor et al, 1965).

(4) *As a chronic progressive disorder - indicating a degenerative disease - affecting the striatum and other sectors of basal ganglia and cerebral cortex.*

The most frequent disease in this category is Huntington's disease. Huntington's disease is of importance beyond its frequency in the population of 5-10/100,000. The disease was originally described in 1872 by George Huntington, based on clinical observations made by his grandfather, his father and himself of an apparent autosomal dominant syndrome of high penetrance that occurred in families living on Long Island in New York state. The syndrome was characterized by the development in mid life of a progressive psychological change (depression, paranoia, psychoses), dementia and involuntary choreiform movements. The disease often affected multiple members of a family over a number of generations. Genetic studies have confirmed that the pattern of inheritance follows an autosomal dominant pattern with extremely high penetrance. In large series, 50% of offspring will manifest the disease that is all who carry the affected gene will manifest the trait. The

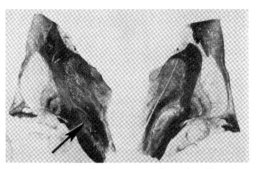

Figure 19-8. Hemiballismus. Myelin stain of basal ganglia demonstrating a discrete hemorrhage into the right subthalamic nucleus. The arrow points to the subthalamic nucleus of the normal left hemisphere. (From Luhan, J.A.: Neurology, Baltimore, Williams & Wilkins, 1968, p.334).

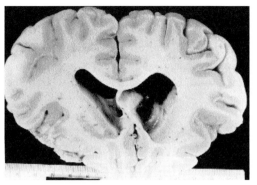

Figure 19-9. Huntington's disease. Marked atrophy of the caudate and putamen with secondary dilatation of the lateral ventricles is evident. Cortical atrophy was less prominent in this case. (Courtesy of Doctor Emanuel Ross, Chicago.)

expression of the gene may vary. Mean age of onset in a large sample as defined by choreiform movements was 42 years. However, behavioral changes, including depression and suicide attempts may precede the movement disorder by ten years or more. Earlier onset of disease is associated with a more rapid course and with the earlier development of features of rigidity. Such patients, for unknown reasons, are more likely to have inherited the gene from the father. Late onset cases (onset between 50 and 70 years of life) have a much more prolonged course and choreiform movements are often the predominant feature (see Sax and Vonsattel, 1992). At times the term senile chorea has been utilized for these late onset cases, which are more likely to have involved cases in the maternal line.

Modern genetic analysis using recombinant DNA techniques and studying families with large numbers of affected individuals or large numbers at risk have allowed the localization of the marker site for the gene on the short arm of chromosome 4. (See Gusella et al, 1983; Martin and Gusella, 1986; Penny and Young, 1988; Roberts, 1990; Wexler, et al, 1991). The specific mutation has now been identified as an unstable trinucleotide repeat in the CAG series coding for poly glutamine tracts at the 4p16.3

locus on this chromosome. Normal individuals have 6-34 repeats; patients with Huntington's disease have 36-121, (Zoghi &Orr, 2000)[10]. The gene product is huntingtin. Patients with onset at an early age or with a more severe form of the disease have a greater number of repeats. As we have already discussed in chapter 9 on the spinal cord, similar mutations involving an excessive number of CAG repeats have also been found in a number of other neurological disorders including several of the spinocerebellar degenerations and spinobulbar muscular atrophy. Other disorders such as myotonic dystrophy and Friedreich's ataxia have an excessive number of repeats in other trinucleotide sequences. Presymptomatic detection is possible but this should be done under circumstances where a full gamut of genetic counseling is available. The underlying gross pathology has been well established *(Fig.19-9)*. Marked atrophy of caudate nucleus and putamen are the pathologic and neuroimaging hallmarks. This gross pathological feature allows for diagnosis and subsequent monitoring of progression by means of CT scan and MRI scan (Fig.19-9) (Vonsattel et al, 1985; and Myers et al, 1985). In addition significant cortical atrophy also occurs as the dis-

---

[10] *The Huntington Disease Working Group1996 however has set the affected range as greater than 38 repeats*

[11]*These neurons are located in the matrix. Matrix neurons that appear late in development are the earliest affected, it is not until very late in the disease that patch neurons are affected.*

ease progresses. The gene product huntingtin is expressed widely throughout the brain. There are high levels in large striatal interneurons and medium spiny neurons, as well as cortical pyramidal cells and cerebellar Purkinje cells.

Additional analysis has indicated that not all cells and transmitters in the caudate and putamen are involved in the process of degeneration. The extensive studies have been reviewed' by Albin et al, 1990; Martin and Gusella, 1986; Penney and Young, 1988; Young et al 1988; and Wexler et al, 1991. The cells affected are classified as medium sized, spiny neurons that project to sites outside the striatum and which constitute approximately 90% of all striatal neurons. All contain the transmitter GABA[11]. However, different subtypes are distinguished based on the specific associated neuropeptide that they contain e.g., enkephalin, substance P. and/or dynorphin and based on the projection destinations. On the other hand other cells with interneuron functions are not affected: (a) The large, spiny acetylcholine containing interneurons and (b) the small aspiny interneurons containing the neuropeptides somatostatin and substance Y. Refer to Table 2 above.

Early in the course of the disease there is a loss of those spiny neurons which contain the GABA and enkephalin transmitters and which project to the lateral globus pallidus. At this point in the disease, the loss of these striatal inhibitory inputs to the lateral globus pallidus results in increased activity (inhibitory) of the lateral globus pallidus neurons. Since the lateral globus pallidus is the main input to the subthalamic nucleus and this input is inhibitory, an increased inhibition of the subthalamic nucleus results. As a result, the excitatory output from the subthalamic nucleus to the medial globus pallidus (MGP) and S. nigra reticularis (SNR) is reduced. The output of neurons in MGP and SNR is reduced. This output is inhibitory and destined for ventrolateral thalamus. The output of ventrolateral thalamus is no longer inhibited - and increased activity occurs in the ventrolateral to supplementary motor cortex circuit. Abnormal excessive movements due to increased activity of SMA then results as discussed above under hemichorea

The following case history provides an example of Huntington's disease.

Case 19-4: This 54-year-old right-handed white female was referred for evaluation of a movement disorder. The patient lived alone and it was difficult to obtain much of any history from the patient. She denied any neurological or psychiatric disorders except for a problem with memory during the last year. She did indicate that she had worked for 10 years as a secretary/computer operator but had been fired 10 years ago because "she did not work fast enough". She had a past history of hypertension but had not been reliable in taking her prescribed medication. The patient's son was aware that changes in emotion, personality, speech, gait and a movement disorder had been present for at least 3 years.

Family history: The patient denied any neurological family history except for a maternal uncle who had the "shakes" (actually a maternal aunt with essential tremor), additional investigation revealed that her father clearly had Huntington's disease with onset of choreiform movements at age 30, and of irritability, paranoia and gait disturbance in his early 50s. The patient's eldest daughter age 34 had been depressed for 5 years and had problems with speech and walking for a number of years.

Neurological examination: *Mental status:* the patient often avoided eye contact. Affect was usually inappropriate with laughter as the response to many questions. The overall mini-mental status score was 27/30. The only deficit related to delayed recall portion of the test 0/3 objects. *Cranial nerves:* the patient had a hyperactive jaw jerk. She had frequent facial grimacing. *Motor system:* the patient had decreased tone at wrists and elbows. As she sat she had frequent restless and at times choreiform movements of hands and feet. Gait was often "bizarre" in terms of occasional sudden movements of a leg as moved down the hall and came to a stop. These same movements occurred, as she remained standing. In addition in standing or walking, there were intermittent dystonic postures of either left or right arm. *Reflexes:* Deep tendon stretch reflexes were everywhere hyperactive but plantar responses were flexor.

Clinical diagnosis: Huntington's disease.

Laboratory data: *MRI scans of head (Fig.19-10)*: The heads of the caudate nuclei were very atrophic. Cortical sulci were wide consistent with cortical atrophy. The lateral and third ventricles were secondarily dilated. PCR testing for the CAG trinucleotide repeats of the huntingtin gene indicated one allele normal at 17; and the second excessive at 42 repeats.

Subsequent course: The patient refused any treatment for the movement disorder. According to the son, the patient's neurological status worsened to a moderate degree over the subsequent 3 years as regards speech, unsteadiness of gait, and mood swings.

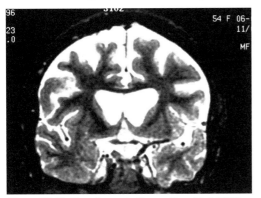

Figure19-10. Huntington's disease. Case 19-4. MRI. (T2) Refer to text. Marked atrophy of cerebral cortex and of caudate nucleus is evident in this patient with a familial history of the disorder and a significant increase in the CAG trinucleotide repeats.

Case 19-5 presented on CD ROM provides an example of a patient with Huntington's disease followed over a longer period of time. Onset occurred at age 32 with definite diagnosis of the disorder in both the father and paternal grandfather.

OTHER MOVEMENT DISORDERS ASSOCIATED WITH DISEASES OF THE BASAL GANGLIA:

1. **Double Athetosis:** Athetosis may be described as instability of posture, a relatively continuous alternation or swing between two positions. For example, in the hand, this would involve a swing from hyperextension of the fingers and thumb with pronation at the wrist to full flexion of the fingers with flexion and supination of the wrist. In a sense, there is an alternation between grasp and avoidance. Occasionally, following cerebral infarction at various age's athetosis may occur as a unilateral phenomenon admixed with hemichorea. In general athetosis occurs as a bilateral congenital phenomenon with involvement not only of the upper extremities but also of the lips, tongue and lower extremities. Double athetosis is one of the varieties of cerebral palsy discussed in chapter 18. The pathology of double athetosis is found in the putamen that may have a marbled or mottled appearance as a result of the presence of excessive numbers of abnormally situated myelin sheaths. To a lesser degree, the thalamus may also be involved. The etiology is not certain: anoxia at birth or other perinatal or prenatal pathology is considered the likely cause. Some of the cases can be related to kernicterus. A rise in bilirubin has occured in the perinatal period owing to hemolysis of red blood cells when an Rh factor or other incompatibility of blood type has been present.

2. **Hepatolenticular Degeneration (Wilson's Disease):** This is a familial disorder (recessive inheritance), which affects, to a variable degree, the liver and the central nervous system and in many cases, the kidney and bones. In the central nervous system, the most severe involvement occurs in the basal ganglia: the putamen and globus pallidus (cavitation, loss of neurons and an increased number of swollen astrocytes). To a lesser extent, the cerebellum and cerebral cortex are involved.

The basic etiology has now been clearly established as a metabolic defect. There is a deficient plasma level of the circulating copper-binding globulin, ceruloplasmin[12]. As a result, there is an excessive serum level of unbound copper. Normally, the binding of copper to ceruloplasmin prevents the passage of copper out of the serum. In Wilson's disease, however, the increased amount of unbound copper results in passage of this metal into the brain[13], liver, and kidney (The total combined level of both bound and unbound serum copper is less than normal). In some cases, usually of early onset, the symptoms of liver involvement predominate. Wilson's disease should always be suspected when hepatocellular disease develops in a child or adolescent. In other cases, the central nervous symptoms predominate and hepatic disease may be only minor. Deposition of copper also occurs in the cornea at the scleral junction. The resultant greenish brown pigmentation is referred to as the Kaiser-Fleischer ring. This ring is almost always present when central nervous system involvement is present. Overall, 54% of cases present with neurological onset, 31% with hepatic dysfunction, 14% with psychiatric symptoms, 2% with eye symptoms, 1 to 2% with hemolytic anemia and 1% with heart disease (see Patten, 1988). In patients with hypersplenism as a complication of the hepatic disease, thrombocytopenia may develop.

The actual symptoms referable to the neurological involvement depend on the age of the patient. In early onset cases (late childhood and early adolescence) rigidity predominates. When cases begin in early adult life, movement disorders predominate (tremor and choreoathetosis). The tremor is coarse and best described as sustained postural. The term "wing beating" is often used. The copper also produces damage to the renal tubules, resulting in glycosuria and aminoaciduria. Diagnosis can be based on the clinical findings, ceruloplasmin levels, (normal levels are 200to400 mg/liter), serum copper (normal total levels are 11 to 24 umol/liter versus 3 to 10 in Wilson's disease). The most definitive study involves analysis of copper content in a liver biopsy. The CT scan and MRI may demonstrate the lesions in the putamen and globus pallidus.

The treatment of Wilson's disease is dependent on a reduction of copper in the diet or the administration of an agent that will bind copper. Penicillamine (B.B-dimethyl- cysteine) is an effective chelator of copper. It is well absorbed on oral administration and promotes the urinary excretion of copper, resulting in a decreased level of copper in the serum, central nervous system, and liver.

The following **Case 19-6** presented on CD ROM illustrates the clinical problem of hepatolenticular degeneration in a 21 year old female who developed progressive neurologic syndrome characterized by mood swings, a wing beating tremor, choreoathetosis, dysarthria, ataxia and a Kayser Fleischer ring. Her brother had died at age 16 with hepatic disease, a Kayser Fleischer ring and choreoathetosis. The patient had serum elevation of free copper and a decrease in protein bound copper and an abnormal liver biopsy. She was successfully treated with penicillamine.

3. Dystonia:

Dystonia is a hyperkinetic disorder dominated by sustained muscle contractions frequently causing twisting and repetitive movements or abnormal postures. (Jankovic & Fahn, 1988) Dystonia may be focal (43%), segmental (30%), generalized (22%)or unilateral (5%). Focal dystonias are more common than the relatively rare generalized form.

Focal and segmental idiopathic dystonias: Focal dystonia affect predominantly a single body part primarily cranial and cervical muscles. Involvement of limbs is less common. Segmental refers to dystonia involving multiple segments usually cranial and cervical. Included within this category of cervical dystonia is the relatively common entity of spasmodic torticol-

---

[12] *In a few cases, ceruloplasmin levels are just within normal limits, but biliary excretion of copper is low. Absorption of copper is normal in all cases.*

[13] *The distribution of copper in the normal brain corresponds in general to the distribution of catecholamine containing neurons. Dopamine B hydroxylase is a copper containing enzyme with considerable localization to the basal ganglia. However, copper is also found in cytochrome C-oxidase that is found throughout the brain.*

lis in which sudden movements of the head to one side occur. The movement may be intermittent or sustained. Focal or segmental dystonia involving the cranial nerves is termed Meige's syndrome.

The most frequent of the cranial dystonias is blepharospasm in which there is forced eye closure. Other aspects of cranial dystonia may involve the tongue and jaw termed lingual and oromandibular. Other overlapping variants may involve the larynx-spasmodic dysphonia, or the pharynx with the latter producing difficulty in swallowing. Other focal dystonias known as occupational dystonias involve the more specific muscles of the hand and arm utilized in specific actions: writers, typists, violinists and pianists. The underlying neuropathology has never been clearly established in idiopathic cases. However, blepharospasm, or dystonia of the foot does occur in Parkinson's disease; retrocollis and oculogyric crises occur in post encephalitis Parkinson's disease. L-dopa and neuroleptic agents may both induce focal dystonias. The most effective treatment of focal /segmental dystonia is botulinum toxin injections that must be repeated every 3-4 months.

Generalized dystonia: The terms "primary generalized idiopathic dystonia" and "idiopathic torsion dystonia" have replaced the older term "Dystonia musculorum deformans". Most cases begin in childhood or adolescence and are familial. Autosomal dominant inheritance is now considered predominant in both Ashkenazi Jewish and non-Jewish families. A mutation in a gene on chromosome 9 that codes for a protein torsin A has been identified for the majority of cases of juvenile limb onset cases. Other cases which predominantly involve the cervical and cranial muscles have a different genetic background. No specific structural neuropathological abnormality has been found. However neurochemical analysis of the brain does suggest abnormalities. Anticholinergic drugs at high dosage levels may improve the symptoms of generalized dystonia in some patients. Approximately 10% of juvenile onset cases beginning in the lower extremities described by Segawa have a remarkable response to low dosage of L-Dopa (termed the L Dopa responsive variant .of progressive generalized dystonia). This is an autosomal dominant disorder coding to chromosome 14 and involving guanosine triphosphate cyclohydrolase It has therefore been recommended that all juvenile onset cases receive a trial of the L-Dopa/carbidopa combination.

Hemidystonia: Approximately 75% of patients in this group do have the history, findings or CT or MRI evidence of a contralateral neuropathology involving the striatum, usually the putamen or striato-pallido-thalamic pathway. The corticospinal pathways are not usually involved. There is then presumably an excessive input into premotor/ supplementary motor cortex resulting in the excessive contralateral movement.

4. Movement Disorders Induced By Dopamine Blocking Agents (Sethi, 2001) has provided a comprehensive review of this topic).

These agents produce frequent acute and chronic complications in psychiatric patients

1) *Acute dystonic reactions*-which occur in 2-5% of patients within hours or days after onset of therapy. Note that this complication occurs not only in psychiatric patients but also in patients receiving agents of this class as antiemetics.

2) *Acute akathisia:* acute restlessness and an inability to sit still, occurs in 20% of patients receiving dopamine-blocking agents

3) *Acute drug induced Parkinsonian syndrome* discussed above. Although symptoms may appear in any patient given sufficient amount of dopamine blocking agents to block 80% of receptors, there does appear to be a predisposition to develop symptoms at lower more therapeutic range in patients with a family history of Parkinsonism.

4). *Tardive dyskinesia and other tardive reactions.* In this context, tardive has been defined as 3 months of exposure to the agent (1 month for patients over the age of 60 years).

---

[14] *In younger individuals, the limbs and trunk are more often involved.*

There may be a genetic predisposition

Tardive dyskinesias are hyperkinetic choreiform movements. Usually these are seen in the older patient involving the mouth. tongue, lips and face[14]: (oral buccolingual facial masticatory syndrome). Symptoms usually develop after 6-24 months of drug therapy and in many patients the symptoms decrease after drug withdrawal.

In another group of patients, symptoms develop only after neuroleptic drugs have been withdrawn or dosage has been reduced again after prolonged use. In one series, such withdrawal or dose reduction resulted in tardive dyskinesia in 40% of patients receiving chronic neuroleptic therapy. The presumed pathophysiology is not entirely clear,

5. **Tics:** Motor tics are sudden, brief involuntary movements resembling jerks or gestures. The patient may experience a poorly defined sensation of an irresistible urge to move prior to the movement. These movements may be simple: an eye blink, a head jerk, or a facial grimace. In some cases, the movement is more complex and patterned. resembling a compulsive act. In some cases (syndrome of Gilles de la Tourette which has autosomal dominant transmission), multiple tics occur accompanied by vocalizations. The vocalization may consist of a simple sound or may involve the use of obscene 4 letter words referring to defecation, genitalia, or sexual acts (coprolalia). In contrast to other movement disorders, voluntary effort accompanied by anxiety may serve to inhibit the occurrence of the tic or tics. Then a flurry of tics may occur after the suppression. Tics may continue to occur in sleep.

The usual idiopathic tics are common in children and may persist into adult life. Lees and Tolosa (1988), indicate one of 10 schoolboys will have idiopathic tics. Males are affected three times as often as females both in idiopathic tics and in Tourette's syndrome. No specific neuropathology has been established. Use of dopaminergic receptor blocking agents such as haloperidol or pimozide or fluphenazine may often produce a significant decrease in the clinical symptoms. In contrast, dopaminergic agents (L-dopa. bromocriptine) and amphetamines may produce an exacerbation of symptoms (Jankovic, 1987, Singer, 2000).

6: **Familial Paroxysmal Dyskinesias:** These disorders although not common are of significance because they undoubtedly occur on the basis of channelopathies. The specific genetic mutation has not yet been identified. In this regard, they are similar to the episodic ataxias discussed in chapter 20, periodic paralysis, and familial hemiplegic migraine in which the specific ion channel mutation has been identified. (See Bhatia, 2001 and Bhatia etal, 2000). In all of the syndromes, the neurological examination is normal between attacks. In the majority autosomal dominance occurs and males predominate. Two major entities may be identified (1) paroxysmal kinesigenic choreoathetosis/dyskinesias (PKC/PKD) and (2) paroxysmal dystonic choreoathetosis/ nonkinesigenic dyskinesias (PDC/PNKD). The first type (PKC/PKD) is the most common. Attacks lasting less than 5 minutes begin in childhood with multiple attacks per day induced by movement of chorea, dystonia, and ballism. The attacks are significantly reduced by the administration of low doses of the anticonvulsant, carbamazepine that has action on the sodium/potassium channel. Several families have mapped to the peri centromeric region of chromosome 16, where a group of ion channel genes are present. The second type (PDC/PNKD) begins in infancy or childhood with infrequent attacks that last from 10 minutes to 6 hours and are predominantly dystonic. The attacks are induced by stress, fatigue, alcohol and caffein. In a number of families (Jarman, etal, 1997) the genetic defect has been mapped to chromosome locus 2q33-q35, an area where a number of ion channel genes are located. Paroxysmal exercise induced dyskinesia (PED) is a third less common variety.

CHAPTER 20
Motor System III: Cerebellum and Movement

ANATOMIC CONSIDERATIONS

Subdivisions of the Cerebellum

A number of schemes for dividing the cerebellum into various lobes and lobules have been devised. From a functional standpoint, it is perhaps best for the student to visualize the cerebellum as composed of various longitudinal and transverse divisions. In order to visualize these subdivisions, it is necessary to unfold and to flatten the cerebellum as shown in *Figure 20-1*.

Longitudinal Divisions

The major longitudinal divisions are the median (vermal cortex), the paramedian (paravermal), and lateral (remainder of the cerebellar hemispheres).

The major projections and functional correlations are outline in Table 20-1. In general, the median (vermal) region is concerned with the medial descending systems and with axial control. The lateral and paramedian regions are concerned with the lateral descending systems and with the appendages (the limbs) (midline to axis; lateral to appendages).

Transverse Divisions

TABLE 20-1 LONGITUDINAL SUBDIVISIONS OF THE CEREBELLUM

Median	Paramedian	Lateral
(Vermal Cortex)	(Paravermal)	(Remainder of cerebellar hemisphere)
Projects to:	Projects to:	Projects to:
Fastigial nucleus (globus and emboliform)	Interpositus nuclei	Dentate nucleus lateral
Role in:	Role in:	Role in:
Posture & movements of body (axial)	Discrete ipsilateral extremity movements (appendicular)	Discrete ipsilateral extremity patterns of movement. Coordinated with cerebral cortex, thalamus, and red nucleus. Initiating, planning and timing of movements.

The transverse subdivisions are essentially phylogenetic divisions.

(1) The archicerebellum is composed of a

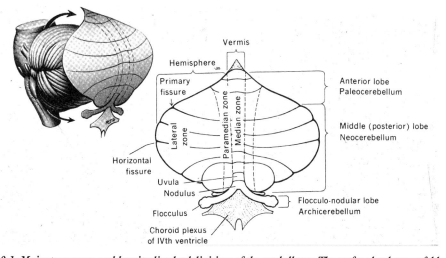

Figure 20-1. Major transverse and longitudinal subdivisions of the cerebellum. The surface has been unfolded and laid out flat. (From Noback, C.R. et al: 1991. The Human Nervous System. 4th edition, Philadelphia. Lea & Febiger p. 282.

flocculus and nodulus and is related primarily to the vestibular nerve and vestibular nuclei. Reflecting this anatomical relation, the archicerebellum has a role in control of equilibrium (balance), axial posture, and eye movement.

(2) The paleocerebellum, or anterior lobe, relates primarily to the spinocerebellar system and to its analogue for the upper extremities, the cuneocerebellar pathway (the lateral cuneate nucleus is the analogue of the dorsal nucleus of Clarke). Its major function, however, appears more related to the lower extremities than to the upper extremities. The primary fissure separates the anterior lobe from the middle posterior lobe.

(3) The middle, or posterior, lobe, the neocerebellum, relates primarily to the neocortex. The function of the neocerebellum is primarily the coordination of discrete movements of the upper and lower extremities (see Table 20-2) Transverse Divisions of the Cerebellum

CYTOARCHITECTURE OF THE CEREBELLUM

In contrast to the cerebral cortex, all areas of the cerebellar cortex are relatively thin and have the same basic three-layered cytoarchitectural pattern, consisting of an outer molecular layer, a Purkinje cell layer and an inner granule cell layer with an underlying medullary layer of white matter. The arrangement of cells and the basic synaptic connections are indicated in *Figure 20-2*. The arrangement of cells and the basic synaptic connections are indicated in *Figure 20-3*. Afferent fibers enter the cerebellar cortex as *mossy fibers*. These fibers are so named because each terminates in a series of moss-like glomeruli, where axodendritic synaptic contacts are made with granule cells. The mossy fibers have already been encountered as spinocerebellar, cuneocerebellar, corticopontocerebellar, vestibulocerebellar, or reticulocerebellar pathways.

The granule-cell axon is sent into the molecular layer where it divides into two long branches. These branches travel parallel to the long axis of the cerebellar folium and are designated as *parallel fibers*. These fibers make excitatory synaptic contact with the extensive dendritic arborizations of Purkinje cells, as well as with stellate cells, Golgi cells, and basket cells. The basket cells inhibit the Purkinje cells via axosomatic synapses as these projections are inhibitory in nature   The outflow of the Purkinje cells is inhibitory in nature   The tar-

TABLE 20-2 TRANSVERSE DIVISIONS OF THE CEREBELLUM

Archicerebellum (Floccular nodular lobe)	Paleocerebellum Anterior Lobe	Neocerebellum (Middle, or posterior, lobe)
CONNECTIONS: Vestibular	CONNECTIONS: Spinocerebellar & Cuneocerebellar & spino-olivary (inferior olive)	CONNECTIONS: Neocortex
ROLE: Axial equilibrium (Trunk primarily) Posture, Muscle tone, Vestibular reflexes, Eye movements	ROLE: Equilibrium Posture Muscle tone in lower extremities Coordinated movements lower extremities eg. heel to shin	ROLE: Limb coordination in phasic posture movements (upper and lower extremities) Possible role in higher executive functions and emotion

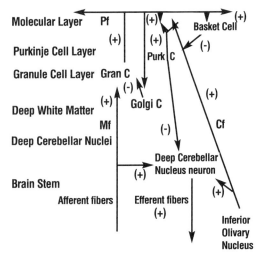

Figure 20-2. The most significant cells, connections, the afferent and efferent fibers in the cerebellar cortex. . Arrows show direction of axonal conduction. (+) = excitatory ;(-) = inhibitory. Cf= climbing fibers; pf = parallel fiber; mf = mossy fiber . Refer to text.

get of the axons of the Purkinje cells is the neurons of the deep cerebellar nuclei. The axons of the neurons in these deep cerebellar nuclei are the major outflow from the cerebellum to the brain stem and thalamus.

Several additional systems must be considered. The *climbing fibers* are excitatory to the Purkinje-cell dendrites. These fibers originate in the contralateral inferior olivary nuclei and represent a long-term latency pathway from the spinal cord or cerebral cortex to the cerebellum (olivocerebellar). These fibers are so named because they climb and wrap around the dendrites and body of the Purkinje-cell neurons, making several hundred synaptic contacts. Stimulation of a single climbing fiber results in large postsynaptic potentials and in large high-frequency bursts of discharges.

The parallel fibers also have excitatory synapses in the molecular layer, but they synapse with the dendrites of the Golgi cells. The Golgi cell, in turn, is inhibitory to the granule cell with a synapse within areas of the molecular layer referred to as glomeruli. The parallel fibers also excite other interneurons (small stellate cells) in the molecular layer, which, in turn, are inhibitory to the dendrites of Purkinje cells.

GABA (gamma aminobutyric acid) is the inhibitory transmitter for the Purkinje, basket, stellate, and Golgi cells. Since the outflow from the cerebellar cortex occurs via the axons of Purkinje cells and since this outflow is inhibitory to the deep cerebellar nuclei, it is logical to ask what excitatory influences drive the neurons of these deep nuclei. The answer is that collaterals of the mossy fibers and climbing fibers serve this function.[1]

AFFERENTS

The inputs to the cerebellum may be summarized as follow:

A. Via the inferior cerebellar peduncle (restiform body):

1. Uncrossed dorsal spinocerebellar and cuneocerebellar

2. Crossed olivocerebellar

3. Uncrossed reticulocerebellar (from lateral reticular and paramedian nuclei)

B. Via the juxta-restiform body

1. Uncrossed from vestibular nuclei are predominantly to the floccular nodular lobe, the archicerebellum, and the midline vermis.

2. Uncrossed from vestibular nerve (note that some fibers from the vestibular nerve bypass the vestibular nuclei and enter the cerebellum directly)

C. Via the middle cerebellar peduncle (brachium pontis):

1. The corticopontocerebellar input (via the pontine nuclei) is primarily to the neocerebellum of the lateral hemisphere. This is crossed from the opposite cerebral hemisphere.

D. Via the superior cerebellar peduncle (brachium conjunctivum):

1. Crossed ventral spinocerebellar is primarily to the anterior lobe (paleocerebellum) and the intermediate area of the paraflocculus.

2. Tectal cerebellar

The superior cerebellar peduncle, then, serves only a minor role as afferent to the cerebellum. Its major role is as the efferent pathway.

EFFERENTS

A. Superior Cerebellar Peduncle (Brachium Conjunctivum).

1. Dentate nuclei and interpositus (globus and emboliform) nuclei, there is a major projection via the crossed brachium conjunctivum (superior cerebellar peduncle) to the ventral lateral (and ventral anterior) nucleus of the thalamus and to the red nucleus (dentatorubral-thalamic pathway).

2. Descending division of the brachium conjunctivum conveys impulses to the paramedian reticular nuclei.

B. Inferior Cerebellar Peduncle.

1. The archicerebellum projects directly from the floccular nodular lobe and via the fastigial nuclei to the vestibular (lateral vestibular nucleus) and reticular areas of the brain stem.

---

[1] *However, the mossy fibers from the pontine and brain stem reticular nuclei apparently do not serve this collateral excitatory function.*

2. The fastigioreticular and fastigiovestibular pathways hook around the superior cerebellar peduncle as the uncinate fasciculus. In addition, impulses are also conveyed via the juxtarestiform body.

C. Middle Cerebellar Peduncle. This contains only afferent connections from the cerebrum via the pons into the neocerebellum.

TOPOGRAPHIC PATTERNS OF REPRESENTATION IN CEREBELLAR CORTEX

Stimulation of tactile receptors or of proprioceptors results in an evoked response in the cerebellar cortex. There is a topographic pattern of representation. Visual and auditory stimuli evoke responses primarily in the midline vermis. Stimulation of the specific areas of the cerebral cortex also evokes responses from the cerebellar cortex in an appropriate topographic manner. Moreover, direct stimulation of the cerebellar cortex in the decerebrate animal will produce, in a topographic manner, movement or changes in tone of flexors or extensors. The general pattern of representation is consistent with the patterns of sensory and cortical representation previously noted.

A possible pattern of somatotrophic representation in the primate cerebellum is shown in *Figure 20-3*. It is important to recall that there is no conscious perception of stimuli arriving at the cerebellar level.

Vestibular stimulation evokes responses not only in the floccular nodular regions but also in the superior and inferior midline vermis.

FUNCTIONS OF THE CEREBELLUM AND CORRELATIONS

The cerebellum acts as a servomechanism, that is, as a feedback loop that dampens movements and motor power to prevent overshoot and oscillation (that is, tremor). In short, it acts to maintain stability of movement and posture. More recently, a role in higher motor function, such as the initiation of planning, the timing of movement, and motor learning, has been suggested for the cerebellum.

REGIONAL FUNCTIONAL CORRELATIONS

We have already discussed the functional relation of the longitudinal and transverse subdivisions of the cerebellum. It is evident that these two classifications overlap from a functional correlation standpoint.

From a functional standpoint, we may specify the following correlations:

1. Vestibular Reflexes and Eye Movement

a. Anatomic Correlation: Floccular nodular lobe

b. Major Input: Vestibular labyrinth (semicircular canals and otolith organs)

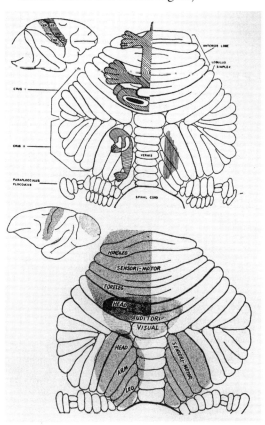

Figure 20-3: Topographic localization in cerebellum. A, Summary of projections from the sensory area and the motor area in the monkey; B, Summary of corticocerebellar projections in the monkey. Note that there is a unilateral representation on the dorsal surface and a representation within each paramedian lobule on the ventral surface. At the lateral margin the division between crus I and crus II is essentially the border between the dorsal and ventral surface. See Figure 20-1 for terminology in human cerebellum. From Snider. R.S.: 1950. Arch.Neurol.Psych. 64:204 (AMA).

c. Major Deep Nucleus: Vestibular nuclei (lateral)

d. Major Connections: Medial descending systems, vestibulospinal and medial longitudinal fasciculus and extraocular motor nuclei

2. Posture and Equilibrium of the Trunk

A. a. Anatomic Correlation: Floccular nodular lobe (vestibulocerebellum)

b. Major Input: Vestibular labyrinth as above

c. Major Deep Nucleus: Vestibular nuclei (lateral)

d. Major Connection: Medial descending systems - vestibulospinal systems (medial and lateral)

B. a. Anatomic Correlation: Midline vermis (spinocerebellum)

b. Major Inputs:
 1. Vestibular labyrinth
 2. Dorsal spinocerebellar and cuneocerebellar from body and proximal limbs
 3. Trigeminal
 4. Visual and auditory

c. Major Deep Nucleus: Fastigial

d. Major Connections: Medial descending motor systems:
 1. Via juxtarestiform body and central tegmental system to pontine and medullary reticular nuclei (origin of medial and lateral reticulospinal tracts) and to vestibular nuclei (origin of vestibulospinal tracts).
 2. Via superior cerebellar peduncle (decussated) to ventral nucleus of thalamus to motor cortex to anterior corticospinal tract

3. Distal Motor Control and Speech/Voice Control

a. Anatomic Correlation: Intermediate lobe (paravermal or paramedian zone-spinocerebellum)

b. Major Input: Dorsal spinocerebellar and cuneocerebellar

c. Major Deep Nuclei: Interpositus (globose and emboliform)

d. Major Connections: Lateral descending motor system, superior cerebellar peduncle via decussation to magnocellular sector of the red nucleus (origin of rubrospinal system) and to ventral lateral nucleus of thalamus and from the thalamus to primary motor area and area 6 of supplementary motor cortex (origins of the crossed lateral corticospinal system)

3A. Speech

a. Anatomic Correlation: Superior paravermal area (more often left than right). Damage produces cerebellar dysarthria which affects prosody, harmony, and rhythm of speech rather than sequential or semantic aspects (see Amarenco et al, 1991; Lechtenberg and Gilman, 1978 for additional discussion).

4. Initiation, Planning and Timing of Movement

a. Anatomic Correlation: Lateral cerebellar hemisphere (cerebrocerebellum)

b. Major Input: Motor and premotor cerebral cortex (areas 4 and 6) and parietal cortex (areas 1, 2, 3, 5) to pontine nuclei to lateral neocerebellum (after decussation) via the middle cerebellar peduncle

c. Deep Nucleus: dentate

d. Major Connection: Via decussation of superior cerebellar peduncle to
 1. Red nucleus (parvocellular) with subsequent rubro-olivary fibers
 2. Ventral lateral nucleus of thalamus, then projecting to motor and premotor cerebral cortex. From these areas, as already discussed, originate the lateral descending systems lateral corticospinal and corticorubrospinal and lateral reticulospinal and the anterior corticospinal systems.

EFFECTS OF DISEASE ON THE CEREBELLUM

The cerebellum, may be viewed as a machine processing information from many sources (sensory receptors, vestibular nuclei, reticular formation, and cerebral cortex) and then acting to smooth out the resultant movements or postures. Sudden displacements or lurches are prevented. The large number of inhibitory feedback circuits in the cerebellum (the major influence of Purkinje cells on deep cerebellar nuclei being inhibitory) qualifies the cerebellum for this role. Lesions may then result in undamped oscillation. As a result, tremor perpendicular to the line of movement may occur, called intention or action tremor. Lesions may also impair the ability to respond to sudden displacements, resulting in instability of posture and gait (ataxia). Release of function of ventral lateral nucleus of thalamus may also occur.

The cerebellum is not designed for the direct, fast-conduction control of movements. Thus, it is not surprising that, in the monkey, total ablation may produce little effect, although some deficit in contact placing may be evident. In humans, extensive destruction of cerebellum may be present with little obvious deficit. Moreover, cerebellar symptoms, if present in a "static" disease process (such as cerebrovascular accidents), often disappear with time. However, more specific tests may still reveal a minor deficit in speed and coordination.

More recent studies suggest a role for the cerebellum in classical conditioning and in motor learning particularly as regards the memory aspect for the timing of movements (Raymond et al, 1996). In addition, a role of the cerebellum in higher order cognitive and emotional functions has been suggested by the studies of Schmahmann&Sherman, 1998, Levinsohn et al, 2000,Riva&Giogi, 2000.In patients with lesions of the posterior lobe and vermis, there was an impairment of executive functions such as planning, set shifting, verbal fluency, abstract reasoning, working memory as well as visual spatial organization and memory. Higher language disturbances also occurred. In addition these patients had a personality change characterized by blunting of affect or disinhibition and inappropriate behavior. This cognitive and affective syndrome suggested a disruption of the modulatory effects of cerebellum on the prefrontal, posterior parietal and limbic circuits discussed in Chapter 18. These changes did not occur with lesions of the anterior lobe. The effects on cognitive and emotional function in patients with prenatal cerebellar hypoplasia may be even more serious and may include autism (Courchesne, et al, 1994,Allin, et, al 2001)

MAJOR SYNDROMES:

While we will discuss three anatomically distinct syndromes of cerebellar disease (the floccular nodular lobe, the anterior lobe, and the lateral hemisphere) it should be pointed out that these strict anatomic borders do not confine many diseases that affect the cerebellum. Moreover, the cerebellum is positioned in a relatively tight compartment with bony walls and a relatively rigid tentorium cerebelli above. An expanding lesion in the posterior fossa, then, may produce a generalized compression of the cerebellum. In such clinical situations, it may be possible to differentiate only between midline (vermal) involvement and lateral (cerebellar hemisphere) involvement. Midline lesions produce disorders of equilibrium and axial ataxia; lateral and lateralized paramedian lesions produce appendicular ataxia and tremor.

SYNDROME OF THE FLOCCULAR NODULAR LOBE AND OTHER MIDLINE CEREBELLAR TUMORS

The major findings in the human, monkey, cat and dog are a loss of equilibrium and an ataxia (unsteadiness) of trunk, gait, and station. Thus, the patient, when standing on a narrow base with eyes open, has a tendency to fall forward, backward, or to one side. The patient may be unable to sit or stand. The patient walks on a broad base, often reeling from side to side, and often falling. Despite the loss of equilibrium, the patient usually does not complain of a rotational vertigo. When recumbent in bed, the patient often fails to show any ataxia or tremor of limbs. Thus, the finger-to-nose and heel-to-shin tests are performed without difficulty. With unilateral disease of the floccular nodular lobe, a head tilt may be present. In addition, spontaneous horizontal nystagmus may be present as a transitory phenomenon.

In humans, the most common cause of this syndrome is neoplastic. The type of neoplasm depends on the age of the patient. The most likely cause in an infant or child[2] is a medulloblastoma, a tumor arising in nests of external granular cells in the nodulus forming the roof of the 4th ventricle. In older children and young adults, the ependymoma, a tumor arising from ependymal cells in the floor of the fourth ventricle, may cause this syndrome by pressing upward against the nodulus. In adults it is more appropriate to speak of midline tumors of the vermis since the tumors do not selectively involve the floccular nodular region. In middle-aged adults, the most common cause is probably the midline hemangioblastoma. In older adults, metastatic tumors may produce this syndrome. At all ages, rare arteriovenous malformations may produce the syndrome.[3]

The course of a medulloblastoma is indicated in the following case history:

Case 20-1: This 27-month old white female had been unsteady in gait, with poor balance, falling frequently since the age of 13 months, when she began to walk. Three months prior to admission, in relation to a viral infection manifested by fever and diarrhea, the patient developed increasing anorexia and lethargy. One-month prior to admission, vomiting increased, and the patient also became more irritable. Perinatal history had been normal, and head circumference had remained normal.

Neurologic examination: *Mental Status:* The child was irritable but cooperative. *Cranial Nerves:* The fundi could not be visualized. Bobbing of the head was present in the sitting position. *Motor System:* A significant ataxia of the trunk was present when sitting or standing. Gait was broad-based and ataxic, requiring assistance. No ataxia or tremor of the extremities was present.

Clinical diagnosis: Probable midline cerebellar tumor, most likely based on age a medulloblastoma

Laboratory Data: *Air-contrast ventriculogram* demonstrated the fourth ventricle was deformed by a mass arising from the nodulus of the cerebellum. There was secondary enlargement of the lateral and third ventricles and marked forward displacement of the cerebral aqueduct.

Hospital Course: Doctor Peter Carney performed a suboccipital craniotomy, which confirmed the presence of a medulloblastoma with many malignant cells present in the cerebrospinal fluid obtained at surgery. Radiation therapy to the entire central nervous axis was begun one week after surgery. There was initial improvement but the patient deteriorated three months after the surgery and expired one month later despite chemotherapy .The findings at autopsy are demonstrated in *Fig. 20-4*.

At present, recommended therapy includes wide resection followed by radiotherapy to the entire CNS axis. Dissemination of tumor cells throughout the cerebrospinal fluid is frequent, because of the friable, cellular, non-stromal nature of the lesion. At present, 5-year survival has been increased to 60%. Recurrences may be treated with chemotherapy although the combination of radiotherapy and chemotherapy may produce significant pathologic alterations in white matter. Refer to Chapter 13 and 27 for additional discussion.

Mid line cerebellar tumor in the adult: the following case 20-2 presents an example of a midline cerebellar tumor metastatic from breast with primary symptoms of gait ataxia, headache, vertigo and vomiting.

Case 20-2: This 67-year-old white female 6 weeks prior to admission developed intermittent vertigo and then two weeks prior to admission a progressive ataxia of gait followed by headaches and vomiting. Past history: This patient had a poorly differentiated highly malignant infiltrating ductal adenocarcinoma of the left breast for which she had undergone lumpectomy 6 years; and a radical mastectomy followed by radiation therapy, 3 years prior to admission.

---

[2] *Medulloblastomas may occasionally occur in adolescents and young adults.*

[3] *Progressive multiple sclerosis may produce in the young or middle-aged adult severe involvement of white matter in the cerebellum and brain stem, resulting in severe truncal ataxia, in which the patient is ataxic in both the sitting and standing positions. Such cases almost always also manifest severe appendicular involvement, that is, the cerebellar syndrome is not selective.*

Neurological examination (in the emergency room): *Cranial Nerves:* There was an absence of venous pulsations on funduscopic examination suggesting a mild increase in intracranial pressure. *Motor System:* She had difficulty standing tending to fall to the right. She was unable to walk because of severe ataxia. There was no limb dysmetria. *Reflexes:* The right plantar response was equivocal.

Clinical diagnosis: Midline cerebellar tumor metastatic from breast.

Laboratory data: *CT scan* demonstrated a large (3-4 cm) enhancing tumor was present involving the midline vermis and right paramedian cerebellum and compressing the fourth ventricle subsequently confirmed by *MRI scans* of the brain *(Fig.20 5)*. *CT scan of abdomen* demonstrated two possible metastatic lesions in the liver.

Subsequent course: The patient received dexamethasone with significant improvement within 12 hours. At the insistence of the patient, Dr. Gerald McGullicuddy resected the solitary central nervous system metastatic lesion. Following surgery she received 3000cGy(rads) whole brain radiation and was begun on the antitumor agent tamoxifen. She expired at home, four months after surgery of systemic complications of the malignancy.

SYNDROME OF THE ANTERIOR LOBE:

Stimulation or ablation of the anterior lobe in the cat or dog may produce significant changes in muscle tone. Thus, stimulation of the middle anterior lobe results in a decrease in spindle discharge and inhibits gamma rigidity of the decerebrate preparation. Stimulation of the intermediate area of the anterior lobe produces an increased spindle discharge and facilitates gamma rigidity in the decerebrate preparation. Actual ablation of the anterior lobe in these animals or functional ablation (cooling) produces an increase in decerebrate rigidity without a change in the gamma system (alpha rigidity). However, in humans, ablation of the anterior lobe does not change tone.[4] The major symptoms are an ataxia of gait with a marked side-to-side ataxia of the lower extremities as tested in the heel-to-shin test. The upper extremities, in contrast, are affected to only a minor degree.

This area corresponds to the representation of the lower extremities and to the area of projection of the spinocerebellar pathways.

Typical of this syndrome are cases of alcoholic cerebellar degeneration *(Figure 20-6)*, where a severe loss of Purkinje cells occurs in the anterior lobe. Although this degeneration occurs primarily in alcoholics, the basic etiology may be a nutritional deficiency (the specific factor is most likely thiamine).

The major findings of a broad-based, staggering gait with truncal ataxia and heel-to-shin ataxia can be related to the anterior lobe of the cerebellum. Since symptoms improve or do not progress after the discontinuation of alcohol, and the administration of multiple B vitamins, a toxic or nutritional factor may be postulated. (The reliability of the history of alcohol intake is always open to question. The amount stated by the patient is usually assumed to be the minimum amount.) Note that a similar acute cerebellar syndrome occurs as one aspect of the Wernicke-Korsakoff's syndrome (refer to Chapter 30). At the present time, the diagnosis of cerebellar atrophy may be confirmed by MRI (midline sagittal section) or by CT scan. As a result, evidence of cerebellar atrophy is now found in many patients presenting with other aspects of chronic alcoholism or other nutritional disease who do not

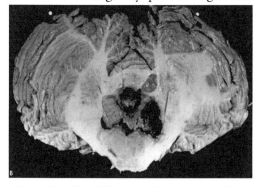

Figure 20-4. Medulloblastoma: Syndrome of the floccular nodular lobe: Case 20-1. Vomiting and truncal ataxia were present without tremor or ataxia of the extremities. (Courtesy of Dr. John Hills). Refer to text.

necessarily present with initial symptoms of ataxia.

Case 20-3 presented on the CD-ROM illustrates alcoholic cerebellar degeneration.

In other cases, this syndrome is the result of a chronic degenerative disease with a genetic basis. As in alcoholic cerebellar degeneration, the predominant histologic change is the loss of Purkinje cells in the anterior lobe. In contrast to alcoholic cerebellar degeneration, the pathology is usually more widespread and not strictly limited to the anterior lobe. In the following case history, the major involvement was primarily of the anterior lobe. There was definite family history, including evidence of consanguinity.

Case 20-4: A 45-year-old single white male carpenter presented with a 4-year history of progressive unsteadiness in walking, present during the daytime as well as at night and a minimal loss of coordinated movements in his hands. He felt that his greatest deficit was unsteadiness. He had some numbness at the toes and minor difficulty in swallowing. His mother and father were second cousins. Several aunts and uncles had "trouble with their legs" in later life, resulting in a difficulty in walking.

Neurologic examination: *Cranial Nerves:* Intact except for minimal dysarthria. Horizontal nystagmus was present on lateral gaze, vertical nystagmus on upward gaze. *Motor System:* The patient walked on a broad base with an ataxia of trunk, as though intoxicated, reeling from side to side, with marked unsteadiness on turns. He was unable to walk a tandem gait with eyes open. A mild intention tremor was present, and there was minimal disorganization of alternating movements. In contrast a marked ataxia and tremor were evident on heel-to-shin test. *Reflexes:* There was a relative decrease in ankle jerks compared to knee jerks. *Sensory System:* There was a minimal decrease in vibration sensation at the toes.

Clinical diagnosis: Hereditary cerebellar degeneration involving predominantly anterior lobe

Laboratory Data: *Pneumoencephalogram* performed at the University of Virginia Hospital (Doctor Stewart) confirmed the significant atrophy of cerebellum.

OTHER CAUSES OF CEREBELLAR ATROPHY:

Prolonged high fever, meningitis, a post-

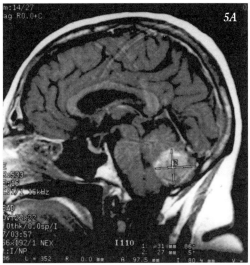

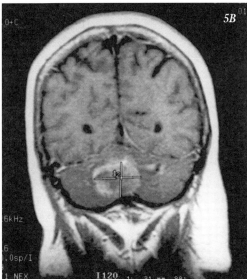

Figure 20-5. Metastatic midline cerebellar tumor: Case 20-2 .MRI (TI) a) sagittal section b) coronal section. Refer to text.

---

[4] *In general chronic lesions of cerebellum in humans do not alter tone or tendon reflexes. Acute lesions of cerebellum resulting from hemorrhage or surgery may produce transient hypotonia (Diener and Dichgans, 1992).*

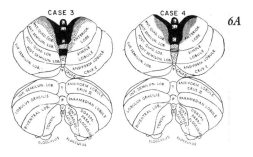

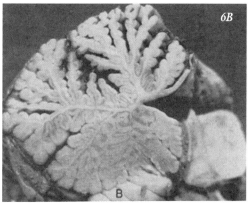

Figure 20-.6 Alcoholic cerebellar degeneration: The anterior lobe syndrome. There is a loss of Purkinje cells with atrophy of the cerebellar folia with relatively selective involvement of the anterior lobe: A) schematic representation of the neuronal loss, B) Atrophy of the anterior superior vermis in sagittal section. From Victor, M., R.D.Adams, and E.L.Mancall: 1959. Arch.Neurol. 1:579, 599, 600 (AMA).

infectious syndrome, and repeated grand mal seizures may all be associated with the development of an ataxia that suggests major involvement of the anterior superior vermis. In some of these cases, there is found on CT or MRI more widespread atrophy. At times, as in the case presented in *Figure 20-7*, other residual neurologic findings may also be present. The topography of cerebellar atrophy, particularly as regards midline structures, can best be studied with MRI. *Figure 20-8* demonstrates the findings in a patient with gait ataxia following high fever and meningitis, in which atrophy, although widespread, was most prominent in the anterior superior vermis.

When the ataxia is of acute or subacute onset in childhood, the possibility of a cerebellitis must be considered. A varicella or other viral infection often precedes this entity.

Mononuclear cells may be present in cerebrospinal fluid. Whether this is a direct viral infection or a post-infectious entity remains unclear. . The major involvement is of the trunk although tremors of the head, trunk and limbs are also present. Recovery usually occurs over 6 months but occasionally recovery is incomplete.

Diener and Dichgans (1992) discuss the localization of ataxia of standing body posture. Their conclusions are as follows:

1. Lesions of the spinocerebellar portion of the anterior lobe (mainly observed in chronic alcoholics) result in body sway along the anterior/posterior axis with a frequency of 3 per second. This sway, or tremor, is provoked by eye closure. Visual stabilization of posture occurs. Since the direction of sway of the trunk is opposite that of the head and legs, the center of gravity is only minimally shifted and the patient does not fall.

2. In contrast, the lesions of the vestibulocerebellum (floccular nodular lobe and lower vermis), which are primarily mass lesions, produce a postural instability of head and trunk during sitting, standing, and walking. Postural sway occurs in all directions often at a frequency of less than 1 per second. Visual stabilization is limited; falls, therefore, are more frequent.

3. Spinal ataxia (as in tabes dorsalis and combined system disease) results predominantly in lateral body sway. Visual stabilization is prominent (positive Romberg test).

SYNDROME OF THE LATERAL CEREBELLAR HEMISPHERES (NEOCEREBELLAR OR MIDDLE-POSTERIOR LOBE SYNDROME):

A number of disease entities may produce symptoms that can be related to the lateral hemisphere. Some of these disease entities are mass lesions such as metastatic tumors *(Fig. 20-9)* or intrinsic tumors *(Figure 20-10)*. The diagnostic problem, moreover, is complicated by the fact that involvement of the cerebellar peduncles may produce many of the same symptoms that result from direct involvement of the cerebellar hemisphere. The student will recall that the lateral medullary infarct produces an ipsilateral intention tremor and an impairment of alternating movements of the ipsilateral upper extremity. Occlusion of the posterior inferior cerebellar artery may produce an infarct of the lateral medulla and/or of the posterior inferior cerebellum as demonstrated in *Figure 13-12*.

Demyelinating disease, that is multiple sclerosis, may involve in a specific case, both the brain stem and cerebellum in a multifocal manner.

Cases presenting relatively pure focal involvement of the lateral hemisphere are those patients surviving after gunshot or shrapnel wounds. Such cases have been studied by Holmes (1939); they are not encountered frequently in civilian practice. Diener and Dichgans (1992) provide a more detailed physiologic analysis of lateral cerebellar lesions affecting the limbs.

The major findings in lateral hemisphere lesions, are the following signs and symptoms in the ipsilateral extremities: intention tremor, disorganization of alternating movements (dysdiadochokinesia), dysmetria, ataxia, and overshoot on rebound. The intention tremor may be described as a tremor perpendicular to the line of motion; often becoming more prominent on slower movements and as the target is approached.[5] This intention tremor may be noted in the finger-to-nose test. One may also demonstrate a similar tremor in the lower extremities as the patient attempts to

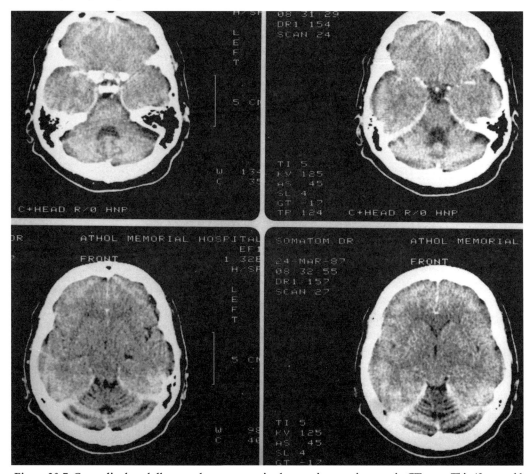

Figure 20-7: Generalized cerebellar atrophy most severe in the anterior superior vermis: CT scan. This 48-year-old female following an acute febrile illness and 3 weeks of coma at age 3 years, developed severe ataxia of stance and gait, with dysmetria on heel-to-shin testing, distal weakness in lower extremities and bilateral Babinski's signs.

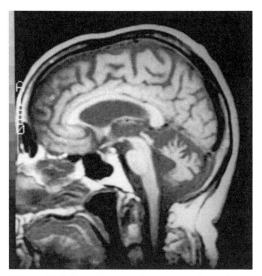

Figure 20-8. Cerebellar atrophy most prominent in anterior superior vermis with predominant anterior lobe syndrome: MRI. This 70-year-old female 23 years previously had prolonged markedly elevated temperature related to meningitis, coma and convulsions with a residual ataxia of stance and gait.

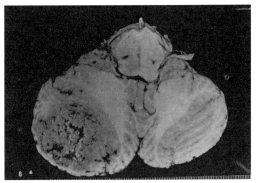

Figure 20-9. Lateral cerebellar syndrome. This patient with bronchiogenic carcinoma had nystagmus and appendicular ataxia. He had this major tumor in the lateral cerebellum and minor metastatic lesions in the cerebral hemispheres. (Courtesy of Dr. John Hills and Dr. Jose Segarra).

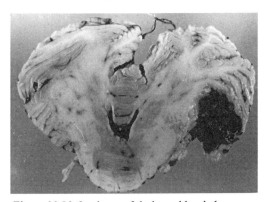

Figure 20-10. Syndrome of the lateral hemisphere: hemangioblastoma of cerebellum. This 72-year-old male had a 2-year history of vertex headache (worse on coughing), blurring of vision, diplopia, papilledema and a lateralized ipsilateral tremor on finger-to-nose testing. (Courtesy of Doctor Jose Segarra).

raise the foot off the bed to touch the examiner's finger or in the heel-to-shin test. As with the other signs of cerebellar disease, intention tremor is usually ipsilateral to the hemisphere involved.

At times, in addition to an intention tremor, a sustained postural "tremor" may be evident, more particularly if the superior cerebellar peduncle or its midbrain connections are involved. Brief jerks at the onset of movement (intention myoclonus) may occur when the dentate nucleus or superior cerebellar peduncle is involved.

The intention tremor represents a defect in the ability to dampen oscillations and to dampen overshoot. In the monkey, cerebellar tremor occurs despite deafferentation (section of the dorsal roots), a finding consistent with the concept that there is a central generator of oscillations. (However, other theories reviewed by Diener and Dichgans (1992) suggest the oscillations arise from dysfunction in long-latency transcortical reflexes.)

Having the patient slap the thigh above the knee with the hand at particular rhythm tests repetitive movements. One can also test for repetitive movement in the lower extremities by having the patient tap toes or heel on the floor. Alternating movements are tested by having the patient slap the hand on the thigh or on the opposite hand, alternating between the palmar and dorsal surface of the hand. In disease of the cerebellar hemisphere a marked disorganization of repetitive and alternating movement occurs. This is almost always ipsilateral to the hemisphere involved.

These patients are also said to demonstrate dysmetria. This may be defined as a defect in the estimation of the force and rate of movement necessary for an extremity to reach a target. The result in the finger-to-nose test is that the patient fails to accurately touch the finger to the nose. The finger may overshoot its goal or fail to reach the goal. Note that in disease involving the older regions of cerebellum, eye movements may also be described as dysmetric. In addition, there is often a failure to properly adjust force when pulling with the arm against an opposing force. As this opposing force suddenly gives way, the patient fails to apply a brake to the action of the arm and tends to have a rebound overshoot. The arm may even strike the body or face.

Finally, patients with cerebellar hemisphere disease are said to have a dyssynergia or disturbance in the synergy of movements, that is a defect of the coordination of multiple sets of agonists and antagonistic muscles when reaching for an object. The result is a decomposition of movement.

Cerebellar Dysarthria may also be present and is characterized by scanning (hesitation) that affects the pattern and intonation of speech. Prosody (rhythm and harmony) is affected rather than word sequence. At times, a tremor of speech is also apparent. Dysarthria may occur in cerebellar atrophy or in focal processes involving the hemisphere or the superior vermis. The crucial area of involvement is the superior paravermal area, as discussed by Amarenco et al, 1991 and Lechtenberg and Gilman, 1978. The left hemisphere is more often involved than the right.

The patient with disease of the cerebellar hemisphere also has a disturbance of gait. This tends to be an ipsilateral disturbance of gait, with a tendency to fall toward the side of the lesion. As discussed above, in lesions of the hemisphere, balance is often well maintained, in contrast to lesions of the floccular nodular lobe, where a disturbance of gait and of sitting balance also occurs. The following case history 20-5 presented on CD ROM illustrates a hemangioblastoma of the right lateral hemisphere with headache, lateralized ataxia, intention tremor and dysmetria of the right upper and lower extremities. At the present time, the diagnosis of these tumors may be readily made by means of MRI and CT scan *(Fig. 20-11)*. Arteriograms may still be of value in defining the vascularity and blood supply of the tumor.

Case 20-6 presented later in this chapter demonstrates a cerebellar infarct with many lateralized appendicular features. A cerebellar hemisphere hemorrhage illustrated later in the chapter also demonstrates many of these lateralized features.

LESIONS OF THE CEREBELLAR PEDUNCLES: (METTLER & ORIOLI, 1958; CARRERA & METTLER, 1947.)

Some patients demonstrating cerebellar symptoms and findings actually have damage to the cerebellar peduncle rather than direct involvement of the cerebellum.

Lesions of the inferior cerebellar peduncle result in an ataxia of the extremities on the side of the lesion, with falling toward the side of the lesion and nystagmus.

Lesions of the superior cerebellar peduncle, the brachium conjunctivum, produce a unilateral ataxia of the limbs, with intention tremor and hypotonia. The symptoms are essentially those of hemisphere lesions. The clinical symptomatology is ipsilateral if the lesion occurs between the dentate nucleus and the decussation of the brachium conjunctivum. If the lesion is above this decussation, the tremor is usually contralateral and often is more than a pure intention tremor. These are components of a sustained postural type of tremor. At times, with involvement of this area within the midbrain, a resting tremor results as well, as discussed in relation to the basal ganglia. It is not unusual to have the tremor evolve from an initial relatively pure intention tremor to a later resting tremor. At times, a minor degree of hemichorea is also present. This is not remarkable when one considers that these areas of the midbrain, the subthalamus and the ventrolateral nucleus of thalamus are close together and all supplied by penetrating branches of the poste-

---

[5] *Diener and Dichgans (1992) suggest that the terms kinetic, goal-directed or terminal better describe the actual tremor. Defect in movement termination is thus referred to as* dysmetria.

rior cerebral artery. Intention myoclonus may also occur at the onset of movement and may merge with hemichorea.

The effects of isolated lesions of the middle cerebellar peduncles are not certain. There is some evidence that an incoordination of fine limb movements and an ataxia of gait may result. In experimental lesions, circling may result.

VASCULAR SYNDROMES OF THE CEREBELLUM

The development of CT and MRI scan has resulted in an increased recognition of infarcts, hemorrhages, and arteriovenous malformations involving the cerebellum. Many have been clinically silent or previously attributed to brain stem or labyrinthine pathology. We have already considered the blood supply of the cerebellum in Chapter 13, which covers vascular syndromes of the brain stem. Essentially, three circumferential arteries of the vertebral basilar circulation supply the several surfaces of the cerebellum, the arteries being named for the surface they supply:

1. The posterior inferior cerebellar artery (PICA), usually originating from the vertebral artery: an initial medial branch of this vessel supplies the lateral tegmental area of the medulla, and the medial and lateral branches supply the cerebellum.

2. The anterior inferior cerebellar artery (AICA), originating from the lower third basilar artery: the initial branches of this vessel supply the lateral tegmental area of the lower pons.

3. The superior cerebellar artery, originating from the rostral basilar artery: proximal branches supply the lateral tegmental area of the rostral pons and caudal mesencephalon.

All three vessels have extensive leptomeningeal anastomoses over the surface of the cerebellum that are similar to the leptomeningeal anastomoses of the arteries supplying the cortical surface of the cerebral hemispheres. As a consequence, vertebral artery occlusion may result in a lateral medullary infarct and syndrome but infarction of cerebellum may be absent or limited. As in the cerebral hemispheres border zone infarctions may

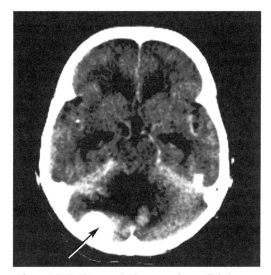

Figure 20-11. Hemangioblastoma of superficial superior and posterior right cerebellar hemisphere. CT scan indicated a densely enhancing mass (arrow) with considerable surrounding edema and cyst formation extending into the opposite cerebellar hemisphere and hydrocephalus. This 60-year-old white female had a several month history of increasing headaches, exacerbated by straining at stool, laughing or any head movement, a sense of dizziness, a sense of instability in walking and a normal neurologic examination. Angiogram demonstrated a vascular mass supplied by the posterior inferior cerebellar and superior cerebellar arteries, as well as meningeal branches. Dr Bernard Stone removed the tumor with relief of symptoms.

occur. On the other hand, embolic occlusion is frequently implicated in cerebellar vascular syndromes.

SYNDROMES OF OCCLUSION AND INFARCTION:

This topic has been well reviewed by Amarenco (1991), in an article that summarizes the detailed studies of his group (Amarenco et al, 1989; Amarenco et al, 1991; Amarenco&Hauw, 1990a and 1990b). The discussion presented here is in large part based on their studies. Refer also to the detailed reviews of Chaves et.al., 1994 and Kase et al, 1993.

Although autopsy series localized only between 1.5 and 4.2% of all infarcts to the cerebellum, in a CT scan series (Shenkin and Zavala, 1982) cerebellar infarcts accounted for

15% of all intracranial infarcts. Cerebellar infarcts account for 85% of all cerebellar strokes and hemorrhages for 15%; therefore, our emphasis will be on infarcts.

The symptoms of infarcts and hemorrhages may be similar, particularly if the infarct is large and associated with considerable edema. The large infarct may have what is referred to as a "pseudotumor" presentation. Essentially, the patient has the acute onset of headaches (usually localized to the occipital or cervical occipital area), severe vertigo, nausea, vomiting and ataxia of gait (often so severe that the patient is unwilling or unable to sit or stand). On examination, nystagmus and ipsilateral dysmetria on finger-to-nose and heel-to-shin tests will be present. Cerebellar type dysarthria is frequent (particularly when the superior cerebellar artery is involved). Coma at the onset or shortly after onset usually indicates severe edema with probable compromise of the fourth ventricle and/or aqueduct of Sylvius or brain stem. Evolving extraocular palsies are also usually indicative of compromise of the brain stem. MRI scan or CT scan allows the following:

a. distinction between infarct and hemorrhage

b. determination of whether hydrocephalus is present

c. determination of the vascular territory of the cerebellum and brain stem involved.

Patients presenting in coma or evolving into coma due to hydrocephalus and/or brain stem compression require immediate shunting to reduce the hydrocephalus and sometimes require evacuation of the hemorrhage or infarct (Shenkin and Zavala, 1982; Heros, 1982).

This acute presentation has long been recognized, although occasional patients are still initially misdiagnosed as suffering from acute labyrinthine vertigo, as in the example presented below. The majority (80 to 90%) of cerebellar infarcts have a more benign course not requiring surgical therapy. Many are recognized only in retrospect when a CT scan or MRI is obtained to analyze another CNS event. The superior cerebellar artery territory or the posterior inferior cerebellar artery territory are most frequently involved. In our own experience, the symptomatic cerebellar infarct is often in the territory of the posterior inferior cerebellar artery (see case history below and Caplan, 1986, and Chaves et. al., 1996). The anterior inferior cerebellar artery is rarely involved in isolation; the posterior inferior cerebellar territory is often also infarcted. One third of patients with superior cerebellar artery infarcts, also have infarcts within the territory of the posterior inferior cerebellar artery suggesting a possible embolus from the vertebral artery to the more distal basilar artery branches. Although some patients with infarcts in the distribution of the posterior inferior cerebellar artery will also infarct the associated brain stem territory and present a lateral medullary infarct syndrome as in case 13-2, the majority have isolated cerebellar infarcts. In contrast most patients with cerebellar infarcts of the anterior cerebellar artery territory also have infarcts of the caudal lateral tegmental pontine territory. Most patients with cerebellar infarcts of the territory of the superior cerebellar artery have associated infarcts in the brainstem territory of the rostral basilar artery. It is therefore easy to see that prior to the era of CT/MRI, the clinical features of the associated brain stem infarction (that included involvement of the cerebellar peduncles) dominated the picture and obscured the infarct of the cerebellum. More limited infarcts of parts of the superior cerebellar (lateral and medial) and posterior inferior cerebellar artery (dorsomedial and lateral) or of the cerebellar border zones are discussed in the various papers of Amarenco. The etiology for most infarcts involving the superior cerebellar artery is embolism of cardiac source. Posterior inferior cerebellar artery infarcts are either embolic of cardiac source or due to atherosclerotic occlusion of the vertebral (or less often) the posterior inferior cerebellar artery.

The following case history presents a typical example of posterior inferior cerebellar artery infarct, the territory most frequently affected selectively.

Case 20-6: This 64-year-old right-handed married white male postmaster, on the morning prior to admission, awoke at 3:00 a.m. with severe posterior headache and severe vertigo, which was exacerbated by any head movement ,nausea, projectile vomiting and slurred speech. Shortly afterwards, he noted left facial paresthesias, clumsiness of the left upper extremity, and diplopia. Two weeks previously, the patient had been seen for a transient episode of difficulty controlling the right arm, a hissing sensation, and impairment of speech *Past history* was significant for hypertension, coronary artery bypass surgery, and six years previously episodes of. transient weakness and numbness of the right arm with transient aphasia for which he had been placed on long term anticoagulation. At that time he had normal aortic arch and carotid angiographic studies but EEG had demonstrated focal left frontal/temporal slow wave activity

Neurologic examination: *Cranial Nerves:* Conjugate lateral gaze to the left was impaired. *Motor System:* Lateralized cerebellar findings were present with marked dysmetria of left upper extremity on finger-to-nose testing and impairment of alternating hand movements. In attempting to stand, the patient was markedly ataxic. *Reflexes:* Deep tendon stretch reflexes were decreased in both lower extremities. Plantar response was extensor on the right and flexor on the left.

Clinical diagnosis: The acute onset of headache, vertigo, vomiting, ataxia of gait, and dysmetria of the left arm were consistent with a left cerebellar hemisphere infarct or hemorrhage with possible minor involvement of brainstem secondary to mass effect.

Laboratory data: *CT scan* demonstrated a large left cerebellar infarct pressing the fourth ventricle to the right. *Magnetic resonance angiography* demonstrated occlusion of the left vertebral artery. *MRI* (2months after onset of symptoms): Infarction of the total cerebellar territory of the left posterior inferior cerebellar artery with no actual infarct of the brain stem (*Fig.20-12*).

Subsequent course: The patient was somewhat vague with some difficulty in memory during the subsequent 8-day hospital course and remained ataxic. He had three weeks of intensive rehabilitation with continuous subsequent improvement. Evaluation 6 months after onset of symptoms demonstrated only a slight slowness of alternating movements of the left hand.

A final note should be made regarding the considerable ability of the human nervous system to recover from extensive cerebellar lesions. Considerable recovery is possible because the cerebellum does not have a direct line role in the control of motor activity but instead, acts as a modulator of the motor control and motor planning circuits.

Causes of Hemorrhage into the Cerebellum.

1. *Hypertensive Hemorrhage Into the Cerebellum.* The cerebellum is the site of hypertensive hemorrhage in 8 to 10% of all cases. In the era before CT scan and MRI, only large hemorrhages or fatal hemorrhages were recognized *(see Figure 20-13a)*. Smaller lesions were not well localized. These are now clearly seen *(Fig.20-13b)*. Either the hemisphere (the hemorrhage often originates in the region of dentate nucleus) or vermis may be involved (Kase and Caplan, 1986).

2. *Arteriovenous Malformations.* The cerebellum is also the site of various types

Before the era of CT and MRI, the precise diagnosis was often not established. Small hemorrhages might have only nonspecific symptoms of headache and dizziness, as in case 20-7, and *Fig. 20-14* presented on CD-ROM.

SPINOCEREBELLAR DEGENERATIONS

In addition to those nutritional and systemic processes that involve primarily the cerebellar cortex, other diseases, usually genetic, involve the cerebellum as part of a degeneration .Other systems may be involved in addition to the cerebellum We can only briefly outline this very extensive topic which has been an area of rapid research advances.

In classifying these disorders ,we will follow

the approach of Wood and Harding (2000), adding more recent genetic information of Klockgether et al (2000) and Fujigasaki et al (2001).

The first step in classification is to separate all of these ataxic disorders into two main categories.

I. The autosomal recessive cerebellar ataxias are usually of early onset, (< 20 years of age).

II. The autosomal dominant cerebellar ataxias (ADCA) are usually of late onset, (>20 years).

I. **Autosomal recessive disorders:** The most common disorder in this category is Freidreich's ataxia which accounts for at least 50% of all cases of hereditary ataxia reported in large series from the United States and Europe.

This disorder which has a prevalence of 1-2 per 100,000, maps to chromosome 9q13. The mutation involves the trinucleotide repeat GAA. The normal number of repeats is 7-22, but in patients with Friedreich's ataxia there are 100-2000 repeats. The marked multiplication undoubtedly explains the early onset and severity of the disease. The gene product is a protein, frataxin. Refer to Chapter 9 for additional discussion.

Other rare disorders in this category include the following.

a) cerebellar ataxia with retained deep tendon reflexes reflexes, in which a mutation in the frataxin gene also occurs. The cases are less severe than Freidreich's ataxia, but with greater cerebellar atrophy on imaging studies.

b) cerebellar ataxia with hypogonadism in which there is predominant involvement of cerebellum and inferior olivary nuclei.

c) cerebellar ataxia with myoclonus (formerly called the Ramsay Hunt syndrome or dyssynergia cerebellaris myoclonica) but now recognized to be a heterogeneous syndrome. Included in this category are such entities as 1) the mitochondrial disorder, myoclonus epilep-

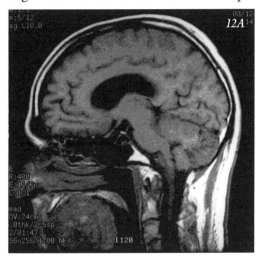

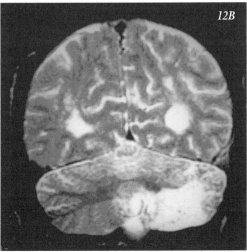

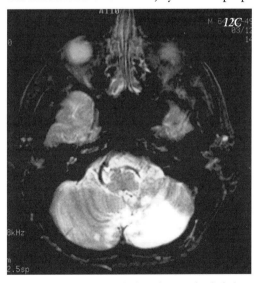

Figure 20-12: Cerebellar infarct in posterior inferior cerebellar artery territory. Case 20-6. Refer to text. MRI: A) T1 weighted sagittal section 10 mm to left of midline. B) T2 weighted posterior coronal section. C) T2 weighted axial sections at level of upper medulla inferior olive.

sy with ragged red fibers (MERRF) and 2)myoclonic epilepsy of Unverricht& Lundborg (Baltic myoclonus or EPM1) .The gene locus which maps to chromosome 21 encodes a small protein cystatin B whose role in the neurological disease is unclear.

d) other rare disorders in which cerebellar ataxia is associated in various combinations with deafness, or optic atrophy or mental retardation or cataracts or retinopathy.

e) There are also a whole host of relatively rare inherited usually autosomal recessive metabolic disorders in which progressive cerebellar ataxia may be a prominent feature. One group of disorders involves deficiencies in vitamin E due to genetic mutations or to malabsorption syndromes such as cystic fibrosis. Some of these disorders might be treatable with administration of vitamin E.

An additional disorder carries the designation ataxia telangiectasia. This is an autosomal recessive disorder, with onset of a progressive cerebellar ataxia beginning in the first 1 to 2 years of life followed by the development of choreoathetosis. One in 40,000-1000,000 live births is affected. An associated finding in all cases, usually appearing by age 6, is the presence of telangiectasis or capillary dilatation of severe and progressive degree, initially involving the conjunctiva but then appearing over the face and neck and the flexor surfaces of the limbs (antecubital and popliteal areas). The neuropathology primarily affects the Purkinje and granule cells of cerebellar cortex. These patients have severe medical problems related to the immune system, affecting both cellular and humoral immunity. The thymus remains in a fetal state. The patients have a marked sensitivity to ionizing radiation and to radiomimetic chemicals, as studied in cultured fibroblasts. The mutation which has been linked in some cases to chromosome 11q results in a defect in repair of DNA. The defective immune state results in a susceptibility to infectious diseases. There is a marked increase in malignancies, particularly leukemia and lymphomas, with rates which are 61 to 184 times normal. More recently, studies of the heterozygotes (who

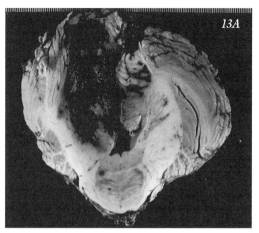

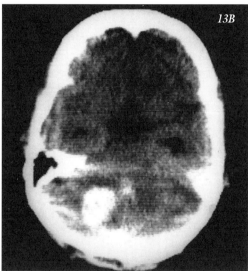

Figure 20-13: A) Large midline-paramedian cerebellar hemorrhage with rupture into the ventricle. This hypertensive (240/110) 61-year-old male had sudden onset of headache, vomiting dizziness, ataxia of gait and trunk, nystagmus on left lateral gaze and minor incoordination of the left hand. (Courtesy of Dr.John Hills and Dr. Jose Segarra.) B) Large hypertensive hemorrhage into right cerebellar hemisphere with hydrocephalus. CT scan (non-enhanced). This 65-year-old white male who had hypertension, diabetes, congestive heart failure and alcoholism, developed over one hour, vomiting, and a marked ataxia of gait. Examination indicated severe ataxia of gait, marked horizontal nystagmus on gaze to the right, dysmetria on right finger-to-nose and heel-to-shin tests, plus a mild right Horner's syndrome and a mild dysarthria, and bilateral Babinski signs. Symptoms resolved over several weeks.

constitute 1% of the general population but do not have neurologic symptoms) have demonstrated a significant increase in cancer rates, with an overall rate approximately 3.5 times that in the general population and breast cancer rates in females 5 times those in the control population. Diagnostic or occupational exposure to ionizing radiation increases the cancer risk of the heterozygotes compared to the controls (Swift et al, 1991).

II. Autosomal dominant disorders: These disorders were a problem in classification. The names of the authors of the various papers describing the clinical features and neuropathology of each of the groups of cases or families were attached as labels and included many of the famous neurologists of the late 1800s and early 1900s such as Holmes, Andre Thomas, Marie, Foix and Alajounine (Greenfield, 1954). Holmes separated these disorders into spinocerebellar degeneration, relatively pure degeneration of the cerebellar cortex and olivopontocerebellar degeneration. More recently Harding (1982), proposed a clinical classification into three major categories of which the first type is the most common

ADCA I: cerebellar ataxia is accompanied at some point in evolution by supranuclear ophthalmoplegia, optic atrophy, basal ganglia symptoms, dementia and amyotrophy.

ADCA II: in addition to the above has retinal degeneration.

ADCA III: presents a relatively pure cerebellar syndrome.

ADCA IV: is a recent addition to the classification cerebellar ataxia is combined with epilepsy.

The problem with all of these classifications from a clinical diagnostic standpoint was that early in the disease, all three groups might have predominantly cerebellar features. With the unraveling of the molecular genetic basis of the diseases, the **term ADCA has been replaced by the term spinocerebellar ataxia (SCA).** To date, 12 different loci have been identified and in many of these the molecular mechanism has been identified (SCA 1-SCA12)*. The gene products of the SCA genes are termed ataxins and are assigned a similar number. In addition 2 channelopathies causing episodic ataxias have been identified as EA 1 and 2.

Gene mechanisms:

1) CAG trinucleotide expansion: SCA 1,2,3,6 ,7, and 12 all involve an expansion of the CAG trinucleotide repeat as in Huntington disease (chromosome 4p16.3) but the chromosome loci are different (SCA 1 @ 6p, SCA 2 @ 12q, SCA 3 @ 14q, SCA 6@19p, SCA 7 @ 3P, SCA 12 @5). This expansion is translated into proteins with expanded polyglutamine tracts. At the ultra structural level, these expanded proteins are found in neuronal intranuclear inclusions.

2) CTG trinucleotide expansion: SCA 8 is a CTG expansion on 13q, the same type of trinucleotide expansion found in myotonic dystrophy.

3) Channelopathy: In EA 1 there is a point mutation in a potassium channel on chromosome 12. In EA 2 there is a point mutation in a calcium channel on chromosome 19q.

Patients with EA 1 have frequent brief attacks lasting minutes of ataxia plus myokymia (a muscle rippling due to peripheral motor nerve hyperexcitability). The patients are normal between attacks. Some cases benefit from acetazolamide or phenytoin. Patients with EA 2, have longer attacks lasting hours to days involving severe truncal ataxia plus vertigo and vomiting. The patients respond well to acetazolamide. However nystagmus is noted between attacks and over the years a slowly progressive ataxia develops. The MRI study at that point will confirm the cerebellar atrophy.

SCA 6 also has features of a channelopathy in that the CAG expansion occurs in a gene that encodes a voltage dependent calcium channel. The chromosome location19 is similar to EA 2.

4). Unknown: The mechanisms involved in SCA 4, SCA 5, SCA 10** and SCA 11 are

---

* As of April, 2002, the number of identified loci has grown to 17. SCA 17 involves a CAG trinucleotide expansion. There is an overlap with basal ganglia disorders for a binding protein.

** SCA 10 has now been associated with a pentaneuclotide expansion.

at the present time unknown. The chromosome locations are SCA 4@16q, SCA 5 @11 centromere, SCA 10 @22q and SCA 11 @ 15q.

Relationship to the clinical classification: SCA 1, 2, 3, and 12 correspond in general to ADCA I. Whether, SCA 2 or SCA 3 is the most frequent mutation depends on the series reviewed (Giunti et al, 1998 and Klockether et al 1998)*

SCA 7 corresponds to ADCA II.

SCA 5, 6 ,8 and 11 correspond to ADCA III. SCA 5 however has some posterior column features .SCA 6 has some sensory and pyramidal features and thus could be placed in the ADCA I group.

SCA 4 has cerebellar features plus a sensory neuropathy and probably would have been classified as ADCA I.

SCA 10** combines cerebellar ataxia with epilepsy (ADCA IV).

Azorean Disease: Machado-Joseph Disease: Refer to the reviews of Rosenberg (1992) and Sudarsky et al (1992).

This autosomal dominant spinocerebellar degeneration is clustered among families who originated in the Portuguese Azorean Islands. This is the region of origin for many families residing in Southeastern Massachusetts and the adjacent area of Rhode Island, and the initial cases were described in this area in 1972. Subsequently, other cases have been described in many areas of the world reached by Portuguese seafarers, explorers, whalers, and fishermen (many of whom were recruited in the Azores). The manifestations depend in part on the age of onset.

I. Early onset (childhood or young adult) cases have predominantly pyramidal and features and basal ganglia dysfunction.

II. Intermediate age of onset (20 to 40 years) Cases have cerebellar deficits and extraocular disturbances as well as the pyramidal features and basal ganglia dysfunction.

III. Later onset (50 to 60 years). Cases are characterized by cerebellar deficits and peripheral neuropathy.

IV. Possible late onset case may have a peripheral neuropathy plus Parkinsonism.

Since the molecular basis of the disorder is now clear, the more specific molecular diagnosis should be utilized when possible: most families carry the SCA 3 mutation. Occasionally the SCA 1 or the SCA 2 mutations have been found. In contrast to OPCA to be discussed below, the cerebellar cortex and the inferior olivary nucleus are not involved. The ataxia instead correlates with a degeneration of the afferent and efferent cerebellar systems (spinocerebellar tracts and Clarke's column dorsal nucleus) and the lid retraction and other extraocular features reflect involvement of the periaqueductal and third nerve nuclear areas.

Olivopontocerebellar Atrophy (OPCA):

This diagnosis once included many of the cases subsequently described as ADCA or SCA. When cases of progressive adult onset ataxia are familial, they should be assigned a diagnosis based on the genetic classification described above. This entity also has been considered in chapter 19 on the basal ganglia and is one of the multisystem atrophies considered in the general category of Parkinsonism plus syndromes. In contrast to the data presented in that chapter, Berciano(1988),when considering both familial and sporadic cases of OPCA.; found cerebellar symptoms (predominantly gait ataxia) to be the most common initial symptom (73%) and the most frequent symptom throughout the course of the disease (88-97%). Parkinsonian symptoms occurred as the initial feature in 8% but eventually in 35-57%. The neuropathology involved in all cases the cerebellar cortex, the pontine nuclei and the inferior olivary nuclei with associated changes in the white matter of cerebellar peduncles (particularly the middle) and of the cerebellar white matter. The substantia nigra is involved in 50% of cases.

Non familial cases of ADCA: Wood and Harding (2000)estimate that approximately two thirds of cases of degenerative ataxia

---

A recent addition to this CAG expansion group is DRPLA- (rubro pallido luysian atrophy) a basal ganglia disorder.

** and SCA 17

beginning after age 20 are single cases without a family history, and suggest that the term idiopathic late onset cerebellar ataxia be utilized rather than the old term OPCA. The majority of cases do demonstrate the pathological findings of OPCA. Some will go on to develop the autonomic features of multisystem atrophy. Some of the patients with onset after age 55 years will demonstrate a relatively pure midline cerebellar syndrome with primarily gait ataxia related to cerebellar atrophy which is most marked in the vermis. With tests for the molecular basis of the disease now available (particularly as regards the trinucletide expansions),the number of such sporadic unclassified cases should be reduced. Note that families have become smaller, full family historical information is not always available. There are however other entities to be considered when faced with a sporadic case.

Alcoholic nutritional cerebellar degeneration: refer to discussion above in relation to anterior lobe.

Paraneoplastic subacute cerebellar degeneration: This degeneration is associated particularly with cancer of the ovary, breast, small-cell cancer of the lung and Hodgkin's disease. The neurologic symptoms may precede the discovery of the primary malignancy in the majority of these patients except in those with Hodgkin's disease where the diagnosis of the malignancy usually has already been established. In general, these patients present with subacute onset of cerebellar symptoms affecting stance, gait, limbs and voice (dysarthria). The neuropathology involves a widespread loss of Purkinje cells and evidence of inflammation. The brain stem and dorsal root ganglia and possibly the limbic system may also show inflammatory changes, as part of a wider "encephalitis". Anti-Purkinje cell antibodies can be identified in many of these patients. Anti-yo antibodies which involve the Purkinje cell cytoplasm are found in patients with ovarian cancer, breast cancer or other gynecological malignancies. Patients with Hodgkin's disease who develop the cerebellar syndrome may have antibodies to a glutamate receptor (Smitt,et al 2000) The paraneoplastic syndromes may improve with control of the primary malignancy. Note, however, that the late cases have evidence on CT or MRI of cerebellar atrophy.

AN OVERVIEW OF TREMORS

Before concluding our discussion of the cerebellum and basal ganglia, we will briefly outline the various types of tremor. This is necessary because many common postural tremors (physiologic and essential) are mistakenly attributed to more serious disease of the basal ganglia or cerebellum.

Tremor is defined as an involuntary movement characterized by rhythmic oscillation of a body part (or parts) that develops when there is a synchronized discharge of many motor units. Many peripheral and central factors enter into this synchronization. Hallett, 1998, Elble, 1998, and Hua discuss the physiology of tremor and the role of mechanical factors and central oscillators, 1998.

Tremor is best classified on the basis of the behavioral situation in which the tremor is observed (Findley, 1988; Hallet, 1991, Deuschl, 1998): **1. tremor at rest 2. action tremor. Within this second category, postural, kinetic, or intention, and task-specific tremors are distinguished.**

1. **Rest Tremor:** This is a tremor that occurs in a body part that is not voluntarily activated and is completely supported against gravity. This tremor particularly when a pill-rolling component is present, is almost always indicative of Parkinson's disease or of related disease of the basal ganglia.

2. **Action tremors:** This is a tremor that is produced by voluntary contraction of muscle. Within this category several subtypes may be specified:

a. **Postural tremors:** These are by far the most common types of tremor and include:

1. *Physiologic Tremor* - usually fine and rapid (6 to 12 Hz), occurs when attempting to sustain a posture. This tremor is significantly increased or "enhanced" by anxiety, excessive caffeine intake, exercise, fatigue. It is also

increased by thyrotoxicosis, hypoglycemia, alcohol withdrawal and beta adrenergic drugs, such as those used in the treatment of asthma and by drugs commonly employed in psychiatry, such as lithium, neuroleptic, or tricyclic medications. Beta adrenergic blockers such as propranolol are often effective against this tremor if specific etiologic factors cannot be corrected (use in asthmatics, however, is contraindicated).

2. *Essential Tremor:* also called familial, benign, or senile tremor: Aside from physiologic tremor, essential tremor is the most common tremor encountered by the physician. Prevalence studies suggest a high frequency, particularly in older age groups. In the over 40 age group, 5% of the population may be affected. When familial, the genetic pattern appears to be autosomal dominant. Onset of familial cases may begin in childhood, adolescence, in mid life, or later in life. The hands are most commonly affected, but in some patients, the head is primarily involved (side-to-side movement). Some also have involvement of the voice or lips. The frequency of tremor ranges from 4 to 12 Hz. The tremor may remain stable or may slowly progress. Although in some patients the tremor is of relatively small amplitude, in other patients it may become relatively coarse, significantly interfering with fine motor activities. Orthostatic tremor is a variant in which a tremor of trunk and legs and to a lesser degree of arms occurs on prolonged standing and is associated with a high frequency of contractions alternating between antagonist muscles.

The underlying pathology of essential tremor is unknown. Functional imaging studies at rest and on action suggest an abnormal bilateral overactivity of cerebellar connections Infarcts (homolateral) of the cerebellum or of contralateral motor cortex or of ventrolateral thalamus may result in the disappearance of the tremor (Dupuis et al,1989). Typical of essential tremor is the temporary reduction in the tremor by ingestion of alcohol. The tremor usually responds to beta adrenergic blockers, such as propranolol. Note that in essential tremor, Parkinsonian and cerebellar features are absent: thus, the patient is able to walk with a good swing of the arms and does not turn en bloc. No pill rolling tremor emerges as the patient sits at rest or walks. No ataxia of gait or stance is present. These distinctions are of importance because postural tremors (of the outstretched hands) can also occur in patients with clear-cut evidence of akinetic rigid variants of Parkinson's disease, Wilson's disease, dystonia and cerebellar disease. A postural tremor may also occur in some forms of peripheral neuropathy or in post traumatic syndromes.

b. **Kinetic tremors** (usually approximately 5 Hz).These are tremors occurring during any voluntary movement. Several types have been specified.

1. *Tremor during target directed movements: intention tremor.* Amplitude increases during visually guide movements towards a target particularly at the termination of movement. This tremor usually is characteristic of cerebellar disease and is described in detail above. Intention tremor is an oscillation perpendicular to the line of movement. Findley (1988), distinguishes the tremor from dysmetria, defining the latter as "the inability to attain the target or normal performance level in a guided, goal-seeking movement. The more severe and violent kinetic tremors are seen in patients with severe multiple sclerosis involving both the brain stem and cerebellum. In these cases, the head and trunk may have a to-and-fro, anterior-to-posterior sway, or tremor, to which the term "titubation" has been applied (Findley, 1988). Cerebellar intention tremor may be decreased by lesions of the contralateral ventral lateral thalamic nucleus as discussed in chapter 19 .

2. *Task specific kinetic tremor: Kinetic tremor which appears or becomes exacerbated during specific activities related to occupation or writing.* It is uncertain whether primary writing tremor and some of the occupational tremors of musicians etc, represent a variant of essential tremor or of dystonia as discussed by Hallet (1991)

CHAPTER 21
Somatosensory Function and The Parietal Lobe

Introduction

The parietal lobe is composed of- post central gyrus, superior and inferior parietal lobules. Each area will be considered in terms of cytoarchitecture and functions. For the superior and inferior parietal lobules, the different effects of disease processes in the dominant compared to nondominant hemispheres will be considered. The major sensory pathways have already been covered in chapters' 7-spinal cord, 11- brain stem and 15- diencephalon. The student may wish to review that material at this time.

POSTCENTRAL GYRUS: SOMATIC SENSORY CORTEX [PRIMARY SENSORY S-I]

The postcentral gyrus is not, from a histological standpoint homogeneous. Four subtypes may be distinguished:

1) Area 3a at the base of central sulcus, and 2) Area 3b along the posterior wall of the central sulcus (anterior surface of the gyrus) both of which are relatively typical granular koniocortex and receive the major projection from the ventral posterior nuclei of the thalamus.

3) Area 1 on the crest of the gyrus and 4) Area 2 on the posterior surface of the postcentral gyrus are modified homotypical cortex. While some of the neurons in areas 1 and 2 receive direct input from the ventral posterior thalamic nuclei, other neurons are dependent on collaterals from area 3.

Postcentral Gyrus Stimulation: Stimulation of the post- central gyrus produces a sensation over the contralateral side of the face, arm, hand, leg, or trunk described by the patient as a tingling or numbness and labeled paresthesias. Less often, a sense of movement is experienced. The patient does not describe the sensation as painful. These various phenomena, occurring at the onset of a focal seizure, may be described as a somatic sensory aura.

There is in the postcentral gyrus a sequence of sensory representation which, in general, is similar to that noted in the precentral gyrus for motor function *(Fig. 21-1)*. The representation of the face occupies the lower 40 percent; the representation of the hand, the middle-upper 40 percent; and the representation of the foot, the paracentral lobule. As in the precentral gyrus, certain areas of the body have a disproportionate area of representation, ie, the thumb, fingers, lips and tongue. Those areas of the skin surface that are most sensitive to touch have not only the greatest area of cortical representation but also the greatest number and density of receptors projecting to the postcentral gyrus[1]. The peripheral field of each of these receptors is also very small (compared, e.g. to the less sensitive skin areas of the trunk. In addition, at the lower end of the postcentral gyrus, extending into the sylvian fissure (the parietal operculum), there is found a representation of the alimentary tract, including taste. Gustatory hallucinations may arise from seizure foci in this area (Hausser-Hauw &Bancaud, 1987). There is also a representation of the genitalia in the paracentral lobule and rare patient with seizures beginning in this area have reported paroxysmal sexual emotions including orgasm and nymphomania (Calleuja et al 1989).

Microelectrode techniques for recording from single cells in the cerebral cortex of the monkey and cat have allowed a considerable elaboration of the functional localization within the sensory cortex. The studies of Mountcastle (1957) indicated a vertical colum-

---

[1] *We will discuss in Chapter 30 the more recent observations that indicate a considerable degree of plasticity may be present in sensory cortex. The maps here illustrated then for an individual human or monkey may be subject to some degree of variability. See also Kass et al, 1983.*

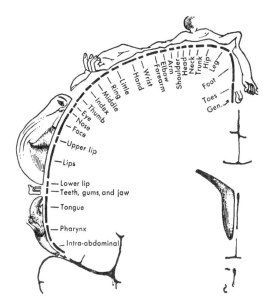

Figure. 21-1 Sensory representation as determined by stimulation studies on the human cerebral cortex at surgery: Note the relatively large area devoted to lips, thumb, and fingers. (From Penfield, W., and Rasmussen, T.: The Cerebral Cortex of Man. New York, Macmillan, 1955, p.214.)

nar organization from cortical surface to white matter. While each column is modality specific, each neuron in a particular column is activated by that specific sensory modality, e.g., touch, movement of a hair, deep pressure, joint position. Jones et al 1982, Jones and Powell, 1970, and Killackey and Ebner, 1973, discuss the specific thalamocortical relationships. The studies of Kass and his associates (1979, 1981, 1983) provide additional elaboration. Within the post central gyrus - four representations of the body and limb surfaces are present in a parallel manner - with each representation relatively modality specific. The cells in area 3a respond primarily to muscle stretch receptors; area 3b to rapidly and slowly adapting skin receptors (as in movement of a hair or skin indentation); area 1 to rapidly adapting skin receptors and area 2 to deep pressures and joint position.

Within areas 1 and 2, there are additional neurons that do not receive direct thalamic input. These neurons respond instead to more complex properties of the stimulus such as the specific direction of movement and have been labeled complex cells.

Studies of cortical potentials evoked by tactile stimulation in the monkey have suggested that there is a secondary somatic sensory projection area (S-II) in addition to the classic postcentral contralateral projection area. This second area has a bilateral representation and is found partially buried in the sylvian fissure at the lower end of the central sulcus). A similar second area of representation has been reported by Penfield & Jasper (1954) Luders et al (1985) and Blume et al (1992) in those seizure patients in whom an abdominal sensation (aura) was followed by a sensation of paresthesias in both sides of the mouth and in both hands *(Fig. 18-11).* Note also that at times stimulation of the precentral gyrus by the neurosurgeon during surgery has at times produced contralateral tingling or numbness, in addition to the more frequent motor responses (Penfield and Jasper, 1954). Additional discussion of seizures originating in parietal lobe will be found in Williamson et al (1992) and Tuxhorn & Kerdar (2001).

Postcentral Gyrus Ablation: Immediately following complete destruction of the postcentral gyrus, there will often be found an almost total loss of awareness of all sensory modalities on the contralateral side of the body. Within a short time there is usually a return of some appreciation of painful stimuli. The patient will, however, often continue to note that the quality of the painful stimulus differs from that on the intact side. An awareness of gross pressure, touch, and temperature also returns. Vibratory sensation may return to a certain degree. Certain modalities of sensation, however, never return or return only to a minor degree. (In partial lesions of the postcentral gyrus, however, these various modalities often return to a variable degree). *These modalities are often referred to as the cortical modalities of sensation or as discriminative modalities of sensation.* The following types of sensory awareness usually included in this category are summarized in table 21-1

In contrast, the sensory modalities of pain, gross touch, pressure, temperature, and vibration

TABLE 21-1: MODALITIES OF SENSATION LOCATED IN POSTCENTRAL GYRUS

Cortical Discriminative Sensory Modalities	Type of Stimulus
1. Position sense	Ability to perceive movement and the direction of movement when the finger or toe is moved passively at the interphalangeal joint, the hand at the wrist or the foot at the ankle
2. Tactile localization:	Ability to accurately localize the specific point on the body or extremity which has been stimulated.
3. Two-point discrimination	Ability to perceive that a double stimulus with a small separation in space has touched a given area on the body or Limb or hand.
4. Stereognosis	Ability to distinguish the shape of an object and thus to recognize objects based on their three-dimensional tactile form.
5. Graphesthesia	Ability to recognize numbers or letters which have been drawn on the fingers, hand, hand face, or leg.
6. Weight discrimination:	Ability to recognize differences in weight placed on the hand or foot.
7. Perception of simultaneous stimuli	Ability to perceive that both sides of the body have been simultaneously stimulated. When bilateral stimuli are presented but only a unilateral stimulus is perceived, "extinction" is said to have occurred.
8. Perception of texture	Ability to perceive the pattern of surface stimuli encountered by the moving tactile receptors.

are referred to as *primary modalities of sensation*. Perception of these modalities continues to occur after ablation of the postcentral gyrus although some alteration in quality of sensation is noted. It has been assumed that the anatomical substrate for such awareness must exist at the thalamic level.

In terms of more specific localization of specific modalities of cortical sensation, the studies of Randolph and Semmes 1974 suggest that in the monkey small lesions restricted to area 3b (hand region) produce deficits in discrimination of texture, size and shape. Lesions in area 1 interfere with the ability to discriminate texture only; lesions in area 2 - size and shape only. The following case histories demonstrate the type of sensory phenomena found in disease involving the postcentral gyrus.

Case 21-1: One week prior to evaluation, this 40-year old right-handed married white male had the onset of repeated 15-21 minute duration episodes of focal numbness (tingling or paresthesias) involving the left side of the face, head, ear, and posterior neck and occasionally spreading into the left hand and fingers, and rarely subsequently spreading into the left leg. Biting or eating would trigger the episodes. There was no associated pain or headache or weakness or associated focal motor phenomena.

Initial neurological examination: Totally intact as regards mental status, cranial nerves, motor system, reflexes, and sensory system.

Clinical diagnosis: Focal seizures originating lower third post central gyrus, of uncertain etiology.

Laboratory data: *EEG*: no focal abnormalities were present. CT Scan *(Fig. 21-2)*: The non-enhanced study was normal. The enhanced study demonstrated a small enhancing lesion just above the right Sylvian fissure. MRI *(Fig.21-3)*: Demonstrated the more extensive nature of this process involving the operculum - above the right sylvian fissure including the lower end of the post central gyrus, the area of representation of the face. A probable infiltrating tumor was suggested as the most likely diagnosis. *Arteriogram* - demonstrated a tumor blush at the right sylvian area, consistent with a glioblastoma, a rapidly growing infiltrating tumor - originating from astrocytic glia.

Subsequent Course: Treatment with an anticonvulsant reduced the frequency of the episodes. Neurosurgical consultation suggested an additional observation period and periodic CT scans. Three months after onset of symptoms, the patient developed a numbness of the left arm and difficulty in coordination of left arm. Exam now demonstrated left central facial weakness, mild weakness of the left hand, and in the left hand a loss of stereognosis and graphesthesia with a relative alteration in pinprick perception. The CT scan showed significant enlargement of the previous lesion. EEG now indicated focal (2-3 Hz delta slow waves in the right frontal central Rolandic area. Doctor Bernard Stone performed a subtotal resection of a glioblastoma - in the right temporal parietal area. Radiation therapy was administered following surgery, 4000 rads total whole brain and 5210 rad total to the tumor area. Reevaluation 3 months after surgery indicated only rare focal seizures involving the face. The neurological examination was normal. Repeat CT scans, however, continued to demonstrate a large area of enhancing tumor in the right temporal-frontal-parietal area. Despite the use of dexamethasone and arterial chemotherapy with cis-Platinum, he continued to progress with the development of a left field defect, left hemiparesis and a loss of all modalities of sensation on the left side. Subsequently he developed significant changes in memory and cognition. CT scan demonstrated progression with extensive involvement of the right side of the brain and spread to the left hemisphere. Death occurred 33 months after onset of symptoms. At postmortem examination of the brain, almost the entire right hemisphere was replaced by necrotic infiltrating tumor. The tumor now had spread into the corpus callosum and temporal and occipital lobes. The internal capsule, thalamus, hypothalamus and basal ganglia were destroyed.

Case 21-2: This 62-year-old white, right-handed housewife, six years prior to admission had undergone a left radical mastectomy for carcinoma of the breast. Five months prior to admission, the patient developed a persistent cough, left pleuritic pain and a collection of fluid in the left pleural space (pleural effusion). She now had the onset of daily headaches in the right orbit, occasionally awakening her from sleep. Four months prior to admission, over a 3 to 4 week period, the patient devel-

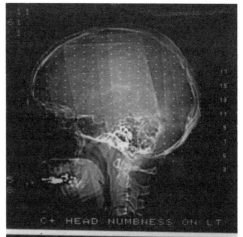

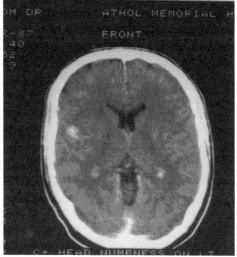

Figure. 21-2. Glioblastoma somatic sensory cortex: focal sensory seizure with onset in left face. Case 21-1.CT scan with contrast demonstrated a small focal enhancing lesion in the right post-central gyrus just above the Sylvian fissure. Upper - reference diagram for plane of section. Lower scan 9 - the enhancing lesion is marked with a white square. (See text).

oped progressive "weakness" and difficulty in control of the right lower extremity. Several weeks later, the patient now experienced, aching pain in the right index finger in addition to a progressive deficit in the use of the right hand. This was more a "stiffness and incoordination" than any actual weakness. One month prior to admission the patient noted episodes of pain in the toes of the right foot and numbness of right foot occurred, lasting 2 to 3 days at a time. She subsequently developed, difficulty in memory and some minor language

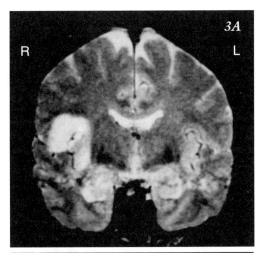

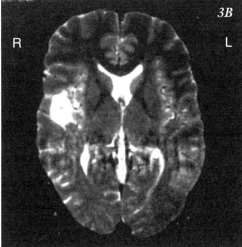

Figure. 21-3. Glioblastoma, somatosensory cortex - focal sensory left facial seizures. Case 21-1. (See text). MRI scans demonstrating the more extensive involvement of cortex and white matter: A) coronal section; B) horizontal section.

deficit, suggesting a nominal aphasia, prompting hospital admission.

Neurological examination: *Mental status:* slowness in naming objects was present although she missed only one item of 6. Delayed recall was slightly reduced to 3-out-of-five in 5 minutes. *Cranial nerves:* early papilledema was noted: absence of venous pulsations, indistinctness of disc margins, and minimal elevation of vessels as they passed over the disc margin. A right central facial weakness was present. *Motor system:* Mild weakness of the right upper limb was present. A more marked weakness was present in the lower limb (most marked distally). Spasticity was present at the right elbow and knee. In walking, the right leg was circumducted; the right arm was held in a flexed posture. *Reflexes:* Deep tendon reflexes were increased on the right. A right Babinski response was present. *Sensation:* An ill-defined alteration in pain perception was present in the right arm and leg - more of a relative difference in quality of the pain than any actual deficit. Repeated stimulation of the right lower extremity (pain or touch) produced dysesthesias (painful sensation). Position sense was markedly defective in the right fingers with errors in perception of fine and medium amplitude movement in the right toes.

Simultaneous stimulation resulted in extinction in the right lower extremity.

Graphesthesia (identification of numbers, e.g., 8, 5, 4 drawn on cutaneous surface) was absent in the right hand and fingers and poor in the right leg. Tactile localization and two-point discrimination were decreased on the right side.

Clinical diagnosis: Metastatic tumor to the left post central and precentral gyri with a cortical pain syndrome (pseudo thalamic pain syndrome).

Laboratory data: *Chest x-ray:* Several metastatic nodules were present in the left lung. *EEG:* almost continuous focal 4 to 7 Hz slow wave activity was present in the posterior frontal central and parietal areas. *Imaging studies* were consistent with a focal metastatic lesion in the parasagittal upper left parietal area.

Subsequent course:: Because there was evidence in this case that the disease had spread to multiple organs, radiation and hormonal therapy were, therefore, administered rather than any attempt at surgical removal of the left parietal metastatic lesion. Temporary improvement occurred in motor function in the arm but the patient eventually expired five months after admission. Autopsy performed by Dr. Humphrey Lloyd of the Beverly Hospital disclosed extensive metastatic disease in the lungs, liver, and lymph nodes and a single necrotic metastatic brain lesion in the upper postcentral gyrus of the left parietal area, 1.0 cm. below the pial surface and measuring 1.2 x 1.0 x 0.6 cm. The occurrence of episodic pain in the involved arms and legs along with the production of an experienced painful sensation on repetitive tactile stimulation (dysesthesias) in patients with sensory pathway lesions is sometimes referred to as a "thalamic" or "pseudo thalamic" syndrome. (Refer to discussion of Wilkins and Brody, 1969). In this case, the anatomical locus for the pseudo thalamic syndrome is apparent.

The effects of deficits in cortical sensation on the total sensory motor function of a limb are apparent in this case. The actual disability and disuse of the right arm and leg were far out of proportion to any actual weakness. Such an extremity is often referred to as a "useless limb." The actual weakness that was present undoubtedly reflected pressure effects on the precentral gyrus and the descending motor fibers in adjacent white matter.

Destructive lesions of the postcentral gyrus during infancy or early childhood often produce a retardation of skeletal growth on the contralateral side of the body. Such a patient examined as an adolescent or adult will be found to have not only cortical sensory deficits in the contralateral arm and leg but also a relative smallness of these extremities (shorter arm or leg, smaller hand and glove size, smaller shoe size).

SUPERIOR AND INFERIOR PARIETAL AREAS

In the monkey these areas are collectively designated the posterior parietal cortex. In both the human and the monkey, superior and inferior parietal lobules can be distinguished divided by the intraparietal sulcus. In man, the superior parietal lobule is composed of Brodmann's cytoarchitectural areas 5 and 7, the inferior parietal lobule of areas 40 and 39, (the supra marginal and angular gyri). In the monkey the superior parietal lobule is designated as area 5; the inferior parietal area as primarily area 7. This results in some confusion as correlations are attempted. There is no clear-cut gross homologue of areas 40 and 39 in the monkey cerebral cortex.

All of these areas may be classified as varieties of homotypical cerebral cortex. The major afferent input of the superior parietal lobule area 5 in the monkey is from the primary sensory areas of the post central gyrus. Area 7 of the monkey cortex receives indirect cortico-cortical connections. The superior parietal lobule receives projections from the posterior lateral nuclei of the thalamus; the inferior parietal lobule from the pulvinar (see Chapter 15). These secondary sensory areas project to an adjacent tertiary sensory area and then to a multimodal sensory association area at the temporal parietal junctional area. This latter area in the posterior parietal cortex then projects to another multimodal sensory association area in the frontal - peri arcuate - principal sulcus area, to the premotor and frontal eye fields. (Discussed earlier in relationship to premotor and prefrontal motor function). Area 7 has connections to the limbic cortex: cingulate gyrus (Mesulum et al, 1977).

The implications of these connections for the integration of complex movements are clear. It is not surprising then that single cell studies as reviewed by Darian-Smith et al (1979) demonstrate responses of neurons in area 5 to manipulation of joints; hand manipulation as in grasping and manipulation objects, or in projecting the hand to obtain a specific object associate with reward. Neurons in area 7 discharged in relationship to complex, eye and limb movements (Lynch et al 1977). The reviews of Andersen et al, 1997 and Colby & Goldberg, 1999 discuss the multiple representations of space in the posterior parietal cortex, particularly the intraparietal sulcus area. When damage to parietal cortex occurs, the patient manifests a variety of spatial deficits. A neglect occurs in relation to multiple sensory modalities but is most prominent with regard to contralateral visual space as we will discuss below. There are corresponding deficits in the generation of spatially directed actions.

Stimulation. The threshold of the superior and inferior parietal lobules for discharge is relatively high. Although Foerster reported the occurrence of some contralateral paresthesias on stimulation in man, Penfield did not confirm these results. Stimulation in the inferior parietal areas of the dominant hemisphere did produce arrest of speech, but this is a nonspecific effect, occurring on stimulation of any of the speech areas of the dominant hemisphere. It must, of course, be noted that space-occupying lesions in the parietal lobules may produce sensory or motor seizures by virtue of their pressure effects on the lower threshold post and precentral gyri.

Ablation: Darian-Smith et al (1979) have summarized the effects of selective lesions of area 5 and or area 7 in the monkey. The most selective studies were those of Stein (1977) using reversible cooling lesions. Cooling of area 5 resulted in a clumsiness of the contralateral arm and hand so that the animal was unable to search for a small object. Cooling of area 7 produced a much more complex clumsiness which was apparent only when the arm was moved into the contralateral visual field. Movements of the hand to the mouth were intact. (Refer to discussion of Colby & Goldberg, 1999, Wise et al, 1997, for a more recent discussion of the topic.)

Earlier clinical studies in humans had suggested possible sensory deficits in relation to ablation of the inferior or superior parietal areas. The detailed studies of Corkin et al, (1964) on patients subjected to limited cortical ablations of the inferior or superior parietal areas (in the treatment of focal epilepsy) clearly indicated that no significant sensory deficits occurred. The postcentral gyrus rather than the parietal lobules is critical for somatic sensory discrimination. The more recent studies of Pause et al (1989) confirmed that anterior parietal (post central lesions) resulted in somatosensory disturbances including surface sensibility, 2 point discrimination, position sense as well as more complex tactile recognition. In contrast, in posterior parietal lesions, there was a preferential impairment of complex somatosensory and motor functions involving exploration and manipulation by the fingers - not explained by any sensory deficit. There was a deficit in the conception and execution in the spatial and temporal patterns of movement. The end result is an impairment of purposive movement as discussed above. The response to visual stimuli in terms of attending to and reaching for objects is impaired (see Baynes et al 1986, Nagel-Leiby et al 1990, Perenin et al 1988, and Pierrott Deseilligny et al 1986). Lesions involving the white matter deep to the inferior parietal lobule may damage the superior portion of the optic (geniculocalcarine) radiation, producing a defect in the inferior half of the contralateral visual field, an inferior quadrantanopsia.

DOMINANT HEMISPHERE PARIETAL LOBULES

Destructive lesions of the parietal lobules produce additional effects on more complex cortical functions. Those lesions involving particularly the supramarginal and angular gyri of the dominant (usually the left hemisphere) inferior parietal lobule may produce one or more of a complex of symptoms and signs known as Gerstmann's Syndrome. These include:

(a) Dysgraphia (a deficit in writing in the presence of intact motor and sensory function in the upper extremities),

b) Dyscalculia (deficits in the performance of calculations),

c) Left-right confusion, and

d) Errors in finger recognition, for example, middle finger, index finger, ring finger, in the presence of intact sensation (finger agnosia).

e) In addition, disturbances in the capacity for reading may be present. Some patients may also manifest problems in performing skilled movements on command (an apraxia) at a time when strength, sensation and coordination are intact.

Usually only partial forms of the syndrome are present. The problem of the dominant parietal lobe in language function will be considered in greater detail and illustrated in the section on language and aphasia. The reader should refer to chapter 24 for an illustrative case history.

NONDOMINANT HEMISPHERE PARIETAL LOBULES

Patients with involvement of the nondominant parietal lobe, particularly the inferior parietal lobule, often demonstrate additional abnormalities in their concepts of body image, in their perception of external space, and in their capacity to construct drawings.

The disturbance in concept of body image may include the following (a) a lack of awareness of the left side of the body, with a neglect of the left side of the body in dressing, undressing and washing. (b) Despite relatively intact cortical and primary sensation, the patient may fail to recognize his arm or leg when this is passively brought into his field of vision. (c) The patient may have a lack of awareness of a hemiparesis despite a relative preservation of cortical sensation (anosognosia) or may have a total denial of illness. At times, this may be carried to the point of attempting to leave the hospital since as far as the patient is concerned; there is no justification for hospitalization. The denial of illness undoubtedly involves more than perception of body image.

The disturbance in perception of external space may take several forms: (a) neglect of the left visual field and of objects, writing, or pictures in the left visual field. At times, this is associated with a dense defect in vision in the left visual field; at other times, there may be no definite defect for single objects in the left visual field, but extinction occurs when objects are presented simultaneously in both visual fields. Again, the problem of possible involvement of occipital cortex or of subcortical optic radiation should be considered as previously noted. (b) An inability to interpret drawings such as a map or to pick out objects from a complex figure. The patient is confused as to figure background relationship and is disoriented in attempting to locate objects in a room. When asked to locate cities on an outline map of the United States, the patient manifests disorientation as the west and east coasts and as to the relationship of one city to the next. Chicago and New Orleans may be placed on the Pacific Ocean; Boston on the Florida peninsula; and New York City somewhere west of the Great Lakes.

The disturbance in capacity for the construction of drawings has been termed a constructional apraxia or dyspraxia. An apraxia may be defined as an inability to perform a previously well-performed act at a time when voluntary movement, sensation, coordination, and understand are otherwise all intact. The following deficits may be present: the patient may be unable to draw a house or the face of a clock; or to copy a complex figure such as a three-dimensional cube, a locomotive, and so forth; in severe disturbances the patient may be unable to copy even a simple square, circle or triangle. The following case demonstrates many of these features.

Case 21-3: This 70-year-old, single, white female, right-handed, retired candy maker underwent a left radical mastectomy for carcinoma of the breast, three years prior to admission. Four months prior to evaluation, the patient became unsteady with a sensation of rocking as though on a boat. She no longer attended to her housekeeping and to dressing. Over a three-week period, prior to evaluation, a relatively rapid progression occurred with deterioration of recent memory. A perseveration occurred in motor activities and speech. The patient was incontinent but was no longer concerned with urinary and fecal incontinence. For 2 weeks, right temporal headaches had been present. During this time, her sister noted the patient to be neglecting the left side of her body. She would fail to put on the left shoe when dressing. In undressing, the stocking on the left would be only half removed.

Family History: The patient's mother died of metastatic carcinoma of the breast.

Neurological examination: *Mental status:* Intact except for the following features: The patient often wandered in her conversation. She often asked irrelevant questions and was often impersistent in motor activities. There was marked disorganization in the drawing of a house or of a clock. A similar marked disorganization was noted in attempts at copying the picture of a railroad engine *(Fig 21-4)*. There was a marked neglect of the left side of space and of the left side of the body. The patient failed to read the left half of a page. When she put her glasses on, she did not put the left bow over the ear. When getting into bed, she did not move the left leg into bed. She had slipped off her dress on the right side, but was lying in bed with the dress still covering the left side. The patient had been reluctant to come for neurological consultation. Although she complained of headache and nausea, she denied any other deficits. Her relatives provided information concerning these problems. Much additional persuasion over a two-week period was required before the patient would agree to be hospitalized. *Cranial nerves:* A dense left homonymous hemianopsia was present. When reading, the patient left off the left side of a page. She bisected a line markedly off center. Disc margins were blurred and venous pulsations were absent, indicating papilledema.

The right pupil was slightly larger than the left. A minimal left central facial weakness was present. *Motor system:* Although strength was intact, there was little spontaneous movement of the left arm and leg. The patient was ataxic on a narrow base with eyes open with a tendency to fall to the left, and was unable to stand with eyes closed even on a broad base. *Sensation:* Although pain, touch and vibration were intact, there was at times a decreased awareness of stimuli on the left side. Errors were made in position sense at toes and fingers on the left. Tactile localization was poor over the left arm and leg. With double simultaneous stimulation, the patient neglected stimuli on the left face, arm and leg.

Clinical diagnosis:: Metastatic breast tumor to right non-dominant parietal cortex.

Laboratory data: *Chest X-ray* indicated a possible metastatic lesion at the right hilum. *EEG* was abnormal because of frequent focal 3 to 4 cps slow waves in the right temporal and parietal areas, suggesting focal damage in these areas *(Fig 21-5 on CD atlas)*. *Imaging studies*

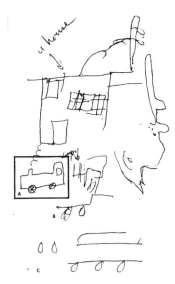

Figure. 21-4. Nondominant parietal constructional apraxia. Case 21-3. The patient's attempts to draw a house are shown on the upper half of the page. Her attempts to copy a drawing of a railroad engine are shown on the lower half of the page (B & C). The examiner's (Dr. Leon Menzer) original is designated (A).

(Brain scan radioisotope Hg197 and arteriograms) were consistent with a large well-defined lesion in the posterior section of the right temporal inferior parietal area *(Fig. 21-6)*.

Subsequent course: Treatment with steroids (dexamethasone and estrogens) resulted in temporary improvement. The patient refused surgery. Her condition soon deteriorated with increasing obtundation of consciousness. She expired two months following her initial neurological consultation.

A CT scan from a more recent case demonstrating many aspects of this syndrome is illustrated in *Figure 21-7*. In many cases, the location of lesion may appear to be predominantly posterior temporal. Such large posterior temporal lesions would certainly compromise the cortex and subcortical white matter of the adjacent inferior parietal area.

In this case, marked deficits in perception of the cortical modalities of sensation were present. In other cases, as in the case demonstrated in *Fig 21-7*, such involvement is much less marked.

In some cases, involvement of the motor

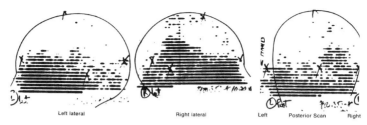

Figure. 21-6 Non dominant parietal lesion. Case 21-3. Brain scanHg197. A large uptake of radioisotope is demonstrated in the right parietal and adjacent posterior temporal area. (Courtesy of Dr. Bertram Selverstone.)

cortex is evident with an actual left hemiparesis accompanied by an increase in deep tendon reflexes and an extensor plantar response. At times in patients with neglect syndromes, there may be several indications in the clinical examination and in the laboratory studies that the involvement of the frontal lobe areas is more prominent than the parietal involvement. We have already indicated that the neglect components of this syndrome may also be noted in lesions of the anterior premotor area (area 8). The prefrontal and premotor areas as discussed above and in chapter 18, receive projection fibers from the multimodal area of the posterior parietal area. It is possible that in some cases the posterior temporal - inferior parietal location of the lesion may also be critical in interrupting these association fibers. For these several reasons, it is perhaps more appropriate to use the term, syndrome of the non-dominant hemisphere, rather than the more localized designation, non-dominant inferior parietal syndrome. Duffner et al, 1990 have presented the concept of a network for directed attention - with right frontal lesions leading to left hemi spatial neglect only for tasks that emphasize exploratory-motor components of directed attention whereas parietal lesions emphasize the perceptual-sensory aspects of neglect.

With lesions of the non-dominant hemisphere, there is a significant alteration of the patient's awareness of his environment. The behavior of an individual is in part determined by his own particular perception of the environment. If that perception is altered or disorganized, the behavioral responses of the patient may appear inappropriate to others. Obviously, not all individuals will respond in the same manner to a given environmental situation; part of the response will be determined by the past experience and personality of the individual. Thus, given the same lesion, one individual may be unaware of a hemiparesis, another may deny the hemiparesis but agree an illness is present, a third may claim to be healthy and claim that people are conspiring to keep him in the hospital.

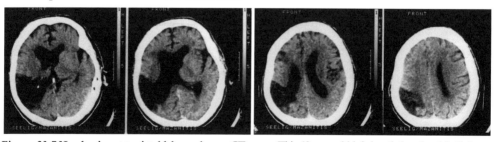

Figure. 21-7 Nondominant parietal lobe syndrome. CT scans. This 68-year-old left-handed male with diabetes mellitus and hypertension had the sudden onset of left arm paralysis, loss of speech and ability to read and write all of which recovered rapidly. On examination 9 months after the acute episode he continued to have the following selective deficits: (1) he was vague in recalling his left hemiparesis (2) in dressing, he reversed trousers and failed to cover himself on the left side. (3) He had extinction in the left visual field on bilateral simultaneous visual stimulation and over the left arm and leg on bilateral simultaneous tactile stimulation. (4) A left Babinski sign was present. CT scan now demonstrated an old cystic area of infarction in the right posterior temporal-parietal (territory of the inferior division of the right middle cerebral artery).

CHAPTER 22
Limbic System

INTRODUCTION

Rhinencephalon

The neurologist Paul Broca in the later half of the 19th century initially designated all of the structures on the medial surface of the cerebral hemisphere the "great limbic lobe." This region, due to its strong olfactory input, was also designated the rhinencephalon.

The olfactory portion of the brain (rhinencephalon, or archipallium) comprises much of the telencephalon in fish, amphibians, and most mammals. In mammals the presence of a large olfactory lobe adjacent to the hippocampus was once considered to be evidence of the important olfactory functions of these regions. However, when a comparative neuroanatomist examined the brains of sea mammals that had rudimentary olfactory apparatus, e.g., dolphins and whales, the presence of a large hippocampus suggested other than olfactory functions for this region.

In 1937, Papez proposed that olfactory input was not the prime input for this region, and the experiments of Kluver and Bucy (1937, 1939) and Kluver (1952 and 1958) demonstrated the behavioral deficits seen after lesions in this zone. More recently, it has been shown that in primates only a small portion of the limbic lobe is purely olfactory: the olfactory bulb, olfactory tract, olfactory tubercle, pyriform cortex of the uncus, and corticoamygdaloid nuclei *(Fig. 22-1)*. The other portions--hippocampal formation, fornix, parahippocampal gyrus, and cingulate gyrus--are now known to be the cortical regions of the limbic system (Fulton, 1953; Green, 1958; Papez, 1958; Scheer, 1963; Isaaccson, 1982).

Emotional Brain

Since the initial observations of Kluver and Bucy (1937) and Papez (1937), which localized emotions in the telencephalon, many other investigators have added information concerning the localization of behavior. We now know that many cortical and subcortical regions are incorporated in the "emotional brain."

Different investigators have coined different terms to succinctly describe the limbic system, particularly the visceral, vital, or emotional brain. The term "visceral brain" would seem appropriate since much of our emotional response is characterized by specific responses in the viscera (Fulton, 1953). On the other hand, the importance of the emotional response for the self-preservation of the individual and the perpetuation of the species has

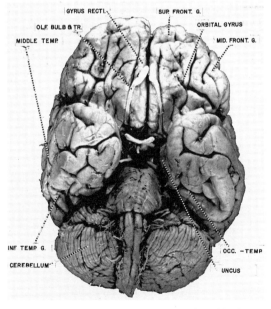

Figure 22-1 Gross view of ventral surface of brain demonstrating olfactory bulb and olfactory tract.

led other investigators to call this region the "vital brain" (MacLean, 1955). The term used most commonly by investigators and the one used in this chapter is limbic lobe (limbus = margin) because the involved region is located on the medial margin of the cerebrum and surrounds the brain stem as it enters the diencephalon.

We can separate the entire central nervous system into a "somatic brain," which controls the external environment through the skeletal muscles, and a " visceral brain," which controls the internal environment through the control of smooth muscles and glands. Our discussion of this region begins with the olfactory system and continues into the Limbic System

I. OLFACTORY SYSTEM

The olfactory system must be included in any discussion of the emotional brain as it sends fibers directly into the medial temporal lobe, and this olfactory information is especially important to the emotional brain.

Olfactory Nerve. The olfactory nerve (cranial nerve I) originates from the uppermost portions of both nasal fossae, which occupy the mucous membrane covering the superior nasal conchae and adjacent septum. The mucous membrane is attached to the walls of the nasal septum that in this region is formed by the ethmoid bone.

Olfactory Receptors. The 100 million or more bipolar receptor cells are embedded in sustentacular cells. The dendrites of the receptor cells are short and ciliated. The cilia are embedded in the odor-absorbing secretion secreted by the Bowman glands that forms the mucosa. New cells from the basal layer are constantly replacing the olfactory receptor cells. All of the axons from the olfactory neuroreceptor cells are unmyelinated. They gather together into about 20 bundles (fila olfactoria), which then pass through openings in the cribriform plate of the ethmoid bone to synapse in the olfactory bulb.

Olfactory Discrimination. Vertebrates with a well-developed sense of smell are called macrosmatic, while those with a poorly developed sense of smell are called microsmatic. Dogs and cats are macrosmatic animals, while humans and all of the great apes are microsmatic *(Fig 22-2)*. In the dog, 15% of the brain

Relates to olfaction while in the human brain, the portion devoted to olfaction is minute. Most dogs and cats have an olfactory system that is infinitely superior to ours both in sensitivity and the ability to discriminate among odors. Nevertheless, we can still distinguish thousands of different odors, and a multimillion-dollar industry has developed to stimulate our olfactory system. In fact, many people have built careers on their olfactory acuity (wine and coffee sniffers and perfumers). Many blind or deaf people have such a refined sense of smell that they can detect subtle changes in their environment.

The olfactory information is distributed from the olfactory bulb through three olfactory stria:

1. Intermediate stria -- olfactory tubercle.
2. Medial olfactory stria -- septal region and via the anterior commissure into the opposite olfactory bulb and,
3. Lateral olfactory stria -- olfactory cortex of the uncus and corticomedial amygdaloid. The terminations in the amygdala produce a strong response in the emotional brain to substances that stimulate the olfactory systems especially smoke, food and pheromones.

Details on the structure of the olfactory bulb, the olfactory connections and the transduction of odors into neuronal signals are included in the CD-ROM section on special senses.

II. LIMBIC REGIONS IN THE BRAIN

The limbic/emotional brain is divided into cortical and subcortical regions.

Cortical Areas are located in: Frontal Lobe, Temporal Lobe and Cingulate Gyrus while Subcortical areas are found in the brain stem, diencephalon and septum.

Table 22-1 identifies the cortical and subcortical limbic structures

SUBCORTICAL STRUCTURES REFER TO CHAPTER 11)

Reticular Formation of the Brain Stem and Spinal Cord. The spinal and cranial nerve roots are the first-order neurons for sensory information to reach the somatic and visceral brain. These peripheral nerves send much sensory information via axon collaterals into the reticular formation of the spinal cord, medulla, pons, and midbrain. The reticular formation is organized longitudinally, with the lateral area being the receptor zone and the medial area being the effector. In the medial zone are the ascending and descending multisynaptic fiber tracts of the reticular formation: the central tegmental tract.

Throughout the levels of the brain stem there are also certain important limbic nuclear groupings in the reticular formation:

1. The mesencephalic portion of the reticular formation seems to be especially important, since this zone provides a direct reciprocal pathway to the hypothalamus, thalamus, and septum.
2. From the locus caeruleus of the upper pons and the raphe of the midbrain the ascending

TABLE 22-1. THE SUBCORTICAL AND CORTICAL NUCLEI OF THE LIMBIC SYSTEM

SUBCORTICAL NUCLEI OF THE LIMBIC SYSTEM:	CORTICAL STRUCTURES OF THE LIMBIC SYSTEM:
1. Reticular formation of the brain stem and spinal cord --Midbrain limbic nuclei (Interpeduncular nucleus, Paramedian nucleus, ventral tegmental area, Ventral half of the periaqueductal gray. 2. Hypothalamus (preoptic, lateral, Lateral mammillary nuclei) 3. Thalamus (midline, intralaminar, anterior, and dorsal medial nuclei) 4. Epithalamus 5. Septum 6. Nucleus accumbens	1. Temporal Lobe: --Amygdala --Hippocampal formation- --Parahippocampal cortex --Rostral portion of temporal lobes 2. Frontal Lobe --Frontal association areas --Supracallosal gyrus and longitudinal stria --Subcallosal gyrus -Orbital frontal cortex 3. Cingulate gyrus and cingulate isthmus

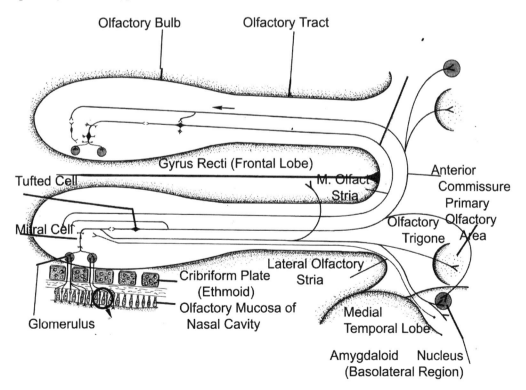

Figure 22-2 Diagrammatic representation of the olfactory system and its connections

serotoninergic and adrenergic systems run into the diencephalon and telencephalon and provide direct input into these regions.

Interpeduncular Nucleus. The interpeduncular nucleus (posterior perforated substance) is found on the anterior surface of the midbrain in the interpeduncular fossa extending from the posterior end of the mammillary body to the anterior end of the pons. It receives fibers from the habenular nuclei (habenulopeduncular tract) and has reciprocal connections with the hypothalamus and the midbrain limbic region. Amygdaloid information reaches this region through connections via the stria terminalis to the septum and then from the septum to the interpeduncular nucleus. Hypothalamic input and septal input are also important parts of the autonomic information to the brain stem passing through this nucleus.

Hypothalamus *(Figs. 22-3, 22-4,* and chapter 16). The hypothalamus is the highest subcortical center of the visceral brain. The basic function of this region is to maintain internal homeostasis (body temperature, appetite, water balance, and pituitary functions) and to establish emotional content. It is a most potent subcortical center due to its control of the autonomic nervous system.

The hypothalamus receives input from all portions of the limbic system, as well as from the reticular formation, basal ganglia, and frontal association cortex. The hypothalamus connects to the thalamus, midbrain, pons, and medulla via the medial forebrain bundle and the dorsal longitudinal fasciculus (Nauta, 1963). Autonomic fibers from the hypothalamus run in the lateral portion of the reticular formation and descend to the cranial nerves and the thoracolumbar (sympathetic) and sacral levels (parasympathetic). The most potent effects result from hypothalamic control of the pituitary gland and the adrenal medulla. The adrenal medulla releases epinephrine and norepinephrine, which produce a decrease in peripheral blood flow, an increase in central blood flow, and an increase in heart rate and force. These agents also stimulate

release of glycogen stores from the liver, providing energy for muscular contractions that

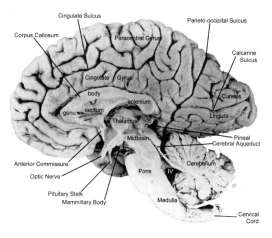

Figure 22-3. Medial surface of a cerebral hemisphere including entire brain stem and cerebellum.

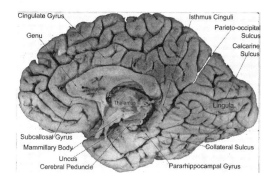

Figure 22-4. Medial surface of a cerebral hemisphere with medulla, pons, and cerebellum removed.

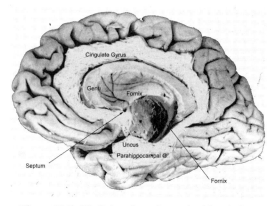

Figure 22-5. Medial surface of a cerebral hemisphere with thalamus removed, demonstrating relationship between fornix and hippocampus.

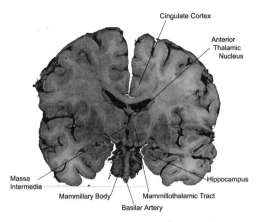

Figure 22-6. Coronal section through diencephalon at level of Massa intermedia showing mammillothalamic tract leaving mamillary body. Weil myelin stain.

accompany the response.

The mammillary nuclei (Figs. 22-3, 22-4, 22-6) of the hypothalamus connect to the anterior thalamic nuclei and habenula, via the mammillothalamic tract, and to nuclei in the brain-stem tegmentum via the mammillotegmental tract. The lateral mammillary nucleus receives axons via the fornix from the hippocampus, axons from other portions of the hypothalamus and septum through the medial forebrain bundle, and input from the midbrain tegmentum and substantia nigra via the mammillary peduncle. Effects of hypothalamic stimulation on the autonomic system are listed in Table 22-2.

Stimulation of the hypothalamus can produce the flight-or-fight response (fear or rage), as well as several visceral responses--sweating, salivation, defecation, and retching. These can be produced in adjacent regions or by moving the electrode site or changing the stimulation parameters (e.g., by varying the current). Another interesting result has been noted in cats made docile with amygdaloid lesions; these animals become savage when a lesion is placed in the ventromedial hypothalamic nuclei. The effects of hypothalamic stimulation and ablation on sleep will be considered in chapter 29.

Thalamus *(Fig. 22-3, 22-4)*. The anterior, medial, midline, and intralaminar dorsal thalamic nuclei receive input from the ascending nociceptive pathways, hypothalamus, reticular system (especially the midbrain reticular formation), cingulate, and frontal association cortex. The intralaminar and midline nuclei connect to the medial and other specific dorsal thalamic nuclei, which then project to the cerebral cortex.

Epithalamus (see chapter 16). The habenular nuclei (epithalamus) give origin to the habenulopeduncular tract, which projects to the midbrain tegmentum and interpeduncular nucleus. The habenular nuclei receive afferents from the septum and preoptic region via the stria medullaris and connect to the intralaminar nuclei.

Septum *(Fig. 22-5)*. The septum forms the medial wall of the frontal horn of the lateral ventricle. The septum consists of two parts: the septum pellucidum containing the septal nuclei (dorsal, lateral and medial) and the caudal velum interpositum. The septum pellucidum is rostral to the interventricular foramen. The septum pellucidum consists of glial membrane and some pia arachnoid (velum interpositum), with the bulk consisting of the column of the fornix. This paired glial membrane, along with the fornix, separates the bodies of the lateral ventricles.

The lower part of the septum pellucidum (septal area) consists of many neurons and glia.

TABLE 22-2 EFFECTS OF HYPOTHALAMIC STIMULATION ON THE AUTONOMIC NERVOUS SYSTEM

STIMULATION OF ANTERIOR HYPOTHALAMUS= PARASYMPATHETIC RESPONSE	STIMULATION OF POSTERIOR HYPOTHALAMUS = SYMPATHETIC RESPONSE OR ANXIETY RESPONSE
Decrease in - heart rate, respiration, and blood pressure; Increase in peristalsis and in gastric and duodenal secretions; Increased salivation; and depending on the internal milieu, even evacuation of the bowels and bladder. Constriction of the pupils; Concomitant with the parasympathetic excitation is sympathetic inhibition.	Increase in- heart rate, respiration, blood pressure, Decrease in peristalsis and in gastric and duodenal secretions, Decreased salivation with sweating, and piloerection. Dilation of Pupils Concomitant with sympathetic excitation there is parasympathetic inhibition.

This zone receives strong input from the amygdala via the stria terminalis and hypothalamus. The septum connects with the:
1. Hypothalamus, interpeduncular nucleus, and the midbrain tegmentum via the medial forebrain bundle,
2. Habenula via the stria medullaris, and
3. Basolateral amygdaloid nuclei through the diagonal band.

Destruction of the septum in cats causes docile animals to become fearful or aggressive, but only for a short time. Complete destruction of the septum may produce coma, probably because it destroys the strong connections between the septum and hypothalamus.

Nucleus Acumbens (see chapter 19. This nucleus lies below the caudate, and receives fibers from the amygdala via the ventral amygdalofugal pathway and from the basal ganglia and thus provides a major link between the limbic and basal nuclei. This nucleus has a high content of acetylcholine. In Alzheimer's disease, there is a significant loss of cholinergic neurons in this nucleus.

ROLE OF CORTICAL STRUCTURES IN EMOTIONS

TEMPORAL LOBE

Parahippcampal Gyrus. The uncus (a hook in Latin) forms the anteriormost region in the parahippocampal gyrus on the medial surface of the temporal lobe. Its surface is the olfactory cortex and the principal nucleus of the amygdala lies internally *(Fig 22-4, 22-5)*.

Amygdaloid Nuclei *(Figs. 22-6, 22-7, 22-8)*. The amygdala is uniquely located to provide the intersection between the primary motivational drives of the hypothalamus and septum and the associative learning that occurs at the hippocampal and neocortical levels.

The amygdala *(Fig. 22-9)* consists of three main groupings of nuclei: corticomedial, basolateral, and central. In addition to these principal nuclei there are extratemporal neurons including the nucleus of the stria terminalis

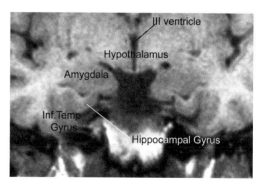

Figure 22-7. Coronal Section through amygdala and hypothalamus. MRI-T1. Hippocampal gyrus = parahippocampal gyrus

and the sublenticular substantia innominata. The amygdala receives extensive projections from many cortical areas and in turn sends projections to these areas. In primates as one expects in comparison to rodents, there has been a significant increase in the projections from and to isocortex (neocortex) as opposed to allocortex and mesocortex. These projections originate from: 1) the multimodality sensory areas, 2) tertiary unimodal sensory association areas, 3) visual association areas which are particularly prominent in the primates, 4) first and second order central sensory neurons of the olfactory system.

From the standpoint of the role of the amygdala in emotion and instinctive behavior, there are important connections with the basal forebrain, medial thalamus (medial dorsal nucleus), hypothalamus (preoptic, anterior and ventromedial and lateral areas), and the tegmentum of the midbrain, pons and medulla (to various nuclei concerned with visceral function such as chewing, licking and the motor components of emotional expression). The more specific connections of the amygdaloid nuclei are listed below

(1) *Olfactory Nuclei.* The corticomedial group receives olfactory information from the lateral olfactory stria and interconnects with the contralateral corticomedial nuclei (via anterior commissure) and ipsilateral basolateral nuclei. The primary efferent pathway of the corticomedial nucleus is the stria terminalis, which projects to the septum, medial hypo-

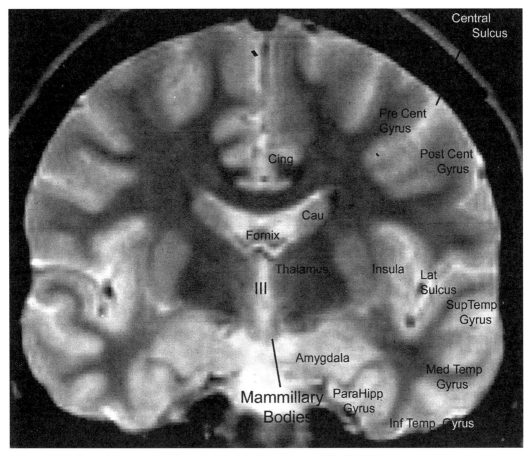

Figure 22-8. Coronal Section through mammillary bodies and amygdala. MRI-T2.

thalamus including preoptic nucleus of the hypothalamus, and to the corticomedial nucleus in the opposite hemisphere.

(2). *Limbic Nuclei*. The central and basolateral nuclear grouping is associated with the limbic brain, and has connections with the parahippocampal cortex, temporal pole, frontal lobe, orbital frontal gyri, cingulate lobe, thalamus (especially dorsomedial nucleus), catecholamine containing nuclei of the reticular formation, and substantia nigra.

(3). The *ventral amygdalofugal pathway* projects from the central nucleus to the brain stem and to the septum, the preoptic, lateral,

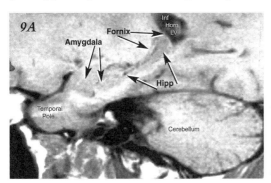 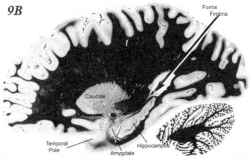

Figure 22-9 A) Sagittal Section. Temporal lobe with amygdala and hippocampus. MRI-T1. B) Sagittal section Weil myelin stain. Inferior horn LV is the junction of posterior and inferior horns of lateral ventricle.

and ventral hypothalamus, and--in the dorsal thalamus--to the dorsomedial, intralaminar, and midline thalamus.

Stimulation of the amygdaloid region. In monkeys, cats, and rats stimulation produces aggressive behavior. The stimulated cats have a sympathetic response of dilated pupils, increased heartbeat, extension of claws, piloerection, and attack behavior. When the stimulus stops, they become friendly. Animals will even fight when the amygdala is stimulated and stop fighting if the stimulus is off. Eating, sniffing, licking, biting, chewing, and gagging may also be stimulated here. In contrast studies in the cat indicate that stimulation of the prefrontal areas will prevent aggressive behavior. In humans, stimulation of the amygdala produces feelings of fear or anger (Cendes et al–1994). A role in sexual behavior has also been postulated, although this may be more prominent in the female.

Ablation of Amygdala. In the studies of Amaral selective bilateral lesions of the amygdala[1] in the adult monkey significantly decreased the fear response to inanimate objects such as an artificial toy snake. This object was now picked up whereas previously, such an object had triggered intense fear responses. The social interactions of the lesioned monkey with other members of a colony were significantly altered. The lesioned animals were more sociable with more sexual and nonsexual friendly contacts with other members of the colony. They were described as socially uninhibited. In the male, aggression was decreased, but occasional females had an increase in aggressive behavior. When the bilateral lesions were produced at two weeks of age, there was a decrease in fear responses to inanimate objects but an increase in fear responses to other monkeys. This latter effect interfered with their social integration into the colony.

In the human, bilateral lesions of the amygdala have been produced for control of aggression. In patients with bilateral lesions of the amygdala there is an impairment of the ability to interpret the emotional aspects of facial expression. (Young et al 1995, Adolph, et al 1998, Anderson et al, 2000). Patients with high functioning autism have a similar disorder (Adolphs et al 2001). On functional MRI studies, these patients failed to activate the left amygdala, as well as the cortical face area or the left cerebellum when implicitly processing facial expressions (Baron-Cohen et al, 2000 Critchley et al 2000). When quantitative MRI is performed in such autistic patients, there is significant enlargement of the volume of the amygdala. Howard et al (2000) suggest that these results may indicate that a developmental malformation of the amygdala (possibly an incomplete neuronal pruning) may underlie the social-cognitive impairments of the autistic patient.

HIPPOCAMPAL FORMATION (22-9, 22-10, 22-13)

This region of the limbic system has been of critical importance in our understanding of both the clinical aspects and underlying biological substrate of memory and of complex partial epilepsy.

Anatomical Correlates:

From an anatomical standpoint, compared to other lobes of the brain, the temporal lobe

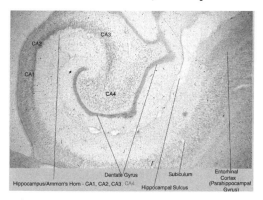

Figure 22-10. The sectors or fields of the hippocampus. The dentate gyrus, subiculum and related structures are demonstrated. Nissl stain.

---

[1] *These selective lesions were produced by injection of ibotenic acid which damages neurons but does not affect fibers of passage*

is a complex structure. It contains four diverse components:

Neocortex- 6 layers: superior, middle and inferior temporal gyri (see chapter 17)

Allocortex – 3-layers: the olfactory cortex, hippocampal formation and subiculum.

Mesocortex- a transitional type of 6-layer cortex (transitional between neocortex & allocortex): entorhinal/parahippocampal gyrus (the large posterior segment of the pyriform region), presubiculum and para subiculum.

Cortical Nucleus - *the amygdala* (discussed above).

The *hippocampus* is phylogenetically the older part of the cerebral cortex termed allocortex and consists of three layers: polymorphic, pyramidal, and molecular. The dentate gyrus[2] fits inside the hippocampus and, like the hippocampus, has three layers: molecular, granular, and polymorphic. The most primitive cortex is the paleocortex of the olfactory bulb.

The fibers of the dentate gyrus are confined to the hippocampal formation while the hippocampal fibers leave the hippocampal formation through the fornix and project to either septum or to the mammillary bodies and mesencephalic tegmentum. The cortex adjacent to the hippocampus changes from three layers to six layers and is classified as transitional mesocortex and includes the parahippocampal gyrus (medial to the collateral sulcus), including entorhinal cortex. The pyriform lobe consists of the lateral olfactory stria, uncus and the anterior part of the parahippocampal gyrus.

---

[2]*In the dentate gyrus in common with the olfactory epithelium and cerebellum, new neurons are formed throughout life. In the vast remainder of the brain there is no differentiation of nerve cells after birth, and when these nerve cells die they are not replaced (Altman 1962, Gage 1994). One must expect considerable research in the future to focus on just which genes permit the continued replacement of nerve cells in these regions.*

Entorhinal Region: The entorhinal/parahippocampal gyrus forms the large posterior segment of the piriform region. This is Brodmann's area 28 and constitutes the bulk of the parahippocampal gyrus. This area has extensive interconnection with the higher association cortex throughout the neocortex and also receives olfactory information from the olfactory stria. This is the major pathway for relating neocortex to the limbic cortex of the hippocampus.[3]

Older classifications have used the terms archicortex for the hippocampus, paleocortex for the piriform area and neocortex for lateral temporal areas based on presumed phylogenetic considerations. Here we will follow in general the terminology as above, employed in the recent monumental work of Gloor (1997). The hippocampus is divided into sectors (referred to as fields) based on cytoarchitectural differences *(Fig.22-10)* CA1-CA4, (or fields of Rose h1-h5). (Note that an older term for the hippocampus is Ammon's {the ram} horn, thus–Cornu Ammonis = CA). CA4, the end folium or end blade merges with the hilus of the dentate gyrus. CA1, the sector closest to the subiculum is referred to as the Sommer's sector. This sector is most severely affected by cell loss following hypoglycemia, anoxia, and status epilepticus *(Fig.22-11)*. However this selective vulnerability may also involve CA3 and CA4 with relative sparing of CA2 and the dentate granule cells. The subsequent gliosis (mesial sclerosis) of the hippocampus is the pathology found at surgery or autopsy in 75% of cases of complex partial seizures arising in the hippocampus.

The hippocampal regions are interconnected by a commissure, the hippocampal commissure. In the primate, the dorsal commissure originates in the entorhinal cortex and presubiculum. The ventral commissure originates

---

[3]*Note that the "entorhinal gyrus" of lower mammals corresponds in humans to a much larger region that includes the middle, inferior, and parahippocampal gyri and the temporal portion of the fusiform gyrus.*

in the CA3 sector and the dentate hilus primarily in their more rostral areas e.g. in relation to the uncus

Cytoarchitecture of the Hippocampus: In the hippocampus, the three layers are as follows:

1. *Molecular* in which the apical dendrites of the pyramidal cells arborize. This layer is usually divided into a more external stratum lacunosum-moleculare and a more internal stratum radiatum. It is continuous with molecular layer of dentate gyrus and adjacent temporal neocortex.
2. *Pyramidal cell layer* (stratum pyramidalis) contains pyramidal cells that are the principal cells of the hippocampus. Dendrites extend into molecular layer and Schaffer axon collaterals arise from pyramidal neurons and synapse in the molecular layer on dendrites of other pyramidal cells.
3. *Polymorphic layer* (stratum oriens)-in, which the basilar dendrites of the pyramidal cells are found. Contains axons, dendrites and interneurons. This layer in CA3 is continuous with hilus of the dentates gyrus. Only the hippocampus sends axons outside the hippocampal formation.

Cytoarchitecture of the Dentate Gyrus: In the dentate gyrus, the following three layers are found:

1. *Molecular layer* contains dendrites of granule cells
2. *Granule cell layer.* These small neurons replace the pyramidal cell layer of hippocampus. The granule cells are unipolar; all of the dendrites emerge from the apical end of the cell into the molecular layer. Efferent neurons from the granule cells are mossy fibers that synapse only with cells of hippocampal areas CA2 and CA 3.
3. *Polymorphic cell.* Also referred to as the hilus, this layer is continuous with CA3 of hippocampus.

The pyramidal and granule cell neurons are excitatory utilizing glutamine as the transmitter. In addition there are inhibitory interneurons (GABA-ergic), basket cells, in the polymorphic layer of both the hippocampus and dentate gyrus. There are also mossy cells in the polymorph layer probably excitatory interneurons. There are also scattered interneurons in the molecular layers. The connections of the hippocampal formation are summarized in Table 22-3.

The **Selective Vulnerability of Hippocampus** *(Fig.2-11)* occurs in diseases such as anoxia and hypoglycemia. In addition the hippocampus appears to be significantly involved in most seizures of temporal lobe origin.

1. *Why does this selective vulnerability occur?* The CA 1 region is rich in NMDA receptors. The dentate hilus and the CA3 sector are rich in kainate receptors. Activation of these receptors by glutamate would allow a considerable entry of calcium ions into the pyramidal neurons beginning a cascade that

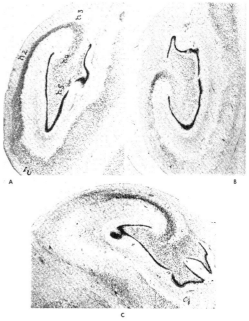

Figure 22-11. Selective vulnerability of the hippocampus demonstrated in Nissl stains. A, Normal Ammon's horn with labels for fields hl-h5 of Rose inserted. B, Anesthetic death: recent necrosis in Sommer sector (hl) and partial loss of neurons in h3 and the end plate (h4, h5). C, Hypoglycemic coma: there has been significant loss of pyramidal cells in section hl. (From Meyer, A. (in) Greenfield, J., et al., Neuropathology 2nd Edition. Baltimore, Williams & Wilkins, 196 p.256).

TABLE 22-3 - CONNECTIONS OF HIPPOCAMPAL FORMATION

AFFERENT INPUT:

1. Perforant pathway from adjacent lateral entorhinal cortex onto granule cells of dentate gyrus and the alvear pathway from the medial entorhinal cortex. The entorhinal areas, in turn, receive their input from many of the highest level of associational cortex in the frontal, orbital, temporal (amygdala), parietal, and cingulate cortex through the cingulum bundle.

2. Septum through the stria terminalis.

3. Contralateral hippocampus, via the hippocampal commissure.

EFFERENT PROJECTIONS:

These fibers, the axons of hippocampal and subiculum pyramidal neurons form the fornix and have connections with the following structures.

1. Lateral mammillary nuclei, habenulae nuclei, anterior midline and intralaminar thalamic nuclei, lateral hypothalamic nuclei, midbrain tegmentum, and periaqueductal grey via a column of the fornix

2. Septum, preoptic region, parolfactory and cingulate cortex via the precommissural fornix and supracallosal fornix

could result in cell death. Moreover the pyramidal neurons of these sectors as compared to CA 2 and the granular cells of the dentate gyrus contain very little calcium buffering protein calbinden.

2. *Seizure Activity.* Most seizures beginning after age 15 are classified as partial, and the majority of these are classified as complex partial. Approximately 75% of the complex partial seizures arise in the temporal lobe, the remainder in the frontal lobe. While these seizures may arise in temporal neocortex, the majority arise in the mesial temporal structures particularly the hippocampus. The hippocampus has a low threshold for seizure discharge; consequently, stimulation of any region that supplies hippocampal afferents or stimulation of the hippocampus itself may produce seizures. Hippocampal stimulation produces respiratory and cardiovascular changes, as well as automatisms (stereotyped movements) involving face, limb, and trunk.

3. *Why do not all of the various pathological processes in the hippocampus produce seizures?* There appears to be a critical age during infancy and early childhood for the acquisition of the pathology that is associated with seizures originating in mesial temporal lobe. There may be an age related remodeling of intrinsic hippocampal connections. Whether a single episode of epileptic status during infancy is sufficient to produce these changes or whether, multiple episodes may be required is still under discussion.

4. *Is the hippocampal pathology alone sufficient to explain the complex partial seizures of temporal lobe origin?* As reviewed by Gloor (1997), many of the specimens examined after surgery or at autopsy also demonstrate extensive changes in the amygdala as well as mesial and lateral isocortex. (Overall however, in the autopsy studies of Margerison and Corsellis (1966) the most frequent site of damage and cell loss was in the hippocampus.) Rather than using the more restrictive term of hippocampal sclerosis, it is more appropriate to use the term mesial temporal sclerosis. Even this term fails to include the involvement of more laterally placed neocortex in some of the symptoms in seizures originating in temporal lobe. (see below)

HIPPOCAMPAL SYSTEM & MEMORY (see also Chapter 30 on Learning and Memory).

The studies of Milner (1972) have shown that bilateral removal or damage of the hippocampus produces great difficulty in learning new information, a condition called anterograde amnesia. The excitatory amino acids, l-glutamate and l-aspartate are important in learning and memory through the mechanism of long-term potentiation. Long-term potentiation is a long lasting facilitation after repeated activation of excitatory amino acid pathways and is most pronounced in the hippocampus and may be related to the initiation of the memory trace

The amygdala and its connections are associated with the conditioning of the emotional response of fear (Refer to review of LeDoux, 2000) and above.

OTHER CORTICAL REGIONS OF THE LIMBIC SYSTEM

Cingulate Cortex. This region *(Fig. 22-5)* receives reciprocal innervation from the anterior thalamic nuclei, contralateral and ipsilateral cingulate cortex, and temporal lobe via the cingulum bundle, as well as projecting to the corpus striatum and most of the subcortical limbic nuclei. The cingulate cortex is continuous with the parahippocampal gyrus at the isthmus behind the splenium of the corpus callosum.

Stimulation of cingulate cortex also produces respiratory, vascular, and visceral changes, but these changes are less than those produced by hypothalamic stimulation. Interruption of the cingulum bundle, which lies deep to the cingulate cortex and the parahippocampal gyrus, has been proposed as a less devastating way to produce the effects of prefrontal lobotomy without a major reduction in intellectual capacity. (For additional discussion of the effects of stimulation and of lesions of the anterior cingulate area on autonomic function and behavior refer to Devinsky et al, 1995)

PRINCIPAL PATHWAYS OF THE LIMBIC SYSTEM

1. *Entorhinal Reverberating Circuit (Fig. 22-12A).* This fiber system is an adjunct to the circuit described by Papez (see below). The entorhinal reverberating circuit functions as follows:

a. Entorhinal cortex receives input from polysensory (multi-model sensory association cortex) cerebral cortex and amygdala

b. Entorhinal fibers project via the perforant pathway to the granule cells of the dentate gyrus via the perforant pathway. There are

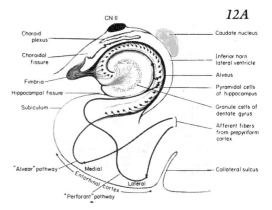

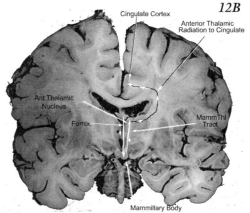

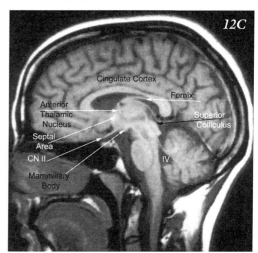

Figure 22-12. A) The perforant pathway modified. From Carpenter MB Core Text of Neuroanatomy, Baltimore William and Wilkins B) The Papez circuit C) Sagittal section showing medial surface of cerebrum demonstrating the limbic structures which surround the brain stem and are on the medial surface of the cerebrum. Portions of Papez Circuit are shown including- fornix, anterior nuclei, mammillary bodies, cingulate cortex and cingulum. MRI-T2.

also direct projections to CA1, CA3 and to subiculum

c. The major projection of the dentate gyrus is to the proximal dendrites of the CA3 hippocampal pyramidal cells. There are also synapses on the neurons of the polymorphic layer.

d. The CA3 pyramidal cells project via Schaffer collaterals to CA1 pyramidal cells.

e. CA1 pyramidal cells collaterals project to the subiculum and entorhinal cortex.

f. The entorhinal cortex projects back to each of the neocortical polysensory projection areas completing a reverberating system.

2. *Papez Circuit*

a. Many hippocampal pyramidal cells synapse on the pyramidal cells of the adjacent subiculum. The pyramidal cells of the subiculum constitute the origin of most of the fibers in the fornix,

b. Fornix projects primarily to the mammillary bodies of the hypothalamus and septum,

c. Mammillary bodies then project via the mammillothalamic tract to the anterior nuclei of the thalamus,

d. Anterior thalamic nuclei project to cingulate gyrus,

e. Cingulate gyrus then projects via the cingulum bundle of fibers to the parahippocampal/entorhinal cortex, which subsequently projects to the hippocampus as outlined above completing the circuit.

3. *Fornix* (*Figs. 22-5, 22-6, 22-7* and chapter 16). Note that the fornix is the efferent pathway from the hippocampus and subiculum and is connected to the hypothalamus, septum, and midbrain. This tract takes a rather circuitous pathway to reach the hypothalamus. The fornix originates from the medial surface of the temporal lobe and runs in the medial wall of the inferior horn of the lateral ventricle, passing onto the undersurface of the corpus callosum at the junction of the inferior horn and body of the ventricles, and running in the medial wall of the body of the lateral ventricle suspended from the corpus callosum. The fornix finally enters the substance of the hypothalamus at the level of the interventricular foramen. Different portions of the fornix have specific names:

- Portio fimbria (fringe) of the fornix found on the medial surface of the hippocampus and consists of fibers from the fornix and hippocampal commissure.

- Portio alveus, band of fornix fibers covering ventricular surface of the hippocampus.

- Portio tenia, connecting the hippocampus to the corpus callosum.

- Portio corpus, located underneath the corpus callosum and entering the hypothalamus

- Portio columnaris, found in the substance of the hypothalamus.

The fornix is also divided into a precommissural portion (in front of the anterior commissure), that enters the septum (from the hippocampus), and the postcommissural portion (behind the anterior commissure), that distributes in the mammillary bodies and midbrain (from subiculum). These relationships should be traced out in the atlas section.

4. *Stria Terminalis.* The pathway of the stria terminalis (the efferent fiber tract of the amygdala) parallels the fornix but it is found adjacent to the body and tail of the caudate nucleus on its medial surface (refer to atlas and chapter 16) The stria terminalis interconnects the medial corticoamygdaloid nuclei, as well as connecting the amygdala to the hypothalamus and septum. There is also a strong termination in the red nucleus of the stria terminalis, which is found above the anterior commissure and lateral to the column of the fornix.

5. *Ventral Amygdalofugal Pathway.* This fiber pathway originates primarily from the basolateral amygdaloid nuclei and to a lesser degree from the olfactory cortex, spreads beneath the lentiform nuclei and enters the lateral hypothalamus and peoptic region, the septum and the diagonal band nucleus, Some of these fibers bypass the hypothalamus and terminate on the magnocellular portions of the dorsomedial thalamic nuclei.

6. *Cingulum.* This fiber bundle is found on the medial surface of the hemisphere and interconnects primarily the medial cortical limbic areas, especially the parahippocampal formation, with one another.
7. *The Intracortical Association Fiber System.* The superior and inferior longitudinal fasciculus, and uncinate fasciculus. The limbic regions on the lateral surface of the hemisphere have strong interconnections with polysensory and third order sensory cortical regions throughout the cerebral cortex through this system
8. *Limbic System and the Corticospinal and Corticobulbar Pathway.* The bulk of the pyramidal pathway originates in the motor/sensory strip. However, most movements begin with a "thought" in the frontal association areas. A possible example of release of the bulbar areas for emotional expression from frontal-lobe control is seen in pseudobulbar palsy. Such patients exhibit inappropriate response to situations because the interrupted corticonuclear/corticobulbar pathway no longer dampens the strong descending autonomic/limbic input to the brain-stem cranial nuclei. The resultant inappropriate behavior is termed emotional lability. A sad story may trigger excessive crying, a funny story excessive laughter.

III. THE NEOCORTEX OF THE TEMPORAL LOBE

Lateral Neocortical Areas. As regards the neocortical areas, (area 41) the primary auditory projection area, is located on the more anterior of the transverse gyri of Heschl *(Fig. 22-13).* This area is a primary special sensory region and has a pronounced layer IV (granular or koniocortex) similar to but considerably thicker than areas 17 and 3. Area 41 receives

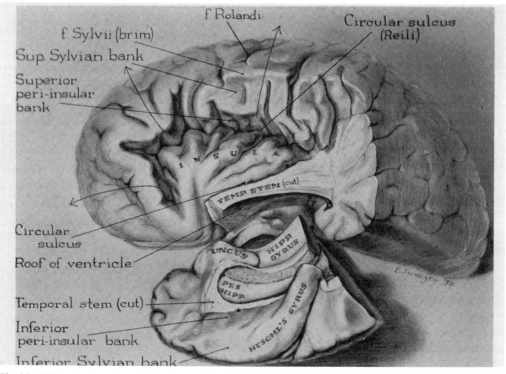

Fig. 22-13 *The superior and medial surfaces of the temporal lobe. The transverse gyrus of Heschl and the relationships of the temporal lobe to the gyri of other structures surrounding the sylvian fissure are demonstrated. The temporal stem refers to the core of white matter relating the temporal lobe to the remainder of the cerebral hemisphere. The hippocampal gyrus of the figure is now termed the parahippocampal gyrus.* (From Penfield, W., and Jasper, H.: Epilepsy and the Functional Anatomy of the Human Brain. Boston, Little, Brown and Company, 1954, p.52).

the main projection from the medial geniculate nucleus of the thalamus. Our understanding of the tonal organization in this region comes from studies in the monkey and chimpanzee where the lowest frequencies project to the more rostral areas. Some investigators limit the term "Heschl's gyrus" to the more anterior of the transverse gyri and localize the primary auditory cortex to the posterior aspects of that gyrus (Liegeois-Chauvel et al., 1991).

The remaining neocortical areas of the temporal lobe have a well-defined six cortical layers and are classified as homotypical.

Auditory and Auditory Association. Area 42 surrounds area 41 and receives association fibers from this area. Area 22, in turn, surrounds area 42 and communicates with areas 41 and 42. Both areas 42 and 22 are often designated as auditory association areas, and in the dominant hemisphere they are important in understanding speech as Wernicke's area.

Visual Perceptions. The inferior temporal areas have significant connections with area 18, providing a pathway by which visual perceptions, processed at the cortical level, may then be related to the limbic areas.

SYMPTOMS OF DISEASE INVOLVING THE TEMPORAL LOBE

In considering the functions of the temporal lobe and the effects of temporal lobe lesions, one must bear in mind as discussed above that the temporal lobe has several structural, phylogenetic, and functional subdivisions. Thus, one may distinguish as discussed above:

1 - the allocortex of the hippocampal formation,

2 - the transitional mesocortex bordering this area,

3 - the neocortex that occupies all of the lateral surface and the inferior temporal areas.

4 - the amygdala already considered above

When one considers disease processes affecting the temporal lobe, signs and symptoms obviously do not follow the precise subdivisions outlined above. There are several

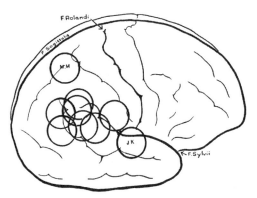

Fig. 22-14 Effects of focal discharge in superior temporal gyrus. The location of the discharging lesion at surgery in nine patients with focal seizures beginning with a sensation of vertigo is demonstrated to be primarily superior temporal gyrus (From Penfield, W., and Kristiansen, K.: Epileptic Seizure Patterns. Springfield, Ill., Charles C. Thomas, 1951, p.49.)

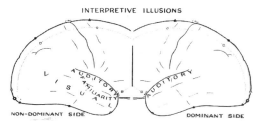

Fig. 22-15. Interpretive illusions (disturbances of perception) produced by stimulation at surgery of temporal lobe cortex in patients with temporal lobe seizures. Visual illusions were produced predominantly in the minor (that is, non-dominant) hemisphere. Auditory illusions were produced from both sides chiefly in the superior temporal gyrus. Illusions of familiarity (déjà-vu) were produced predominantly from the non-dominant hemisphere. Sensations of fear, unreality, loneliness, or of detachment were produced from stimulation in either temporal region. From Penfield, W., and Perot, P.: Brain, 86:599, 1963 (Oxford Univ. Press). After Mullan and Penfield: Archives of Neurology & Psychiatry, 81:269, 1959.)

reasons for this:

1. The basic lesion (such as a glioma) often involves both neocortical and non-neocortical areas of the temporal lobe.

2. The threshold of the hippocampus for seizure discharge is relatively low, whereas that of the neocortical areas on the lateral surface is relatively high. Since connections from the

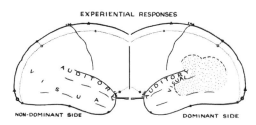

Fig. 22-16 Experiential responses (hallucinations) to stimulation at surgery of temporal lobe and adjacent inferior parietal area in patients with temporal lobe seizures. Auditory responses occur from stimulation in either dominant or nondominant hemisphere, (primarily superior temporal gyri). Visual responses occur primarily from stimulation of the nondominant hemisphere. The stippled area within the interrupted lines indicates the posterior speech area of the dominant hemisphere. No experiential responses occur on stimulation of this speech area in the dominant hemisphere. From Penfield, W., and Perot, P., Brain, 86:676, 1963 (Oxford Univ. Press).

lateral temporal and inferior temporal area to the hippocampus exist, seizure discharges beginning in the lateral temporal neocortex often activate discharge in the hippocampus.

3. Lesions in the more posterior temporal areas often involve the adjacent posterior parietal areas, that is, the inferior parietal lobule. In contrast anteriorly placed lesions of the temporal lobe often involve the adjacent inferior frontal gyrus. Lesions spreading into the deeper white matter of the temporal lobe (particularly in its middle and posterior thirds) often involve part or all of the optic radiation.

Symptoms following Stimulation of the Temporal Lobe.:

Seizures involving the temporal lobe are frequent and produce a variety of symptoms. These may be classified as simple partial if awareness is retained or complex partial if awareness is not retained and the patient is amnestic for the symptoms. A simple partial seizure may progress to a complex partial seizure and may subsequently generalize. These symptoms are outlined in Table 22-4 with the most likely anatomical correlate.

A detailed discussion of the correlation between symptoms in simple and complex

TABLE 22-4 CORRELATION OF TEMPORAL LOBE SEIZURE /LIMITED STIMULATION SYMPTOMS WITH ANATOMY

SYMPTOMS	ANATOMICAL CORRELATION
Autonomic phenomena	Amygdala. Less often cingulate gyrus or orbital frontal
Fear, less often anger or other emotion	Amygdala
Crude auditory sensation: tinnitus	Heschl's transverse gyrus (primary auditory projection)
Visual and auditory illusions: perceptual distortions (22-15)	Higher order visual and auditory association cortex. Superior temporal gyrus in studies of Penfield. For visual predominantly non dominant hemisphere in studies of Penfield
Vestibular sensations: dizziness/vertigo (22-14)	Superior temporal gyrus posterior to auditory cortex
Arrest of speech	Wernicke's area and posterior speech area (posterior temporal/ parietal (angular-supramarginal)- dominant hemisphere
Olfactory hallucinations	Olfactory cortex of the uncus (termination of lateral olfactory stria)."Uncinate epilepsy of Jackson"
Experiential phenomena/dreamy states: More complex illusions such as déjà vu*, déjà vécu*, jamais vu*, other illusions of recognitions, visual and auditory hallucinations (often of past experience) (22-15, 22-16)	Lateral temporal cortex: primarily superior temporal gyrus. Controversial. However ictal phenomena is abolished by ablation limited to lateral temporal neocortex (Blume et al 1993, Mullan &Penfield, 1959) Refer to CD ROM
Automatisms= repetitive simple or complex often stereotyped motor acts, most commonly involving mouth, lips etc. Speech or limbs may be involved.	Primary or secondary bilateral involvement of amygdala / hippocampus. Invariably accompanied by confusion and amnesia for the acts
Defects in memory recording followed by amnesia for the event	Hippocampus

*Déjà vu=sensation of familiarity, déjà vécu =sensation of strangeness, jamais vu =perception is dream like

partial epilepsy with temporal structures is included on the CD for this chapter.

In general, as we have indicated, there is mixed symptomatology during partial seizures of temporal lobe origin. The following case history illustrates many of the points just discussed regarding seizures of temporal lobe origin.

Case 22-1: Three months before admission, this 56-year-old right-handed male baker had the onset of 4-minute episodes of vertigo and tinnitus, unrelated to position, followed by increasing forgetfulness. Later that month he had a generalized convulsive seizure that occurred without any warning. The patient then developed episodes of confusion and unresponsiveness, followed by a left frontal headache. Several studies; EEG, pneumoencephalogram, and carotid arteriogram were all negative. One month before admission, the patient began to have minor episodes, characterized by lip smacking and a vertiginous sensation, during which he reported seeing several well-formed, colorful scenes. At times, he had hallucinations of "loaves of bread being laid out on the wall." In addition, he would have a perceptual disturbance (e.g., objects would appear larger than normal). He also had colorful visions and terrifying nightmares "terrifying dreams, crazy things".

Neurological examination: *Seizures observed:* The patient had frequent transient episodes of distress characterized by saying, "Oh, oh, oh, my" and at times accompanied by automatisms: fluttering of the eyelids, smacking of the lips, and repetitive picking at bedclothes with his right hand. Consciousness was not completely impaired during these episodes, which lasted from 30 seconds to 3 minutes. The patient reported afterward that at the onset of the seizure he had seen loaves of bread on the wall and smelled a poorly described unpleasant odor. At other times, the olfactory hallucination was described as pleasant, resembling the aroma of freshly baked bread. *Mental status:* The patient was disoriented to time, could not recall his street address and could not pronounce the name of the hospital. *Cranial nerves:* A possible deficit in the periphery of the right visual field and a minor right central facial weakness were present. *Reflexes:* A right Babinski sign was present

Clinical diagnosis: Simple and complex partial seizures originating left temporal lobe probably involving at various times left lateral superior temporal gyrus, uncus, amygdala and hippocampus, with tumor the most likely etiology in view of age and the focal neurological findings.

Laboratory data: *EEG:* Frequent focal spike discharge was present throughout the left temporal and parietal areas consistent with a focal seizure disorder (Fig. 29-1 and CD atlas). A recording 6 days later indicated almost continuous 3 to 5 (Hz) focal slow-wave activity (focal damage) in the left temporal area (fig. 29-1 and CD Atlas). *Imaging studies:* Left carotid arteriogram indicated a possible avascular mass lesion left posterior lateral frontal.

Subsequent course: These episodes were eventually controlled with anticonvulsant medication. Seizures recurred 7 months after onset of symptoms. An aura of unpleasant odor was followed by a generalized convulsion followed by four or five subsequent seizures of a somewhat different character (deviation of the head and eyes to the right, then tonic and clonic movements of the right hand spreading to the arm, foot, and leg lasting approximately 1 to 2 minutes, followed by a post-ictal right hemiparesis). He also experienced minor seizures characterized by sensory phenomena on the right side of the body. Neurologic examination now indicated progression with a marked expressive aphasia, with little spontaneous speech and difficulty in naming objects. There was a dense right homonymous hemianopia, a flattening of the right nasolabial fold, and a right hemiparesis, with a right Babinski sign. The symptoms and findings suggested that the basic disease process might well have spread to involve the adjacent areas across the sylvian fissure--the speech areas of the inferior frontal convolution, premotor areas, and sen-

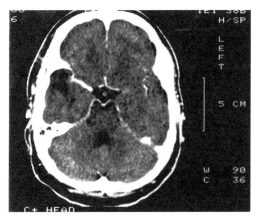

Figure. 22-17. Complex and simple partial seizures: cystic-mixed glial tumor and cortical dysplasia right temporal lobe. Case 22-2. CT scan demonstrated an area of atrophy or cyst right anterior temporal lobe with possible minimum enhancement at the border. This 17-year-old right-handed male, at age 4 years began to have "staring spells and bizarre behavior" with some proceeded by the "sound a musical rhythm". An EEG spike discharge was present in the right anterior-middle temporal area. (Refer to text).

sory motor cortex. An arteriogram indicated a large space-occupying tumor of the left temporal lobe, and with additional progression, craniotomy was performed by Dr. Robert Yuan 16 months after the onset of symptoms. A necrotic glioblastoma was found involving the superior temporal gyrus, the deeper temporal and extending superficially under the Sylvian fissure to involve the adjacent posterior portion of the inferior frontal gyrus. A temporal lobectomy was performed (from the anterior temporal pole posteriorly for a distance of 6 cm).

Today, CT scan and MRI would be employed for early diagnosis in patients with this type of seizure disorder as in the following cases and surgery might be limited to a stereotaxic biopsy.

Case 22-2. This 17-year-old right-handed male, at age 4 years began to have "staring spells and bizarre behavior." Some episodes had been preceded by the "sound of a musical rhythm". EEG at that time indicated a right temporal spike discharge Seizures were controlled for ten years with anticonvulsants but then recurred and were poorly controlled

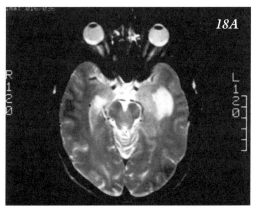

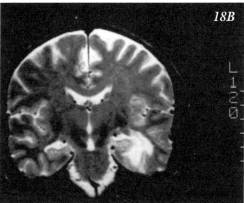

Figure. 22-18. Glioblastoma left temporal lobe. Complex partial seizures with secondary generalization in a 47 year old. Case 22-3. MRI T2 non-enhanced. A) Horizontal section; B) coronal section. (Refer to text.)

occurring several times per day.

Neurological examination: *Observed 3-minute seizure:* The seizure began with loss of contact and a stare and then automatisms of the hands. He stood up, walked around the room, went to the physician's desk and attempted to pull open a nonexistent middle drawer. He answered questions vaguely with one or two word answers. Confusion was present for 1-2 minutes after the end of the episode. *Cranial nerves:* a left central facial weakness was present.

Clinical diagnosis: **Focal seizures originating** right temporal lobe. Observed seizure was complex partial. The seizures beginning with the musical sound might be classified as simple partial with secondary complex partial.

Laboratory data: *EEG:* Intermittent focal

spike discharge anterior-middle temporal area. *CT scan (Fig.22-17):* An area of focal atrophy was present in the right anterior temporal lobe with possible enhancement at the border. MRI: AT the Yale Epilepsy Center was more consistent with a tumor in the right anterior temporal lobe.

Subsequent course: Dr. Dennis Spencer at the Yale Epilepsy Center performed a temporal lobectomy. This demonstrated a cystic glial tumor with components of astrocytes and oligodendrocytes. Cerebral cortex also demonstrated abnormal lamination and dysplastic features. Seizures were fully controlled over the next three years.

Case 22-3. This 47-year-old ambidextrous male experienced his first seizure one month prior to admission. He felt light-headed and warm. Then he was observed to walk 100 feet down a hallway, appearing "dazed" and not recognizing people. He fell to the floor, with a loss of consciousness, bit his tongue and was confused afterwards for several minutes.

Neurological examination at 2, 16 and 48 hours after the episode a persistent right Babinski sign, and slight right hyperreflexia were present.

Clinical diagnosis: Complex partial seizures. In view of age of onset and persistent focal signs a brain tumor should be suspected as the etiology.

Laboratory diagnosis: *Sleep deprived EEG:* normal. *CT and MRI scans (Fig. 22-18)* demonstrated an extensive infiltrating tumor of the left temporal lobe most likely a glioblastoma.

Subsequent course: Despite Anticonvulsant therapy (phenytoin) additional episodes occurred over the next month: loss of train of thought while talking and several minutes of loss of memory. A subtotal resection by Dr. Bernard Stone (St. Vincent Hospital) demonstrated a Grade III-IV astrocytoma (glioblastoma). Despite radiotherapy and chemotherapy, the disease pursued a relentless course producing early and severe disability prior to death, approximately one year after onset of symptoms.

Symptoms from Ablation of or damage to the Temporal Lobe.

1. *Effects on Hearing.* Unilateral lesion of auditory projection areaway result in an inability to localize a sound. Bilateral lesions may produce cortical deafness.
2. *Aphasia.* Destruction of area 22 in the dominant hemisphere produces a Wernicke's receptive aphasia. Such a patient not only has difficulty in interpreting speech but also, in a sense, has lost the ability to use previous auditory associations. Lesions that deprive the receptive aphasia area of Wernicke in the dominant temporal lobe (area 22) of information from the auditory projection areas of the right and left hemispheres result in pure word deafness. Such a patient can hear sounds and words but is unable to interpret them.
3. *Visual Defects.* Unilateral lesions of the temporal lobe that involve the subcortical white matter often produce damage to the geniculocalcarine radiations. Since the most inferior fibers of the radiation that swing forward around the temporal horn (representing the inferior parts of the retina and referred to as "Meyer's loop") are often the first to be involved, the initial field defect may be a contralateral superior field quadrantanopia.
4. *Klüver-Bucy Syndrome.* This syndrome results from bilateral ablation of the temporal pole, amygdaloid nuclei, and hippocampus in the monkey (Klüver and Bucy 1937). The animals could see and find objects, but they could not identify objects (visual agnosia). The animal showed marked deficits in visual discrimination, particularly in regard to visual stimuli related to various motivations. They also had a release of very strong oral automatisms and compulsively placed objects in their mouths, which, if not edible, were dropped. They had a tendency to mouth and touch all visible objects and manifested indiscriminate sexual practice. They willingly ate food not normally a part of their diet (such as corn-beef sandwiches). They showed a lack of response to aversive

stimuli and had no recollection or judgment. The animals also lost their fear (release phenomenon) and, in the case of wild monkeys, became tame and docile creatures. They had a marked absence of the fight-or-flight response and also lost fears of unknown objects or objects that previously had frightened them. (See amygdala above). The problems in visual discrimination may reflect a disconnection of the visual association areas from the amygdala/ hippocampal areas.

5. *Memory:* Bilateral damage to the hippocampus produces a marked impairment of the ability to form new associations, an inability to establish new memories at a time when remote memory is well preserved.
6. *Unilateral effects on memory:* Patients with resection of the left temporal lobe may lose the ability to retain verbally related material but gain the ability to retain visually related material relevant to the right hemisphere. The reverse is true for right temporal lobectomy patients.
7. *Psychiatric disturbances:* Waxman and Geschwind (1975) have described an interictal personality disorder characterized by hyposexuality, hyper religiosity, hypergraphia, and a so-called stickiness or viscosity in interpersonal relationships. In some patients a psychosis may be apparent in periods between seizures (the interictal period) that is often most severe during periods when the seizure disorder is well under control. In other patients a transient psychosis may be related to the actual seizure discharge or the postictal period. There is controversy regarding this issue.
8. *Aggressive behavior:* aggressive behavior and episodic dyscontrol may be related to dysfunction of the amygdala or to damage to prefrontal areas, which inhibit the amygdala. This remains an area of considerable controversy (Mark and Ervin, 1970; Geschwind, 1983 Ferguson et al, 1986).
9. Complex partial seizures as discussed above may follow mesial temporal sclerosis.

IV. FRONTAL LOBE: PREFRONTAL CORTEX AND EMOTIONS (Areas 9, 10, 11, 12, 46, 13, 14)

The term prefrontal refers to those portions of the frontal lobe anterior to the agranular motor and premotor areas. The prefrontal region is in a unique position in the brain as it receives a vast stream of information from polysensory, third order sensory and from cortical and subcortical limbic areas including especially dorsomedial nucleus, hypothalamus and indirectly the amygdala.

In the prefrontal cortex three regions are usually delineated:
1) Lateral - dorsal convexity,
2) Medial, and
3) Orbital.

All three are concerned with executive function and relate to the dorsomedial thalamic nuclei. In general the lateral-dorsal relates to motor association executive function, the orbital –ventral and the medial to the limbic –emotional control executive system. Recent studies have questioned this simplistic approach. Thus certain functions previously considered to be associated with dorsolateral location have been now localized to a medial location (Stuss et al 2000). In addition, there is evidence that dorsal lateral lesions also alter affect and motivation. Moreover as the clinical examples will demonstrate, most pathological processes involve the functions associated with more than one division either directly or through pressure effects.

Although the term prefrontal traditionally included areas 9, 10, 11 and 12, most studies of stimulation or of ablative lesions involving the prefrontal areas have included, within the general meaning of the term prefrontal, areas beyond these Brodmann designations. Thus the posterior orbital cortex (area 13) and the posteromedial orbital cortex (area 14) have been added to the orbital frontal group (Fig. 18-28). These areas all share a common relationship to the medial dorsal nucleus of the thalamus. Areas 46 and 47, located in man on the lateral surface of the hemisphere, are also often grouped with areas 9, 10, 11, and 12.

Another area, the anterior cingulate gyrus (area 24) is also included in the prefrontal area, although it relates to the anterior thalamic nuclei. Some authors (particularly in discussing frontal epilepsy) also include area 8 and the supplementary motor cortex already discussed above in relationship to premotor cortex.

The multiple connections of the prefrontal areas are discussed by Damasio 1985, Goldman-Rakic 1987, Jacobson and Trojanowski 1977, Nauta 1964, Pandya et al 1971 and Stuss and Benson 1986. Essentially, all sensory association and polysensory (multimodal) areas and the olfactory cortex project to the prefrontal areas. The prefrontal areas in turn have reciprocal connections to the premotor, temporal, inferior parietal and limbic cortex. The connections to the limbic system include (a) cingulate gyrus and then via cingulum to hippocampus and (b) uncinate fasciculus. Subcortical bidirectional connections are prominent in relationship to the medial dorsal nucleus of the thalamus, amygdala, and hippocampus. Significant projections to the caudate nucleus - putamen and the premotor areas provide the anatomical substrate for influencing motor function. Projections to the superior colliculus provide the substrate for modifying eye movements - e.g., "look to the direction opposite to the target."

Here we will briefly outline the major connections of the major subdivisions

(1) *Orbitofrontal Cortex.* The orbitofrontal cortex (areas 11 and 12) receives input from the cingulate regions, other sections of the prefrontal cortex, the magnocellular division of the dorsomedial nucleus, the amygdala and temporal areas. The orbitofrontal cortex in turn has strong projections to the autonomic regions of the hypothalamus and magnocellular dorsomedial nucleus of the thalamus in addition to the amygdala and temporal structures. Stimulation of the posterior orbital surface, like stimulation of the cingulate, parahippocampal gyrus, and amygdala changes the respiration, cardiac rhythm, and visceral contractions. These responses are believed to be the somatic basis of the emotional changes that originate in the limbic cortex.

(2) *Frontal Association Areas/Lateral Prefrontal Granular Frontal Cortex.* This region that is important in controlling behavior lays rostral to the premotor cortex, area 6 & 8 and has already been discussed in chapter 18. The frontal association areas (areas 9, 10, 11, and 12) receive their input from the cingulate regions and cortical association areas in the temporal, parietal and occipital lobes. They also have a strong reciprocal connection with the parvicellular portions of the dorsal medial thalamic nucleus via the anterior thalamic radiation. The dorsal medial thalamic nucleus, in turn, receives most of its input from the hypothalamus and the midbrain limbic nuclei. The frontal association region has strong projections to the hypothalamus and thalamus.

Overview of the role of the prefrontal area in motor and cognitive function: this topic has already been discussed in chapter 18.

Stimulation complex partial seizures. These seizures, partial seizures with impairment of consciousness, were formerly considered to be entirely temporal lobe in origin. It is now recognized that 20-30% of such seizures are extratemporal in origin - predominantly frontal lobe (Schwartz et al 1989, Williamson et al 1988). Compared to complex partial seizures of temporal lobe origin, complex partial seizures of frontal lobe origin are usually briefer, occur more frequently and tend to have a much shorter post ictal period of confusion. Recent studies of St.Hilaire at al; Wada and Weiser (see Chauvel et al 1992) suggest that these automatisms of orbital or medial parasagittal origin are often complex involving both arms or both legs or trunk or pelvis, at times in organized kicking, struggling, running or screaming. In contrast, automatisms of temporal lobe origin are more often oral or alimentary (licking, chewing, swallowing, etc). Such frontal lobe seizures must also be differ-

TABLE 22-5: SUMMARY OF THE LOCALIZATION TYPES OF FRONTAL EPILEPSY

LOCATION OF SEIZURE	BEHAVIORAL EFFECTS
Supplementary motor	Tonic, postural and speech arrest;
Cingulate	Complex partial
Anterior frontal polar	Forced thinking or initial loss of consciousness plus adversive components
Orbital frontal	Complex partial plus olfactory hallucinations plus illusions;
Dorsolateral	Tonic plus or minus aversion
Opercular	Mastication, salivation, swallowing plus speech arrest plus or minus epigastric and gustatory symptoms (see under parietal) plus or minus secondary spread to other simple partial.

entiated from seizures of psychogenic origin and from the "petit" absence seizures of generalized epilepsy (see Chapter 29). Thus, orbital frontal seizures present many features previously associated with pseudo or psychogenic seizures. Frontal polar seizures or dorsolateral convexity may present as forced thinking or apparent absence seizures. Quesney et al, 1990 make a distinction between discharges originating in the dorsolateral convexity, which are more likely to present with automatisms and affective components compared to parasagittal discharges that are more likely to present with motor or somatosensory components.

The present International Classification (see Commission - 1989) provides a summary of the localization types of frontal epilepsy that includes more than the prefrontal areas and is presented in Table 22-5:

Additional information concerning frontal lobe seizures/epilepsy will be found on the CD ROM.

Prefrontal cortex damage in humans: In general, most studies of prefrontal function have dealt with bilateral damage or with a unilateral lesion which, because of its parasagittal location in relation to the medial aspect of the prefrontal area, has produced essentially bilateral effects.

The first well-authenticated case of the frontal lobe syndrome is the crowbar case of Mr. Phineas P. Gage reported by Harlow in 1868. In an explosion in 1848, a pointed tamping iron shot through the skull of the patient, an efficient, well-balanced, shrewd, and energetic railroad foreman. The bar, 3.5 feet in length and 1.25 inches in greatest diameter, entered below the left orbit and merged in the midline vertex, anterior to the coronal suture, lacerating the superior sagittal sinus in the process. Following the injury, a marked personality change was noted. The balance "between his intellectual faculties and animal propensities" had been destroyed. He was "impatient of restraint or advice which conflicts with his desires; at times - obstinate yet capricious and vacillating, devising many plans of future operations which are no sooner arranged than they are abandoned in turn for others appearing more feasible."

Additional studies of such patients with bilateral damage to prefrontal areas have been reviewed by Damasio (1985), Damasio et al, 1994 and by Stuss and Benson, (1986). In humans, gross correlation of location of pathology (trauma and mass lesions) with major behavioral syndromes has been suggested.

(1) A *syndrome of "frontal retardation" or "pseudo depression" (abulia)* manifested by apathy, non-concern, lack of motivational drive and a lack of motivational drive and a lack of emotional reactivity has been associated with lesions involving the *frontal poles and/or the medial aspects of both hemispheres*. In its most severe form a syndrome of akinetic mutism occurs. Some have also noted aspects of this syndrome with dorsolateral frontal lesions.

(2) In contrast, the *"pseudopsychopathic" syndrome* is distinguished by a lack of inhibi-

tion including: facetiousness ("Witzelsucht"), sexual and personal hedonism, and lack of concern for others has been associated with *orbital frontal* pathology. Some of these patients with ventromedial orbital lesions have been described as manifesting confabulation, a particular disorder of memory found in patients with the Korsakoff syndrome; (a syndrome in which lesions of the dorsal medial nucleus of the thalamus have been implicated (chapter 30)). Patients with lesions of lateral orbital and lateral convexity have been described as restless, hyperkinetic, explosive and impulsive. As indicated above, damage in the dorsolateral sector appears to result in impairment of that high level cognitive ability which allows for abstraction, and creative activities. A rigidity or concreteness of response often is present. Additional discussion of the effects of frontal lobe lesions on emotional regulation will be found in Grafman et al 1986. Most patients do not fall into these discrete groups as regards the effects on emotion and personality and the correlation with tumor location is often not precise. Large prefrontal tumors often involve several prefrontal regions.

Prefrontal Lobotomy. Based on the initial reports of Jacobsen regarding the effects of prefrontal lobotomy on the emotional responses of the Chimpanzee, Moniz, a neuropsychiatrist and Lima a neurosurgeon introduced in 1936 the surgical procedures of prefrontal lobotomy to modify the behavior and affect of psychotic patients. Subsequently, Moniz introduced the procedure of prefrontal leukotomy: bilateral disconnection of prefrontal areas from subcortical (thalamic and basal ganglia) and other cortical areas *(Fig. 22-19)*. A more specific approach to problems of severe anxiety, manic behavior, and chronic pain was the procedure of stereotaxic anterior cingulotomy introduced by Ballantine et al, 1967. A number of lobotomies were performed in the 1940s and early 1950s for psychiatric reasons or to modify the emotional

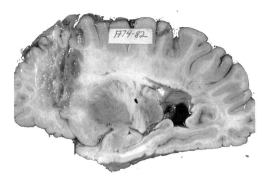

Figure 22-19. Prefrontal leukotomy. A surgical section has separated the prefrontal connections with the thalamus. Courtesy of Dr. Thomas Sabin and Dr. Thomas Kemper.

response of patient with chronic pain previously requiring large doses of narcotics. Such studies must be interpreted with a certain degree of caution. In the psychiatric patients, the lesions are produced in individuals with preoperative abnormalities of personality function. The effects are not necessarily those that the same lesion would produce in otherwise normal individuals. There is, however, a considerable resemblance in these cases to the effects produced by trauma (to the prefrontal areas) in relatively normal individuals.

The early results did suggest that these procedures did produce an alteration in the emotional response with a reduction of anxiety generated in conflict and painful situations. Emotional response was often detached from the pain and conflicts.

However, these effects on emotion can now be produced by the use of the tranquilizing drugs that were developed beginning in the middle 1950s. The development of these drugs, in addition to the frequent postoperative complications including seizures and personality alterations, led to the discontinuation of the procedure. (Valenstein, 1986). The personality changes are those that have been elaborated above. The personality change results from the isolation of the limbic subcortical structures from input provided by the prefrontal areas. Following prefrontal lobotomy or leukotomy, these patients were often impulsive and distractible. Their emotional responses were often uninhibited with an apparent lack of concern over the consequences of their actions. A related finding was an inability to plan ahead for future goals; at times, the patients were unable to postpone gratification, responding to their motivations of the moment. (In a sense, a loss of the reality principle had occurred). Although distractible, a certain perseveration of response was noted with an inability to shift responses to meet a change in environmental stimuli or cues. A rigidity and concreteness of response was apparent with deficits in abstract reasoning. A somewhat similar but less drastic personality change can be produced by bilateral severance of the anterior thalamic radiation from the medial nucleus or by direct destruction of the medial nuclei or orbital cortex. This less massive resection of cortex lessens anxiety with fewer personality changes. Destruction of the anterior cingulate regions also produces this less drastic change. Current psychosurgical or functional neurosurgical procedures have been reviewed by Diering and Bell(1991).

The results of partial or complete prefrontal lobotomy have shown that this region is important in motivation, intellect, judgment, abstract reasoning, and emotional control.

Parasagittal Lesions. It must be noted that lesions (e.g., meningiomas) which were initially parasagittal or subfrontal in relation to the prefrontal areas may, as they progress, compromise the function of adjacent premotor areas (8 and 6 and the supplementary motor cortex). The resultant clinical picture may then include not only the changes in personality but also the release of an instinctive tactile grasp and of a suck reflex. An incontinence of urine and feces may be present in addition to an apraxia of gait - an unsteadiness of gait, which is apparent as the patient attempts to stand and begins to walk but clears up once the act has been initiated. The following case histories illustrate many of the features of focal disease involving the prefrontal areas.

Case 22-4: Two years before admission,

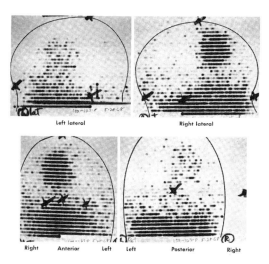

Figure 22-21. Frontal parasagittal meningioma. Case 22-4 refer to text. Radioactive brain scan (Hg^{197}) demonstrates dense uptake in right frontal parasagittal area.

this 69-year-old right handed white housewife; visited a relative (a physician) in Israel whom she had not seen in a number of years. He noted a change in the patient's personality. The patient was described as apathetic with silly immature reactions. In retrospect, the patient's husband felt that these alternations had begun insidiously a number of years previously. Her letters became incoherent .Ten months prior to evaluation, a left central facial weakness was first noted followed two months later by a decreased left arm swing. Six months prior to evaluation, her husband noted, she was purchasing items for which she had no need: 24 pairs of shoes, 15 brassieres. Subsequently she became careless in her housework. On occasions she lost her purse. Her ability to play bridge had decreased over the last year, although her golf game was unchanged.

Neurological examination: *Mental status:* There was a marked impersistence in motor activities. She was often inattentive. There was inappropriate joking. Her answers were often irrelevant. (According to her husband this had been present for many years). Digit span was decreased to 5 forward and 3 in reverses. There were marked deficits in serial 7 substractions. At times she began to subtract other numbers, at times she added rather than subtracted. There was a significant spatial disorientation. There was a marked impairment in the ability to copy a cube and deficits in drawing a house. There was disorientation in locating cities on a map. *Cranial nerves:* There was a significant neglect of single stimulus objects in the left visual field and a greater neglect when bilateral simultaneous stimuli were utilized. At times there was a limitation of voluntary gaze to the left. A left central facial weakness was present. *Motor system*: There was minimal weakness of the left arm and leg. Variable resistance was present on passive motion at left wrist and elbow suggesting the "gegenhalten"of frontal lobe disease. There was a decreased swing of the left arm in walking. The gait was initially apraxic but improved as the patient picked up speed. *Reflexes* Deep tendon stretch reflexes were increased on the left, with a left Babinski sign and a bilateral release of the grasp reflex.

Clinical diagnosis: Right frontal lobe tumor: probably a meningioma based on the long history.

Laboratory data: EEG (Fig 22-20 on CD Atlas), and *imaging studies (Fig. 22-21)* all indicated a right frontal parasagittal lesion. A meningioma was considered most likely since vascular supply was derived from the left and right middle meningeal and left anterior meningeal arteries.

Hospital course: At craniotomy, Dr. Samuel Brendler found a large right posterior parasagittal frontal fibrous, meningothelial meningioma measuring 8x8cm attached to the right side of the superior sagittal sinus.

Case history 22-5 presented on CD ROM provides an example of a patient with a slowly progressive subfrontal meningioma manifested by symptoms of apathy, loss of ambition, "depression" of mood, emotional lability, with a tendency to introspection and a slowness of mental processes. All of these symptoms were initially attributed to depression and a hypothyroid state. As gait and memory problems developed, these symptoms were attributed to Alzheimer's type senile dementia. When increased intracranial pressure and coma suddenly developed, neurological and neurosurgical intervention occurred.

Limbic Brain as a Functional System.

The emotional brain is organized into a hierarchy of function proceeding from the reticular formation, including mesencephalic midbrain nuclei to the hypothalamus and thalamus to the limbic and neocortical regions.

Reticular Formation. The reticular formation is the site where information is received from the peripheral nerves. This system is so organized that only certain stimuli trigger it to alert the brain. If it were possible for any response to trigger this system, then the individual's survival would be threatened. The response, however, is selective because throughout our lives we have evolved a set of emotional responses that determine whether we will respond to situations calmly or with rage or fear. What happens is that the reticular system, based on the sensory information with probably some subconscious cortical assistance, focuses the attention by sorting out the relevant information, thus enabling the central nervous system to continue functioning efficiently throughout a crisis.

Hypothalamus. By the time the data reaches the hypothalamus there are already distinct, well-organized emotional responses. The hypothalamus, with some assistance from the thalamus, sets the level of arousal needed for the emotional state and organizes and mobilizes the cortical and subcortical centers (especially the autonomic nervous system).

Pleasure/Punishment Areas. Throughout the limbic brain are found pleasure or punishment centers. These were located by implanting electrodes at various subcortical sites in an animal and training it to press a bar that connects the electrode to an electrical current (Olds, 1958). If the electrode is in certain pleasure centers, the animal will self-stimulate until it is exhausted. In fact, the animal would rather press the lever than eat. The pleasure centers are located throughout the limbic system, but especially in the septum and preoptic

TABLE 22-6: THE NEUROLOGICAL SUBSTRATES OF PSYCHIATRIC DISORDERS

DISORDER	DEFINING FEATURES	NEUROLOGIC CORRELATE
Schizophrenia	Progressive psychotic* disturbance with deterioration -Positive symptoms: thought disorder: delusions and hallucinations -Negative symptoms: poverty of speech, decreased movements poverty of affect, withdrawal from interpersonal relations Rx Respond to neuroleptics**	Based on neuropathology, MRI, PET -Decreased gray matter in temporal areas- left posterior superior temporal and mesial temporal -Prefrontal, left parietal and bilateral temporal
Affective disorders	Majority is not psychotic. Disorder is not progressive and no deterioration occurs. -Bipolar –Unipolar-manic -Unipolar depression ***	In post stroke depression, variable location of infarct: left lateral frontal &basal ganglia, or right frontal or right temporal/parietal - Metabolites norepinephrine (+) in csf - Norepinephrine &serotonin (-), Also Hypothalamic with (+) ACTH
Anxiety/panic	Autonomic activation is similar to pattern with fear responses	Brain stem adrenergic and serotonergic centers are implicated in anxiety. In panic attack patients blood flow to right limbic system increased between attacks
Obsessive/compulsive	Recurring thoughts trigger repetitive motor acts.	Increased metabolic activity in dorsolateral prefrontal, anterior cingulate and caudate nucleus
Personality disorders	Various types: hysterical, passive aggressive, antisocial (overlaps with criminal), passive dependent, schizoid, obsessive/compulsive, paranoid	Pseudopsychopathic syndrome may follow prefrontal damage. Studies of violent antisocial individuals indicate smaller prefrontal areas
Autism	Retarded development of social/emotional interactions with development of stereotyped motor automatisms	Cerebellar hypoplasia plus alterations in many limbic structures including amygdala, hippocampus, entorhinal cortex, septal nuclei, mammillary
Asperger's	Milder form of autism	Neocortical migration disorders

* Psychosis=gross impairment in reality testing affecting perception, thought and insight.

** Neuroleptics=dopamine antagonists

*** Anti depressants such as tricyclic agents and monoamine oxidase inhibitors increase norepinephrine and serotonin at selective receptor sites in hypothalamus and the limbic system by inhibition of reuptake of biogenic amines. Selective serotonin reuptake inhibitors (SSRI's) are effective for less severe cases of depression. For mania and bipolar disorder, lithium carbonate is effective. Neuroleptics may decrease manic behavior. For severe affective disorders electroshock therapy may be utilized.

region of the hypothalamus.

Other brain centers when stimulated produce fear responses: pupil dilation, piloerection, and sweating. In these "punishment regions," located in the amygdala, hypothalamus, thalamus, and midbrain tegmentum, the animal quickly stops pressing the bar. If the situation is non-threatening, the normal operations of the viscera continue. In a mildly stressful situation, such as one involving anger or heavy work, some of the digestive processes slow down, and the heart rate and blood flow increase. If the situation is threatening--triggering the reactions of fear, pain, intense hunger or thirst, or sexual arousal--most of the digestive processes slow down and heart rate, blood flow, and respiration increase. Once the threatening situation passes, conditions quickly return to a normal balance between the sympathetic and parasympathetic nervous systems.

Limbic Cortical Regions. The limbic cortical regions are strongly influenced by the emotional patterns set by the hypothalamus and amygdala which are then transmitted with only a very few synaptic interruptions into the limbic cortex where the emotional pattern is elaborated and efficiently organized. The neocortex (prefrontal cortex and, to a lesser degree, the temporal lobe), based on past experience, examines the situation, sorts out the emotional responses from the intellectual, and inhibits or controls the situation based on what past experience has proven to be expedient for individual survival.

Consider the following examples: a mother responds to her baby's crying, while the father sleeps on; in the middle of the night a jet airplane thundering overhead causes no response, while a whiff of smoke or breaking glass quickly arouses the central nervous system and keeps it focused.

Various types of dysfunction in this system will produce those significant alterations in behavior that represent the spectrum of psychiatric disease, briefly summarized in the next section

The Role Of The Limbic System In Psychiatric Disorders

The involvement of the limbic system either from a structural or functional standpoint is central to the psychiatric disorders which affect large numbers of patients. These disorders may affect perception, cognition and affect.

The neurological substrate is presented in Table 22-6. The epidemiology and genetics of the major disorders are presented in Table 22-7. Additional information will be found on the CD ROM.

TABLE 22-7: EPIDEMIOLOGY AND GENETIC ASPECTS OF THE MAJOR PSYCHIATRIC DISORDERS*

Disorder	Population Frequency %	Concordance in Monozygotic Twins %	Concordance in Dizygotic twins, Siblings, Parents %
Schizophrenia	0.5-1	68**	11
Affective disorders	10-20		14-25**
-Unipolar	18-19	40	11
-Bipolar	1-2	72	14

* Anxiety disorders and personality disorders affect large numbers of the population
** Risk is unchanged when adopted at birth and removed from biologic family

CHAPTER 23
Visual System

INTRODUCTION

Of all our senses, vision is the most important: we perceive the world mostly through our eyes. Even though light intensity varies by a factor of 10 million between the brightest snowy day and a starlit night, our eyes and visual system adapt to these intensity changes. We can discriminate between thousands of hues and shades of color. Our eyes are set in our heads in such a way that each eye sees almost the same visual field, making depth perception possible.

In the visual system the primary, secondary, and tertiary neurons are in the retinae and are all part of the central nervous system. The right field of vision projects to the left cerebral hemisphere; the left field of vision projects to the right cerebral hemisphere.

STRUCTURE OF THE EYE

The anatomy of the receptor organ, the eye is shown in *Figure 23-1*. It has three layers, or tunics:

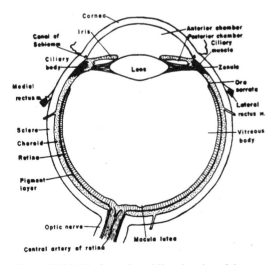

Figure 23-1A. Horizontal meridional section of the eye. (From Leeson and Lesson. 1970. Histology, Philadelphia, W.B.Saunders.)

1. The outer fibrous tunic - cornea and sclera.
2. The middle vascular and pigmented tunic - choroid, ciliary body, iris and pupil.
3. The inner neural tunic - retina with pigmented epithelium and neuronal layers.

The **outer fibrous tunic** consists of two parts: the anterior transparent cornea and the posterior fibrous white sclera.

The **cornea,** the window to the world, allows light rays to enter the eye. Most of the refraction needed to focus light rays on the retina occurs at the air-cornea junction. The sclera forms the white of the eye and the rest of the outer covering.

Middle Tunic- the choroid, the ciliary body, and the iris. The middle tunic is richly vascular and provides oxygen and nutrients to the inner, photoreceptor layer, and the retina.

The choroid is vascular and pigmented and forms the posterior portion of this tunic. Its inner portion is attached to the pigmented layer of the retina. The ciliary body is found in the anterior portion of the middle tunic and consists of a vascular tunic and the ciliary muscle. The ciliary body surrounds the lens and consists of a vascular tunic (the ciliary muscle) and the suspensory ligaments (the zonule), which suspend the lens *(Fig 23-1B)*. The ciliary muscles consist of meridional and circular fibers. The meridional fibers are external to the circular fibers. The ciliary muscles are the muscles used in accommodation, focusing on near objects.

Lens. The lens separates the anterior chamber from the vitreous body and completes the refraction of the entering light. A fibrous network, the zonule, suspends the lens. For distance vision the fibers are taut and the anterior surface of the lens is pulled flat. For close vision the ciliary muscle contracts,

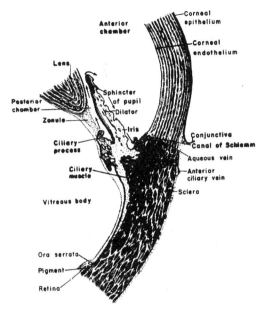

Figure 23-1B. Enlargement of a potion of the meridional section to show the angle of the eye. The letter P indicates the pectinate ligament or the trabecular meshwork. The arrows indicate the course of circulation of aqueous fluid. (From Leeson and Leeson, Histology, Philadelphia, WB Saunders, 1970.)

the fibers go slack, and the anterior surface becomes more convex; the eye accommodates. A decreased ability to focus on close objects is a normal consequence of aging. The term "visual acuity" refers to the resolving power of the eye in terms of the ability to focus on near or distant objects. For example, 23/40 means the patient's eye sees at 23 feet what the normal eye sees at 40 feet.

The pigmented iris on the anterior portion of the vascular tunic divides the space between the lens and the cornea into an anterior chamber (between the lens and the cornea) and a narrow posterior chamber (between the suspensory ligaments of the lens and the iris). The iris is a circular structure that acts as a diaphragm to control the amount of light falling on the retina. The opening is analogous to the f-stop of a camera. The eye has a range equivalent to the range from f2 to f22.

PUPILLARY REFLEXES

Pupillary Muscles. Two sets of muscles in the iris control the size of the pupil, sphincter and the dilator pupillae. When the circular muscles, the sphincter pupillae contract, the iris is drawn together and the pupil constricts, such as purse strings close the mouth of the purse. The second set of muscles, the dilator pupillae, are radial and draw the iris back toward the sclera.

The sphincter pupillae are supplied by the parasympathetic nervous system via the ciliary nerves, the fibers of which run together with nerve III. Transmission by this pathway is cholinergic, the dilator pupilla are supplied by the sympathetic nervous system via the superior cervical ganglion. Neurotransmission in this pathway is alpha-adrenergic.

Pupillary Reflexes. When light is flashed into cither eye, the pupil dilates, the light reflex. When the eye focuses on close objects, accommodation occurs

1. *Light Reflex.* In dim light the aperture of the pupil increases (dilator muscle) while in bright light the aperture of the pupil decreases (sphincter muscle). The afferent fibers arise in the retina and travel with the optic nerve and tract, these fibers pass through the lateral geniculate and synapse in the pretectal region of the midbrain. Neurons of the pretectal nuclei project bilaterally to the Edinger-Westphal nucleus.

2. *Accommodation.* This reflex changes the refractive power of the lens by the ciliary muscle contracting and decreasing the force on the suspensory ligament of the lens; the lens assumes a more rounded appearance with a shorter focal length. The neural pathway arises in the occipital lobe and the final efferent fibers run together with nerve III. In looking at distant objects the pupil dilates while in focusing on nearby objects the pupil constricts.

3. *Fixed Pupil.* In the absence of trauma to the eye, a dilated pupil that does not respond to light, fixed, is usually a sign of pressure on the third nerve. This is typically from supratentorial mass lesions that have produced herniation of the mesial segments of the temporal lobe through the tentorium. Compression of the third nerve and of the

midbrain occurs. Progressive rostral- caudal damage to the brain stem then evolves. This problem when first detected must be treated as a neurosurgical emergency.

Inner Tunic, Retina - pars optica, pars ciliaris, and pars iridica. The inner tunic contains photoreceptor cells and nerve cells. They are organized so that photoreceptor cells containing the photopigments are closet to the sclera and the nerve cells are above them. The nerve cells send fibers to the optic nerve through a pigment-free area, the optic disc. Light must traverse the nerve network to reach the photoreceptors.

The pars ciliaris and iridica are primarily a pigmented region on respectively the ciliary body and iris. The pars optica contains the photoreceptor cells and will be the focus of our discussion.

The pars optica contains:
- *outer pigmented layer,*
- *inner photoreceptor cells closest to the sclera,*
- nerve cells that are internal to photoreceptors and form the optic nerve that leaves the retinae through a pigment-free area, the optic disk. Light must traverse the nerve network to reach the photoreceptors.

The pars optica of the retina consists of the following layers *(Fig.23-2A):*
1. Pigmented epithelium (most external),
2. Receptor cell layer - rods and cones,
3. External limiting membrane,
4. Outer nuclear layer containing the nuclei of rods and cones,
5. Outer plexiform layer- containing the synapses of rods and cones with the dendrites of bipolar and horizontal cells, and the cell bodies of horizontal cells,
6. Inner nuclear layer- containing the cell bodies of bipolar, and amacrine cells,
7. Inner plexiform layer- containing the synapses of bipolar and amacrine cells with ganglion cells,
8. Ganglion cell layer,
9. Optic nerve layer, and
10. Internal limiting membrane.

Photoreceptors - Rods and Cones.

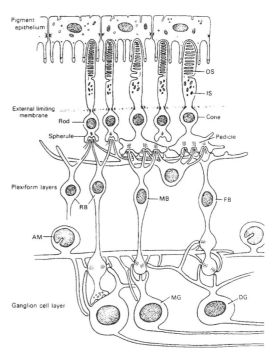

Figure 23-2 A. Schematic diagram of the ultrastructural organization of the retina. Rods and cones are composed of outer (OS) and inner segments (IS), cell bodies, and synaptic bases. Photopigments are present in laminated discs in the outer segments. The synaptic base of a rod is called a spherule; the synaptic base of a cone is called a pedicle. Abbreviations: RB = bipolar cells, MB=midget bipolar cell, FB a flat bipolar cell, AM = amacrine cell, MG = midget ganglion cell, DG = diffuse ganglion cell. (From Carpenter, M.B.A Core Text of Neuroanatomy Baltimore, Williams and Wilkins).

There are two types of photoreceptor cells: rods and cones.

The *rods* are sensitive in dim light as they contain more of the photosensitive pigment rhodopsin than the cones. This system has poor resolution; newspaper headlines are the smallest letters that can be recognized.

The *cones* are sensitive to color. There are there separate groups of cones each of which contains photopigments that are primarily sensitive 1) to blue - short wave lengths, 2) to green - middle wave lengths, or 3) to red longer wave lengths. Color vision requires light levels greater than bright moonlight and has high resolution-fine detail can be seen.

The rods and cones are not uniformly distrib-

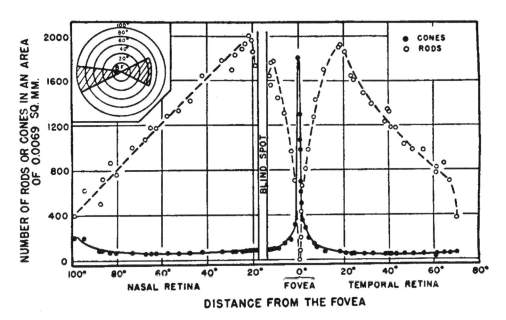

Figure 23-2B. The densities of cones and rods on or near the horizontal meridian through a human retina. The insert is a schematic map of the retina showing the fovea (F) and the blind spot (B). The striped area represents the regions of the retina, which were sampled in obtaining the counts plotted here. From Chapanis after Osterberg, Acta Opth, 1935, Suppl.6 (Blackwell).

uted in the retina. Most of the 6 million cone cells are located in an area 2 mm in diameter, the macula lutea (Fig. 23-1,), which can be seen through the ophthalmoscope. In the center of the macula lutea lies a zone of pure cones, the fovea *(Fig. 23-2B)*. The rest of the macula lutea is composed of both rods and cones, and most of the peripheral retina contains only rods. There are about 123 million rods.

Rods - Vision in Dim Light and Night Vision

The rods permit vision, *scotopic* (dark vision), in the dark-adapted eye. Rods contain a single pigment, *rhodopsin*, which is related to vitamin A_1 (retinol). The spectral sensitivity of night or scotopic vision is identical to the absorption spectra of rhodopsin (Fig 23-3) Light that can be absorbed by rhodopsin is seen; light that cannot be absorbed such as red light is not seen at these low light intensities. Scotopic vision has quite poor definition; two light sources must be quite far apart to be distinguished as two sources rather than one. Peripheral vision has progressively poor definition. The modern human uses this system very little since we have bright, portable light sources or radar. Several modern inventions,

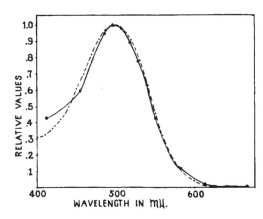

Figure 23-3. Comparison of the absorption curve of pure visual purple (interrupted line) with the intensity of light just visible in a dark-adapted human subject (continuous curve). From Ludvigh, Arch.Ophthal.20: 713.1938 (AMA).

such as night goggles with infrared sensors, are designed to compensate for these limitations.

The visual pigment, rhodopsin absorbs photons. Each rhodopsin molecule consists of retinal and opsin. Retinal absorbs the light while opsin is the protein in the plasma membrane of the rods. The photon converts retinal from the 11-cis form to the all-trans form. The all-trans retinal has to be re isomerized to 11-cis retinal to reform rhodopsin and to begin the process again. The photons result in hyperpolarization of the plasma membrane that is the light dependent part of visual excitation.

The photoreceptors are depolarized in the dark due to the sodium channels remaining open, called the dark currents. The dark current is turned off in light. Daylight prevents the regeneration of rhodopsin in the rods.

After 5-10 minutes in the dark, scotopic vision begins to return as the rhodopsin is regenerated, and reaches maximal sensitivity after 15 to 23 minutes. Since rhodopsin does not absorb red light, using red goggles can preserve night vision.

Cones - Color Vision

Color vision requires much higher light intensities and occurs primarily when the image is focused upon the macula. Each cone contains one of the three-color pigments whose absorption spectra are shown in *Figure 23-4*. The visual pigment of the cones consists of opsin and retinal.

The cones are divided into three separate groups that contain photopigments primarily sensitive to a different part of the visible spectrum as follows:

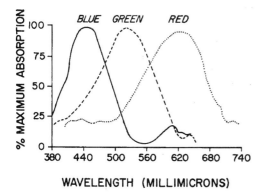

Figure 23-4. The absorption spectra of three types of cones. (Drawn from records in Marks, Dobelle, and MacNichol. 1964. Science 143:1181; and Brown and Wald. 1964. Science 144:45.)

1. Blue -423 nm (short wavelengths),
2. Green -530 nm (middle wave lengths), and
3. Red - 560 nm (longer wavelengths).

Color vision requires light levels greater than bright moonlight and has high resolution so that fine detail can be seen. Any color that does not excite two pigments, such as deep purple (400 nm) and deep red (650 nm) will be hard to discriminate. For this reason, a color of 660 nm in bright light can mimic a color of 640 nm in weaker light. To distinguish the color orange, 600 nm, the visual system compares the relative absorption by two visual pigments, in this case red and green.

Comparing the absorption by one visual pigment to the overall brightness apparently makes these color discriminations. For this reason a color of 660nm, in bright light, mimics a color of 640 nm in weaker light. Only when the two types of cones are excited can color and brightness be determined independently.

Electrophysiology of Retinal Photoreceptors and Neurons.

Synaptic Organization of the Rods, Cones and Neurons in the Retinae (Figure 23-2A)

1. Rods and cones synapse on the dendrites of bipolar cells and horizontal cells. In the peripheral portion of the retina, 60 degrees from the center, as many as 600 rods may converge on single bipolar cell, while in the fovea, only one or two cones may do so. In the macula lutea both rods and cones may innervate a single bipolar cell.

2. Axons of bipolar cells and amacrine cells synapse on dendrites of ganglion cells. Once again, the degree of convergence on the ganglion cell depends on the region of the retina. Ganglion cells in the fovea connect with only one bipolar cell.

3. Axons of ganglion cells leave the retina; form the II cranial nerve that synapses in the lateral geniculate, hypothalamus, and midbrain. There are about a million optic nerve fibers. Clearly, considerable data reduction has occurred between the 100 million rods and

cones and the one million optic nerve fibers.

Two other types of neurons are found in the retina: horizontal and amacrine cells. Horizontal cells are found in the outer plexiform layer (next to the rods and cones); their dendrites make contact with numerous rods and cones over a 230 to 400-µM field. Their axons run transversely along the retina and branch to make contact with many other rods or cones at the bipolar-cell junction. The horizontal cells connect groups of visual receptors in one area with groups in another area. Amacrine cells lie in the inner plexiform layer (next to the ganglion cells). The dendrites form postsynaptic contacts with bipolar cells and presynaptic contacts with ganglion cells; no axon has been described.

Response of the cells to light *(Fig. 23-5)*

Rods and Cones. Both rods and cones show a hyperpolarizing response to light. This very surprising result has been obtained in all vertebrate visual receptors studied to date. Contrary to what one would expect, they act as if dark were the stimulus. During darkness, high concentrations of cyclic GMP increase membrane conductance and give rise to a dark current that flows from the outer to inner segment, depolarizing the cell. After absorbing a photon, rhodopsin activates cyclic-GMP-phosphodiesterase, which cleaves cyclic GMP, lowers cyclic GMP levels, and by reducing the membrane conductance, hyperpolarizes the rod. Also contrary to expectation, the rod and cone cells release transmitter when it is dark, but this release ceases when it is light.

Horizontal Cell of retinae - hyperpolarizes in response to light, but it has a large receptive field. Even when the light is shining on receptor cells 250µm away, there is a strong hyperpolarizing response. Potential spread through the horizontal layer is by slow potentials; no action potentials have been observed.

Bipolar Cell: Both the direct hyperpolarizing receptor response and the horizontal-cell output impinge on the bipolar cell. Here we see the first type of data reduction carried out in the retina: center surround contrast enhancement. The bipolar cell has one type of output--hyperpolarization--for a bright spot on a dark field and another type of output--depolarization--for a large bright field.

Amacrine Cell. Recordings from the amacrine cell show that it responds primarily to sudden changes in light intensity. Note that it is active primarily when the light turns on or off. This is the second type of processing, or data reduction, that takes place in the retina: dynamic detection.

Ganglion Cells. There are two types of ganglion-cell responses –amacrine or bipolar cell.

Amacrine Cell Response. This response is, primarily a dynamic response to turning the light on or off. Some cells react more strongly to the light coming on than off; others have the opposite response. These cells are inhibited by illumination of both a spot and the surrounding area.

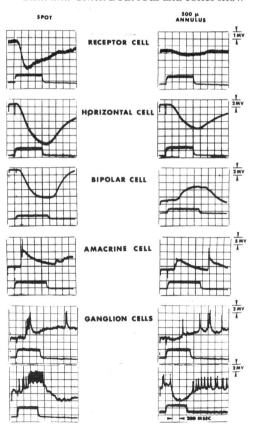

Figure 23-5. Responses of the cell types in the Necturus retina. From F.S. Werblin and J.Dowling. 1969. J. Neurophysiol. 32:339 (Amer. Physiol. Soc.)

Bipolar Cell Response. The second type of ganglion-cell response reflects bipolar-cell activity. When only the central portion of its receptive fields is illuminated, the ganglion cell responds with an intense train of action potentials during the illumination that abruptly stops when the light ceases. The frequency of the action potentials during illumination is proportional to the logarithm of the brightness of the light. When both the spot and the surrounding area are illuminated, there is less activity than occurs in the dark (the background level). When only the surround is illuminated, there is no specific response.

A third type of ganglion cell response (not shown in figure 23 –5)- is a center off type. When the center of the receptive field is illuminated, action potentials frequency decreases. When the surround is illuminated activity is greater than the background level, but not as intense as in the second type of response to a spot noted above.

To summarize ganglion cells signal the onset and cessation of light flashes and are effectively stimulated only by edges or abrupt transitions between light and darkness in their fields. Diffuse illumination does not stimulate ganglion cells very well.

The axons from the ganglion cells traverse the inner surface of the retina to the optic disk, where they turn and form the optic nerve, which runs for about 1 inch before entering the optic foramen on its way toward the diencephalon.

Optic Nerve. Electrical recordings from optic nerve fibers also show the types of data reduction occurring in the retina. First there is convergence; many receptors must fire one bipolar cell in the fovea Consequently the "grain" of the visual image depends on the location on the retinae. Second brightness can be detected. Third there is a center surround contrast enhancement. Lastly there is a dynamic or motion detection.

A spot of light on the retinae will be coded in terms of its brightness and its edges. As it is turned on and off, the dynamic receptors fire to alert higher centers. The spot of light will receive much more attention in the optic nerve bandwidth if it is moved around because each tiny movement will set of a new set of dynamic or movement receptors and these receptors make up about half the total optic nerve fibers. For example the rotating flashing lights of emergency vehicles attracts more attention than a fixed light.

VISUAL PATHWAY *(Fig.23-6)*

1. Origin of Cranial Nerve II from gan-

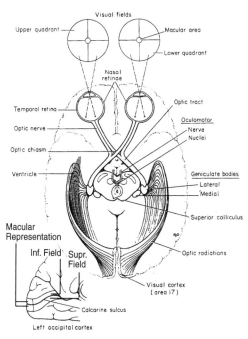

Figure 23-6. The visual pathways viewed from the ventral surface of the brain. Light from the upper half of the visual field falls on the inferior half of the retina. Light from the temporal half of the visual field falls on the nasal half of the retina; light from the nasal half of the visual field falls on the temporal half of the retina. The visual pathways from the retina to the striate cortex are shown. The plane of the visual fields has been rotated 90 degrees toward the reader. The inset (lower left) shows the projection of the quadrants of the right visual field on the left calcarine (striate) cortex. The macular area of the retina is represented nearest the occipital pole. Fibers mediating the pupillary light reflex leave the optic tract and project to the pretectal region; fibers from the pretectal olivary nucleus relay signals bilaterally to the visceral nuclei of the oculomotor complex. (Modified from Carpenter and Sutin. 1983. Human Neuroanatomy. Courtesy of Williams and Wilkins.)

glion cell layer in the retina,

2. Optic nerve,

3. Optic chiasm,

4. Optic tract,

5. Synapse in lateral geniculate nucleus of the thalamus,

6. Visual radiation from lateral geniculate nucleus to the striate (visual) cortex - area 17 of the occipital lobe,

7. Visual association cortex - areas 18 and 19, and

8. Multimodal associations with cortical regions in the posterior temporal-parietal, and frontal lobes in both hemispheres (Fig. 18-11).

VISUAL FIELDS

Retina.

Each retina is divided into a temporal or nasal half by a vertical line passing through the fovea centralis. This also serves to divide the retina of each eye into a left and right half. A horizontal line passing through the fovea would divide each half of the retina and macula into an upper and lower quadrant. Each area of the retina corresponds to a particular sector of the visual fields, which are named as viewed by the patient. For each eye tested alone, there is a left and right visual field corresponding to a nasal and temporal field. Because of the effect of the lens, the patient's left visual field projects onto the nasal retina of the left eye and the temporal retina of the right eye. Similarly, the upper visual quadrants project to the inferior retinal quadrants.

Optic Nerve and Optic Chiasm.

At the chiasm, the fibers in the optic nerve from the left nasal retina cross the midline and join with the fibers from the right temporal retina to form the right optic tract. This dense bundle of over a million fibers runs across the base of the cerebrum to the lateral geniculate nucleus of the thalamus. Fibers from the inferior nasal retina (the superior, temporal visual field) cross in the inferior portion of the chiasm, which usually lies just above the dorsum sellae.

By this crossing, all the information from the left visual field is brought to the right lateral geniculate nucleus and subsequently to the right calcarine cortex. The fibers of the right nasal retina also cross in the chiasm to join the fibers from the left temporal retina. Thus, the visual pathway repeats the pattern in other sensory systems; the right side of the visual field is represented on the left cerebrum.

Lesions in the Optic Pathway:

1. In the retina or the optic nerve before the optic chiasm (2 in *Figure 23-7*) there is no sight in that eye - monocular blindness.

2. In the chiasm, the fibers from both temporal visual fields are cut; the result is bitemporal hemianopsia (3 in Figure 23-7)

3. Lesions behind the optic chiasm produce blindness in one visual field, a homonymous hemianopsia from injury to the optic tract, the lateral geniculate body or the genicu-

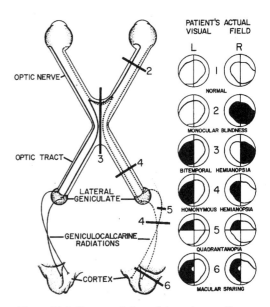

Figure 23-7. Common lesions of the optic tract. 1) Normal, 2) optic nerve, 3) optic chiasm, 4) optic tract or complete geniculate or complete geniculocalcarine radiation or complete cortical 5) temporal segment of geniculocalcarine radiation (Meyer's loop) producing a superior quadrantanopia. If the parietal segment of this radiation were involved, an inferior quadrantanopia would result. 6) Calcarine or posterior cerebral artery occlusion producing a homonymous hemianopsia with macular sparing.

localcarine radiation (4 in Figure 23-7).

4. Partial lesions of the geniculocalcarine radiations produce a quadrantanopsia (5 in figure 23-7).

5. Macular sparing is seen with lesions in the visual cortex following calcarine artery infarcts (6 in Figure 23-7). The macula is represented close to the occipital pole and this area receives anastomotic flow from the middle cerebral artery.

The following case is an example of a lesion in the optic nerve, anterior to the optic chiasm.

Case History 23-1: This 53-year-old white right-handed housewife had a progressive 23-year history of right-sided supraorbital headache with decreasing acuity and now almost total blindness in the right eye. During the 3 years before admission, intermittent tingling paresthesia had been noted in the left face, arm, or leg. About 1 year before admission, the patient had a sudden loss of consciousness and was amnestic for the events of the next 48 hours. No explanation for the episode was clearly established. The cerebrospinal fluid protein was reported to be elevated (230 mg/dl). The patient and her family reported some personality changes over a period of several years, including a loss of spontaneity and increasing apathy.

General Physical Examination: There was a minor degree of proptosis (downward protrusion of the right eye).

Neurologic Examination: *Mental Status:* The patient was, in general, alert but at times would become lethargic. Her affect was flat. At times, she would laugh or joke in an inappropriate manner. *Cranial Nerves:* There was anosmia for odors, such as cloves, on the right and a reduced sensitivity on the left. Marked papilledema (increased intracranial pressure with elevation of the optic disk and venous engorgement was present in the left eye. In contrast, there was pallor of the right optic disk indicating optic atrophy. Visual acuity in the right eye was markedly reduced. This combination of funduscopic findings is termed the Foster Kennedy syndrome. The patient had only a small crescent of vision in the temporal field of the right eye; only vague outlines of objects could be seen. A slight left central facial weakness was present. *Motor System:* movements on the left side were slow. *Reflexes:* A release of grasp reflex was present on the left side.

Clinical diagnosis: Subfrontal meningioma rising from the inner third sphenoid wing. Alternative location would be olfactory groove.

Laboratory data: *Imaging studies* demonstrated a tumor blush in the right subfrontal region extending back to the right optic nerve groove consistent with a meningioma arising from the olfactory groove or inner third sphenoid wing, *(Fig. 23-8)*

Hospital Course: A bifrontal craniotomy performed by DR. Sam Brendler exposed a well-encapsulated smooth tumor a meningioma attached to the inner third of the sphenoid wing. Approximately 90 to 95% of the tumor was removed exposing the right optic nerve. Examination 4 months after surgery indicated right anosmia and right optic atro-

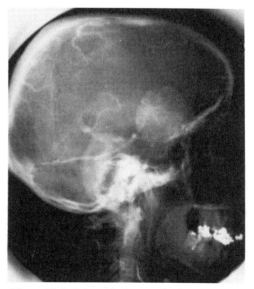

Figure 23-8. Sphenoid-wing meningioma producing compression of the right optic and olfactory nerves. Case 23-1 Refer to text. Right carotid arteriogram, venous phase demonstrating tumor blush in the subfrontal area of the anterior fossa, extending into the middle fossa. (Courtesy of Dr. Samuel Wolpert. New England Medical Center Hospitals.)

phy were present.

Optic Chiasm: A bitemporal defect results (*Fig. 23-9, 23-10, 23-11*). This may be a

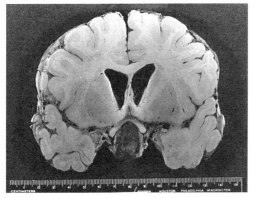

Figure 23-9. A large pituitary adenoma with secondary hemorrhage has extended upward out of the sella to compress the optic chiasm and optic nerves. This 53-year-old white male 11 months before death experienced difficulty reading small print, with greater involvement of the right eye than the left. This progressed to total blindness in the right eye and shortly thereafter-sudden loss of vision in the left eye, an absence of pupillary responses, and a bilateral anosmia. Vision improved after partial surgical removal and irradiation, but fatal meningitis developed. (Courtesy of Drs. John Hills and Jose Segarra).

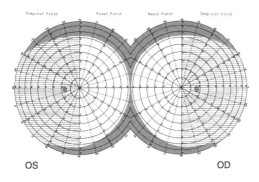

Figure 23-10. Pituitary adenoma. Visual fields demonstrating a bitemporal hemianopia. O.S. = left eye; O.D. = right eye. This 51-year-old obese male with large puffy hands and a prominent jaw, declining sexual interest for 18 years, an 8-year history of progressive loss of visual acuity, a 6- to 7-month history of diplopia due to a bilateral medial rectus palsy and headache had stopped driving because he was unable to see the sides of the road. Urinary adrenal and gonadal steroids and thyroid functions were low, with no follicle-stimulating hormone. Imaging studies demonstrated a large pituitary adenoma with significant suprasellar extent.

bitemporal hemianopsia or an incomplete bitemporal field defect. The usual cause is a pituitary adenoma or a supra sellar tumor such as a craniopharyngioma.

Case 23-2. Patient of Dr. Martha Fehr. This 45-year-old man had an 18-month history of a progressive alteration in vision. He was unable to see objects in the right or left periphery of vision. He was concerned that in the process of driving he might hit pedestrians stepping off the sidewalks in the periphery of vision. He had also experienced over 6-8 months, progressive headaches, loss of energy and libido.

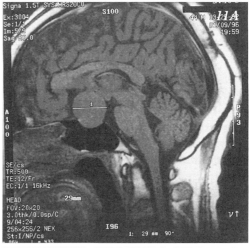

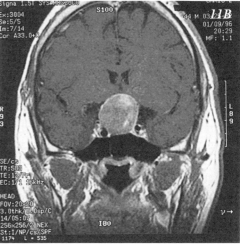

Figure 23-11. A large pituitary adenoma in a 44-year-old man with a bitemporal hemianopia. Case 23-2. Significant extrasellar extension compresses the T1 optic chiasm. MRI scans: A) midline sagittal section B) coronal section. (T1 with contrast.)

Neurologic exam: Normal except for a bitemporal hemianopia.

Clinical diagnosis: Pituitary adenoma compressing the optic chiasm.

Laboratory data: *Endocrine studies:* all were normal. *MRI:* A large macroadenoma measuring 3 cm in diameter extended outside of the sella turcica to compress the optic chiasm (23-11).

Subsequent course: Dr. Gerald McGuillicuddy performed a gross total trans sphenoidal resection of the tumor, with a significant improvement in vision. Three years later, the patient again had visual symptoms and eye pain. MRI scans indicated regrowth of tumor with possible impression on the right optic nerve. He was treated with radiotherapy.

Note that males are more likely to present with large pituitary adenoma compressing the optic chiasm. Women are more likely to present with microadenomas or small macroadenomas since they are more likely to be seen at an earlier stage due to an initial complaint of amenorrhea.

Lateral Geniculate Nucleus (LGN).

The LGN is primarily a relay station in the visual pathway. Some optic fibers for pupillary light reflexes bypass the LGN and descend into the pretectal region of the midbrain. Others, for coordinating eye and head or neck movements, descend to the superior colliculus. Also some of the retinal fibers synapse in the suprachiasmatic nucleus of the hypothalamus and then project to the pineal by way of the sympathetic fibers where they influence circadian rhythms and endocrine functions.

The lateral geniculate nucleus is a six-layer horseshoe-shaped structure. Visual processing begins in the large and small cells in the LGN. The small cells are in layers 1 to 4, and the large cells in layers 5 and 6. Each LGN cell receives direct input from a specific area of the retina, and each area of the LGN is driven from a specific area of the visual field. Each receptive field has a center that is either on or off and a surround that has the opposite polarity.

The optic nerve fibers from the temporal visual field, the crossed fibers, end in layers 1, 4, and 6. The fibers from the temporal retina (the nasal visual field), the uncrossed fibers, end in layers 2, 3, and 5. Each optic nerve fiber ends in one layer on 5 or 6 cells. The total number of cells and fibers is roughly equal. LGN cells respond well to small spots of light and to short, narrow bars of light, as well as to on-off stimuli. Diffuse light is a very poor stimulus.

Cells in the LGN grow rapidly in the first 6 to 12 months after birth, under the influence of visual stimuli. Visual sensory deprivation during this period leads to atrophy and life-long blindness. Two major streams of visual information leave the retina to the LGN (1) The magnocellular stream starts in the large cells of the retinal ganglion (Y-cells) that project to the magnocellular layers (layers 1 and 2) which subsequently project to the striate cortex. The cortical neurons of the magnocellular stream respond to movements, orientation and contrast but respond poorly to color. (2) The parvocellular stream begins with small cells of the retinal ganglion layer (X-cells) that project to parvocellular layer of LGN layers 3 and 6. These cells respond either to shape and orientation or color red/green and blue/yellow.

Optic Radiation.

The fibers to the visual cortex leave the lateral geniculate nucleus and form the optic radiation, which sweeps around the posterior horn of the lateral ventricle, through the parietal lobe to the occipital lobe and the calcarine cortex (Fig. 23-6). Some fibers from the lateral geniculate nucleus also enter the inferior pulvinar nucleus.

The visual radiation is so extensive that not all the fibers can pass directly posterior to the calcarine cortex. Only the most dorsally placed fibers pass back deep into the parietal lobe. The more ventrally placed fibers pass forward (Myer's loop) and deep into the temporal lobe, swing posteriorly around the anterior portion of the inferior horn of the lateral ventricle, and continue posteriorly lateral to the ventricle wall to the calcarine cortex. In contrast to the densely packed optic tract, the optic radiation fans out widely on its passage through the parietal and temporal lobes.

OCCIPITAL LOBE

The occipital lobe in Brodmann's numerical scheme consists of areas 17, 18, and 19. From a cytoarchitectural standpoint, area 17 represents a classic example of specialized granular cortex, or koniocortex. Areas 18 and 19 represent progressive modifications from koniocortex toward homotypical cortex (Refer to chapter 17 for a discussion of cytoarchitecture).

Area 17, the striate cortex, the principal visual projection area in humans is found primarily on the medial surface of the hemisphere, occupying those portions of the cuneus (above) and lingual gyrus (below) that border the calcarine sulcus. For this reason, it is often termed the calcarine cortex. Much of this cortex is located on the walls and in the depths of this sulcus.

The extrastriate visual cortex, including parastriate areas 18 and 19 forms concentric bands about area 17 and is found on both the medial and lateral surfaces. Area 19 in humans extends onto the adjacent mid and inferior temporal gyri.

Most of our understanding of occipital-lobe function comes from simian studies. In non-human primates five visual areas have been identified:
V1 - corresponding to area 17 in humans,
V2 and V3 - corresponding to area 18, and
V4 and V5 - corresponding to area 19.

Area 17 is the primary visual area, and receives the termination of the geniculocalcarine (or optic) radiation. This projection is arranged in a topographic manner, with the superior quadrant of the contralateral visual field represented on the inferior bank of the calcarine fissure, and the inferior quadrant of the contralateral visual field represented on the superior bank. The macula has a large area of representation, which occupies the posterior third of the calcarine cortex and extends onto the occipital pole.

Area 17 receives fibers from and sends fibers to area 18 but does not have any direct callosal or long-association-fiber connections to other cortical areas. The columnar and horizontal organization of the primary visual cortex has been discussed previously in chapter 17. Area 17 is more mature at birth and has the most precise map. The other visual areas develop through maturation and experience, although experience modifies V1 as well (see Chapter 29).

Area 18 has extensive connections with areas 18 and 19 of the ipsilateral and contralateral hemispheres, which were demonstrated by both the earlier strychnine neuronography studies and the more recent horseradish peroxidase studies.

Callosal fibers enter the opposite hemisphere, and association fibers communicate with the premotor and inferior temporal areas and area 7 of the adjacent parietal cortex.

As discussed in greater detail below, Zeki and associates (1992) studied each of these visual areas to create a map of the retina. Parallel pathways process various aspects of visual information--color (wavelength), motion, stereopsis, and form (line orientation)--and extend from the retina to the lateral geniculate nucleus, striate cortex, and finally, extrastriate cortex.

Ocular dominance Columns. Ocular dominance columns are seen in the striate cortex. They are seen as an alternating series of parallel stripes that represent a column of neurons in the striate cortex innervated by either the ipsilateral or contralateral eye. Ocular dominance columns extend in alternating bands through all cortical areas and layers, and are absent in only the cortical region representing the blind spot and the cortical area representing the monocular temporal crescent of the visual field. The mosaic appearance of these ocular dominance columns is demonstrated with autoradiography using 2-deoxyglucose and stimulation of only one eye. Simple cells are driven monocularly while the

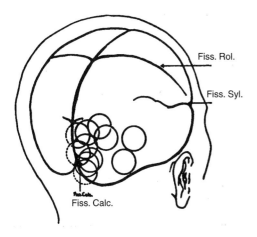

Figure 23-12 Focal discharge in occipital lobe. Location of the lesions at surgery in 11 patients who had focal seizures beginning with visual sensations. (From Penfield, W., and Kristiansen, K.: Epileptic Seizure Patterns, Springfield, Charles C. Thomas, 1951, p.47.)

complex and hyper complex cells are stimulated from both eyes.

Stimulation of Areas 17, 18, and 19

Direct electrical stimulation of areas 17, 18, and 19 in conscious people produces visual sensations (Pollen, 1975) *(Fig. 23-12).*

These images are not elaborate hallucinations (as in complex partial temporal seizures) but rather are described as flickering lights, stars, lines, spots, and so forth. Often the images are described as colored or moving around. The images are usually localized to the contralateral field, at times to the contralateral eye. An illustrative case, 23-4, is presented below.

At times, the patient cannot determine laterality. In addition, stimulation of areas 18 and 19 (and sometimes 17) produces conjugate deviation of the eyes to the contralateral field and, at times, vertical conjugate movements. Discharge of the occipital cortex may be followed by transient visual defects similar to the transient post-ictal hemiparesis that may follow focal seizures beginning in the motor cortex (see Aldrich et al., 1989). More complete discussion of seizures beginning in occipital cortex can be found in Salanova et al., 1993, and Williamson et al., 1992.

Physiology

The optic (geniculocalcarine) radiation terminates on the cells in the calcarine cortex, which respond like retinal ganglion cells. Their receptive fields consist of an excitatory region surrounded by an inhibitory region. Fields with the opposite pattern are also seen in higher mammals. These fields are excited by an annulus (donut) of light.

Hubel and Wiesel have described three types of higher-order cells in the visual cortex:
1. Simple cell.
2. Complex cell and,
3. Hypercomplex cell.

Simple cell. This cell type responds best to a bar of light with a critical orientation and location. The receptive fields of several simple cells are shown in Figure 23-13. The excitatory field of each simple cell is bordered on one or both sides by an inhibitory field as shown for simple cell a. The small portion of the visual field outlined in Figure 23-13 drives thousands of simple cells, each with a discrete location and orientation. Each simple cell appears to receive input from a number of ganglion cells, which have their excitatory and inhibitory fields in a straight line.

Every simple cell can be driven by input from either eye, but they usually show a preference for one eye or the other. For example, a simple cell may be strongly excited by a bar seen with the right eye and only weakly stimulated by the same bar seen with the left eye. This differential sensitivity may form the basis of binocular vision.

Complex Cell. A number of simple cells having the same orientation activate a complex cell. The complex cell has a definite orientational preference, but a much larger receptive field. An example of this is complex cell alpha in Figure 23-13. Any horizontal line of any length within the outlined visual field excites complex cell alpha. Lines with different orientations in the same visual field will drive other complex cells.

Hyper complex cell. The final cell type is a

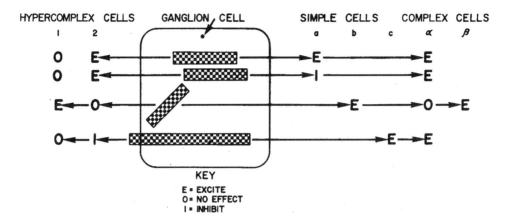

Figure 23-13. Types of visual stimuli that excite each of the four types of cells found in the calcarine cortex. The square represents a box 2 inch by 2 inch, 5 feet away from the monkey's eye. The actual stimuli are bright bars of light on a dark field. (Data from Hubel and Wiesel. 1968. J. Physiol. 195:215.)

hyper complex cell. It has characteristics very similar to the complex cell except it discriminates lines of different lengths. The bottom light bar in *Figure 23-13* will inhibit hyper complex cell 2. If its left-hand border were moved into the visual field, it would strongly excite hyper complex cell 2. Once again, it is excited by lines of a critical orientation anywhere within a large portion of a visual field, but they must not be too long. Corners and angles also excite these cells.

About half a million optic-tract fibers enter each occipital lobe. It should be clear that there are many more cells processing this information, and indeed, the banks of the calcarine cortex contain many millions of cells.

The response to a large rectangle of light varies with the type of cell. Only if the edge falls on the center of a cell's receptive field will it be excited. Ganglion cells are very sensitive to position, contrast, and dynamics. Ganglion cells, in turn, excite simple cells that outline the edges of the rectangle of light. Simple cells have a larger receptive field than ganglion cells, are less sensitive to position, but are sensitive to the direction of the edge. Simple cells do not respond to continuous light or dark stimuli in their fields or to bars of light with an orientation other than their preferred orientation. Complex cells have much larger receptive fields than simple cells but are heavily influenced by the direction of the edge. Lastly, a hyper complex cell must have a corner, an angle, or the end of the edge within its field to be activated.

When a rectangle of light is flashed upon the retina, the cells described are initially activated but adapt fairly quickly and then are hardly stimulated. However, if the pattern flashes on and off, the cells are repeatedly stimulated. The cortex is also stimulated if the pattern moves a little on the retina. A different group of ganglion and simple cells will react, but because of their larger receptive fields, most of the complex and hyper complex cells continue to be stimulated. The nervous system has overcome the problem of stationary images by continually producing rapid, saccadic eye movements that constantly shift the image on the retina. Indeed, when the image is stabilized by special contact lenses, it disappears.

In the process of data reduction, the visual system loses the ability to judge absolute light intensity for the most part.

Differences in intensity within a field can be determined easily but not the overall intensity of the field; photographers who estimate exposures usually waste a lot of film. Only the pupillary reflex system retains the ability to measure absolute intensity.

Different calcarine cells within a narrow

column perpendicular to the surface repeat this processing for each of the primary-color receptors to give information on form, color, motion, and contrast. Lesions as small as l mm result in a detectable loss of all of these modalities in a small area of the visual field.

Zeki and associates (1992) demonstrated how the four identified perceptual pathways *(color, form with color, dynamic form, and motion)* relate to the specific visual areas.

The neurons of area V5 (area 19) are responsive to *motion and directionally selective,* but they are nonselective for color. In contrast, the majority of cells in V4 (area 19) are selective for specific wavelengths of light (color sensitive), and many are selective for form (line orientation) as well.

The cells of the adjacent areas V3 and V3A (area 18) are selective for *form (line orientation) but are indifferent to wavelength (color).* V1 (area 17) and V2 (area 18) distribute information to these specialized fields. V1 has column and intercolumnar areas. The columns stain heavily for the energy-related enzyme cytochrome oxidase. The neurons within the columns are wavelength sensitive, tend to concentrate in layers 2 and 3, and receive input primarily from the parvocellular layers of the lateral geniculate nucleus. Information concerning color is then projected either directly or through thin-column stripes in area V2 to V4.

By contrast, *form-selective neurons* are located in the intercolumnar areas in V1. Form-related-to-color derives from the parvocellular-neuron layers of the lateral geniculate nucleus, which project first to the intercolumnar areas of V1 and then to V4.

The second *form system is independent of color* and depends on inputs from the magnocellular layers of the lateral geniculate nucleus to layer 4B of V1 projecting to V3. Extrastriate areas send projections back to the dorsal lateral geniculate nucleus and the pulvinar of the thalamus.

The *motion-sensitive system* consists of inputs from the magnocellular layers of lateral geniculate nucleus to layer 4B of area V1, which then projects to area V5, either directly or through area V2. Each of the prestriate areas V5, V4, and V3 sends information back to V1 and V2, as well as to the parietal and temporal areas. This provides for more integrated visual perception. Striate

To summarize, let us consider the example of a slowly moving colored ball. Visual information is fractionated into the modalities of form, color, and motion in the calcarine cortex. That information is reassembled elsewhere to give us a unitary view of the ball.

We are just beginning to understand where and how this parallel processing is done.

If the object is the letter A, then further processing takes place in the dominant (usually left) lateral occipital cortex (Brodmann's areas 18 and 19), the visual language association area. Letters seen in the left visual field and represented in the right calcarine cortex must project across the posterior corpus callosum before they can be recognized in the left visual association cortex.

Information needed to reconstruct complex forms flows through the inferior occipital/temporal region, primarily on the nondominant side. Located here are cells that can discriminate between human faces, for example. Similarly, sheep have cells that respond only to the image of a wolf (not of a sheep), and monkeys have cells that respond only to a human face but not to a snake or a banana. Many of these cells are sensitive to gaze, being strongly stimulated by direct eye contact but only weakly stimulated when the eyes are averted.

Extra-occipital higher-level visual processing then occurs in area 7 of the parietal lobe, areas 23 and 21 of the temporal lobe, and area 8 of the frontal lobe (see Pandya and Kuypers, 1969).

Occipital Lobe and Eye Movements.

Areas 18 and 19 send fibers to the tectal area of the superior colliculus. Such connections are necessary for visual fixation and accurate following of a moving object. The slow and smooth conjugate eye movements that occur when the eyes are following a moving visual stimulus (pursuit movements) should be distinguished from the independent phenomenon of voluntary (saccadic) conjugate eye movements.

Saccadic movements are rapid and shorter in latency and duration. They do not require a visual stimulus and do not depend on any connections to area 18. Rather, they are dependent on area 8, which sends fibers to the lateral gaze center of the pons.

This discussion of cortical control of eye movements is clearly simplified (see also chapter 18). Both areas send some fibers to the superior colliculus, and ultimately, both the saccadic and pursuit movements involve the pontine gaze centers. In addition, vestibular and cerebellar influences also determine eye movement.

Lesion in Area 17

Complete unilateral ablation of area 17 (V1) produces a complete homonymous hemianopia in the contralateral visual field (the same half field in each eye has no conscious visual perception). Such patients may have some crude visual function such as spatial localization, presumably because of optic-tract connections to the superior colliculus. A report by Barbur and colleagues (1993) suggests that conscious perception of the type of visual stimulus and its direction of motion can continue to occur after selective V1 lesions have been made, due to activation of V5 (possibly by subcortical input to V5). Vision is clearly abnormal in this blind field. V1 to V5 input is the dominant input in intact individuals.

With partial lesions, a partial defect results. For example, if only the superior bank of the calcarine fissure is involved, the visual field defect is limited to the inferior quadrant of the contralateral field. A bilateral infarct of either the upper or lower bank of the calcarine fissure may produce a homonymous altitudinal hemianopia in which apparent blindness exists in either the entire lower or upper field of vision (as appropriate).

With such calcarine-cortex lesions, the field defects are usually similar (congruous) in both eyes. By contrast, incomplete optic-tract lesions may produce somewhat unequal (non congruous) homonymous field defects in both eyes.

Vascular lesions, in particular occlusion of the posterior cerebral artery, often result in a homonymous hemianopia with macular sparing. That is, vision in the macular area of the involved field remains intact. Such preservation of macular vision probably occurs because the macular area has a large representation in the most posterior third of the calcarine cortex and is the area nearest the occipital pole. This area, then, is best situated to receive leptomeningeal anastomotic blood supply from the middle cerebral and anterior cerebral arteries. Occasionally, with occipital infarcts, there may be preservation of vision in a small peripheral unpaired portion of the visual field, called the temporal crescent (see Benton et al., 1980).

In vascular insufficiency or occlusion of the basilar artery, infarction may occur in the distribution of both posterior cerebral arteries, producing a bilateral homonymous hemianopia and a syndrome of "cortical blindness." Such patients lose all visual sensation, but pupillary constriction in response to light is preserved.

Lesions of Extrastriate Areas 18 and 19

Lesions in areas 18 and 19 often referred to as the visual association areas, produce deficits in visual association, including defects in visual recognition and reading. The problem is that in humans with a unilateral lesion, one is almost never dealing with disease limited to areas 18 and 19 (however, see discussion below of more recent studies).

Rather, one finds that adjacent portions of either the inferior temporal areas or of the inferior parietal lobule (the angular and supramarginal gyri) are involved or that the lesion extends into the deeper white matter of the lateral occipital area and involves association and callosal fibers. Such more-extensive lesions will, of course, produce the deficits previously noted in the discussion of parietal function. A limited unilateral ablation might produce defects in visual following, as tested by evoking opticokinetic nystagmus (e.g., moving vertical black lines on a white background or looking at telephone poles from a moving train).

Bilateral lesions limited to areas 18 and 19 occur only rarely in humans. In humans such a lesion would deprive the speech areas of the dominant hemispheres of all visual information. The patient would presumably see objects but be unable to recognize them or to place visual sensations in the context of previous experience. The patient would be, moreover, unable to relate these visual stimuli to tactile and auditory stimuli. In monkeys, restricted lesions in the "visual" areas of the temporal and parietal lobes also interfere with pattern discrimination (Mishkin, 1972).

Horton and Hoyt (1991) have reported that unilateral lesions specific to the extrastriate V2/V3 cortex produce quadrantal visual field defects. As discussed by Zeki (1992), rare patients with restricted lesions in V4 present with achromatopia. The patients see the world only in shades of gray but perception of form, depth, and motion are intact. Rare patients with lesions limited to V5 have akinetopia.

Stationary objects are perceived, but if they are in motion, the objects appear to vanish. Plant and coworkers (1993) have reviewed such impaired motion perception. Selective deficits in form perception are even less frequent. Destruction of both form systems and therefore of V3 and V4 (areas 18 and 19) would be required and, as discussed above, such a lesion would also destroy V1 (area 17), resulting in total blindness.

The following case illustrates the effects of a space-occupying lesion in the occipital lobe. One should compare these findings to those reported earlier in Case 23-1, a lesion in the visual system anterior to the optic chiasm.

Case History 23-3 *(Fig 23-14)* This 47-year-old white, right-handed, real estate salesman developed a cough with some blood present in the sputum (hemoptysis), one month before admission. Five days before admission the patient developed a generalized headache, which was precipitated by coughing or straining and which awakened him or prevented him from sleeping. Two days before admission, the patient noted blurring in the left inferior quadrant of his field of vision. The day before admission, he noted complete loss of vision in this quadrant. On the day of admission, the headache increased, and the patient was unable to see anything in the left visual field. Past history was significant. The patient had multiple pulmonary infections, treated with antibiotics.

Neurologic examination: *Cranial nerves:* Early papilledema was present. A non-congruous left homonymous hemianopia was present *(Fig. 23-14A)*.

Clinical diagnosis: Mass lesion right occipital? Tumor, abscess.

Laboratory data: *Imaging* demonstrated an enhanced lesion in the posterior and medial aspect of the right hemisphere (occipital and adjacent parietal lobes). *EEG* was abnormal because of frequent focal 3 to 4 cps, slow waves in the right occipital area and to a lesser degree, the right posterior parietal area. *Cerebrospinal fluid* pressure was elevated to 210 mm CSF, 97 lymphocytes were present (upper limit is 5 to 7 lymphocytes). Protein was increased to 75 mg/dl. Glucose was 75 mg/dl (normal when compared to blood sugar of 95 mg/dl).

Subsequent course: The patient was treated with antibiotics (cephalothin sodium, penicillin, and streptomycin), and visual fields, EEG, and spinal fluid findings improved. Within 3 weeks the field defect had resolved to a non congruous left inferior quadrantanopia *(Fig. 23-14B)*.

Ten days later the patient was readmitted to the hospital with a three-day history of right eye pain, sweats, and chills. Neurologic examination now revealed a recurrence of blurred optic-disc margins, a left homonymous hemianopia and a slight increase in deep tendon reflexes on the left. Imaging studies now revealed a large space-occupying lesion in the right parietal/occipital region, displacing the right lateral ventricle forward and downward. After 10 days treatment with antibiotics (penicillin and streptomycin), a craniotomy was performed by Dr. Bertram Selverstone revealing an abscess and a large surrounding area of hard granulomatous cortex and white matter that were removed. The etiologic organism was subsequently found to be a microaerophilic streptococcus.

Follow-up examination 6 months after surgery was normal, except for the left homonymous hemianopia *(Fig. 23-14C)*.

The case that follows is another example of the effects of a lesion in the occipital lobe with very different consequences then those seen in the previous case.

Case History 23-4. This 18-year-old left-handed single white female restaurant employee had the onset of her seizures at age 14 when she had a sequence of five seizures in less than 12 hours. Each began with flashing lights, "like Christmas tree lights," all over her visual field, plus the sensation" that peoples' faces were moving. She then would have an apparent generalized convulsive seizure. Neurologic examination, and CT scan were all reported as normal. An electroencephalogram reported occasional focal sharp and slow waves in the left hemisphere. The patient was treated with carbamazepine (Tegretol), 400 mg in the morning, and 230 mg at hour of sleep with no additional

She apparently had done quite well with no additional grand mal seizures. She had rare episodes of "fear attacks," which would last 23 to 25 minutes. The last attack had occurred 2 years ago, but as recently as two months ago, she had had one episode of flashing lights. A recent EEG was normal.

Neurologic examination: normal.
Clinical diagnosis: Seizures of focal origin left occipital.
Subsequent course: The patient did well

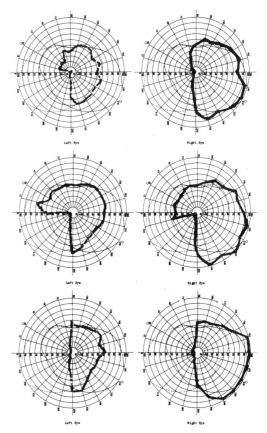

Figure 23-14. Case History 23-3, Brain abscess, right occipital area: Perimetric examination of visual fields. A, Initial examination demonstrated a somewhat asymmetrical (non congruous) left homonymous hemianopia less in the left eye than the right. The fields are shown from the patient's point of view. B, 16 days later; with antibiotic therapy, an improvement had occurred. A non congruous, quadrantanopia is now present. C, Fields after an additional 24 days. A relatively complete homonymous hemianopia, present at the time of readmission, persisted following surgery as shown above.

for 3 years then had a recurrence of a generalized convulsive seizure possibly related to omission of medication. Six weeks later, she reported two additional episodes characterized by flashing lights, then movement of the lights away from a center circle, then dizziness, then a sensation of unreality. She also had other episodes of feeling unreal accompanied by fear.

The neurologic examination demonstrated a minor right central facial weakness not present previously.

Laboratory data: *EEG:* normal. *MRI*

scan now revealed a small tumor at the left occipital pole with surrounding edema *(Fig. 23-15)*. *CT scan*, this tumor appeared partially calcified but did show enhancement. Review of the CT scan obtained in Arizona at age 14 indicated similar findings. Angiograms suggested tumor vascularity of a type seen with meningiomas.

Dr. Bernard Stone removed a discrete encapsulated tumor that appeared to be a meningioma. Subsequent microscopic examination, however, indicated a rare type of indolent follicular adenocarcinoma of the thyroid, which sometimes spreads as a single lesion to brain and remains quiescent for many years. In the postoperative period, a non-congruous right-inferior-field defect was present (partial quadrantanopia). The blood level of thyroid-stimulating hormone (TSH) was elevated. A thyroid nodule was found. A thyroidectomy was performed, and thyroid replacement medication was prescribed. No additional seizures were observed over the next 18 months. The patient subsequently developed a lumbar vertebral metastasis.

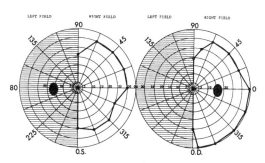

Figure 23-16. Occlusion of right posterior cerebral artery. Case 26-6 on CD ROM. Tangent screen examination of visual fields, 3 months after the acute event. This 75-year-old woman had the acute onset of headache, bilateral blindness, confusion, vomiting, mild ataxia and bilateral Babinski signs. All findings cleared except a residual left homonymous hemianopsia with macular sparing. O.S. = left eye; O.D. = right eye.

Vascular lesions of the calcarine cortex: Various types of visual field defects may occur including a homonymous hemianopsia, with *(23-16)* or without macular sparing, or a quadrantanopsia. The CT scan correlations are presented in *Figures 23-17 and 23-18*.

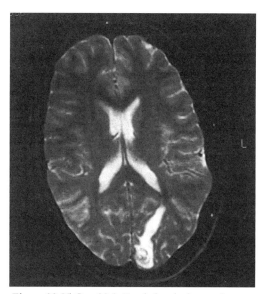

Figure 23-15 Case 23-2. Focal visual seizures characterized by "flashing lights" and the sensation of movement of the visual field with secondary generalization beginning at age 14 due to metastatic (thyroid) tumor of the left occipital lobe. MRI, T2-weighted, non-enhanced demonstrated a small tumor at the left occipital pole with surrounding edema. (See text for details.)

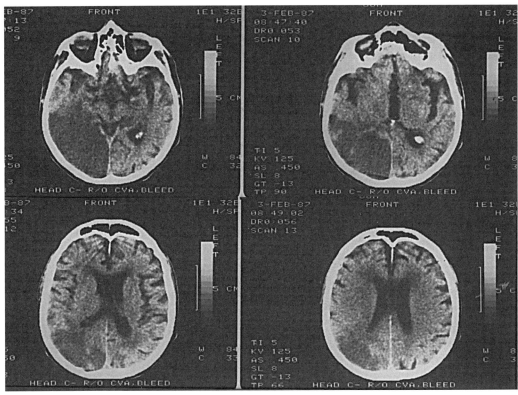

Figure 23-17 CT scan of a total infarct, presumably embolic in the right posterior cerebral artery cortical territory (occipital and posterior temporal lobes) obtained 5 days after the acute onset of confusion and possible visual hallucinations in an 86-year-old right-handed male. As confusion cleared examination indicated a dense left homonymous hemianopia with no evidence of macular sparing. The patient would look to the right and to the midline but would not follow objects to the left beyond the midline. Contrast to Fig. 23-18.

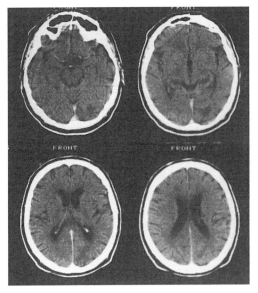

Figure 23-18. CT scans of a partial infarct of the left posterior cerebral artery territory (inferior calcarine) with superior quadrantanopia probable basilar vertebral ischemia with artery-to-artery embolic infarction of the occipital (calcarine artery). This 71-year-old right-handed male with a history of diabetes mellitus and hypertension had acute onset of vomiting, ataxia, minor confusion, and a persistent problem with peripheral vision in the right visual field, followed by episodes of vertical diplopia. Examination the following month demonstrated a homonymous visual field defect involving the right superior field, minor ataxia, and a right Babinski sign. Compare to figure 23-17.

CHAPTER 24
Speech, Language, Cerebral Dominance and the Aphasias

INTRODUCTION

In the preceding sections, we have alluded to various areas in the dominant hemisphere concerned with speech and language. The reader may well have been confused by the introduction of such terms as aphasia, apraxia, agnosia and dyslexia. It is well to warn the student beginning the study of language function that prior to the development of modern neuroimaging this had been an area of much confusion, with much disagreement and multiple hypotheses. This discussion will be limited to the more practical problems of anatomical localization.

DYSARTHRIA

We should indicate at the onset that we are not concerned here with the problem of dysarthria. Dysarthria refers to a difficulty in articulation of speech from weakness or paralysis or from mechanical difficulties. Dysarthria may have many causes at many levels of the peripheral or central nervous system as indicated in Table 24-1.

CEREBRAL CORTEX AND COMPLEX DISTURBANCES OF VERBAL EXPRESSION

There are, however, more complex disturbances of regards verbal expression, that occur at a time when the basic motor and sensory systems for articulation are intact. Similarly there are complex disturbances in language function, as regards comprehension of written and spoken symbolic forms that occur at a time when the basic auditory and visual receptor apparatus is intact. We refer to these more complex acquired disturbances of language function as aphasias. In general, it is possible to relate these language disturbances to disease of the dominant cerebral hemisphere, involving the cerebral cortex or cortical association fiber systems. The term dysphasia, which is used in England interchangeably with aphasias, is used

TABLE 24-1 CAUSES OF DYSARTHRIA

LOCATION OF PATHOLOGY	EXAMPLES
Local diseases of the larynx, tongue or lips	Laryngeal carcinoma
Disease of the muscles affecting the tongue, the lips or pharynx,	Oculopharyngeal dystrophy
Diseases affecting the neuromuscular junction in the muscles of the-tongue, lips and pharynx	Myasthenia gravis, botulinum toxin
Diseases affecting the peripheral nerve supply of the tongue, lips, or pharynx,	A) Hypoglossal nerve (carotid endarterectomy). B) Recurrent laryngeal nerve (thyroid surgery). C) Metastatic or regional cancer involving the lymphatics in the neck. D) Diphtheritic neuropathy
Diseases of the brain stem motor nuclei particularly nucleus ambiguus.	Lateral medullary infarcts, amyotrophic lateral sclerosis (so called progressive bulbar palsy), or bulbar poliomyelitis.
Bilateral damage to the corticobulbar fibers (cerebral cortex, internal capsule, or upper brain stem).	Bilateral infarcts, multiple sclerosis producing a pseudobulbar palsy.

in the United States in reference to developmental language disorders as opposed to the acquired aphasias.

CEREBRAL DOMINANCE

It is perhaps appropriate at this point to consider the question of cerebral dominance. Most individuals are right-handed (93 per cent of the adult population), and such individuals almost always (>99%) are left-hemisphere dominant for language functions. A minority of individuals (some on a hereditary basis) is left-hand dominant. Baseball appears to have collected a high percentage of these individuals as left-handed pitchers. Some left-handers are right-hemisphere dominant (50%) but a certain proportion is left-hemisphere dominant. It has been estimated that 96 percent of the adult population are left-hemisphere dominant for speech.

Hand, eye and foot preference occurs in monkeys and limb preference occurs in cats (Cole 1957). One might inquire as to why the majority of humans are right-handed. Although hand preference does not become apparent until one year of age, the studies of Yakolev and Rakic (1966) would suggest that even before this age, there is an underlying anatomical basis for the dominance of the right hand. In the study of the medullae and spinal cords of a large number of human fetuses and neonates, the fibers of the left pyramid were found to cross to the right side in the medullary pyramidal decussation at a higher level in the decussation than fibers of the right pyramid. Moreover, more fibers of the left pyramid crossed to the right side than vice versa. Although the majority of pyramidal fibers decussated, a minority of fibers remained uncrossed. It was more common for fibers from the left pyramid to decussate completely to the right side of the lower medulla (and eventually spinal cord. The minority of fibers that remained uncrossed was more often those descending from the right pyramid into the right side of the lower medulla. The end result, in the cervical region at least, was for the right side of the spinal cord and presumably the anterior horn cells of the right side of the cervical cord to receive the greater corticospinal innervation. There is then an anatomical basis for the preference or dominance of the right hand. Since the majority of fibers supplying the right cervical area have originated in the left hemisphere, on this basis alone one could refer to the dominance of the left hemisphere. Study of the adult brain results in similar conclusions (Kertesz & Geschwind, 1971).

The studies of Geschwind and Levitsky (1968) established that an actual anatomical asymmetry was present in the adult brain between the two hemispheres in an area significant for language function. That area of the auditory association cortex, posterior to Heschl's gyrus on the superior-lateral surface of the temporal lobe (areas 22 and 42) bordering the Sylvian fissure and including the speech reception area of Wernicke, was found to be larger in the left hemisphere. Similar differences have been found in the brain of the fetus and newborn infant (Wada et al 1975). The studies of LeMay and her associates (1972, 1978, and 1978), Galaburda et al, (1978), Chui and Damasio (1980) elaborated on the asymmetries of the Sylvian fissure and occipital lobe demonstrated on arteriography and CT scan during life. Similar asymmetries were demonstrated in the brains of the great apes, e.g., chimpanzee, but not in the Rhesus monkey (LeMay and Geschwind 1978). LeMay and Culebras (1972) studied the endocranial casts of human fossil skulls in which imprints of the major fissure could be noted and demonstrated similar differences in Neanderthal man. Damasio and Geschwind (1984) provide additional review of this topic.

Development Aspects:

It is clear that from a functional standpoint a considerable degree of flexibility exists in the child as regards dominance for language function. Thus, in a child under the age of 5 years, destruction of the speech areas of the dominant left hemisphere during pre or postnatal life produces only a minor long-term language disturbance (Varqha-Khademe at al 1985). If the damage to the speech areas in the dominant hemisphere occurs between the ages of 5 and 12 years a more severe aphasia occurs; however some limited recovery of language may occur within a year. These cases suggest an equipotentiality of the two hemispheres for language function during the early years of development. In these cases under the age of 5 years the right hemisphere must assume a dominant role in mediating those learned associations important in language and in the use of symbols either through process compensation and/or reorganization. If total destruction of the speech area of the dominant hemisphere occurs in adult life only a minor recovery of language functions will occur. Patients with left hemisphere damage sustained under the age of 5, develop a strong left-hand preference. Patients with damage sustained after the age of 5 have only weak left-hand preferences or are ambidextrous. Additional discussion of the role of genetics in dominance and of the effects of early damage to the left hemisphere can be found in Dellatollas et al, (1993).

Normal development of speech in the child: There is considerable variability. Milestones may be outlined as follows: 9-12 months—babbles, "mama, dada". At 12 months single words utilized and echoes sounds. At 15 months has several words, at 18 months has vocabulary of 6 words and possible hand dominance. At 24 months puts 2-3 words together into a sentence.

APHASIA

In approaching the patient with aphasia it is well to keep in mind that textbook discussions of this problem are often artificial, in the sense that such discussions tend to deal with relatively isolated pure types of language disturbance, which have relatively specific localization. The actual patient with aphasia more often presents a mixed disturbance with damage to several areas; some have damage to all major areas in a global aphasia. This is not unexpected when one realizes that all of the major speech areas of the dominant hemisphere are within the cortical vascular territory of the middle cerebral artery. Other areas involved in related disorders are supplied by the anterior cerebral artery (the superior speech area) or by the posterior cerebral artery, (those areas involved in pure dyslexia and agnosias). Willmes and Poeck (1993) and Kirshner (2000) provide additional reviews.

Three cortical areas of the dominant hemisphere are of major importance in language disturbances *(Fig.24-1):*

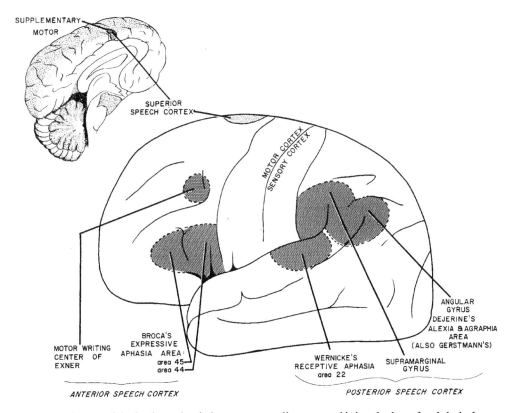

Figure 24-1. Speech areas of the dominant hemisphere, summary diagram combining the data of pathological lesions and stimulation studies. Precise sharp borders are not implied. The designations of Penfield and Roberts 1959 the terms "anterior speech area", "posterior speech area", and "superior speech cortex".

(1) *Broca's motor aphasia or expressive speech center:* The opercular and triangular portions of the inferior frontal convolution (areas 44 and 45).

(2) Wernicke's receptive aphasia area: The auditory association area on the superior and lateral surface of the posterior portion of the superior temporal gyrus (area 22 and adjacent parts of area 42).

(3) Inferior parietal lobule: The angular and supra- marginal gyri (area 39 and 40), at times associated with Gerstmann's syndrome of dysgraphia, dyscalculia, left-right confusion and finger agnosia.

Other cortical areas: the supplementary motor cortex (superior speech area) may also play a role in language function. Stimulation studies by Luders et al, 1988, have also identified a *basal temporal speech area.*

In addition to these cortical areas, the *association fiber systems* relating these areas to each other and to other cortical areas are of considerable importance for certain types of language disturbance, as we will indicate later.

From a practical localization standpoint we can, in a general sense, speak of patients as presenting an anterior or posterior type of aphasia (anterior or posterior to the central sulcus) *(Fig.24-1).*

1. The anterior type of aphasia relates to Broca's motor aphasia area. The patient's speech may be generally described as nonfluent with little spontaneous verbalization.

2. The patients with the posterior types of aphasia are fluent and do speak spontaneously. The lesions in such cases involve Wernicke's area in the posterior portion of the superior temporal gyrus or the inferior parietal lobule. The fluent aphasias include (a) Wernicke's receptive aphasia, (b) disconnection syndromes, (c) Gerstmann's syndrome, (d) dyslexia plus, e) dyslexia without agraphia, (f) the agnosias. However since the posterior areas are relatively close to one another it is not unusual for a single disease process to involve both areas.

Now let us consider in greater detail language functions and the postulated anatomical correlations of these various language centers.

In these correlations, it is important to make a distinction between acute and chronic effects

It is necessary first to review those components of language function that are evaluated in the neurological examination (adapted from Benson 1985).

(1) Conversational speech:

a. Amount: Nonfluent versus fluent. Mute refers to total absence.

b. Articulation, rhythm and melody, defective in nonfluent, normal in fluent. However note that the nonverbal aspects of language prosody, (the affective, intonation and melodic aspects of speech) and gesture often involve the right hemisphere (refer to Ross, 1993)

c. Content:

1. Grammar - (referred to as syntax): Nonfluent speech is often telegraphic - lacking the words or word parts and endings needed to express grammatical relations; whereas, patients with fluent aphasia can produce long sentences of phrases with a normal grammatical structure.

2. Meaning and substance - (referred to as semantics): Patients with fluent aphasia often demonstrate a number of abnormalities in terms of the choice of words.

a. The precise word may be replaced by a nonspecific work or phrase.

b. The correct word or phrase answer may be replaced by a phrase that describes the use of an object (circumlocution). Because such patients are fluent, casual examination may fail to detect such glib features.

c. The correct word may be replaced by an inappropriate word or phrase substitution (verbal or semantic paraphasias), e.g., "clocks for wristwatch".

d. The correct word may contain replacement sounds or syllables, which are incorrect (literal or phonemic paraphasias) e.g., "note for nose".

e. The correct word may be replaced by a totally new word (neologism), e.g. "Zingbam".

(2) Repetition: the ability to repeat words or phrases

a. This may be absent in a mute individual, in a severe nonfluent patient or in a patient with a severe type of Wernicke's aphasia.

b. This may be impaired as an isolated symptom (so-called conduction aphasia)

c. This may be preserved in an individual with other wise intact language

Or preserved in isolation (so-called Sylvian isolation syndrome).

(3) Comprehension of spoken language is tested by providing spoken or written commands that do not require a verbal response, such as "hold up your left hand, close your eyes, and stick out your tongue".

4. Word finding and selection: Naming of objects or selection of the proper name of an object. Defect is defined as anomia or nominal aphasia.

5. Reading: Defect is defined as alexia or dyslexia.

6. Writing: Defect is defined as agraphia or dysgraphia.

7. Related functions:

a. Calculations: Defect defined as acalculia or dyscalculia.

b. Perception in visual, auditory or tactile sphere: defects in recognition when the sensory modality and naming are otherwise intact are defined as agnosia. The main sensory modality involved is vision and this merges with optic aphasia, dyslexia and nominal aphasia. The inability to recognize faces (prosopagnosia) and color agnosia are special examples (De Renzi, 2000).

c. Apraxias: Defects in use of objects when motor and sensory functions are otherwise intact.

STIMULATION OF THE SPEECH AREAS

In general, as demonstrated in *Figure 24-2*, stimulation of the dominant hemisphere - speech areas - produces arrest of speech. ("Aphasic Arrest" in the terminology of Penfield and Rasmussen). Paroxysmal arrest of speech often occurs as one aspect of seizures involving the dominant hemisphere. More selective aspects of aphasia such as ictal alexia or agraphia or anomia without speech arrest have been reported in rare patients - (Ardila and Lopez, 1988, Warner, 1988) with the appro-

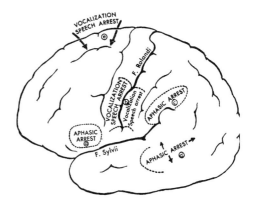

Figure 24-2. Arrest of speech on stimulation of dominant hemisphere at surgery. (From Penfield, W., and Rasmussen, T.: The Cerebral Cortex of Man. New York, The Macmillan Company, 1955, p.107.)

priate localization as noted below. Luders' et al 1988 and Dinner and Luders (2000) have recently reviewed this subject with the following conclusions:

(1) Speech arrest or interference was elicited in five areas:

a) Primary motor cortex when stimulating the representation of muscles involved in speech of the dominant or non-dominant hemisphere;

b) Negative motor areas (suppressor areas), that is the supplementary motor areas, and inferior frontal gyrus of the dominant hemisphere;

c) Broca's area - dominant hemisphere only; (note that Broca's area overlaps the motor suppressor area of the inferior frontal gyrus,

d) Wernicke's area - dominant hemisphere only;

e) Basal temporal language area - dominant hemisphere only. Dinner and Luders concluded that any effect of stimulating the superior speech area was related to inhibition of motor activity. This area did not appear to have speech function per se. (Ablation type studies do suggest possible speech functions).

(2) In stimulation of Broca's, Wernicke's and the basal temporal areas at high stimulus intensities, complete speech arrest occurred with a global receptive and expressive aphasia; however, the patient could still perform non-verbal tasks. At lower intensities of stimulation,

the effects were incomplete: speech was slow, simple but not complex tasks could be performed (both motor sequences and calculations). Simple but not complex material could be repeated. A severe naming defect (anomia) still occurred.

In children some seizures may be associated with long-lasting expressive aphasia (Landau-Kleffner Syndrome) (Sawhney et al, 1988).

ABLATION OF SPEECH AREAS: ANATOMICAL CORRELATION OF SPECIFIC SYNDROMES

The nonfluent aphasias involving Broca's area: In 1861, Broca described patients with nonfluent aphasia: little spontaneous speech with poor repetition but with considerable preservation of auditory comprehension. The spontaneous speech that remained was agrammatic and telegraphic. Based on his examination of the lateral surface of the cerebral hemispheres at autopsy, he emphasized the damage to the third left frontal gyrus, the frontal operculum of the inferior frontal gyrus. Although considerable controversy was generated at the time, Broca's name has continued to be associated both with the disorder, nonfluent aphasia, and with the postulated anatomical area. As discussed by Damasio and Geschwind, (1984) subsequent CT scans of the museum specimen of Broca's case indicated much more extensive damage than that described by Broca. Damasio, (1992) in an extensive recent review of aphasia discusses in detail the differences between true chronic Broca's aphasia and the more limited and transient effects of lesions restricted to Broca's area alone.

Broca's area is essentially a continuation of premotor cortex and may be considered a specialized motor association area with regard to the tongue, lips, pharynx, and larynx. This area is adjacent to the motor cortex representation of the face, lips, tongue and pharyngeal muscles. It is not, therefore, unusual for these patients to have also a supranuclear type weakness of the right side of the face and a right hemiparesis. This by itself is not sufficient to explain the problems in speech, which these patients manifest, since lesions involving the face area of the nondominant motor cortex do not produce the speech disturbance. Moreover, tongue and pharynx receive a bilateral corticobulbar supply, so that a unilateral cortical lesion involving the motor cortex in this area is not a sufficient explanation. These patients appear to have lost the motor memories for the sequencing of skilled coordinated movements of tongue, lips and pharynx that are required for the vocalization of understandable single words, phrases and sentences. They are usually able to vocalize sounds. At times, they may be able to vocalize single words into a proper grammatical sequence for telegraphic sentences.

In the patient with a relatively pure form of Broca's motor aphasia, the capacity to formulate language and to select words in a mental sense is intact. The student might logically inquire as to whether the patient has also lost his motor memories or motor associations for making a nonverbal sequence of movements of the tongue and lips. Frequently this is the case, as will be demonstrated in the illustrative case history. We may refer to this defect in skilled sequential motor function (at a time when motor, sensory and cerebellar function are intact, and when the patient is alert and understands what movements are to be performed) as a motor apraxia. There are several types of apraxia with different localization. In general, we may relate the more purely motor forms of apraxia to the motor association areas of the premotor cortex. One could then refer to patients with Broca's motor aphasia as patients with a motor apraxia of speech.

The following presents an example of this type of aphasia. An embolus to the superior division of the left middle cerebral artery produced marked weakness of the face tongue and distal right upper extremity and a severe selective nonfluent aphasia. The hand weakness rapidly disappeared, the right central facial weakness, an apraxia of tongue movements and some expressive difficulties persisted. Most likely, the embolus fragmented and passed into the cortical branches supplying Broca's area.

Case 24-1: This 55-year-old, right-handed white housewife while working in her garden at 10:00 a.m. on the day of admission suddenly developed a weakness of the right side of face and the right arm and was unable to speak. Nine years previously the patient had been hospitalized with a three-year history of progressive congestive heart failure secondary to rheumatic heart disease (mitral stenosis with atrial fibrillation). An open cardiotomy was performed with a mitral valvuloplasty (the valve opening was enlarged) and removal of a thrombus found in the left atrial appendage. Postoperatively, the patient was given an anticoagulant, bishydroxycoumarin (Dicumarol), to prevent additional emboli since atrial fibrillation continued until the present admission. She had done well in the interim. However, prothrombin time at the time of admission was close to normal; that is, the patient was not in a therapeutic range for anticoagulation.

General physical examination: Blood pressure was moderately elevated to 170/80. Pulse was 110 and irregular (atrial fibrillation). Examination of the heart revealed the findings of mitral stenosis: a loud first sound, a diastolic rumble and an opening snap at the apex as well as the atrial fibrillation.

Neurological examination: *Mental status and language function:* She had no spontaneous speech, could not use speech to answer questions and could not repeat words. She was able to indicate answers to questions by nodding (yes) or shaking her head (no), if questions were presented in a multiple-choice format. In this manner it was possible to determine that she was grossly oriented for time, place and person. She was able to carry out spoken commands and simple written commands, such as, "hold up your hand, close your eyes." However, she had significant difficulty in performing voluntary tongue moments (such as "wiggle your tongue"; "stick out your tongue") on command. *Cranial nerves:* She tended to neglect stimuli in the right visual field and was unable to look to the right on command although the head and eyes were not grossly deviated to the left at rest. A marked right supranuclear (central) type facial weakness was present. The patient had difficulty in tongue protrusion and was unable to wiggle the tongue. When the tongue was protruded it deviated to the right. *Motor system:* there was a marked weakness without spasticity in the right upper extremity, most prominent distally and a minor degree of weakness involving the right lower extremity. *Reflexes:* deep tendon stretch reflexes were increased on the right in the arm and leg. Plantar responses were both extensor, more prominent on the right. *Sensory system:* Intact within limits of testing.

Clinical diagnosis: Broca's: anterior aphasia embolus from heart to left middle cerebral artery, superior division.

Laboratory data: *EEG* was normal (48 hours after admission). *Electrocardiogram* revealed atrial fibrillation. *CSF* was normal

Subsequent course: Within 24 hours a significant return of strength in the right hand had occurred; independent finger movements could be made. Within 48 hours after admission the patient was able to repeat single words but still had almost no spontaneous speech. She appeared aware of her speech disability and would manifest some frustration. She was able to carry out two- or three-stage commands although tongue movements and perseveration remained a problem. Within 6 days of admission, the patient used words, phrases and occasional short sentences spontaneously and was better able to repeat short sentences. At this time strength in the right arm had returned to normal, but a right central facial weakness was still present. Two weeks after admission, she still had expressive disabilities consisting of word-finding and apraxic components in tongue placement and alternating tongue movements. Although complete sentences were used, sentence formulation in spontaneous speech was slow and labored with word finding difficulties. Repetition was better performed. In reading sentences aloud, substitutions or word omissions were made. The patient did well in naming common pictures and in matching printed words to spoken words or printed words to pictures. She could write from dictation and would often respond preferentially in writing when difficulty in speaking was encountered. A right central facial weakness was still present.

With the passage of time, this type of patient would continue to show some degree of improvement (Mohr, 1973). Improvement could continue to occur over a two-year period. Although lesions of Broca's area may produce the acute onset of a nonfluent aphasia, rapid amelioration of the deficit occurs even when the acute lesion involves underlying white matter, as well as the superficial cortex.

In order to produce a more persistent severe nonfluent aphasia with good comprehension but poor repetition, more extensive lesions are required. In the CT scan correlation study of Naesar, (1983), this larger lesion extended from Broca's area to the anterior parietal lobe usually including the deep structures such as caudate, putamen and/or internal capsule. Ludlow et al (1986) compared persistent versus nonpersistent nonfluent aphasia fifteen years after penetrating head injuries of the left hemisphere in the Vietnam War. Both groups had nonfluent aphasia still present at six months after injury with equal involvement of Broca's area. The group that failed to recover at 15 years had a more extensive left hemisphere lesion with posterior extension of the CT scan lesion into Wernicke's area and some involvement of the underlying white matter and basal ganglia. The additional study of Naeser et al (1989 compared severe nonfluency to less severe Broca's aphasia - 6 months to 9 years following an infarct of the left cerebral hemisphere. Those patients with severe nonfluency had more extensive lesions of the subcortical white matter involving: a) the subcallosal fasciculus deep to Broca's area containing projections from the cingulate and supplementary motor areas to the caudate nucleus plus b) the periventricular white matter near the body of the left ventricle deep to the lower motor-sensory cortical area for the mouth. The MRI of a case (24-2) with chronic residuals of a more persistent nonfluent aphasia is presented in *Figure 24-3*.

Case 24-2: On May 30, 1985, at midday, this 37-year-old right-handed woman with a past history of rheumatic fever, rheumatic heart

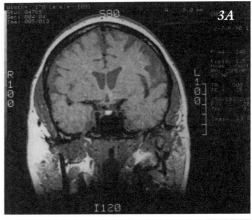

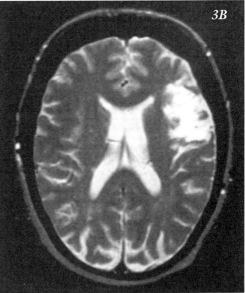

Figure. 24-3. Anterior aphasia: embolic infarct (secondary to rheumatic atrial fibrillation. Case 24-2 (refer to text) MRI: A) T1 weighted, coronal section. B) T2 weighted, horizontal section.

disease and "an irregular pulse" (atrial fibrillation) had the sudden onset of loss of speech, central weakness of right face, an inability to protrude the tongue and a right hemiparesis. Although hemiparesis, face and tongue problems disappeared within a week, recovery of speech was slower and limited. At one week, she could produce some two-syllable words.

Neurological examination at 17 months: *Language function:* Spontaneous speech was slow and relatively scanty. She could name objects slowly without difficulty, could do some simple repetitions and could write from dictation. She could carry out two and three stage commands. She could read slowly aloud but had little comprehension of what she read. *Reflexes:* A residual right Babinski sign and right-sided hyperreflexia were present.

Clinical diagnosis Anterior aphasia: embolic infarct of superior division of middle cerebral artery.

Laboratory data: Although initial CT scans in May 1985 had been reported as negative; the MRIs in November 1986 and June 1990 demonstrated the infarct, which included Broca's area, as well as adjacent frontal areas with predominant involvement of frontal operculum (inferior frontal gyrus) and adjacent middle frontal gyrus. In addition, a minor independent infarct was present in right occipital area (presumably embolic to the calcarine artery).

Subsequent course: Similar findings were present at 5 years, June 1990).

A brain section from a patient (Case 24-3) with a persistent nonfluent aphasia (is presented in *Figure 24-4*.

Case 24-3: This 69-year-old, right-handed male expired in 1965. In 1961 and 1963, the patient, had experienced transient episodes of right hemiparesis and aphasia, both followed by complete recovery. In 1964, a more persistent right hemiparesis and aphasia had developed.

General physical examination: No left carotid pulsation was present.

Neurologic examination (1965): *Language functions:* Speech was nonfluent with only a small verbal output. A significant nominal aphasia with paraphasias was present. Reading was limited. Comprehension of spoken language was good. A marked apraxia was present for movements of tongue, lips and hands. *Motor system:* Right hemiparesis with associated reflex findings.

Pathologic diagnosis: Left carotid occlusion with old infarction, predominantly of left middle cerebral territory with lesser involvement of anterior cerebral territory Note that minor cortical infarcts are also present involving cortex of right hemisphere. (Courtesy of Dr. John Hills,; and Dr. Jose Segarra.)

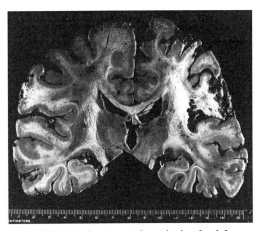

Figure 24-4. Persistent anterior aphasia after left carotid occlusion with old infarction, predominantly of left middle cerebral territory with lesser involvement of anterior cerebral territory. Case 24-3(refer to text). (Courtesy of Dr. John Hills and Dr. Jose Segarra)

The Fluent Aphasias

Wernicke's aphasia and Wernicke's area: In 1874, Wernicke described a type of aphasia that differed significantly from the nonfluent aphasia described by Broca and that had a more posterior localization. This type of patient has fluent spontaneous speech with poor comprehension and poor repetition. The patient has difficulty in understanding symbolic sounds, i.e., words that have been heard and is unable to carry out verbal commands. The patient is unable to use those same auditory associations in the formulation of speech. Moreover, the patient is essentially unable to monitor his own spoken words as he talks, since he lacks the ability to interpret and to compare the sounds, which he himself is producing to previous auditory associations. The end result is a patient who is fluent but who uses words combined into sentences that lack any meaning to the listeners. Word substitution (paraphasia) is frequent. In severe cases, not only are phrases and words combined into meaningless sentences but also the patient combines syllables into words that have no meaning (jargon aphasia). The patient, however, is usually unaware of his errors. The patient usually fails to show the "rationale" frustration, which is characteristic of patients who have

Broca's motor aphasia and who are aware of their errors. At times, the patient shows some awareness that his verbal responses are failing to deal with the environment. At times, the problems in communication are so severe that the patient becomes agitated and is mistaken for psychotic, as demonstrated in the case history below. The anatomical localization for Wernicke's aphasia based on early autopsy studies has been confirmed by subsequent CT scan and MRI studies: the posterior temporal-parietal area. In the studies of Naeser (1983) the essential lesions involved Brodmann's Area 22 (Wernicke area) of the posterior-superior temporal gyrus and the adjacent supramarginal gyrus of the inferior parietal area.

In general, the responsible lesion is usually an infarct within the territory of the inferior division of the left middle cerebral artery often embolic. There is usually very little evidence of a hemiparesis. In contrast, Broca's aphasia usually reflects infarction within the territory of the superior division of the left middle cerebral artery and a right facial weakness and right hemiparesis are often present. (See Chapter 26 for additional discussion of vascular anatomy.)

Case 24-4 provides an example of a patient with a Wernicke's type aphasia

Case 24-4: This 40-year-old, right-handed white male had experienced a series of myocardial infarctions beginning 10 months prior to consultation. Angiographic studies indicated occlusion of the right coronary artery and the left anterior descending coronary artery with stenosis of the circumflex artery and a coronary artery bypass procedure was performed using extracorporeal circulation. The anesthetist noted that the patient was nonreactive 15minutes after the last administration of halothane anesthesia. When the patient did wake up, he was quite agitated, combative and in retrospect had transient difficulty using the right arm. and he continued to behave despite medications throughout the postoperative period, in a disoriented and combative manner. Transfer to the psychiatry service was actively planned but a neurology consultation was obtained.

Neurological examination: *Mental status:* He was alert but agitated. Speech was fluent and usually grammatical. He made use of many paraphasias and neologisms (nonsense words). He often appeared unaware of his errors. He was unable to name single objects such as pencil, cup, and spoon though he could recognize and demonstrate their use. He could read occasional individual words; for example, cat, house. He could not comprehend or carry out, either spoken or written commands. His attempts at spontaneous writing produced only a few letters. He was unable to write words or phrases from dictation. He did significantly better when asked to copy a printed sentence or phrase. Repetition of spoken phrases or sentences was poor. Calculations of even a simple nature were poorly performed. *Reflexes:* A minor increase in deep tendon stretch reflexes was present on the right side. The plantar response on the right was equivocal. The left was flexor.

Clinical diagnosis: Wernicke's type of fluent aphasia probably secondary to an embolus to the inferior division of the middle cerebral artery.

Pure word deafness: We have indicated previously in discussing the auditory cortex, that this is a related relatively rare problem (Auerbach et al 1982 and Coslett et al 1984). In cases of pure word deafness, the hypothetical center of auditory word association (Wernicke's area) appear to be intact but auditory information is unable to reach this center from the auditory projection cortex (of either hemisphere); in a sense disconnection has occurred. The auditory word association center is intact. Previous auditory associations may be used in the formulation of the patient's own speech. Input of visual information to the center may also be intact. It should be evident that the distinctions that have been made here between Wernicke's receptive aphasia and word deafness are to be some extent artificial. In general, lesions in this area do not respect such artificial distinctions and often destroy not only the center for auditory word associations but also the inputs into this area from the auditory projection areas.

Both Wernicke's receptive aphasia and word deafness are often grouped under the general term auditory agnosia.

Conduction or repetition type fluent aphasia: This type of aphasia is characterized by a relatively fluent spontaneous speech but with marked deficits in repetition. Literal paraphasias are present and word-finding problems may be present. The patient is usually aware of the errors made. In contrast to the Wernicke's type of aphasia, auditory comprehension is usually relatively well preserved. The anatomical location of the lesion is somewhat variable (Geschwind, 1965, Benson et al 1973, Damasio and Damasio 1980, Demasio 1992, Nausea 1983, Damasio and Geschwind 1984). The lesion essentially serves to disconnect Wernicke's area from Broca's area (the posterior from the anterior speech areas). The essential fiber system is the arcuate fasciculus *(Fig.24-5).* This fiber system arches around the posterior end of the Sylvian fissure to join the superior longitudinal fasciculus. In the study of Naeser (1983) this system was involved deep to the supramarginal gyrus and/or deep to Wernicke's area. In other cases the white matter deep to insular cortex has been involved (Damasio and Geschwind 1984). In the resultant conduction aphasia, the patient understands spoken commands and is usually able to carry out complex instructions that do not require a repetition of language. Although able to speak spontaneously with little or no evidence of a Broca's aphasia, the patient is unable to repeat test phrases and sentences, e.g., "The rain in Spain falls mainly on the plane" or "no ifs, ands or buts" and is unable to write from dictation. Often, the syndrome is seen in a less than pure form as in the case history that accompanies Figure 24-6. This is not surprising since the involved territory is that of the middle cerebral artery predominantly the inferior division. The essential lesion must spare Broca's and Wernicke's area to a considerable degree but significantly damage the interconnection of these areas. If there is significant damage to either Wernicke's or Broca's areas, repetition will be defective on either the input or output side of the loop. A case demonstrating conduction aphasia will be presented as part of the next section.

Fluent aphasias associated with lesions of the dominant inferior parietal areas: angular and supramarginal gyri: Geschwind (1965) indicated the critical location of the inferior parietal lobule in man situated between the visual, auditory and tactile association areas. As such, it may act as a higher association area between these adjacent sensory association areas. In earlier chapters; we have already identified this area as the posterior parietal, multimodal sensory association area. Geschwind has suggested that correlated with the development of this area in the human (the angular and supramarginal gyri cannot be recognized as such in the monkey and are present only in rudimentary form in the higher apes), there has developed the capacity for cross modality sensory-sensory associations without reference to the limbic system. He contrasts this situation with that in subhuman forms where the only readily established sensory-sensory associations are those between a non limbic (visual, tactile or auditory) stimulus and a limbic stimulus: those stimuli related to primary motivations such as hunger, thirst, sex. (Such considerations of course do not rule out the establishment of limbic-non limbic stimuli associations in man, particularly during infancy and childhood.) These theoretical concepts would suggest the underlying basis for man's language functions. It is not surprising then

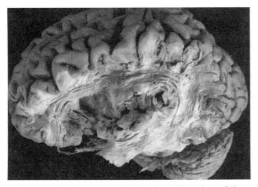

Figure 24-5. The arcuate fasciculus: dissection of the long fiber system, the superior longitudinal fasciculus demonstrating the arcuate fasciculus, passing beneath the cortex of inferior parietal, and posterior temporal areas. A small bundle of U fibers are also demonstrated incidentally at posterior, inferior margin of the temporal lobe interconnecting adjacent gyri.

that those aspects of language function which are most dependent on association between auditory and visual stimuli and auditory-visual-tactile stimuli (reading and writing) are most disturbed by lesions of angular and supramarginal gyri. In the words of Geschwind, "It is a region which turns written language into spoken language and vice versa."

The component parts of the resultant syndrome are not constant. The patient will have a varying degree of difficulty in writing (dysgraphia), particularly as regards spontaneous writing. Often the ability to copy letters, words and sentences will be relatively well preserved. In contrast the patient will usually have difficulty writing from dictation, although at times, this function may be less affected than spontaneous writing. There is often an associated deficit in reading (dyslexia) although this is not invariable.

Significant defects in spelling and in calculations (dyscalculia) may be present. There may be an associated left-right confusion. Errors may be made in finger identification, with confusion as to index, ring and middle fingers (finger agnosia). The combination of dysgraphia, dyscalculia, finger agnosia and left right confusion is identified as Gerstmann's Syndrome.

From the standpoint of localization, several points of caution should be indicated. Dysgraphia may also result from disease involving the premotor cortex anterior to the motor cortex representation of the hand. It is likely that information is conveyed from the inferior parietal area to the premotor motor association cortex and then to the precentral motor cortex. This area of premotor cortex (sometimes termed the writing center of Exner) must function in a manner analogous to Broca's area. It would be reasonable to consider dysgraphia from a lesion in this premotor location as essentially a motor apraxia due to destruction of an area concerned in the motor association memories for writing.

Pure dyslexias may occur without actual involvement of the angular supramarginal areas. The basic lesion as we will indicate, is a lesion that deprives the inferior parietal cortex of information from the visual association areas.

Many lesions in the inferior parietal area extend into the subcortical white matter involving the long association fiber system, the arcuate fasciculus - as discussed above.

Patients with dominant inferior parietal lobule lesions often demonstrate problems in drawing (constructional drawing). Whether this reflects damage to the cortex of the angular and supramarginal gyri or a disconnection of these areas from the more anterior motor areas because of involvement of subcortical association fibers is usually uncertain.

As we have already discussed, many patients have involvement of all of the components of the posterior speech areas, the inferior parietal, the arcuate fasciculus, and Wernicke's area. The following case history demonstrates such a sequence of involvement in a tumor infiltrating this area with greatest difficulty in repetition as the most prominent initial feature.

Case 24-5: This 42-year-old right-handed truck driver noted fatigue and irritability 6-8 weeks prior to admission. Three to four weeks prior to admission, he noted he was using words which he did not mean to use. Certain words would not come to him. During the week prior to admission, he had two generalized convulsive seizures; each proceeded by a ringing sensation in the ears.

Neurological examination: *Mental status:* The patient was alert and oriented to time, place and person cooperative and able to carry out the four-step command: "stick out your tongue, close your eyes, hold up your hand and touch the thumb to your ear". There was however evidence of a significant left-right confusion when laterality was introduced e.g. "left hand to right ear". The ability to do even simple calculations was markedly impaired with or without paper e.g.100-9=99. There was little evidence of an expressive aphasia. Flow of speech was slow with only minor mispronunciations. Reading was slow but with few errors. The patient's did have minor difficulty in naming objects-a mild nominal aphasia. The patient's greatest difficulty was in repetition of simple test phrases. There were moderate defects in drawings of a house and a clock but few errors in copying simple figures. There was a significant dysgraphia with marked difficulty in writing a simple sentence spontaneously or in writing from dictation. On the other hand, the patient was much better able to copy a simple sentence. Significant errors were made in spelling. Memory was impaired; with immediate recall object recall limited to 2/5 objects and delayed recall in 5 minutes limited to 0/5. Digit span was limited to 2 forward and 0 in reverse. *Cranial nerves:* A mild right central facial weakness was present. *Sensory system:* Tactile localization was impaired on the right side.

Clinical Diagnosis: Fluent posterior aphasia, disconnection syndrome plus elements of Gerstmann's syndrome (left right confusion, dyscalculia, dysgraphia) plus seizures of focal origin (Heschl's gyrus posterior temporal lobe) with secondary generalization. All suggested a lesion in the left posterior temporal-posterior parietal area

Etiology was uncertain but tumor (glioma) was considered most likely in view of the evolving course.

Laboratory data: *EEG:* Focal 4-5 Hz slow wave activity was present in the left parietal and posterior temporal areas. *Brain scan (radioactive Hg 197):* Increased up take of isotope in the left posterior temporal –parietal area measuring 4x5 cm.

Left carotid arteriogram: A vascular mass was present in the region of the angular gyrus with tumor stain in the area.

Subsequent course: Papilledema soon developed. Dr.Robert Yuan performed a craniotomy which revealed a palpable firm mass slightly above and posterior to the angular gyrus. At 2 cm below the surface, firm yellow tumor tissue was encountered. All visible tumor and adjacent cortex of the angular and supramarginal gyri were removed and a partial left temporal lobectomy was performed. Histologic examination indicated a highly malignant glial tumor with active mitosis and necrosis (a glioblastoma). Post operatively, the patient had a marked expressive and receptive aphasia. By 30 days after surgery, the patient had a significant improvement in fluency, he continued to have a severe posterior type of aphasia. He was unable to carry out simple commands. His speech was incoherent, with nonsense words and neologisms. He appeared unaware of his errors and did not appear frustrated by his failure to follow commands. In addition to this Wernicke's type aphasia, the patient was unable to repeat words of two syllables. He could do no calculations, had severe difficulties in copying drawings and was unable to even write his name. The patient received radiotherapy but his condition continued to worsen with the development of a progressive expressive and receptive aphasia and hemiparesis all suggesting the spread of a rapidly growing glioblastoma. Coma, pupillary changes and respiratory changes then intervened suggesting the effects of herniation of the residual temporal lobe. Death occurred 7 months after onset of symptoms, and 5 months following surgery.

The neurosurgical approach to the management of a glioblastoma particularly those involving the speech areas of the dominant hemisphere has changed in the 35 years since this patient was seen. A limited stereotaxic biopsy under CT scan guidance would be undertaken and the patient would then be treated with dexamethasone and radiotherapy. The end results would have been similar and actually may even have occurred earlier. The postoperative quality of language function may have been transiently better. Unfortunately this patient had a very malignant and aggressive tumor. The median survival of glioblastomas with surgery alone is 26 weeks, with the addition of radiotherapy 52-60 weeks.

Case 24-6: This 57-year-old female, with a long history of hypertension in the summer of 1985, had a brief episode of difficulty in word finding and a more prolonged episode in December 1985 with confusion, "visual problems", "inappropriate speech and difficulty finding the appropriate words". Following an additional episode in March 1986, she had persistent problems in reading and apraxia of the right hand.

Neurological examination (May 1986): *Language function:* Spontaneous speech was usually quite fluent. She described the room, the environment and her friend. In contrast, capacity for repetition was very poor both as regards spoken and written speech. Spontaneous writing was very poor except for name and address Reading to herself aloud was poor with little comprehension. She was able, however, to follow single one stage spoken or written commands but unable to do a sequence of commands. Naming of objects was well performed. *Cranial nerves:* Extinction occurred in the right visual field for bilateral simultaneous stimuli. *Motor system:* No hemiparesis was present. *Reflexes:* Right-sided hyperreflexia and a right Babinski sign were present.

Clinical diagnosis: Posterior aphasia, infarction left posterior temporal and parietal areas, probably due to embolic occlusion of inferior division of the left middle cerebral artery.

Laboratory data: *CT scan:* Infarct left posterior temporal and parietal areas *(Fig. 24-6). Angiograms:* Occlusion of the left internal carotid artery. All left anterior and middle cerebral artery branches filled from the right carotid. An artery-to-artery embolus was presumed.

Global aphasia: These patients have little verbal output either spontaneous or on repetition, no comprehension and no ability to read or write. CT scans demonstrate massive infarction of the entire middle cerebral artery territory of the dominant hemisphere involving presylvian areas of the frontal, temporal and parietal areas (see Naeser 1983).

Anomic or amnesic or nominal aphasia: There remains a variety of aphasia - nominal aphasia - which is commonly encountered but which is not as readily localized to a particular cortical area. The patient recognizes objects when these are present in the visual or tactile sphere but is unable to name the object or is unable to access the internal dictionary to find the name, Nominal aphasia is often noted as part of other aphasic syndromes. When encountered in relatively pure form, speech output is usually fluent, comprehension is normal and repetition is normal. The patient will

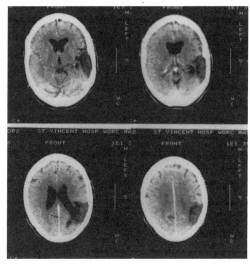

Figure 24-6. Posterior aphasia, infarction left posterior temporal and parietal areas, inferior division, left middle cerebral, (probably embolic following left carotid occlusion) CT scan. Case 24-6 (Refer to text).

substitute other names, but more often will describe the use of the object as a means of naming it, e.g., for a light switch - "That's the thing that makes the light go on and off" (circumlocution). At times, associated findings suggest localization to the dominant posterior temporal parietal area; at times, there is some association with a Broca's area motor aphasia.

Mixed transcortical aphasia or isolation of the speech areas: In contrast to the patient with conduction aphasia and to patients with Broca or Wernicke's aphasia, there are patients who have excellent repetition but little spontaneous speech. Broca's area, Wernicke's area and the arcuate fasciculus are intact. The peri sylvian areas supplied by the middle cerebral artery remain intact but are isolated from other cortical areas by infarction at the borders between the middle and anterior cerebral arteries or between the middle and posterior cerebral arteries. (See Geschwind et al 1968, Geschwind 1971, Damasio and Geschwind 1984, Benson 1985 and Bogousslavsky et al 1988). As discussed in greater detail in chapter 26, carotid artery stenosis may produce a perfusion defect at the border zone of the middle and anterior cerebral artery (watershed infarct). On the other hand, a fall in blood pressure may produce perfusion-related infarction at the border zone between the middle and posterior cerebral artery. The patient can repeat without difficulty (fluent for repetitions) but has little spontaneous speech and little comprehension. At times, the patient echos (echolalia) everything stated by the examiner.

Transcortical motor aphasia: This relatively rare syndrome reflects isolation of Broca's area from the more anterior frontal areas and from the supplementary motor area. The lesion is within the territory of the anterior cerebral artery. The speech pattern resembles that found in Broca's aphasia except that repetition is intact.

Transcortical sensory aphasia: This relatively rare syndrome reflects isolation of Wernicke's area from more posterior temporal occipital areas. The lesion is within the territory of the posterior cerebral artery. The resultant syndrome resembles Wernicke's aphasia except that repetition is intact.

Damasio and Geschwind (1984), Benson (1985) and Kirshner (2000) discuss the different types of transcortical aphasia (motor sensory, mixed) with anatomical correlation.

Visual agnosia and selective dyslexia (alexia without agraphia): Visual agnosia may be defined as a failure to recognize visually presented material. Dyslexia or alexia represents a more selective deficit: a failure to recognize visually presented words (this is sometimes also called word blindness). The student may conceptualize the basic disturbance as a disconnection of the visual areas from the speech areas of the posterior temporal and inferior parietal areas. This disturbance could result from bilateral lesions or from a unilateral lesion. As we have previously indicated, bilateral destruction of the visual association cortex (areas 18 and 19) would theoretically produce this effect. An alternate combination of lesions would involve destruction of the primary visual cortex (area 17) of the dominant left hemisphere and the visual association cortex of the right hemisphere.

A large unilateral lesion producing essentially this same effect has been implicated in dyslexia. This lesion has usually destroyed the visual cortex of the dominant left hemisphere and at the same time damaged the posterior portion or, at least, the splenium of the corpus callosum (or the fibers radiating from the splenium). The fibers conveying visual data from the right hemisphere must presumably reach the speech areas of the dominant hemisphere by passing through the posterior segment of the corpus callosum. The following case 24-4 would suggest this type of unilateral lesion producing a transient and partial but selective type of dyslexia.

Case 24-7: Seven days prior to admission, following a period of athletic activity, this 15-year-old, right-handed white male high school student had the onset of a 48-hour period of sharp pain in the left jaw and supraorbital area. The next day the patient noted blurring of vision and the following day had the onset of vomiting. Four days prior to admission, the patient noted that he was unable to read and to translate his Latin lessons. No definite aphasia was apparent. Two days prior to admission, the patient had the sudden onset of severe left facial pain, accompanied by tingling paresthesias of the right leg. The patient became stuporous and then unresponsive for approximately 90 minutes. An examination of spinal fluid at that time revealed pressure elevated to 290 mm of water with 250 fresh blood cells.

Neurological examination: *Mental status and language function:* Intact. *Cranial nerves:* small flame shaped hemorrhages were present on funduscopic examination's slight, right, supranuclear-type (central) facial weakness was present. *Neck:* A slight degree of resistance was present on flexion of the neck (nuchal rigidity) consistent with the presence of blood in the subarachnoid space.

Clinical diagnosis: *Subarachnoid hemorrhage of uncertain etiology.* In view of age and the transient dyslexia, and the transient parerethises of the right leg, an arteriovenous malformation in the left parasagittal parietal-occipital area might be suspected.

Laboratory data: *Arteriograms:* An arteriovenous malformation was present in the left occipital region. The main arterial supply was the posterior temporal branch of the left posterior cerebral artery. The malformation drained into the lateral sinus. The posterior portion of the anterior cerebral artery was shifted across the midline, indicating a hematoma as well as the malformation.

Subsequent course: On the evening after admission, the patient had the sudden onset of a bifrontal headache and was unconscious for a 2- to 3-minute period with deviation (repetitive driving) of both eyes to the right, and with sweating and slowing of the pulse rate. Over the next two days an increase in blurring of the optic disc margins was noted (early papilledema). Four days after admission, the patient suddenly complained of being unable to see from his right eye. He became restless and agitated. Additional headache and neck pain with numbness of right arm and leg were reported. Examination now disclosed a dense right visual field defect (homonymous hemianopsia). The deep tendon reflexes were now slightly more active on the right. Funduscopic examination suggested that recent additional subarachnoid bleeding had occurred since new retinal hemorrhages were present. The next day the patient was noted to have a significant reading disability (dyslexia), although other language functions were intact. Because of the progressive evolution of neurological findings due to an expanding intracerebral hematoma, Doctor Robert Yuan performed a craniotomy, which revealed blood present in the subarachnoid space. The left lateral occipital cortical surface had numerous areas of bluish discoloration indicating an intracerebral clot. The main arterial vessel was a lateral and inferior branch of the posterior cerebral artery, bearing no relationship to the calcarine region. The draining vein was located in the occipital-temporal area, close to the tentorium. At a depth of 1 cm, a hematoma of 50 cc of clot was found. The clot extended down to the occipital horn of the lateral ventricle but did not enter the ventricle. The malformation, hematoma and related cerebral tissue of the lateral occipital area were removed. The calcarine area remained intact at the end of the procedure. Over a period of several days, the visual field deficits disappeared.

Detailed language and psychological testing approximately two weeks after surgery indicated that language function was intact except for reading. Oral reading was very slow (4 times slower than normal for a test paragraph), halting and stumbling. Comprehension of the material read was good. Minor errors were made in the visual recognition of letters. This was reflected in occasional errors in spelling. Writing and drawing were intact but slow. Visual, auditory and tactile recognition and naming of objects was intact. Calculations were intact. No right-left confusion or finger agnosia was present. Repetition of speech and spontaneous speech were intact.

Follow-up evaluation at 20 and 30 months indicated that, although the patient was doing well in school, receiving A's and B's with excellent grades in mathematics, he still was described as slow in reading compared to his level prior to illness. No actual errors were made in reading when this was tested.

Geschwind (1965) has reviewed many of the previously reported cases of selective dyslexia. It is uncertain whether the information from area 18 of the non-dominant hemisphere passes to the non-dominant angular gyrus and then crosses to the dominant temporal parietal areas or whether these fibers pass instead directly to area 18 of the dominant hemisphere with information then conveyed to the dominant temporal parietal areas. He finally concludes that both pathways are probably operative. The relatively minor nature of the final residual deficit in the present case would be consistent with such multiple pathways. It addition, non-fatal hemorrhages always produce less damage than infarcts of comparable size. Finally, Coslett (2000) has reviewed various recovered pure cases of dyslexia where the right hemisphere may have played a role in reading. Such patients do better with concrete nouns than abstract nouns, verbs and functors (pronouns, prepositions, conjunctions and interrogatives). It should be noted that a relatively selective dyslexia in the absence of a severe general visual agnosia might occur. Geschwind has suggested that whether a general visual agnosia or dyslexia occurs may depend on the degree of disconnection of the dominant speech areas from the parietal and occipital areas of the non-dominant hemisphere. In the patient who already has left occipital cortex damage, dyslexia may require only damage to the splenium, whereas visual agnosia may require more extensive damage to the posterior body of the corpus callosum as well. As noted bilateral lesions of visual association cortex (areas 18 and 19) would also produce a visual agnosia. (See also Coslett and Saffran 1989 and Friedman and Albert 1985).

In addition there are patients who have acquired dyslexia, which is unrelated to the processing of the visual information input. These patients have a central (rather than the peripheral dyslexia of the pure alexia syndrome discussed above). The central dyslexic patient has an impairment of the "deeper" or "higher" reading functions by which visual word forms mediate access to meaning or speech production mechanisms. This problem has been reviewed in detail by Coslett (2000). Feinberg and Farah (2000) have reviewed the general topic of agnosias

APRAXIA

As we have previously indicated this term may be defined as impairment in motor performance in the absence of a paralysis or sensory receptive deficit and at a time when cerebellar and cognitive functions are otherwise intact. In the carrying out of a skilled movement on command there are several stages, which are, similar to those considered under language function. The command, if auditory, must be received at the cortical sensory projection area of the dominant hemisphere. The information must then be relayed to the auditory association areas for the words of the command to be comprehended, that is, to evoke the appropriate auditory associations. From these area information must be relayed, to the multimodal association area of the dominant inferior parietal association area. Information is then conveyed as discussed previously to the premotor and supplementary motor association areas of the dominant hemisphere. Information must then be sent to the dominant motor cortex and via the corpus callosum to the premotor and motor areas of the non-dominant hand so that hand can also perform skilled movements on command.

Apraxia then could result from damage at any point in this series of association centers and their interconnections. Depending on the point of disruption in the circuit, the type of impairment will vary. Thus, a lesion of the anterior one-half of the corpus callosum will result in an apraxia limited to the nondominant hand. In contrast, a lesion of the dominant premotor association areas would be more likely to produce a bilateral impairment in certain tongue and hand movements.

The concept that many varieties of apraxia represent a disconnection between various cortical areas, as reintroduced by Geschwind 1965, provides a more anatomically based approach than the use of terms such as motor or limb kinetic (motor association areas) or ideomotor apraxia (parietal lobe apraxia). Heilman et al (2000) and Ochipa & Rothi (2000) have provided a more recent analysis of this topic.

The student will also encounter the term ideational apraxia. This term that has received multiple definitions over the years is now employed to indicate a defect in performing a series of acts in their proper sequence. This problem may occur in diffuse disorders such as Alzheimer's disease or frontal lobe disorders or left parietal disorders. Some would suggest that this type of dysfunction really represents not an apraxia but a deficit in recent memory and would require among the defining conditions of apraxia that dementia not be present. (The topic of apraxia remains an area of controversy).

NONDOMINANT HEMISPHERE FUNCTIONS

As we have previously indicated, certain symptoms appear to follow damage to the non-dominant parietal areas such as lack of awareness of hemiplegia, neglect syndrome, defects in spatial construction and in perception of three-dimensional space. Some have questioned whether a true dominance of these functions actually exists in the hemisphere that is non-dominant for speech. Certainly, neglect syndromes may also follow lesions of the dominant parietal area or of either premotor area. As regards concepts of visual space, there is some evidence from the studies of Gazzaniga et al (1965), on section of the corpus callosum that the dominant left hemisphere is dependent on information that must be conveyed from the right hemisphere. Moreover, the studies of Penfield discussed in chapter 22, suggest that in patients with temporal lobe seizures, visual hallucinations are more likely to arise from stimulation of the right hemisphere than of the left, The stimulation studies of Fried et al (1982) suggest short-term memory for visuospatial functions may be localized to the right superior temporal gyrus. Perception for visual spatial material may be localized to the right parietal-occipital and to right frontal operculum. A defect in the recognition of faces (prosopagnosia) has been related to lesions of the right parietal-occipital or inferior temporal-occipital junction (see Luders et al 1988, Fried et al 1982, Sergent and Villemore 1989, Landis et al 1988).

ROLE OF CORPUS CALLOSUM IN TRANSFER OF INFORMATION

The corpus callosum allows the non-dominant hemisphere access to the special language centers of the dominant hemisphere. Thus, the intact right-handed individual can name an object, such as a key; independent of whether the object is placed in the left or right hand. Following section of the corpus callosum, the object can be named when placed in the right hand but not when placed in the left hand. In a similar manner, the corpus callosum allows the dominant hemisphere access to those specialized areas of the non-dominant hemisphere concerned with concepts of visual space.

There is also evidence that the corpus callosum is involved in the transfer of information concerned with learned sensory discriminations. Thus, the intact monkey or man, who has been trained to press a key with the right hand only when a particular visual pattern appears in the right visual field, has no difficulty in performing the same discrimination without additional learning when the left visual field and left hand are employed. Section of the corpus callosum interferes with such transfer of learned information. There is evidence, however, that some sensory information and learned sensory discrimination is transferred between hemispheres via non-callosal pathways. (Refer to Ettlinger 1965 and Sergent 1986)

In considering the functions of the corpus callosum based on studies of patients subjected to callosotomy for control of intractable epilepsy, much caution must be exercised. (1) The patients also have focal multifocal or diffuse cortical damage sustained at variable age. At times, some reorganization of function has occurred. (2) The acute disconnection syndrome which has been described after corpus callosum surgery (see Reeves 1985, Sass 1988, Engel 1989) consisting of mutism, agnosia, apraxia or the non-dominant limbs, apathy, bilateral grasp release and gait apraxia may, in part, relate to traction on non-dominant parasagittal frontal, supplementary motor and premotor areas. (3) Ontogenetic effects and effects of staging surgery have been noted. (4) The same deficits do not necessarily occur in-patients with agenesis of the corpus callosum.

CHAPTER 25
Case History Problem Solving: Part III Cerebral Cortex: Cortical Localization

These cases represent the gamut of disease affecting the cerebral cortex. The pathology represented may be tumor, infarction, or hemorrhage. Some of the lesions are intrinsic; others are extrinsic. The nature of the pathology may be uncertain; the location of the pathology, however, should be evident to you.

CASE 25-1A: Three months prior to admission, this 39-year-old, right-handed, white, male mechanic had the first of repetitive episodes characterized by "vertical wavy lines" in his left visual field. The initial episode lasted 30 minutes; subsequent episodes, 1 to 2 minutes. The patient had been seen by an ophthalmologist during the first of these episodes and a left visual field defect, which would disappear at the end of the episode, was detected. An electroencephalogram revealed no definite abnormality. Approximately 1 month later, the left visual field defect remained as a permanent deficit. At this time the patient also noted the onset of numbness in his left hand. Two months later the patient noted the onset of bitemporal and bifrontal headaches precipitated by any motion of the head. One week prior to admission there was an onset of vomiting and these symptoms prompted his admission.

Past History: Not remarkable.

General Physical Examination: Not remarkable.

NEUROLOGICAL EXAMINATION:

1. *Mental status:* This was intact except for some vagueness in the chronology for events in the present illness. The patient was oriented as to time, place and person. Delayed recall was intact. No aphasia was present; reading and writing were intact.
2. *Cranial nerves:*
   a. A complete left homonymous hemianopsia was present.
   b. Fundoscopic examination indicated bilateral papilledema with venous engorgement, arteriovenous nicking, and a recent hemorrhage in relation to the right disc. Visual acuity, however, was well preserved.
   c. Both pupils demonstrated a sluggish response to light.
   d. A left central (supranuclear) type of facial weakness was present with a droop of the left corner of the mouth.
3. *Motor system:* Strength was intact and cerebellar tests were negative. However the patient tended to lean slightly to the left when walking.
4. *Reflexes:*
   a. Deep tendon reflexes were asymmetrical, being slightly more active on the left than on the right.
   b. Plantar responses were flexor. Abdominal reflexes were present bilaterally.
   c. No release of grasp and suck reflexes had occurred.
5. *Sensory system:*
   a. Pain, touch, and vibration were intact.
   b. Position sense for fine amplitude movements at the fingers and toes was decreased on the left.
   c. On simultaneous tactile stimulation of the left and right side, extinction of stimuli occurred on the left.
   d. Errors were made in object identification in the left hand.
   e. Occasional errors were made in the identification of numbers drawn on the fingers of the left hand.

LABORATORY DATA:

The *electroencephalogram and imaging studies* now suggested a single lesion.

QUESTIONS:

1. Localize the initial symptoms? In your con-

sideration as the location of the lesion, you will wish to consider the localizing significance of the following: the initial episodes of "vertical wavy lines" b. the homonymous hemianopsia. What is the nature of these initial symptoms?

2. Which areas were subsequently involved by the lesion? Take into account the pattern of findings on sensory examination.

3. What is the pathological nature of the lesion? Take into account the short duration of symptoms and the age of the patient.

4. Which diagnostic study would you perform?

5. How would you treat the seizures?

6. How would you manage the basic lesion?

CASE 25-2: This 35-year-old, right-handed, white male had the onset of a sudden pins-and-needles sensation which began in the left foot and then spread within a few seconds to involve the entire left side of the body including the arm and the face. The total episode lasted for 8 minutes. Similar episodes then recurred once or twice weekly beginning most frequently on the left side of the face or tongue and less often on the left foot or hand. Four months later the patient had a similar episode but fell to the floor with a loss of consciousness of several minutes duration during which observers reported some clonic movements of the left arm and leg. Neurological examination, lumbar puncture, and electroencephalogram at that time were reported as normal. The episodes were temporarily controlled with anticonvulsant medication, phenytoin (Dilantin), but then recurred with increasing frequency. At the time of hospital admission, 19 months after onset of symptoms, these episodes of numbness were occurring 3 to 4 times per day.

Past History: There was no history of significant head trauma. There was no family history of seizures.

NEUROLOGICAL EXAMINATION:

1. *Mental status:* Orientation and memory were normal.

2. *Cranial nerves:* No abnormalities were present.

3. *Motor system:* No relevant findings were present. The patient had fallen during the secondarily generalized seizure, sustaining a fracture of the left humerus and dislocation at the left shoulder. Since that time he had a poor swing of the left arm due to limitation of left shoulder movement. A slight atrophy of the hypothenar eminence with some weakness in the ulnar distribution had also been present since that injury.

4. *Reflexes:* Deep tendon reflexes were symmetrical; plantar responses were flexor.

5. *Sensory system:* All modalities were intact except for errors in graphesthesia in the left hand.

LABORATORY DATA:

1. *Skull x-rays:* The calcified pineal was shifted 4 to 5 mm compared to a shift of 2 mm. in films obtained 1 year previously.

2. *Electroencephalogram:* Focal slow wave activity of 4 to 6 Hz was present, consistent with focal damage.

3. *Cerebrospinal fluid:* Normal pressure and protein (21 mg./100 ml.) were present.

4. The *imaging studies* were consistent with a single focal lesion.

QUESTIONS:

1. Where is the lesion in terms of the origin of the seizure activity?

2. What are the possible pathological diagnoses? In answering this question you should keep in mind that the patient is of middle age (35), the seizures are clearly focal, and there is a 19-month history. These facts alone, even without the confirmatory laboratory data, should lead to the most likely diagnosis.

3. Which diagnostic study would you perform?

4. How would you treat the seizure?

5. How would you treat the basic lesion?

CASE 25-3: A 66-year-old, right-handed, white widow developed changes in affect and lack of interest in her surroundings over a 1-month period. Although previously she had been alert, caring for herself, doing her own shopping, and interested in her relatives, she now took to her bed and became disheveled in appearance. She would talk little if at all. She complained of a vague weakness in her legs, was unwilling to walk, but was able to walk to the bathroom without assistance.

Past History: A myocardial infarct had occurred 10 months previously. There was a 10-year history of Paget's disease involving the bones of the lower extremities.

NEUROLOGICAL EXAMINATION:

1. *Mental status:*
   a. The patient was apathetic with a dull appearance.
   b. The patient was disoriented for person and time. She knew that she was in a hospital in the state of Massachusetts but did not know the city.
   c. She was unable to recall her birth date or to name any president. She could give no account of her illness.
   d. She was unable to add 5 and 5 and could not do multiplication or subtractions.
   e. She repeated 4 digits forward and none in reverse.
   f. Spontaneous speech was scanty but without a definite dysarthria or aphasia. She could name 3-out-of-6 objects, could repeat 6-out-of-6, and could read aloud 4-out-of-6. She was able to perform slowly a two-step command. She was apparently unable to write.
2. *Cranial nerves:* II-XII were intact except that the left disc margin was blurred on funduscopic examination.
3. *Motor system:*
   a. No focal weakness was present.
   b. There was a variable resistance in both lower extremities described as "gegenhalten."
   c. The patient was very hesitant and very fearful of sitting, standing, or walking. There was some retropulsion (a tendency to fall backward when attempting to sit or walk). When she walked her right foot appeared to be glued to the floor.
4. *Reflexes:*
   a. Deep tendon reflexes were symmetrical and active.
   b. An equivocal plantar response was present on the right; that on the left was flexor.
   c. Strong grasp reflexes were released at hands and feet. Visual and tactile suck reflexes were also released.
5. *Sensory system:*
   a. Pain, touch, and position sense were intact.
   b. There was a minor decrease in vibratory sensation at the toes, which may well have related to the patient's general nutritional status.

LABORATORY DATA:

1. *Sedimentation rate* elevated to 98 mm. in one hour.
2. *Skull x-rays:* An osteolytic lesion (area of bone destruction) was present.
3. *Lumbar spine x-ray:* A suspicious, probably osteolytic lesion was seen in the body of the L1 vertebra.
4. *Stool Guaiac:* A small amount of occult blood was present. Hematocrit and hemoglobin were normal. Rectal and proctoscopic examination revealed a firm mass at approximately 9 cm. above the anus.
5. *Electroencephalogram* indicated a focal lesion having the quality of damage (continuous focal 2 to 4 Hz slow waves).
6. *Imaging studies* were consistent with a single focal lesion.
7. A *lumbar puncture* 3 weeks prior to admission had revealed pressure increased to 220

mm. of CSF and protein elevated to 59-mg./100 ml.

QUESTIONS:

1. Where is the lesion? (Assume a single central nervous system lesion).
2. What is the most likely pathology?
3. How would you manage this problem as regards more specific diagnosis and therapy? Which neuroimaging studies would you perform?

In considering these questions you will wish to keep in mind the following points:

1. As regards the location of the lesion:
   a. The personality change, accompanied by impairment of all areas of mental function,
   b. The release of strong grasp reflexes in both upper and lower extremities, in addition to suck reflexes,
   c. The apparent inconsistent findings as regards resistance to passive motion,
   d. The disturbance of sitting posture and of gait.
2. As regards the type of pathology:
   a. The elevation of sedimentation rate,
   b. The apparently rapid onset of symptoms,
   c. The presence of occult blood in the stool,
   d. The presence of osteolytic lesions,
   e. The mass noted on examination of the rectum.
3. Which diagnostic studies would you perform?

CASE 25-4: On the evening prior to admission, his wife noted that this 62-year-old right-handed, white, male civil engineer had the relatively sudden onset of slurred speech associated with a drooping of the left side of the face. The patient denied that these symptoms were present. The following morning he was examined by his brother, a physician, who confirmed the presence of slurred speech and weakness of the lower one-half of the left side of the face but also noted a weakness of the left arm and leg.

Past History: Gout had been present since age 20. He had sustained an apparent coronary artery occlusion in the past.

NEUROLOGICAL EXAMINATION:

1. *General:* blood pressure: 130/80; pulse 85 and regular.
2. *Mental status:*
   a. The patient was oriented for time, place, and person.
   b. Memory for recent and remote events was intact. Immediate and delayed recall was intact.
   c. The patient could do simple calculations but had difficulty with more complex problems.
   d. These results may have been influenced by the fact that the patient demonstrated a marked impersistence in following commands. The patient was somewhat apathetic and yawned frequently during the examination. There was a flattening of mood.
   e. The patient denied that he had any particular problems requiring hospitalization. He denied illness and stated he was as healthy now as he had been one week prior to hospitalization.
   f. Abstract reasoning was intact, as tested in proverb interpretation (used abstract concepts).
   g. No aphasia was present.
   h. A distinct deficit in the construction of three-dimensional drawings was present.
3. *Cranial nerves:* All were intact except for:
   a. A significant left facial weakness of supranuclear type.
   b. An inconstant neglect of left field on simultaneous stimuli to left and right visual fields.
4. *Motor system:* Strength and gait were intact and cerebellar tests were negative.
5. *Reflexes:* Deep tendon reflexes were symmetrical, plantar responses were flexor and no pathological grasp reflex was present.

6. *Sensory system*
   a. Pain and touch were intact.
   b. Stereognosis, two-point discrimination, and tactile localization were normal. However, position sense was decreased at toes and fingers on the left. There was a tendency to ignore tactile stimuli on the left face, arm, and leg when simultaneous left and right stimulation was carried out. Graphesthesia was intact in the right hand, but defective on the left.

LABORATORY DATA:

1. Blood *serological test* for syphilis was negative.
2. *Skull x-rays* were normal.
3. *Electroencephalogram*: abnormal because of focal 2-5 Hz slow waves.
4. *Brain scan (radioactive Hg 197):* A diffuse focal uptake was present.

SUBSEQUENT COURSE:

Over the next 3 months, significant improvement occurred. The patient was able to do complex calculations. He had no difficulty in drawing or in locating points on a map. Sensory examination was normal. The patient still had difficulty in maintaining his concentration. Psychological testing now indicated verbal IQ of 123, performance IQ of 104, with full scale IQ of 116, compared to much lower scores of 111, 87, and 101 respectively at the time of his acute illness.

QUESTIONS:

1. Where is the lesion (be specific)?
2. What is the nature of the lesion? If vascular discuss the type of lesion and the vascular territory
3. Which studies should be performed (in terms of neuroimaging and other studies)?

CASE 25-5: This 52-year-old housewife was referred for evaluation of episodic "twitching of the left thumb and forefinger" beginning 2 weeks prior to admission. The patient had noted the sudden onset of a numbness of the left thumb, followed in seconds by twitching of the left eyelid and then almost immediately by a twitching of the distal segment of the left thumb. This soon spread to the left forefinger, then to the middle finger, and then involved the hand in repetitive clonic movements. According to observers, the clonic movements then spread into the left arm. The twitching about the eyelid spread to the entire face. This initial episode lasted a total of 15 minutes; there was, however, a residual numbness of the left thumb and index finger: "as though they had fallen asleep." There was also a transient weakness of those areas first involved. Since the first episode, the patient had experienced minor recurrences of "tingling" of the left thumb and index finger.

Past History: Six years prior to admission the patient underwent a right radical mastectomy for carcinoma of the breast. Regional lymph nodes were reported as negative. The patient's mother and sister also had carcinoma of the breast.

NEUROLOGICAL EXAMINATION:

1. *Mental status:* intact with no evidence of aphasia. No deficits were noted in drawing.
2. *Cranial nerves:* A minimal left supranuclear central facial weakness was present.
3. *Motor system:* A mild weakness was present in the left upper extremity and rapid alternating movements of the left hand were slowly performed
4. *Reflexes:*
   a. Deep tendon stretch reflexes were increased on the left.
   b. An equivocal plantar response was present on the left; that on the right was flexor.
   c. No grasp reflex was present.
5. *Sensory System:*
   a. Pain, light touch, position, and vibration were normal.
   b. There was a slight disturbance in graphesthesia over the left hand and face. There was minor impairment of two-point discrimination over the left thumb and forefinger.

QUESTIONS:

1. Indicate the location of the lesion in this case.
2. Indicate the most likely pathology.
3. (a) Attach a label to the twitching which begin in the left thumb, then spread to the other fingers, then to left hand, arm, and face.
   (b) Attach a label to the transient weakness, which followed these episodes of twitching.
4. Which diagnostic studies would you perform?
5. Discuss management of this patient.

CASE 25-6: This 19-year-old, right-handed, white, male pharmacy student was referred for evaluation of frequent minor seizures. At age 18 months, the patient had fallen striking his head against a concrete floor and within hours had lapsed into coma. Later that day he required emergency neurosurgical treatment at the Children's Hospital for a "skull fracture and a bleeding vessel which the neurosurgeon had to tie." The patient apparently made an excellent recovery. However, from 3 years of age until 5 years of age he experienced "large convulsions" (apparently generalized convulsive seizures). During this time he also experienced minor episodes characterized by the sensation that objects were becoming smaller or larger. The patient recalled that at times, he had the sensation that bright geometric patterns were present and shrinking in size. Almost invariably the patient experienced the sensation of fear with these episodes. The patient had no seizures from age 5 years until age 16 years. Since age 16, the patient had frequent minor episodes of several types occurring as frequently as 15 to 30 times a day.

One type of episode was accompanied by dreamy sensations lasting a few seconds; at times things looked unreal. At the same time the patient would feel that the whole episode was somehow very familiar, that objects looked very familiar in a manner that could not be described. There was an accompanying sensation of fright. During such an episode the patient would be able to continue driving or walking.

A second type of episode, lasting 30 to 120 seconds, was characterized by an inability to talk properly, a babbling of speech, and an inability to understand speech. During this time a sensation of vague familiarity and fatigue and at times a sense of reincarnation would exist.

A third variety was characterized by the visual sensation of looking down a long dark corridor with a light at the other end. There was no concept of time: at times the corridor seemed to move from right to left. At the end of such an episode things would not look as they should; objects appeared distorted, and color altered. Objects were seen as black and white. The whole scene would seem unworldly. Sounds would be perceived but would fail to be registered and would not make sense.

NEUROLOGICAL EXAMINATION:

Mental status, cranial nerves, motor-system reflexes, and sensory system were all intact except for a minimal lag of the left side of the face in smiling. During the examination, the patient had a minor episode; his head dropped and he appeared out of contact for 20 to 30 seconds. He did not answer questions directly but replied by saying, "What was then, what was that then, what was that then?" He was confused for several minutes, after the episode. He was subsequently amnestic for the events of the episode.

QUESTIONS:

1. The electroencephalogram indicated a stable focus of discharge. Indicate the location of that focus.
2. Classify the seizure according to the International Classification of Seizures.
3. What is the most likely explanation for the acute events, which occurred at age 18 months?
4. Why did seizures recur at age 16 years?

CASE 25-7: This 20-year-old white male (Mr. K.B.) had been admitted to Boston City Hospital at age 9 months with acute right hemiplegia and acute right-sided hemi convulsions associated with a fever. A right hemiplegia was present for several days and he continued to have a mild right hemiparesis. Several additional seizure episodes occurred between ages 1-3 years. No additional episodes occurred until age 17 when the patient had a focal seizure characterized by tonic posture of the right arm – abducted at the shoulder and flexed at the elbow – followed by turning of the head and eyes to the right. Consciousness was impaired but not completely lost. For several minutes the patient was unable to speak. Upon regaining speech, he was able to follow commands.

Treatment with anticonvulsants (Eskabarb – long-release form of phenobarbital) was relatively effective in preventing seizure recurrence. Once a month the patient did experience a few seconds of "weakness or limpness" of the right hand with a possible small jerk of the right hand. Approximately two weeks before his initial office visit in January 1968, the patient discontinued his medications. Four days prior to this visit, at a time of sleep deprivation, the patient had a focal seizure with tonic movement of the right arm and deviation of the head and eyes to the right followed by a generalized convulsive seizure. Since age 9 months, residual weakness of the right arm had been present. However language functions had developed and he had done well in school

NEUROLOGICAL EXAMINATION:

1. *General:* The patient was left-handed with long-standing smallness of the right side of the face, arm, and leg.
2. *Mental status:* All areas were intact with no aphasia.
3. *Cranial nerves:* All were intact except for slight lag on the right in smiling.
4. *Motor system:*
   a. There was a mild weakness of the right upper extremity, most marked distally and particularly marked in the thumb abductor.
   b. Gait: A minimal circumduction of the right leg was noted.
5. *Reflexes:*
   a. Deep tendon reflexes were increased on the right.
   b. Plantar response was equivocal on the right.
6. *Sensory system:* Minor deficits were present in the right upper extremity characterized by errors in fine-movement perception of the right fingers and errors in graphesthesia of the right hand. Pain and touch were intact.

SUBSEQUENT COURSE:

With adjustment of anticonvulsant medication, excellent seizure control was attained. Only rare limited focal (partial) seizures occurred over the next 30 years.

QUESTIONS:

1. Locate the lesion.
2. What is the origin of focal (partial) seizures characterized by tonic posture of the right arm with deviation of the head and eyes to the right and with arrest of speech?
3. The patient has a right hemiplegia and yet he does not have aphasia. Explain.
4. In view of the acute onset of the hemiplegia at age 9 months and the subsequent lack of progression, speculate

Concerning the etiology.

5. Why was the weakness most prominent in the distal muscles of the upper extremity (particularly the thumb)?
6. Why did seizures recur?
7. Why has the smallness of the right side occurred?
8. What is the nature of the sensory deficit?
9. How would you manage the seizure problem?
10. Does the recurrence of seizures at age 17, after no seizures since age 3, indicate that a progressive pathology is present?

CASE 25-8: This 23-year-old, right-handed, white male was admitted with chief complaints of frequent and increasing headaches and "seizures". The problem apparently began 15 months prior to initial evaluation, when several months after their marriage, the patient's wife noted a change in his personality. In particular, the patient no longer had any desire for sexual intercourse. He became more irritable, at times with only minimal provocation and he was verbally and physically abusive.

Eight months later, (7 months prior to initial evaluation), he began to awake from sleep with numb feeling of right side of the lips which quickly spread to the right side of face; would last a few minutes and resolve. When the patient awoke during such episodes, he found himself sitting upright. These would occur weekly and only as the patient was going to sleep or after he had fallen asleep.

After a few weeks the seizures changed in character. While sleeping, the patient would suddenly be thrown toward his right side, his head would arch back and he would have difficulty breathing. He would be aware during these episodes and could understand what people said, but could not speak. Sometimes the patient would not be thrown to the side, but would simply have tonic stiffening and claw-like formation of the right hand, lasting 20 to 30 seconds.

Two months later, (5 months prior to evaluation) he underwent a right inguinal herniorrhaphy and had several attacks in the recovery room. Since that time, the patient has experienced one or two seizures per day, some of which occurred while awake.

Because of these episodes, he was admitted to a local hospital for evaluation. EEG and LP were reported to be normal. Following the lumbar puncture he developed a bifrontal headache. While sitting in a wheelchair, he began to complain of headaches. A nurse placed her hand on his forehead and his wife told the nurse to get her hand off. According to other observers, he ran down the corridor pushing the nurses aside and ran into a wall. He had to be restrained by several people, and was given some injections. Psychiatric treatment was begun with no significant improvement. Two months prior to evaluation, the patient's wife inquired of him whether the psychiatrist had been able to tell him anything about the cause of his episodes. He threw a jar of mustard at her, then the plates off of the dinner table. He rushed at her. As she retreated, he kept coming after her. He seemed unaware of his actions, and did not seem to recognize her. He placed his hands about her throat and began to choke her. With difficulty, she managed to break free. She noted that he appeared dazed and sleepy. At this point, fearing for her life, the wife decided upon a divorce. The patient denied any memory for this episode. Intermittent bitemporal and bifrontal headaches began to increase in frequency. Occasionally, he had incontinence of urine and tongue biting during his severe nocturnal seizures, which now were occurring, on a daily basis.

Fifteen days prior to admission, the patient became angry at a dog that was barking outside the house. He ran outside to chase the dog, but could not catch him and he kicked the fence gate. This is the last he remembers. His girlfriend relates that the patient began punching the brick wall of the house and then began to bang his head forcibly against the wall, stopped, and fell on his back staring, whereupon his girlfriend helped him walk into the house. The patient next recollects the trip to the hospital where lacerations of his hands were sutured and he was released.

NEUROLOGICAL EXAMINATION:

On admission evaluation, this was entirely normal. Bilateral paresthesias of the hands were reproduced by hyperventilation.

SUBSEQUENT COURSE:

8 months later, minor findings were present:

1. *Mental status:* He was anxious but affect was indifferent and passive.

2. *Cranial nerves:* A slight right central facial weakness was present
3. *Reflexes:* Deep tendon reflexes were slightly increased at right biceps, patellar and Achilles. The right plantar response was equivocal.

QUESTIONS

1. What is the most likely explanation and anatomical basis for the episodes of aggression in this case?
2. How often does directed aggression occur as an ictal event as opposed to non-directed aggressive behavior during ictal or post ictal confusional states?
3. From the anatomical standpoint what is the explanation for the change in personality?
4. What is the localization for the episodes of seizures while sleeping characterized by the following description "the patient would suddenly be thrown toward his right side, his head would arch back and he would note difficulty in breathing"?
5. What is the anatomical; localization for the episodes of tonic stiffening and claw-like formation of the right hand? What is the explanation for the initial episodes of tingling of the right side of face and lips?
6. How can you tie all of this together into a comprehensive diagnosis?
7. Discuss the possible underlying pathology and your diagnostic and therapeutic management.

CHAPTER 26
Cerebral Hemispheres: Neuropathology and Clinical Correlation I. Vascular Syndromes

The course and nature of some of the more common diseases affecting the cerebral hemispheres are already familiar to the student on the basis of those case histories presented in the previous chapter on cortical localization. We will refer back to some of those case histories. However, we will not limit our discussion simply to the neocortex but will discuss the cerebral hemispheres in a more general sense. It will also be convenient at this point to consider several broad categories of diseases, not previously discussed, which do not necessarily restrict themselves to the cerebral hemispheres.

Diseases affecting the cerebral hemispheres account for the largest proportion of central nervous system diseases encountered in medical practice. These diseases may be focal or diffuse in nature. There is also a third or intermediate category of multifocal (that is, comprised of many focal events). Our major concern in this chapter will be with focal vascular disease.

VASCULAR DISEASES OF THE CEREBRAL HEMISPHERE

Vascular problems account for the largest category of diseases affecting the cerebral hemispheres. Cerebrovascular disease remains the third leading cause of death in the United States. The annual incidence is close to 300,000. Approximately 261,000 fall into the ischemic/occlusive/category and 36,000 into the category of intracerebral and subarachnoid hemorrhage. The overall mortality due to this cause has been falling due to: (1) better control of hypertension; (2) better control of cardiac arrhythmias; (3) decreased incidence of rheumatic heart disease; (4) decreased incidence of meningovascular syphilis. Several types must be considered: (1) atherosclerosis with ischemia due to stenosis and occlusion (2) small infarcts (Lacunar infarcts) due to hypertension, (3) embolism, (4) intracerebral hemorrhage, and (5) subarachnoid hemorrhage. These various categories are not mutually exclusive. Thus occlusive vascular disease usually occurs in relation to the process of atherosclerosis - the deposition of fatty material (cholesterol) in the walls of arteries. However, emboli produce their effect essentially by occlusion of cerebral arteries. Such emboli generally originate in relation to clots within the chambers of the heart. At times, however, such embolic material has originated at an area of occlusive disease when atheromatous material or an overlying thrombus (clot) has broken loose to be carried into the more distal circulation. There is, on the other hand, an overlap between ischemic occlusive disease and intracerebral hemorrhage. Thus certain areas of tissue damage resulting from ischemia may become hemorrhagic secondarily. The symptoms and signs, which then evolve, may resemble those of intracerebral hemorrhage. The use of angiography, CT scan, and MRI have allowed a more precise delineation of infarct vs hemorrhage but have left a large group of patients where a specific cause of the infarct remains uncertain.

ISCHEMIC-OCCLUSIVE DISEASE:

Within this category the subtypes indicated in Table 26-1 may be specified (based on Kistler, 2000, Sacco et al 1989, and Wolf et al. 1987.

Earlier clinical and autopsy studies suggested a higher incidence of large or small vessel disease compared to embolic disease. (See Adams et al 1997).

Atherosclerosis involves the large- and medium-sized extracranial and intracranial portions of the cerebral arteries. With increasing deposition of fatty material at points of bifurcation or angulation, a progressive narrowing (stenosis) of the lumen occurs *(Fig. 26-1)*. Eventually, at approximately 70-75% stenosis, a decrease in cerebral blood flow through this point of narrowing in the vessel will occur (the

TABLE 26-1 TYPES OF ISCHEMIC-OCCLUSIVE DISEASE

Cause of Ischemic-Occlusive Disease	Percent of Total Cases
1. Large vessel atherosclerosis:	15 % (internal carotid artery constitutes 9% of all ischemic)
2. Small vessel (lacunar)	9%
3. Embolic	Overall 60%(secondary to atrial fibrillation constitutes 15% of all ischemic)
4. Dissection and other causes	3%

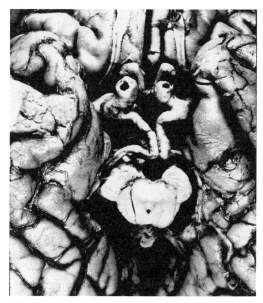

Figure 26-1. Atherosclerosis of arteries in the circle of Willis at the base of the brain. (Courtesy of Drs. John Hills and Jose Segarra).

actual relationship may be calculated according to Bernoulli's principle). The area supplied by this vessel may then at some point receive less than the required amount of blood (oxygen, glucose, and so forth). When this occurs, we may speak of cerebral ischemia. Kalimo et al (1997) outline the effects of decreased perfusion as follows: with a blood flow>40% of the normal value, normal spontaneous and evoked activities of nerve cells is still present. Between 30-40% of normal flow, neurons are unable to produce sufficient energy to continue transmission of impulses. At < 30 % of perfusion, transmission by the neuron no longer occurs although the neuron is still viable. When perfusion falls below 15% of normal, membrane failure begins to occur. The transmembrane ion gradients are no longer maintained and extracellular potassium rises. Unless perfusion and energy production are restored, irreversible nerve cell damage will occur. Note that there is a zone adjacent to this focal area of damage referred to as the penumbra where total failure has not yet occurred. This area represents tissue that could be saved by therapeutic measures.

The clinical correlates of this sequence of focal ischemia are as follows. A dysfunction of the involved ischemic area occurs with focal symptoms of weakness, numbness, and the like. This dysfunction may be only temporary in nature or it may be prolonged with actual tissue death (infarction) and residual neurological deficit. As progressive narrowing of the lumen occurs, eventually a point of total occlusion is reached. The narrowing of the lumen is moreover accentuated by several additional processes: (a) thrombus formation - the formation of a clot of platelets, fibrin, and blood cells due to stasis, change in intimal surface, and so forth, (b) ulceration of the intimal surface at the locus of the atherosclerotic surface at the locus of the atherosclerotic plaque, and (c) subintimal hemorrhage in relation to this area of deposition. The process of ulceration of the intimal wall may produce an originating site for emboli composed of the atheromatous material. Such an ulcerated intimal wall is a likely site of thrombus formation producing additional stenosis. The study of Eliasziw et al (1994) confirmed that plaque ulceration more than doubles the risk of stroke at higher degrees of stenosis. Fragments of the clot may also travel as emboli into the distal circulation.

We have used the joint term, atherosclerotic ischemic occlusive vascular disease, since there is no one-to-one relationship between actual occlusion of a vessel and tissue death (infarction or encephalomalacia) in the region supplied by that artery. Thus, total occlusion of a vessel may occur with or without the development of significant neurological symptoms and signs. There may be at times persistent symptoms and signs correlated with significant tissue destruction. At other times, with stenosis, there may be only episodes of transient symptomatology without actual tissue destruction. These transient episodes, related to vascular insufficiency, are referred to as transient ischemic attacks.

The several factors which account for this lack of close correspondence between the state of a particular artery and the vascular status of the region supplied by that vessel have already been covered in the earlier discussion of basilar vertebral ischemic/occlusive disease. (Refer also to figure 26-7 below)

1. There are significant variations in the capacity of the circle of Willis to provide collateral circulation.

2. There are significant leptomeningeal anastomoses over the surface of the cerebral cortex between the anterior, middle, and posterior cerebral arteries of a given hemisphere. With regard to each cerebral artery one may then speak of its area of central supply and its area of peripheral supply. This latter peripheral area usually receives also the peripheral supply of adjacent vessels and is in a sense a border zone or watershed.

3. To a variable degree, anastomoses are present between the external and internal carotid arteries.

4. Finally, a considerable variability occurs because of certain general systemic and metabolic factors. Thus a narrowed vessel may still deliver sufficient blood when the systemic blood pressure is 190/100 but fails to do so when the blood pressure falls to 120/70 during sleep, on sudden standing after a long period of recumbency, and in relation to antihypertensive medications. Similarly, stenotic blood vessels may deliver a sufficient blood flow when metabolic conditions such as temperature serum sodium and glucose are normal but this blood flow may fail to meet essential requirements when metabolic conditions have changed.

Infarction (encephalomalacia): It is appropriate at this point to briefly summarize the gross and microscopic changes, which occur when ischemia has been of a sufficient degree to produce tissue death. The neuropathologist generally distinguishes between pale and hemorrhagic infarction. Most infarcts following ischemia are "pale" and are not complicated by hemorrhage at a gross level. During the first 4 to 6 hours no gross or microscopic changes are apparent. Between 8 and 48 hours swelling of gray and white matter is noted; the white matter may appear somewhat granular. After 48 hours the infarcted area feels soft and mushy; the descriptive term necrosis may be used. This state may be generally recognized by visual inspection. Over the next 10 days a decrease in swelling occurs. Through enzymatic processes, liquefaction occurs. By 3 weeks, in larger lesions a gross cavity begins to appear. Within a period of several months, all of the necrotic tissue is replaced by a fluid-filled cavity.

Advances in imaging studies have allowed for earlier detection of ischemia and infarction. Refer to Shaeblitz and Fisher, 1999. Perfusion imaging MRI studies will demonstrate decreased perfusion and presumably ischemia within 15 seconds. This ischemia may be reversible. This is similar to the older neurophysiologic test; the electroencephalogram which when employed in the operating room during carotid endarterectomy will begin to detect the changes of ischemia within 10-15 seconds. Diffusion weighted MRI studies may demonstrated differences between normal and ischemic presumably infarcted tissue within minutes to one hour.* Standard MRI will detect the increased water content of infarcted tissue at 6-12 hours. CT scans may also begin to detect early subtle changes related to

---

The dicrepancy between perfusion imaging and diffusion weighted imaging may allow the identification of the ischemic not yet infarcted penumbra around an infarct.

changes in water content (some effacement of sulci and gyri) at 3-4 hours although the clear cut appearance of infarction is usually delayed until 18-24 hours (Warach & Edelman, 1993, Kalimo et al, 1997).

From the microscopic standpoint indistinct staining of the neurons (which are more sensitive to anoxia than other elements, e.g., capillaries, and glia) may be noted as early as 12 hours. Within 1 to 3 days, swelling or shrinkage or alteration in the distribution of Nissl substance (chromatolysis) may be noted. Within 2 to 4 days disintegration of cells occurs. As these neuronal changes are developing, alterations are also occurring in the appearance of axons (swelling of myelin sheaths, poor staining, and disintegration) and of glial cells (swelling of astrocytes with fragmentation of processes).

While this process of necrosis is underway, various histological responses to the infarction are also taking place. Within 24 to 36 hours, the area of infarction is infiltrated by neutrophilic polymorphic leukocytes (the acute response cells). Within 24 hours of infarction, phagocytic cells (macrophages) begin to appear. These macrophages arise from various sources: microglia, blood vessel walls, circulating blood cells. By 48 hours of infarction, macrophages, filled with fatty debris (from myelin and cell breakdown), are noted. These cells remain the predominant repair cell from the period 5 days to 30 days *(Fig. 26-2)*. Some of these cells may be found years later at the edge of the infarct. If some hemorrhagic component has also been present, products from the breakdown of the blood may also be present in these cells. This removal of debris by the macrophages will eventually result in the appearance of the cavity that may be seen grossly.

At the same time that macrophages are removing the debris, the capillaries and astrocytes are beginning to engage in repair processes. Within several days to weeks after the infarct, new capillaries are noted at the edge of the lesion. There is also a thickening of the capillary walls with increased cellularity and the

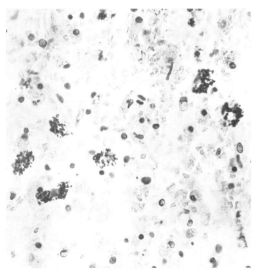

Figure 26-2. Macrophage stage of reaction in an area of necrosis. Approximately 28 days after a hemorrhagic infarct of the brain stem. Many macrophages are filled with heme pigment. (H & E x 400). (Courtesy of Dr. John Hills).

appearance of mesodermal fibrils in the adjacent tissue. There also occurs in this adjacent tissue a proliferation of astrocytes (first noted at 3 days) with the formation of reaction astrocytes containing large amounts of pink cytoplasm in H & E stains. There is a correlated proliferation of astroglial fibers. By 4 to 5 weeks a meshwork wall has been formed about the cavity by these mesodermal and glial fibers.

Hemorrhagic infarcts occur in 18-48 % of autopsy studies. Infarcts due to emboli are more likely to be hemorrhagic (51-71 % compared to non embolic infarcts, 2-21%). Venous infarcts are often hemorrhagic.

The syndromes associated with ischemic-occlusive disease of each of the arteries supplying the cerebral hemispheres will now be considered in greater detail[1]. The student, however, should interpret this material keeping in

---

[1] *At present most cerebral vascular disease involves the arterial circulation. Disease of the venous circulation is now less common. In the pre-antibiotic era, infectious processes involving the face, paranasal sinuses the mastoid and middle ear could secondarily involve the venous sinuses and cortical veins.*

mind those limitations of correlation that have been indicated above.

INTERNAL CAROTID ARTERY

The carotid artery may be involved by atherosclerosis in its extracranial or intracranial portions. The most common location for stenosis in this system is at the origin of the vessel at the bifurcation of the common carotid artery into the internal and external carotid branches *(Fig. 2-27)*. Another favorite location is at the carotid siphon. The branches of the internal carotid artery are in order as follows: the ophthalmic, posterior communicating, anterior choroidal, anterior cerebral, and middle cerebral. With stenosis of the internal carotid artery, a variable symptomatology and anatomical pattern of ischemia and infarction may develop *(Fig. 26-3)*. The pattern and time course of the syndrome are dependent on the availability of collateral circulation. One may distinguish transient ischemic attacks (TIAs) which last less than 24 hours with total recovery[2], progressing strokes, incomplete or partial strokes and completed total territory strokes. Transient ischemic attacks may be retinal or hemispheric and may be related to alteration in perfusion or to microemboli originating at the carotid bifurcation. This discussion will emphasize the perfusion etiology.

One syndrome involves the occurrence of repeated episodes of monocular blindness on the side of the involved carotid artery with or without the later development of hemispheric symptoms e.g. motor and sensory symptoms involving the contralateral extremities. The patient may describe a curtain descending over the field of vision of the involved eye. The episodes of monocular blindness are referred to as *amaurosis fugax*. In such cases, when symptoms are limited to monocular blindness we

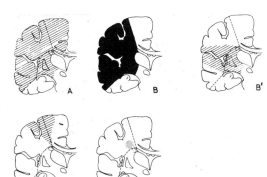

Figure 26-3. Various types of the 40 infarcts in 52 cases of occlusion of carotid artery at autopsy. A) Combined MCA and ACA Territories- 37.5%. B) Total MCA- 13%. B') Proximal MCA -21%. C) Watershed 19%. D) Terminal or deep border zone MCA/ACA- 5% (modified from Torvik, A., and Jorgensen, L.: J. Neurol. Sci., 3:415, 1966). MCA=middle cerebral artery territory, ACA=anterior cerebral artery territory.

may presume that there is ischemia predominantly in the ophthalmic artery distribution with the more distal carotid branches temporarily receiving adequate collateral supply from the posterior communicating and anterior communicating arteries and via leptomeningeal anastomosis over the cortical surface. With progression of the occlusive process even this collateral supply may become inadequate and additional symptoms develop.

One of the common patterns of carotid artery insufficiency is the carotid border zone syndrome *(Fig. 26-3, 26-4)*. With decreased perfusion by the carotid artery, symptoms will initially develop in the upper extremity, which corresponds to the cortical border zone between the middle and anterior cerebral arteries[3]. As such it receives some overlap supply from both the anterior and middle cerebral arteries. As perfusion pressure drops in the

---

[2] *In reality, most TIAs are minutes in duration and rarely greater than one hour in duration. This has significance when the use of the thrombolytic agent recombinant tissue type plasminogen activator (t-PA) is considered requiring a more realistic definition (see below). This agent must be administered within three hours of stroke onset.*

[3] *For additional discussion refer to Bogousslavsky & Regli (1986). Note also that although the carotid (anterior border zone syndrome may be relatively common as an insufficiency syndrome, actual infarction limited to the border zone alone is less frequent (Mounier-Vehier, 1995)*

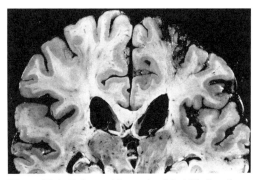

Figure 26-4. Left carotid stenosis with old infarction predominantly within the border zone between anterior and middle cerebral arteries. This 72-year-old male had a persistent paralysis of the right upper extremity with defective position sense at fingers. (Courtesy Drs. John Hills, and. Jose Segarra.)

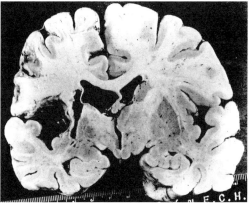

Figure 26-5. Total MCA territory infarct after left carotid clamping and excision 8 months prior to death in a 62 year old left handed male during a radical neck dissection. Patient progressed from a weakness of right hand grip, 8 hours after surgery to a complete right hemiparesis with right central facial weakness and a mixed aphasia. Subsequently recovery occurred in the lower extremity He wrote jargon with his left hand, could copy but was unable to comprehend. (Courtesy of Drs. John Hills and Dr. Jose Segarra.)

carotid artery due to severe stenosis, this border zone will be first affected and the patient will experience numbness and weakness involving the arm and hand.

With increasing stenosis and increasing ischemia there will be involvement of the face with numbness and a supranuclear type weakness. If the ischemia involves the dominant hemisphere an expressive aphasia will develop due to involvement of Broca's motor speech area. These symptoms indicate the progression to involvement of the more central cortical supply area of the middle cerebral artery. As a general rule, one may assume that carotid ischemia and infarction will occur predominantly over the areas of cerebral cortex within the peripheral and central supply area of the middle cerebral artery *(Fig. 26-5)*. Involvement of the area supplied by the anterior cerebral artery is usually less marked since the anterior cerebral artery is more likely to receive a collateral supply via the anterior communicating artery and via leptomeningeal anastomosis over the corpus callosum from the opposite anterior cerebral artery. At times, however, both the middle and anterior cerebral areas will be involved and infarcted as in case 26-1. At times artery to artery embolization will occur involving primarily the middle cerebral artery *(Fig 26-14)*.

Case 26-1: This 23 year old right handed female on birth control pills, smoking one pack of cigarettes per day, on June 15, 1985 shortly after intranasal "snorting" of cocaine complained of gradually increasing right sided headache and then fell to the floor with a flaccid left hemiplegia with the head and eyes deviated to the right.

Neurological examination: *Mental status:* The patient denied any illness and could not explain why she was in the hospital. *Cranial nerves:* There was a left homonymous hemianopsia, deviation of head and eyes to the left and a left central facial weakness. *Motor system:* A flaccid left hemiparesis was present. *Reflexes:* deep tendon reflexes were initially depressed on the left side. Plantar responses were extensor bilaterally. *Sensory system:* a severe left hemisensory deficit, with a neglect of the left side of space and body was present.

Initial clinical diagnosis: Total right middle cerebral artery territory infarct.

Laboratory data: *Initial CT Scan* demonstrated hypodensity (infarction in right frontal and parietal areas). *Right carotid angiogram* demonstrated severe stenosis of the supraclinoid segment of the right internal carotid artery with severe stenosis of the M-1 segment of the right middle cerebral artery and multiple areas of narrowing of the remainder of the right MCA. *Left carotid angiogram* indicated only limited cross filling of the distal right middle and anterior cerebral arteries.

On 6-21-85, 6 days after admission the previously alert patient suddenly developed a decreased level of consciousness and a fixed, dilated right pupil suggesting that tentorial herniation had been produced by severe edema or hemorrhagic transformation of the infarct. CT Scan now demonstrated massive infarction and edema-right anterior and middle cerebral artery territories plus-transtentorial and sub falx herniation *(Fig.26-6)*. With emergency management of the increased intracranial pressure- (monitor and osmotic agents) the patient became more alert and was transferred to a rehabilitation facility.

The following case histories indicate additional patterns of disease in patients with stenotic-occlusive disease of the carotid artery. In both cases the disease was in the extracranial portion of the vessel; thus an opportunity for surgical correction of the stenotic lesion was presented. Of course, similar symptoms could have occurred with disease of the intracranial portion of the carotid artery, e.g., with stenosis at the siphon. However, in evaluating patients presenting these symptoms, the student should remain alert to the possibility that surgically remedial disease is present. The aim should be recognition of such disease when insufficiency

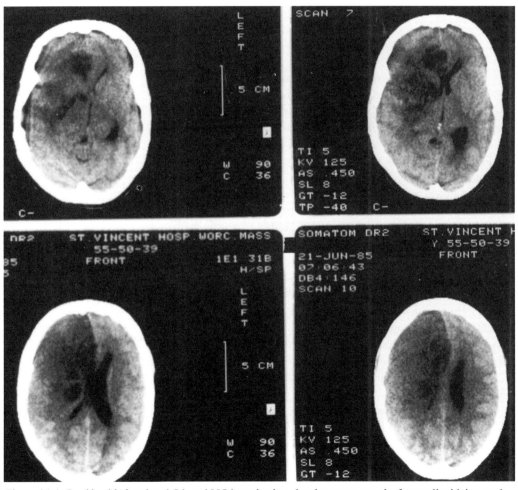

Figure 26-6. Combined infarction ACA and MCA territories related to severe stenosis of supraclinoid, internal carotid artery and proximal (M1) segment of MCA, with infarction territories of right anterior and middle cerebral arteries. CT scan. Case 26-1. Refer to text.

symptoms alone are present, that is, prior to the development of actual infarction. It also should be noted that in some cases no premonitory symptoms will be present but the patient will present with a completed stroke instead of a transient ischemic attack. A patient with carotid TIA's is presented in Case 26-2.

Case 26-2: This 54-year-old, right-handed, white male for 7 years had experienced twice per month intermittent 30-60 second episodes of blurring and blacking out of vision in the left eye. Ten days prior to admission, the patient had a 45-minute episode of minor weakness of the right face, arm, and leg plus numbness of the right side of the face accompanied a transient difficulty in speech (possibly dysarthria, possibly difficulty in word finding). The patient's mother died of a "stroke" at age 66 years, and his father, of heart disease at age 57.

General physical examination: Blood pressure elevated to 160/100 in both arms. Bruits (murmurs) were present over each carotid artery. The retinal artery pulsation was easily obliterated by pressure on the globe of the left eye. Peripheral pulses in the lower extremities were poor.

Neurological examination: Intact except for a minor right central facial weakness.

Laboratory data: Cholesterol was elevated to 284 mg%. *Arteriography (aortic arch study)* demonstrated a complete occlusion at the origin of the left internal carotid artery. A *right brachial arteriogram* demonstrated excellent collateral circulation with filling of the left anterior and middle cerebral arteries by cross flow through the anterior communicating artery from the right side. The intracranial portion of the left internal carotid artery filled from the left posterior communicating artery.

Subsequent course: A left carotid endarterectomy performed by Dr. Allan Callow restored blood flow following removal of the occlusive lesion (atherosclerosis with recent thrombosis) at the carotid bifurcation.

In general, carotid transient ischemic attacks (TIA) are more likely than vertebral-basilar TIA's to proceed to completed strokes. Patients with hemispheric attacks that present with hemiplegia are more likely to proceed to completed strokes than those patients with transient monocular blindness. In the study of Hurwitz et al (1985) cumulative risk for infarct after hemispheric carotid TIAs was 27% compared to 14% for monocular blindness-carotid TIAs. In general, most patients will experience the stroke after one or two transient ischemic attacks. Multiple TIA's over a long period of time are much less likely to result in completed strokes (Kase et al 1987).

Not all patients with carotid TIA's are found to have significant (>75%) stenosis or occlusion of the internal carotid artery. Pessin et al (1977) found that only 52% of 95 patients with carotid TIA's had significant stenosis (42% if hemispheric, 58% if transient monocular blindness and 80% if both types). The explanation for TIA's in patients with less than significant stenosis of the carotid artery is not certain. Emboli from ulcerated plaques; emboli from other sources such as the heart (or aortic arch) or micro platelet and fibrin emboli have all been suggested as possible explanations. (See also Ringelstein et al 1983 and Amarenco et al, 1992; Marshall 1971.)

The use of the antiplatelet agent aspirin has been effective in reducing both strokes and myocardial infarctions in patients with TIA's by 22%.

A current approach to the management of carotid TIAs is to promptly image the extracranial carotid arteries with ultrasound -doppler (Duplex scan)[4]. If significant stenosis (>70-75%) is found at the carotid bifurcation in the neck, then magnetic resonance angiography (MRA) or classical angiography followed by endarterectomy is undertaken. In many centers, the initial noninvasive study is the MRA eliminating the need for both the duplex scan and the angiography. The role of carotid endarterectomy in patients with severe carotid stenosis of 70-99% and TIA's or non disabling strokes was clearly established by the random-

---

[4]*Refer to Hennerici et al(1992) for a discussion of the technique.*

ized North American Symptomatic Carotid Endarterectomy Trial (1991), Kistler, et al (1991), Moore et al (1995) and emphasized by the long term follow up of these patients (Barnett et al 1998).

If significant stenosis is not present, then the patient is treated with aspirin. If this is not effective, then anticoagulation with Warfarin derivatives may be considered (see Brust 1977, Whisnant et al 1978). Such therapy, however, is not without risks. The risk of bleeding complications with anticoagulation is greater than with the use of antiplatelet agents.

In Caucasians, most symptomatic lesions of the carotid artery occur at the bifurcation of the carotid artery in the neck. In non-Caucasians, intracranial lesions increase in frequency[5]. If attacks persist, angiography may be undertaken to document intracranial lesions before embarking on long-term anticoagulation.

In addition, the absence of a significant extracranial lesion should raise the possibility of a source of multiple emboli from cardiac or other sources (see below).

If total occlusion of the artery is demonstrated, several courses of action have been suggested:

a) Emergency removal of an acute thrombus; for example, if progression is occurring (see Walters et al 1987); This may be complicated by intracerebral hemorrhage particularly if infarction has already occurred and blood flow is restored into infarcted tissue (Piepgras et al 1988).

b) External carotid to internal carotid artery bypass. This has been found to be of little value (EC-IC Bypass Study Group, 1985).

Controversy exists as to the management of the patient with the asymptomatic carotid artery bruit or with asymptomatic carotid artery stenosis. In the past, many of these patients with no neurological symptoms or with symptoms unrelated to the specific carotid artery were unnecessarily subjected to carotid artery surgery with accompanying morbidity and mortality of that procedure (see Winslow et al 1988, Dyken, 1986, Hennerici et al, 1986, Levin et al, 1980, Bornstein and Norris, 1989). The study of Freischlag et al, 1992 does suggest a definite role for the procedure where the degree of stenosis is over 75% and where perioperative morbidity and mortality are low (<3%). The study of Chambers and Norris (1986) had previously demonstrated that asymptomatic patients with a stenosis of >75% had an ipsilateral stroke rate of more than 10% per year. The editorial of Kistler and Furie (2000) discusses the effects on both low flow and the formation of thrombus leading to embolic phenomena that follows upon stenosis. The results of randomized trials of patients with asymptomatic carotid artery stenosis suggest that severe stenosis (>90%) may benefit from surgery but leave unsettled the question of whether lesser degrees of stenosis (50-90%) may benefit (Barnett, 1993, Barnett & Haines 1993, The CASANOVA Study Group, 1991, Hobson, et al 1993, Mayberg & Winn, 1995).

The following case demonstrates a case in which the patient began with carotid TIAs progressed to a carotid border zone syndrome of infarction and then involved additionally all of the middle cerebral artery territory.

Case 26-3 (Patient of Dr. Thomas Mullins): This 72-year old right-handed white male had two episodes of right upper extremity weakness, each of two hours duration, beginning 2 days prior to admission. Following the first episode, the patient had transient slurring of speech, persistent left -right disorientation, dysgraphia, and could not understand what he had read.

Neurological examination: *The only findings related to language function:* speech was fluent with intact repetition, and object naming; however, reading comprehension, writing, and left-right orientation were all impaired. *Carotids:* a bruit was present on the left.

Clinical diagnosis: Carotid artery stenosis with TIA's reflecting decreased perfusion in

---
[5]*The effects of race and ethnicity on ischemic stroke are discussed by Zwefler et al (1995) and Selva & Chimowitz(1995).*

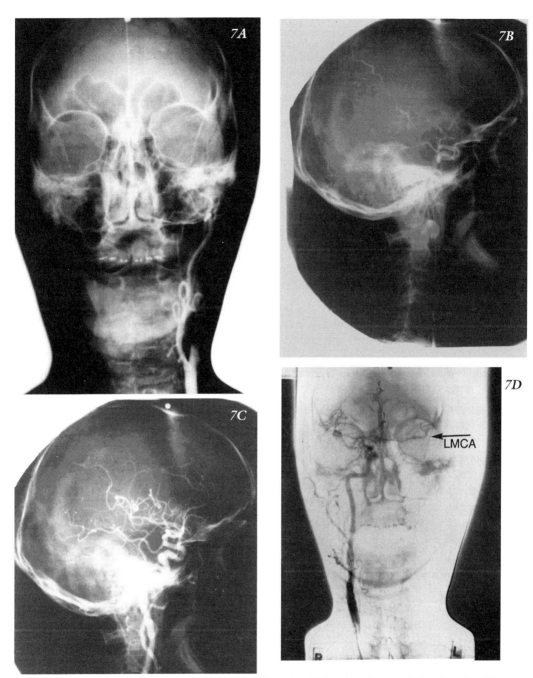

Figure 26-7. Internal carotid artery occlusion at the bifurcation. TIAs with subsequent infarction of middle and posterior cerebral artery territories. Case 26-3 (refer to text.) A and B) Left Carotid injection A) occlusion of left internal carotid just above the bifurcation. B) External carotid anastomotic branches (periorbital etc) allow late filling of the left internal carotid artery at the siphon and its branches. C and D) Right carotid injection C) Lateral view: posterior cerebral artery originates from carotid artery, D) AP view: cross filling of the left anterior cerebral artery and limited filling of the left middle cerebral artery (arrow).

the border zone followed by thrombosis and passage of emboli to the inferior division of the middle cerebral artery.

Laboratory data: *Arteriograms:* (a) the left internal carotid artery was completely occluded just above the bifurcation *(Fig. 26-*

7A). (b) The left external carotid artery filled the left internal carotid artery at the siphon through anastomosis about the orbit *(Fig. 26-7B)*. However, there was a cutoff just below the siphon suggesting that the thrombus (clot) at the bifurcation had extended to this level. Moreover, the left middle and anterior cerebral arteries failed to fill in this manner. (c) The left posterior cerebral artery did originate from this segment of the otherwise occluded carotid siphon segment but with no retrograde filling into the basilar artery *(Fig. 26-7B)*. (d) The right carotid artery filled not only the right middle cerebral and anterior cerebral artery but also was the source of the right posterior cerebral artery *(Fig. 26-7C)*. (e) The right carotid system also supplied the left anterior cerebral artery and to a limited extent the left middle cerebral artery *(Fig. 26-7D)*.

Subsequent course: On the day following admission, weakness in the right hand at wrist and finger extensors recurred. During the arteriogram, the patient was noted to have additional problems with speech. Weakness in the right upper extremity increased. Despite subsequent anticoagulation, additional progression occurred with total paralysis of the right upper extremity, aphasia and a right homonymous hemianopsia. The CT scan, which had been normal on admission now, 72 hours after these events, demonstrated acute infarction with severe edema involving the entire cortical and deep territory of the middle cerebral artery and the cortical territory of the posterior cerebral artery but with relative sparing of the anterior cerebral artery territory and most of the thalamic territory of the posterior cerebral artery (Fig. 26-8). Coma developed secondary to the herniation effects on the brainstem and the patient expired 8 days after admission.

Such massive swelling with herniation of temporal lobe through tentorium and secondary brainstem compression is the major acute cause of death in patients dying after acute carotid artery or massive middle cerebral artery infarctions (31%). Other causes of death are pneumonia (29%), cardiac arrest (17%), pulmonary embolus (13%) (Bounds et al, 1981).

Not all patients with acute carotid infarcts continue to progress. In the series of Jones and Milliken (1976), 39% of patients remained stable. Thirty-five percent had gradual improvement and 19% had a progressing deficit over 48 hours.

MIDDLE CEREBRAL ARTERY

The middle cerebral artery is the final branch and, in a sense the direct continuation of the internal carotid artery. One of the classic internal carotid artery syndromes is essentially a syndrome of the cortical branches of the middle cerebral artery.

Depending on the point of occlusion of the middle cerebral artery and its branches several syndromes are possible. Essentially these syndromes reflect disease of the lenticulostriate penetrating branches or the cortical branches. In some cases, the clinical picture reflects involvement of the territory of both penetrating and cortical branches due to occlusion at the stem or main trunk of the artery.

Syndrome of the Lenticulostriate penetrating branches: These branches supply the putamen, most of the caudate nucleus, the outer segments of the globus, pallidus, and most of the adjacent internal capsule (particularly the anterior half of the posterior limb). These vessels ascend with no anastomotic oops between adjacent branches. The classic clinical picture associated with occlusion of these branches is that of the pure motor syndrome, relatively rapid in onset, involving equally the face, arm, and leg in a spastic hemiplegia (so-called capsular hemiplegia). As discussed by Fisher and Curry (1965), there are often no sensory symptoms and no aphasia. These hemiplegic symptoms, reflecting capsular damage, tend to predominate initially over any symptoms related to the basal ganglia, although portions of the basal ganglia may also be infarcted. If the lesion is small (<1.5 cm.) termed a lacune, the patient may demonstrate a rapid improvement, so that within weeks or months, there is little if any residual motor disability. However with multiple lacunes, deficits may persist *(Fig. 26-*

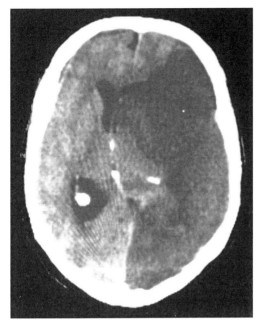

Figure 26-8: Case 26-3 as in figure 26-7. CT Scan obtained 72 hours after acute massive progression demonstrating infarction and massive edema of the left MCA territory and the cortical territory of the posterior cerebral artery, with sparing of ACA.

9). As indicated in *Fig. 26-10*, however, larger infarcts involving the lenticulostriate territory may result in a persistent spastic hemiplegia with considerable motor deficits and symptoms may include sensory and other features.

The basic pathological process has been termed segmental fibrinoid arterial degeneration (Fisher 1969). The walls of these small vessels become thickened and converted into a hyalinized material. Usually but not invariably, this occurs on a background of long-standing hypertension. At times, diabetes mellitus is present. The lenticulostriate branches rising as the initial branches of the middle cerebral artery are under relatively high pressure; not only may these vessels become occluded, they may be the site of small or large intracerebral hemorrhages also complicating the process of hypertension. We shall have cause to discuss these vessels again later in this chapter. However, not all patients with lacunar infarcts have hypertension or atherosclerosis or diabetes mellitus, and at autopsy not all patients demonstrate the specific changes in penetrating

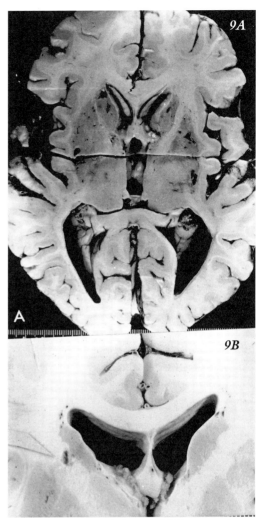

Figure 26-9. Lacunar Infarcts in MCA penetrating branch territories. A) Multiple small infarctions in a 71 year old hypertensive (210/116) who after multiple episodes of right hemiparesis and dysarthria and "dizzy spells" became wheelchair bound, with apraxia of lips, limbs and gait, pseudo bulbar, bilateral pyramidal and frontal release signs. B) Detail of a small lacunae from a 48-year-old white male, with severe hypertension (250/140). (Courtesy of Drs. John Hills and Dr. Jose Segarra.)

vessels noted by Fisher (1969). Other disease processes may certainly affect the penetrating vessels and these include: a) small emboli originating in the heart or arteries, b) the arteritis of meningovascular syphilis (Johns et al 1987); c) the arteritis of tuberculous meningitis; d) the vasculitis of autoimmune diseases such as polyarteritis, and lupus erythematosus (see

Millikan and Futrell 1990). Mast et al (1995) suggest that hypertension and diabetes are more related to multiple as opposed to single lacunar infarcts

Other types of penetrating branch infarct, so called "lacunar syndromes": A number of additional syndromes have been identified in the subsequent studies of Fisher and other investigators (Fisher 1982, Mohr 1982). These syndromes of penetrating branches of various blood vessels are outlined in Table 26-2.

Although many of these above lesions can be demonstrated on CT scans, MRI is the most effective technique for confirming localization *(Fig. 26-11)* (see Larson et al 1988, Rothcock et al 1987).

Lacunar State: Over time some patients have multiple infarcts involving multiple penetrating branches. The patient then begins to manifest a typical clinical syndrome that has been assigned the diagnosis of "lacunar state", based on the findings at the time of pathological examination of multiple small cavities *(Fig. 26-9, 26-11)*. The patient begins to manifest

TABLE 26-2: SYNDROMES OF PENETRATING BRANCHES OF CEREBRAL VESSELS:

PENETRATING ARTERY SYNDROME	LOCATION OF OCCLUSION
a) Pure motor syndrome of hemiplegia	Penetrating middle cerebral artery supply to internal capsule, less often basilar pontine paramedian arteries
b) Pure hemisensory	Posterior cerebral penetrating branches to ventral posterior lateral nucleus of thalamus or penetrating middle cerebral arteries to medial border of the posterior limb of the internal capsule
c) Dysarthria & Clumsy Hand Syndrome	Paramedian penetrating branches supplying the pons
d) Ataxic Hemiparesis:	Infarcts of the corona radiata and less often the posterior limb of the internal capsule-posterior limb (both middle cerebral artery penetrators) and to a minor degree, the pons (basilar-paramedian-penetrating)
e) Hypesthetic-ataxic hemiparesis	Infarction of the posterior limb (posterior medial segment) of the internal capsule and lateral thalamus (anterior choroidal artery). Less often the posterior cerebral artery penetrating branches are involved.
f) Hemichorea-hemiballism	Infarction of the subthalamus or thalamus penetrating branches post. cerebral artery, or lateral putamen of the striatum (penetrating branches of the middle cerebral artery (See chapter 19)
g) "Silent strokes"	Found on CT or MRI scan unrelated to clinical history.
h) Caudate infarcts producing slow apathetic (abulic) state or hyperactive restless state	Lesion in caudate with extension to anterior limb internal capsule - perforators proximal anterior or middle cerebral artery

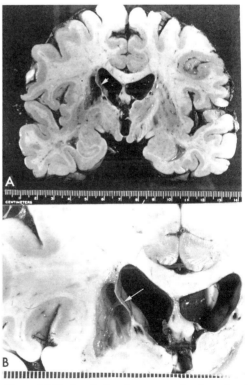

Figure 26-10. Lenticulostriate penetrating branches of the middle cerebral artery. This 73 year old hypertensive had at age 53 sustained a right hemiplegia with persistent and dense motor deficits involving the right central face, arm, and leg. A) Coronal section, B) close up of the cavity. (Courtesy of Drs. John Hills and Jose Segarra).

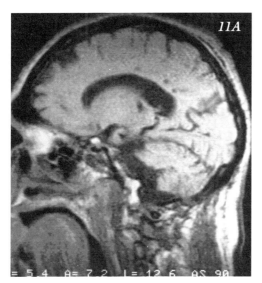

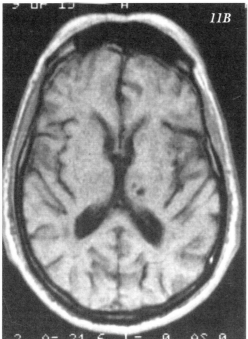

Figure 26-11. Multiple lacunar infarcts most prominent in penetrating branches left posterior cerebral artery: thalamus, posterior one third corpus callosum and adjacent white matter of left posterior cerebral hemisphere in a 70 year old right handed hypertensive male with multiple episodes of "collapse", possibly related to severe orthostatic changes. A) MRI - T1 weighted sagittal section- 12.6 mm left of midline) MRI-horizontal section.

an apraxia of gait, and a release of grasp reflex occurs as a result of damage to the descending fibers from the premotor areas. Clinical findings associated with disease of the basal ganglia may develop: rigidity and a slowness of movement and at times, a minor degree of tremor. These symptoms may reflect the damage to the putamen and globus pallidus or, perhaps, damage to the descending fibers from the premotor areas. This aspect of the syndrome has sometimes been designated as "arteriosclerotic Parkinson's disease" and can usually be distinguished from the classical Parkinson's disease (refer to chapters 18, &19). There may be a variable degree of cognitive dysfunction with considerable rigidity in problem solving, a loss of inhibition of emotional responses and impairment of executive function. Often the patients appear apathetic and are misdiagnosed as manifesting dementia or depression. Subcortical infarcts involving descending frontal systems are responsible (Wolfe et al 1990).

Syndromes of the Cortical Branches. (Figs. 26-12, 26-13). After providing a series of lenticulostriate branches, the main stem of the middle cerebral artery divides into two main cortical divisions: a superior division and an inferior division. At the point of division, the smaller

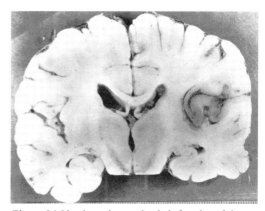

Figure 26-12. Acute hemorrhagic infarction of the central cortical territory of the MCA: calcific embolus occluded MCA following atheromatous occlusion of the carotid artery at the bifurcation. This 74-year-old, right-handed, white male, 5 days prior to death, had the sudden onset of pain in the left supraorbital region and lost consciousness. Upon regaining consciousness, the patient had a right central facial weakness, a right hemiparesis and was mute but was transiently able to follow commands. (Courtesy of Drs. John Hills and Jose Segarra.)

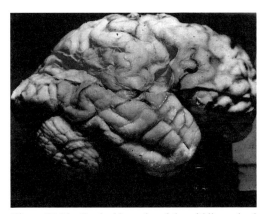

Figure 26-13. Cortical branches of the middle cerebral artery. Old infarcts involving much of the total cortical territory of the right middle cerebral artery. This 72-year-old, right-handed male with severe hypertension and auricular fibrillation expired after a 3 year history of least 4 episodes of left central facial weakness and left hemiplegia involving primarily the arm and a focal seizure involving the left arm. (Courtesy of Drs. John Hills and Jose Segarra.).

lateral orbital frontal and temporal polar arteries originate. The superior division has pre-Rolandic branches (to inferior frontal gyrus and premotor area including Broca's area in dominant hemisphere), Rolandic branches (to pre- and post-central gyri), and anterior parietal branches (to post-central gyrus). The inferior division includes anterior temporal, posterior parietal, angular and posterior temporal branches, named after the areas supplied. The syndrome that results from occlusion will depend on whether the main cortical stem, the superior or inferior division, or the branches thereof have been involved. In general the MCA cortical syndromes are assumed to be embolic in origin particularly in the Caucasian population. The stem, divisions and the cortical branches of the left middle cerebral artery are a frequent site for the lodgment of embolic material because there is a relatively direct vertical takeoff of the left common carotid artery from the arch of the aorta compared to the right carotid artery. In addition the middle cerebral artery is the direct continuation of the internal carotid artery; the cortical branches are the direct continuation of the main middle cerebral trunk.

As noted above, infarcts may occur due to carotid disease often due to artery-to-artery emboli.

Occlusion of the main cortical stem produces a mixed cortical motor -sensory syndrome involving the contralateral face (supranuclear) and arm; this is termed a faciobrachial paralysis. If this occurred in the dominant hemisphere a mixed type of aphasia would also be present with anterior (nonfluent) and posterior (fluent) components. Face and language involvement would be greater than hand involvement. The degree to which the hand is involved would depend on the degree of collateral circulation. Thus only the central cortical territory or the entire cortical territory of the middle cerebral artery could be involved.

Occlusion of the superior division in the dominant hemisphere would produce a mixed cortical motor- sensory syndrome involving again predominantly the face and hand plus an expressive Broca's type aphasia. This syndrome has already been illustrated in Chapter 24.

The less frequently encountered occlusion of the inferior division branches, producing dominant or non-dominant inferior parietal-post temporal syndromes including the syndromes of posterior type aphasias and parietal neglect already discussed in chapter 24. Refer also to Caplan et al, 1985, 1986 and Mohr et al, 1986.

Combined Syndrome: Total Occlusion of Initial Segment of Middle Cerebral Artery (Fig. 26-5). The resultant syndrome in this case will be the summation of the lenticulostriate and cortical branches syndrome. The following case history illustrates such an occlusion with massive infarction of the entire territory of the middle cerebral artery.

Case 26-4: This 61-year-old, right-handed hypertensive black housewife one week prior to admission, suddenly fell to the floor while taking a bath and lost consciousness. She was found by her son who noted that she was unable to move her left arm and leg. Her speech had been thick, but no aphasia had been present. No headache had been noted. Both parents had died of heart disease and hypertension.

Physical examination: Blood pressure was 150/100 and pulse was regular at 84.

Neurological examination: *Mental status:* The patient was obtunded and slow to respond but grossly oriented to time, place, and person. *Cranial nerves:* Papilledema was present bilaterally particularly in the right fundus where a recent hemorrhage was present. The head and eyes were deviated to the right. The eyes did not move to the left on command but did move to the left on vestibular stimulation. The patient neglected stimuli in the left visual field. Pain sensation was decreased or neglected on the left side of the face. A marked left central facial weakness was present. *Motor system:* A complete flaccid paralysis of the left arm and leg was present. *Reflexes:* Deep tendon reflexes were increased on the left compared to the right. The left plantar response was extensor. *Sensory system:* All modalities of sensation were decreased on the left side of the body.

Clinical diagnosis: Acute occlusion of the stem of the middle cerebral artery, possibly embolic, although primary occlusions of intracranial arteries may occur in the non-Caucasian population. Hypertensive intracerebral hemorrhage was also possible.

Laboratory data: The EEG indicated severe focal damage with a relative absence of electrical activity in the right temporal area and focal 3 to 5 Hz slow waves most prominent in the right frontal area (Fig.2-23 and CD EEG atlas). Radioactive brain scan demonstrated a marked right hemisphere uptake of isotope (Hg^{197}) measuring 11x6x6 cm., extending from the right frontal area to the posterior parietal area extending from the surface to the deep midline *(Fig. 26-14)*. Right brachial arteriogram revealed a complete occlusion of the right middle cerebral artery at its origin *(Fig.26-15)*. Cerebrospinal fluid contained no significant cells.

Hospital course: The patient showed no significant improvement during a four-week hospital course.

Prognosis for survival during the acute state of the first week depends on the degree of

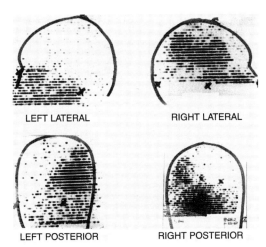

Figure 26-14. Right middle cerebral artery occlusion Case 26-4. Brain scan. A marked uptake of isotope (Hg197) is found throughout the territory of the right middle cerebral artery (22 days after the event). This is demonstrated in the posterior, anterior, and right lateral scans. (Courtesy of Dr. Bertram Selverstone).

edema. In patients with massive middle cerebral artery territory infarcts due to MCA occlusion who already had early CT scan evidence of brain swelling within 24 hours of the ictus, fifty five percent of these patients expired, over-

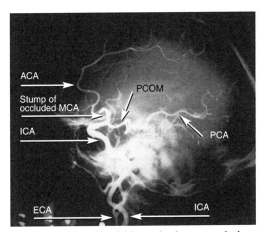

Figure 26-15. Right middle cerebral artery occlusion. Case 26-4. A) Selective common carotid injection reveals total right middle cerebral artery occlusion. (Courtesy of Dr. Samuel Wolpert). B) Reference diagram of the major branches of the internal carotid artery. ICA= internal carotid artery, ACA= anterior cerebral artery, PCA= posterior cerebral artery (supplied directly from the IC as a continuation of the PCOM posterior communicating artery), MCA= middle cerebral artery. (Refer to Fig. 26-7C and 2-27c for normal reference).

whelmingly of herniation effects. Patients older than 45 years had a poorer prognosis than younger patients. Infection of the lungs is a serious complication of large infarcts and a major cause of death during the second week. If the patient survives, prognosis for recovery is related to the size of infarct. Less than 3 cm. infarcts may be associated with a good recovery, greater than 3 cm. with severe disability (Olson, 1991).

ANTERIOR CEREBRAL ARTERY

In our clinical experience occlusive disease of the anterior cerebral artery is much less common than that involving the internal carotid or middle cerebral arteries accounting for approximately 1.8% of all ischemic infarcts - Bogousslavsky and Regli 1990). The anatomical explanation for this is evident. Each anterior cerebral receives collateral circulation from several possible sources: (a) leptomeningeal anastomotic end-to-end loops from the middle cerebral artery of the same side, (b) leptomeningeal anastomotic loops from the contralateral anterior cerebral artery over the corpus callosum, and (c) anterior communicating artery from the contralateral anterior cerebral artery. Finally, at postmortem examination atherosclerotic plaques are found less frequently in the anterior cerebral than in the larger internal carotid, middle cerebral, basilar, and vertebral arteries. In the series of Bogousslavsky and Regli (1990) 63% of the 27 patients had infarcts secondary to an embolus from the heart or carotid artery.

Penetrating and cortical branches may be distinguished. *The penetrating branches* including a larger vessel, the recurrent artery of Heubner supply the anterior limb of the internal capsule and the anterior head of the caudate nucleus. *The cortical branches:* orbital frontal, frontal polar, callosal marginal and pericallosal supply much of the corpus callosum (genu and body), and the orbital frontal cortex, the medial aspects of the frontal and parietal lobes including the sensory motor areas of the paracentral lobule and the supplementary motor cortex. Symptoms then when they occur will be most severe in the contralateral lower extremity with minor involvement of the contralateral supper extremity and minor or no involvement of the face. Most patients will have involvement of both the cortical and penetrating branch-subcortical territories. Infarcts limited to the penetrating branch territories may occur (Caplan et al, 1990). Note that not all vascular syndromes with predominant leg involvement are due to occlusion of the anterior cerebral artery. Occlusion of the superior sagittal sinus or basilar-vertebral disease affecting the brain stem may also produce predominant leg weakness (refer to Schneider &Gautier, 1994 for additional discussion).

At times both anterior cerebral arteries may be essentially branches of the same proximal segment. Occlusion of this proximal anterior cerebral artery segment will then result in infarction of the medial frontal-parietal areas of

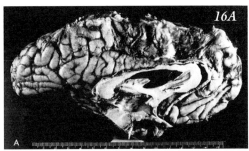

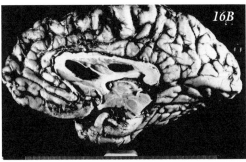

Figure 26-16. Anterior cerebral arteries: old bilateral infarcts with due to severe bilateral stenosis ACA (complete occlusion on right). Left hemisphere involvement was marked (A), with a lesser involvement of the right hemisphere (B). This 61-year-old, right-handed female 16 months prior to death developed difficulty in speech, right-sided weakness, apraxia of hand movements and progressed 2 days later to an akinetic mute state with urinary incontinence.(Courtesy of Drs. John Hills, and Jose Segarra.

both hemispheres *(Fig .26-16)*. The resultant syndrome will include a more prolonged change in personality and affect (due to involvement of the prefrontal and anterior cingulate areas), an akinetic and mute state (Freeman, 1971) (due to involvement of the prefrontal and premotor areas), urinary incontinence (due to involvement of the para- central lobule, medial premotor, and supplementary motor areas), and a sensory motor syndrome involving both lower extremities (due to involvement of paracentral lobule). There may also be a significant apraxia of the nondominant hand from damage to the corpus callosum depriving the premotor and motor areas in the nondominant motor hemisphere of information from the dominant hemisphere. A bilateral release of the grasp reflex is present because of the bilateral involvement of the premotor and supplementary motor areas. Many of these features may follow unilateral infarction in lesser degree. Transient mutism and transcortical motor aphasia are more prominent with dominant hemisphere infarcts, neglect syndromes with nondominant infarcts. Release of instinctive grasp is usually contralateral to the infarct. An example of infarction in the distribution of the anterior cerebral arteries complicating aneurysm surgery is provided in case 26-9.

ANTERIOR COMMUNICATING ARTERY

Penetrating branches from this vessel supply the anterior hypothalamus, the optic chiasm and the suprachiasmatic and paraolfactory areas.

ANTERIOR CHOROIDAL ARTERY

This is the only branch of the internal carotid artery that has not yet been discussed. This vessel supplies the inner portion of the globus pallidus, a small adjacent section of the posterior limb of the internal capsule (including the area occupied by the ansa and fasciculus lenticularis) and to a variable degree the adjacent sector of the ventral posterior lateral nucleus of the thalamus and of the lateral geniculate or geniculocalcarine tract. Various clinical results may follow occlusion of this vessel. There is no single characteristic syndrome. The initial surgical procedures of Cooper indicated that occlusion of this vessel in the patient with Parkinson's disease as discussed in chapter 19 resulted in a relief of tremor and rigidity in the contralateral limbs without the development of hemiparesis or other focal deficits. Foix in 1925 first described a triad of hemiplegia, hemianesthesia, and homonymous hemianopsia associated with infarction of the territory of the anterior choroidal artery (Bruno et. al. 1989 and Decroix et. al. 1986). With the development of CT scan and MRI scan, the syndrome is now more frequently recognized. Essentially this is a variable penetrating or lacunar syndrome ranging from a pure motor syndrome to the full-blown triad.

POSTERIOR CEREBRAL ARTERY

The posterior cerebral arteries usually originate as the direct continuation of the basilar artery following its bifurcation. At times, then, occlusive disease of the vertebral or basilar artery will be manifested by the development of symptoms within the posterior cerebral territory. At times, a mixture of brain stem and cerebral hemisphere symptoms will be present. At times, bilateral posterior cerebral artery symptoms will be present. Moreover, since the cortical and penetrating vessels of the posterior cerebral artery are the distal branches of the basilar vertebral circulation, they will at times be subject to embolization from extracerebral sources (cardiac) or from occlusive disease in the more proximal sections of the basilar or vertebral arteries. As demonstrated previously in some patients (25%), the cortical divisions of the posterior cerebral arteries may originate as a direct continuation of the posterior communicating artery.

Essentially two categories of posterior cerebral symptomatology may be recognized: (a) syndromes of the penetrating branches, and (b) syndromes of the cortical branches. With main trunk occlusions, a mixture of these two syndromes may be present.

Overall, posterior cerebral artery cortical

TABLE 26-3. COMPARISON OF SYNDROMES FROM PENETRATING VERSUS CORTICAL BRANCHES OF PCA.

VESSEL	SYNDROME
The Penetrating Branches. -supply the rostral portion of the midbrain and the thalamus (often involved in lacunar infarcts)	-Occlusion of penetrators to midbrain damages the cerebral peduncle, the third cranial nerve, and the red nucleus (Weber's or Benedikt's Syndrome -Occlusion of penetrators to subthalamus & VL nucleus of thalamus produces contralateral hemichorea or hemiballismus Damage to VPM/VPL of thalamus produces contralateral loss of sensation and possible pain syndrome (thalamogeniculate artery-thalamic syndrome).
The Cortical Branches (Fig. 26-17,26-18). -anterior temporal and posterior temporal branches to the inferior and medial surfaces of the temporal lobe, posterior cerebral artery then divides into a parieto-occipital branch and the calcarine artery to the visual (calcarine) cortex.	-In bilateral posterior cerebral disease impairment of recent memory may occur, accompanied by a significant degree of confusion and disorientation ±bilateral blindness -If the dominant temporal lobe is involved, a confusional state may follow unilateral posterior cerebral disease. -Ischemia or infarction within the calcarine artery territory produces a contralateral homonymous hemianopsia, often with sparing of the macular area due to collateral circulation from MCA (see above if bilateral) (See Chapter 23).

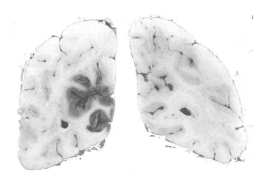

Figure 26-17. Posterior cerebral artery: acute hemorrhagic infarction predominantly within the distribution of the right calcarine branch. This 53-year-old male with a right frontal-anterior temporal glioma had tentorial herniation with compression of the right posterior cerebral artery (in addition to brain stem structures). (Courtesy of Drs. John Hills and Jose Segarra.)

infarcts account for 5% of all infarcts. Usually the infarcts reflect embolic occlusion (see Fisher 1986, Pessin et al 1987, Koroshetz and Ropper 1987). Among the 54 cases presented by Kinkel et.al. (1984), 11% demonstrated total involvement of the cortical territory. Most had partial involvement primarily of calcarine and/or the posterior temporal artery territory. Nine percent had bilateral infarcts.

The following case history illustrates many aspects of ischemia within the cortical distribution of the posterior cerebral artery. This patient demonstrates the general rule that posterior cerebral cortical infarcts are usually embolic.

Case 26-5: This 55 year old ambidextrous married white female research coordinator at approximately 11:30 AM on the day of admission suddenly developed blurring of vision in her left visual field, possibly in her left eye and was found to have a left visual field defect. Shortly thereafter, she developed tingling paresthesias of the left face arm and leg. Hypertension was under treatment with hydrochlorothiazide and Lasix.

Physical examination: Blood pressure was elevated to 160/70. Weight was elevated to 264 pounds. A small ecchymosis was present under the toenail of the left second digit.

Neurological examination, (3hours after onset): *Cranial nerves:* A non-congruous left homonymous hemianopsia was present greater in the left temporal field than the right nasal field. *Reflexes:* Deep tendon stretch reflexes were decreased in the lower extremities (patellar, 1+ and Achilles, 0). Plantar responses were equivocal, with the left probably extensor. *Sensory system:* pain and graphesthesia were decreased over the left foot.

Clinical diagnosis: Posterior cerebral artery ischemia and possible infarct, (possibly embolic), involving the right calcarine cortex and thalamus or a lacunar event involving the right thalamus.

Laboratory data: *MRI,* 4days after onset *(Fig.26-18)* demonstrated a small infarct in the right occipital visual cortex area in the distribu-

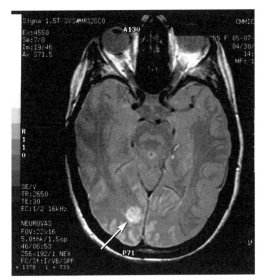

Figure 26-18. Infarct right occipital area, calcarine artery branch territory (Embolus to posterior cerebral artery from heart). MRI. Case 26-5.

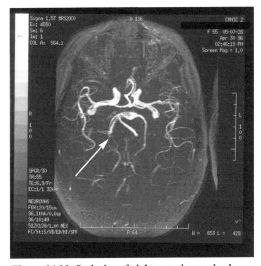

Figure 26-19. Occlusion of right posterior cerebral artery(arrow). MRA. Case 26-5 as above.

tion of the calcarine artery. *MRA (Fig.26-19)* demonstrated decreased flow in the right posterior cerebral artery. There was filling of more distal branches via anastomatic flow. *Holter monitoring* demonstrated multiple brief episodes of paroxysmal atrial fibrillation. The *transesophageal echocardiogram* demonstrated a definite patent foramen ovale with right to left shunting.

Subsequent course: Re-evaluation of the patient 22 hours after the onset of symptoms demonstrated clearing of all neurologic findings. When seen 2 weeks after the event her only complaint related to some problems in reading. This patient already been receiving aspirin at the time of the event and this was discontinued. She was begun on lifelong systemic anticoagulation with coumadin by the consultant cardiologist Dr. Filiberti since she had two conditions associated with embolization: paroxysmal atrial fibrillation and patent foramen ovale with right to left shunting. The patient did well over the subsequent 5 years.

Case 26-6 presented on CD ROM provides an example of a patient with bilateral posterior cerebral artery ischemia, manifested by acute onset of blindness and confusion.

CEREBRAL EMBOLISM

This topic has already been discussed and illustrated above in cases 26-3, 26-4, 26-5 and in the chapter on aphasia. The middle cerebral and posterior cerebral artery cortical branches are primary targets.

In general, cerebral embolism is a complication of cardiovascular disease. However, as we have indicated, artery to artery (the carotid, vertebral or basilar) embolization of thrombus or atheromatous material is also frequent. Rarely emboli may be composed of tumor cells or, following trauma, of fat or air (Jacobson et al, 1986; Knoppee et al, 1988). The embolus in most cases is a fragment of thrombus (clot, platelets, fibrin, and blood cells) that has become detached from a larger thrombus within the heart or proximal artery. The causes of such a thrombus within the heart are several:

a. A clot will often form in the left auricular appendage when atrial fibrillation is present. This is often the case when a relative stasis of blood in this area occurs as in mitral stenosis (Wolf et al, 1978, 1991, Peterson 1990).

b. A "mural thrombus" may form on the endocardium in relation to an area of myocardial infarction (Foster and Halperin, 1989, Konnrad, 1984).

c. Thrombus may collect on heart valves (so called valvular vegetations), usually the mitral valve, when these valves are the subject of infection and inflammation as in bacterial endocarditis with embolic material breaking off and entering into the circulation at times of change in cardiac rhythm. (Hart et al 1990, Jones and Siekert, 1989, Salgado et al 1989).

d. The aorta may be a source of embolic material (Amarenco et al, 1994, Jones et al 1995, Kistler 1994)

e. In patients with a patent foramen ovale, within the heart, embolic material may originate in the veins of the legs or pelvis. Such patients may have both pulmonary and "paradoxical" cerebral emboli. (Lechat et. al. 1988, Jones et al, 1983).

Multiple small emboli of cardiac origin may occur to many cerebral vessels to produce a syndrome of multiple small vessel occlusions with a resultant clinical picture of dementia, alteration in personality, and so forth. Patients with multiple cerebral emboli also have embolization to extracerebral areas as well, such as the femoral artery, the kidney, the spleen, and the skin. Emboli to the kidney will often be detected following the sudden appearance of red blood cells in the urine (hematuria).

Differentiating embolic from non-embolic ischemic-occlusive disease. Several points should be considered: (see also Ramirez-Lassepas et al, 1987):

a. Perhaps most important is the fact that embolic occlusion of a vessel is sudden. In general there are no preceding transient ischemic events. At times, the event may be so sudden that the patient stops speaking in mid sentence with a complete loss of speech. On the other hand, infarction due to ischemic-occlusive disease is often preceded by one or more transient episodes of ischemia.

b. Since embolic occlusions are sudden, the event may be accompanied by a focal or generalized seizure.

c. Embolic occlusions may occur during any time of the day and often during periods of activity. Ischemic-occlusive infarctions tend to occur during sleep when blood pressure is relatively lower or shortly after arising in relation perhaps to transient falls in blood pressure.

d. Infarction due to embolic occlusion of a vessel is more likely to be hemorrhagic than a non-embolic infarction. This is related to reperfusion of an area of recent necrosis due to disintegration of the embolus or to the development of collateral (leptomeningeal anastomotic flow). (Fisher and Adams 1951, Okada, 1989, Ogata, et al, 1989.) In such an area, the walls of blood vessels have also been ischemic and recently damaged. Such vessels when again perfused under a normal head of pressure will then leak blood into the surrounding tissues. In the study of Honig et al (1993), MRI at 3 weeks demonstrated hemorrhagic transformation in 68% of cardioembolic infarcts but always without clinical deterioration.

e. Infarctions due to embolic occlusions are also often subject to rapid improvement because with the disintegration of embolic material, blood flow is restored.

Treatment is discussed below.

MANAGEMENT OF ISCHEMIC AND OF EMBOLIC DISEASE

As regards management of patients with atherosclerotic ischemic-occlusive disease the student should consult Biller & Love (2000) and Brott &Bogousslavsky (2000). For patients seen within 3 hours of an occlusion usually embolic of the middle cerebral artery, the use of intravenous t-PA may be considered if the CT scan does not demonstrate hemorrhage or more than one third of the middle cerebral artery territory involved by infarction. For patients seen within 6 hours direct intra-arterial thrombolysis may be considered. For most patients who have suffered infarcts, there are no specific curative therapies. Considerable recovery will occur in a supportive setting for many patients. Some patients with transient ischemic episodes or infarcts with minimal residual within the carotid distribution will have treatable extracranial vascular disease, which will benefit from surgical correction of the stenosis or occlusion. In addition to direct surgical carotid endarterectomy, the use of balloon angioplasty with or without stents is being developed as for coronary artery disease. Many patients with transient ischemic attacks or with a prior stroke will have a significant reduction in stroke risk with antiplatelet therapy (aspirin). Anticoagulation has a significant role in the prevention of cerebral emboli in patients with atrial fibrillation of both rheumatic and nonrheumatic origin but has little value in other types of occlusive disease. The best treatment is prevention. The major factors which can be controlled are hypertension, the risk factors for cardiac disease including the prevention of rheumatic and coronary artery heart disease, cigarette smoking, diabetes mellitus, dietary factors and illicit drug use (Refer also to Bronner et al, 1995 and White et al, 2000, Inzitari et al, 2000). A major risk factor that unfortunately cannot be well controlled is the family predisposition for cerebrovascular or coronary artery disease (Graffagnino et al, 1994). For a discussion of ischemic-occlusive disease involving the basilar-vertebral circulation, the student should refer to the chapters on the brain stem and cerebellum.

INTRACRANIAL HEMORRHAGE:

There are two types of intracranial hemorrhage: *intracerebral hemorrhage and subarachnoid hemorrhage*. In the Framingham study (Sacco et al 1984), intracerebral hemorrhages constituted 5% of all strokes and subarachnoid hemorrhage; 9% of all strokes. Both have a high mortality. The young and middle age Black population in the United States has an increased risk of both intracranial and subarachnoid hemorrhage compared to the white population, (Broderick, et al 1992).

INTRACEREBRAL HEMORRHAGE

Intracerebral hemorrhage is primary in 83 % of cases and secondary in the remaining 15%. Hemorrhage into the brain may occur in patients receiving anticoagulants, Kase et al 1985, Franke et al 1990, or as a complication of various hematological problems which are characterized by bleeding disorders, e.g., leukemia (Fig. 26-20). Bleeding into the brain may also occur in relation to arteriovenous malformation or to intracranial tumors (such as glioblastomas or metastatic malignancy, Little et al, 1979) or as the complication of the rupture of a saccular aneurysm where there is extension of blood from the subarachnoid space into the substance of the cerebral hemisphere. In general, however, the term intracerebral hemorrhage is applied to those primary massive and medium-sized hemorrhages into the parenchyma of the brain. The major risk factor is hypertension. However in approximately 40% of patients, blood pressure is normal (Brott et al 1986). Other risk factors include excessive alcohol intake, and cerebral amyloid angiopathy with deposition of beta amyloid in cerebral blood vessels. The latter condition is associated with the presence of the e-2 and e-4 alleles of the apolipoprotein E gene. Recurrent hemorrhages may occur at different sites. Patients with intracerebral hemorrhages are seen less commonly on general neu-

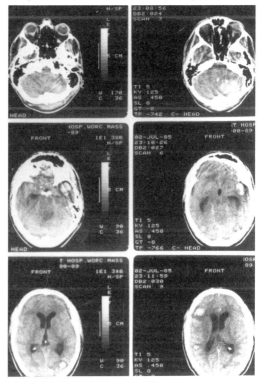

Figure 26-20. Multiple cerebral hemorrhages secondary to acute thrombocytopenia (platelet count of 17,000) with acute lymphatic leukemia. CT scan without contrast in a 24 year old with headache and acute onset of coma, clinical brain death and a flat EEG.

TABLE 26-4. INTRACEREBRAL HEMORRHAGE

Location	Vessel	Frequency*/Figure	Clinical Manifestations
Putamen-lateral ganglionic mass-internal capsule	Lenticulostriate penetrators: middle cerebral artery	33-40% Fig.26-21, 26-22	Progressive headache, hemiparesis hemianesthesia, confusion, herniation and coma
Temporal lobe stem and lobar (Lobar often secondary to amyloid angiopathy)	Superficial small and medium arteries	23% Fig. 2-29	Progressive temporal lobe or other cortical mass with focal seizures. Recurrent or multiple.
Thalamus	Penetrators: posterior cerebral artery	20% Fig. 26-24	Unilateral numbness, Hemianesthesia, and vertical gaze impairment
Pons	Paramedian penetrators of the basilar artery	7% Fig 13-26	Sudden quadriplegia, coma, respiratory impairment, death Small lateral tegmental lesions may survive.
Cerebellum	Branches posterior inferior or superiorcerebellar arteries	8% Fig. 20-13	Sudden headache, vomiting, vertigo, ataxia, then extra-ocular findings and coma

* Based on series of Hier et al, 1977, Kase et al, 1982, Kase and Caplan, 1986

rological and neurosurgical services than patients with occlusive vascular disease and those with subarachnoid hemorrhage. Patients with intracerebral hemorrhage are, however, relatively common when acute general medical admissions and autopsies are considered. There is certainly a predilection for certain areas of the brain as demonstrated in Table 26-4.

The basic pathogenesis involved in intracerebral hemorrhage is not certain. The fact that penetrating arteries such as the lenticulostriate arteries, are also involved in occlusion with lacunar infarction in hypertension has led to several hypotheses concerning the etiology of the hemorrhages:

a. The vessel wall may be weakened by this process and may dilate and rupture (the miliary aneurysm of Bouchard and Charcot).

b. The vessel may occlude with infarction. When collateral flow is then introduced into the acutely necrotic area, hemorrhage occurs.

c. Related to this hypothesis is the concept of fluctuations in blood pressure. Certainly such wide fluctuations may occur in some hypertensive patients. With a fall in blood pressure, infarctions would occur. With restoration of elevated blood pressure, these necrotic vessel walls would be unable to take the stress of the increased pressure and would bleed.

d. Another explanation suggests that in some cases venous occlusion may have occurred. Venous occlusions are usually hemorrhagic.

In both explanations (b) and (c) there is implied a concept of hemorrhagic infarction. In some cases of premonitory ischemia, symptoms will be noted.

e. It has been suggested that in younger patients when hypertension is not present, bleeding has occurred from an arteriovenous malformation, the evidence of which has been destroyed by the massive hemorrhage (Case 26-7 on CD).

f. The use of sympathomimetic drugs utilized in nasal decongestants and cough syrups in addition to amphetamines and the use of illicit street drugs such as "crack" cocaine have also been implicated in younger adults

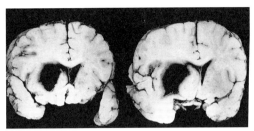

Figure 26-21. Intracerebral hemorrhage from the penetrating (lenticulostriate) branches of the right middle cerebral artery, into the putamen with displacement of adjacent structures, and ruptured into the lateral ventricle. This 46-year-old, black female with severe hypertension (230/130) had the sudden onset of left face, arm, leg paralysis, left visual field defect, and then 72 hours later tentorial herniation with progressive brain stem deterioration. (See Fig. 13-8) (Courtesy of Tufts Pathology Department.)

(Harrington et al 1983, Levine et al 1990).

g. In elderly patients when hypertension may or may not be present and when the location of the hemorrhage is lobar involving the superficial small and medium sized arteries rather than small deep penetrating branch vessel, a process of "amyloid angiopathy" has been identified. Deposits of amyloid occur in the media and adventitia of these vessels without the presence of systemic amyloidosis. (See discussion above and Finelli et al, 1984, Molinari, 1993).

The pathological features to be found with an intracerebral hemorrhage may be briefly summarized. The hematoma may continue to expand during the first three hours resulting in continued deterioration. Within a few hours of cessation of the hemorrhage, clotting occurs. In general with hypertensive hemorrhages, bleeding does not again occur into the same area. The blood is not quickly removed from the brain parenchyma. If this hemorrhage is actually a complication of infarction, then in the early stages, the changes of ischemic infarction will be noted with infiltration of adjacent ischemic tissue by polymorphonuclear leukocytes.

The actual clot of a hemorrhage will remain as a red or reddish black mass for a matter of several weeks. Macrophages that may have been already abundant in adjacent necrotic tissue begin to digest red blood cells at the periphery of the hemorrhage producing the yellow-brown iron pigment hemosiderin. At three weeks, this produces at the periphery, a rim of orange. The more central area of hemorrhage never undergoes phagocytosis but is converted to a semiliquid mass after several additional weeks or months.

When the brain is examined many months after a hemorrhage, there will be found a residual cleft or cavity with orange-stained walls. The staining is due to the persistence of hemosiderin-containing macrophages. The wall, in general, is similar to that surrounding the cavity, which results from an ischemic infarct. A proliferation of capillaries has occurred with the formation of connective tissue fibers. In adjacent tissue, proliferation of astrocytes has occurred with the formation of glial fibers.

The gross changes over time may be demonstrated with sequential MRI or CT scans *(Fig. 26-22, 26-23)*.

The clinical manifestations will depend on the location and size of the hemorrhage and are summarized in Table 26-4. In general with the most common variety - those into the putamen - there is the sudden onset of progressive headache, followed by progressive hemiparesis, and then within minutes to hours by hemianesthesia, confusion, and coma. The progression of symptoms is due to the progressive enlargement of the hematoma and to dissection of the hemorrhage along fiber pathways. Displacement of midline brain stem structures and tentorial herniation often occurs. With these brain stem complications and with extension of the hemorrhage into the ventricular system, a decerebrate state and a compromise of respiratory functions develop. Death occurs in a high percentage of such patients (75 per cent) within hours to days. Overall, 50% of patients with large hemorrhages expire. The prognosis for smaller hemorrhages is more favorable (21%). Prognosis is directly related to a) size of hematoma on CT scan and b) level of consciousness (Hier et al 1977, Broderick et al 1993). Patients with small hemorrhages may have only a minimal degree of disability since

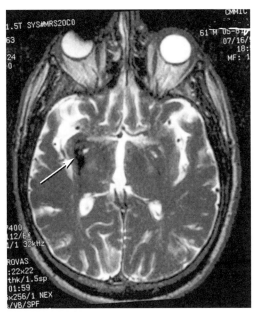

Figure 26-22. Resolving hematoma right putamen, hypodensity of a cystic cavity with residual hemosiderin and dystrophic calcification (arrow), on MRI T2, 16 months after the acute hemorrhage. This 61 year old right handed white male had the acute onset of a left hemiparesis, had an excellent recovery of leg function but continued to have severe spastic weakness in the left upper extremity with tonic neck effects, a hemiplegic gait, decreased pain sensation in left face and arm and a dressing apraxia left arm.

the hemorrhage displaces structures rather than destroying structures as in an infarct.

The diagnosis of intracerebral hemorrhage may be made on the basis of the sudden onset of symptoms. The onset occurs during a period of activity as opposed to the pattern for arteriosclerotic ischemic-occlusive disease where onset during sleep or shortly after awakening is more characteristic. Occasionally warning prodromal symptoms are present; such warning symptoms, however, are more characteristic of ischemic-occlusive disease.

In general, the cerebrospinal fluid is under increased pressure and usually but not invariably contains red blood cells. The number of such cells, however, is usually less than that found in subarachnoid hemorrhage.

At the present time, the major aids in diagnosis are the CT scan and the MRI scans. There is then no indication for lumbar puncture such a procedure in any case may be

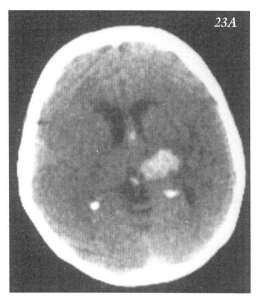

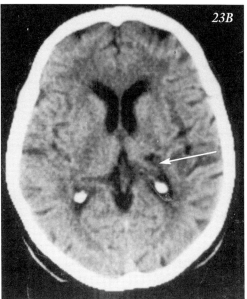

Figure 26-23. A) Acute hemorrhage left thalamus on CT Scan of a 72 year old woman with clinical findings of severe hypertension (210/130), dense right sided hemiparesis and sensory deficits plus fluent aphasia. B) Old cystic cavity (arrow) on CT Scan, 2.5 years later. She had significant sensory symptoms and cortical sensory deficits in right arm and leg, ataxia of gait, intention tremor of right hand and leg and occasional choreiform movements of the right hand.

dangerous when increased intracranial pressure is present.

As regards treatment, the deep location within the putamen, thalamus, and pons means that surgical evacuation of these massive hemorrhages is not feasible. Moreover, with most large pontine hemorrhages, death rapidly ensues. With extension of hemorrhages of the putamen or thalamus into the ventricle, death usually occurs within a short time. However, expanding masses of more superficial lobar hematoma within temporal or parietal lobe, offer the possibility of surgical therapy. Cerebellar hematomas when diagnosed prior to brainstem compromise are often evacuated with a high degree of recovery. Qureshi et al (2001) have reviewed current approaches to management.

SUBARACHNOID HEMORRHAGE

The diagnosis of primary subarachnoid hemorrhage is made on the basis of gross blood present in the cerebrospinal fluid in the absence of primary intracerebral hemorrhage or trauma. Both of these conditions may have secondary leakage of some blood into the subarachnoid space. In 85% of cases of primary subarachnoid hemorrhage in the adult, the source of bleeding is a ruptured saccular aneurysm. (Refer to table 26-5). The basic pathology in the saccular or berry aneurysm is a defect in the media and internal elastic membrane of the vessel wall. Since the media of cerebral arteries develops in a multicentric manner, adjacent sections meet at arterial bifurcations where clefts or gaps in the media are common. These developmental defects in the media probably occur to some extent in all individuals. When later in life thinning or loss of the internal elastic membrane is superimposed at this point of defect, the thin layer of intima bulges out and is covered only by the loose connective tissue of the adventitia. The ballooned protrusion is referred to as a saccular aneurysm.

The unruptured saccular aneurysm is usually asymptomatic. Unruptured saccular aneurysms occur as an incidental finding in 1-2 percent of routine autopsies[6]. Most do not rupture and do not cause symptoms. Occasionally, however, symptoms may be noted prior to rupture. Thus, an enlarging aneurysm of the posterior communicating artery may compress the third nerve. At times an aneurysm of the cavernous portion of the carotid artery may invade the pituitary fossa or compress the optic chiasm (table 26-5) or rupture into the cavernous sinus. In general, the initial symptoms are those related to rupture. The usual symptoms are those of sudden severe headache, (the most severe ever experienced by the patient), vomiting, neck pain, and stiffness. The latter symptoms are due to meningeal irritation. Sudden straining, exercise, or sexual activity often precipitates the headache. In massive rupture, sudden onset of coma may occur. In a less severe bleed, consciousness is sometimes well preserved. Often no specific focal symptoms are present, and it may then be difficult to determine particularly in the comatose patient, on the basis of the clinical findings alone, the actual site of rupture. At times, the presence of minor focal signs will allow the localization of the aneurysm. Those cases where consciousness has been preserved and where few focal signs are present can of course be readily distinguished from cases of primary intracerebral hemorrhage. In intracerebral hemorrhage, secondary leakage of blood into the subarachnoid space may occur, but in the early stages although consciousness may be preserved or clouded well-developed focal signs are present, e.g., hemiparesis, hemianesthesia, hemianopsia, and quadriparesis.

At times in cases of ruptured aneurysm, significant focal or bilateral signs of cerebral involvement will be present. In some cases coma will be an early sign. In some of these cases extension of the hemorrhage into the substance of the cerebral hemisphere has

---

[6]*More recent large autopsy studies suggest an overall frequency of intracranial aneurysms of 5% with a population based incidence of subarachnoid hemorrhage secondary to aneurysms of 10 per 10,000 per year (see Phillips et al, 1980). Patients with autosomal dominant polycystic kidney disease or a family history are at increased risk for aneurysms and require non-invasive screening (Wiebers and Torres, 1992).*

TABLE 26-5: SUMMARY OF THE CAUSES OF SUBARACHNOID HEMORRHAGE (SAH), ANEURYSM LOCATION AND MANIFESTATIONS.

LOCATION OR TYPE	FREQUENCY ALL ANEURYSMS FIGURE #, CASE#	FREQUENCY AS CAUSE OF SAH	EARLY MANIFESTATIONS OR SYMPTOMS OF COMPRESSION PRIOR TO RUPTURE
I. ANEURYSMS		85%	**SUDDEN ONSET OF WORST EVER HEADACHE PLUS STIFF NECK**
Junction of post Communicating & internal carotid	30% Fig .26-24 Cases 12-1, 26-8 (On CD)		Third nerve paralysis since third cranial nerve runs close to and parallel to the posterior communicating artery
Bifurcation MCA	20% Fig.26-25 Case 26-9(CD)		Focal symptoms in MCA territory: focal weakness or seizure face, speech etc
Junction anterior communicating-anterior cerebral	30% Fig. 26-26, 26-27, 26-28 Case 26-10(CD)		Compression optic chiasm, or bilateral prefrontal or bilateral lower extremity or coma or mute state or if giant ,dementia. Often non localized SAH
Basilar vertebral system	5-10%		Variable
Multiple aneurysms	15%		Variable, usually only one is the site of SAH
II. OTHER SAH CAUSES		15%	
Perimesencephalic		10%	Non localized: worst headache, stiff neck, drowsiness (Fig. 26-29)
Other: AVM, mycotic aneurysms		5%	Variable depending on location (often distal cortical branches in mycotic). Seizures are frequent in both.

occurred. At times, rupture from the base of the brain into the third ventricle has occurred. At times, the vessel beyond the point of rupture has been deprived of blood, and infarction has occurred within its territory. Embolism from clot in the dome of the aneurysm may occur. At times, spasm of the artery distal to the point of aneurysm will occur, related perhaps to the presence of blood in the subarachnoid space. Such arterial spasm may produce sufficient vasoconstriction, apparently via adrenergic mechanisms, to result in ischemia and infarction.

Those red blood cells that have entered the subarachnoid space undergo a series of changes. Within a few hours, disruption of red blood cells has begun with a yellow discoloration (so called xanthochromia) of the spinal fluid. At this time oxyhemoglobin and methemoglobin may also be detected in the spinal fluid. After 48 hours, increasing amounts of bilirubin pigment may be noted with an increase in the degree of xanthochromia. The red blood cells have usually been completely disrupted by 10 days after the hemorrhage. The bilirubin product of this disruption disappears more slowly. The spinal fluid does not become colorless for 15 to 30 days after the hemorrhage. In tissues adjacent to the subarachnoid blood, hemosiderin-filled macrophages may be seen for several weeks after the acute episode of bleeding. Initially in the spinal fluid obtained by lumbar puncture, the ratio of white blood cells to red blood cells is the same as in the peripheral blood. With the passage of time, after the acute hemorrhage, as red blood cells are disrupted and as meningeal reaction occurs, a relative increase in white

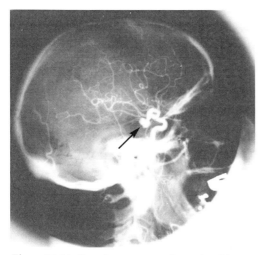

Figure 26-24. Saccular aneurysm, (arrow) arising from junction right internal carotid artery and posterior communicating artery demonstrated by right carotid arteriogram. This 70-year-old widow, 18 months previously, had the sudden onset of severe pain in the right eye and complete third nerve paralysis which had partially improved in the interim. (Courtesy of Dr. Samuel Wolpert.). (Compare to Fig. 26-7C and 2-27 for non-aneurysm reference).

Figure 26-25. Giant aneurysm arising from the bifurcation of the right middle cerebral artery. (Courtesy of Dr. C. W. Watson.)

blood cells will be noted.

Evaluation and Management of the Patient with Subarachnoid Hemorrhage: Early recognition of subarachnoid hemorrhage is important since the overall mortality of ruptured aneurysms is at least 50% and the survivors have a significant morbidity. The recent reviews of Schievink (1997, and van Gijn & Rinkel (2001) are excellent sources of information. Prognostic factors include early recognition at the time of the initial hemorrhage, age, level of consciousness and neurological condition.

Patients presenting at the emergency room with the sudden onset of the worst headache ever should undergo an immediate CT scan searching for subarachnoid blood *(Fig. 26-29)*. The CT Scan will be negative in 5-10% of patients, usually those with a minor bleed and these patients should undergo a lumbar puncture searching for subarachnoid blood. If either of these studies demonstrates subarachnoid blood, then immediate angiography should be performed, followed by immediate surgery procedure designed to clip the aneurysm. Surgical therapy has as its primary objective the

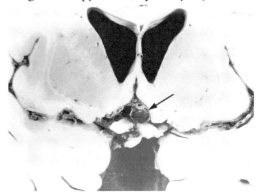

Figure 26-26. Saccular aneurysm at junction of anterior communicating and left anterior cerebral artery situated close to the optic chiasm. This 52-year-old, white male had the sudden onset one evening of headache, nausea, and vomiting. On examination, the next morning he was stuporous, inattentive, disoriented but could move all of his extremities, had bilateral Babinski sign and release of strong grasp reflexes plus. urinary and fecal incontinence. CSF had > 100,000 red blood cells per cu.mm; the supernatant fluid was xanthochromic.(Courtesy of Drs. John Hills, Dr. Jose Segarra)

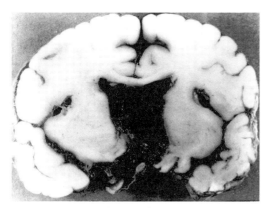

Figure 26-27. Saccular anterior communicating aneurysm with rupture into ventricular system. This 65-year-old white male, 14 days after a confirmed subarachnoid hemorrhage without definite localizing findings, had a generalized seizure followed by quadriplegia and coma. (Courtesy of Drs. John Hills, and Jose Segarra)

prevention of additional subarachnoid hemorrhage. Modern advances now under development include the use of spiral CT scan angiography or high resolution MRA. In some patients the direct clipping procedures are being replaced by endoscopic obliteration of the aneurysm. In patients who survive an initial bleed, a second bleed is associated with a very high mortality of 50%. Complications are summarized in Table 26-6.

In considering the matter of therapy, it is important to examine the mortality of the disease. In the study of Sacco et al 1984, one third of the 36 total cases were in coma at the onset and this group had a 30-day mortality of 83%. Two thirds were conscious at the onset with a 29% mortality at 30 days. Of this conscious group of 24 patients, 15 developed no deficits and 13 of 15 survived with a 30-day mortality of 13%. Of the 9 conscious patients who developed a delayed deficit, the mortality was 56%. The overall 30-day mortality was 47%.

Many patients (18%) die acutely and may never reach a neurosurgical center. In the Cooperative Study of Subarachnoid Hemorrhage (Locksley 1969), 27% of patients who reached a neurosurgical center died within the first week. Of patients dying within 72 hours, 90% had an associated intracerebral hemorrhage. Re-bleeding is a major cause of morbidity and mortality in these patients. Approximately 33% of patients re bleed in the first 30 days with a mortality of 42%. Most episodes of rebleeding occur within the first 14 days particularly between day 5 and day 9. However, rebleeding continues to occur even 10-20 years after the initial hemorrhage in patients not subjected to surgical therapy. In the series of Winn et al (1977), rebleeding in such patients was estimated at 3.5% per year during the first ten years with an associated mortality of 67%.

Although direct intracranial surgery was at one time attended by a considerable mortality and morbidity, major advances including microsurgery, neuroanesthesia with more effective control of blood pressure and

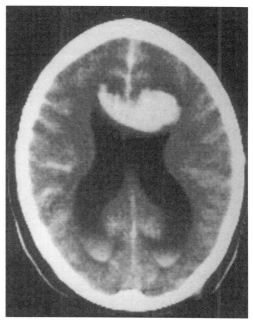

Figure 26-28. Aneurysm anterior communicating-anterior cerebral junction: Sub arachnoid and intracerebral hemorrhage with rupture into ventricular system and hydrocephalus. CT Scan (nonenhanced). This 51 year old male was found in a confused state and soon lapsed into unconsciousness with mild nuchal rigidity, miotic pupils, decerebrate movements on the left side, flaccidity on the right side and bilateral Babinski signs. He improved with osmotic shrinking agents (Mannitol), steroids, ventricular drainage, and subsequent clipping of the aneurysm, evacuation of clot and eventually a shunt procedure.

TABLE 26-6: COMPLICATIONS OF THE INITIAL HEMORRHAGE OR OF A SECOND HEMORRHAGE

COMPLICATION	PREVENTION OR TREATMENT
Recurrence of SAH	Prevent by immediate arteriography and clipping of aneurysm after initial SAH
Rupture into the ventricle with brain stem compression Massive SAH with tentorial and tonsillar herniation	Prognosis poor, coma then death, possible use of ventricular drainage and mannitol
Dissection into brain with hematoma formation	Evacuate hematoma and clip aneurysm to prevent herniation and rebleeding
Vasospasm secondary to aminergic substances in blood	Calcium channel blockers (Nimodipine) and increase blood pressure after clipping aneurysm to increase perfusion
Acute or chronic hydrocephalus due to blockage of cisterns, ventricles and subarachnoid space	For acute shunt after ventricular drainage For chronic shunt
Seizures	Treat with anticonvulsants: phenytoin or phenobarbital
Compressive effects of large and giant aneurysms	Clip and decompress aneurysm
Cardiac arrhythmias	Treat with pacemaker or appropriate anti-arrhythmia agents

intracranial pressure plus the use of CT scanning have resulted in a marked improvement in results (see Adams et al 1987, 1983, Biller et al 1988, Ohman and Heiskanen, 1989, Ropper and Zervas, 1984). The use of calcium channel blockers has decreased vasospasm-induced infarction.

Patient selection is of considerable importance and this is dependent on grading of the patient as regards level of consciousness and the presence or absence of focal neurological deficit. The Classification of Hunt and Hess, 1968, (Table 26-7) utilizes a 5-grade scale. The World Federation of Neurological Surgeons grading scale is discussed by van Gijn and Rinkel (2001).

The Unruptured Intracranial Aneurysm: These aneurysms may be detected because of a) compression of cranial nerves, b) angiogram performed in patients with transient ischemic attacks, or ruptured aneurysm at another site c) the increasing use of CT and MRI scan (Ojemann, 1981, Wiebers et al 1981, and Zacks et al 1984). The most extensive study is the International Study of Unruptured Intracranial Aneurysms (Wiebers et al, 1998) demonstrating a clear-cut relationship of aneurysm size and risk of hemorrhage. Essentially those smaller than 10 mm in diameter on angiographic study at the time of diagnosis are unlikely to subsequently rupture, cumulative rate of rupture was less than 0.05 % per year. If the patient already had a repaired aneurysm at another site after subarachnoid hemorrhage the rate of rupture was 0.5% per year. For aneurysms of 10 mm or more in diameter, the rate of rupture approached 1.0 % per year for both types of patient the previously untreated and the previously treated at another site (about 20X-11 X the rate for the smaller aneurysms). The rate of rupture for giant aneurysms (> than 25mm in diameter) was 6% in the first year. Currently, MRI angiography can reliably detect unruptured aneurysms greater than 2-3 mm in diameter, although many earlier studies have utilized 5mm in diameter as the sensitivity cut off (see Wiebers and Torres). When two or more members of the same family have aneurysms, MRA of asymptomatic members should be considered.

NEUROLOGIC COMPLICATIONS OF BACTERIAL ENDOCARDITIS (see CD ROM for, case 26-11 and Figures 26-30, 26-31)

Osler in 1885 first recognized the neurological complications of endocarditis (association of fever, heart murmur and hemiplegia). The neurological effects of infections of the heart valves remain a major problem. Patients with rheumatic heart disease were once the

TABLE 26-7 HUNT AND HESS CLASSIFICATION-WITH CORRELATIONS

GRADE	DESCRIPTION	PROGNOSIS
1	Asymptomatic or minimal headache and slight nuchal rigidity	Excellent if early surgery
2	Moderate to severe headache and nuchal rigidity but no neurologic deficit	Excellent if early surgery
3	Drowsiness, confusion, or mild focal deficit	Excellent if early surgery
4	Stupor, moderate to severe hemiparesis, early decerebrate rigidity and vegetative disturbance	Poor unless on CT, treatable hydrocephalus or hematoma
5	Deep coma; decerebrate rigidity and moribund appearance.	Very poor unless as above

major patients at risk. At present, other major etiologic factors have emerged including intravenous drug addiction, prosthetic valves, and the use of various intravenous and intracardiac lines.

The streptococcus (particularly viridans) remains the predominant organism, although increasingly Staphylococcus aureus and fungi must also be considered. The overall incidence of neurological complication in large series from the Mass. General Hospital] amounts to 39% (Pruitt et al, 1978). In 16% of endocarditis patients, the neurological complication was the initial complaint. The development of neurological complications had a significant effect on prognosis. If neurological complications were present, mortality was 58%, if not present, 20%. The major neurological complications are summarized in table 26-8.

Durack (1995) discusses the prevention of infective endocarditis.

ARTERIOVENOUS MALFORMATIONS (Refer to CD ROM for Fig. 26-32, 26-33, 26-34 and case 26-12 in which a long standing seizure disorder was secondary to an AVM)

Included within this diagnostic category of vascular malformations are a variety of pathological entities, (based on frequency) such as 1) arteriovenous anomalies, 2) cavernous angiomas, 3) venous malformation, 4) telangiectasis 5) arteriovenous fistulas and 6) dual vascular malformations. All are congenital. (However, arteriovenous fistulas may occur with trauma e.g. carotid cavernous). In general these represent developmental (at times on a hereditary basis) tangles of abnormal vessels within the substance of the brain (less often involving meningeal vessels). These abnormal vessels provide arteriovenous shunting of blood. Since the vessels are abnormal with defects in their walls, bleeding into the substance of the brain (producing an intracerebral hematoma) or into the subarachnoid space

TABLE 26-8: NEUROLOGIC COMPLICATIONS OF BACTERIAL ENDOCARDITIS

COMPLICATION	FREQUENCY	MANIFESTATIONS
Cerebral embolism	17% (6-31%), 3% of all cerebral emboli	Usually MCA cortical territory or stem (see above)
Multiple micro-emboli	11%	Confusional state or personality change or seizures
Mycotic aneurysm	2-10% (2.5-6 % of all aneurysms)	Subarachnoid hemorrhage often more distal branches of the MCA
Intra cerebral hemorrhage	7-8%	Usually lenticulostriate
Meningitis	16%	In 50% of purulent usually staphyloccus was identified
Brain abscess	Single large is not common	Usually multiple and microscopic- possible confusional state or seizure
Infection of vertebrae or disc space with compression syndromes	Not a common syndrome, although osteomyelitis may occur	Spinal cord or root compression

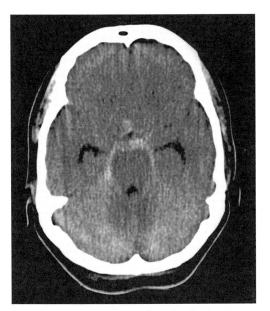

Figure 26-29. Perimesencephalic, unlocalized subarachnoid hemorrhage. Non enhanced CT scan demonstrating blood in the interpeduncular, prepontine, suprasellar and ambient cisterns. This 46 year old woman experienced the acute onset of "the worst headache in her life" lethargy and nuchal rigidity but had an otherwise normal neurological examination. Two series of angiograms and all hematologic studies were normal. She has done well over the subsequent 7 years.

sinus is primarily a disease of the infant less than one month of age and reflects dehydration and other systemic and perinatal factors. Refer to the recent report of DeVeber et al (2001) for a more complete review.

may occur. Moreover, these abnormal vessels are also prone to occlusion with the development of local infarction of cerebral tissue .In general; these malformations produce symptoms earlier in life than do saccular aneurysms. They are moreover less often fatal. Thus several episodes of subarachnoid hemorrhage may occur without loss of life. The initial manifestations are as follows: seizures, often focal in 50%, intracerebral hemorrhage in 20 %, and subarachnoid hemorrhage in 20%. Although involving primarily the cerebral hemisphere, the brain stem, spinal cord or cerebellum may be involved.

DISORDERS OF VEINS AND SINUSES.

Cavernous sinus thrombosis or cortical vein thrombosis may occur in adults due to infection and has already been discussed. Occlusion of the superior sagittal and lateral and straight

CHAPTER 27
Cerebral Hemispheres: Neuropathology and Clinical Correlation II. Non-Vascular Syndromes

PART I: TRAUMA

In the era of the automobile, bicycle, and motorcycle, trauma to the head constitutes one of the most frequent neurologi-cal problems presented to emergency room physicians. Not all head injuries involve the brain. Significant scalp lacerations and simple skull fractures may be present without the development of significant neurological deficits. On the other hand, closed head injuries may be present with significant neurological signs but without major external signs of head trauma (i.e., scalp and skull appear intact).

Concussion: Perhaps the most common neurological syndrome occurring in relation to head trauma is the cerebral concussion, a syndrome characterized by a transient alteration of consciousness and of higher cerebral functions lasting a matter of minutes to hours, following a blow to the head. Consciousness may or may not be transiently lost at the onset. The patient complains of "being in a dazed state" with blurring of vision and a sense of unsteadi-ness. Memory is often impaired for events during this period following the injury: post-traumat-ic amnesia. The transient inability to record and recall events that occur after the traumatic event is called anterograde amnesia. In addition, memory for events which occurred minutes, hours, or even several days prior to that injury may be impaired: retrograde amnesia. As recovery occurs, this period of retrograde amnesia may shrink.

No definite pathology has been associated with cerebral concussion; the pathophysiology remains unclear. There is some indication that differential acceleration or deceleration of the brain and skull may force the cerebral cortex against the hard surface of the skull producing a transient neuronal dysfunction. Rotational forces may also set up stresses that result in the stretching (shearing) of white matter systems at a subcortical and brain stem level.

Contusions and Lacerations (Fig 27-1): More serious injuries of the cerebral hemisphere involve contusions, lacerations, intracerebral hemorrhages and edema.

Contusions (local areas of swelling and capillary hemorrhage resembling bruises) are found particularly at the anterior temporal poles and the under surfaces of the frontal and temporal lobes. These contusions of the cerebral cortex occur as a result of the sudden impact of the cortex against the bony wall of the skull. The anterior portions of the anterior and middle fossa provide relatively constricting compartments favoring the development of such contusions. Contusions may directly underlie the site of the blow to the skull (so-called coup injuries) or may be across from the site of injury (so-called contra coup injuries). When examined at autopsy years later, the brain may show small orange-yellow colored areas of depression on the orbital-frontal and anterior-temporal areas (plaques jaunes).

Intracerebral hemorrhages may also be present. These are usually multiple and small.

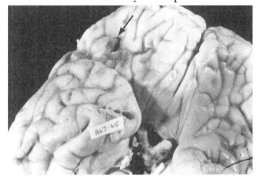

Figure 27-1. Old contusion and laceration of the right orbital frontal area (arrow). This 59-year-old white male had a long history of heavy alcohol intake In addition to the traumatic lesion shown above, the patient had clinical and pathological evidence of an old Wernicke's encephalopathy involving the mammillary bodies and of alcoholic cerebellar cortical degeneration. (Courtesy of Dr. Jose Segarra)

Occasionally, a single large hematoma may be present requiring surgical therapy if mass effects are present.

Lacerations of the brain involve actual tears in the cortical surface. Both lacerations and contusions may result in the appearance of some red blood cells in the subarachnoid space. In general, patients with contusion and laceration have a loss of consciousness in relation to the injury. Residual alterations in mental function, memory and personality are often noted. Such alterations are uncommon with a simple concussion. Focal neurological findings may be present during the acute stage, at times remaining as residual deficits. Severe focal deficits such as hemiparesis or aphasia are, however, usually absent.

Complications of Skull Fractures: Certain types of skull injuries provide serious problems as regards the brain and require surgical attention. Thus, depressed skull fracture may compress the cerebral cortex, with bony splinters lacerating the dura and cortex.

Penetrating head injuries (the object causing trauma has penetrated the skin, bone and meninges) may introduce contaminated material into the brain, producing abscess formation and meningitis. Fractures involving the cribriform plate with laceration of the overlying dura and arachnoid may result in the loss of olfactory sensation (anosmia) and cerebrospinal rhinorrhea: the drainage of spinal fluid through the nose. The presence of such a pathway, of course, allows for the ready entry of bacteria from the nasopharynx into the cerebrospinal fluid spaces, producing recurrent episodes of meningitis. Air may also enter the intracranial cavity through this passageway. Both air and infection may also enter when skull fractures extend into any of the paranasal sinuses or the middle ear. In the latter instance, a cerebrospinal fluid otorrhea may occur: drainage of spinal fluid from the external ear canal. Such basilar skull fractures may extend across the petrous pyramid or tear the tympanic membrane. Blood may be present in the external ear canal or behind the tympanic membrane. If the basilar skull fracture extends into the sigmoid sinus, the tissue over the mastoid may show delayed hematoma with discoloration of the skin and subcutaneous tissue - ("Battle sign").

EXTRADURAL HEMATOMAS AND SUBDURAL HEMATOMAS

Of great importance from the neurological standpoint are these two traumatic conditions in which progressive compres-sion of the cerebral hemispheres occurs. These are conditions where early recognition and treatment will produce excellent results. The failure of recognition will result in progressive deterioration of neurological status with death as the eventual outcome. Note that large traumatic intracerebral hematomas will also produce a progressive course and these lesions may also benefit from surgical therapy (see also Zervas and Hedley-White 1972). The use of CT scan has revolutionized the diagnostic approach to these problems: (see Cordobes et al 1981, Markwalder 1981, Masters et al 1987).

Extradural Hematoma: Extradural hematoma is a complication of skull fractures that extend across a groove containing a meningeal artery usually the middle meningeal artery. Less often the skull fracture has torn a large venous sinus. Since the bleeding is usually arterial and brisk and the extradural space is narrow (normally the dura is closely applied to the skull), compression of the cerebral hemisphere soon results with the rapid development and progression of neurologic symptoms. The classic description is usually that of a short period of unconsciousness related to the acute trauma, then a short lucid interval, which is followed by progressive confusion and coma. A rapidly progressive hemiparesis is often noted with the development of a fixed dilated pupil on the side of the hematoma indicating uncal herniation by the mass that may be placed laterally over the temporal lobe. At other times, as in the following case history, these lateralized findings may be overshadowed by the rapid progression to a stage of functional midbrain transection.

Case 27-1: This twenty-two-year-old, white airman, during a winter storm, was involved in an auto accident at 11:30 p.m., in which he struck his head against the windshield. The patient apparently was dazed, perhaps unconscious, for a matter of seconds to minutes. He was taken by ambulance to the emergency room of the nearby army hospital where a brief evaluation indicated that the patient was alert, without definite neurological findings, but skull X-rays did indicate a linear fracture over the right temporal area

The patient was apparently alert upon his arrival on the ward. However, by 5 a.m., the patient was reported by the nurses to be agitated and confused.

Neurological examination: (8 a.m.) *Mental Status:* The patient was agitated and unable to cooperate. He was moving all limbs sitting up in bed, holding his head, which was tender to palpation moaning and hyperventilating to a marked degree. He answered only occasion-al questions and then with a yes or no. *Cranial Nerves:* Pupillary responses were sluggish. The right pupil was perhaps slightly larger and slightly more sluggish than the left. *Reflexes:* Deep tendon reflexes were increased bilaterally. Plantar responses were bilaterally equivocal.

Neurological diagnosis: Possible evolving acute bilateral epidural or subdural hematoma

Subsequent course: Progression occurred. By 4 pm, the patient was now deeply comatose with little response to stimulation. The pupils were now bilaterally fixed and dilated. The four limbs were extended in a decerebrate posture with significant spasticity on passive motion. The degree of spasticity in the upper limbs could be modified by tonic neck maneuvers. The plantar responses were bilaterally extensor. Spontaneous respirations were now irregular and infrequent, and the patient required the assistance of a mechanical respirator.

Revised Neurological diagnosis: Acute tentorial herniation with brain stem compression secondary to epidural hematoma.

Subsequent course: A general surgeon immediately placed burr holes and evacuated an epidural hematoma over the right temporal-parietal area. He coagulated and ligated a bleeding middle meningeal artery branch. There was a rapid improvement in the patient's status. The patient returned to active duty but was experiencing some problems related to changes in recent memory, motivation and personality.

Subdural Hematomas: Subdural hematomas *(Fig. 27-2, 27-3)* represent the accumulation of blood within the subdural space overlying the cerebral convexities. Such hematomas may be acute, subacute or chronic. The acute and subacute types are clearly associated with trauma. The acute type is usually associated with other significant injuries of the brain such as cerebral contusion, laceration and intracerebral hematomas. With tears in the arachnoid, blood from the lacerations passes into the subdural space. Moreover, small bridging veins from the pia arachnoid to the superior sagittal sinus are also likely to be torn. The clinical picture of the acute subdural hematoma is often then modified by the clinical manifestations of these associated injuries. At times, however, the typical picture of a rapidly progressive supratentorial space-occupying lesion with progressive obtundation of consciousness and the signs of tentorial herniation will develop in acute (24 to 48 hours) relationship to the head injury. The chronic variety may appear after closed-head trauma, often apparently trivial. At times, there may be no definite history of trauma.

The subacute and chronic subdural hematomas develop after a variable latent period of days, weeks, or months following the injury. The bleeding is venous from small bridging veins passing from the pia arachnoid to the superior sagittal sinus. The resultant accumulation of blood then is relatively slow. Associated injuries to the brain are usually not present. The blood in the subdural space is not readily removed and remains adherent to the dura. Fibroblasts and blood vessels grow from the dura at the edges of the clot, producing a thin membrane that separates the clot from the arachnoid. At the same time, a thicker layer of connective tissue and blood vessels is forming a membrane where the clot is adherent to the

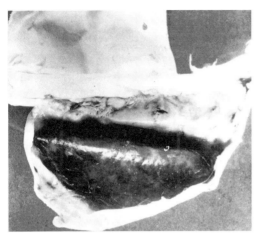

Figure 27-2. Subdural hematoma. This 69-year-old patient with Parkinson's disease had many dizzy spells and had fallen frequently. (Courtesy of Drs. John Hills and Dr. Jose Segarra)

inner layer of the dura. The hematoma often continues to increase in size apparently for several reasons: continued venous oozing may occur; moreover, bleeding may occur from the blood vessels that are growing into the hematoma from the dural surface. Finally, it has been suggested that as some breakdown occurs within the clot, a fluid with high protein content is produced. The resultant fluid of high osmotic pressure then attracts more fluid into the hematoma.

The clinical manifestations are those of a progressive supratentorial mass lesion. At times, progressive focal symptoms and signs are present; for example, hemiparesis, aphasia and focal seizures. Eventually, alterations in consciousness occur, often of a fluctuating nature. With eventual tentorial herniation, a fixed dilated pupil and the signs of a progressive functional midbrain transection are noted. Often, however, a syndrome is present in which focal symptoms and signs are not prominent, perhaps reflecting the fact that bilateral subdural hematomas are present in a high percentage of cases. The resultant picture then is that of a progressive state of confusion with fluctuating alterations in consciousness. A bilateral grasp reflex is often present. Eventually, if untreated, coma and the signs of tentorial herniation develop. In both types of

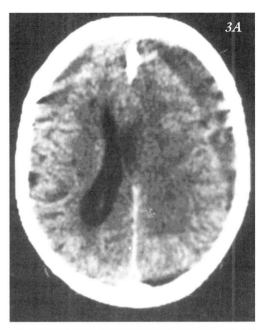

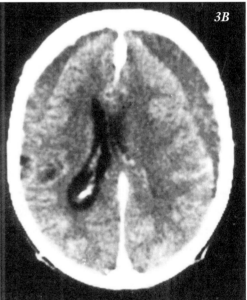

Figure 27-3. Chronic subdural hematoma, bilateral CT Scans. Case History 27-2. A) Non-contrast enhanced. B) Contrast Enhanced: note enhancement at membrane margins. This 83-year-old female - 3 months after a motor vehicle accident in which she struck her head, developed fluctuating confusion and lethargy. (See text for details).

syndrome, progressive headache may be present although this complaint as well as the past history of head trauma may be difficult to elicit when the patient is admitted to a hospital in a confused state.

Case 27-2. (Patient of Dr. Alex Danylevich). This 83-year-old white female struck her head in a motor vehicle accident 2 months prior to admission. She had sustained fractures of her ribs and left clavicle. She had a scalp hematoma in the right frontal area. She was admitted to her local hospital at that time for several days. She was described as having post-traumatic antero-grade and retrograde amnesia and obtundation, which cleared over her several day hospital stay. She returned to her usual activities but then was readmitted with a three-day history of increasing confusion, lethargy followed by a mute state.

Neurological examination: *Mental status:* The patient was lethargic, mute and did not follow commands even when relatively alert. *Cranial nerves:* Pupils were 2 mm. and reactive. There were sudden paroxysmal nystagmoid deviations of the eyes to the right. *Motor system:* Tone was increased in the right leg. *Reflexes:* Plantar responses were extensor bilaterally.

Clinical diagnosis: Chronic subdural hematomas, probably bilateral.

Laboratory data: The *CT scan (Fig. 27-3)* demonstrated significant bilateral chronic subdural collections much denser on the left where the lateral ventricle was obscured and shifted to the right. *Serum sodium* was low at 117 mEq/liter

Hospital course: Serum sodium was corrected. Then, bilateral subdural hematomas were removed via burr holes on the right and a bone flap on the left where substantial chronic membrane and fluid were found. Neurological examination eight hours after conclusion of the procedure, indicated an alert patient who had no apparent language disorder and no weakness. She did develop transient right face and hand focal motor seizures that were treated with anti-convulsants.

There are patients who are at increased risk for subdural hematomas: patients receiving anticoagulant medications, or with bleeding disorders, (e.g., as in leukemia. The elderly, chronic alcoholics, and patients with Parkinson's disease are also at increased risk perhaps because such population groups are more prone to unsteadiness and falls (See discussion of chapter 18). In the patient with cortical atrophy as in Alzheimer's disease, the bridging veins may be stretched and thus more prone to tear. Because of the atrophy, a large amount of blood may accumulate in the subdural space without producing early symptoms of headache and other symptoms.

The problem also exists in the infant during the first year of life. The picture presented, however, differs significantly from that seen in the adult because in the infant, separation of the sutures may occur, allowing the intracranial cavity to expand. The onset is then often non-specific and nonfocal: an enlargement of the head, a failure to thrive, a failure to gain weight, and a failure to reach developmental landmarks. At times, generalized seizures, vomiting, and papilledema may be noted. The etiology often relates to presumed birth trauma.

A related problem is the occurrence of subdural effusions in infants following H.influenza meningitis.

MANAGEMENT OF SEVERE HEAD TRAUMA

Severe head injuries represent major problems requiring intensive care. Such patients generally are brought to the emergency room in coma or an obtunded state. As opposed to the patient with a simple concussion (where any loss of consciousness is brief - minutes), these patients continue with an altered level of consciousness for hours, days, weeks or months. From a pathological standpoint, these patients are found to have contusions, laceration, and multiple hemorrhages within the substance of brain and associated edema and infarction. These findings may be diffuse or concentrated in the diencephalic and midbrain areas involved in maintaining consciousness. At times, the direct effects of trauma have been complicated by cardiopulmonary arrest due to chest injuries or on a central basis producing anoxic encephalopathy. Most do not have epidural or subdural hematomas (epidural in 15%, subdural in 15%). The details of management are considered in Jennett and Teasdale 1981, Bullock and Teasdale 1990, Jaffe and Wesson 1991, White and Likavec 1992, and Ropper, et al 1988. Essentially, the emergency critical care management of the traumatic comatose patient involves the following major steps.

1. ABC -Stabilization of vital functions including Airway, Breathing and Circulation (control of bleeding and blood pressure).

2. Rapid neurological evaluation with rapid rating of the level of consciousness (see chapter 29).

3. Emergency CT scan should be obtained .A simple cervical spine series at the same time should serve to rule out cervical fractures and dislocations.

4. Epidural hematomas, significant subdural hematomas and large focal intracerebral hematomas require emergency neurosurgi-cal intervention.

5. Monitoring and management of increased intracranial pressure are necessary where contusions, edema, small intracerebral hematomas, and small ventricles are present on CT scan).

6. Temperature, electrolytes and fluid balance and nutrition must be maintained to avoid the metabolic causes of coma.

7. Seizures must be treated.

8. Prevention of deep vein thrombosis in the lower extremities, and of the subsequent complication of pulmonary embolus (Geerts et al, 1994).

Late Effects Of Head Trauma 1) *Post-Traumatic Epilepsy:* The seizures may be focal or generalized. Penetrating injuries of the head, as in wartime wounds from bullets or shrapnel, are likely to be associated with post-traumatic seizures. In the World War II series of Watson, 41.6 percent of such patients developed seizures within 3 years of the injury, the majority within 6 to 12 months. There is some evidence that factors such as the location of the wound within the cerebral hemisphere and the complication of abscess may influence the incidence of such post-traumatic seizures. Most head injuries in civilian life, however, are not penetrating missile wounds but rather blunt and non-penetrating injuries (closed head trauma), and under these circumstances, seizures are much less common. Thus, in the series of Jennett, the overall incidence of seizures in 1000 consecutive blunt head injuries was approximately 5 percent over a four-year period. In summary, the post-traumatic seizures are more likely to occur in those head injuries where actual structural damage of the cerebral cortex has occurred.

2) *Post-Concussion Syndrome:* Another relatively common late complication of head trauma is the post-traumatic or postconcussion syndrome consisting of recurrent episodes of headache and dizziness (vertigo or light-headedness). The episodes often appear to be precipitated by exercise or change in posture from the recumbent to the upright. The severity and duration of the syndrome bears no relationship to the severity of the trauma; the episodes may continue for years after a trivial injury. In a few cases, the vertigo may be related to a traumatic disturbance of the labyrinth. Patients may develop benign positional vertigo. In many cases, the dizziness is actually a light-headedness related to anxiety and hyperventilation. In general, no actual pathology is found. As a general rule, stable, well-motivated individuals who are anxious to return to their previous occupation or studies do not experience this syndrome in severe degree. On the other hand, individuals with disorders of personality or psychoneurosis or with problems in adjusting to studies or employment, appear to be severely affected (Evans 1992).

In recent years, there has been recognition of possible long duration changes in memory and cognitive function following concussion or other mild head trauma primarily on the basis of neuropsychological testing (see Capruso and Levin 1992, Packard 1994 for additional discussion of cognitive effects of head trauma).

PREVENTION

Trauma is the most frequent cause of death

and disability in children and young adults. Neurological trauma to brain and spinal cord, constitutes the major reason for prolonged costly hospitalizations in young adults since such patients require prolonged intensive critical unit and low term rehabitative care. The long-term effects of interruption of what were once hopeful productive careers cannot be calculated simply in terms of the billions of dollars involved but must also take into account the personal and family miseries, which are incurred. Many of the problems are preventable. Table 27-1 summarizes the measures available for prevention of head injuries.

Other traumatic problems include a.) anosmia as in case 30-3 on CDROM (shearing of the olfactory filaments passing through cribriform plate), b.) carotid/cavernous fistula and c.) dissections of the carotid and vertebral arteries.

PART II: NEOPLASMS

Based on the previous chapters, it is evident that particular tumors occur at particular ages. In childhood most intracranial tumors are infratentorial; in adult life, the larger proportion are supratentorial. Meningiomas, glioblastomas, metastatic carcinomas, pituitary adenomas, vestibular Schwannomas (or neurinomas), are tumors of the adult. On the other hand, medulloblastomas and ependymomas are primarily tumors of childhood, while brain stem and spinal cord gliomas are found primarily in older children, adolescents and adults

It is also evident that knowledge of the cell type of a tumor taken in isolation does not necessarily enable the observer to predict the future biological and clinical behavior of the tumor. Identity of the cell type and knowledge of the location of the tumor does allow for a higher order of prediction.

Thus, glial tumors may arise in the cerebral hemisphere, the brain stem, and the cerebellum. The biological behavior of the tumor in each location is quite different. Thus, most glial tumors of the cerebral hemispheres in the adult are highly malignant. They are rapidly

TABLE 27-1: PREVENTION OF HEAD INJURIES

PREVENTATIVE MEASURES	RESULTS OF PREVENTATIVE MEASURE
1. Tri-corner seat belts	Mandatory use has been demonstrated to significantly reduce head, face, chest and spinal cord injuries.
2. Strict enforcement of driving regulations regarding alcohol and drugs	The majority of severe accidents involve the disregard of these regulations. Often the use of these agents is combined with excessive speed.
3. Mandatory use of safety helmets by bicycle and motorcycle riders	Reduces the frequency of: a. Head and neck injury deaths, b. Severe head trauma morbidity c. Minor head and neck injuries, d. Post-traumatic epilepsy - one of the few preventable causes of one of the major diseases of the nervous system.
4. Strict regulation of boxing – professional and amateur with mandatory use of helmets. Compulsory suspensions or revocations after knockouts should be required Based on long-term neurological, EEG and CT scan studies of professional boxers (Ross et al 1983). Many medical organizations including the AMA and the American Academy of Neurology have gone so far as to call for the elimination of all professional boxing	Reduce the incidence of multiple small contusions and lacerations of the orbital frontal and temporal areas etc, which result in a chronic frontal lobe syndrome, "the pinch drunk" state and possible post traumatic Parkinsonian state.
5. Effective regulation of sales of arms.	Weapons constitute a major cause of injuries and deaths in adolescents and young adults.

growing invasive necrotic tumors with variable cell morphology and frequent nuclear mitotic figures. The brain stem and spinal cord astrocytoma, usually seen in children and young adults, on the other hand, is usually slower growing and has a more constant cellular and nuclear pattern. The cerebellar astrocytoma is a limited cystic and encapsulated tumor of low histological grade, which can be removed in its entirety with essential cure of the patient.

The clinical manifestations of a particular tumor are determined in part by the cell type of the tumor (astrocytoma, meningioma) and its biological activity (benign meningioma, low-grade astrocytoma, or highly malignant astrocytoma). Much more important in the determination of clinical signs and symptoms is the location of the tumor. This applies to both the local symptoms and signs (focal cortical symptoms and signs or posterior fossa symptoms and signs) and to the general symptoms and signs such as headache, vomiting, drowsiness and papilledema that reflect the increased intracranial pressure resulting from the tumor mass, cerebral edema, and blockage of the ventricular system. Focal seizures are more likely to occur with low-grade gliomas and meningiomas and less likely to occur with highly malignant gliomas.

Headaches: Headache in patients with brain tumors may reflect not only increased intracranial pressure but also the direct or indirect compression of pain sensitive intracranial structures such as blood vessels, dural sinuses, meninges or cranial nerves at the base of the brain. At the present time, less than 50% of patients with brain tumors complain of headaches. Headache as an early symptom is more frequent in patients with highly malignant gliomas, and less frequent in low-grade gliomas and meningiomas. Supratentorial tumors tend to produce frontal, temporal, orbital or parietal located headaches. Infratentorial tumors tend to produce occipital or upper cervical located headaches.

In general, patients with headaches alone have another (non tumor) explanation for the headaches. This includes tension headaches, migraine headaches, or disease of the nose, nasal sinuses, orbits or acute or chronic disease of the cervical spine. Headaches that awaken the patient from sleep, or are produced by positional changes or by coughing or similar maneuvers should raise questions about a possible non-benign pathology. Even under these circumstances, the majority of patients will fall into a benign pathology group. Acute onset of headache, "the worst ever" may indicate a subarachnoid hemorrhage or meningitis.

1. *Migraine Headaches.* In general migraine (vascular headaches) occur intermittently, are throbbing, accompanied by nausea and vomiting and may be unilateral or bilateral. They are classified as migraine with aura if focal symptoms such as visual phenomena lateralized tingling or other cortical phenomena precede the headache. They are classified as migraine without aura if the headache occurs without these phenomena.

2. *Periodic cluster headache.* This variety of unilateral vascular headache often awakens the patient from sleep each night over a period of 1 to 2 weeks. They are throbbing but also involve acute sharp pain localized to the eye. Usually, the headache is accompanied by unilateral tearing of the involved eye and unilateral nasal congestion. They are usually relieved by inhalations of oxygen. Such headaches must be differentiated from the unilateral pain of glaucoma. Acute maxillary sinusitis may also produce unilateral pain in the maxillary and ophthalmic distribution accompanied by nasal congestion, local tenderness, swelling and erythema.

3. *Muscle tension headaches* tend to occur on a daily basis, and tend to occur as the problems of the day build up. They are described as a pressure or aching type pain located in the frontal areas but often spreading to the temporal and occipital areas. Often increased tension may be palpated in the temporalis, frontalis or cervical muscles. Degenerative disease of the cervical spine is frequent and such patients may have chronic muscle tension and aching pain in the cervical and occipital areas.

4. Another common cause of headache is

occipital neuralgia. This occurs in patients with prior whiplash injuries who have unilateral sharp pains in the distribution of the occipital nerves or the cervical 2-nerve root. These patients will have significant local tenderness on palpation of the occipital nerve at the occipital notch. The pain is relieved by cervical traction or local injection of the occipital nerve.

Brain Tumor Epidemiology: The overall relative frequency of the various types of tumors affecting the central nervous system varies from series to series depending on whether a surgical or autopsy series is being reported. In addition, the particular interests of the neurosurgeon have often resulted in a specialized tumor type being referred to a particular neurosurgeon. Table 27-2 presents the data of two neurosurgeons with extensive experience in the surgery of brain tumors. The autopsy data of an earlier period when the postmortem examination was more frequently performed is also presented.

Metastatic tumors are under-represented in the operative series since patients with multiple metastatic tumors are not considered surgical candidates. Pituitary adenomas were of particular interest to Cushing and, thus, are over-represented in the operative series. The true clinical frequencies are somewhere between the operative and autopsy incidences. It is important to realize that silent meningiomas may be found on CT scans or MRI scans or at autopsy in older patients. Small pituitary adenomas are found frequently at autopsy.

It is also important to place in perspective, the overall frequency of brain tumors.

1. Primary intracranial tumors diagnosed prior to death occur in 11.8 patients per 100,000 per year (Radhakrishman et al, 1995).

2. Asymptomatic tumors. In the same series 7.3 patients per 100,000 per year are found to have asymptomatic tumors.

3. Pediatric age group: Brain tumors are the most frequent form of all solid neoplasm (medulloblastomas, brain stem and cerebellar astrocytomas, ependymomas). Among all malignancies in the pediatric age group, brain tumors are out ranked only by leukemia. In the age group 15-34 years, malignant gliomas are the third leading cause of death from cancer.

4. Overall malignant gliomas alone account for 2.5% of deaths due to cancer. Among adults, more die of primary brain tumors than of Hodgkin's disease or of multiple sclerosis.

5. Moreover, 13% of the overall 385,000 cancer deaths per year have some clinical evidence of central nervous system involve-ment. Autopsy studies suggest that 25% of patients dying of cancer have intracranial metastases. With cancer of the lung, the percentage is even higher.

TABLE 27-2 RELATIVE FREQUENCY OF VERIFIED INTRACRANIAL TUMORS (AFTER MERRITT, 1967)

TYPE OF TUMOR	COMBINED NEUROSURGICAL SERIES OF 4349 CASES (Grant and Cushing)	AUTOPSY SERIES OF 3010 CASES (Courville)
Type of Tumor	Percent of Total	Percent of Total
Gliomas	43.0	41.5
Meningiomas	15.0	11.6
Metastatic Tumors	6.5	23.7
Pituitary Adenomas	13.0	3.4
Acoustic Neuromas	6.5	2.5
Congenital Tumors	4.0	3.5
Blood Vessel Tumors*	3.0	7.8
Miscellaneous: papillomas, Granulomas** and sarcomas	9.0	5.7

* There are very few true tumors of blood vessels (hemangioblastomas and angioblastic meningiomas). Today many of the "tumors" of this group would be considered malformations.

** Today infectious granulomas would not be considered in this series. In some parts of the world tuberculomas and parasitic granulomas still constitute a significant percentage of mass lesions coming to surgery or autopsy.

Etiology Of Primary Brain Tumors

For the majority of cases, no specific factors can be identified.

1. Radiation of the head early in life for unrelated reasons, certainly does significantly increase the risks, for all types of tumors involving the nervous system including, gliomas (2.6X), meningiomas (9.5X), and nerve sheath tumors (18.8X), compared to a matched control group. Overall risk for 1.5Gy neural tissue dose was 6.9X control and 20X control for 2.5Gy tissue dosage. (Ron et al 1988).

2. A possible increased risk for the development of gliomas following occupational exposures in the petrochemical, chemical and rubber industries has been discussed by a number of authors (Selikoff and Hammond 1982).

3. Head injury of a significant nature, e.g., depressed skull fractures, may be associated with meningiomas.

4. In a small number of patients, certain familial syndromes may be present as predisposing factors, e.g., *neurofibromatosis* associated with acoustic neuromas, meningiomas and gliomas (of various types - including optic nerve), and *tuberous sclerosis* associated with astrocytomas.

5. Black (1991) and Shapiro and Shapiro 1992 have reviewed concepts of oncogenesis, as applied to primary brain tumors, Oncogenes that potentiate or initiate cell mitosis may behave inappropriately or excessively in neoplastic cells. Alternatively, or in addition, neoplastic cells may have lost tumor suppressor sequences. Chromosomal analysis has identified deletion loci on chromosome 22 in meningiomas, chromosome 17 and 22 in acoustic neuromas and chromosomes 10 and 17 in astrocytomas. Chromosome 7 has been implicated with increased frequency in the cells of malignant gliomas.

5. Other Factors. The role of various other growth factors and of immune mechanisms remains to be clarified. Hormonal factors are undoubtedly of importance in meningiomas. These tumors are more common in women than men (2:1 ratio), may undergo growth during pregnancy and are more common in women with carcinoma of the breast. Estrogen and progesterone receptors have been demonstrated in meningiomas (Lesch et al 1987).

PRIMARY INTRINSIC TUMORS OF NEUROEPITHELIAL ORIGIN

The most frequent tumors affecting the cerebral hemispheres are those arising from glia. Those of the astrocytic series are by far the most frequent whether operative or autopsy series are considered (Table 27-3). Within this group the glioblastoma is the most frequently encountered, based on autopsy statistics. In the era of Cushing, glioblastomas were often not subjected to surgery. (See below for more recent data and approaches). Oligodendrogliomas are relatively infrequent and ependymomas involving the cerebral hemispheres are rare.

Astrocytic tumors. In general these tumors infiltrate the cerebral hemisphere. The astrocytomas have been subdivided into various types. According to the classification of Kernohan and his associates (1952), these tumors were subdivided into four grades based on their histological degree of malignancy. This classification to a variable degree, replaced the older nomenclature of Bailey and Cushing (1926), which classified the glial tumors on an embryological basis (The tumor was classified after the actual or hypothetical embryological cell of closet resemblance). More recent classifications have utilized features, which correlate with prognosis (see Burger et al 1985, 1987; Daumais-Duport et al 1988). Features of importance in grading are degree of (1) hypercellularity; (2) pleomorphism, including nuclear atypia and mitosis; (3) vascular proliferation; (4) necrosis.

In the discussion, which follows, we will utilize the several classifications as well as the most recent classification of the World Health Organization as modified by DeAngelis (2001)

From a prognostic standpoint, the following grades of astrocytoma are recognized.

A. The low grade Astrocytoma Tumor (grade I or II astrocytoma). This is a mildly hypercellular tumor with a variable degree of pleomorphism but with no vascular proliferation and no necrosis There are increased number of mature astrocytes of relatively normal appearance with no evidence of mitosis. Those arising in the cortex are composed predominantly of protoplasmic astrocytes and were referred to as "gemistocytic"; the cell resembles reactive astrocytes. Those arising in the white matter are composed predominantly of fibrillary astrocytes and sometimes referred to as "piloid or pilocytic". Grossly, the lesion is non-encapsulated, firm, and granular and gray. Cysts may be present. At times, the only hint of the lesion may be a lack of distinction between gray and white matter. On CT the cyst may be evident. The lesion, however, is usually best seen on MRI scan, which initially does not demonstrate tumor enhancement.

Grade I - Mild, hypercellularity with little pleomorphism or mitosis. The major example is the pilocytic astrocytoma that is predominantly a tumor of children and adolescents. The predominant cell type is usually the fibrillary astrocyte, a relatively uniform elongated mature astrocyte infiltrating along white matter tracts and often found in the pons, optic chiasm and diencephalon. Ther term polar spongioblast was at one time applied to this type of tumor.

Grade II – The major example in this group is the low grade fibrillary astrocytoma demonstrating a greater hypercellularity with greater pleomorphism (mitosis present) but with no vascular proliferation or necrosis. Most (65-95%) present with seizures either focal or secondarily generalized and initially the neurological examination may be without focal features. Most eventually progress to high-grade malignant gliomas. The appearance of progressive neurological findings and of tumor enhancement in the MRI or CT scan may signify such a progression. The median survival despite therapy is 5 years but some survive for 10-15 years. Case 27-3 presented on CD ROM provides an example of the course of such a tumor with onset of generalized seizures at age 30 and followed for over 11 years prior to surgery and radiotherapy.

B. Malignant Astrocytoma: These tumors

TABLE 27-3 THE OVERALL INCIDENCE OF VARIOUS TYPES OF ALL VERIFIED NEUROEPITHELIAL TUMORS*

TYPE OF GLIOMA	NEUROSURGICAL SERIES (CUSHING) Total Number= 862 PERCENT	AUTOPSY SERIES (COURVILLE) Total Number= 1259 PERCENT	AVERAGE SURVIVAL: MONTHS AFTER ONSET OF SYMPTOMS**
Astrocytoma grade I	30.0	23.0	76
Astrocytoma grade II	4.0	1.0	28
Glioblastoma (Astrocytoma grades III &IV)	24.0	52.0	12
Polar Spongioblastoma	4.0	2.0	46
Ganglioglioma	0.3	2.0	-
Oligodendroglioma	3.0	1.0	66
Ependymoma	3.0	6.5	32
Medulloblastoma	10.0	6.0	17
Pinealoma	2.0	1.0	-
Unclassified	20.0	4.0	-

* The series are not restricted to the cerebral hemisphere but include all gliomas.
** After Merritt 1967

correspond to histologic grades III and IV and unfortunately are the most common glial tumors with an annual incidence of 3 to 4 per 100,000 populations. Two tumors are included: the anaplastic astrocytoma and the very malignant glioblastoma. Both types of tumor have a significant contrast enhancement on CT and MRI scans which is often an irregular ring like area at the apparent margin. In actuality the tumor extends beyond this border into the adjacent tissue.

Grade III: The anaplastic astrocytoma tends to begin between the ages of 30 to 50 years. Characteristics include a significant degree of hypercellularity, a significant degree of pleomorphism and a moderate degree of vascular proliferation *(Fig 27-4)*. Median survival with aggressive treatment is approximately three years. Treatment is similar to that discussed for glioblastoma below.

Grade IV: The glioblastoma multiforme accounts for 80% of all malignant gliomas, with the highest frequency of onset of symptoms between the ages of 50 to 70 years. However there are two categories within this group: primary glioblastomas that tend to begin in older individuals (mean age 55years) and secondary glioblastomas, occurring in somewhat younger individuals (mean age 45 years). These latter tumors have begun as a lower grade of tumor and then have secondarily evolved so that younger patients often appear to have a longer duration of disease From the histological standpoint a marked hypercellularity is present with marked variability of cellular and nuclear appearance *(Fig. 27-5)* Multinucleated cells and mitotic figures are frequent (4-5 per high power field). Hyperplasia of blood vessels is very evident (endothelial proliferation). The additional feature of necrosis serves to distinguish this tumor from the anaplastic astrocytoma. The vascular proliferation often results in hemorrhage into the tumor. The vascular hyperplasia of adventitia and endothelium may be so prominent as to suggest actual sarcomatous alteration. Both the hyperplasia of the blood vessel wall and the rapid growth of the tumor often result in a tumor that out grows the blood supply with resultant necrosis within the tumor. Grossly, the presence of hemorrhage and necrosis is evident *(Fig. 27-6)*. These features are also evident on CT scans *(Fig. 27-7)*, which will demonstrate area of hemor-

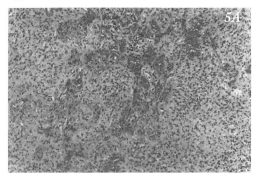

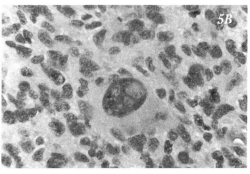

Figure 27-5. Glioblastoma surgical histological features. Case27-4. Refer to text and Figure 27-7 A) Low powered field demonstrating marked cellularity and marked vascular proliferation with areas of micro-hemorrhage (approximately 100x in original). B) High powered field demonstrating marked pleomorphism; giant cells and mitotic figures (approximately 400x in original).

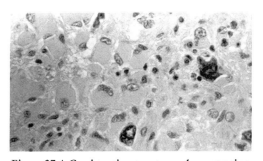

Figure 27-4. Gemistocytic astrocytoma, demonstrating a relatively high grade of malignancy: swollen astrocytes, with nuclei often bizarre and multinucleated displaced to the periphery. (H & E x 400). (Courtesy of Dr. David Cowen, Columbia-Presbyterian, Neuropathology)

rhage, necrosis, and edema and with contrast, a considerable degree of ring enhancement.

Prognosis and Treatment: For malignant astrocytomas, surgery may be employed for purposes of decompression and for debulking the tumor mass so that fewer malignant cells remain to be destroyed by radiotherapy. For the glioblastoma, median survival with surgery only is 6 months, with surgery and radiotherapy, 12 months. The role of chemotherapy is uncertain, the percentage of long-term survivors maybe moderately increased. (Refer to DeAngelis, 2001 for additional discussion). In the study of Shapiro et al (1989), 29-37% of patients with glioblastoma receiving radiotherapy survived for 18months Chemotherapy produced minor additional survival (approximately 2-3 months).

In general younger patients with polar lesions who are able to tolerate extensive resection, radiotherapy and perhaps chemotherapy have the longer survivals (Scott et al, 1999). Why do these patients do poorly? Although at times these tumors may arise superficially in the cerebral cortex, more often they seem to originate in the subcortical white matter. They then appear to infiltrate widely through the cerebral hemisphere, often along white matter systems to involve other areas of the brain (e.g., via the corpus callosum to involve the opposite hemisphere). At times, a multi-centric origin may be suspected. Necrosis and hemorrhage are frequent; and progression of neurological deficit is often rapid. At times, the episodes of necrosis and hemorrhage may suggest vascular accidents.

The variable clinical course of a glioblastoma in younger (40 year old) patients has already been presented in chapters 21 and 24. Those cases should be reviewed at this time. The clinical course in an older patient with a 4-5 day course prior to diagnosis, which initially suggested a possible cerebrovascular event, is presented as case 27-4 and in Figures 27-5 &7. In retrospect, it was evident on the CT scan and at surgery that the acute onset reflected the hemorrhage and necrosis, which had occurred

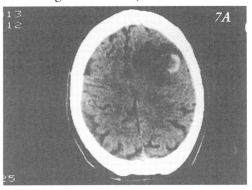

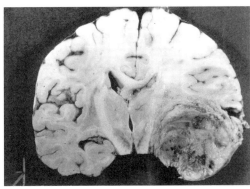

Figure 27-6. Glioblastoma multiforme of the left temporal lobe compressing brain stem and ventricular system. This 43-year-old white male had the onset of short "dreamy" states preceded by a distinctive taste and a sensation of familiarity. Despite a craniotomy and radiation therapy, the patient had a 14-month course characterized by a progressive right hemiparesis, a nominal aphasia, a severe receptive aphasia, increasing disorientation, lethargy, headache, vomiting and papilledema. (Courtesy of Drs. John Hills and Jose Segarra)

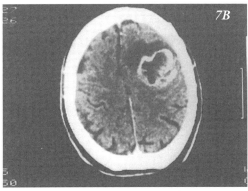

Figure 27-7. Glioblastoma: case 27-4.Refer to text. CT scans A) Non-enhanced CT scan demonstrated a low-density mass in the left frontal parietal parasagittal region, with areas of hemorrhage. B). CT scan (with contrast) demonstrated an irregular rim of enhancement.

within the tumor.

Case 27-4 (patient of Dr. Tom Mullins): This 68 year old right handed white male was admitted to St Vincent Hospital with an apparent 4 to 5 day history of difficulty in doing calculations and in speaking, "he could think of the word, but had difficulty getting it out".

Neurological examination: *Mental status:* Speech was limited to short phrases with loss of connectors and sentence structure. Writing was impaired but repetition, comprehension, naming and reading were intact. *Motor system:* Fine motor movements of the right hand were impaired. Tone was increased on the right. *Reflexes:* The right patellar reflex was increased. Bilateral Babinski signs were present.

Initial Clinical Diagnosis: Diagnosis was uncertain. The abrupt onset suggested possible vascular event such as embolus to the middle cerebral artery but overall pattern including bilateral Babinski sign without significant hemiparesis was unusual.

Laboratory data: *EEG:* Focal left hemisphere delta and theta slow wave activity. *CT scan* (Fig.27-7): A) the *non-enhanced study* demonstrated a low-density mass in the left frontal-parietal parasagittal region with areas of hemorrhage. B) The *contrast-enhanced study* demonstrated an irregular rim of enhancement. *Angiograms* indicated an avascular mass in the left frontal-parietal area

Subsequent course: Dr. Alex Danylevich performed a subtotal resection of this tumor that was found to contain blood clot and cystic fluid. The histology of the tumor was glioblastoma multiforme (Fig. 27-5). The patient subsequently received 4000cGy whole brain radiation and an additional 2000cGy to the tumor site. Three months following surgery, the patient developed increasing obtundation. CT scan revealed a large left frontal necrotic mass. Fever, hypocalcemia and hypernatremia developed and he expired one month later.

Comment: This patient presented with what appeared to be an abrupt onset of language problems raising the question of a cerebrovascular accident. In retrospect, it was evident on the CT scan and at surgery that the

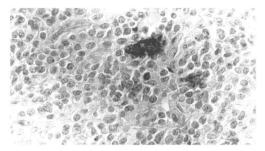

Figure 27-8. Oligodendroglioma. A relatively uniform mosaic pattern of cells with a small nucleus surrounded by a clear halo of cytoplasm is evident. A small area of calcification is present in this field; much larger areas of calcification were evident in other fields. (H & E x 400) (Courtesy of Dr. David Cowen).

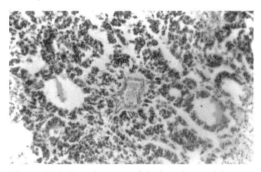

Figure 27-9. Ependymoma of the fourth ventricle. Clusters of ependymal cells are arranged as rosettes or as perivascular pseudorosettes. (H & E x 125) (Courtesy of Dr. John Hills).

acute onset reflected the hemorrhage and necrosis that had occurred within the tumor. This was a rapidly growing, extremely malignant tumor, which apparently had outstripped its blood supply despite the neovascularity.

Other types of glial tumors: oligodendrogliomas: this variety of glioma is found in the cerebral hemisphere of young and middle aged adults constituting 5 –20 % of all glial tumors. The majority is low grade with a long course of 5-15 years; refer to case 18-2. A small percentage are anaplastic or mixed tumors containing anaplastic astrocytoma elements. (Fig 27-8 demonstrates the histology)

Other tumors of neuroepithelial origin: This group contains two very common tumors of the pediatric patient: (1) *Primitive neuroectodermal tumor: medulloblastomas* (*Fig.27-9* and case 20-1), and (2) *ependymomas (Fig.27-*

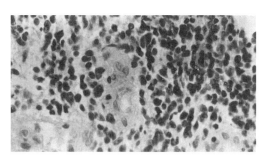

Figure 27-10. Medulloblastoma. Clusters of small cells with densely staining nuclei are present. (H & E x 400) Refer Chapter 20 for illustration of a gross specimen. (Courtesy of Dr. David Cowen).

10). These brain tumors of the posterior fossa constitute the most common solid malignant tumors found in the pediatric population. These tumors have been discussed and illustrate in chapters 13 and 20.

The other 2 tumors in this group are rare (1) *Ganglioglioma or neuroastrocytoma:* now considered to represent the diffuse infiltration by a low-grade astrocytoma among neurons of the cerebral cortex - basal ganglia or diencephalic nuclei. At times, this type of histological picture has been noted in the lesion of tuberous sclerosis. (2) *Pinealomas:* These relatively rare tumors of the pineal have been considered already in relation to the brain stem (Chapter 13). The most common tumor of the pineal region is the germinoma.

OTHER PRIMARY INTRINSIC TUMORS OF NON NEUROEPITHELIAL ORIGIN

(1) *Primary central nervous system lymphomas:* Prior to 1980, primary intrinsic tumors of the nervous system of non-neuroepithelial origin were considered rare. Primary lymphomas of the central nervous system have now assumed increasing importance and are no longer rare. In general these malignant tumors have B cell surface markers and are of the diffuse large cell subtype. They were formerly designated reticulum cell sarcoma or microgliomas. As with glioblastomas, the peak incidence occurs in the 50 to 70 year age range. The increased incidence has occurred in both the immune suppressed (AIDS) and the immune competent population.

(2) *Tumors of blood vessel* origin: Probably the only true vascular neoplasm intrinsic to the central nervous system is the hemangioblastoma of the cerebellum composed of embryonic vascular elements. Giant aneurysms angiomas or arteriovenous malformations of the brain and meninges are composed of adult vascular elements and are not neoplasms but may have mass effect. Hemangioblastomas are discussed in chapter 20

EXTRINSIC TUMORS

Meningiomas: The most common extrinsic tumor above the tentorium is the meningioma. Meningiomas constitute 20% of all types of brain tumors. Based on the epidemiological data from Rochester Minnesota (Radhakrishnan et al, 1995) meningiomas have a total annual incidence of 7.8 per 100,000 populations but the true incidence of symptomatic meningiomas is 2 per 100,000. Most meningiomas (74%) are asymptomatic and are discovered at autopsy or on imaging studies performed for unrelated problems *(Fig. 27-11)*. Meningiomas are tumors of the adult population 30 to 70 years of age. They are more frequent in females due to the fact that many meningiomas contain estrogen receptors. There is an additional increased incidence in patients with breast cancer, at times providing a diagnostic dilemma. Meningiomas arise from arachnoidal cells embedded in the dura. Normally, arachnoidal villi invaginate into the venous sinuses as arachnoidal or pacchionian granulations but arach-noidal cells may be embedded in the dura. In addition, fibrous and vascular elements are, to a variable degree, also included in these tumors. The variable composition has resulted in a histologic classification according to the following main types:

1. A *syncytial or meningothelial* variety composed of clusters of cells similar to the outer layer of arachnoid arranged in nests or whorls and surrounded by layers of elongated or flattened cells.

2. A *fibrous* variety in which interlacing bundles of fibro-blastic elements predominate.

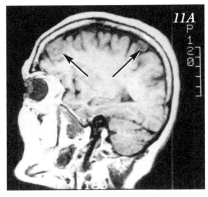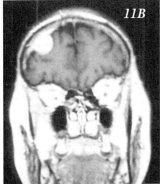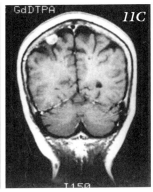

Figure 27-11. Meningiomas: incidental discovery multiple asymptomatic in a 71-year-old ambidextrous female with a normal neurological examination had a history of a transient left posterior thalamic lacunar infarct which was confirmed on subsequent MRI studies. Those studies also disclosed right prefrontal and right parietal asymptomatic meningiomas A) non-enhanced - T1 - right parasagittal section, B) enhanced - T1 - frontal coronal section, C) enhanced T1 parietal coronal section. No changes occurred over the next 5 years.

3. A *psammomatous* variety (probably the most common type) consisting of the same type of arachnoidal cells, arranged in a whorl. In the center of the whorl is an area of hyalinization arranged in concentric lamellae and calcified (psammomatous granules). The calcification is often visible on plain X- rays of the skull.

4. A less common *angioblastic* variety, that appears spongy both on gross and microscopic examination, because it contains multiple small vascular channels. (At times, the term hemangiopericytoma has been applied to this type of meningioma.)

5. A *mixed or transitional* variety. Considerable overlap is apparent as demonstrated in *Figure 27-12*.

It is unclear that these different histologic types have prognostic significance. However a small percentage (7%) from a histologic standpoint have atypical or frankly malignant features of a sarcoma. Regardless of the histologic type, all meningiomas on molecular analysis have a loss of chromosome 22q. A similar defect occurs in patients with neurofibromatosis type 2. These latter patients often have multiple meningiomas in addition to bilateral Schwannomas of the vestibular nerve

These tumors are in general benign and slow growing. Almost all occur external to the brain with a dural attachment - compressing, displacing, and at times invaginating into the brain. Rarely, the tumors may be found without a dural attachment within the ventricle or within the Sylvian fissure. Although not invading the brain the otherwise benign meningioma may invade the overlying bone of the skull. Others excite an osteoblastic reaction in the overlying bone with a marked local thickening of the bone (hyperostosis). What is important about meningiomas is not their histological cell type (sarcomatous degeneration is rare) but their location; in general they are readily amenable to neurosurgical removal. In asymptomatic cases, the risks of surgery may outweigh any conceivable benefit to the patient.

Meningiomas tend to occur in certain specific locations outlined in Table 27-4.

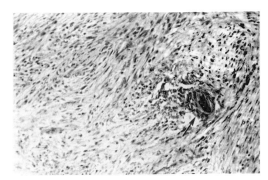

Figure 27-12. Meningioma. Histological appearance of a mixed or transitional type meningioma. Interlacing fibrous bundles and whorls of meningothelial cells are present in addition to a calcified psammoma body (arrow). (H & E x 100) (Courtesy of Dr. David Cowen).

TABLE 27-4 MENINGIOMAS

LOCATION	FREQUENCY*	MANIFESTATIONS
Parasagittal a) Rolandic b) Prefrontal	21-23% (or 33-50%)	Focal motor or sensory seizures or leg weakness (Cases 1-1,18-1), Fig. 27-20 on CD Personality change and dementia (Cases 22- 3)
Lateral convexity	17%	Focal motor seizures hand or face
Sphenoid wing a) Inner third b) Middle third c) Outer third	9-17%	Unilateral exophthalmos and optic nerve atrophy + contralateral papilledema** Bilateral papilledema then inner or outer third findings Temporal lobe seizures (Fig 27-13).
Anterior fossa a) Olfactory groove b) Tuberculum sellae	8-18%	Unilateral anosmia, prefrontal syndrome, dementia (see case 22-4) May mimic pituitary tumor, with compression of optic chiasm, prefrontal syndrome
Posterior fossa	7%	See chapter 13,CP angle, foramen magnum and tentorial syndromes
Other sites a) Intraventricular b) Spinal cord	Not common	Hydrocephalus Spinal cord compression see chapter 9

* Series of Cushing & Eisenhardt (1938), of Courville (1967) and of Taveras and Wood (1964)
** Foster –Kennedy syndrome (See case 23-1)

The course of several meningiomas has been presented already in several of the case histories of Chapter1, 18, 22 and 23.

TUMORS OF PITUTARY GLAND – see chapter 16 and below.

Other Extrinsic Tumors Involving the Brain Refer to Table 27-5

SECONDARY TUMORS -
Metastatic Tumors:

The greatest number of secondary tumors of the central nervous system is spread from a distant site via the blood stream. The majority spread to the cerebral hemisphere or to the cerebellar hemisphere. Metastatic lesions may be solitary or multiple. (Fig. 27-18, 27-19, 27-20). In the series of Delattre et al, 1988, 47% of patients with metastatic lesions to the brain on CT scan had single lesions. These secondary metastatic tumors represent a significant percentage of patients seen with intracranial tumors in a general hospital population. As already noted, 25% of cancer patients have cerebral metastases at autopsy. With increasing length of survival after the diagnosis of cancer, metastatic spread to both brain and leptomeninges has increased in frequency.

1. The most frequent site of primary tumor is the lung accounting for almost a third of the cases. It has been estimated that almost half of all patients with small cell and non-squamous, non-small cell carcinoma of the lung, at autopsy will have cerebral metastasis (Cox et al 1979).

2. The second most frequent site of primary lesion is the breast. Several examples have been presented in Chapter 21.

3. Certain tumors, which have a lower overall frequency of occurrence than carcinoma of the lung and breast, have a particularly high frequency of metastatic spread to the brain:(a) malignant melanoma (usually multiple), b) hypernephromas (often solitary), and (c) choriocarcinoma. (In the surgical series of Delattre et al, (1988), lung malignancies accounted for 40%; melanoma for 11% and kidney cancer for 11%.)

Treatment Solitary metastatic lesions may be subjected to surgical removal, since median survival is significantly improved in series where surgery plus radiation is compared to radiation alone: 40 weeks versus 15 weeks (Patchell et al 1990). At other times, when multiple metastasis occur; radiation therapy or hormonal therapy may be employed, with a significant reduction in symptoms (for additional discussion see, Posner1990, 1992).

Carcinomatous meningitis: Malignant cells are present in the CSF implanting on nerve roots, cranial nerves, hemispheres and meninges (Fig. 27-20, 27-21). Breast cancer, lymphoma, lung cancer and melanoma are the most frequent primary lesions. Hydro-cephalus is frequent. Additional discussion and illustrative case histories will be found in Hedges et al 1988. Winkler and DeLaMonte 1987, Gruber and Sobel 1992, Little et al 1974, Olson et al 1974.

Local Invasive Tumors: Direct extension of tumors to involve the brain is less common than metastatic disease. Common primary sites are carcinomas originating in the nasopharynx and nasal sinuses. These tumors may erode the base of the skull or spread through the foramina at the base of the skull to involve cranial nerves such as nerves V and VI. Treatment involves local radiation; results are dependent on the capacity for response of the primary lesion.

III. INFECTIONS:

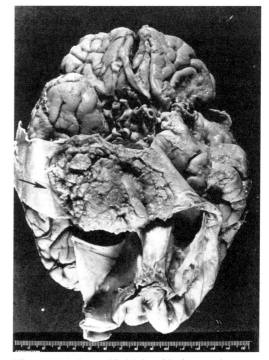

Figure27-13. Outer third Sphenoid wing meningioma. The dura overlying the sphenoid has been reflected revealing a large meningioma (arrow) compressing the inferior surface of the anterior temporal lobe and adjacent orbital frontal area. This 73-year-old black male had a 14-year history of complex partial and generalized tonic clonic convulsions. (Courtesy Drs. John Hills and Jose Segarra).

The central nervous system may be involved by all of the general types of organisms that infect other systems of the body: bac-

TABLE 27-5-OTHER EXTRINSIC TUMORS INVOLVING THE BRAIN

TYPE	LOCATION	STRUCTURES INVOLVED OR MANIFESTATION
Craniopharyngioma	Suprasellar	Pituitary, hypothalamus, optic chiasm (Fig. 27-14)
Epidermoid,	Bone of skull, dura or cisterns	C-P angle or suprasellar
Dermoids/Teratomas	Usually midline	Pineal, pituitary, 4th ventricle
Chordomas	Clivus	Compression of ventral brain stem, basilar artery
Colloid cysts*	F. Monro of Third ventricle,	Acute hydrocephalus with acute headache and coma, often intermittent due to ball valve mechanism (Fig.27-15, 27-16)
Schwannomas**	Posterior fossa:	C-P angle, vestibular nerve most common (see Chap.12)

* The colloid cyst of the third ventricle although uncommon must be recognized in the emergency room, since death may result.
** Except for Schwannomas at CP angle (Fig.27-17), which may also arise from cranial nerves (jugular foramen) and from the sensory nerve roots, all of these tumors are relatively rare. Refer to chapter 13.

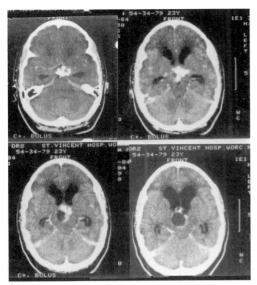

14A

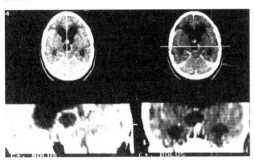

14B

Figure 27-14 Craniopharyngioma. A 23-year-old female with a 6-year history of headaches and amenorrhea followed by diplopia and papilledema. Contrast enhanced scan demonstrates suprasellar calcification with large cyst which projects upward into the third ventricle A) thin sections through suprasellar region; B) sagittal and coronal reconstructions.

teria, viruses, rickettsiae, fungi, protozoa, and helminths. Certain organisms do have a predilection for the central nervous system and often for particular segments or certain systems of the neuraxis. Thus, we have already seen that the virus of poliomyelitis involves predominantly the motor neurons of the anterior horn and brain stem. Bacteria such as meningococcus and haemophilus influenzae involve predominantly the meninges. In most instances, when infectious agents involve the central nervous system, they also involve to a variable

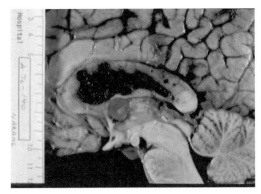

Figure 27-15. Colloid cyst of the third ventricle. This 20-year-old male had sudden attacks of headache; with the last such attack he presented to the emergency room with rapid onset of coma and death.

degree other systems of the body in a general or focal manner. The student should bear in mind that the clinical and pathological manifestations of an infectious disease reflect not only the direct effects of the invading organism but also effects that indicate the body's defensive responses to invasion. Moreover, when immune mechanisms have been suppressed, as in AIDS (acquired immune deficiency disease), organisms that do not usually invade the brain or other organs of the adult patient, may assume major invasive roles. Examples include toxoplasmosis, fungi, cytomegalovirus, etc.

Infections involving the nervous system may be considered under two broad categories:

1. Those infections which involve the central nervous system or subdivisions of the central nervous system in a general or diffuse manner, and

2. Those infections which involve the central nervous system in a focal (or multi-focal) manner

FOCAL INFECTIONS OF THE NERVOUS SYSTEM

The problem of focal infections of the nervous system has an importance far out of proportion to the actual frequency of occurrence of such cases. These focal infections represent, in general, rapidly progressive and compressive space-occupying lesions and they often require emergency neurosurgical intervention.

We have already seen that at the level of

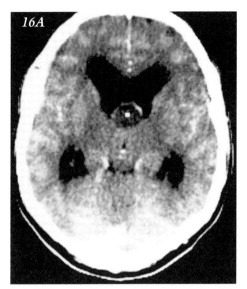

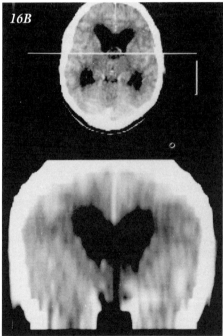

Figure 27-16 Colloid cyst of third ventricle. This 23-year-old right-handed male had a 4-6 week history of generalized throbbing headache and episodic visual obscuration with bilateral papilledema and enlargement of the blind spots. CT scan demonstrated a third ventricular colloid cyst at the level of the foramen of Monro with secondary dilation of the lateral ventricles. A) Contrast enhanced CT at horizontal level just above foramen of Monro. B) Coronal reconstruction. All symptoms disappeared after Dr. Alex Danylevich drained and removed this cyst and shunted the ventricular system.

spinal cord, focal infections are essentially extradural in location. The favorite site of acute focal infection is the fatty tissue of the epidural space. The resultant clinical picture is that of a

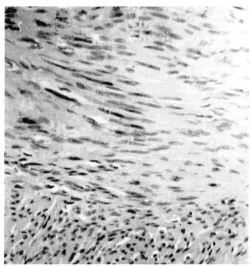

Figure 27-17 Vestibular (acoustic) neuroma or Schwannoma. Microscopic examination of these tumors usually indicates two tissue types: interwoven bundles of spindle-shaped cells with alignment of nuclei in the form of palisades; and looser, somewhat cystic areas. (H & E x 100) (Courtesy of Dr. David Cowen, Columbia-Presbyterian, Neuropathology).

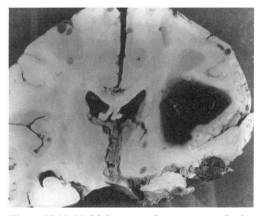

Figure 27-18. Multiple metastatic tumors to cerebral hemisphere from malignant melanoma, with secondary hemorrhage into several of the pigmented lesions in a 40-year-old white female with excision of primary melanoma of the left knee with dissection of the left groin 5 years prior to death. She had multiple cutaneous metastatic nodules despite chemotherapy and then, right temporal headache, nausea and vomiting sudden coma with papilledema and sluggish pupils (Courtesy of Dr. John Hills

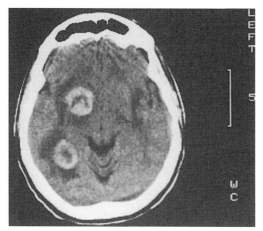

Figure 27-19. Multiple metastatic tumors to brain. CT scans demonstrate multiple high-density lesions with low-density centers, and surrounding areas of edema. Minor enhancement occurred in some of the lesions with contrast. This 43-year-old white male with a 30-pack -year smoking history and a poorly differentiated adenocarcinoma with invasion of peribronchial lymph nodes and of parietal pleura initially presented with hemoptysis. Four months later he presented with a one-day history of confusion, lethargy and frontal release signs and died one month later.

rapidly progressive extradural compressive lesion of the spinal cord. The symptoms and signs are due both to the direct damage from compression and the indirect destruction of tissue produced by vascular compression and occlusion. At the level of the spinal cord also, spinal cord compression, may result from the involvement of and collapse of vertebral bodies by chronic tuberculosis.

ACUTE FOCAL INFECTIONS INVOLVING THE BRAIN

Three syndromes should be considered: 1. Subdural empyema, 2.Purulent brain abscess, and 3. Focal viral cerebritis - h. simplex.

1. *Subdural empyema:* In general, subdural empyema occurs as a result of the direct extension of purulent infection from an adjacent focus of infection in the nasal sinuses or middle ear or following compound skull fractures. The nasal sinuses are now implicated as the major source of infection. Osteomyelitis of the intervening bone is a frequent finding; throm-

bophlebitis of the venous sinuses and of the feeding cortical veins is a not unusual complication. The pus is present in a space that offers

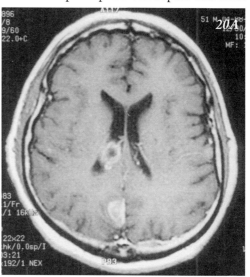

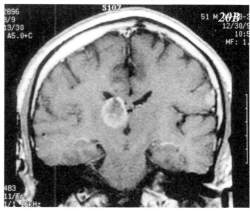

Figure 27-20. Multiple metastatic tumors to brain adjacent to ventricular system and subarachnoid space with carcinomatous meningitis. Which process was the primary one—periventricular solid metastatic lesions or the carcinomatous meninsitis remains uncertain. This 51 year old white male teamster had progressive headaches for 6-7 weeks and a 2 week history of a progressive third nerve paralysis. CT scans and MRI scans demonstrated at least 5 distinct areas of contrast enhancing metastatic tumors. CSF contained 90 cells consistent with non small cell carcinoma. Chest x-ray demonstrated a lesion in the upper lobe of the right lung. He received radiotherapy and dexamethasone with improvement in headache. Three weeks later he developed weakness, tingling and depression of reflexes in both lower extremities, plus urinary and fecal incontinence.

A) T1 horizontal section. B) T1 coronal section.

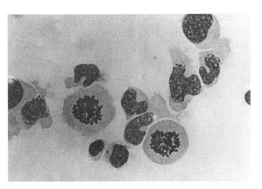

Figure 27-21. "Meningeal Carcinomatosis". Five years after a complete remission from diffuse histiocytic lymphoma (Stage III-B non-Hodgkin's lymphoma), which had presented with symptoms of fever, night sweats, abdominal pain and lymphadenopathy, this 77-year-old female had a biopsy proven recurrence of disease (lymph nodes and bone marrow). She had drowsiness; poor recall, right eyelid ptosis and diffuse weakness of the left lower extremity with tenderness over left femoral and sciatic nerves. Over 2 days she progressed to a complete pupil sparing right third nerve paralysis, incomplete left CN III paralysis (down gaze was preserved), left CN VI paralysis, bilateral peripheral facial paralysis, decreased hearing on left, and bilateral toe extensor weakness. CSF cytology (approximately 1000x) demonstrated large pleomorphic neoplastic cells. The patient received intrathecal chemotherapy (Methotrexate). (Courtesy pathology department St. Vincent Hospital)

very little resistance to the expansion of this purulent mass. The mass is apparently never well encapsulated although in older cases after the removal of the pus, a fibrinous exudate will be found on the inner surface of the dura and the inner surface of the arachnoid. The responsible organism is most commonly the Staphylococcus aureus. A sterile secondary inflammation is usually present in the subarachnoid space. This results in spinal fluid findings on lumbar puncture of moderate increases in cells and protein but a normal sugar and negative culture.

Case history 27-5 on the CD presents an example of subdural empyema.

2. *Purulent brain abscess*: A brain abscess (Fig. 27-22) reflects the spread of purulent infection to the brain by several pathways. Direct extension may occur from a nearby focus of infection in the nasal sinuses or mastoid or middle ear. For this reason many solitary brain abscesses tend to be located in the frontal or temporal lobe of the cerebral hemisphere or in the cerebellar hemisphere. Direct introduction of infected material may occur in relation to trauma, e.g., compound skull fractures, or in relation to neurosurgical procedures. Finally, hematogenous spread may occur in patients with a primary source in the lung, in patients with bacterial endocarditis and in patients with cyanotic congenital heart disease where the blood may bypass the screening system of the lungs. Patients with pulmonary arteriovenous fistula are also at special risk. Such hematogenous spread is more likely to result in multiple abscesses rather than a solitary abscess.

Since the introduction of antibiotics, there has been a significant decrease in the number of patients presenting with brain abscess particularly in the number of cases where the abscess is secondary to sinusitis and mastoiditis. Moreover, among the remaining patients there has been a decrease in the percentage of cases with an identifiable primary focus of infection and an increase in cases without an identifiable primary focus. These latter cases may reflect instances where partial antibiotic therapy has eradicated the primary focus but not the brain abscess.

As might be expected, the most common organism is Staphylococcus aureus. Streptococci and pneumococci are also encountered as responsible organisms. Rarely, under special circumstances, a fungus infection may be implicated, e.g., mucormycosis in severely debilitated diabetics and aspergillosis in patients receiving immunosup-pressive therapy after renal transplantation or in patients with AIDS.

When the cerebral hemisphere is involved the clinical symptoms will correspond to the specific area impacted with focal seizures and rapidly evolving focal neurological deficits as the prominent features.

The electro-encephalogram is likely to show continuous focal 0.5 to 2 cps slow wave activity indicating severe focal damage (refer to Fig. 2-22). The CT scan *(Fig.27-23)* will usu-

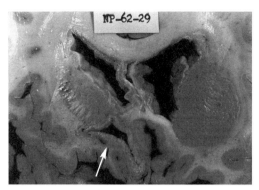

Figure 27-22. Subacute brain abscess (arrow) with associated ventriculitis. This 41-year-old white male, as a child, had a fibroma of the right maxillary sinus that had been treated with surgery and radon implants. Four months before death he developed the first of 4 episodes of meningitis with headache, nuchal rigidity, fever, and chills; cerebrospinal fluid contained 495 white blood cells (75% polymorphonuclear) and a CSF-sugar of 40 mg./100 ml. He had chronic sinusitis of right maxillary and ethmoid sinuses on x-rays and defect in the bone and dura in relation to the right ethmoid sinus. (Courtesy of Drs. John Hills and Jose Segarra).

ally demonstrate a contrast ring-enhancing lesion.

In general the outcome in the untreated brain abscess is fatal. Treatment consists of surgical drainage and total excision of the abscess cavity. In addition appropriate antibiotic therapy must be administered. In the series of Loeser and Scheinberg (1957), surgical therapy in the period 1940-1944 was associated with a 47 percent mortality. By the period 1949-1956, an era of specific antibiotics, this mortality had dropped to 19 percent.

A brain abscess involving the right parietal-occipital area has already been presented in Chapter 23. In reviewing that case, the student should keep in mind that events in a temporal lobe abscess may proceed at a much more rapid pace as regards tentorial herniation with third nerve and brain stem compression.

CHRONIC FOCAL OR MULTIFOCAL INFECTIONS: GRANULOMAS: The followings infections may produce granulomas: tuberculosis, cryptoccus, larva of the pork tape worm (cerebral cysticercosis), toxoplasmosis and schisto-somiasis. In some parts of the world these infections are a common cause of mass lesions, seizures or hydrocephalus. (See also discussion of AIDS)

GENERAL OR DIFFUSE INFECTIONS: We will consider this section in detail on the CD ROM

This category of diseases is usually presented in detail in microbiology courses and in the infectious disease sections of internal medicine.

a. Those involving primarily the leptomeninges (pia and arachnoid) producing leptomeningitis. Generalized infection of the dura (pachymeningitis) is uncommon. *(Fig.27-24, 27-25)*

b. Those involving primarily the parenchyma of the brain, producing encephalitis. (Fig.23-26, 27-27)

There are some infections, often viral, that involve both structures producing, a meningoencephalitis.

MENINGITIS (Leptomeningitis)

This is the most common form of infection of the nervous system. The manifestations of meningitis depend on the organism involved, the age of the patient, and the underlying physical status of the patient. Acute, subacute and chronic forms may be considered.

Acute Purulent (Bacterial or Septic) Meningitis): This is the most common form of infection of the central nervous system among patients requiring hospitalization . In each case, a specific bacterial organism may be isolated from the cerebrospinal fluid (CSF) in the sub-arachnoid space. In addition, the spinal fluid shows evidence of an acute inflammatory reaction. The fluid is cloudy and under increased pressure with large numbers of white blood cells, predominantly polymorphonuclear leukocytes, e.g., 90 to 98 percent of 500 to 40,000 white blood cells per cubic mm. The CSF sugar content is markedly reduced, relative to the blood sugar. Normal spinal fluid sugar is >50% of the blood sugar. Gram stain of the spinal fluid, latex agglutination tests of CSF and cultures will allow identification of the specific organism. The specific organism and the recommended treatment are indicated in Table

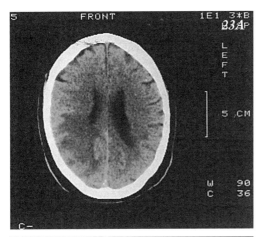

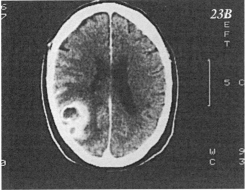

Figure 27-23. Brain Abscess. CT Scan: A) Non-enhanced demonstrates an edematous mass in white matter of right posterior temporal parietal-occipital areas with compression of the right lateral ventricle. B) With contrast a 3-4 cm ring-like enhancement occurs. This 53-year-old left handed white male with borderline diabetes mellitus had been treated for an abscess in the muscle of the abdominal wall due to anaerobic Bacteroides with incision drainage and antibiotics. Seven days later he developed severe "blinding" right sided headaches, confusion, difficulty in reading, word finding, following commands, repetition of short phrases plus flat affect, a neglect of the left visual field and a mild left hemiparesis. All neurological findings resolved over the next 2 days after treatment with antibiotics, dexamethasone and mannitol (to reduce cerebral edema) and immediate evacuation of the abscess. (Courtesy of Drs. Ralph Sama and Bernard Stone)

27-6.

Clinical Presentation: The clinical picture of acute purulent meningitis depends on the age of the patient. An infant under 6 months of age may be febrile, listless and drowsy and may vomit and fail to take feedings. The anterior fontanelle of an infant will be under increased pressure and bulging. An older infant or a child is likely to have nuchal rigidity (resistance to passive flexion of the neck) in addition to fever, convulsions, and coma. A posture of

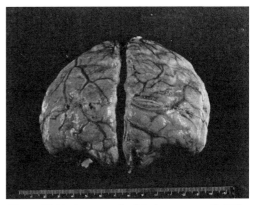

Figure 27-24. Acute meningitis (pneumococcal). This 57-year- old white male had been living in a nursing home with a diagnosis of chronic undifferentiated schizophrenia. Two weeks previously he had become less responsive than usual, then developed a temperature of 104degrees for which he received penicillin. On the day of admission, the patient fell out of bed and became comatose. Cerebrospinal fluid was cloudy with 11,861 with blood cells (80% polymorphonuclears) and a sugar of 36-mg./100 ml. Culture of the cerebrospinal fluid subsequently revealed S. pneumoniae. A left and then bilateral fixed dilated pupils developed followed by Cheyne-Stokes periodic respirations, decerebrate posture, and death. At autopsy, the brain was swollen with a yellow-green purulent exudate present in the subarachnoid space. (Courtesy of (Drs. John Hills and Jose Segarra)

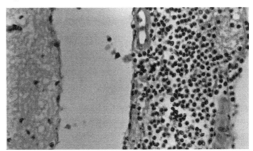

Figure 27-25. Acute meningitis (pneumococcal). The subarachnoid space is filled with inflammatory cells, predominantly polymorphonuclear leukocytes. There has been artifactual detachment of the pia from the cortical surface. (H & E x 400). (Courtesy of Pathology Department, Tufts University School of Medicine).

opisthotonos may be present (extreme extension of head, neck, trunk and limbs).

The older child and adult with meningococcal meningitis will usually have a prodromal period characterized by symptoms of an upper respiratory infection, low-grade fever, and various body aches. With septicemia and, then, the subsequent involvement of the meninges, the symptoms of chills, vomiting, severe headache, and nuchal rigidity occur, often followed by an alteration in consciousness. The signs of central nervous system involvement often appear relatively abruptly. A skin rash is common in the stage of bacteremia. Small or large areas of skin hemorrhage (petechiae and purpura) due to involvement and occlusion of the skin capillaries occur during the process of septicemia and are caused either directly by the organism or indirectly as a result of disseminated aggregation of platelets into small thrombi.

The early recognition of acute purulent meningitis is of importance. If it is untreated, serious intracranial complic-ations develop and death is the usual outcome. Prior to the era of specific antimicrobial therapy, the mortality was close to 90%. With specific therapy, this figure has been markedly reduced so that the actual mortality in meningococcal meningitis is now 3%, although the mortality in patients with overwhelming meningoccemia may still approach 40%. For S. pneumoniae the case fatality rate remains high at 21% due to the age of the patients. The case fatality rate for H.influenzae is 6 % and for group B streptococcus 7% percent. Eleven to 19% of survivors of meningococcal disease have sequelae usually minor if treated early (peripheral facial weakness, hearing loss), at times major: focal neurological deficit or focal seizures or loss of a limb.

There are several possible complications of meningitis. While these are more likely to occur in delayed or untreated cases, some may occur in those patients receiving adequate treatment. With a marked increase in intracranial pressure (at times, lumbar subarachnoid cerebrospinal fluid pressure at lumbar puncture may be 600 mm. of CSF compared to a normal pressure of 150 mm.) herniation of medial temporal areas through the tentorium may occur with compression of the third cranial nerve and midbrain. A vasculitis may involve blood vessel walls, resulting in occlusion of cortical arteries and veins. The occlusion of corti-

TABLE 27-6: ACUTE BACTERIAL MENINGITIS: COMMON ORGANISMS AT SPECIFIC AGES AND RECOMMENDED TREATMENT

AGE OF PATIENT	BACTERIAL ORGANISMS (on culture)	RECOMMENDED TREATMENT (All are intravenous)
0-1 Month	B. Streptococcus*	Penicillin G (plus in neonates initially gentamicin for 72 hours)
1-23 months	S. Pneumoniae-45%, N.meningitidis-31%	Vancomycin plus a broad-spectrum cephalosporin for S. Pneumonia (see below for N. meningitidis)
2-18 years	N. Meningitidis	Penicillin G**
19-59 years	S. Pneumoniae-60%; N.meningitidis-20%	See above
>60 years	S. Pneumoniae	See above

* Prior to introduction in 1990 of H.influenzae conjugate vaccine immunization; most cases were due to this organism. Recommended treatment for H.influenzae is ceftriaxone

** For epidemics of N. meningitidis in developing countries, a single intramuscular injection of a suspension of chloramphenicol in oil has proven to be effective.

*** Less common agents: L. monocytogenes ampicillin + gentamicin,

*** Enterobacteriaceae-broad spectrum cephalosporin +aminoglycoside.

cal veins and superior sagittal sinus may lead to hemorrhagic infarction of the parasagittal frontal parietal areas with a resultant weakness of the lower extremities and focal seizures. In addition, a further increase in intracranial pressure may occur.

As we have already indicated, a thick exudate at the base of the brain (somewhat more likely to occur in pneumococcal meningitis) may damage cranial nerves and also obliterate the subarachnoid cisterns. The end result may be a significant degree of hydrocephalus since cerebrospinal fluid will be unable to pass up over the hemispheres to the areas of absorption. In the child, a progressive enlargement of the head occurs.

Overall, in the recent series of Pomeroy et al (1990), 14% of children had neurological deficits that persisted beyond a year after bacterial meningitis. Of these 10% had only sensoryneural hearing deficits and 4% had multiple neurologic deficits. Seven percent had late, non-febrile seizures. (See also Taylor et al 1990 and Smith 1990).

Treatment: The present recommendations Table 27-7. (Quagliarello &Scheld, 1997, Rosenstein et al, 2001) as to choice of intravenous antibiotics which are to be started based on the Gram's stain of the CSF when the patient is first seen but before csf or blood cultures are available are presented in Table 27-7.

Once the organism has been identified on cultures or agglutination studies of CSF or blood the recommendation of Table 27-6 should be followed.

Acute Aseptic Meningitis: In these cases, the clinical signs and symptoms of meningitis are present in the sense that the patient complains of headache and stiffness of the neck. Vomiting and nuchal rigidity are present. However these findings are usually less fulminating than those in acute purulent meningitis, e.g., sudden coma and purpura in the adult is unlikely; consciousness is usually well-preserved. The spinal fluid, moreover, is often clear or only minimally cloudy (often described

TABLE 27-7 TREATMENT RECOMMENDATIONS BASED ON CSF STAINS PRIOR TO CULTURE RESULTS

Identification of Agent in CSF (Gram stain) before Culture	Treatment Recommendations (Intravenous)
COCCI	
Gram positive cocci:	Vancomycin plus a broad spectrum antibiotic
Gram negative cocci	Penicillin G
BACILLUS	
Gram positive bacilli	Ampicillin (or penicillin G) plus aminoglycoside
Gram negative bacillus	Broad spectrum cephalosporin plus aminoglycoside
EMPIRICAL APPROACH	
Children <1 month	Ampicillin + broad spectrum cephalosporin and vancomycin
Children >1 month	Vancomycin +cefotaxime or ceftriaxone

as opalescent). A relatively small number of white blood cells is present (5 to 2000 per cubic mm. and usually less than 500 per cubic mm.) in comparison to acute purulent meningitis. Moreover, the white blood cells are predominantly mononuclear (lymphocytes and monocytes). The sugar content of the spinal fluid is normal. Moreover, smears, agglutination studies and bacteriological cultures of the spinal fluid fail to reveal a responsible organism. In general, the causative organism is a virus: ECHO, Coxsackie, non-paralytic poliomyelitis, mumps and lymphocytic choriomeningitis.

The student should note, however, that a similar cerebrospinal fluid reaction may also characterize certain diseases where a secondary meningeal reaction occurs: subdural empyema, brain abscess, and venous sinus thrombosis. At times, a similar spinal fluid formula may be noted in a viral encephalitis. The aforementioned viral agents may, of course, at times present a combined syndrome of meningoencephalitis.

Moreover, at times several more significant subacute infections in their early stages may present a predominately mononuclear reaction in the spinal fluid: tuberculous and cryptococcal (fungal) meningitis. Bacterial organisms will be absent on routine smears and cultures of the cerebrospinal fluid. However, both of these infections are characterized by a low spinal fluid sugar, and appropriate stains and cultures (and in the case of cryptococcal infection specific antigen studies) will eventually disclose the organism. These infections, although not common at the present time, in the non-immunosuppressed population, are of importance because a) specific therapy is required, b) with the increase in cases of acquired immune deficiency syndrome; the incidence of these diseases has increased.

Differentiation of the various types of viral meningitis may be made by specialized immunological techniques for the measurement of neutralizing antibodies. Viral meningitis is a self-limited disease. No specific treatment is available (see below, however, under H.simplex). Recovery is in general complete - unless, the meningitis is an epiphenomenon in the midst of an encephalitis.

Other types such as aseptic, tuberculous, fungal, meningovascular neurosyphilis and other spirochetal infections are considered on the CD ROM

ENCEPHALITIS: This topic is considered in detail on the CD Rom and in general medical texts. See also figures 27-26 and 27-27. A brief introduction is provided here.

The term encephalitis refers to a diffuse invasion of the parenchyma of the brain by an infectious agent.

At the present time, clinical presentation of encephalitis is probably less common than those diseases considered under the category of meningitis. During particular epidemics certain infectious agents involving the central nervous system have produced a large number of cases with encephalitis. However, subclinical or minor diffuse involvement of the central nervous system probably occurs in the course of a number of common viral diseases, primarily as a mild aseptic meningitis or meningo-encephalitis. Even in epidemics of encephalitis or poliomyelitis, there may be many subclinical cases. For example, for Japanese encephalitis, the ratio of subclinical to overt clinical cases has been estimated at 200 or 300:1 (Monath 1988).

In general, most cases falling into this category reflect viral infection. Evidence of viral particles will be found as inclusions within the cytoplasm or nuclei of neurons. With the exception of the spirochete, bacteria do not produce a diffuse encephalitis, although bacteria may produce multiple areas of abscess formation. Other infectious agents may produce encephalitic involvement of the nervous system. The rickettsiae which are intermediate in size between bacteria and viruses, usually produce signs of a meningoencephalitis in addition to a characteristic skin rash. Examples are typhus, and Rocky Mountain spotted fever). Moreover, certain protozoal parasites may invade the central nervous system producing a subacute encephalitis Examples include toxoplasmosis and trypanosomiasis (African Sleeping Sickness).

Encephalitis may be acute or chronic. Most cases fall into the acute category; most viral and rickettsial diseases are acute. On the other hand, the spirochetal infection and protozoal infections of the central nervous system are, in general, chronic or subacute processes. Recently, however, several viruses and prions have been implicated in chronic diffuse progressive processes (previously considered to be of an unknown etiology) involving the nervous system.

ACUTE ENCEPHALITIS: ACUTE VIRAL INFECTIONS

The pathology of viral infections of the neural parenchyma involves a viral invasion of neurons with the production of intranuclear or intracytoplasmic inclusions *(Fig 27-27)* and the acute degeneration and destruction of nerve cells. There is a cellular infiltration of neural tissue and an accumulation of inflammatory cells about the degenerating nerve cells, in a perivascular location (Fig. 27-26). The cells are usual-

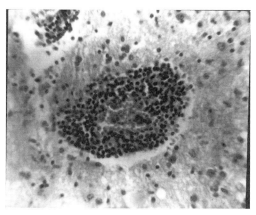

Figure 27-26. Encephalitis (herpes simplex). A significant collection of mononuclear cells is present in the perivascular space with infiltration of adjacent cerebral cortex. This 32-year-old patient had the acute onset of fever, headache, and generalized convulsions. Over a 30-day period, he developed increasing stupor, bilateral extensor plantar responses, and dilated pupils. (H & E x 320). (Courtesy of Dr. John Hills).

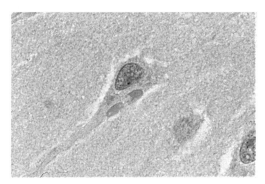

Figure 27-27. Encephalitis. Rabies. Negri bodies are present as cytoplasmic inclusions in neurons of the hippocampus. Intracellular inclusions are characteristic of viral diseases. (H&E 100x) (Courtesy of Dr. Thomas Smith)

ly mononuclear although in some processes, e.g., acute poliomyelitis or in a severe case of encephalitis, there are often noted a significant number of polymorphonuclear leukocytes at times involved in neuronophagia. A microglia proliferation is often noted in the involved neural parenchyma. To a variable degree adjacent meninges may be infiltrated by inflammatory cells. Depending on the degree of meningeal infiltration and on the severity to which the underlying parenchyma is involved, a variable increase in cells (predominantly lymphocytes) will be noted in the spinal fluid. Occasionally a normal cell count may be present, more often 100 to 500 cells per cubic mm. are present. The protein content is usually increased; the sugar and chloride content are normal.

The differentiation of the particular virus involved depends on:

a. The specific clinical pattern of the disease: some viruses involve particular areas of the central nervous system; some have associated involvement of other organ systems.

b. Virus isolation studies: inoculation of blood, nasal washings, excretions, cerebrospinal fluid, or fresh post-mortem brain tissue into susceptible animals. In some specialized centers, tissue culture techniques may be employed.

c. Application of acute and chronic serological tests to measure antibody levels (complement fixation or antibody neutralization tests).

It had been traditional to divide these viral infections into *neurotropic group* (primary involvement of the central nervous system) and a *non-neurotropic group* (involvement of the central nervous system in humans is usually secondary to or less prominent than involvement of other organ systems). As we will discuss with the herpes group of virus infection, the distinction is to some degree artificial. The infections with the neurotropic may be epidemic (the arthropod bourne viruses such as equine encephalitis or West Nile) or non epidemic such as rabies.

A more modern classification divides viruses into those containing RNA and those containing DNA.

Examples of RNA containing viruses include: the enteroviruses (polio, Coxsackie, ECHO), the arboviruses (Eastern and Western Equine, Japanese, etc.) rubella, mumps and measles, rabies and HIV.

Examples of DNA containing viruses include Herpesvirus, papovavirus and pox-rusvirus. An overview of viral encephalitis is provided in the review of Whitley (1990).

Neurotropic Group: *Epidemic group:* Within the epidemic group several infections may be grouped together because they share common characteristics: a) Equine encephalomyelitis with Eastern U.S.A., Western U.S.A., and Venezuelan subtypes(Deresiawicz et al,1997). b) Japanese B encephalitis (probably the most common form. World wide in recent years). c) Russian tick encephalitis. d) St. Louis encephalitis. e)West Nile encephalitis a recent problem in the eastern United States (Nash et al 2001, Tyler, 2001) f) LaCrosse virus encephalitis ,probably the most frequent variety in the U.S. in recent years (McJunkin et al, 2001).

In each of these infections a relatively diffuse involvement of the cerebral cortex, diencephalon, brain stem and cerebellum occurs. In each, an arthropod, usually a mosquito, has been implicated as a vector of transmission. A species of bird or mammal has often served as a reservoir of infection. For these reasons, cases have occurred predominantly in a seasonal manner. (In northeastern United States, for example, most cases of Eastern equine and West Nile encephalitis have occurred in the late summer or early fall months, a time when mosquitoes are frequent). Vaccines are available for prevention of some of the more common varieties (Hoke et al, 1988).

The clinical syndrome in most cases is characterized by the sudden onset of headache, vomiting, drowsiness, confusion, convulsions, coma, fever, plus or minus, stiffness of the neck. Evidence is often present on examination of the patient of diffuse and, at times, multifocal involvement at multiple levels of the neural axis: coma, confusion, myoclonic jerks, status epilepticus, decorticate or decerebrate rigidity and cerebellar findings. In some cases, cranial nerve findings, aphasia and hemiplegia, will be present. The duration of the disease is a matter of days to several weeks. *The state presented by the patient may be described as an encephalopathy.* The major differential diagnosis is that of toxic metabolic encephalopathy. A variable mortality occurs in part related to the age of the affected patients. The mortality of Eastern equine encephalitis, which has affected children predominantly, has approached 75 percent, with significant residuals in those recovering. *Encephalitis lethargica (Von Economo's encephalitis)* occurred as an epidemic in the period 1916-1926. More recent case have been described by Howard and Lees(1987). A viral etiology was suspected but never confirmed: the means of transmission remained uncertain. This infection differed from the previously described varieties of encephalitis in the sense that although diffuse, the pathology tended to be concentrated in a periventricular location with severe involvement of structures bordering the aqueduct and third and fourth ventricles. The periaqueductal region and other structures of the midbrain were particularly affected. Reflecting the significant involvement of the midbrain and of the associated structures of the extrapyramidal system, disorders of eye movement and movement disorders of various types were prominent; in addition to the more general symptoms of lethargy, headache and fever. Since the involvement of the cortex was less severe, seizures and focal cortical deficits were less frequent. The mortality approached 25%. Among those who survived, post encephalitic residuals were common. In contrast to other forms of encephalitis, Parkinson's disease and other disorders of extrapyramidal function were frequently noted; the Parkinson's disease usually appeared to emerge or evolve as a symptom after the clearing of the acute symptoms (see chapter 19).

Acute anterior poliomyelitis, the other member of this epidemic group of neurotropic viruses, has already been discussed in relation to the spinal cord. The portal of entry is the gastrointestinal tract. The central nervous system is involved either by spread along the axis cylinders or via a generalized viremia. The virus involves predominantly the large motor neurons of the anterior horn and of the brain stem with lesser involvement of other neurons in the spinal cord.

Non epidemic neurotropic virus infections
The two most familiar members of this group - Herpes zoster (discussed previously) and rabies - do not have their major effects at

the level of cerebral cortex.

Rabies: The virus of rabies is present in the saliva of rabid animals and invades the central nervous system several weeks to months after a bite from an infected animal (dog, cat, wolf, fox, raccoon, squirrel, or bat). The virus travels to the central nervous system along peripheral nerves. The symptoms relate to the characteristic involvement of nuclei of the brain stem, Purkinje cells of the cerebellum, and pyramidal cells of the hippocampus (Ammon's horn). In addition, there is a significant inflammation and necrosis of the spinal cord or brain stem at the segmental site corresponding to the radicular dermatome involved by the bite. The cerebral cortex is relatively intact and consciousness is preserved.

The initial symptoms consist of numbness and tingling in the distribution of the involved peripheral nerve, followed by headache, vomiting and a stage of agitation. This latter stage is characterized by restlessness, generalized convulsions, and at times, visual and auditory hallucinations. Marked alteration in emotions occurs: unreasonable fear, rage and depression. In this stage laryngeal and pharyngeal spasm (with fear of water, "hydrophobia") is prominent. In a later stage of flaccid paralysis, impairment of vocal cords and of respiratory centers develops. In approximately 20 percent of cases, an ascending flaccid paralysis dominates the acute stage (Plotkin and Koprowski, 1978).

Death occurs within 2 to 5 days of onset of central nervous system symptoms. At autopsy, the diagnosis may be established in man or other infected animals by the findings of characteristic acidophilic (eosinophilic) inclusions (Negri bodies) within the cytoplasm of hippocampal pyramidal cells (Fig.27-27).

Because of the long incubation between exposure to the virus and the development of neurological symptoms, prophylactic treatment is possible. Thus, the administration of a vaccine containing the attenuated virus (first introduced by Pasteur), induces the production of anti-bodies, and may prevent the development of neurological symptoms. The combination of this active immunization with passive immunization with antiserum containing antibodies to rabies has been demonstrated to be much more effective. The current vaccine is the human diploid cell vaccine, the current antisera - human rabies immune globulin. Previous treatment protocols involving nervous system derived vaccines resulted in post vaccinal demyelinating reactions. Once symptoms have developed, no effective treatment is available. In under developed areas of the world a high mortality occurs. (see Baer & Fishbein 1987 and Fishbein &Robins,1993 for a review of epidemiology and current concepts of treatment.).

Non-Neurotropic Viral Infections:

These viruses were traditionally considered as not producing significant symptoms of central nervous system involvement. Indirect evidence suggests that some of these viruses (mumps, measles, enterovirus) do invade the central nervous system in a much greater percentage of otherwise asymptomatic cases. In some of these diseases, e.g., herpes simplex, there is evidence of direct invasion of the central nervous system when the syndrome of acute encephalitis occurs. In other viral infections, e.g., mumps and measles, it is often unclear when central nervous system symptoms develop, whether one is dealing with an actual viral invasion of the central nervous system producing an acute encephalitis (direct invasion of neurons), or whether one is observing an immunological reaction to infection elsewhere pro-ducing an acute allergic post-infectious encephalomyelitis, (white matter predominantly involved with perivenous demyelina-tion).

Types Of Non-Neurotropic Viral Infections:

1. *The herpes virus family:* included herpes simplex, herpes zoster/varicella, cytomegalovirus and Epstein Barr.

The *Herpes simplex virus (HSV)* has two types - *Type 1* associated with the common cold sore of the lip or oral cavity, often activated by fever and *Type 2* genital herpes with labial vaginal and penile infection. In both instances, the virus remains in a latent form in neural ganglia. (Baringer and Swoveland, 1973), particularly the trigeminal ganglion in the case of Type 1 HSV and the sacral ganglia in the case of Type 2 HSV.

Type 1 HSV is the most common cause of non-epidemic; sporadic acute encephalitis in the United States. The early signs of headache and fever are followed by the development of signs that suggest involvement of limbic structures one or both temporal lobes, and at times orbital frontal *(Fig.30-1)*. The virus apparently gains access to the central nervous system by infecting structures that are related to the olfactory system, perhaps by axonal transport along the olfactory bulb and its anatomical connections. In some cases spread may be hematogenous (see review of Picard et al ,1993). Personality and behavioral changes may be present for days or a week; followed by seizures , hemiparesis and aphasia. Rapid or subacute progression to a stage of increasing coma then occurs. In a small number of cases, the patient may present with an overwhelming rapidly fatal state characterized by repeated generalized convulsions; myoclonic jerks of the extremities, decerebrate rigidity, coma and periodic complexes in the EEG. In contrast in a small number of cases, the clinical picture is that of a progressive subacute dementia and confusional state. The CSF demonstrates increased pressure, and a variable number of mononuclear cells plus or minus a variable small number of red blood cells. The electroencephalogram may provide evidence of focal or bilateral temporal lobe abnormalities, including slow waves. At variable points in the disease course, the radioactive brain scan, contrast enhanced CT scan or MRI will demonstrate focal or bilateral temporal (and frontal) abnormalities : enhancing necrotic lesions with considerable edema. Sequential serological studies of blood and CSF may be useful in the diagnosis. The treatment of choice is intravenous acyclovir which can be administered without serious side effects. The response to treatment is directly related to how long the symptoms have been present and to the level of consciousness. Therefore, if the diagnosis is strongly suspected on clinical grounds - treatment is begun as soon as possible, even though the ancillary laboratory studies are not yet confirmatory. In most instances, a brain biopsy is not obtained unless the patient fails to respond to therapy. Even with acyclovir therapy, 53% of patients will die or will be severely impaired. (See discussions of Hanley, 1990; Whitley, 1992).

Type 2 HSV in adults is associated with an aseptic meningitis or with severe radicular symptoms involving the sacral segments. The latter may include urinary retention (Caplan et al, 1977).

Type 2 HSV infection in the infant or fetus results in severe disseminated disease with multi-system involvement of brain and viscera.

2. *Paramyxovirus* group includes mumps and measles. In the child and adult mumps encephalitis is rare. In the fetus and neonate, mumps virus infection has been implicated as a possible cause of aqueductal stenosis producing hydrocephalus (see Johnson and Johnson 1969).

Measles virus infection may be associated with a variety of central nervous system syndromes including:

a. an acute or subacute encephalitis
b. a post infectious encephalomyelitis
c. subacute sclerosing pan-encephalitis: (SSPE) a form of progressive chronic encephalitis manifested by dementia, ataxia, seizures and myoclonus. The disease usually develops - in a delayed manner after an early childhood measles infection. There is a reactivation of the measles virus within the CNS.

3. *Retrovirus - Human Immunodeficiency Virus (HIV-1)*: This virus infects cells of the immune system (e.g., T helper lympho-cytes) - producing an acquired-immune deficiency state (AIDS). Carriers who may be in an early or asymptomatic stage of the disease outnumber actual cases of AIDS or AIDS-related complex (ARC). In the United States during the 1980s specific groups were at high risk: (a) homosexual or bisexual men (66%); (b) intravenous drug users (17%); (c) heterosexual -

sexual partners of patients with AIDS (4%); (d) recipients of blood and blood products (3%); and (e) combined risk - homosexual or bisexual male - also drug user (8%)[1]. By the year 2001, Sekowitz quoted figures of 36 million people world wide infected with HIV, with an additional 21.8 dead of AIDS. Approximately 70% of cases occur in sub-Saharan Africa where there are two types of epidemics : 1) a horizontal transmission in adults spread primarily by heterosexual contact and to a lesser degree by shared needles and 2) a vertical transmission, infected mothers give birth to infants who have been infected in utero. Significant numbers of cases are also occurring in the Caribbean, southeast Asia, and South America.

*The nervous system is involved in at least 60% of patients with clinical disease:*The infections may be generalized or focal due to the immunodeficiency state. Thus, there is an increased incidence of cryptococcal fungal meningitis(10%) and of toxoplasmosis granuloma (5%). A high percentage of patients (>30%) have a direct invasion of the CNS by the virus producing a subacute encephalitis referred to as AIDS dementia or HIV, meningitis. In patients with acute HIV-1 infection 24% have an aseptic meningitis.

Other infections of the nervous system (spirochetes, fungi, protozoa, and tuberculosis) will be considered on the CD ROM.

IV. SYSTEM DISORDERS: The following section is included in detail on the CD ROM and in general medical texts. The following subsections are included:

A. Degenerations: These disorders are considered in the chapters on muscle, peripheral nerve, spinal cord, basal ganglia, cerebellum and memory systems. Refer to CD ROM for an approach to classification.

B. Nutritional disorders: Refer to the CD ROM Table 27-8 and to general medical texts

C. Toxic Disorders: These are considered in detail on the CD ROM and in general medical texts. Heavy metals effects are summarized in Table 27-9 Various central nervous systems pharmacologic agents taken in excess may all produce effects on CNS function. Some such as sedatives, anticonvulsants, benzodiazepines or narcotics may produce depression of cognitive function and or ataxia and eventually in severe overdose coma. Neuroleptics may produce extrapyramidal effects. Hallucinogens and amphetamines may produce a psychosis. Use of the latter may be associated with hemorrhagic strokes Cocaine may produce arrhythmias and stroke. Carbon monoxide may produce headaches or coma.

C. Complications of Endocrine disorders: see CD ROM regarding diabetes, thyroid disease, pituitary hormones etc.

D. Remote effects of malignancy: see CD ROM and chapter 21

V. DISORDERS OF MYELIN: Refer to CD ROM. See also Chapters 9 and 13. Figures 27-28 and 27-29 demonstrate cases with clinical involvement of the cerebral hemispheres.

---

[1] *These percentages vary markedly in different geographic areas and in different population groups. Heterosexual cases are increasing significantly among drug users or their contacts and in Africa, South America and Southeast Asia. New cases among homosexual males are decreasing.*

TABLE 27-8: NUTRITIONAL AND RELATED DISORDERS OF THE NERVOUS SYSTEM*

STRUCTURES INVOLVED	VITAMIN (S) DEFICENT	MANIFESTATIONS
Cerebral Cortex	-Niacin	-Pellagra: dementia, dermatitis, diarrhea
	-Pyridoxine	-Seizures (possibly of hippocampal origin)
	-B12	-Dementia
Corpus callosum	Specific unknown, possibly toxic, described initially with excess of Italian red wine. Possible role of thiamine.	Marchiafava –Bignami disease: dementia and apraxia. May be seen in patients who also have Wernicke-Korsakoff syndrome.
Diencephalon and brainstem and cerebellum-Periventricular	-Thiamine	Wernicke-Korsakoff syndrome (Ch. 30)
Pons	Specific unclear but does follow too rapid correction of hypo-natremia	Central pontine myelinolysis. Patient is mute, spastic and akinetic (Chapter 13)
Cerebellum	Thiamine or multiple B	Alcoholic cerebellar degeneration (Ch. 20)
Spinal cord	B12	Combined system disease (Ch.9)
Peripheral nerves	Thiamine, or pyridoxine or B12 or multiple B	Nutritional mixed sensory neuropathy or with thiamine beri-beri.
Optic nerves	Thiamine, B12, riboflavin	Optic neuropathy (retrobulbar neuropathy usually bilateral)

- Muscle may also be involved in an alcoholic myopathy-specific unknown but multiple B suspected.
- In infants and children deficiencies of proteins and fat may retard development of nervous system.

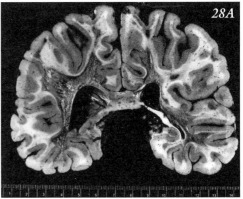

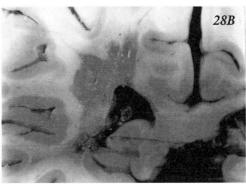

Figure 27-28. Multiple sclerosis: cerebral involvement. A) Severe involvement of the cerebral white matter has occurred, with multifocal gray sclerotic lesions as well as large confluent lesions in a periventricular location. The large confluent lesions and the relative sparing of the arcuate fibers is reminiscent of Schilder's diffuse sclerosis. Destruction of axons, as well as of myelin, has occurred in the larger lesions. This patient had the onset of multiple sclerosis at age 26 and had a progressive course so that by age 30, the patient was essentially bedridden with paralysis of all four extremities, marked impairment of vision, and marked internuclear ophthalmoplegia, and by age 33 had a severe dementia. B) A more typical area of demyelination in the periventricular subcortical white matter is demonstrated in this coronal section from another case of multiple sclerosis. (Courtesy of Dr. John Hills).

TABLE 27-9: EFFECTS OF HEAVY METALS ON THE NERVOUS SYSTEM

METAL	ACUTE AND HIGH LEVEL	CHRONIC LOW LEVEL
Lead	Encephalopathy: acute cerebral edema	In adults: motor peripheral neuropathy In children long term retardation of cognitive function/intelligence quotient
Inorganic mercury	Acute gastrointestinal, renal and hematologic effects	Cerebellar degeneration and the "mad hatter syndrome": dementia and psychosis
Organic mercury	Blindness, cerebral, cerebellar, pyramidal, and anterior horn cell degeneration primarily affecting fetus infants and children (Minimata Bay in Japan etc)	
Arsenic	Acute hemorrhagic encephalopathy	Peripheral neuropathy: sensory/motor
Copper		Hepatolenticular degeneration: Wilson's Disease –genetic defect- (Chap.19)
Manganese		Parkinson's disease

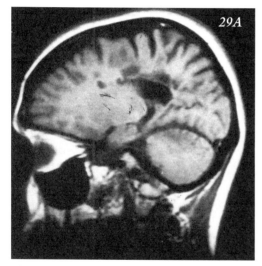 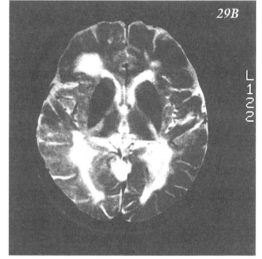

Figure 27-29. Multiple sclerosis. MRI scans. This 38 year old white female bank employee had a 15 year history of relapsing-remitting secondarily progressive multiple sclerosis initially affecting spinal cord Lhermitte's sign), then optic nerves ,then cerebellum and cognitive function. CSF indicated an IgG index of 2.42 (normal 0.45 + 0.1) with 3 oligoclonal bands (normal < 1). In the 5 years following this MRI, she developed dementia and a bedridden state. A) T1 sagittal section. B) T2 horizontal section.

CHAPTER 28
Problem Solving: Part IV
Diseases of the Cerebral Hemispheres

LESION DIAGRAMS: For each of the following diagrams indicate the structures involved by the cross hatched lesion. For each structure involved, indicate the expected clinical symptoms or signs with appropriate lateralization. Where appropriate, indicate the vascular territory involved or the designation of the most likely type of pathology. (Diagrams to be inserted).

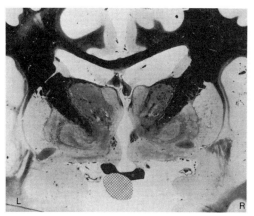

Figure 28-1.

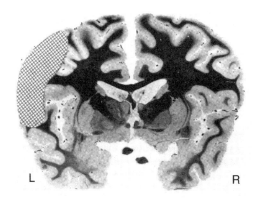

Figure 28-2.

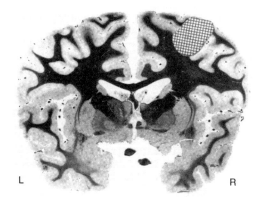

Figure 28-3.

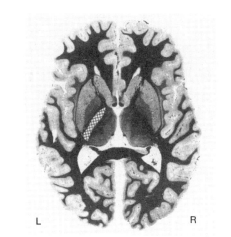

Figure 28-4.

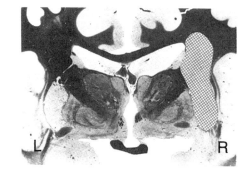

Figure 28-5.

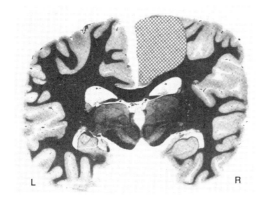

Figure 28-6.

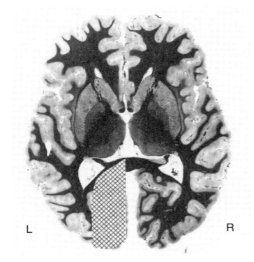

Figure 28-7.

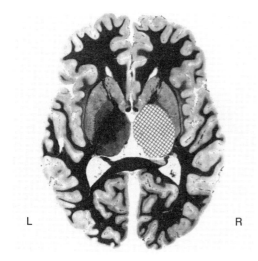

Figure 28-8

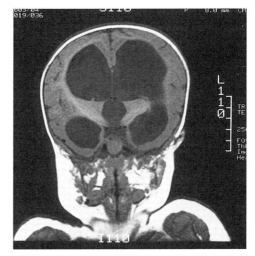

Figure 28-9A

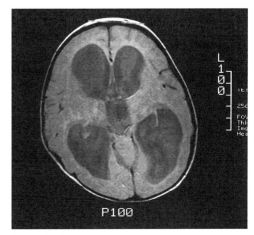

Figure 28-9B

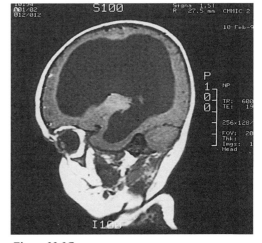

Figure 28-9C

CASE HISTORY PROBLEM SOLVING PART IV: CEREBRAL HEMISPHERES

These cases represent the gamut of disease affecting the cerebral hemispheres. The pathology represented may be tumor, infarction or hemorrhage. Some of the lesions are intrinsic; others are extrinsic. The nature of the pathology may be uncertain; the location of the pathology, however, should be evident to you.

Diagram each lesion. If the etiology appears to be vascular, attempt to identify the vessel involved.

Case 28-1: This 53-year-old, right-handed, white housewife had experienced intermittent right-sided headaches and retro-orbital pain for approximately 2 months. Five days prior to admission to the neurological center, the patient awoke from a sound sleep at 7:00 a.m. complaining of the most severe headache she had ever experience. "It felt like something bursting inside my head". This was accompanied by nausea and vomiting. Patient was admitted to her local community hospital. Two days prior to transfer to the neurological center the patient reported ptosis and diplopia. Severe headaches persisted, requiring the administration of meperidine hydrochloride (Demerol hydrochloride).

NEUROLOGIC EXAMINATION

The patient was a well-developed white female with blood pressure of 160/90. Cardiac status was normal.

1. *Mental status:* The patient was lethargic but easily aroused. When aroused, she then responded with alert, appropriate responses.
2. *Cranial nerves:*
   a. The right pupil was dilated and sluggish in response to direct or consensual stimulation with light. The left pupil was of normal size. Ptosis of the right lid was present. There was partial weakness of right medial rectus, inferior rectus, and inferior oblique and superior rectus.
   b. All other cranial nerves were intact.

3. *Motor system:*
There was minor weakness of the left arm and left leg with an increase in tone in the left arm and left leg (minor spasticity).

4. *Reflexes:*
   a. Deep tendon reflexes were everywhere increased on the left side.
   b. Plantar responses were extensor bilaterally but much more prominently so on the left.

5. *Sensory system:* No definite abnormalities were present.

6. *Head & Neck:*
   a. There was resistance to flexion at neck.
   b. No bruits were present over neck, orbits and head.

QUESTION:

What is the most likely clinical diagnosis at this point?

LABORATORY DATA:

1. *Skull and chest X-rays* were normal.
2. *Lumbar puncture:*
   a. Opening pressure was 300 mm of CSF; closing pressure was 150 mm of CSF.
   b. The fluid was grossly bloody. First tube demonstrated 55,000 red blood cells per cubic millimeter and 60 white blood cells per cubic millimeter. Third tube demonstrated 72,000 red blood cells per cubic millimeter.
   c. The CSF protein was 84 mg/100 ml. The spinal fluid sugar was 60-mg/100 ml with a simultaneous blood sugar of 90-mg/100 ml.

HOSPITAL COURSE:

The patient remained relatively stable and alert until definitive procedures were performed.

QUESTIONS:

1. Based on the symptoms and signs in this case what is your diagnosis?
2. How do you interpret the results of the lumbar puncture?
3. If you had seen this patient in the emergency room directly at the onset of symptoms, which study would you have obtained? Would the lumbar puncture have been your initial study?

4. Where is the basic pathologic lesion located? Be specific.
5. Indicate the single most definitive diagnostic procedure that should be performed in this case prior to definitive therapy. When should this be performed?
6. Define definitive therapy in this case. When should this therapeutic procedure be performed?
7. Discuss the complications of this disorder.
8. Discuss prognosis of this clinical problem.

Case 28-2: This 56-year-old right-handed white male truck driver, on the day prior to admission, had the sudden onset of a transient monocular blindness involving the right eye.

On the morning of admission the patient developed a weak, numb and useless left arm. These symptoms had shown considerable improvement by the time of his initial admission that afternoon.

PAST HISTORY:

1. The patient smoked two packs of cigarettes/day.
2. Occasional symptoms of intermittent claudication had been present.

PHYSICAL EXAMINATION:

1. Blood pressure was 150/80; pulse was 100 and regular.
2. Bruits were present over the right carotid, left subclavian, and both femoral arteries.
3. Posterior tibial pulses were absent.
4. There were no cardiac murmurs.

NEUROLOGIC EXAMINATION:

1. *Mental status:* Intact.
2. *Cranial nerves:* Intact.
3. *Motor system:* There was minor weakness and incoordination of the left hand.
4. *Reflexes:* Intact.
5. *Sensory system:* Intact.

SUBSEQUENT COURSE:

Six days later, the patient had a focal motor seizure involving his left arm. Twenty-three days later, the patient experienced another transient episode of weakness and numbness involving the left arm. Approximately four weeks later definitive diagnostic studies were performed.

QUESTIONS:

1. What is the most likely diagnosis?
2. Outline your approach to effective management.
3. Which non-invasive studies would you obtain and when would you obtain those studies?
4. You have obtained these studies and the results confirm your clinical diagnosis. Which physician would you now consult?
5. Which surgical procedure would now be appropriate?
6. Indicate the classical definition of a transient ischemic attack (TIA). Now indicate the more modern operational definition that takes account of new developments in the acute treatment of ischemic stroke. What is the definition of the term reversible ischemic neurological deficit (RIND)? What is the definition of the term completed stroke? What is the meaning of the term stroke in evolution?

Case 28-3: This 32-year-old white male was admitted to the hospital because of severe headache, confusion and aphasia. The patient had been in good health except for frequent headaches for the 10 years prior to admission. Six days prior to admission at approximately 2:00 p.m., the patient returned from a hunting trip, and went into the bathroom. His wife heard a noise, a thud, as though the patient had fallen to the floor. She found the patient lying on the floor, moaning and groaning, as if in pain. The patient showed evidence of significant confusion; he was, however, moving all his extremities and she surmised that he had no actual paralysis. The patient was admitted to his local hospital, where a lumbar puncture demonstrated the presence of blood in the spinal fluid. Over the next few days, the patient complained more and more of headaches, became more and more agitated, and increasingly restless. He was subsequently transferred to a neurosurgical center.

GENERAL PHYSICAL EXAMINATION:

There were no remarkable features. Blood pressure was 125/70.

NEUROLOGIC EXAMINATION:

1. *Mental status:*
   a. Patient was oriented for time and place
   b. The patient was sleepy but agitated when roused from sleep
   c. The patient had difficulty with word selection and would not cooperate in repeating simple words or identifying simple objects
   d. The patient often spoke words that were unrelated to questions he had been asked or to the conversation.
2. *Cranial nerves:*
   a. A minor right visual field defect was suggested
   b. A right central facial weakness was present
3. *Motor system:* A minor weakness of the right upper extremity was present. When outstretched, the right hand rapidly fell downward.
4. *Reflexes:*
   a. Deep tendon reflexes were symmetric.
   b. Plantar response was extensor on the right, flexor on the left
5. *Sensory system:* All modalities were intact.
6. *Neck:* Marked nuchal rigidity was present. The patient complained of severe pain on neck motion.

HOSPITAL COURSE:

Over the next four hospital days, the patient developed increasing expressive aphasia, increasing right central facial weakness and increasing weakness of right upper extremity. The degree of right upper extremity weakness was now marked, with total paresis of movements of right hand and partial paresis of movement at the shoulder and elbow. The patient continued drowsy and intermittently agitated. Some observers felt that the patient now had a slightly larger pupil on the left side than on the right. Specific diagnostic and therapeutic procedures were performed.

QUESTIONS:

1. Indicate the original pathology and location of pathology.
2. Indicate the complications of this pathology that occurred prior to surgery and within 2 weeks of the initial episode.
3. Indicate the specific diagnostic procedures that were performed to localize the lesion prior to surgery.
4. What primary indications for surgery were present in this case?

When should surgery be undertaken? Which surgical procedure should be performed?

SUBSEQUENT COURSE:

The patient then underwent a surgical procedure 12 days after the initial episode. Postoperatively the patient had a marked right hemiplegia, with increase in the degree of aphasia. This was a mixed type of aphasia; when the patient did speak, he used jargon and he also appeared unable to understand language.

Eleven days postoperatively, the patient developed fever of 104° and a recurrence of nuchal rigidity. Spinal fluid examination now indicated 1200 white blood cells (92% polymorphonuclear); spinal fluid sugar was 50-mg/100 ml with blood sugar of 130-mg/100 ml. Specific treatment was administered.

QUESTIONS:

5. Indicate the diagnosis that explains this complication of fever and nuchal rigidity. Be as specific as possible in terms of the agent producing fever.
6. How would you treat this complication?

SUBSEQUENT COURSE

At the time of discharge, the patient was alert and oriented, but had a significant expressive aphasia and a moderate hemiparesis (arm greater than leg) in addition to a right central facial weakness. Four months after the initial episode, the patient had a generalized convulsive seizure. Two to three weeks later, he had a focal seizure characterized by clonic movements of right upper extremity. One year after the initial episode, patient had focal seizures consisting of clonic contraction of the right face accompanied by arrest of speech. These seizures were preceded by a sensation of tightness in the face. Twenty-one months later, the patient was reported as having frequent focal seizures characterized by numbness of the right hand and face, followed by transient paralysis of the right lower extremity. Follow-up 33 months after that indicated several minor seizures every day involving numbness of the right hand and perioral area of face.

Neurologic examination almost 6 years after the initial episode, indicated the following significant residual deficits:

1. Hesitancy in speech with mild expressive aphasia and a more significant nominal aphasia. In addition, left right confusion was present.
2. Right upper quadrantanopia.
3. Mild right central facial weakness.
4. Minimal weakness, right upper extremity in a right-handed man.
5. Deep tendon reflexes were increased on the right side with a right extensor plantar response.

The patient was treated with specific medications resulting in a reduction of the frequency of his focal seizures.

QUESTIONS:

7. In the years following surgery, the patient had focal sensory and motor seizures involving hand and face. Indicate the medications to be used in the treatment of these episodes.
8. The patient also had secondarily generalized seizures in the post-operative years. Indicate the recommended medical treatment.

Case 28-4 (patient of Dr. George Robertson and Dr. Huntington Porter): This 73 year old right handed white businessman from Maine was involved in an auto accident approximately 10 days prior to admission, in which his auto was struck by a heavy logging truck. The patient walked away from his car but was dazed and amnestic for the events of the next 15 minutes. This condition however then cleared. At the emergency room of his local hospital skull x-rays were normal and the patient was found to have no significant injuries except for ecchymosis about the right eye and knee. The patient subsequently however did note a steady bilateral frontal temporal headache, which would on occasion awaken him from sleep and which gradually, became worse. Five days prior to admission, some fogginess in thinking was noted and the patient's family reported some personality change. The patient reported some weakness of both legs in getting out of a bathtub perhaps somewhat greater on the right side. Two days prior to admission, the patient was noted to be sleeping longer. On the day of admission, the patient and his physician noted a more persistent left sided weakness. The patient was admitted to the neurology service.

PAST HISTORY:

Hyperthyroidism had been under treatment for three years. Three years prior to admission, the patient had been evaluated for a minor episode of dizziness, which had occurred in relation to bradycardia and a transient fall in blood pressure. The neurological consultant at that time had noted a totally normal examination. The right carotid pulse however was noted to be decreased compared to the left.

NEUROLOGIC EXAMINATION
(day of admission):

1. *Mental status:* The patient was alert, well oriented in all three spheres with an excellent knowledge of current events. He could remember 4/5 objects in 5 minutes. His digit span was 8 forward and 5 in reverse.
2. *Cranial nerves:* All were intact except for a questionable left central facial weakness.

3. *Motor system:*
   a. There was a significant decrease in strength predominantly distal in the left leg. To a lesser degree, a mild weakness of the left upper extremity was present.
   b. The patient dragged the left leg in walking.
4. *Reflexes:* Deep tendon stretch reflexes were increased bilaterally in a symmetrical manner. Plantar responses were flexor.
5. *Sensory system:* All modalities were intact
6. *Carotids:* The right carotid pulse was decreased as previously but no bruits were present.

QUESTIONS:

1. It is an often quoted axiom in neurology that 75% of neurological diagnosis is based on the history. This patient was seen before the era of the CT scan and yet the most likely diagnosis was clear. What is your diagnosis as regards location of the lesion and the nature of the pathology? Be specific.
2. Is the apparent decrease on palpation of the right carotid pulse of any significance?
3. How would you evaluate this patient at this point in time utilizing current technology?
4. An EEG was performed demonstrating focal 3-5 Hz slow wave activity. Predict the location of that abnormality.
5. The patient had some transient symptoms after the initial trauma; he was dazed and amnestic for a short period of time. What term is applied to those symptoms?

HOSPITAL COURSE:

A progressive change in neurological status was soon evident. Twenty-four hours after admission, the patient was found to be intermittently lethargic, if roused, he was able to answer questions but it was evident that he was disoriented for time. In addition deep tendon reflexes were now increased on the left side, and a left Babinski sign was now present. Later that day, increasing weakness of the left arm and leg developed. The patient was unable to raise the arm at the shoulder or the leg at the hip. The patient became increasingly more lethargic. Thirty hours after admission, the right pupil was found to be sluggish in response to light. The pupils which had been previously symmetrical were now asymmetrical, the right measured 3.5 mm, the left 2.5 mm. An emergency carotid arteriogram was performed. Based on that study, a surgical procedure was immediately undertaken by the neurosurgeon, Dr. Bertram Selverstone.

The patient made a rapid recovery. Examination, 10 hours after surgery, demonstrated an alert and well-oriented patient who was able to lift his leg off the bed and had only a minimal weakness in the left upper extremity. Pupils were now equal. The deep tendon reflexes were now symmetrical and the plantar responses were now flexor. At the time of hospital discharge, 13 days following surgery, the neurological examination was now normal.

QUESTIONS:

6. What change if any would you now make in your diagnosis?
7. Which complication of the original disorder has now occurred?
8. How would you manage that complication?
9. What did the carotid arteriogram demonstrate?
10. Which diagnostic procedure would you perform today? Describe the expected results of that study?
11. Which surgical procedure was performed? Describe the most likely surgical findings.

Case 28-5: This 62-year-old white right-handed male research associate, 18 months prior to admission, had the onset of focal motor seizures, which would begin in the left side of the tongue, and then spread to the face, shoulder, arm, and hand.

At the same time, he would note spread of tingling paraesthesias over the same parts of the left side of the tongue, face, and hand. Neurologic examination at that time was not remarkable.

Anti-convulsant medication produced a significant reduction in focal seizures. However, 1 month prior to admission, the patient had a focal seizure that was followed by transient residual weakness of left face and hand.

Three weeks prior to admission, a minor degree of weakness of the left hand developed. This progressed in degree and also began to involve the left leg and left side of face, as well. Two weeks prior to admission, right-sided headache developed.

NEUROLOGIC EXAMINATION:

1. *Mental Status:* Intact
2. *Cranial Nerves:* Intact except for a severe left central facial weakness.
3. *Motor System:*
   a. Weakness of the left hand was present, with thumb opposition most severely involved. Minor weakness was present in the left lower extremity.
   b. In walking, the patient dragged the left lower extremity to a minor degree.
4. *Reflexes:*
   a. Deep tendon reflexes were increased on the left.
   b. Instinctive grasp reflex was present on the left.
5. *Sensory System:*
   a. Graphesthesia was decreased over left face and hand.
   b. Tactile localization and two-point discrimination were decreased over the left hand.

QUESTIONS:

1. Where is the lesion? Be specific.
2. In terms of the pathological diagnosis, list the three most likely possibilities.
3. Indicate the most definitive laboratory test for establishing the diagnosis and the expected findings.
4. Indicate the other studies that would be useful and the expected findings.
5. Indicate a plan of management for each of the possible diagnoses.

Case 28-6: This 58-year-old white right-handed female with a past history of hypertension and angina pectoris had the acute onset of a right-sided weakness while at work. She had no change in consciousness, no sensory symptoms, no impairment of language function and no headache. She had experienced no vertigo, no diplopia, no vomiting and no visual symptoms. She had no complaints of chest pain or of shortness of breath.

PHYSICAL EXAMINATION:

Blood pressure was elevated to 180/120

NEUROLOGIC EXAM:

1. *Mental status:* This was entirely intact, with no evidence of aphasia.
2. Cranial nerves: A marked right central facial weakness was present with minimal deviation of the tongue to the right.
3. *Motor system:* A dense right hemiparesis was present.
4. *Reflexes:*
   a. Deep tendon reflexes were increased on the right.
   b. The plantar response was extensor on the right (sign of Babinski).
5. *Sensory system:* All modalities were intact.

LABORATORY DATA:

Assume the EKG was not relevant and the cerebrospinal fluid studies were normal.

HOSPITAL COURSE:

The patient was stable and alert during the next 2 days and at her request, was transferred to her local hospital.

QUESTIONS:

1. What is the most likely diagnosis? How would you confirm the diagnosis?
2. *Contrast Case #1:* Assume that this patient presented with a 1-month history of 5 transient episodes of numbness and weakness of the right hand. In addition, the patient had experienced two or three episodes of sudden transient blindness of the left eye. Then on the morning of admission, she awoke with numbness and weakness of the right hand and arm, with increased deep stretch

reflexes at the right biceps, triceps, patellar and Achilles tendons and a right sign of Babinski on plantar stimulation. You also find errors in position sense and in graphesthesia involving the right fingers and hand. The patient has moderate but incomplete recovery over the next week. At no time does the patient complain of headache. What is your conclusion as to the most likely diagnosis? How do you establish the diagnosis and manage the case?

3. *Contrast Case #2:* Assume that this patient has a history of rheumatic heart disease with mitral stenosis and atrial fibrillation. The patient suddenly, while at work at 11:00 a.m., stopped speaking in mid-sentence fall to the floor with a focal seizure involving the right side, with subsequent generalization. On recovery from the post-ictal confusion, the patient is found to manifest:

1. *Mental status:* Patient appears alert but has a dense global aphasia.
2. *Cranial nerves:*
   a. A dense right homonymous hemianopsia.
   b. Deviation of head and eyes to the left.
   c. A dense right central facial weakness.
3. *Motor system:* A dense flaccid right hemiparesis
4. *Reflexes:*
   a. A depression of the deep tendon reflexes at the right arm and right leg.
   b. A right Babinski sign.
5. *Sensory system:* Absence of all sensory modalities on the right side.

Assume that the initial CT scan at one hour after onset of symptoms is interpreted as normal and that the cerebrospinal fluid is clear with no cells.

QUESTIONS:

4. After seeing this patient at 12 noon in the emergency room, what conclusions do you reach as to diagnosis and what course of action do you recommend?
5. Unfortunately, the patient is found to have contraindications to the therapy, which you recommended. An alternate course of action is followed. No significant change occurred over the next 10 days except for transient lethargy and the minimal return of deep tendon reflexes. Discuss these issues.
6. Why were the deep tendon reflexes acutely depressed on the side of the hemiparesis?

Case 28-7: This 14-year-old right-handed white female was admitted with a chief complaint of "draining left ear" and left frontal headache. Approximately 4 weeks prior to admission, the patient had the onset of progressive pain in the left ear accompanied by fever up to 104 degrees F. Four days later a discharge from the left ear developed. Three days later a total left sided facial paralysis developed. The patient was evaluated at the Massachusetts Eye and Ear infirmary where drainage of the left middle ear was performed yielding purulent material. Culture of this drainage revealed Staphylococcus aureus .the patient was treated with antibiotics with a complete resolution of symptoms. However one week prior to admission, the patient developed a constant severe left frontal headache with swelling about the left eye. There was now a recurrence of discharge of a thick green purulent material from the left ear. Two days prior to admission, the patient developed confusion and delirium. She was also noted to have marked difficulties in language function and in calculations.

GENERAL PHYSICAL EXAMINATION:

Temperature was normal at 98.6 degrees F. The patient appeared chronically ill. Gauze soaked with purulent drainage was present in the left external auditory canal.

NEUROLOGIC EXAMINATION:

1. *Mental status:*
   a. The patient was oriented for the day and year but not for the month. There was considerable confusion as to past and recent events.
   b. She had a marked deficit in naming objects. Even when the name of the object was

supplied in a multiple-choice scenario, she was unable to recognize the correct name. Paraphrasing was evident. She was able to carry out spoken commands. Simple writing and spelling were intact but reading of isolated words was defective. She had no left-right confusion. She was able to identify and to point out the various parts of her body with the exception of the thumb.

c. At times, the patient acted in a "silly" manner and would giggle inappropriately in response to questions

2. *Cranial nerves:*

a. Fundoscopy demonstrated papilledema with elevation of the disk on the right. A recent hemorrhage was present just below the disk of the right eye.

b. A minor peripheral left facial weakness was present, most prominent in frontalis muscle.

c. Hearing was decreased on the left.

3. *Motor system:* Strength, gait and cerebellar functions were intact.

4. *Reflexes:* Deep tendon reflexes were intact. The right plantar response was equivocal; the left was flexor.

5. Sensory system: All modalities were intact.

QUESTIONS:

1. What is your diagnosis as regards localization and pathology?
2. What laboratory studies would you request?
3. Would you perform a lumbar puncture on this patient? Justify your answer.
4. What is your explanation for the left peripheral facial weakness?

LABORATORY DATA:

1. White blood count and differential were normal.
2. Skull and mastoid X-rays indicated an acute mastoiditis on the left with an increased density throughout the left mastoid extending throughout the medial aspect of the petrous ridge.
3. EEG was abnormal because of focal and hemispheric 2-3 Hz slow wave activity.

QUESTIONS:

5. Do you wish to alter your diagnosis or diagnostic approach?
6. Outline your approach to therapy. Which categories of physicians would you consult?

SUBSEQUENT COURSE:

Dr. Daniel Miller of the ENT service performed a left mastoidectomy and epidural exploration. Pus was present in the softened mastoid cells. Purulent granulation tissue extended into the adjacent epidural space. Cultures were obtained and antibiotics were re instituted. Over the next 4 day improvement occurred in the clinical and EEG findings but then findings worsened. Dr. Sam Brendler of the neurosurgical service then performed more definitive procedures. Following surgery, the patient had a nominal aphasia and a noncongruous right visual field deficit with some sparing of the inferior quadrant. The patient had a single generalized convulsive seizure, 6 years after surgery and anti convulsants were instituted. At that point in time, the nominal aphasia had shown marked improvement; she hesitated and stumbled over words. The right visual field defect was unchanged.

QUESTIONS:

7. Which procedures were performed by the neurosurgeon?
8. Explain the visual field deficits.

Case 28-8: This 29 year old right-handed single white female schoolteacher was about to testify in court in the late afternoon on the day of admission, when she suddenly dropped to the floor. She began to babble but did not lose consciousness. She was taken to the emergency room of the local hospital at 4:15 p.m., where she was initially described as "confused, disoriented, unresponsive to questions, and not verbalizing". At 7:50 p.m., she was subsequently described as agitated with garbled speech. Later that evening, she was transferred to this hospital.

PAST HISTORY:

Patient was receiving birth control pills.

GENERAL PHYSICAL EXAMINATION:

1. Blood pressure 120/70. Pulse 90 and regular
2. No heart murmurs were present.
3. No bruits were present over the carotid arteries and carotid pulses were strong.
4. Neck was supple with no nuchal rigidity.

NEUROLOGICAL EXAMINATION:

1. *Mental Status:* The patient was anxious and alert with intermittent agitation. She appeared to understand but only intermittently followed verbal commands. She used only a few words. She attempted to name objects but produced only paraphasia and jargon speech.

2. *Cranial Nerves:*

   a. A right homonymous hemianopsia was present

   b. There was decreased pain and touch over the right side of the face.

   c. A right central facial weakness was present.

3. *Motor System:* A dense right hemiplegia was present.

4. *Reflexes:*

   a. Deep tendon stretch reflexes were increased in the right upper and lower extremities with clonus at the right patellar and Achilles tendon reflexes.

   b. The plantar response was extensor on right and flexor on left.

5. *Sensory System:* An apparent right hemisensory deficit was present to all modalities.

LABORATORY DATA:

1. CSF pressure was normal at 160, with clear fluid (cells = 0, protein =11mg%, sugar = 75mg% with blood sugar of 104).
2. Skull X-rays were normal.
3. EKG - normal.

HOSPITAL COURSE:

By 48 hours after onset, speech had shown some improvement. The patient was using more recognizable words. Occasional appropriate two-word phrases were employed. The dense right hemiparesis had persisted but the right homonymous field cut and the sensory pain deficit were no longer present. By 72 hours after onset, slight movement was present in right lower extremity. However at 5 days after onset, the patient had the sudden development of a cold, pulseless left arm.

QUESTIONS

1. Localize the lesion.
2. Characterize the speech deficit at various stages. Did this patient have a dysarthria or aphasia? If aphasia was present was this primarily a fluent or non fluent.
3. Discuss the use of the term "confusion". Discuss the use of the term "babbling".
4. Discuss the pathology and likely etiology and pathophysiology. If vascular indicate the vascular territory.
5. What other information would you wish to have?
6. This patient was seen before the era of the CT scan. Assume that this patient had arrived in the emergency room today 30 minutes after the acute events, and you are the emergency room physician or the consulting neurologist how would you manage this problem.

CHAPTER 29
Alterations in Consciousness: Seizures, Coma and Sleep

BASIC DEFINITIONS

Consciousness: The waking state in which the individual has awareness, with " the ability to perceive, interact and communicate with the environment and others in the integrated manner which wakefulness normally implies" (Zeman, 2001).

Sleep: A normal periodic alteration in consciousness with various stages. Sleep is an active physiologic process dependent on a complex interaction of cerebral cortex, hypothalamic and brain stem areas. Metabolic activities such as oxygen use remain normal. The level of consciousness may be fully restored by an appropriate stimulus

Sleep disorders: This term includes such major disorders as insomnia, hyposomnia, narcolepsy and sleep apnea as well as sleep walking (somnambulism) and abnormal movements during sleep.

Abnormalities of consciousness: occur in a brief paroxysmal manner in seizures and syncope and in a more prolonged manner in coma.

Seizure: A sudden (paroxysmal) alteration in behavior due to an excessive discharge of cerebral cortical neurons .The alteration in behavior may be limited or may result in a generalized convulsive seizure. The state of consciousness may be unaffected, may be altered or may be lost. Seizures represent an alteration in the usual balance between excitation and inhibition. Seizures and epilepsy are most frequent in the infant and in the elderly.

Epilepsy: A chronic condition in which seizures recur at variable frequency over time. The period of seizure activity is termed the ictus or ictal event. The time between seizures is termed the interictal period

Syncope: a condition in which consciousness and usually posture are briefly impaired in a relatively sudden, manner due to a general decrease in cerebral perfusion. This decrease in perfusion may be due to a general peripheral vasodilation (vasovagal reaction) or hypotension or a decrease in cardiac output. The loss of consciousness almost always occurs when the patient is in an upright position. The actual loss of consciousness is preceded by prodromal symptoms that last from seconds to minutes: light-headedness, weakness, "dizziness", sweating and blurring and dimming of vision. The patient is observed to slump limply to the around (as opposed to the patient with epilepsy who usually stiffens and then falls violently to the around). Tonic or clonic movements are rare but may occur. The period of unconsciousness is brief, usually a matter of seconds, occasionally a minute. During this time, the patient is noted to be pale and sweaty. The pulse is usually weak and the blood pressure often reduced. The return to consciousness is rapid occurring usually when the patient assumes a recumbent position. There is no significant period of mental confusion following the episode

Coma: an abnormal state in which various degrees of prolonged alteration of consciousness occur. In contrast to sleep, coma is a passive process in which a loss of function has occurred. Metabolic activities are depressed. Stimulation will not restore a full and persistent state of consciousness. Various gradations of coma exist. In describing the degree of coma, it is best to use operational terms describing brain stem reflexes, response to stimuli etc. The Glasgow coma scale provides such an approach. Various terms continue in use, in deep coma the responses to stimuli are completely lost. In moderately deep coma, rudimentary responses of a reflex nature only are present (e.g.. corneal reflex. withdrawal of a limb to painful stimulus), but psychologically understandable responses to external stimuli and inner need are absent. Various terms have been employed to describe lesser degrees of impairment of consciousness: semicoma, stupor, obtundation or lethargy

Confusion: an abnormal state characterized by an impaired capacity to think clearly and with customary speed to perceived stimuli and to remember current stimuli. Confusion includes disorientation and memory impairment. In addition to confusion associated with an altered state of consciousness, confusion may also occur with an alert state in patients with dementia (chapter 30)

Delirium: an abnormal state in which agitation (of motor activity) is present in addition to confusion.

Akinetic mutism: a state in which the patient lies in bed with eyes open but fails to move or to speak. There are two types. (1) In the mesencephalic/diencephalic variety, usually due to occlusion of a penetrating branch of the posterior cerebral artery or of the proximal/mesencephalic artery segment of the paramedian branch at the top of the basilar artery bifurcation, there is a state of apathetic or somnolent akinetic mutism in which the patient remains in a lethargic state most of the time. The patient may open his eyes in response to strong stimulation but soon closes them and returns to a lethargic state. Paralysis of vertical gaze and of other extraocular movements is usually found. Infarction at the diencephalic-mesencephalic junction is bilateral and includes rostral midbrain tegmentum, the adjacent pretectal area, subthalamus, periventricular gray matter at the posterior end of the third ventricle: midline thalamic nuclei: reticular nucleus of the thalamus and intralaminar nuclei of the thalamus (centrum medianum). (2) In the bilateral frontal type of hyperalert akinetic mutism (vigilant coma) the patient will follow with intact eye movements and will cycle (alternation periods of agitation). The medial, orbital and anterior surfaces of the frontal lobes plus adjacent cingulate gyri or the septal area are involved. Although the patient lies immobile in bed, he provides some suggestions of alertness in the sense of appearance. We should distinguish several other conditions: acute massive infarct of the dominant hemisphere may be mute and hemiparetic. (3) in late Parkinson's disease marked rigidity, akinesia and mutism my be present but the patient is alert. (4) a similar state may follow bilateral destruction of the putamen or globus pallidus (apallic state). Refer to Segarra (1970).

Locked in syndrome (pseudo akinetic mutism or pseudocoma): a state in which the patient is aware and conscious but because of paramedian pontine infarction usually related to thrombosis of the basilar artery, is unable to provide verbal responses. When the midbrain is intact, the patient may provide responses by opening or closing the eyes or by vertical eye movements.

Brain death: Up until the late 1960s, patients were declared dead when cardiac (and respiratory) activity ceased. With the cessation of cardiac respiratory activity, neurological function ceased within 3-4 minutes. With the development of cardiopulmonary resuscitation and of life support equipment, an era dawned in which cardiorespiratory function might be continued or restored but function of the brain could not be restored. Thus, the advance of technology required an expansion of the prior definition of death to include brain death. In most jurisdictions, either by legislative law or by court decision case law, brain death is recognized as legal death.

Brain death is defined by the following criteria:

1. An absence of consciousness.
2. No cerebral responsivity is present.
3. No brain stem reflexes are present.
4. No spontaneous respirations are present.
5. If obtained, the EEG will show no evidence of cortical electrical activity.
6. Spinal cord reflexes may be present.
7. The state is irreversible: reversible metabolic and toxic conditions have been eliminated, a sufficient period of time has elapsed and appropriate studies have been performed to establish a *diagnosis*.

It is important to note that in those patients who meet the criteria of brain death but where respiratory and blood pressure support is continued, cardiac arrest usually occurs within days, rarely in weeks.

Persistent vegetative state or neocortical death: this term refers to a state in which neocortical death or disconnection has occurred but brain stem reflexes and spontaneous respiration are preserved. Although this condition had long been recognized within the larger group of chronic coma, following cardiac respiratory arrest and trauma, Jennett and Plum popularized the specific term in 1972. Such patients have periods of "wakeful" eye opening and movement. The EEG is usually isoelectric (or occasionally demonstrates a burst suppression pattern). The major cause of the syndrome is cardiac pulmonary arrest with subsequent resuscitation. There is diffuse damage to neocortex and hippocampus. At times, laminar necrosis is present. The patient may live for weeks, months or years. Such patients never regain consciousness. The management of this state and the ethical dilemmas involved in the continued care of this patient has been a major area of medical and legal controversy.

THE BASIS OF THE ELECTROENCEPHALOGRAM:
Review chapter 17 (Cerebral cortex) for a discussion of this subject.

ABNORMALITIES OF THE EEG AND CORRELATION WITH SEIZURE DISORDERS: Refer to chapter 2 for examples of normal and non-seizure abnormalities.

Kershman and Jasper (1941) delineated the two major categories of EEG abnormalities to be found in seizure disorders: (1) those findings which were focal and (2) those findings which were bilateral and relatively symmetrical and synchronous. The international classification of seizures and of the epilepsies has continued to follow this long established distinction.

There are two general rules for the normal electroencephalogram (1) the left and right hemisphere are relatively similar in appearance (2) for a given age and level of alertness: awake vs. drowsy vs. asleep, the frequency and rhythm will fall within certain parameters.

All abnormalities of the electroencephalogram then will represent deviations from these two general rules.

Type 1 abnormality: termed focal. The following sub types may be listed:

a. Focal slow waves correlated with a focal cortical dysfunction with the quality of damage (Fig. 2-22 and CD ROM Atlas)

b. Focal spikes and sharp waves correlate with a focal dysfunction having the quality of excessive neuronal discharge that is partial epilepsy *(Fig.29-1)*. Spikes have a duration of less than 75 milliseconds; sharp waves have a duration of the greater than 75 milliseconds. Both are named from their sharply contour-rapidly rising slope These abnormalities are not to be confused with the spike of the short duration action potential of the axon (a matter of 1-2 msec) for they represent summated post synaptic potentials. Single focal spikes may occur in interictal (interseizure) periods without the occurrence of the actual focal seizure. In general, the clinical focal seizure is likely to occur when frequent repetitive focal spikes occur or when the focal spike is followed by an after discharge. Any focal seizure may secondarily generalize *(Fig.29-2)* Additional examples of focal spike discharges with specific correlation have already been presented. Focal (partial) epilepsy is the predominant seizure disorder in those patients with onset of seizures after the age of 15 years and occurs in 45% of patients with seizure onset prior to age 15.

c. Focal voltage suppression correlates with a reduced electrical activity in the tissue underlying the electrodes for example the presence of a subdural or intracerebral hematoma, or massive cerebral infarction (Fig. 2-23).

The 2nd type of abnormality is termed generalized and the following subtypes are recognized:

a. Continuous generalized slow wave activity or dominant activity slower than the

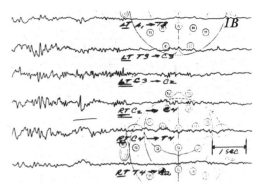

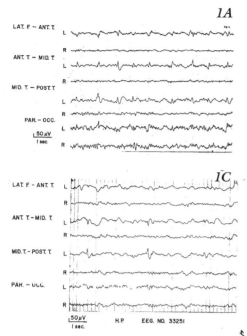

Figure 29- 1A. Focal spike discharges in the left temporal lobe with phase reversal at the left anterior temporal electrode. Case 22-1-Refer to text. Left temporal glioma, with frequent clinical seizures originating in temporal lobe LAT.F. = Lateral frontal, ANT.T. = Anterior temporal, MID.T. = Mid temporal, POST.T. = Posterior temporal, PAR. = parietal, OCC. = occipital. 1B. Focal spike discharge in the right parasagittal Rolandic area (Cz-C4) in a 28 year old woman with focal seizures beginning with clonic movements of the left thigh and numbness of the left leg. Transverse bipolar montage. Case 18-1.Refer to text. 1C. Focal spike discharges with phase reversal at left mid temporal electrode in a 46 year old man with subarachnoid hemorrhage secondary to a middle cerebral artery aneurysm,The patient had focal sensory seizures beginning in the right face tongue or thumb. Case 26-9 on CD ROM

alpha rhythm correlated with a generalized dysfunction having the quality of damage (Fig. 2-24, 2-25)

b. Generalized rhythmic paroxysmal electrical discharges correlated with generalized epilepsy: Spike-slow wave complexes, polyspike -slow wave complexes and polyspike (repetitive fast spike discharges). Depending on the clinical findings, and frequency of discharge, (e.g. slow 1-2hz vs, 2.5-4 Hz vs >4hz spike-wave or polyspike-wave complexes), these EEG findings may be associated with idiopathic (primary) generalized or secondary generalized epilepsy. As explained below, this latter term is to be distinguished from partial epilepsy with secondary generalization.

c. the generalized burst suppression pattern discussed in chapter 17.

d. Generalized suppression of all electrical activity correlated with neocortical death or with the deepest stage of anesthesia.

UNDERLYING BASIS OF THE FOCAL SEIZURE AND FOCAL SPIKE DISCHARGE:

Focal seizures indicate a focal cortical dysfunction and imply an underlying focal cortical pathology. In some cases the actual pathology is in the white matter just below the cortex and the acute or chronic effects of cortical isolation or deafferentation are being seen. Seizures do not arise from an area of total necrosis since dead neurons do not discharge. Surviving neurons must be present (for example at the margin of an infarct or hemorrhage).

The pathological causes of focal (partial) seizures are variable, dependent in part on the age of the patient and those pathological processes affecting cerebral cortex that are common at that particular age (Table 29-1).

The incidence of seizures following specific pathological processes: Essentially seizures will occur in approximately 50% of meningiomas ,low grade gliomas, oligodendrogliomas and brain abscesses involving the cerebral cortex. High-grade gliomas have a lower frequency of approximately 35%. Seizures will occur in 25% of all cerebral infarcts with a higher % in embolic MCA infarcts of 45%. Cortical migration disorders have a significant association with focal

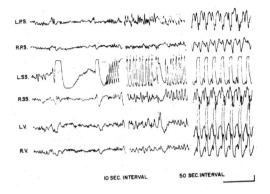

Figure 29-2. Generalization from an experimental focal spike discharge in the left suprasylvian area (SS) of the cat. After discharge of repetitive spikes occurred at the focus with subsequent spread to other recording areas of the same and contralateral cerebral hemispheres. P.S= posterior sigmoid gyrus (sensory cortex); V. = visual projection area of marginal gyrus. (Calibrations: 1 second and 100 u volts) From Marcus, E.M., Watson, C.W., and Goldman, P.L.: Arch. Neurol. 15:525, 1966 (AMA).

seizure disorders (exact % unclear) possibly because of aberrant circuits (Duchowney, 2000) Refer to Table 29-2 on CD ROM

Regional Cortical Variations in the Capacity For Seizure Discharge: A given pathological process is more likely to cause seizure discharge if it involves an area of low threshold (motor cortex or mesial temporal areas) rather than an area of high relative threshold (prefrontal or occipital areas). (Refer to chapter 17.)

The Epileptic Focus And The Epileptic Neuron: At times, focal seizures may occur in acute or subacute relationship to penetrating head injuries or to the cortical ischemia that follows a cerebral embolus and such patients are more likely to have recurrent seizures. In many cases however the spike discharge and the focal seizure begin as a late effect. It is customary to distinguish in all cases between the interictal spike and the ictal discharge. The ictal discharge accompanies the clinical seizure. The spike discharge recorded at the surface represents summated excitatory postsynaptic potentials. Spike discharges at the cortical surface, then, are recorded as predominantly surface negative waves. The more recent studies of Conners (1998) utilizing isolated columns of horizontal layers of rat neocortex in vitro have indicated that the discharges are primarily initiated in the large pyramidal neurons of layer 5.Discharges may then propagate both vertically and horizontally. Schmidt and Wilder (1968) define the focus as a group of abnormal cells ("epileptic neurons") with certain unique properties. Engel (1989) has summarized many of the studies utilizing the best-studied acute model of partial epilepsy: topical application of penicillin. Similar changes are observed in chronic models.

(1) In the acute and the chronic focus there are bursting units discharging in a paroxysmal manner at high rates of discharge up to 1000 Hz with high amplitude *paroxysmal depolarization shifts* (PDS) as the intracellular correlate of the bursting unit discharge. The PDS corresponds to activation of glutamate-mediated synapses.

(2) In addition abnormally hypersynchronous activity of large numbers of adjacent neurons occurs; the PDSs of most of the neurons within the focus summate to produce the interictal spikes recorded at the surface by the EEG. The PDs are followed by prolonged after hyperpolarizations (AHPs). The AHP reflects

TABLE 29-1. CAUSE OF PARTIAL SEIZURES:

Age	Cause
Within the first 48 hours of life -	Acute birth trauma (poor prognosis),
Within the first 2 years -	Birth trauma and congenital malformations (e.g.. Porencephalic Cyst)
From 2 to 10 Yrs.	Birth injury and head trauma
From 10 to 20 years -	Head trauma and late effects of pathology in infancy and childhood (complex partial seizures),
From 20 to 50 years-	Head trauma and neoplasms and late effects of pathology acquired in infancy and childhood (complex partial seizures),
Over 50 years	Tumor and vascular infarcts and degenerations, (Alzheimer's disease).

activation of calcium and voltage dependent potassium channels in addition to GABA/A and GABA/B synapses. These AHPs summate to produce the EEG surface recorded slow wave that follows the EEG spike discharge, as a spike slow wave complex.

(3) Neurons in the cerebral cortex surrounding the focus develop large membrane hyperpolarization. This "inhibitory surround" limits the spread along the surface from the focus but also may interfere with normal activities of those surrounding neurons (Prince and Wilder (1967). The inhibitory surround is in a sense built into the columnar organization of the neocortex.

As long as discharge is restricted to the pool of neurons within the seizure focus, in general there are no clinical manifestations.

When however surround inhibition is surmounted and after-hyperpolarization decreases, propagation to other regions (cortical and subcortical) occurs and clinical seizure phenomena results.

Propagation from the focus: Propagation of discharge from the focus, occurs via several pathways: (1) short intracortical pathways involving synapses and neurons within the adjacent neuropil, (2) long fiber systems to other cortical areas within the same cerebral hemisphere, (3) callosal and other commissural pathways to the contralateral cerebral hemisphere, and (4) descending pathways to basal ganglia, thalamus, reticular formation of the brain stem, and to the spinal cord. Obviously, the specific pattern of spread will depend on the location of the focus and on the anatomical connections or projections from the particular cortical area. Refer to chapter 18 for additional discussion. It is *important to realize that any focal spike discharge may become secondarily generalized (Fig.29-2). From a clinical stand- point, any focal seizure may become a secondarily generalized seizure.*

Histological changes at the focus: (Babb&Pretorius, 1997) There are multiple changes at the chronic focus: damage to capillaries, fibroblastic and capillary ingrowth from meningeal adhesions (meningocerebral cicatrix), gliosis, mechanical deformation of neuronal processes a progressive loss of dendritic spines, the development of recurrent collateral excitatory synapses and partial loss of afferent input (possibly inhibitory) to the neuron. Within the hippocampus, neuronal loss and gliosis are accompanied by sprouting of mossy fibers. It is uncertain whether the changes at the chronic focus in the human focus relates to a decrease in the inhibitory transmitter (GABA) or an increase in the excitatory transmitter (glutamate). See Ribak 1986, McDonald et al, 1991, Selzer &Dichter, 1992

Secondary epileptogenesis: Two experimental models involving the chronic focus suggest that persistent spread of discharge from a focus may produce persistent alterations in excitability in those cortical and subcortical areas that are the targets of the propagated discharge. These projection areas then eventually take on the capacity for independent generation of focal or multifocal discharge.

The first model was described by Morrell (1960,1991) - *the mirror focus.* The second model was described shortly afterwards by Goddard (1969) *the kindling model.* (Refer also to Mayersdorf&Schmidt, 1982 and to chapter 30).

MULTIFOCAL ABNORMALITIES

Many traumatic injuries and other pathologic processes do not produce limited focal damage. Thus, a blow to the occipital area of the skull may not only produce injury to the occipital cortex at the site of the blow, but may also produce contusions of the cortex at a distance from the site of injury: the anterior temporal poles and frontal poles and orbital frontal surfaces. The end result may be multifocal slow waves, implying multifocal damage or multifocal spikes, which in turn implies the evolution of multifocal areas of excessive neuronal discharge. Another possible explanation for multifocal spikes when a unilateral pathology may be present is discussed above under secondary epileptogenesis. Note also that multiple foci may interact to produce bilateral discharges as discussed below.

GENERALIZED EPILEPSY OR GENERALIZED SEIZURES:

In the International Classification of the Epilepsies, this category is defined as those seizures that are generalized from their onset. Other terms that have employed in the past are essential, nonsymptomatic and cryptogenic or idiopathic—but all of these are now used in a somewhat different context. *The term idiopathic is now used again for this category.*

In this category of epilepsy and seizures no focal origin is evident for the electrical discharge and the clinical, seizure is characterized from its onset by the loss of consciousness and or by motor manifestations that are bilateral and symmetrical. In general, these seizures begin in childhood or adolescence.

This type of electrical abnormality and the correlated clinical seizure disorder should be contrasted to the focal spike and the focal (or partial according to the International Classification) seizure disorder where a focal origin is clearly evident and where total loss of consciousness is usually not the initial sign of the seizure. Such a distinction becomes a little blurry when complex partial seizure, particularly those arising in the frontal lobes are considered.

Within the overall category of generalized epilepsy or generalized seizures, the classification has two divisions (A). The major category of: primary generalized or idiopathic epilepsy. (B) secondary or symptomatic generalized epilepsy.

A. Primary generalized or idiopathic epilepsy has the following features:

1. No specific underlying gross neuropathology is present. However as we will discuss below there is now evidence that an underlying cortical micro dysgenesis may be present.

2. Neurological and psychological - intellectual status are usually normal

3. The background EEG activity is relatively normal for the age

4. A familial genetic background is often present. Early twin studies by Lennox and his associates (1960) indicated a concordance rate for monozygotic twins of 70-100% and for dizygotic twins of 5-15%. Seminal studies by Metrakos and Metrakos (1961) demonstrated that the 2.5-3.5 discharge associated with idiopathic epilepsy was inherited as an autosomal dominant. Other studies have suggested a general predisposition gene with other modifying genes determining whether and what type of clinical seizures will occur. The specific altered genetic process is unknown, but studies by Janqua and Andermann (1989) suggested significant abnormalities of amino acids in the serum of both patients and first-degree relatives–particularly increased levels of the excitatory amino acid glutamic acid.

5. The seizures are usually very responsive to medication

6. The overall course is benign

Included within this division are the following that shortly will be discussed in greater detail:

1. childhood absence epilepsy (CA)
2. juvenile absence epilepsy (JA)
3. juvenile myoclonic epilepsy (JME)
4. generalized tonic clonic seizures on awakening
5. isolated generalized tonic clonic seizure (GTCS)
5. benign myoclonic epilepsy of childhood

B. Secondary or symptomatic generalized epilepsy has the following features.

1. A diffuse or multifocal pathology is usually present

2. The neurological examination is abnormal

3. Mental retardation is present or psychological deterioration develops

4. The background EEG is abnormal

5. There is in some cases a progressive course; deterioration may occur

6. Response to medication is usually poor

7. A familial genetic background may or may not be present

Included in this category are the following syndromes:

1. The various progressive myoclonic

epilepsies beginning in infancy, childhood, adolescence and adult life - many due to specific metabolic disorders; lipidosis, lysosomal storage diseases and mitochondrial disorders

2. Infantile spasms: with the associated EEG pattern of hypsarrhythmia (high amplitude irregular asynchronous spikes and slow waves) and often as well a decremental pattern resembling burst suppression during sleep (Fig29-3).

3. Lennox Gastaut Syndrome: childhood encephalopathy with associated EEG pattern of "slow"(1-2Hz.) spike-slow wave complexes. This syndrome may be a residual of infantile spasms. The term "petit mal variant" has been employed in the past. In addition to atypical absence seizures, atonic drop attack seizures, tonic seizures and generalized tonic/clonic seizures may occur. Multiple foci are often present and section of the corpus callosum may decrease the interaction of these foci (see below).

A. PRIMARY/IDIOPATHIC GENERALIZED EPILEPSY: MAJOR SYNDROMES:

1. Childhood absence epilepsy (CA) that usually begins between 5-10 years of age has the correlated EEG pattern of generalized relatively symmetrical and synchronous bursts of 3 Hz (2-1/2- 4 Hz) spike slow wave complexes *(Fig.29-4)*. The spike components of the bursts are most prominent in the frontal-Rolandic parasagittal recording areas. In some cases, the bursts may contain considerable polyspike-slow wave components, particularly during drowsiness and sleep. Short bursts of spike slow wave complexes (1-2 seconds) in general, will have only limited or no definite clinical accompaniment. Bursts of 3-30 seconds duration are in general accompanied by some interruption of consciousness. The absence seizure consists of a short (3-30 seconds) interruption of consciousness in which general postural tone is relatively well preserved. During this brief period, the patient is observed to stare, oblivious to stimuli introduced into the environment. Ongoing activities are interrupted. For example, if in the midst of eating, the patient may stop with his spoon in mid air. The interruption is relatively abrupt in onset and relatively abrupt in cessation. The patient's awareness and motor activities return promptly at the end of the episode. Memory is defective only for the period of the seizure. It is as though the patient, although physically present, has been absent as regards his higher cortical functions for the brief interval of the seizure. In actuality, the impairment of awareness, the impairment of response capacity and the impairment of motor activity is a relative phenomenon, occurring in variable degree. Thus, some patients may be unaware of phrases or numbers that are provided as auditory or visual stimuli during the episode. Other patients are aware of these same stimuli and are able to repeat them during questioning at the end of the episode. Some patients are able to continue a familiar recitation during the absence seizure. Whether a response to a stimulus occurs during these brief seizures may, in part, depend on the intensity of the stimulus and the motivation and past experience of the patient. It is the degree of interruption that is variable. There are minor motor phenomena that may often accompany the seizure: eye opening, eyelid and myoclonus of facial extraocular muscles. Automatisms of mouth, jaw and hands may also occur. Some minor loss of postural tone in the neck muscles may occur, causing the head to fall onto the chest. Usually many attacks occur per day interfering with school work. In most patients the attacks may be triggered by hyperventilation.

In many families there is strong evidence of a genetic (autosomal dominant) predisposition to primary generalized epilepsy-based on EEG studies. In such families the genetic predisposition, is apparently for idiopathic epilepsy and not absence seizures per se. There is a predominance of females affected. For most families a polygenic pattern rather than a monogenic pattern of inheritance is suspected. The high concordance rate in identical twins was noted by Lennox cited above. In a minority of cases there is little evidence of a genetic background

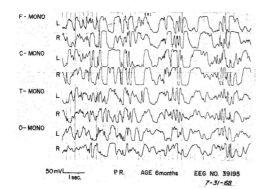

Figure 29-3. Hypsarrhythmia in an infant with myoclonic spasms Refer to Text: Referential (Monopolar) recordings from frontal (F), rolandic (C), temporal (T), and occipital (O) areas.

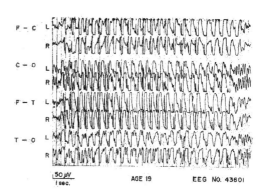

Figure 29-4. Generalized 3 Hz spike wave discharge. This 19 year old male had a long history of frequent absence seizures characterized by a blank stare flickering of the eyelids and a failure to respond to instructions. This accompanying EEG discharge in this episode was of 8 seconds duration. Other episodes were of 20--40 seconds duration.

and the seizures apparently reflect perinatal or other multifocal and diffuse pathological processes beginning in early childhood.

The seizures and the EEG pattern are particularly responsive to medications such as ethosuximide, trimethadione and valproic acid. Since approximately 50% of theses patients will also have generalized convulsive tonic clonic seizures, valproic acid may be the preferred medication since it has significant therapeutic actions against both types of seizures. Note that the terms pyknolepsy and petit mal absence seizure have been employed in the past.

2. Juvenile absence epilepsy (JA) has a similar EEG pattern. The attacks are less frequent and there is a higher incidence of generalized tonic clonic seizures. There is no male/female difference.

3. Juvenile myoclonic epilepsy (JME) was identified in 1957, by Janz and Christian, as a syndrome which incorporated features of juvenile onset tonic-clonic and myoclonic seizures with a bilateral symmetrical and synchronous 4-5 Hz polyspike-slow wave pattern (Fig.29-5). Myoclonus may be defined as a sudden rapid jerk or twitch resulting from the sudden contraction of one or more muscle groups. The jerks may be bilateral and symmetrical or multifocal. Facial muscles, extraocular muscles, and head or limbs or body may be involved. The intensity of the contraction may be strong enough to throw the patient to the floor or isolated and weak enough to result in the twitch of a finger. In 40% of patients, the discharges and often the myoclonus are triggered by intermittent photic stimulation with a stroboscope *(Fig.29-6)*. In some series, associated with absence seizures the 3 -4Hz spike wave or polyspike slow wave EEG are more common. This syndrome of juvenile myoclonic epilepsy is now recognized as the most common variety of primary generalized epilepsy

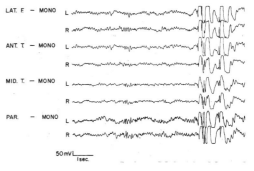

Figure 29-5. Generalized bursts of poly spikes and slow waves. Juvenile myoclonic epilepsy in a 13 year old who had the onset of sudden single generalized myoclonic jerks, which initially were sufficiently violent to throw the patient to the floor but did not impair consciousness. Neurological examination has remained normal and the patient has never (now age 22 years) experienced generalized convulsive seizures or petit mal absences.

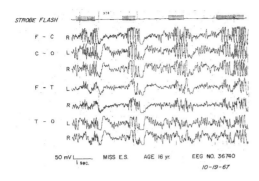

Figure 29-6. Generalized bursts of poly spikes and slow waves triggered by photic stimulation. Juvenile myoclonic epilepsy with photosensitivity. This young woman at age 13 had the onset of generalized convulsive seizures and of frequent myoclonic seizures involving eyelids and arms. The myoclonic seizures involving eyelids, facial muscles and limbs were clearly triggered by bright flickering light.

beginning in the adolescent or young adult years (12-18), accounting for 7-12% of all seizure patients depending on the series (see Dreifuss, 1989; Delgado-Escueta, 1989). Almost all patients have both generalized tonic-clonic and myoclonic seizures, the latter often precipitated by sudden awakening, sleep deprivation, alcohol consumption or stress. This syndrome has considerable significance

In the early genetic studies of EEG photosensitivity- this particular trait appeared to be inherited as an autosomal dominant *(Fig 29-7)*, although the majority of affected relatives did not have clinical seizures (Watson &Marcus, 1962). In more recent studies, of JME (Delgado-Escueta et al, 1989), 50% of probands had a 1st or 2nd degree relative with epileptic seizures. Among 1st degree relatives of the proband, 80% of symptomatic siblings and 6% of asymptomatic siblings had diffuse 4-6hz poly spike slow wave discharges. Linkage analysis suggested a relationship to chromosome 6,although subsequent studies have suggested linkage to 15q in other families (Delgado-Escueta etal, 1999) Therapy is very specific: Valproic acid will totally control seizures in 80% of patients but must be continued on a relatively life- long basis. In the series of Janz (1985), 91% of patients had recurrent seizures when anticonvulsant was discontinued after two seizure-free years.

Anatomical differential diagnosis of myoclonus: The anatomical substrate for myoclonus is variable. In the monkey, myoclonus may occur as a result of bilateral epileptogenic foci in the motor cortex or the adjacent premotor cortex. A similar cortical onset of discharge has been noted in the light sensitive baboon (papio, papio). Myoclonus may also occur in various extrapyramidal diseases involving the basal ganglia or ventro-lateral nuclei of the thalamus. In patients with Parkinson's disease, myoclonic jerks may complicate the use of L-dopa (in addition to the more common choreiform dyskinesia). Segmental myoclonus has been described in viral diseases involving the spinal cord.

In the non-cortical varieties of myoclonus, it is possible to have the behavioral phenomenon without the occurrence of accompanying spike or polyspike-slow wave discharges in the electroencephalogram. In the cortical variety of myoclonus, conduction of discharges to the brain stem and spinal cord occurs along both pyramidal and non-pyramidal pathways.

Stimulus sensitive myoclonus refers to the precipitation of myoclonic jerks by sound, light flash, proprioceptive or tactile stimuli. Usually, the underlying disease is a diffusely distributed process. In some cases, it is primarily cortical; in others it is primarily non-cortical. The disease process involves neurons primarily rather than white matter.

Etiological and prognostic classification of myoclonus:

1. Benign-nonprogressive myoclonus –sudden jerks on falling asleep or on awakening as a common normal phenomenon. . A familial variety of essential myoclonus-may also occur.

2. Non-progressive myoclonus as a manifestation of idiopathic epilepsy: the JME syndrome.

3. Symptomatic Generalized epilepsy, usually with mental retardation

   a. Infantile spasms

   b Lennox-Gastaut syndrome.

4. Progressive myoclonic disorders (Berkovic et al,1991,Serratosa et al,1999).

ALTERATIONS IN CONSCIOUSNESS: SEIZURES, COMA AND SLEEP

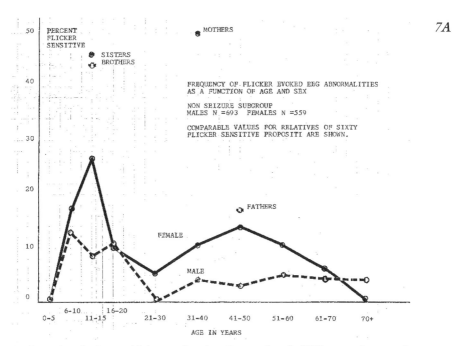

Figure 29-7A. The EEG trait:photosensitivity as a function of age and sex in 1256 consecutive non seizure patients. The role of genetics is also demonstrated in the relatives of 60 photosensitive propositi matched for age. Almost all of the propositi had idiopathic epilepsy. Almost all of the photosensitive relatives did not have epilepsy.

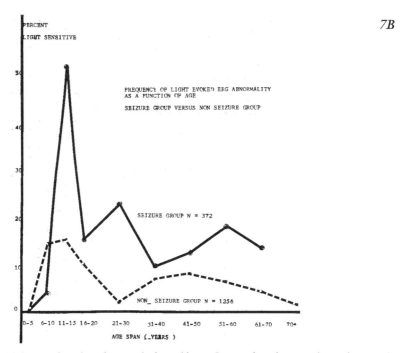

Figure 29-7B. Photosensitivity as a function of age and seizure history. Consecutive seizure and nonseizure patients. The frequency of this trait in those patients with focal (partial) epilepsy did not differ from the nonseizure patients. The excess of photosensitivity in the 10-20 age seizure group reflected the occurrence of idiopathic epilepsy in this age group. The excess in the 50-70 age seizure group may have reflected the occurrence of diffuse disorders in this age group.

a. mitochondrial disorders,
b. lipidosis-lysosomal storage diseases
c. myoclonus epilepsy of Unverricht and Lundborg
d. Lafora's disease
5. Infectious-viral related
a. acute encephalitis e.g. acute herpes simplex.
b. subacute inclusion body encephalitis-reactivation of the measles virus
c. Spongiform encephalopathy-Creutzfeldt-Jakob disease-prion disorders
6. Metabolic disorders
a. uremia
b. hypocalcemia
c. CO_2 narcosis
d. anoxic encephalopathy: acute or chronic effect
e. pyridoxine deficiency state of infancy
f. L-Dopa toxicity

4. Generalized tonic clonic seizures (GTCS): The stereotype of the generalized convulsion, the grand mal seizure, consists of two phases: the tonic and clonic. Occasionally myoclonic jerks of increasing frequency may precede the actual loss of consciousness and the tonic phase. Unless the tonic clonic convulsion represents a secondarily generalization of a partial seizure, the onset is abrupt, with loss of consciousness the initial sign. The patient stiffens and, if standing erect, is thrown to the floor. Respiration is arrested and the pupils dilate. After a short but variable period of 10 to 30 seconds, the tonic stage then, passes into the clonic stage. The symmetrical jerks of clonus then slow in frequency and cease. The total duration of the actual uncomplicated tonic/clonic phase is usually a matter of 1 minute. This state is followed by a short but variable period of coma, followed by, stupor, confusion and drowsiness. The total duration of confusion and drowsiness is usually less than 30 to 60 minutes. The patient will be amnesic about the onset of the attack, about subsequent events of the seizure and about the postictal period of confusion. If the bladder has been full, incontinence of urine will have occurred during the seizure. The plantar responses are usually extensor during the attack and will remain extensor during the early part of the postictal period. The following EEG features are correlated with the clinical sequence: fast spike discharge at 8-20 Hz with the tonic phase *(Fig.29-8A)*, repetitive spike wave or polyspike wave generalized discharge of gradually decreasing frequency with the clonic phase *(Fig.29-8B, C, D)*, a short period of relative electrical silence in the electroencephalogram *(Fig.29-8E)* with the initial post ictal stage. Slow wave activity of 1-3 Hz is then apparent and is correlated with the period of depression of consciousness. With clearing of confusion, the electroencephalogram gradually returns to a normal frequency range. During interictal periods, the electroencephalogram in patients who are subject to generalized convulsive seizures may present a variable appearance. The record often is normal. Generalized bursts of polyspike and slow waves may occur. In some cases, focal or multifocal abnormalities may be present even though a clinical focus of seizure onset was not apparent. The causes of GTCS at various ages is indicated in Table 29-3. The preferred treatment of idiopathic epilepsy characterized by generalized tonic clonic seizures is valproic acid. For some patients phenytoin may be utilized.

Other conditions must be distinguished from generalized tonic clonic seizures: several conditions in which a loss of consciousness and postural tone occur must be differentiated from generalized tonic clonic seizures.

(1) Syncope

(2) Akinetic (atonic) seizures are seizures in which the patient falls suddenly to the floor but then stands up again almost immediately. The loss of consciousness is brief. In some cases, there may be no definite evidence of an actual loss of consciousness. In some cases, the fall represents a sudden loss of postural tone; in other cases, the fall represents a massive myoclonic jerk that throws the patient violently to the floor. Atonic or akinetic seizures are frequently seen in the Lennox Gastaut syndrome in which other seizure types: atypical absence seizures (termed previously petit mal variant) generalized convulsive seizures and in many cases, tonic seizures are also present.

(3) **Drop attacks in the adult** may occur in relation to ischemia in the paramedian branches of the basilar artery with transient dysfunction in descending motor pathways and of the motor centers of the pontine or medullary reticular formation. In such cases, consciousness is usually preserved but patients often have other symptoms suggesting that basilar vertebral insufficiency may be present (see Chapter 13).

OVERVIEW OF PRIMARY/ IDIOPATHIC GENERALIZED EPILEPSY

(1) **The Overlap Of Syndromes:** As demonstrated in Table 29-4 from Genton (1995) although clinical syndromes and their EEG patterns have been delineated, it is important to emphasize the considerable overlap among the various subcategories.

Many of the models to be discussed below illustrate an overlap of syndromes. In addition, both in the human cases and in the experimental models, there is not a 1 to 1 correlation of EEG and behavior.

(2) **The Genetic Molecular Biology Of Idiopathic Epilepsy:** For most cases this has not yet been established. However a number of specific genetic mutations involving channels and receptors have been identified. Recent papers by McNamara (1999), Prasad et al (1999) and Steinlein, (1999) have reviewed this topic. Specific defects have now been identified in several varieties of presumably monogenic seizure disorders.

(a) **Partial idiopathic epilepsy: Autosomal dominant nocturnal frontal lobe epilepsy (ADNFLE).** This type of nocturnal motor seizure disorder (at times with secondary generalization) was originally misdiagnosed as a sleep disorder. In a large pedigree reported from Australia, linkage to the 20-q 13.3 chromosome site has been established. The a4 subunit of the neuronal acetylcholine nicotinic receptor (CHRNA4) maps to the

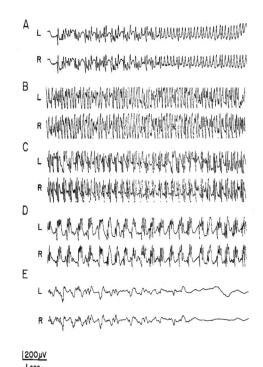

Figure 29-8. Experimental generalized tonic clonic seizure with out focal onset in a monkey. The animal was tonic during the fast 8 Hz spike discharge of segment A. The animal had repetitive clonic jerks, during the poly spike –slow wave discharges of segments B, C and D, gradually decreasing in frequency. The period of relative electrical silence shown at the end of segment E and the subsequent period of 1-2 cps slow wave activity (not shown) corresponded to the postictal depression of consciousness. Bipolar recordings from left and right premotor - precentral areas.

same region. A missense mutation occurs with the addition of serine to the 2nd transmembrane amino acid sequence. This mutation is present in all affected individuals and in the "carriers" (there is a penetrance of 70 - 80 %). In a Norwegian family with the same syndrome, an extra leucine was inserted in to the terminal end of the 2nd trans membrane domain on the same receptor subunit.

The overall result of this mutation is a significant reduction of calcium permeability and a hypoactive receptor. The additional step necessary to produce seizures is unclear. However it is possible that since some acetylcholine receptors are in a presynaptic location a

TABLE 29-3: CAUSES OF NON FOCAL GENERALIZED TONIC CLONIC CONVULSIONS

AGE	ETIOLOGY
NEW BORN	Metabolic: Na/K, alkalosis, hypoglcemia, hypocalcemia, pyridoxinedeficiency, dependency, aminoacidurias (phenylketonuria), respiratory /anoxia, perinatal pathology
INFANT	Metabolic disturbance, febrile convulsions*, meningitis /encephalitis, Residuals of perinatal pathology
CHILD	Febrile (<5 years), idiopathic epilepsy, meningitis/encephalitis
ADOLESCENT	Idiopathic epilepsy***, drug use and withdrawal:
YOUNG ADULT	Alcohol/drug withdrawal**, cocaine use, idiopathic epilepsy
OLDER ADULT	Alzheimer's disease and other degenerative disorders, anoxic encephalopathy

** Abrupt withdrawal or reduction of intake after chronic intake of alcohol may trigger seizures within 7-48 hours of the withdrawal. During this period the EEG may demonstrate precipitation of myoclonus and of electrical discharges by photic stimulation. There are in these cases of alcohol withdrawal, correlations with low serum magnesium, increased pH and low pCO2 levels (see Victor, 1990). More recent cellular studies of sedative drug withdrawal are discussed in Carlen et al. 1990. In these cases the post-ictal state may then merge into a more prolonged agitated and confusional state of delirium tremens. This same clinical state may also follow withdrawal from short-acting barbiturates, meprobamate, diazepam (Valium) and other benzodiazepines.[1]

*Febrile convulsions are frequent in the age group under 5 years and occur predominantly between 6 months and 3 years of age. It has been estimated that 4 to 7 percent of all children under the age of 5 years will have at least one generalized convulsive seizure (the majority febrile). Children with febrile seizures have a high familial incidence of seizure disorder suggesting a factor of genetic predisposition. A significant proportion (but not the majority) of these children will be found later to also have non-febrile convulsions.

*** Note that 86% of JME patients also have generalized convulsive tonic clonic seizures. A related syndrome is that of epilepsy with generalized tonic clonic seizures on awakening which differs from JME - only in the absence of myoclonic seizures (see Dreifuss, 1989).

decrease in calcium flux might reduce the release of GABA at this location. More recently a 2nd locus on chromosome 15 q 24 has been identified; this site also contains multiple subunits of nicotinic cholinergic receptors.

(b) **Generalized idiopathic epilepsy: Benign, familial neonatal convulsions (BFNC).** This is also a rare autosomal dominant disorder in which partial or generalized clonic convulsions start around the 3rd postnatal day and persist until week 6. Approximately 12% of patients have seizures again later in life. In most families linkage to chromosome site 20 q 13.3 occurs with a missense mutation in the gene for the voltage gated potassium channel-KCNQ2. There is a loss of amino acids in the sequence at the pore of the channel. *The mutation abolishes the potassium currents that are activated by depolarization. As a result excitability is increased.* In a few families, the mutation occurs at the KCNQ3 site on chromosome 8 with a similar effect on excitability. Both KCNQ2 and 3 are predominantly expressed in the brain. Neurological development is otherwise normal.

A similar mutation in another potassium channel, KCNQ1, produces the cardiac arrhythmia of the long QT syndrome.

(c) **Generalized idiopathic epilepsy: Generalized epilepsy with febrile seizures plus (GEFS+).** In this autosomal dominant syndrome, febrile seizures occur before 3 months or after 3 years of age. In addition however and this is of particular interest, multiple forms of classical idiopathic seizures occur without fever including absence, myoclonic, atonic, and generalized tonic-clonic. Different members of the same family may have different types of these seizures despite the presence of the same genetic mutation. The mutant gene

---

[1] *It is important to realize that in patients with underlying predisposition to either partial seizures (as in post-traumatic epilepsy) or to primary generalized epilepsy withdrawal effects from acute use of relatively moderate amounts of alcohol may precipitate seizures. These patients then are more sensitive to withdrawal from alcohol and other sedative agents (see Mattson. 1990).*

maps to chromosome locus 19 q 13.1. A mutation in the beta subunit of a voltage gated sodium channel, (SCN1B) occurs as a result of amino acid substitution of tryptophane for cysteine.

This is a rapidly evolving area; additional discoveries are to be expected. Note however, that in most seizure disorders in the human patient, penetrance is often incomplete and transmission may be polygenic.

Later, in our discussion of the idiopathic epilepsies, we will discuss other genetic defects in rodent models

(3) The Underlying Neuropathology Of Idiopathic/Primary Generalized Epilepsy: Although a longstanding general criteria cited for this disorder was that no structural neuropathology be present, it is now evident that cortical micro dysgenesis is present in those cases that come to autopsy. In most cases, neuons are found in the molecular layer which is normally relatively devoid of neurons. In the series of Meencke (1996), 100% of 12 patients with childhood absence epilepsy and 100% of 3 patients with JME had significant micro dysgenesis; but none had severe migration disorders in the deeper cortical layers. In contrast 50% of 24 patients with infantile spasms had micro dysgenesis and 13% had severe migration disorders. Among 30 patients with the Lennox-Gastaut syndrome, 50% had micro dysgenesis and 16% had severe migration disorders. Among 27 patients with temporal lobe epilepsy, only 7% had micro dysgenesis but 11% had severe migration disorders. In terms of increased neuron density in the molecular layer, (a prominent feature in idiopathic epilepsy), this was most severe in the frontal lobes, but also present in other cortical areas. Recent quantitative MRI studies have also raised the question of cerebral grey matter structural changes in all types of patients with idiopathic epilepsy most prominent in juvenile absence epilepsy (Woermann et al, 1998) In addition among patients with JME, more specific MRI and PET scan changes have been noted in superior frontal and mesial frontal gyri (Woermann et al, 1999) These patients may also have a correlated deficit in visual working memory.

(4) Underlying Pathophysiology: A number of models of idiopathic epilepsy are available, which reproduce the behavioral and electrical (EEG), and pharmacologic features of these disorders. If the genetic and biochemical pathophysiology is known in man, this should be reproduced. Unfortunately, except for a small number of the clinical syndromes that we have earlier reviewed, in relation to genetic disorders of channels, these latter features are often unclear. In general, because of phylogenetic differences in cortical and callosal development, observations in the monkey may be more difficult but have greater validity in terms of generalization to man. In the models of absence seizures there is the question of whether the process recorded in the EEG reflects the activity of cerebral cortex alone or of the effects of an interaction of thalamus with cerebral cortex. Studies in the rat are con-

TABLE 29-4: SEIZURE TYPES IN IDIOPATHIC EPILEPSY SYNDROMES IN 253 CONSECUTIVE CASES, MODIFIED FROM GENTON

MAJOR SYNDROME	# OF CASES	SEIZURE TYPE(S)			
		Absences	GTCS	Myoclonic	Other
Childhood absence	73	100%	44%	3%	7%
Juvenile absence	28	100%	75%	11%	11%
Juvenile myoclonic	59	22%	86%	95%	---
GTCS on awakening	30	7%	100%	13%	-
Isolated GTCS	39	-	100%	11%	-

sistent with the thalamo cortical interaction. Studies in the monkey suggest a cortical basis As regards the JME syndrome, the baboon with light induced myoclonus provides a model in which the discharges begin in frontal rolandic cortex. As regards GTCS the EEG discharge can be reproduced in a preparation composed of cerebral cortex and corpus callosum. Table 29-5 below provides a summary of pathophysiology.

TABLE 29-5: OUTLINE OF MODELS OF IDIOPATHIC EPILEPSY PATHOPHYSIOLOGY: REFER TO CD ROM FOR ADDITIONAL DISCUSSION.

TYPE OF SEIZURE TYPE OF MODEL	SUBSTRATE OF EEG SEIZURE	MECHANISM FOR BILATERAL SYNCHRONY	BEHAVIOR
ABSENCE			
1. midline/intralaminar thalamic stimulation Fig.29-9	Recruiting response thalamo-cortical Stimulus related 3Hz SW Refer to Chapter 17.	Unclear probably corpus callosum interaction.	Arrest of movement
2. Bilateral symmetrical premotor foci (monkey) Fig.29-10,11,12	Cerebral cortex. Bilateral 2.5-3.5 Hz SW persists in bilateral cortical callosal isolate preparation.	Corpus callosum if posterior premotor.	Absence: eyeopening, staring, arrest of movement if anterior premotor. Greater myoclonus
3. Feline general penicillin epilepsy (cat) Fig. 29-13	Onset in the cerebral cortex, later in the specific thalamic nuclei later in nonspecific thalamic nuclei. Thalamic nuclei are recruited into a cortical/thalamic-cortical oscillation. 3.5-6 Hz SW	Corpus callosum	Impairment of responsivity during bursts plus. myoclonus of face/ extremities
4. GAERS(rat) (Generalized absence epilepsy of the rat- Strasbuurg)	Discharges are predominant in frontoparietal cortex and d posterolateral thalamus, Thalamo-cortical interaction is required . 7-11 Hz SW	Corpus callosum	Animals are immobilized with myoclonus face/extremities and impairment of responsivity during the bursts.
MYOCLONUS			
1. Subthreshold penty - lenetetrazol (monkey)	Motor cortex: bilateral 3Hz SW	Corpus callosum	Bilateral myoclonic jerks of extremities
2. Bilateral motor cortex foci	Motor cortex: initially bilateral spike discharges	Corpus callosum	Myoclonus of extremities then GTCS**
3. Photosensitive baboon	Origin in frontal/rolandic Cortex (areas 4,6). Bilateral 4Hz Polyspike slow wave. Blocked by DOPA agonists	Corpus callosum	bilateral myoclonus of eyelids, face and limbs. At times, progression to GTCS
GTCS Generalized Tonic Clonic Seizure			
Threshold dose of Pentylenetetrazol (monkey/cat). Refer to chapter 17 for illustrations	Cerebral cortex: onset at low threshold areas e.g. motor cortex. Pattern is similar in bilateral cortical callosal isolate	Corpus callosum	GTCS often preceded by bilateral myoclonic jerks

ALTERATIONS IN CONSCIOUSNESS: SEIZURES, COMA AND SLEEP   29-17

Genetic Models in the Mouse-(Table 29-6 on the CD ROM): At present 6 inbred mouse strains have been reported with spike wave discharges and absence seizures. In each of these strains the bilateral bursts of spike wave discharge are accompanied by behavioral

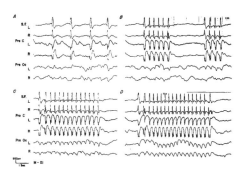

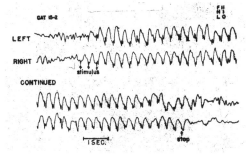

Figure 29-9. *Experimental 3 Hz spike wave discharges following midline or intralaminar thalamic stimulation. A stimulus-related bilaterally synchronous spike and wave discharge is produced in cat cerebral cortex by repetitive 3 cps stimulation of the Massa intermedia. (From Jasper, H., and Droogleever-Fortuyn, J.: Res.Publ.Ass.Nerv.Ment.Dis. 26:272-298, 1947).*

Figure 29-10. *Experimental 3 Hz spike wave discharge following bilateral symmetrical epileptogenic foci in anterior premotor cortex of the monkey. Effects of hyperventilation as in patients with absence seizures .A= just prior to hyperventilation; B, C, D., Two to three minutes after beginning of hyperventilation. SF= superior frontal gyrus, Pre C = precentral gyrus Pre 0 = preoccipital area. From Marcus, E.M., and Watson, C.W.: Arch.Neurol.19: 103, 1968 (AMA). Refer also to chapter 17 for an additional example of a 3-4 Hz spike wave discharges in this preparation.*

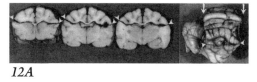

12A

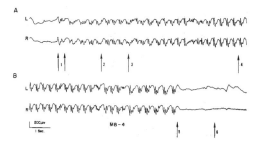

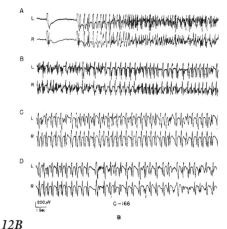

12B

Figure 29-11. *Bilateral symmetrical foci: anterior premotor cortex. Correlation of behavior (absence seizure) with bilateral 3 Hz spike-slow wave discharge. Continuous bipolar recordings: premotor to precentral. Segments A and B are continuous.1) Head moves to midline and eyes are wide open. 2) Start of testing with pencil moved back and forth across visual fields. The animal failed to follow. 3) Light tap on nose produced only partial blink but no other response of hands or head. 4) Hands were tapped; without response although during non-seizure periods, animal had vigorously grasped the pencil.5) Animal now blinked fully to approaching object complete eye closure then opening; no tactile stimulus to nose was required. 6) Animal now followed pencil with head and eye, turning as pencil moved towards the animal's left peripheral field. Licking also occurred (had not occurred during seizure). Reproduced with permission from Marcus, E.M., et al Epilepsia, 9:2371968b (Blackwell).*

Figure 29-12. *Prolonged bilateral synchronous 2-3.5 Hz spike-slow wave and poly spike slow wave discharges in the cortical callosal preparation following the production of bilateral epileptogenic foci. A) The isolated bilateral cortical callosal preparation in the cat (From Marcus, E.M., and Watson, C.W.: Ach.Neurol. 14:632, 1966). B. The pattern of discharge. Similar results may be obtained in the monkey. From Marcus, E.M., Watson, C.W., and Simon, S.A.: Epilepsia, 9:243, 1968 (Blackwell).*

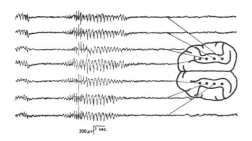

Figure 29-13. Generalized penicillin epilepsy in the cat: IM administration. Generalized bursts of polyspike and 3-5 Hz spike –wave discharges occur. From Gloor, P.Epilepsia 20:571-588, 1979 (Blackwell).

arrest and the discharges are blocked by drugs that are effective against absence seizures (Noebels et al 1997).However almost all of these strains have other neurological deficits usually related to cerebellar degeneration and the discharge consists of a 6-7 HZ spike and wave. The importance of these strains relates to the monogenic inheritance of actual or presumed ion channel mutations (Ca++ or Na+/H+). Table 19-6 on CD-ROM modified from Barclay and Rees(1999) summarizes the major findings. Other genetic seizure disorders in rats are discussed by Noebels et al (1997) and Sarkisian et al (1999)

CONCLUSIONS REGARDING THE BILATERAL SYNCHRONOUS DISCHARGES OF IDIOPATHIC EPILEPSY:

1. All of the models leaving aside the mouse models suggest that the basic pathology involves an increased hyperexcitability of neurons at the cortical level. This is consistent with the now available neuropathological data and MRI data in man.

2. The bilateral synchrony in the various models in the monkey, baboon, cat and rat is dependent on the corpus callosum and the related major commissures. There is no direct anatomical substrate in the thalamus of the human providing the basis for the bilateral synchrony found in idiopathic epilepsy.

3. There may be multiple possible mechanisms providing the substrate for the oscillations of excitation and inhibition responsible for the spike-slow wave discharge. The minimal substrate is built into the design of cerebral cortex and of the cortical –callosal circuits. There is little doubt that discharges having begun in cerebral cortex do subsequently spread to the thalamus, basal ganglia, the brain stem and spinal cord and these structures do reverberate back to cerebral cortex. However to what extent, the cortico-thalamo-cortical oscillation is necessary to produce and sustain the spike –wave complex in the human is uncertain. The idiopathic epilepsies and the spike wave discharge may represent a biological spectrum, rather than a unitary disease as noted by Berkovic et al (1988)

4.Behavior is a more complex matter. There is a representation and re representation of patterns of behavior at various levels of the neuraxis.

WHY ARE AWARENESS AND RESPONSIVITY ALTERED DURING THE ABSENCE SEIZURE? SEVERAL EXPLANATIONS ARE POSSIBLE.

1) There is widespread involvement of cerebral cortex in seizure discharge; particularly of the multimodality sensory projection areas and of other frontal and temporal association areas involved in working memory etc., so that these areas can not participate in their normal activities

2) There is impairment in the capacity of the thalamus to transmit information.

3) Both of these explanations are operational in the intact individual since discharge of cerebral cortex secondarily spreading to thalamus via the cortico –thalamic system will also alter the capacity of the thalamus to transmit information.

WHAT ARE THE MECHANISMS FOR ENDING SEIZURE DISCHARGE?

The total explanation for the cessation of the seizures is not entirely clear but presumably could include various inhibitory processes involving the cortex, subcortical nuclei, thalamus and brain stem. In this regard, we should note that the activity of isolated cerebral cortex is characterized by bursts of activity with intervening periods of relative electrical silence. There are then intrinsic mechanisms within cerebral cortex for cessation of discharge. This is then the most likely overriding consideration. Additional discussion of the role of structures such as the red nucleus, basal forebrain, pontine and medullary reticular formation and the pars reticularis of the S. Nigra will be found in the cited references (Gale, 1990, Moshe &Sperber, 1990).

GUIDELINES FOR MANAGEMENT OF SEIZURE DISORDERS:

Evaluation: Over 75% of diagnosis in neurology is dependent on the history, this is even more important with regard to seizure disorders. However in this case, information must be obtained from witnesses regarding events for which the patient may have little recollection. Information from the patient and witnesses regarding focal manifestations, aura etc is vital. The neurological examination is critical in terms of the presence or absence of signs suggesting focal cortical pathology. Laboratory studies should include a complete blood count, blood sugar, calcium, electrolytes and creatinine/bun. The electroencephalogram performed during awake and sleep states may provide information regarding focal or generalized electrical seizure discharges or slow wave activity. The yield of the EEG may be increased by activation procedures such as sleep deprivation (SD), hyperventilation (HV) and photic stimulation (PS). All patients should have an imaging study. Based on the information considered above, whenever possible this should be an MRI. When this is not possible, then a CT scan with and without contrast should be obtained.

Specific aspects of seizure management are considered in Table 29-7. The use and properties of anticonvulsants are considered in Tables 29-8, 29-9, 29-10

The upper limit of therapeutic range for all of these agents is variable related to protein binding etc. For valproic acid, the upper limit of therapeutic range is 100-150 ug/ml. The half-life of phenytoin is constant only up to 12ug/ml. above that level the half-life increases with blood level. The time to achieve steady state is approximately 4-5X the half life. Dosing schedule relates to half-life. Note that primidone is in part converted to phenobarbital, both metabolites have anticonvulsant action.

The molecular basis of anticonvulant drug action: Refer to table 29-9.

In general anticonvulsants that are effective against partial and secondarily generalized seizures are effective in the experimental maximal electroshock test, (a model of seizure spread) and produce Na+ channel blockade. In general those drugs that are effective against spike wave epilepsies, particularly absences are effective against experimental pentylenetetrazol induced seizures, elevate threshold and produce T type Ca++ channel blockade. However benzodiazepines and barbiturates are also effective against this type of experimental seizure.

Teratogenic and side effects of anticonvulsant medications

1. teratogenic effects in the offspring of the pregnant seizure female. In general, the non-seizure population has a 2.5% risk of malformation, seizure patients not on anticonvulsants - a 3.5% risk. Seizure patients on anticonvulsants - 5.5% risk. Teratogenic effects occur primarily in the first trimester and consist of cleft lip, neural tube closure defects and congenital heart disease. The risk of malformation is increased if any of the following factors is present: (1) poor seizure control (2) polypharmacy (3) high blood levels (4) use of trimethadione. Risk of neural tube defects can be significantly reduced by administration of folic acid.

2. Other side effects of anticonvulsants fall into general groups:

a. Allergic - dermatological, hepatic, immune for which the drug must be discontinued

b. Dose related (ataxia, drowsiness, dizziness, diplopia)- for which the dose must be lowered

c. Chronic - hematological, lymphatic (lymphoma, lupus, etc.

TABLE 29-7: MANAGEMENT OF SEIZURE DISORDERS

PROBLEM	MANAGEMENT	EXPECTED RESULTS
SINGLE GTCS	No specific anti convulsant treatment if (1) History indicates, no focal features, no prior minor seizures, normal growth and development, no mental retardation, and no family history of seizure disorder (2) Neurological examination is normal (3) EEG with sleep deprivation, HV and PS is normal (4) Imaging studies are normal (5) Conditions such as fever, sleep deprivation, alcohol and drug withdrawal or use of drugs such as cocaine, isoniazid or other agents or metabolic conditions lowering threshold are present.	Annegers et al (1986) Found recurrence rate of 16% @1 year, 22%@3 years, 26%@5 years.
RECURRENT SEIZURES (EPILEPSY)	(1) Rule out metabolic abnormality and mass lesion, Prevent additional seizures with anticonvulsants (See Tables 7, 8, 9, 10). (2) Start with monotherapy push dose to seizure control or toxicity. (3) If still no control use second or third major agent then newer add on agents. (4) Treat for 3-5 years seizure free, If at that point, EEG indicates no seizure discharges, consider gradual withdrawal from medications	In generalized /idiopathic epilepsy complete control In 63% and significant reduction in an additional 25%. In complex partial, complete control in 20-35%. If seizure free 5 years and discontinue medication 30% relapse over 10 years
RELAPSES	**If on anticonvulsants:** (1) Compliance issues, omission of medication. Check levels. (2)Insufficient daily dosage.Check levels and adjust dose. With phenytoin check the free unbound fraction (usual therapeutic range is 1- 2 ug/ml). Alternatively push dosage until therapeutic effect is attained or dose related toxicity occurs. (3) Wrong medication for the type of seizure-reevaluate in terms of history and EEG. Do long term video-EEG monitoring. Whether on or off medications consider following factors: (1) Sleep deprivation (2) Alcohol or drug withdrawal (3) Fever or metabolic abnormalities or use of antihistamines. (4) Falls with frontal or temporal injuries =additional foci. (5) Epileptic status producing hippocampal sclerosis so that additional foci of seizures resulting in complex partial seizures. Re-evaluate in terms of history, EEG & monitor. (6) Secondary epileptogenesis. (7) Underlying progressive disorder is present. (8) Possible pseudo seizures + actual seizure disorder.	(1) Recurrences are more likely if focal rather than idiopathic, or if long time before initial seizure control or if seizure discharges are still present in EEG. (2) Intensive re-evaluation at an epilepsy center with video EEG monitoring will result in a significant number of patients reclassified with treatment modified. (3) Combination of epilepsy +alcoholism or drug addiction difficult to treat because off compliance issues and withdrawal effects triggering seizures.
REFRACTORY PATIENT DESPITE USE OF 3 TRADITIONAL MEDICATIONS	(1) Consider all factors listed above. (2) If focus is well defined, and psychiatric issues are not present consider surgery after intensive non-invasive ± invasive monitoring, and determination of dominance. Because of high incidence of complex partial seizures usual procedure is anterior/mesial temporal resection. Less often, resections in frontal lobe. (3) If poorly controlled secondary generalized epilepsy (e.g.Lennox Gastaut) or multiple foci consider vagal stimulator or corpus callosotomy.	In patients with complex partial epilepsy due to mesial temporal sclerosis, 78% had complete seizure control at one year after anterior medial temporal resection (Doyle&Spencer(1998) (see also Wiebe et al,2001). With frontal lobe resections,68% became seizure free.Vagal stimulator will reduce seizure frequency. Callosotomy will decrease GTCS and drop attacks but focal seizures may increase.
EPILEPTIC STATUS	Discussed below	

TABLE 29-8: PHARMACOLOGIC PROPERTIES OF THE MAJOR ANTICONVULSANT DRUGS

AGENT	USUAL DAILY DOSE CHILDREN (MG/KG)	USUAL DAILY DOSE ADULTS (mg) (TOTAL)	DOSING	HALF-LIFE	BLOOD LEVEL (µg/ml) EFFECTIVE	BLOOD LEVEL (ug/ml) TOXIC
Phenobarbital	3-5	90-180	1X/day	96+/- 12	>15 (4-13)	>40
Phenytoin	4-7	300	Split 2X/day	24± 12	>10	>20
Primidone	10-25	750	Split 3X/day	>6	>12	
Ethosuximide	20-30	1000	Split 2X/day	30 ± 6	>40	>100
Carbamazepine	8-20	600-1400	Split 3X/day	12±3	>8 (4-13)	>13
Valproic Acid	15(initial) 15-30 (maintain)	1000 2600	Split3-4X/day	8	>50	>125

TABLE 29-9: MOLECULAR BASIS OF ANTICONVULSANT DRUG ACTION
MODIFIED FROM RHO & SHANKAR (1999) ON CD ROM

TYPE OF ACTION	DRUGS* TRADITIONAL	DRUGS* NEWER AGENTS
(1) Inhibit voltage gated sodium channel	PHT, CBZ, VPA	FBM, LTG, TPM, ZNA
(2) Enhance GABA/A receptor Cl-channel:	PB, BZD	FBM, TPM
(3) Increase brain GABA	VPA	VGB, GBP, TPM, TGB
(4) Inhibits Na+currents through non NMDA receptor		TPM
(5) Inhibits Ca++ currents through NMDA receptor		FBM
6) Inhibit slow threshold T type volt gated Ca++ channel	ESM (?VPA)	ZNA
7) Inhibits brain carbonic anhydrase	AZM	

Drugs* PHT, phenytoin; CBZ, carbamazepine; VPA, valproicacid: henobarbital; BZD,benzodiazepines;
ESM, ethosuximide; AZM,acetazolamide; FBM,felbamate; GBP, gabapentin; LTG, lamotrigine;
TPM, topiramate; TGB, tiagabine; VGB, vigabatrin; ZNA, zonisamide

STATUS EPILEPTICUS - This is a frequent major neurological emergency. Overall incidence is at least 41-61 cases per 100,000 population. Less than 1 year of age the incidence is 156 per 100,000, in the over 65 year group, the incidence is 86 per 100,000 population. Overall 30% of cases are generalized from the onset and 32-43% are secondarily generalized after focal onset. This acute condition must be recognized and treated as rapidly as possible. The strict terminology of the International Classification of Epileptic Seizures defines epileptic status as any seizure lasting > 30 minutes or intermittent seizures lasting >30 minutes without regaining consciousness. Since the usual tonic clonic seizure is usually <1-2 minutes in duration, most experts would now shorten the time before treating as epileptic status to 5-10 minutes. The earlier the treatment, the lower the morbidity and mortality. The epidemiologic data of DeLorenzo et al (1995) as presented in table 29-11 indicates a current overall mortality of 22% in cases meeting the 30 minutes criteria. The greatest risk of mortality occurs in those cases that are generalized from the onset or are secondarily generalized after focal onset and in the over 65 age group where mortality is 38%. The causes of morbidity and mortality are cerebral anoxia, cerebral edema, aspiration pneumonitis, hyperthermia, hypotension and cardiac arrest as possible complications. Due to damage to hippocampal areas prolonged grand mal status may be followed by prolonged confusion and a residual impairment of recent memory as well as temporal lobe seizures.

TABLE 29-10 SEIZURE TYPES AND DRUG CHOICE*:

Seizure Type	Drug Choice -Traditional	Drug Choice - Newer* **
Partial and secondarily generalized	PHT, CBZ, VPA	LTG, TPM, ZNA, GBP, TGB**
Absence	ESM, VPA	
Myoclonic	VPA, BZD	
Primary generalized tonic/clonic	VPA, PHT, CBZ	
Infantile spasms	ACTH, BZD	
Status epilepticus: generalized, partial)	BZD (intravenous)	

* Drugs= PHT, phenytoin; CBZ, carbamazepine; VPA, valproicacid: henobarbilul; BZD, benzodiazepines; ESM, ethosuximide; AZM, acetazolamide; FBM, felbamate; GBP, gabapentin; LTG,lamotrigine; TPM, topiramate; TGB, tiagabine; VGB, vigabatrin; ZNA, zonisamide ACTH = adrenocorticotrophic hormone
** With the exception of LTG all of these newer agents are adjunctive add on agents
*** Additional new agent for partial seizures - levetiraetam
**** LTG may also be utilized in the Lennox-Gastaut syndrome

TABLE 29-11: EPIDEMIOLOGY OF STATUS EPILEPTICUS (SE) BASED ON DATA OF DELORENZO ET AL (1995)

Age Range	Incidence* Per 100,000	Mortality %	Major Etiology	Past History Of Epilepsy %	Type Seizure In SE: Partial %	Type Seizure In SE: Partial To GTCS	Type Of Seizure In SE: GTCS %
<1 year	156						
1month-15 year (pediatric)	38	2.5	Non CNS infection (52%) Low levels (21%)	38	28	32	32
16-59 years (Adult, non elderly)	27	14	CVA: acute or Remote (>41%) Low Level (34%)	54	24	43	27
>65 –80+	86	38	CVA: acute or Remote (61%)	30	Included in Adult	Included in Adult	Included In adult
Total	41-61	22		42			

* These are minimum figures. Extrapolating from the data for Richmond Virginia to the total U.S. population results in an estimate of 152,000 cases and 42,000 deaths per year.

The etiology of status epilepticus can be divided into three major groups:

I. Status in patients with prior seizures due to (a) acute withdrawal or omission of medication (b) new metabolic factors and fever. (Group I constitutes 42% of all cases).

II. Status as an acute or remote effect of a focal process most frequently a cerebrovascular accident (CVA). However, frontal lobe tumor or abscess may also be the underlying focal process.

III. Status as the acute manifestation of systemic disease such as non-CNS infection in infants or children (52% of all cases in this age group), and alcohol withdrawal or metabolic factors or hypoxia in the adult.

Treatment has 4 essential components:

(1) Assure stability of vital functions: (a) Airway (b) Breathing (c) Circulation: maintain blood pressure in normal range (d) maintain temperature in normal range (generalized seizures will produce hyperthermia due to central mechanisms). Draw blood levels and chemistries. Administer intravenous 50% glucose.

(2) Stop the seizures: intravenous benzodiazepines: lorazepam, (0.1mg/kg at rate of 1-2 mg/min to maximum dose of 8 mg) or diazepam (5-20 mg at rate of 1-2 mg/min), or midazolam.

(3) Prevent additional seizures: load with intravenous fosphenytoin (18-20 mg/Kg at rate of 50 mg/min).

(4) Determine the cause: this will require EEG and CT scan or and /or MRI, blood levels, etc as well as neurological examination and observation of the seizures. Must still make distinction - focal versus generalized.

(5) If status does not end rapidly at the emergency room level continued intensive care unit monitoring, including continuous EEG will be required. General anesthesia including intravenous pentobarbital may be required.

(6) Non-convulsive status: For acute onset of confusional states, consider the diagnosis of complex partial and absence status: The diagnosis will depend on immediate neurological examination and immediate EEG as well as trials of intravenous therapy with benzodiazepines.

This is a skeleton outline of management of status for greater details see Delgado-Escueta, 1982 and Pellock & Delorenzo, 1997.

SLEEP STAGES: ANATOMICAL BACKGROUND

Sleep is characterized by a sequence of clinical and EEG stages *(figure 29-14)*: **Stage W:** Awake –alpha rhythm of 8-13 Hz with suppression of alpha rhythm and emergence of low voltage fast activity on eye opening or intense arousal. **Stage 1**- drowsiness –decease in alpha amplitude and emergence of intermittent 4-7 Hz theta activity. **Stage 2**- early sleep –high amplitude diphasic or triphasic sharp waves are present over the vertex areas-termed K complexes, or vertex sharp waves intermixed with 14-15 Hz spindles. **Stage 3 & 4** are characterized by increasing amounts of 0.5-3 Hz delta slow wave activity (3 =delta 20-50% of the time, 4=delta >50% of the time). *Stages 2,3,4 are referred to as slow wave or NREM (non REM) sleep.* **Stage Rem** After 90 minutes in the normal human, there then emerges a different stage of sleep characterized by low amplitude alpha rhythm and fast activity, rapid eye movement and a loss of muscle tone. This stage is also called paradoxical sleep since the individual is deeply asleep but the EEG corresponds to a stage of relative wakefulness. This stage is associated with dreaming.

BRAIN STEM, DIENCEPHALIC AND CORTICAL STRUCTURES INVOLVED IN CONSCIOUSNESS AND SLEEP.

REM sleep: recent studies have indicated an interaction of several areas in the brain stem. *Cholinergic neurons* in two nuclei located in the pontomesencephalic tegmental area of the reticular formation: the pedunculopontine tegmental (PPT) and the lateral dorsal tegmental (LDT), discharge during REM sleep. These neurons send fibers to the thalamic relay and reticular nuclei. In contrast, during REM stage of sleep there is an inhibition of activity in the *serotoninergic neurons* in the midline raphe nuclei of the pons and medulla, in the *nora-*

TABLE 29-12: ROLE OF FOREBRAIN, DIENCEPHALIC AND BRAIN STEM STRUCTURES IN SLEEP AND COMA

STRUCTURE	BEHAVIORAL CHANGE	EEG CHANGE
Basal forebrain, preoptic, orbital frontal: stimulation	Induction of sleep,	Slow wave stage of sleep
Rostral (anterior) hypothalamus: preoptic, suprachiasmatic -Lesion* -Stimulation (low frequency)	-Insomnia -Decreased spontaneous activity and of muscle tone	-Desynchronized. Aroused
Caudal hypothalamus +upper midbrain -Lesion* -Stimulation (low frequency)	-Hypersomnolent -Behavioral activation.	-Slow wave pattern
Lateral hypothalamus-lesion*	Periodic narcolepsy/cataplexy	
Thalamus -Anterior/dorsomedial ---Lesion ---Stimulation -Intralaminar system ---Lesion ---Stimulation (high frequency)	---Insomnia ---Sleep induced ---Lethargy ---Wakefulness	---Slow wave ---Desynchronized aroused
Mesencephalic reticular formation -Lesion -Stimulation (high frequency)	-Coma -Shifts animal from sleep to wake	-Slow waves with no arousal -Shifts from slow to low volt fast
Midline raphe nuclei pons and medulla serotonergic neurons: lesion	A marked increase in the total duration of wakefulness. Sleep decreased by 80-90 %	Persistent awake pattern

* Hypothalamic lesions based on neuropathologic correlation studies of Von Economo in encephalitis lethargica. otherwise studies of Hess (1949)

drenergic neurons of the locus ceruleus in the pontine tegmentum and in the *histaminergic nuclei* in the hypothalamus. Thus it is possible to speak of "REM on", as opposed to "REM off neurons". (See reviews of Steriade, 1992 and Chokoverty, 2000 and table 12). Both the cholinergic and the monoamine areas appear to be controlled by the ventrolateral preoptic nuclear complex (Lu et al, 2000). Table 29-13 summarizes the relationship between these centers in relation to sleep stages.

Circadian Timing System And Anatomical Basis Of Sleep Wake Cycles: Sleep-wake behavior, occurs on an alternating cyclic basis, referred to as *circadian*. Other endocrine and hypothalamic functions also are subject to rhythmic variations over the period of approximately 24 hours. e.g. secretion of growth hormone and of cortisol. Most non-primate mammals sleep during the day and are awake at night. Primates, in general, sleep at night and are awake during the day. It is also evident that, these cyclic rhythms are also altered by the light dark cycle. Experimental studies reviewed by Moore (1990), indicate that in the presence of constant light (referred to as free running), the cycle still occurs and

TABLE 29-13: RELATIONSHIP OF BRAIN STEM NEURONS AND SLEEP STAGES (AFTER SAPER ET AL 2001)

Transmitters/ neurons	Awake Stage	Slow Wave Stages	REM Stage
Cholinergic	2+	0	2+
Monoamines	2+	1+	0

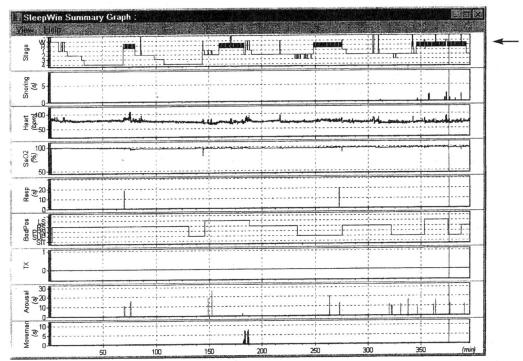

that the pacemaker for this rhythmic cycling is located in the suprachiasmatic nucleus of the hypothalamus. The major projections of the suprachiasmatic nucleus are to other nuclei of the hypothalamus - primarily to the anterior hypothalamus. The hypothalamus has reciprocal connections to the brain stem reticular formation and projects to the nucleus basalis and to the cerebral cortex.

Additional studies have demonstrated a retinal hypothalamic pathway and a lateral geniculate hypothalamic pathway to this nucleus allowing for the entrainment of this nucleus by visual, light dark influences. An additional input is also present from the serotonin secreting neurons of the midbrain raphe nucleus. The secretion of melatonin by the pineal gland is regulated by the suprachiasmatic nucleus. This hormone, in turn, can significantly modify the cyclic activity of neurons within the suprachiasmatic nucleus.

DISORDERS OF SLEEP: I. Insomnia, II. Excessive Daytime Sleepiness.

I. Insomnia is defined as an inability to enter a state of sleep. **Hyposomnia** is defined as a reduction in sleeping time.

Figure 29-14. Polysomnogram. An all night 7-hour sleep study demonstrating the periodicity of sleep with the rapid eye movement (REM) stage (arrow on interrupted block line) occurring at 60-90 minute intervals. Note that in addition to sleep stages, snoring, respiration, EKG, movements, arousals and oxygen saturation are all recorded. (Courtesy of Sleep Disorders Institute of Central New England /DR. Jay Phadke) Refer also to Fig. 29-16.

- Both conditions may occur rarely on a neurological basis following lesions of the rostral hypothalamus and preoptic area or anterior or dorsomedial thalamus (Lugaresi – 1992).

- An infrequent cause of fragmented sleep with frequent awakenings is *central sleep apnea* which may be due to the following conditions: a. autonomic system lesions, b. high spinal cord lesions, e.g., bilateral cordotomy for control of pain, c. medullary lesions, d. abnormalities at the craniocervical junction, e. muscular disorders, f. congestive heart failure. The changes in blood gases may result in significant effects on cardiac rhythm, sudden death may occur.

- *The most common medical condition associated with insomnia relates to anatomical prob-*

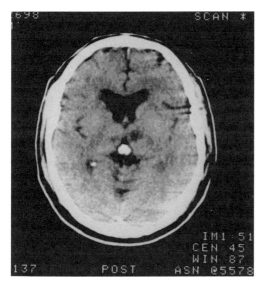

Figure 29-15. Excessive daytime sleepiness. CT scans which had been normal 7 years previously now. Demonstrated bilateral thalamic paramedian infarcts confirmed on MRI. Angiograms, echocardiogram and CSF were normal and EKG nonspecific. This 64-year-old right handed usually extroverted male had the acute onset of excessive nocturnal snoring and the following day fell asleep on multiple occasions even while driving. He also had a change in personality, failure to complete his thoughts, and problems in recognition of familiar photographs. Examination, 6 days later, demonstrated only hypertension (180/100), a subdued appearance a decrease in spontaneous speech and finger agnosia. He returned to baseline within 3 weeks and had no additional episodes over the next 4 years.

lems of the airway. For example, patients with sleep apnea syndrome, particularly obstructive sleep apnea syndrome (who as discussed below, also manifest excessive daytime sleepiness), will have frequent nocturnal awakenings. These patients who may be obese or have abnormalities of the pharynx, or short crowded necks may have collapse of the pharynx and cessation of airflow, while diaphragmatic efforts continue, often resulting in snoring. More important from a physiological standpoint, hypoxemia often occurs (arterial pO_2 actually falls as measured by pulse oximetry) and CO_2 retention also occurs. With or without the fall in pO_2, the decrease in airway results in increased diaphragmatic effort and frequent arousals with fragmentation of sleep. As a result, the patient has excessive daytime sleepiness.[2]

- The most frequent causes of disorders affecting the initiation and maintenance of sleep (the insomnias) relate to non-neurological factors (1) poor sleep hygiene including excessive daytime sleep in elderly patients, (2) Underlying psychiatric disorder, (3) pharmacologic factors, (4) Alterations in sleep wake cycle, jet lag, etc

II. Excessive daytime sleepiness: As with insomnia - the major causes of excessive somnolence, particularly daytime somnolence, are not associated with specific neuropathological causes.

-In most patients excessive daytime sleepiness is a reflection of poor nighttime sleep. This includes sleep apnea such cases may be studied with all night sleep studies utilizing polysomnography.

-Specific neuropathological lesions producing somnolence are not common, Tumors or encephalitis or infarcts (Figure 29-15) involving the areas cited in table 29-12 may produce somnolence (Culebras, 1992).

-**Narcolepsy:** Narcolepsy is a disorder characterized by paroxysmal attacks of sleep and of a desire to sleep. Following a period of sleep that varies from minutes to hours, the patient awakens refreshed. As in normal sleep, the patient can often be aroused from sleep by strong stimulation. Most cases have no known structural pathology and the neurological examination is normal. At times, a family history of a similar disorder is present. The specific genetic aspects are not clear, whether multifactorial or autosomal dominant remains uncertain. There is a relationship to the HLA (HLA = Human leukocyte antigen - histo compatibility complex) located on chromosome 6, linked gene - DW-2 (see Honda and Matsuki, 1990). There is no relation to epilepsy.

---

[2]*The effects of hypoxemia, of CO_2 retention and of the apneas - are much more complex. The responses of the respiratory centers to hypoxia and hypercapnia may be altered (see Grunstein and Sullivan, 1990). Changes in heart rate and rhythms may occur with fatal arrhythmias - resulting in sudden death during sleep (see Krieger, 1990).*

As regards the underlying basis for narcolepsy, recent studies summarized by Silber and Rye (2001) indicate a marked decrease in the content of a peptide, hypocretin-1 (Hert-1) synthesized by the cells of the ventral and lateral hypothalamus in the brains and CSF of patients with narcolepsy. These neurons innervate all of the cholinergic and monoaminergic arousal centers discussed above. *Many of the patients with narcolepsy manifest other periodic disorders that occur frequently enough to be grouped within the tetrad of the narcolepsy syndrome: cataplexy, sleep paralysis and hypnogenic hallucinations.*

Cataplexy refers to a sudden loss of muscle tone and of postural reflexes, usually occurring in relation to laughter or sudden emotional stimulation. *Sleep paralysis* refers to episodes of inability to move and occurs in relation to the process of falling asleep or awakening. *Hypnogenic hallucinations* are vivid sensory, dream-like experiences that occur upon awakening. Many aspects of the narcolepsy syndrome have been related to the desynchronized, rapid eye movement (REM) stage of sleep. Thus, the multiple sleep latency tests will demonstrate the early occurrence of periods of desynchronized sleep and rapid eye movements in these patients shortly after onset of sleep (as contrasted to normal where such

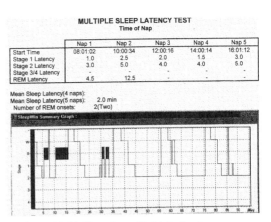

Figure 29-16. Narcolepsy. Multiple sleep latency tests demonstrating mean sleep latency for 5 naps of 2 minutes with REM onsets in the first 2 naps at 4.5 and 12.5 minutes. In contrast the all night sleep study demonstrated sleep onset at 6.7 minutes and REM onset at 60-90 minutes. This 32-year-old woman had a 7-year history of falling asleep during the day, if sitting at the table or in a car. She also had episodes of sleep paralysis. Her mother had narcolepsy. (Courtesy of Sleep Disorders Institute of Central New England /DR. Jay Phadke) Stages = stages of sleep, R = REM.

episodes do not occur for 90 minutes). In the multiple sleep latency tests, five daytime, 20-minute naps are offered at 2-hour intervals. Mean sleep latency and mean REM sleep latency after onset of sleep are calculated. Patients with narcolepsy have daytime sleep latencies of approximately 2-3 minutes, control subjects 7-12 minutes. Within the 20-minute nap peri-

TABLE 29-14: ETIOLOGY OF COMA OF ACUTE ONSET. ADAPTED FROM PLUM & POSNER (1966)

MAJOR CATEGORY	FREQUENCY	MECHANISM	EVALUATION
Toxic/metabolic/diffuse	68%	Diffuse biochemical disorder, e.g. alcohol intoxication hypoglycemia, overdoses, ketoacidosis, anoxia, encephalitis,	Early depression of consciousness without focal findings
Supratentorial	18%	Herniation due to mass lesions with compression of brain stem e.g. tumor or abscess of temporal lobe, intracerebral hemorrhage, massive MCA infarct	Lateralized hemisphere findings before brain stem findings then herniation lateral (ipsilateral fixed dilated pupil then bilateral findings) or central (Cheyne Stokes respiration then bilateral midbrain with bilateral fixed pupils) with progressive rostral-caudal deterioration of brain stem function
Infratentorial	10%	Direct damage to brain stem: basilar artery thrombosis, pontine hemorrhage	Onset with brain stem findings usually bilateral e.g.. Bilateral pyramidal tract and pontine findings

ods, all patients with narcolepsy will have two or more REM periods. (REM within 15 minutes of sleep onset is referred to as REM onset sleep as demonstrated in *Fig.29-16*). In the human, the most effective agents in preventing the narcolepsy are those drugs that enhance the presynaptic release of norepinephrine. Agents include methylphenidate and pemoline. The tricyclic antidepressants, such as Protriptyline are effective in preventing the cataplexy, through blockage of norepinephrine and serotonin re-uptake.

COMA IN MAN

In discussing depression of consciousness in man, we will consider separately two aspects: (1) the patient who presents at the emergency room or on the general wards of the hospital with acute of subacute onset of coma; (2) the patient with prolonged depression of consciousness.

Emergency Onset Of Coma: In the emergency room evaluation of the patient in coma of relatively acute onset, the physician is concerned with the differentiation of three general etiologic categories as presented in table 29-14:

Prolonged Disturbances Of Consciousness: There are two major causes (1) *intrinsic diencephalic and mesencephalic disorders* as in infarcts or infiltrating gliomas (As discussed above von Economo's encephalitis was once included in this category) (2) *disorders producing a diffuse loss of neurons in cerebral cortex* as in cardiac arrest with anoxic encephalopathy and laminar necrosis.

An apathetic-drowsy state may occur in relation to bilateral infraction of the frontal lobe and basal forebrain territory within the distribution of the anterior cerebral arteries, as in anterior communicating-anterior cerebral aneurysms.

It is evident then that from an anatomical standpoint, the conscious state reflects a complex interaction between brain stem structures, diencephalic structures, and the basal forebrain areas on the one hand, and the cerebral cortex on the other. *Without functional neocortex, consciousness does not exist. Thus patients with cerebral anoxia resulting in wide spread laminar necrosis of the cerebral cortex are in a state of irreversible coma.*

Care Of The Comatose Patient: Several general rules must be followed if life is to be preserved in the comatose patient until such time as recovery has occurred. The ABC (airway, breathing, circulation) approach has already been presented in relation to the management of status epilepticus. The patient must be placed on the side and turned every two hours to avoid aspiration and passive congestion of the lungs. This will also serve to avoid decubitus ulcers. Joints must be manipulated passively through their full range of motion several times per day to avoid contractions. The legs should be protected from the weight of bedclothes by cradle. Sand bags should maintain physiological ankle extension. Pressure areas should be protected by foam rubber to avoid decubitus ulcers.

Suctioning is required at period intervals to avoid aspiration of nasal and oral secretions. While the intravenous route may provide fluids and glucose temporarily, this will not allow administration of adequate protein, lipids and carbohydrates. Careful nasogastric feeding will be required to maintain nutritional status, in the short term. After 7-10 days, if swallowing reflexes have not returned, the placement of a percutaneous gastrostomy or jejunostomy should be considered. Nutritional intake and urinary output must be recorded. A condom catheter or indwelling catheter will be required with periodic replacement. Urinary tract and respiratory infections must be treated with antibiotics but prophylactic antibiotics should not be administered. Constipation and fecal impaction should be avoided by the periodic use of mild cathartics. Enemas may be required. Low dosage cutaneous heparin may be administered to prevent thrombophlebitis. For more complete discussion see Ropper and Kennedy 1988, and Weiner 1992.

Prognosis In Coma: Management of the neurological patient includes not only diagnosis of the underlying location of lesion and of the underlying neuropathology but also assessment of the prognostic implications of the diagnosis. From the standpoint of advice to the family of the comatose patient, and from the standpoint of allocation of limited, costly intensive care resources, early assessment must be made as to the likelihood of survival and recovery. This is best done in a uniform, reproducible and if possible, quantitative manner. Considering coma in general, prognosis also relates in general to age of patient and etiology.

The prognosis for patients with coma secondary to head injuries or intracranial hemorrhage, often is better than for patients with coma secondary to cardiac arrest. For such patients, the standard method utilized is the Glasgow Coma Scale. Values are assigned for (a) degrees of eye opening (E) Spontaneous = 4; nil = 1: (b) best motor response (M) obeys = 6; nil = 1: (c) verbal response (V) oriented = 5. Nil = 1. The total coma scale is then computed E + M + V and may range from 3-15. There is a clear-cut correlation: patients with a low scale are unlikely to recover; those with a high score are very likely to survive with a good quality of survival.

Essentially, most patients with cardiac arrest outside of the cardiac intensive care unit have a very poor prognosis. For assessing recovery after hypoxic ischemic coma or other forms of non-traumatic coma, Levy et al (1985) have utilized evaluation of brain stem reflexes: (including eye movements, corneal and pupillary responses), motor responses and verbal responses. Patients who, at first assessment, within the first 24 hours (usually at 6 or 12 hours after onset), had no corneal, pupillary or oculovestibular (caloric) responses, did not recover or remained in a chronic vegetative state, in 97% of cases. Only 1% had a good recover or moderate disability. By 24 hours, if three of the following (corneal, pupillary or oculovestibular, or motor responses) were not reactive, 98% remained in vegetative state or had no recovery.

Absence of pupillary responses at initial examination was associated with no recovery or vegetative state in 94%, severe disability in 6%. Failure to regain brain stem reflexes by 24 hours, is an extremely poor prognostic sign, essentially no patient recovers. At one day, the absence of motor withdrawal and of spontaneous eye movements was associated with 95% in recovery or vegetative state and severe disability in 4%. If brain stem reflexes have recovered by 24 hours, failure to regain consciousness by day three, has a poor prognosis, failure to regain signs of consciousness by day 7, an even bleaker prognosis. At 7 days, absence of motor response to commands and absence of spontaneous eye movements - was associated with no recovery or vegetative state outcome in 100% of cases. (See Maiese and Caronna, 1988).

The use of the EEG and of evoked may also be of help in assessing prognosis. (Marcus and Stone, 1984).

CHAPTER 30
Learning, Memory, Amnesia, Dementia, Instinctive Behavior and the Effects of early Experience

DEFINITIONS

(1) **Learning:** a relatively permanent change in behavior that results from practice or experience. Learning is the process of acquiring knowledge about the world (Kupperman, 1991). This definition implies plasticity in central nervous system function and it also implies plasticity in the formation of stimulus response connections. The definition excludes changes in behavior result-ing from maturation, sensory adaptation and fatigue.

(2) **Memory:** the processes or neural mechanisms involved in the encoding and storage or repre-sentation of an experience and the retrieval of that information. At times the definition is restricted to the "read in" consolidation and storage stages and the terms remembering and retrieval are defined as the "read out phase" of learning.

(3) **Instinctive behavior:** complex behavior occurs on the first presentation of the triggering stimulus. Such behavior occurs in all species even those raised in isolation. In birds, complex behavior may be triggered by hypothalamic or midbrain stimulation.

(4) **Imprinting:** Not all behavioral patterns are learned or instinctive. *Certain behavioral patterns require exposure to a specific stimulus at a critical period in development.* The concepts of imprinting evolved from the study of "following" behavior in newly hatched chicks. Thus, chicks will usually follow their natural mother since during the specific critical period after hatching they are usually exposed to the mother. However, if chicks do not encounter the mother but rather encounter a man, they will tend to follow the man. Critical periods in human postnatal development have also been demonstrated.

(5) **Disorders of memory:** This problem may reflect focal or more generalized disease of the nervous system. The process may be static or progressive. Included in the definition in addition to the focal disorders involving hippocampus and diencephalon are more generalized disorders such as dementia, amentia and mental retardation.

(6) **Dementia:** *A progressive impairment* of previously intact mental faculties without loss of consciousness. In general, in the most common type, Alzheimer's disease, the loss involves initially and most severely recent memory and the ability to learn and retain new memories. To some extent, particularly as time passes, other areas of mental capability are also affected: remote memory, abstract reasoning, insight, arithmetic abilities, language function, personality mood and social behavior. In other types personality is initially affected, followed later by memory.

(7) **Amentia:** The term amentia implies a *non-progressive congenital absence or relative deficiency of mental faculties.* At times, in the absence of an adequate history, the distinction may be difficult to make. The term **mental retardation** is somewhat less specific, referring to retardation in the development of mental abilities or retardation in the development of intelligence.

LEARNING IN MAN AND RELEVANT ANATOMICAL SUBSTRATE

Classification of learning and the anatomical substrate: (Squire & Zola-Morgan, 1991, Desimone, 1992, Rolls, 2000). Essentially two broad categories are considered:

I. **Declarative (Explicit) learning** - This refers to the conscious recollection of facts or events. This system is rapid; one trial may be sufficient.

II. **Non-Declarative (Implicit or "Reflexive")** - This refers to a non-conscious alteration of behavior by experience. This type of learning is slow and requires multiple trials. Table 30-1 presents the neural substrate for these types of learning

STAGES OF HUMAN MEMORY: DECLARATIVE LEARNING

The terminology employed has changed with time particularly as regard the terms short term and long-term memory. As presented in table 30-1, there are essentially three major stages.

Immediate or short term working memory– This stage has been discussed in chapter 18. This is a matter of seconds (estimated as <10 seconds. Total capacity has been estimated as less than 12 items. It is best exemplified in simple digit repetition as in digit span testing or the immediate repetition of three or 5 objects or the **delayed response test** utilized in monkeys and children. Monkeys with prefrontal lesions and infants less then 8 months (presumably with immaturity of the frontal function) are unable to consistently perform the task. The neural substrate undoubtedly involves the reception in the appropriate primary sensory projection area with relay to the adjacent sensory association cortex; for example, areas 18, 22, 5/7 then transfer to the multimodal posterior parietal cortex, then relay to prefrontal area for very short term storage and then relay to motor association cortex and then to motor cortex if the information is to be recited to the examiner (as in the digit span or immediate recall tests).

Long-term memory. Labile stage - This stage may be considered a transcription and transduction stage. The duration of this intermediate process has been variously estimated in infrahuman species as a matter of 20 to 180 minutes. The time required is probably shorter as one descends the phylogenetic scale. Theoretically, this stage involves the transcription from a relatively localized reverberating neuronal circuit, indicated in the immediate memory stage, into a more permanent macromolecular form of recording. It has been hypothesized that RNA and protein synthesis

TABLE 30-1 TYPES OF LEARNING AND NEURAL SUBSTRATE

TYPE	ANATOMICAL SUBSTRATE
I Declarative: conscious, facts or events	
-Stage 1: short term/ working/immediate memory	Prefrontal
-Stage2: long term memory: labile stage (Transcription/ transduction and consolidation)	Limbic system: hippocampus-thalamus
-Stage 3: long term memory: stage of remote memory	Diffuse cerebral cortex
II Non Declarative: non conscious, reflexive	
- Motor habit and skill learning	–Cerebellum, striatum and motor cortex.
-Classical and operant conditioning	-Amygdala for emotional responses, -Cerebellum for motor responses
-Non associative learning: Sensitization*/ habituation)**	Reflex mechanisms of spinal cord and brain stem
-Priming: recall of words/ objects improved by prior exposure to these stimuli	Neocortex

*Sensitization=increased response to non noxious stimuli after presentation of noxious stimulus

**Habituation=decreased response to a benign stimulus when this is presented repeatedly

are involved in this stage. It should be noted that RNA turnover does increase in tissues undergoing learning-like experiences. RNA turnover also increases in areas of neural tissue subjected to repetitive stimulation. It is clear that interference with RNA synthesis or with protein synthesis (by the use of the antibiotic puromycin) interferes with this stage of memory. Presumably, as we will discuss below, an accompanying change in synaptic connection is beginning to occur.

The neural structures involved are the entorhinal cortex, parahippocampal gyri, the hippocampus, the fornix, and the anterior and dorsal medial nuclei of the thalamus. (see

Zola-Morgan and Squires, 1990, and below). These structures are inter-related as the "limbic system" previously discussed. From a clinical standpoint, we evaluate this stage of memory with delayed recall test (list of 5 objects to be remembered for 5 minutes)[1] and by asking the patient to recite a paragraph ("cowboy"or "gilded boy story") that he has just read or heard. This ability to learn and to retain new experiences and new material is often referred to as retentive memory. Disturbance of this stage in memory will be considered later under the topic of the amnestic-confabulatory syndrome.

Long-term Memory: stage of Remote Memory - This phrase is not discretely localized. Rather, it represents a diffuse storage throughout the cerebral cortex and possibly other areas of the central nervous system. Long-term memory would appear to relate more to the actual volume of cortex remaining intact than to any specific localized process (consistent with the concept of Lashley). Although not discretely localized, it must be recalled that the read out mechanism for such remote memories may be triggered by stimulation of the lateral aspect of temporal lobe in the particular abnormal situation of patients who are subject to complex partial seizures of temporal lobe origin (refer to Chapter 22). In such patients, ablation of the area that on stimulation produces the remote memory does not abolish this remote memory. From a clinical standpoint, we evaluate this stage of memory asking the patient to indicate his date of birth, date of marriage, dates of World War I and II, and so forth.

FACTORS INFLUENCING LEARNING AND MEMORY

In the consideration of the learning process, it is important to realize the influence of several additional factors.

(1) **Motivational drives** alter the rate of learning. There are primary or instinctive drives such as hunger thirst, sex present in all members of a species, irrespective of cultural influences and in general present in the infant. There are learned, secondary drives or motivational forces such as achievement, anxiety, and dependency. These are learned drives in the sense that motivations, such as desire for acquiring money and fame, or particular fears or guilt are not present in the infant and are not present in all members of a species. During the process of development, previously neutral cues may become attached to primary drives and take on the capacity to motivate performance. Some drives such as severe anxiety may trigger responses that are not goal-directed, thereby interfering with learning. In contrast, lesser degrees of anxiety may motivate performance.

2. Motivation also affects the central process of interpretation of stimuli (perception). In a sense, what we see depends upon what we wish to see and upon our frame of reference particularly where the stimulus is ambiguous.

3. Whether stimuli are perceived will also depend on the general state of attention or alertness of the individual mediated by the reticular formation. In addition prefrontal and parietal cortex influence attention.

4. At times retention may be intact but the search and read out mechanism may be temporarily defective. This may account for the phenomena of shrinking retrograde amnesia following head trauma.

NEUROBIOLOGICAL MECHANISMS IN MEMORY AND LEARNING -

The psychologist, Hebb, in 1949 had proposed that when in associative learning the axon of neuron A - fires synapses on neuron B in a persistent manner, a growth process or metabolic change occurs in one or both neurons such that the probability or efficacy of neuron A firing neuron B is significantly increased. Subsequent studies (reviews of Kandel & Hawkins, 1992, Kandel & O'Dell, 1992) have confirmed these observations and demonstrated that the synaptic connection between neuron A and B could be strengthened by a third (modulatory neuron) – acting

---

[1] *Other tests utilize 3 objects for 3 minutes.*

to enhance - release of transmitter from the terminals of the presynaptic neuron. If neuron A and the modulatory neuron -discharged simultaneously, then the connection between neuron A and B would also be strengthened in an associative manner in a variety of classical conditioning and in the hippocampus in relationship to spatial learning. Subsequent studies indicated that eventually the number of presynaptic terminals was increased providing an anatomical basis for this type of learning.

Long-term potentiation: Bliss and Lomo (1973) demonstrated that brief, high frequency trains of action potentials within each of the three major pathways of the hippocampus would produce a long lasting increase in synaptic strength in that pathway lasting for hours in the anesthetized animal but for days and weeks in the alert, freely moving animal. The CA3 to CA1 Shaffer collateral pathway and the perforant-afferent pathway to the dentate gyrus have associative learning properties and the model of Hebb is followed. The pre and postsynaptic neurons must fire simultaneously. The transmitter involved is glutamate. Both NMDA and non-NMDA receptors are involved. Similar long-term potentiation has been subsequently demonstrated in other areas of cerebral cortex. The basic point to be made is that long-term potentiation provides a possible mechanism for the labile period of memory - the transition period which maintains memories for hours, days and weeks, while more permanent transduction is occurring in terms of protein, RNA, etc.

PLASTICITY IN THE NERVOUS SYSTEM

(1) **Mirror focus:** Repeated experimental focal seizure discharge with repeated transcallosal response will produce changes in the excitability of the contralateral hemisphere so that eventually an independent focal disharges develops.

(2) **Kindling:** Repeated subthreshold stimulation of the amygdala or hippocampus, or of the frontal cortex in many species will over time alter the excitability of widespread cortical area so that eventually spontaneous seizures occur. Both the mirror focus and kindling have been proposed as possible demonstration models of long-term synaptic changes that may be involved in learning. That is, the alteration of neuronal function as a result of experience.

(3) Modification of cortical motor and sensory maps. (See chapter 17). These maps are not fixed but instead have a plasticity, which can be modified by experience.

(4) **The corpus callosum also has a role in learning.** Thus, a monkey or human trained to carry out a task with one hand, e.g., how to follow a maze or somatosensory discrimination, for roughness, etc., does not have to relearn that task when the other hand is utilized a rapid transfer of the learned skill or discrimination has occurred.

(5) **Early Experience And Postnatal Brain Development** (Refer to Spitz, 1945, Harlow, Lewis et al 1990, Siegel et al, 1993, Martine al 1991, Sirevaag& Greenough, 1985, 1987, Hubel &Wiesel, 1963). A number of studies have demonstrated a significant effect of early environment of the infant animal or human at critical periods on later brain development and function. These observations may be related to the phenomena of imprinting. Beyond a certain critical age of 15 months in the human infant, and 6-12 months in the infant monkey these changes related to environmental deprivation are irreversible. Cultural deprivation at a somewhat later stage of childhood may also produce long-term effects on psychological development. These studies also demonstrate correlated changes in the microscopic structure and biochemistry of the cortex, or basal ganglia or hippocampus. These considerations provide a biological foundation for the use of "head start" programs in preschool and primary grade students from culturally deprived backgrounds. (Refer to Hellmuth, 1968.)

DISORDERS OF RECENT MEMORY: THE AMNESTIC-CONFABULATORY SYNDROME: WERNICKE-KORSAKOFF ENCEPHALOPATHY:

Wernicke, in 1881, described a syndrome that is of relatively acute onset, occurring in alcoholics or nutritionally deficient patients and consisting of a triad: mental disturbance (confusion and drowsiness), paralysis of eye movements, and an ataxia of gait. The basic cause of the syndrome is a dietary deficiency of thiamine. As a deficiency of thiamine and of the other B-complex vitamins also results in a peripheral neuropathy, symptoms relevant to this degeneration of peripheral nerves often will be present as an associated finding. The basic pathological process consists of a necrosis of neural parenchyma and a prominence of blood vessels due to a proliferation of adventitial and endothelial cells. Petechial hemorrhages also occur about these vessels. The pathological findings involve the gray matter surrounding the third ventricle, aqueduct and the fourth ventricle. Lesions in general are usually most prominent in the mammillary bodies and the medial thalamic areas. Table 30-2 correlates the symptoms/signs with lesion location

The Memory Disturbance: As drowsiness clears it will often be noted that a severe deficit in recent memory is present, particularly if the patient has delayed seeking medical diagnosis and treatment. The patient will be unable to learn new material. He will have little memory of events surrounding his illness but will have little difficulty recalling events in the distant past. The patient may demonstrate confabulation supplying imaginary answers for questions concerning the recent past and for questions involving new material he has been requested to learn by the examiner. A significant disorientation for time is usually present. *The memory disturbance involves both retrograde amnesia (events prior to the illness) and anterograde amnesia (events since the onset of the illness).* The deficit in memory for recent events may be transient, clearing with continued treatment, or may be more persistent. The persistence of these disorders of recent memory and of the state of confusion as a chronic phenomenon is referred to as Korsakoff's psychosis. In the studies of Victor et al, pathological examination of the brain in these latter cases revealed persistent lesions in the dorsal medial and anterior nuclei of the thalamus in contrast to cases of Wernicke's encephalopathy without the persistent memory deficits. The following case history illustrates the problem of Wernicke's encephalopathy.

Case 30-1. Patient of Doctor John Sullivan and Doctor John Hills): This 62-year-old, white, right-handed stonecutter had been a known heavy alcoholic 6-8 week spree drinker for many years. Two years previously, the patient had been admitted to the Boston City Hospital because of delirium tremens (tremor and visual hallucinations). Two months prior to admission shortly following the death of a brother-in-law, the patient began his most recent drinking spree. Apparently, he had drifted aimlessly for 5 weeks with no definite food intake for a month. He unaccountably found himself in Florida, not knowing where he was and why he was there. The patient was brought back by his family and hospitalized at his local commu-

TABLE 30-2: WERNICKE'ENCEPHALOPATHY

SYMPTOM/SIGN	LESION LOCATION	EFFECT OF THIAMINE
Ophthalmoplegia (bilateral)	Predominantly CN VI, less of IIII	Rapid, reversal hours to days
Nystagmus	Vestibular nuclei	Clears over days
Drowsiness	Periaqueductal grey/ Periventricular diencephalon	Clears over days
Ataxia: gait &heel to shin	Anterior superior vermis cerebellum	If severe may persist as alcoholic cerebellar degeneration
Confusion/memory deficit	Dorsal medial/anterior thalamic nuclei	If persists: Korsakoff psychosis

nity hospital with diplopia, ataxia, marked impairment of memory and complaints of numbness of his fingertips and unsteadiness of gait. After beginning treatment, he was transferred to a neurological center

General physical examination: enlarged liver with the edge palpated approximately 2-1/2 finger breadths below the costal margin.

Neurological examination: *Mental status:* The patient was markedly disoriented for time and place Confabulation was also evident when it was suggested that he had recently seen various fictitious persons or was in particular locations. The patient was unable to state his age but could provide his birth date. He was confused with little insight as to his disorientation, or his condition. At times, the patient often indicated to visitors that his mother and father were still alive, though both parents had been dead for over 20 years. The patient's digit span was normal at 7 forward and 6 in reverse. The patient could name various objects correctly when these were presented to him and yet he was unable to retain any memory of which objects had been presented to him five minutes previously. He was unable to retain any information concerning a story that he had been requested to learn. Calculations reading and writing were intact. *Cranial Nerves:* There was horizontal diplopia on right lateral gaze. A minor weakness of the right lateral rectus was suspected[3]. Horizontal nystagmus was present on lateral gaze, bilaterally and vertical nystagmus on vertical gaze. *Motor System:* a minor degree of weakness was present in the distal portions of the lower extremities. There was no longer an ataxia of gait but a positive Romberg test was present. *Reflexes:* Deep tendon reflexes were absent at patellar and Achilles even with reinforcement. *Sensory System:* Pain and touch were decreased in the lower extremities below the mid calf. Vibratory sensation was absent at the toes and decreased over the tibia to a marked degree and to a lesser degree over the knees and at fingertips and wrists. Position sense was decreased at fingers and toes.

Clinical diagnosis:
1) Wernicke' encephalopathy
2) nutritional poly neuropathy

Hospital course: The patient was treated with thiamine. There was a significant improvement in extraocular functions. The patient had no diplopia after the day of admission. There was no significant change in his mental condition or peripheral neuropathy. Evaluation 3 months later indicated persistent disorientation for time and place and severe selective deficits in memory (delayed recall was still grossly defective) suggesting a residual Korsakoff psychosis.

In many cases, the state of confusion in Wernicke's encephalopathy is preceded by or accompanied by a period of delirium tremens indicating alcohol withdrawal. The use of intravenous glucose feedings without supplemental intravenous thiamine during such a withdrawal state may actually increase the requirements for thiamine, thus exacerbating the thiamine deficiency state. For this reason, all patients under treatment for alcohol withdrawal (or admitted to the hospital with a recent past history of alcoholism) should be treated with high dosage B vitamin therapy as well, on the presumption that they are candidates for nutritional deficiency.

In general, as we have indicated, the majority of patients with Wernicke's encephalopathy progress to a more persistent memory disturbance. Victor and Adams reported that 75 percent of their 86 cases progressed to a permanent amnestic confabulatory syndrome. The Korsakoff syndrome with its particular deficits in memory may also occur in other disease states involving the diencephalon:

(1) Tumors involving the posterior but not the anterior hypothalamus affect recent memory.

(2) Infarcts of the medial or anterior but not the posterior thalamic areas may also produce defects in recent memory. Refer to case 30-7 and Fig. 30-10 below. In the study of von Cramon et al (1985) a combined lesion of both the mamillothalamic tract and the ventral portion of the lamina medullaris interna were

most effective in producing amnesia. (See also Graff-Radford, et al, 1990 and Tatemichi, et al, 1992)

(3) Lesions of the fornix may also be associated with significant problems in memory recording. Such damage is likely to occur with colloid cysts or with the surgical procedure necessary to remove this potentially life-threatening nonmalignant tumor (see–Fig 27-11). The effects are most prominent if bilateral but may also occur with unilateral left-sided damage. Recall that the fornix is the major outflow pathway from the hippocampus. Defects in recent memory have also been reported in tumors involving the posterior or anterior portion of the corpus callosum.

(4) Lesions of the basal forebrain as in anterior communicating artery rupture. (Irle et al, 1992 Morris et al 1992) may produce persistent anterograde and retrograde amnesia.

BILATERAL LESIONS OF HIPPOCAMPUS (BILATERAL MESIAL TEMPORAL LOBE LESIONS):

The anatomy of the hippocampus and its connections has been considered in detail in chapter 22, and should be reviewed at this time.

Bilateral Ablation: Scoville and Milner (1957) were the first to observe the effects of bilateral mesial temporal lesions following bilateral ablation of the anterior two-thirds of the hippocampus and the parahippocampal gyrus with removal of the uncus and amygdala (Corkin et al, 1997). This surgical procedure had been performed for treatment of temporal lobe epilepsy in a patient who had bilateral temporal lobe epileptic spike foci. Following surgery, a gross loss of the ability to retain current experiences and to learn new material was noted, in addition to a significant retrograde amnesia. Remote events were well recalled. It should be noted that a similar defect might follow unilateral ablation of the temporal lobe structures if disease were present in the contralateral temporal lobe (Penfield and Milner (1958). These observations stimulated considerable clinical and experimental investigation. Extensive studies in the monkey by Squire and Zola-Morgan (1991) have now clarified the relative roles of these structures.

Essentially, monkeys with selective bilateral lesions of the hippocampus are impaired in memory tasks that involve the ability to acquire new information (declarative memory) but are not impaired in their capacity for skill and habit learning (that is, non-declarative memory). The type of task employed by Squire and Zola-Morgan is the delayed non-matching to sample: A single object is presented; after a delay, two objects are presented - the original object and a novel object. The animal is rewarded for choosing the novel object and rapidly learns to select the novel object. Monkeys with selective bilateral lesions of hippocampus, parahippocampal and entorhinal cortex were even more impaired than monkeys with a selective hippocampal lesion. The entorhinal/parahippocampal areas receive input from all of the higher sensory association neocortical areas and then project via the perforant and alvear pathways to the hippocampal formation. There are however other inputs to the hippocampus. Monkeys with bilateral lesions of the perirhinal cortex and parahippocampal cortex sparing the hippocampus also demonstrated severe impairment on these same memory tasks. Combined bilateral lesions of hippocampus, parahippocampal gyrus and perirhinal cortex produce greater impairment of memory than the selective lesions of hippocampus and parahippocampal gyrus. Monkeys with selective bilateral lesions of entorhinal cortex demonstrated impairment but recovered when retested at 9-14 months (Leonard et al, 1995). All of these monkeys had no impairment of remote memory. Monkeys with lesions of the amygdala demonstrated a marked alteration in emotional behavior but had no impairment on memory tests (see chapter 22 and Zola-Morgan et al, 1991).

While most patients with bilateral medial temporal lesions have both retrograde and anterior grade amnesia, several recent studies have allowed a differentiation of these two aspects of amnesia. Zola-Morgan et al, 1986, reported that a patient with selective bilateral lesions of the entire rostral caudal extent of field CA1 of the hippocampus (due to global ischemia) had enduring anterograde amnesia but minimal retrograde amnesia. Rempel-Clower et al, 1996, confirmed this in an additional group of three patients. On the other hand, Kapur et al (1992) described a patient with a closed head injury who had severe post-traumatic retrograde amnesia (extending back to childhood) but only mild patchy anterograde amnesia. On MRI scan, there was bilateral damage to anterior temporal cortex and to a lesser degree prefrontal areas. There was no damage to hippocampus, thalamus or other limbic structures of the diencephalon.

Disease involving the medial temporal areas may occur under a variety of circumstances and may result in an inability to retain current experiences and to learn new material. Material learned and already stored in long term remote memory will be intact. In the clinical discussion that will follow, we will often be considering the medial temporal areas in a more general sense. As already noted early cases of destruction of medial temporal areas were nonselective. Damage to hippocampus also included damage to amygdala and the adjacent cortex.

Bilateral Infarction - Bilateral infarction of the hippocampal areas may occur with disease of the posterior cerebral artery (Victor et al, 1961 and De Jong et al 1969). Occlusions of the posterior cerebral arteries usually reflect embolic events with the embolus originating in the heart or at a lower level in the vertebral basilar circulation. The ischemia or infarction of the mesial temporal areas may be accompanied by ischemia or infarction of the territories of the calcarine arteries. In other cases stenosis of the basilar artery may produce decreased blood flow in both posterior cerebral arteries. In still other instances, the ischemia apparently may be limited to the medial temporal area, that is, hippocampal areas, with little involvement of the occipital cortex. In these cases, transient episodes may occur and may be difficult to distinguish from the syndrome of transient global amnesia.

We should note at this point, that the posterior cerebral arteries via their penetrating branches also supply the medial diencephalic areas. Such ischemia in the posterior cerebral circulation, then, might well produce ischemia of the dorsal median nucleus, the anterior thalamic nuclei, and the mammillary bodies. Thus, in the case of persistent memory deficit following posterior cerebral artery lesions reported by Victor et al (1961), the areas of infarction involved not only the hippocampal formation and fornix but also the mammillary bodies. An example of bilateral medial thalamic lesions producing alterations in recent memory has been provided in Chapter 29.

Unilateral Infarcts: The study of Ott and Savez (1992) demonstrated the occurrence of the amnestic syndrome following infarcts within the territory of the posterior cerebral artery involving the hippocampus or the thalamus. In 85% of unilateral infarcts, the lesion was left sided.

Transient Global Amnesia Syndrome: This syndrome involves the sudden onset of a defect in memory. The memory deficit is similar to that which occurs in the amnestic confabulatory syndrome seen in Wernicke's encephalopathy. There is a significant retrograde amnesia for the events of the preceding days, weeks, and months. This retrograde amnesia slowly clears over a period of hours. The patient has a persistent amnesia for the period between the onset of the attack and the point of complete recovery. During the period of the episode, the patient is unable to learn new material, that is, the patient is unable to register new memories. Apart from this defect, the patient usually shows no other abnormality of behavior during the episodes. Forsometime, a possible vascular etiology involving the posterior cerebral arteries (or basilar vertebral system) has been postulated. Whether the selective transient ischemia involves the hippocampal formation, rather than the diencephalic areas remains unclear. As discussed, the hippocampus is more likely to manifest a selective vulnerability to anoxia and would be expected to show functional impairment prior to the medial thalamic areas. Mathew and Meyer (1974) demonstrated that elderly patients with transient global amnesia had one or more risk factors for vascular disease and 4-vessel angiography demonstrated significant lesions in the vertebral basilar and posterior cerebral systems. Patients with single episodes had no permanent impairment of memory. Those with recurrent episodes had some degree of permanent memory impairment, as well as mild visual spatial or visual motor dyspraxia. Several recent studies confirm the previously postulated ischemic etiology but leave undecided the location of the ischia: thalamus versus hippocampus (Goldenberg et al, 1991 Stillhard, et al, 1990 Lin et al, 1993). The majority of patients makes an apparent total recovery and do not have significant infarcts (see Melo, 1992). The following case history illustrates this syndrome. A transient disease process involving the temporal areas is suggested but not proven.

Case 30-2: This 55-year-old, right-handed, white male, college professor, awoke at 3:00 a.m. on the morning of admission in an uneasy and restless state. He was confused as to time, kept repeating himself and asking the same questions. His wife arranged for him to be seen early in the morning by his family physician who lived a short distance away. At approximately 8:00 he became lost driving a familiar route to his doctor's office and when he arrived with the assistance of his wife, he was unable to explain why he had come. He could remember no significant events from the 3-week period prior to the onset of his illness. The patient's more remote recall and other intellectual capacities remained intact.

Neurological examination: When examined in the early afternoon, findings were essentially limited to the mental status examination: the patient was beginning to regain some of his ability to retain new information he was still disoriented for the day and month. Store of information was quite intact.

The patient had marked difficulty with delayed recall. He could recall none of the 4 objects after 5 minutes. He could not remember any of the three test phrases given to him when asked about these 5 minutes later. He did recall in a vague manner that a memory test had been given him. was unable to remember his visit to his family physician earlier in the day. Digit retention, however, was relatively well -preserved; 6 forward and 5 backward. The patient had no defects in calculation. There was no evidence of a constructional apraxia. Language function was entirely intact.

Clinical diagnosis: Probable transient global amnesia

Laboratory data: EEG and CSF were all within normal limits.

The brain scan demonstrated a small area of increased uptake of radioisotope (Hg197) in the left temporal region.

Hospital course: Over the several hours following admission, the patient gradually regained his ability to retain new information and to recall the events of the preceding 3 weeks, the more remote events being recalled first. The memory for the events of the day prior to admission was regained last. This pattern is referred to as a shrinking retrograde amnesia. The specific events that occurred on the morning of admission were never recalled. The following morning delayed recall was four-out-of-four objects after 8 minutes. The patient's mental status and neurological examination were otherwise within normal limits. All brain scan findings had resolved at 5 months. No additional episodes occurred during the one-year after the acute episode.

Stimulation Of The Temporal Lobe:

Bilateral stimulation of the temporal lobe or unilateral stimulation with secondary spread to the contralateral hemisphere may occur during limited electric stimulation of the medial temporal structures during epilepsy surgery or unilateral electroshock therapy, during general stimulation of the brain (electrical shock therapy or a spontaneous generalized convulsive seizure), or during a temporal lobe seizure. Thus, it may be demonstrated that electrical stimulation of the depths of the temporal lobe in patients susceptible to temporal lobe seizures can produce a defect in memory for recent events without disturbing the patient's ability to recall remote memories. In general, the longer the stimulation, the longer the duration of retrograde amnesia and the longer the recovery time before the new memories can be recorded. Table 30-3 summarizes the example cited by Doty, (1967)

A primary or secondarily generalized convulsive seizure is followed by a period of confusion during which the patient is confused, unable to record new information and uncertain of memories prior to the seizure. In general, on recovery from this period of confusion, the patient is amnestic for the entire period from the beginning of the seizure until the clearing of the postictal confusion. The patient usually, however, has no impairment of retrograde memories up to the point of the seizure. Thus any "aura" may be recalled. *The effects produced by electrical stimulation of the hippocampus may also occur as relatively selective ictal and postictal phenomena in a complex partial seizure originating in mesial temporal lobe structures* as illustrated in the **case history 30-3 presented on the CD ROM**. Focal temporal lobe seizure phenomena were followed by a period of impaired recent memory with an inability to record new memories.

Selective Vulnerability of the Hippocampus to Anoxia Or Hypoglycemia or Status epilepticus: As discussed in chapter 22, **transient** or permanent defects in memory may occur. The effects of repeated seizures on the hippocampus may involve more than simply the effects of hypoxia. Excessive discharge of neurons may also damage the neurons due to excitotoxic effects of the glutamate transmitter.

Toxic effects on the hippocampus: A recent accidental intoxication of nature provides additional information about selective damage to the hippocampus resulting in an amnestic syndrome. In 1987, an outbreak characterized by gastrointestinal symptoms and neurologic symptoms occurred in Canada in 107 persons who had eaten cultivated mussels from Prince Edward Island. The acute phase symptoms consisted of headaches, seizures and hemiparesis. Twenty-five percent of patients had persistent severe anterograde amnesia. PET scanning demonstrated decreased glucose metabolism in the medial temporal areas. In the four patients who died, necrosis and loss of neurons occurred predominantly in the hippocampus and amygdala, (Teitelbaum, et al, 1990). Two patients also had involvement of the dorsal medial thalamic nucleus. Not all patients with the persis-

TABLE 30-3: STIMULATION OF "DEEP" TEMPORAL STRUCTURES (AFTER DATA OF DOTY, 1967)

Duration of Stimulation	Length of Retrograde Amnesia	Length of Anterograde Amnesia
2 seconds	Just prior to stimulation	1-2 minutes
5 seconds	Current day	5-10 minutes
10 seconds	Previous 3 weeks	1-3 hours

tent memory deficits had uncontrolled status epilepticus - which may also produce hippocampal damage. The mussels were found to be contaminated by domoic acid produced by a form of marine vegetation; Nitzchia pungens. Domoic acid has structural similarities to the natural excitatory transmitter glutamic acid and to kainic acid. Both kainic acid and domoic acid bind strongly to the glutamate receptor. Domoic acid is 30 - 100 X more potent than glutamate and 2 - 3 X more potent than kainic acid. Experimental administration of domoic acid or of kainic acid to rats produces limbic seizures, memory disorders and degeneration of the hippocampus.

Herpes Simplex Encephalitis: As discussed in chapter 27, the virus; herpes simplex type 1, the agent responsible for the common cold sore, may on occasion invade the central nervous system *(Fig 30-1)* and produce several syndromes In addition to the more common acute generalized or focal encephalitis a third rare syndrome results from a more localized involvement of the medial temporal structures and is characterized by a subacute but progressive dementing process and temporal lobe seizures.

The disturbance of mental status is characterized by confusion and disorientation with a defect in memory for recent events and often for remote events as well. There is a marked inability to form new associations. The selectivity of the memory disturbance is less evident in these cases than in those cases of hippocampal disease previously discussed in detail.

TRAUMATIC AMNESIA:

Perhaps the most common transient impairment of memory occurs in relation to

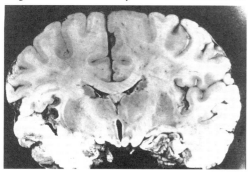

Figure 30-1. Herpes simplex encephalitis. Predominant involvement of medial and inferior temporal structures (and of insula) by an acute necrotic process is evident, although a generalized encephalitis was also present. Same case as Figure 27-26. (Courtesy of Dr. John Hills)

head injury. When blunt trauma to the head occurs, consciousness is lost. As the individual recovers consciousness, there is a period of confusion before a return to normal behavior. The loss of consciousness and the subsequent period of confusion in such patients are often referred to as concussion. Following this period of confusion, there is often a period during which behavior is otherwise normal but the ability to form new memories is defective. Examination of the patient at that time will indicate that not only is there no memory of the actual period of unconscious-ness and the following confusional state but also that there is defective memory for a period of time preceding the period of injury. The patient has then both an anterograde and retro-grade amnesia. In general, the longer the period of retrograde amnesia the more severe the head injury and in general, the longer the period of time before memory will be regained. As recovery occurs, the period of retrograde amnesia is gradually reduced. The underlying pathophysiology has been discussed previously in the section on trauma.

PROGRESSIVE DEMENTING PROCESSES

Introduction

As already defined dementia refers to a progressive impairment of previously intact mental faculties. In general, in the most common type, Alzheimer's disease, the loss involves initially and most severely recent memory and the ability to learn and retain new memories. To some extent, particularly as time passes, other areas of mental capability are also affected: remote memory, abstract reasoning, insight, and arithmetic abilities, language function, personality mood and social behavior. In other types, such as the "frontal temporal dementias", personality and behavior are involved earlier than memory .In still other types, Lewy body dementia, motor function is early involved followed by memory. In general, these processes involve the older adult population. However, certain rare disorders producing a progressive dementia affect infants, children and adolescents and the young or middle-aged adult. Some of these pediatric disorders reflect known or suspected inborn and genetic errors of metabolism: aminoaciduria, lipidosis, galactosemia, Hurler's disease (accumulation of mucopolysaccharides), and leukodystrophies. Other disorders are placed in the degenerative category because in general the etiology remains unknown; e.g., spongy degeneration. A detailed consideration of these various early onset problems is beyond the scope of this text. Dementia is a major medical problem because of its high frequency in the elderly and because of the significant aging of the population. Life span has been significantly extended but the capacity to prevent the dementing disorders is limited. The prevalence of dementia and of Alzheimer's disease in the general population is presented in Table 30-4.

The frequency of various types of dementia is indicated in Table 30-5.

At one point the dementias were classified as cortical (Alzheimer's was cited as the most common example), or subcortical (Parkinson's, Huntington's, PSP, were cited as subcortical examples). However, degeneration of the subcortical basal forebrain nucleus of Meynert is prominent in Alzheimer's disease and cortical pathology or indirect cortical effects can be demonstrated in most of the so-called subcortical models. Although rare thalamic dementia may occur. An example is provided in case history 30-7 below.

ALZHEIMER'S DISEASE:

The most common cause of a progressive impairment of mental faculties in the older adult population is the degenerative disease known as presenile or senile dementia (Alzheimer's disease). Whether the process is called presenile or senile is arbitrary, based on the age of the patient. When the process begins before the age of 65 years, the designation presenile is used; when the process begins after the age of 65 years, the designation senile is employed. The basic pathological process (Perl, 2000) is, however, the same. Grossly *(Fig. 30-2)*, there is an atrophy of cerebral cortex, involving primarily in a diffuse manner the frontal and temporal areas but sparing the motor cortex, sensory and visual areas. In some cases, (more often presenile) involvement of the parietal association areas is evident as well. There is a thinning of gyri and widening of sulci. There is usually a secondary dilatation of the lateral ventricles. These findings in patients with senile dementia overlap with the patients in the same age group without dementia. In presenile patients, there is less overlap with non-demented patients and the atrophy and the microscopic changes discussed below are also more widespread. There is no one to one relationship of the degree of neocortical atrophy with the severity of the dementia. These overall findings are reflected in the neuroimaging studies. Thus, patients with a significant degree of dementia may manifest only a minimal degree of neocortical atrophy as noted on the CT or MRI scan. On the other hand, it is not unusual to find elderly patients with a significant degree of neocortical atrophy on such radiological studies that demonstrate relatively little actual cognitive impairment. There is, however a significant correlation in Alzheimer's disease between the degree of atrophy of the hippocampus and the presence of dementia. This hippocampal atrophy is evident on the CT scan but is best seen in MRI studies that utilize measurements of hippocampal volume *(Fig.30-3)*. **It is evident then that the earliest changes occur in the hippocampus and entorhinal cortex.**

The microscopic changes probably have a higher correlation with Alzheimer's disease than the gross changes in neocortex. These microscopic changes may be outlined as follows:

TABLE 30-4: PREVALENCE OF DEMENTIA AND ALZHEIMER'S DISEASE

AGE RANGE YEARS	60-64	65-69	70-74	75-79	80-84	85-93 ****
Overall all Types/100**	0.4	0.9	1.8	3.6	10.5	23.8
Alzheimer's Disease/100***	0.3*		3.2*		10.8*	

* For Alzheimer's disease, age ranges are 60-69,70-79,80-89

** Derived from Bachman et al, 1992 (Framingham).   *** Derived from Rocca et al (1991)

**** Skoog et al, 1993, in 494patients at age 85 (Sweden) overall prevalence 29.8%: Severe 11%, moderate, 10%, mild 8%.

TABLE 30-5: CAUSES OF DEMENTIA

NEUROPATHOLOGICAL & CLINICAL SYNDROME*	PERCENTAGE OF TOTAL
Alzheimer's disease	55%
Dementia with Lewy bodies	15-20%
Multi-infarct and other vascular dementia**	15%
Combined multi-infarct/vascular plus Alzheimer's disease	12%
Other degenerative causes***	<3-5%

* Derived from Tomlinson et al, 1970, McKeith & Burn, 2000 and other sources

** Vascular dementia includes a variety of syndromes: 1. A single infarct in areas critical for memory. 2) Infarcts (single or multiple) in which total volume of cortex destroyed is >100ml. 3) Small vessel disease producing multiple lacunar infarcts) senile leukoencephalopathy: involving periventricular white matter in association with hypertension (Roman et al, 1993)

*** Frontal-temporal dementia including Pick's disease, Huntington's disease, Progressive Supranuclear Palsy (PSP)

1. **Loss of neurons in the cerebral cortex:** The large pyramidal cells of the frontal, temporal and parietal association areas neocortex and particularly the hippocampal and related medial temporal areas. Some loss of neurons occurs in all normal aging individuals but the degree in Alzheimer's is markedly greater.

2. **Loss of neurons** in certain subcortical nuclei that project to cerebral cortex: The basal forebrain nucleus of Meynert (cholinergic), the locus ceruleus (nonadrenergic), and the amygdala.

3. **Loss of dendritic spines** and branches affects the pyramidal neurons of the temporal and frontal cortex and the limbic cortex. Normal aging individuals actually have an increase in dendritic trees until the 80s and 90s are attained.

4. **Neurofibrillary tangles,** develop within the cytoplasm of the surviving large pyramidal neurons of these neocortical areas and the hippocampus *(Fig. 30-4, 30-5)*. These tangles are composed of the spiral (helical) winding of paired protein filaments probably derived from the protein skeleton of the cytoplasm. They contain tau proteins, a normal constituent of microtubules. Normally these tau proteins are highly soluble .In Alzheimer's disease, the tau protein found in the paired helical filaments is hyper phosphorylated and highly insoluble. These insoluble tau aggregates in the tangles are usually complexed with another protein ubiquitin. The anterior frontal, temporal neocortex, and particularly the medial temporal area (hippocampus and entorhinal cortex) are primarily affected. The amygdala, basal forebrain, cholinergic nuclei, thalamus, substantia nigra and locus ceruleus are also affected. Betz cells of the motor cortex and the Purkinje cells of the cerebellum are resistant to the degeneration. Neurofibrillary tangles also occur in a number of other CNS diseases that have been grouped together as tauopathies. It is likely that the change in tau protein represents a response of neurons to a variety of pathological insults of a genetic and non-genetic nature. These include such entities as Down's syndrome, post encephalitic Parkinson's disease, amyotrophic lateral sclerosis/Parkinson-dementia complex observed primarily on Guam, progressive supranuclear palsy, corticobasal degeneration, and hereditary frontal temporal dementia (linked to chromosome 17), as well as dementia associated with repeated head trauma in boxers.

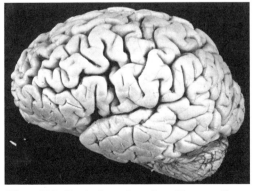

Figure 30-2. Cortical atrophy as found in presenile and senile dementia of Alzheimer type. There is widening of sulci and narrowing of the gyri, particularly in frontal areas.

Some neurofibrillary tangles will be found in the brain of intellectually normal individuals primarily in the anterior medial temporal areas,

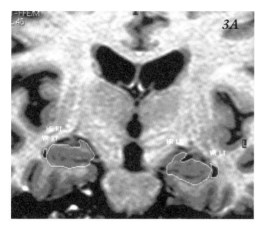

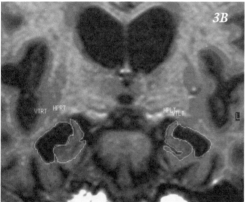

Figure 30-3. Atrophy of the hippocampus in Alzheimer's disease. MRI: area of hippocampus circled for quantitative analysis. A) Normal hippocampus in a 92-year-old patient with Parkinson's disease but normal cognitive function. B) Severe atrophy of hippocampus in a 72-year-old patient with well developed Alzheimer's disease. Neocortical atrophy is also present. (Courtesy of Dr. Daniel Sax).

but in comparatively small numbers. In the studies of Tomlinson, 5% of normal patients of 40 years had some minor involvement. By the seventh decade, 50% of patients had some involvement and all 90-year-old patients were affected to some degree. In Alzheimer's disease not only is the anteromedial temporal lobe affected to a much greater degree but also there is widespread involvement of the remainder of the hippocampus and marked involvement of the neocortex.

In some patients with clinical Alzheimer's disease, an alternate form of neuronal inclusion is found in the cortical pyramidal cells; the Lewy body composed of a-synuclein protein already discussed in relation to Parkinson's disease. These cases are labeled as the Lewy body variant of Alzheimer's disease.

What distinguishes the patient with Alzheimer's disease from the normal elderly patient and from the other tauopathies is the presence of the senile neuritic plaque to be discussed below.

5. **Dystrophic neurites** - These are altered neuronal processes axons, dendrites and/or synaptic terminals - found free in the neuropil as well as surrounding senile plaques *(Fig. 30-4)* The study of McKee et al, 1991, noted a significant increase in these dystrophic neurites in Alzheimer's disease and a high correlation of the number of both dystrophic neurites and neurofibrillary tangles with the severity of dementia.

6. **Extracellular Plaques (senile or neuritic plaques) containing insoluble fibrils of amyloid –beta proteins** *(Fig.30-4, 30-6, 30-7)*. In Alzheimer's disease fragments of abnor-

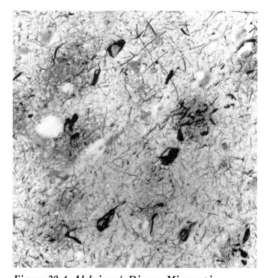

Figure 30-4. Alzheimer's Disease: Microscopic Features: This 87-year-old female had a 7 year history of a progressive dementia. This 100 x magnification section of the hippocampus CA-1 sector, stained with the Bielschowsky silver stain demonstrates the following features: (1) Loss of neurons, (2) argyrophilic neurofibrillary tangles in surviving pyramidal cells, (3) senile plaques containing fragments of silver staining neuronal and glial processes, (4) dystrophic neurites, silver staining processes surrounding the plaques (Courtesy Dr. Tom Smith).

mal appearing neuronal processes (axons and dendrites) surround the amyloid core of these plaques. There is also evidence of surrounding altered astrocytes and microglia. The plaques have a predilection for the frontal temporal and parietal association neocortex and particularly accumulate in the medial temporal areas. A "diffuse" deposit of amyloid in the neuropil, in a non-fibrillar form and without the altered glia and without the surrounding dystrophic neurites may occur in otherwise normal individuals. Similar amyloid may also accumulate in the walls of cerebral blood vessels.

The primary constituent of the amyloid core of the plaque is the B (beta) amyloid peptide of 28-43 amino acids. This peptide is generated by proteolytic cleavage of a larger transmembrane glycoprotein containing 695, 751 or 770 amino-acid residues: the amyloid precursor protein (APP). This cleavage is mediated by a series of proteolytic enzymes alpha, beta and gamma secretase. The B. amyloid in the plaque is the internal fragment of the amyloid precursor protein. Selkoe (2001) provides additional discussion of the mechanisms of cleavage.

A secondary area of controversy involves the relationship between the B.amyloid (and the senile plaque) and the neurofibrillary degenerative changes in the neurons and the loss of neurons. One hypothesis suggests that the local microglial activation, reactive astrocytosis and cytokine release surrounding the plaque may produce a cascade of effects that damage the neuron. An alternative hypothesis is based on the demonstration by Yankner and Mesulam, (1991) of a direct toxic effect when B.amyloid is added to tissue cultures of mature neurons. In contrast, a neurotrophic effect is evident when B.amyloid is added to a tissue culture of immature neurons. (For a review of the controversy surrounding these studies and the studies of the effects of direct injection of B.amyloid into the brain, see Marx, 1992). That there is some relationship of the senile plaques to the total process is evident, from several standpoints.

a. The plaques tend to be concentrated in those areas with severe neuronal loss and with a high density of neurofibrillary changes.

b. The relationship of Alzheimer's disease to Down's syndrome. Patients with Down's syndrome have trisomy 21. The gene for the amyloid precursor protein localizes to chromosome 21. Patients with Down's syndrome who have in a sense a triple dose of this chromosome develop a severe dementia once they reach age 40-50 (almost 100% after age 50). The dementia from a clinical and neuropathologic standpoint is identical to Alzheimer's disease. Moreover, patients with Down's syndrome dying in their teens, 20s or 30s before clinical dementia has developed will demonstrate a significant accumulation of plaques and changes surrounding the plaque prior to the development of neurofibrillary changes. Thus, the plaques may be the earliest change.

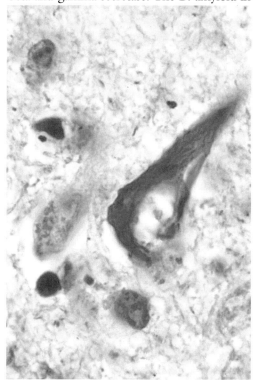

Figure 30-5. Neurfibrillary tangles in pyramidal neurons of the hippocampus. Bielschowsky silver stain. 100X magnification. (Courtesy of Dr. Thomas Smith)

c. In some families with hereditary

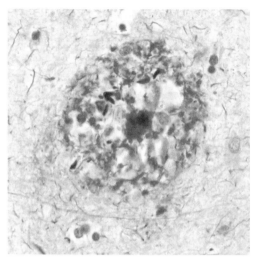

Figure 30-6. Senile plaque. Silver stain 100x magnification. Courtesy of Dr. Thomas Smith).

Alzheimer's disease of relatively early onset - a linkage to chromosome 21 could be demonstrated. In many of these families, missense mutations in the amyloid protein precursor have been demonstrated. These tend to cluster at the beta secretase cleavage site, others occur just after the gamma secretase cleavage site. All of these various mutations would result in enhanced production of increased amounts of beta amyloid. However, other familial cases have localized to chromosomes 14 or 1. In both of these latter sites, there has been localization to mutations in presenilin 1 or 2. These are membrane proteins that apparently regulate gamma secretase activity. Other late onset familial cases localize to chromosome 19. Other familial cases "The Volga Germans" localize to none of these chromosomes. The monograph of Pollen (1993) provides a review of the search for the genetic basis of the disease.

Late onset cases: All of these various mutations do not explain the majority of late onset non-familial Alzheimer's disease cases. Chromosome 19 however does provide a locus for the apolipoprotein E (ApoE) gene. Three variants (alleles) are found in the general population: E2, E3 and E 4. The E3 allele is the most common with a frequency of 0.73 .E4 has a frequency of 0.14. In families with late onset Alzheimer's disease, and individuals with late onset disease, the frequency of the E4 allele is increased to 0.40. Inheritance of one or two alleles of the E4 variant increases the density of senile plaques, increases the risk and lowers the age of onset of the late onset disease. Patients with two E4 alleles had an average age of onset 68 years; with one E 4 allele; 79 years and with no E4 allele, 84 years. By age 80, over 90% of patients with two E 4 alleles developed Alzheimer's disease, whereas only 50% of patients with no E 4 allele developed the disease by this age. ApoE 4 is a risk factor but not an invariant cause of late onset Alzheimer's disease. The E-2 variant may provide some protection from the disease. As regards the mechanism, this is not entirely clear. High concentrations of Apo E are found in liver and brain. with production in brain by glial cells. In the Alzheimer brain, ApoE has been localized to the senile plaque and to neurons containing neurofibrillary tangles. The Beta amyloid peptide binds more rapidly to

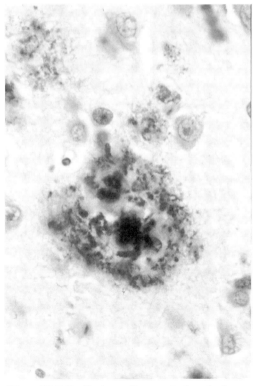

Figure 30-7. Senile plaque. Immunologic stain for beta amyloid. 100 x magnification. (Courtesy of Dr. Thomas Smith).

the E 4 form than to the E 3 form. In contrast the E 3 form binds to tau protein forming a stable molecular structure, whereas the E 4 form does not bind to the tau protein. Additional discussion of this topic will be found in Strittmatter et al (1993), Sanders et al (1993) and Corder et al (1993). Additional discussion of the molecular biology will be found in Selkoe (2001).

Clinical findings and course: The clinical symptomatology in Alzheimer's disease relates to a progressive impairment of mental faculties, usually beginning with recent (retentive) memory. Initially, the social graces, remote memory and rational reasoning capacities are well preserved. As the disease progresses, these aspects of mental function are also affected. In general, during the early stages, focal motor findings do not develop, deep tendon reflexes remain symmetrical and plantar responses are flexor. It is not unusual to have a significant degree of nominal aphasia present as the memory impairment becomes more severe. (The patient is unable to recall the names of objects.) Apraxias are also not unusual in these cases. As the disease progresses a release of a grasp reflex is also often noted. In late stages, frontal lobe gait apraxia becomes prominent and eventually, bilateral Babinski signs emerge. Eventually, the patient becomes bed ridden and unable to care for daily needs. A terminal state of paraplegia in flexion or a fetal position may be the final posture. Not all cases have the same initial appearance and not all have the same rate of progression (see Mayeau, 1985). In some cases, focal aphasia with apparent focal atrophy may be noted as an early presentation (Mesulam, 1982) but these cases are more likely to fall into the diagnostic category of frontal temporal dementia. In a considerable percentage of patients late in the disease course, myoclonus or generalized seizures or partial seizures may develop and account along with post infarct seizures for a significant increase in the frequency of seizure disorders in the elderly. Criteria for the clinical diagnosis of Alzheimer's disease have been outlined in the report of an N.I.H. work group (McKhann, et al, 1984). See also Markesbery (1992), Katzman (1993).

The following case history presents an example of presenile dementia.

Case 30-4: This 64 year old right handed white male formerly an administrative assistant for the veteran's administration and newspaper distributor was evaluated for progressive impairment of recent memory of 10 year's duration. At age 60, delayed recall was limited to 0/4 and there was minor time disorientation. During last year there were now personality changes and word finding difficulties. Family history was negative.

Neurological examination: *Mental status:* The mini mental status exam indicated a total score of 19 out of 30. He had particular problems in time orientation. He was able to do the immediate recitation of three objects and of a test phrase but could remember none of these on a delayed recall test. He also had difficulty copying a test figure. The patient was often tangential in his answers and demonstrated inappropriate joking. *Motor system:* premotor / frontal lobe functions were abnormal with a release of the instinctive grasp reflex, and impairment of motor sequences.

Clinical diagnosis: Alzheimer's disease

Laboratory data: All studies were normal except a CT scan demonstrated significant dilatation of the temporal horns suggesting hippocampal atrophy, in addition to a general increase in lateral ventricular size with blunting of the angles of the frontal horns. A SPECT scan demonstrated slight decrease in perfusion in the left parietal region.

Subsequent course: After 10 weeks of treatment with 5mg per day of donepezil (Aricept) a centrally acting acetylcholinesterase inhibitor and high dosage of vitamin E (a possible antioxidant), the test score had increased to 25 out of 30 with particular improvement in the delayed recall section of the exam. There was however no change in personality. Administration of the acetylcholinesterase inhibitor temporarily improved memory function for approximately 18 months. However by age 67, behavioral disturbance (agitation,

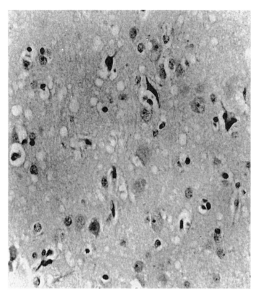

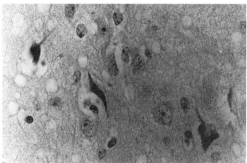

Figure 30-8 Creutzfeldt-Jakob Disease: Histological features: (A) low power H&E - approx. 40x (B) high power H&E - approx.150x. Diffuse micro vacuolization of the neuropil and neuronal cell bodies is present in this biopsy of frontal lobe. This 45-year--old female presented in November 1992 with a 3-4 week history of progressive ataxia and mental deterioration. EEG evolved into the characteristic periodic pattern. Patient expired after a total course of approximately 5 weeks. (Courtesy of Dr. Tom Mullins and Dr. Tom Smith)

aggression nocturnal wandering and sexual disinhibition) and urinary incontinence were becoming major problems. He could no longer be managed in his home and day care setting despite the use of haloperidol and he was placed in a nursing home At age 69, he was now described as relatively nonfluent, and very confused and restricted to a wheel chair. He was "stiff and afraid" when requested to walk.

The use of the donepezil has been estimated to postpone nursing home placement by approximately 15 months. The rational for the use of acetylcholinesterase inhibitors is based on the demonstration by Drachman and Leavitt (1974) that administration of scopolamine, an anticholinergic drug that crosses the blood brain barrier, would produce a transient syndrome similar in many respects to the memory and cognitive changes of Alzheimer's disease. Subsequent studies demonstrated a marked loss of neurons in the basal forebrain nucleus of Meynert, a major source of cholinergic fibers to the cerebral cortex. (Whitehouse et al 1981 Rogers et al, 1985 editorial of Growden 1992). However, a simplistic cholinergic hypothesis is insufficient to explain all of the features of Alzheimer's disease. Administration of acetylcholine esterase inhibitors in mild cases (mini mental status score of 10-25) has only a limited value as in case 30-4. Other neuronal systems are involved (noradrenergic synapses are also involved. The hippocampal system is severely involved).

The **Evaluation of the Patient with Alzheimer's disease**: The primary purpose of a complete evaluation in these cases is to rule out the treatable causes of dementia. In addition, several less frequent causes of dementia must be differentiated from Alzheimer's disease. The treatable causes of dementia are outlined in table 30-6. The usual screening tests are the following: TSH, B12, and folate, RPR or VDRL and CT scan. Other causes of progressive dementia are outlined in table 30-7.

Case 30-7: (Patient of Dr. Thomas Mullins). This 74-year-old right-handed widow presented to the emergency room at St Vincent hospital on 2-19-02 with an 8-10 week history of worsening memory problems. In December 2001, accompanied by her boyfriend she made her annual trip to Florida. Although she had been in her usual state of independent living prior to her departure, she was found to have serious problems with memory on arrival. She was unaware as to where she was. In a telephone conversation with her daughter, "she was talking non-

TABLE 30-6: TREATABLE AND RELATED CAUSES OF DEMENTIA

CATEGORY	SPECIFIC ETIOLOGY	TEST
Infections	-Tertiary syphilis (general paresis)* -AIDS dementia	-RPR, or specific FTA or CSF VDRL -If indicated HIV
Nutritional	B12, thiamin, niacin	B12 and folate levels
Intoxications	Various (Refer to chapte r27)	Toxicology screen +if suspected Pb,
Endocrine	Hypothyroid	TSH, then if +, T4, T3
Tumors	Frontal meningiomas, and gliomas**	CT or MRI scan
Post traumatic	Chronic subdural hematomas	CT scan
Hydrocephalus	NPH, etc, (ataxia>dementia)	CT scan, improve after CSF removal (30cc)
Vascular	Various types of vascular dementia	Multiple strokes+ multifocal signs, CT/MRI
Pseudodementia	Depression>dementia + variability	Trial of antidepressants
Other meningeal	Carcinomatosis of meninges	MRI, CSF cytology

* An example of general paresis case 30-5 is presented on the CD ROM   **See case 30-7 and *Fig 30-10* below.
Other progressive disorders producing dementia are summarized in Table30-7.

sense". During the next 2 months, memory and word finding continued to deteriorate. She was taken to the emergency room of the local hospital in Florida where a CT scan without contrast on 02/07/02 demonstrated mild cortical atrophy but no acute process. Her son drove to Florida and found that her memory was poor and her conversation did not make sense. "My husband is in the service: my mother died last year." However she could play cards with her grand daughter. Her gait was slow and stooped. She had urinary incontinence on her way to the bathroom.

She had no history of alcoholism and took various vitamins and herbs.

Family history: her brother had died of a brain tumor of unknown type.

Neurologic examination: The positive findings were the following: *Mental status:* The patient was awake and alert. Overall Mini-Mental Status Sore was 19/30. The major problems were in orientation and in delayed recall. She was able to indicate that she was in a hospital in Massachusetts and the date was February. She registered 3 objects but could not recall any in 3 minutes. She was however able to read, name objects, follow instructions, spell "world" backwards and copy a figure. She was unable to draw a clock with the hands set at 11:10. *Motor system:* Although strength and tone were normal, her gait was broad-based and shuffling. *Reflexes:* Patellar and Achilles reflexes were absent but plantar responses were flexor.

Clinical diagnosis: 1) Subacute dementia etiology uncertain possibly of Alzheimer's or Lewy body type. Various entities such as chronic subdurals, tumor etc to be ruled out 2) Peripheral neuropathy, most likely etiology at this age diabetes mellitus.

Laboratory data: *Basic CBC and chemistries:* Normal except for an elevated fasting glucose. *CT scan (Fig 30-10):* An enhancing 2.75 cm tumor with a necrotic center was present in the thalamus at the level of the upper third ventricle. There was bilateral involvement of the anterior and medial thalamus but with no involvement of the posterior thalamus or hypothalamus. The study suggested a lymphoma or glioblastoma.

Subsequent course: *Stereotaxic biopsy of the right thalamus:* Frozen section raised the question of a lymphoma or glioblastoma but special immunological stains were consistent with a glioblastoma. Necrosis and endothelial hyperplasia were present. The patient was

TABLE 30-7: OTHER PROGRESSIVE CAUSES OF DEMENTIA

DISORDER	DISTINGUISHING CLINICAL	PATHOLOGY
Diffuse Lewy* Body Disease	Parkinsonism and psychotic features with relative sparing of memory at onset	Diffuse Lewy body involvement of cortical & subcortical neurons + senile plaques
Huntington's Disease	Chorea and personality changes at onset Autosomal dominant, high penetrance	Atrophy caudate/putamen + cortex, seen CAG trinucleotide repeats
Frontal/temporal Dementia (several types)	1) Personality changes early, memory changes late. 2) progressive nonfluent aphasia at onset. 3) semantic dementia In 40% autosomal dominant tauopathies are found linked to chromosome 17.	1) Bilateral prefrontal atrophy early, hippocampus late 2) Lobar left temporal-frontal frontal atrophy, 3) Bilateral temporal atrophy with gliosis and/or spongy changes. In some Pick's bodies.
Creutzfeldt – Jakob	Rapidly progressive dementia, myoclonus, seizures, pyramidal, basal ganglia signs (+cerebellar in 50%). Familial **and new variant (bovine) cases have a younger onset, a slower course and greater cerebellar involvement.	Transmissible spongiform encephalopathy - Prion disorder (Fig 30-8). Periodic EEG complexes in sporadic (Fig.30-9). In new variant, diffuse amyloid plaques +spongiform changes.

* CD ROM Case 30-6. **10-15% of Creutzfeldt Jakob cases have an autosomal dominant mutation on chromosome 20 (prion protein)

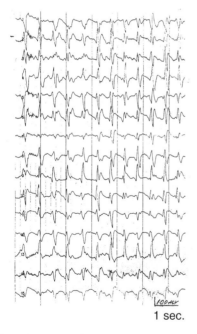

1 sec.

Figure 30-9 Creutzfeldt-Jakob disease: Biopsy proven spongiform encephalopathy EEG Features: Frequent periodic discharges of generalized triphasic blunt spikes are present at a late stage of the disease. This 58-year-old male was admitted on 01/18/93 with a brief subacute onset of ataxia and mild cognitive deficits and a mildly abnormal EEG. He evolved rapidly over a 4-week period, (from onset to death) both from the clinical and EEG standpoint. (Courtesy of Dr. Tom Mullins and Dr. Sandra Horowitz.)

treated with dexamethasone and radiotherapy with improvement in gait.

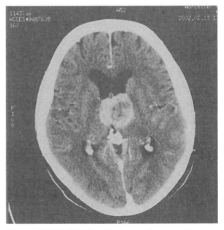

Figure 30-10: Thalamic dementia secondary to a glioblastoma. Case 30-7: CT scan with contrast. An enhancing tumor with a necrotic center involves the anterior and medial thalamus without involvement of the posterior thalamus.

CHAPTER 31
Case History Problem Solving: Part V
General

Case 31-1: This 59-year-old, white male, married plumbing contractor, approximately 7 years prior to admission, had the onset of a tremor of the right hand occurring mainly at rest. One year prior to admission, the patient had noted progression. The tremor was now present at the wrist and shoulder in addition to the fingers. During this time, the patient had also noted a stiffness of the right leg when walking. For one year prior to admission, since an alleged back injury, he had been unable to straighten up when walking and had had to walk in a very stooped position. During the 6 months prior to admission, the patient had experienced short episodes of subjective vertigo, resulting in a tendency to fall forward.

Past History:

1. Poliomyelitis at age 2, with residual atrophy of left gastrocnemius muscle.
2. Influenza at age 18.

NEUROLOGICAL EXAMINATION:

1. *Mental status:*

   a. Patient was depressed and anxious.

   b. There was mild impairment of digit span and of delayed recall, and an inability to retell stories. Some evidence of confabulation was present.

2. *Cranial nerves:* A fixed facies with minor facial asymmetry was present.

3. *Motor system:*

   a. Strength was intact, except for plantar flexion at left ankle, with associated atrophy of gastrocnemius and shortening of the left Achilles tendon.

   b. Gait: The patient walked stooped over, with slow short steps. He turned *en bloc*. There was a lack of associated movements of the arms, with greater impairment on the right side.

   c. There was a resting tremor of the right upper extremity at the shoulder, wrist, and fingers. The tremor was rhythmical at 3 to 4/sec. and pill rolling in type.

   d. There was a plastic resistance to passive motion with an intermittent cogwheel component in the right upper extremity at the shoulder, elbow, and wrist. To a lesser degree, rigidity was present in the right lower extremity.

   e. Alternating movements were impaired bilaterally (right more so than left).

4. *Reflexes:*

   a. Deep Tendon Reflexes: Biceps and triceps were more active on the left; patellar and Achilles were more active on the right.

   b. Plantar responses were extensor bilaterally.

5. *Sensory system:* Intact.

LABORATORY DATA:

1. Thyroid function normal.
2. Cerebrospinal Fluid: Normal pressure, cells, protein, and Hinton.
3. EEG: Normal

QUESTIONS:

1. Indicate the diagnosis: provide a differential diagnosis.
2. Where is the lesion?
3. What significance do you attach to the bilateral Babinski signs?
4. What significance do you attach to the changes in mental status?
5. Indicate the pathophysiology and etiology.
6. Outline current concepts of diagnostic and therapeutic management.
7. Indicate the prognosis.

Case 31-2: This 42 year-old woman reported to the doctor with a complaint of "trouble with my eyes". At around the age of 20, her menses ceased and never returned. The patient stated she had had an underactive thyroid for 20 years. Her menstrual symptoms did not respond to thyroid medication. During the past few years, the patient had had some automobile accidents and had noted that these accidents occurred when there were objects to her right. During the month before entry, the patient felt excessively tired and was sleeping more than usual but was able to go to work. Her skin had become dry; she had gradually put on weight over the years; her hair was scant with loss of underarm hair and scant pubic hair. A bifrontal dull headache had been present for one week.

PHYSICAL EXAMINATION:

There was puffiness of hands and face and dryness of the skin. Blood pressure was 108/72, pulse 82 and regular. Skin and mucous membranes were pale, and there was bilateral pallor of the optic discs.

NEUROLOGICAL EXAMINATION:

1. *Mental status:* Intact.
2. *Cranial nerves:* Intact except the visual fields showed a bitemporal upper quadrantic field defect that was almost complete in the right eye and less marked on the left. There was also bilateral pallor of the optic discs. Pupils and ocular movements were normal.
3. *Motor system:* No abnormalities noted.
4. *Reflexes:* Within normal limits.
5. *Sensory System:* Normal.

SUBSEQUENT COURSE:

Patient returned the next day, having had an emotional upset at work with the onset of a severe pressure headache. She complained of sudden worsening of vision in the left eye and a film over the temporal fields. On examination, she was pale and sweating. Blood pressure was 100/70; pulse was 100. There was more marked cutting of visual fields, particularly on the left, with a true hemianoptic defect (both eyes now demonstrated a full temporal hemianopia). The patient was drowsy but could provide concise answers. Neurologic examination was otherwise normal. The patient was immediately admitted to the hospital. Laboratory studies in the hospital confirmed the clinical impression of hypothyroidism. In addition, there was no FSH detected on an assay of a 24-hour urine specimen. Plain skull films were positive and bilateral carotid arteriograms were performed. Patient was started on endocrine therapy, and an operation was performed. Three days following surgery, it was noted that the patient was complaining of constant and excessive thirst and was consuming huge quantities of fluids.

QUESTIONS:

1. In this case, neurological evaluation many years previously would have established the diagnosis. Indicate the location of this lesion. Discuss the differences between this lesion at onset and the lesion at the present time.
2. Indicate the most likely pathology. Speculate as to the other possible causes of this syndrome.
3. What happened the day after the initial office evaluation?
4. Today, which neurodiagnostic study would have best demonstrate the location and nature of the pathology?
5. This patient had 24-hour urine FSH levels performed. Which additional endocrine blood test would be performed today? What are the causes of abnormalities in that additional test?
6. What is the localizing significance of the thyroid dysfunction?
7. What is the localizing significance of the ovarian dysfunction?
8. Why did the patient have excessive thirst and excessive consumption of fluids (polydipsia) after surgery?
9. Why was the patient sleeping more than usual and why was she markedly drowsy at

the time of admission? (Several explanations are possible in this case.)

Case 31-3: Three weeks prior to admission, this 60 year-old right-handed white male had the acute onset of clumsiness of his right hand. Soon thereafter, he noticed weakness and numbness involving the right hand. He then noticed that his right leg was weak, and he fell to the floor. The weakness and numbness of the hand and leg cleared in a matter of minutes. The patient, however, continued to have clumsiness of the right hand. On two subsequent occasions, he had a recurrence of weakness of the right hand.

The patient's history was also of significance in that over a 2-month period, he had had 3 episodes, each 3 minutes in duration, of a monocular blindness of the left eye. Diabetes mellitus had been present for approximately 20 years. Six months before admission, the patient had had a sudden onset of weakness and coldness of the right lower extremity, due to sudden occlusion of the right femoral artery. Symptoms improved when a right femoral endarterectomy with bypass was performed. The patient had had 5 to 10 minute episodes of chest pain (angina pectoris) for a number of years.

Family history was significant in that the patient's mother, father, daughter and son had all experienced severe diabetes mellitus.

GENERAL PHYSICAL EXAMINATION:

1. Blood pressure was 130/80 in both the right and left arms.
2. Light pressure on the left eyeball led to a blackout of vision; similar symptoms were not produced on the right side.

NEUROLOGIC EXAMINATION:

1. *Mental status:* The patient was oriented for time, place, and person.
2. *Cranial nerves:* No significant findings.
3. *Motor system:* Strength was intact, except for a minimal drift downward of the outstretched right arm when both arms were outstretched. Cerebellar tests and gait were intact.
4. *Reflexes:* Deep tendon reflexes were slightly increased in the right arm compared to the left. The Achilles' tendon reflexes were absent bilaterally.
4. *Sensory System:*

   a. Vibration sensation was decreased bilaterally at the toes.

   b. Occasional errors were made in letter and number recognition in the right hand

QUESTIONS:

1. What is the diagnosis? Be specific as to the vessel and area involved, if vascular in nature.
2. Outline your diagnostic approach to this problem.
3. Indicate possible therapeutic measures.
4. What is the explanation for the absent Achilles' deep tendon reflexes and the decreased perception of vibration at toes?

Case 31-4: The patient was a 20 year-old, right-handed white housewife who was seen initially in the 23rd week of her third pregnancy. The patient presented with a history of generalized convulsive (tonic clonic) seizures beginning at age 7. These apparently would occur once a month without warning. She was begun on treatment with phenytoin and phenobarbital about age 8. She had had recurrences of seizures two years prior to the visit and one year prior to the visit. On each occasion, these seizures related to her omission of medication for several days. Her medications at the time of the initial visit were phenytoin, 100 mg a day and phenobarbital, 15 mg a day.

FAMILY HISTORY:

A maternal grandfather had convulsions and staring spells. Two cousins also had a seizure disorder, manifested by staring spells.

NEUROLOGICAL EXAMINATION:

In detail, the entire examination was normal.

LABORATORY DATA:

1. *Electroencephalogram:* The patient had bilat-

eral, relatively synchronous bursts of poly spikes -slow waves complexes most prominent in the frontal recording areas. At times, the bursts would continue for 2 to 3 seconds. No clinical phenomena were noted during the discharges.

2. *Blood Levels:* Dilantin blood level was less than 2.5 uG/ml.; Phenobarbital was 5 uG/ml.

SUBSEQUENT COURSE:

The patient agreed to increase her intake of Dilantin to 200 mg a day and to increase her intake of Phenobarbital to 15 mgs. two times a day. The blood levels were checked again approximately 3.5 weeks later. Her phonation level was still low at 2.8 ug/ml., her phenobarbital was 4.5 ug/ml. The patient did not return again for follow up until 17 months later. In the interim, she had had no additional seizures of either a major or minor type. Her infant and previous offspring were normal.

1. Classify these seizures. If possible be specific.
2. What other types of seizures might be expected in this syndrome?
3. What would be the most appropriate anticonvulsant?
4. What would neuro imaging studies demonstrate?
5. Discuss the relationship between epilepsy and pregnancy.
6. Discuss the relationship between anticonvulsants and pregnancy.
7. Discuss the role of anticonvulsants and related drugs in teratogenesis.
8. Discuss the interaction of anticonvulsants and birth control pills.

Case 31-5: This 23 year-old white right-handed, army recruit private, on the day prior to admission, was evaluated on sick call for an apparent upper respiratory infection. The patient was given a day of barracks rest. Early on the morning of admission, the patient apparently vomited. At reveille, the patient was found in his bed, unresponsive and he was found to have been incontinent of urine and feces.

GENERAL PHYSICAL EXAMINATION:

1. The patient was febrile with a temperature of 102 to 103∞F. Blood pressure was 130/70; pulse was 96.
2. There was a marked degree of nuchal rigidity.
3. Examination of the skin revealed a diffuse petechial rash with a number of ecchymotic areas on the extremities. Some of these areas had apparent necrotic centers.

NEUROLOGICAL EXAMINATION:

1. *Mental status:* The patient was in a coma; he responded to painful stimuli by withdrawing his extremities.
2. *Cranial nerves:*

   a. The pupils were midposition and responded very poorly to light.

   b. Funduscopic examination was not remarkable.
3. *Motor system* There was no definite lateralized weakness. Cerebellar system and gait could not be tested.
4. *Reflexes:*

   a. Deep tendon reflexes were equal but hyperactive throughout.

   b. An extensor plantar response was present on the left side.
5. *Sensory System:* Could not be adequately tested.

QUESTIONS:

1. What is your diagnosis?
2. Which studies must be performed and when must they be performed?
3. When must treatment be started? Which treatment would be most appropriate?

LABORATORY DATA:

1. *CBC:* White blood count was markedly elevated to 42,000, with a differential count of

64% neutrophils, 31% bands, and 5% lymphocytes. Platelets were decreased to 95,000 per mm3. Blood serology was negative.

2. *Spinal fluid examination* revealed a marked increase in pressure (>600mm of CSF) The fluid was grossly cloudy. There were 7250 white blood cells per cubic mm; 100% of these were polymorphonuclear leukocytes. Spinal fluid sugar was 6-mg/100 ml (blood sugar was within normal range); spinal fluid protein was increased to 314-mg/100 ml.

4. *Smears and eventually cultures* of spinal fluid and blood were consistent with the clinical diagnosis.

SUBSEQUENT COURSE:

The patient was begun on specific therapy. As the patient's level of consciousness began to improve, on the second hospital day it was apparent that a left facial weakness was present. On the fourth hospital day, as the patient began to respond to questions by opening his eyes and moving his right hand, it became apparent that he had a left facial weakness, a weakness of the left arm and leg, and a left Babinski response. On the sixth hospital day, even though the patient was continuing to improve, he had the onset of focal seizures beginning on the left side of the body. Some remained localized to the left side; some became generalized. Following readjustment of therapy, seizures disappeared. With physiotherapy, a significant improvement in function of the left arm and left leg occurred.

QUESTIONS:

4. Do you wish to modify your diagnosis and proposed treatment based on the CSF findings? What do you expect the cultures of blood and CSF to indicate?

5. Indicate the pathophysiology involved in the complications that developed during the patient's hospitalization.

Case 31-6: This 62 year-old white divorcee was referred for evaluation of memory subsequently of personality change. The patient had a high school education and had been described in the past as a bright, alert and intelligent lady. The patient was unable to provide a history. She showed little insight into her problem and seemed unaware of any reason for her neurological evaluation. The patient's son indicated that she had begun to show some decline in intellectual capacity as long as 3 or 4 years prior to admission. During the preceding 12 months, a more marked change had occurred, particularly with regard to memory. In more recent months, a lack of concern for her personal appearance and a lack of spontaneity had developed. She had lost interest in her home, friends, and activities.

Past history was unremarkable as regards cardiovascular disease, diabetes or alcohol or drug and ingestion.

GENERAL PHYSICAL EXAMINATION:

Blood pressure was 140/96; cardiac status was normal.

NEUROLOGICAL EXAMINATION:

1. *Mental status:*

    a. The patient was pleasant and well mannered.

    b. She was disoriented for the day of the week, for the day of the month, for the month and for the year.

    c. She was unable to remember any of four test objects after a 5-minute period.

    d. The patient was unable to remember the year she was born.

    e. She was unable to recall any president but Kennedy.

    f. She was able to follow very simple directions, but was unable to remember any instructions beyond these.

    g. She was unable to comprehend subtraction of serial 7s and could not do even simple additions. She was unable to repeat numbers in reverse.

    h. She was unable to draw a triangle but did recognize a circle.

    i. There was a severe dysnomia for even

common objects.

2. *Cranial nerves:*

   a. Pupillary reactions were normal.

   b. At times, the patient had extinction of simultaneously presented visual objects in the left or right visual field.

3. *Motor system:*

   a. Strength and gait were intact.

   b. Minor tremulousness was present in the performance of hand and arm movements.

   c. Minor variable resistance on passive motion existed and was described as Gegenhalten.

4. *Reflexes:*

   a. Deep tendon reflexes were symmetric and physiologic.

   b. Plantar responses were flexor bilaterally.

   c. Grasp was absent.

5. *Sensory system:* Normal.

QUESTIONS:

1. Does this patient have focal disease or a more generalized disease? If the latter, which areas are predominantly affected? What is the significance of the dysnomia in this case? Which diagnosis would you assign to this case?

2. Describe the most likely pathology in this case. Present a differential diagnosis.

3. What treatable causes of this progressive syndrome much be considered? Discuss in terms of:

   a. Infections

   b. Nutritional diseases

   c. Metabolic diseases

   d. Intoxications

   e. Other treatable etiologies For each entity indicate the appropriate diagnostic tests.

4. How would you manage this patient as regards diagnosis and treatment?

5. What would an electroencephalogram demonstrate?

6. What would a CT scan demonstrate?

7. What cerebrospinal fluid findings are to be expected?

8. What is the natural history and prognosis?

9. What more restricted mental status changes might have been present early in the disease course? What would be a more likely diagnosis if personality changes had been the primary early symptoms? If basal ganglia symptoms had been present early in the course of the disease, which diagnoses might have been considered.

Case 31-7 (Patient of Doctor John Sullivan and Doctor Huntington Porter): This 67-year-old white right-handed retried schoolteacher awoke one morning with numbness over the entire right side of her body. The right corner of her mouth drooped. She was able to go to a family physician but was unsteady on her feet. She had no diplopia or aphasia. Numbness on the right side disappeared gradually over a two-week period, but then appeared in the left foot. The left hand became restless and would more in a jerky up-and-down movement. The hand and arm would take on abnormal postures and tend to trail behind her. One week later, this involuntary movement began to affect the left foot, which became restless. Its constant motion seriously interfered with her ability to walk. Over the next several weeks, the movement disorder in the left hand disappeared.

Past History was not remarkable except for menopause at age 30. Family history indicated that her mother died at age 63 of a "heart attack". Father died at age 72 of a "shock".

GENERAL PHYSICAL EXAMINATION:

Blood pressure was 160/100, with a normal cardiac examination.

NEUROLOGICAL EXAMINATION:

1. *Mental Status:* All areas were intact, with no evidence of aphasia.

2. *Cranial Nerves:*

   a. A slight droop of the left corner of the mouth was present.

b. There was an unsustained nystagmus on far lateral gaze.

3. *Motor System:*

   a. Strength was intact, with good retention of skilled movements of the hands.

   b. In the left leg, there was an almost constant play of movement, smooth and rhythmical, consisting of eversion/inversion of foot and ankle with flexion/extension movements of the toes. In the left arm, the same type of movement was present. This movement was not synchronized with that in the foot and was often absent for periods of time. As the patient used the right arm, the movement disorder became apparent in the left hand. When the patient performed repetitive movements of the left fingers, mirrored movements of the right fingers occurred.

   c. Gait: The body was twisted to the left. The left hand trailed behind, with the index finger pointing down and the other fingers clenched. The gait was unsteady. Walking did not dampen the movement in the left leg. At times, the left leg would suddenly buckle.

4. *Reflexes:*

   a. Deep tendon reflexes were symmetric and physiologic.

   b. Plantar responses were flexor bilaterally.

   c. There was no release of the grasp reflex and no repellent apraxia.

5. *Sensory System:* Pain, touch, position, and other discriminative modalities were intact. There was a slight decrease in vibration sensation at the toes.

LABORATORY DATA:

1. *Fasting blood sugar* was elevated to 250-mg/100 ml.
2. *Cerebrospinal fluid:* Opening pressure was 60; no cells; protein was 45-mg/100 ml.
3. *Chest and skull X-rays* were negative, except for calcification of the carotid siphon.
4. *Sedimentation rate* was 17 mm in one hour.

SUBSEQUENT COURSE:

The patient did well for approximately one month. Then she began to have sudden "jerky movements" of the left hand, causing the arm to strike her in the face. The left arm also began to posture in a position of relative flexion at the elbow, wrist, and fingers. The patient also reported occasional violent flinging movements of the arm. To some extent, numbness had returned to the right upper extremity.

When she was re-examined 4 months after the initial episode, certain changes had occurred in the movement disorder. A distal fluid, flowing 2 to 4 Hz movement was still noted at the fingers, toes, hand, and foot, alternating between flexion/supination and extension/pronation. This movement had now appeared at more proximal joints, and some spread of movement into the face was reported. In addition, there were sudden pronation and posterior rotatory movements at the shoulder. At times, this was described as a wild swinging or flinging movement of the left upper extremity. At the time of a visit 5 months later, significant improvement had begun to occur. By the time of a follow-up 2 months later, approximately one year after onset of symptoms, there was little evidence of any movement disorder. Occasional distal movements occurred only when the patient was emotionally excited.

QUESTIONS:

1. Provide a diagnostic label for the abnormality of movement noted on the initial neurologic examination.
2. Provide a diagnostic label for the movement disorder that subsequently evolved.
3. Discuss the neuroanatomic basis of these disorders.
4. Discuss diagnostic and management approaches.

Case 31-8: One week prior to admission, this 51-year-old white housewife, developed episodic posterior headaches beginning in the neck and radiating to the vertex of the head. The headaches were usually related to and were definitely exacerbated by straining or coughing. At the same time, the patient noted clumsiness of the left hand, and a tendency to fall to the left side. During the week prior to admission, a progressive gait ataxia also had developed.

PAST HISTORY:

Eight months previously, the patient had been admitted to the gynecology service before with a 10-week history of postmenopausal vaginal bleeding. She was found to have an infiltrative anaplastic carcinoma of the uterine cervix. The patient was treated with radiotherapy (4000R to the pelvis), followed by vaginal insertion of radium. Re-evaluation one month prior to the present admission had revealed that this anaplastic epidermoid carcinoma of the cervix had spread to the anterior vaginal wall. In addition, metastatic nodules were now present in the lung.

NEUROLOGIC EXAMINATION:

1. *Mental status:* The patient was an anxious white female who was cooperative, alert and well oriented.

2. *Cranial nerves:*

   a. There was coarse nystagmus on gaze to the left.

   b. There was a minor dysarthria for lingual and guttural sounds.

3. *Motor system:*

   a. Strength was intact.

   b. Gait was unsteady, with tendency to fall to the left, especially when turning to the left. On a narrow base, the patient tended to fall backwards.

   c. Cerebellar tests revealed no definite truncal ataxia. There was a marked Appendicular ataxia on the finger-to-nose test in the left upper extremity and the heel-to-shin test in the left lower extremity.

4. *Reflexes:*

   a. Deep tendon reflexes were active bilaterally. This, however, was felt to be consistent with the patient's degree of anxiety.

   b. Plantar responses were flexor.

5. *Sensory system:* All modalities were intact, except for a minimal decrease in vibratory sensation at the toes.

LABORATORY DATA:

The results of specialized studies were consistent with the clinical diagnosis.

QUESTIONS:

1. Where is the lesion? Be specific.
2. What pathological process is to be expected?
3. Outline your diagnostic and therapeutic approach for this case.
4. Which is the appropriate neuroimaging study?
5. Should a lumbar puncture be performed? Indicate why or why not.

Case 31-9: A 21-year-old white right-handed wife of an air force enlisted man had been married for two months and had just moved from her home in the south to an air force base in New Jersey. She had been complaining of increasingly severe right frontal temporal or bifrontal headaches for three to four weeks. During the 10 days prior to admission, she had been treated on several occasions at the base dispensary for these headaches, which were relatively constant and were now accompanied by blurring of vision and frequent vomiting. Various medications had been prescribed for an upper respiratory infection or sinusitis. In the 48 hours prior to admission, the patient had become increasingly confused and ataxic. Her past history was otherwise negative as regards head injury or pulmonary and ear infection.

GENERAL PHYSICAL EXAMINATION:

1. The patient was dehydrated and somewhat lethargic.
2. She was complaining of severe headache, particularly on head movement.
3. There was a moderate degree of hyperventilation.
4. No otitis media was present.

NEUROLOGICAL EXAMINATION:

1. *Mental status:* The patient was disoriented for time and place. She was unable to provide a history and was unable to cooperate for the examination. She was impersistent in fixing gaze or in maintaining eyes in open position.

2. *Cranial nerves:*

   a. On funduscopy, bilateral papilledema was present, with 2 to 3 diopters of disc elevation, accompanied by fresh hemorrhages and venous engorgement.

   b. The right pupil was dilated and fixed in response to light. Extraocular movements were otherwise intact.

   c. A left central facial weakness was present to a marked degree.

3. *Motor system:*

   a. All limbs were moved spontaneously, but there was a downward drift of the outstretched left arm.

   b. The patient was ataxic in a sitting position and in attempting to stand, even on a broad base.

   c. There was a tremor of the outstretched hands, increasing to a minor degree on movement and intention.

4. *Reflexes:*

   a. Deep tendon reflexes were increased on the left in the upper and lower extremities.

   b. Plantar response was extensor on the left, equivocal on the right.

   c. Grasp reflex was present bilaterally.

5. *Sensory system:* Pain sensation was intact; modalities could not be tested.

6. *Neck:* Minor resistance to passive motion.

LABORATORY DATA:

1. *Skull and chest X-rays* were negative.

3. *EEG* demonstrated almost continuous focal 1 to 2 Hz slow-wave activity.

QUESTIONS:

1. This patient was seen relatively late in her disease course. You should be able to localize the primary location of the pathology.

2. Indicate what secondary complications have occurred.

3. The nature of the pathology may be somewhat uncertain to you. However, you should be able to present a differential diagnosis and then to indicate the most likely diagnosis. Keep in mind that the headaches had been present for only 3-4 weeks and the additional symptoms for only 2-10 days.

4. Was the papilledema long-standing?

5. What is the significance of the pupillary findings?

6. Assuming a cerebral hemisphere lesion, why was the patient ataxic in sitting and standing when examined late in her course?

7. Which diagnostic studies would you perform in this case, and when would you perform these?

8. Which diagnostic studies would you not perform? State reasons.

9. How would you manage this problem from a therapeutic standpoint? Indicate when you would institute this therapy.

Case 31-10: This 55-year-old white widow, 6 to 8 years prior to admission, had the onset of a peculiar reeling gait. At the same time, she was noted to be "very nervous", having many peculiar restless movements. At other times, she was described as constantly fidgeting. The patient's children felt that in recent years, she had not been thinking as clearly as she once did. A personality change had also occurred.

FAMILY HISTORY:

The patient's mother died in her sixties with a disorder that had been labeled as Parkinsonism. From the description of relatives, however, it was evident that the mother had restless movements that were similar to the patient's.

NEUROLOGICAL EXAMINATION

1. *Mental status:* The patient was oriented and alert with an intact general store of information. Object recall was three out of four in 10 minutes. Digit span was six forward and

three in reverse

2. *Cranial nerves:* Intact.

3. *Motor system:*

   a. Strength was intact with no significant spasticity or rigidity.

   b. Gait showed a swooping quality of variable degree; there was also a decrease in associated arm movements.

   c. As the patient sat, she would move her head, neck, and feet almost constantly.

   d. The patient was able to walk a tandem gait, and there were no abnormalities of cerebellar function.

4. *Reflexes:* Deep tendon reflexes were symmetric. Plantar responses were flexor. No grasp reflex was present.

5. *Sensory system:* Intact.

LABORATORY DATA:

1. Routine laboratory studies were not remarkable.

2. *Specialized neuroradiology studies* were consistent with the clinical diagnosis.

QUESTIONS:

1. Present a differential diagnosis of this problem and indicate the most likely diagnosis.

2. Outline your diagnostic and therapeutic approach to this problem.

3. What did the specialized neuroradiology studies reveal?

4. Which neurodiagnostic study would be most cost-effective and specific in establishing the diagnosis today?

5. Discuss the underlying molecular basis of this disorder.

Case 31-11 (patient of Dr. Thomas Sabin): A 50-year-old white man was admitted to the hospital because of drowsiness and confusion of 5 days duration. His wife stated that he was in good health until that time. For 4 days prior to admission, he merely slept more than usual and napped during the day, but while awake, was lucid. The day before admission, he was confused and this confusion increased on the day of admission.

NEUROLOGIC EXAMINATION:

This was a thin man with normal vital signs who answered questions quickly but often incorrectly. In addition to the confusion, drowsiness and problems in memory, pertinent findings included the following: Extraocular movements were limited in all directions, and there was horizontal and vertical nystagmus. The gag reflex was diminished bilaterally. Finger-to-nose and heel-to-shin tests showed a mild deficit. The patient had an unsteady, broad-based gait. No other motor findings were present

QUESTIONS:

1. Do you require more historical information? If so, indicate what is required.

2. Present a differential diagnosis.

3. What is the most likely diagnosis?

4. In a general sense what is the localization of lesions in this disease. What is the localization for each specific symptom or sign (confusion and impairment of new learning, lethargy, impairment of extraocular movement and ataxia?

5.

6. How could your suspected diagnosis be confirmed?

7. How and when would you treat this patient? Which findings would rapidly resolve over the next 1-2 days?

8. You administer specific treatment but the difficulties with memory fail to clear and one year later the patient is still confused and confabulating. How do you modify your diagnosis?

9. You administer specific treatment but one year later, the patient is still ataxic. How do you modify your diagnosis?

Case 31-12: This 60-year-old right-handed retired priest was referred for evaluation of problems in walking. The history was not precise. Two and a half years previously, he had been admitted to his local hospital, for severe diabetic coma. Although he recovered, he had some difficulty with memory thereafter. He began to fall. Problems in walking had progressed, particularly in the last 6 to 8 months, so that he was no longer able to walk without assistance. His memory problems had progressed over the last year. He denied numbness in his extremities but was aware of "twitching" of his extremities. He denied any significant family history. He indicated he had been a very heavy drinker and had been hospitalized once in the previous year for treatment of alcoholism. Admitted intake was at least half a liter (500 cc) of gin per day.

NEUROLOGIC EXAMINATION:

1. *Mental status:* He was very vague about the date. Delayed recall was limited to two out of five objects in five minutes.

2. *Cranial nerves:* All were intact, except for a tremulous voice.

3. *Motor system:*

   a. The patient stood on a broad base and was very unsteady standing on a narrow base, requiring support. This unsteadiness slightly worsened with eye closure.

   b. He had been brought into the office in a wheelchair. He walked with assistance with small steps but was very ataxic.

   c. There was no impairment on finger-to-nose testing but there was dysmetria on heel-to-shin test.

4. *Reflexes:*

   a. Deep tendon reflexes were everywhere absent.

   b. Plantar responses were not responsive.

5. *Sensory system:*

   a. Pain sensation was absent to the level of the knees.

   b. Position sense was intact.

   c. Vibration sensation was markedly decreased at the toes and moderately decreased at the ankles.

LABORATORY DATA:

1. *Blood studies:* Folic acid, B12 levels, serum and urine immuno-electrophoresis, ESR, and ANA were normal. Serological tests (RPR and FTA) were negative. Liver function studies had been previously elevated but now, except for alkaline phosphatase, were normal.

2. *Electromyogram and nerve conduction* studies indicated a severe sensory-motor peripheral neuropathy, primarily axonal in type, as well as left median nerve (carpal tunnel) and right ulnar nerve (olecranon groove) entrapment neuropathies.

6. *MRI* demonstrated:

   a. Prominence of cerebellar folia, particularly of the superior vermis;

   b. Moderate cerebral cortical atrophy;

   c. Periventricular white matter involvement (leukoariosis).

SUBSEQUENT COURSE:

An alcoholism program with discontinuation of all alcohol intake and administration of multiple B vitamins was recommended. When last seen in follow-up, the patient was consuming at least 500 cc of vodka per day. His neurologic examination was unchanged except that pain sensation was now intact. He refused additional neurologic followup.

QUESTIONS:

This patient had several neurologic problems: At least five can be specified.

Discuss in terms of:

   a. Ataxia
   b. Memory changes
   c. Peripheral nerve disorders
   d. Toxic and metabolic diseases

CHAPTER 31A
Case History Problem Solving: Part II General Case History Review with Correlation to Illustrations

PHOTO SERIES

Each series of photos is derived from a particular case history. Indicate the most likely diagnosis in terms of localization and pathology.

Then use the photos for the following problem solving case histories.

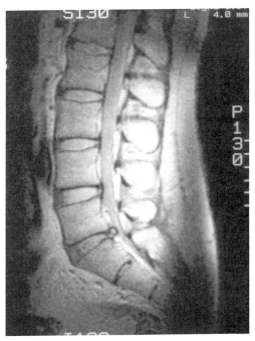

Figure 31-1.

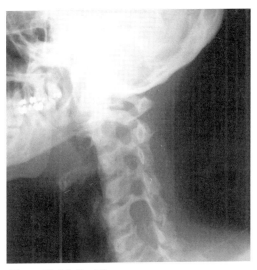

Figure 31-2A. Lt oblique

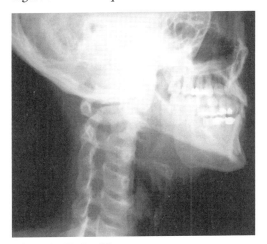

Figure 31-2B. Rt oblique

31A-2 CHAPTER 31A

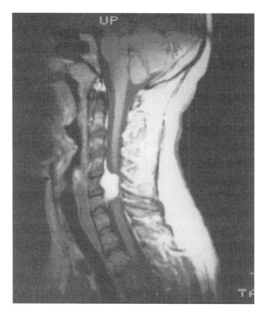

Figure 31-3A.

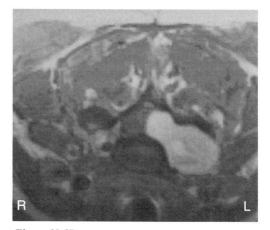

Figure 31-3B.

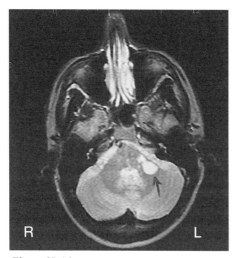

Figure 31-4A

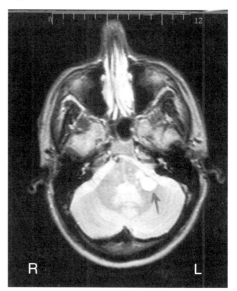

Figure 31-4B.

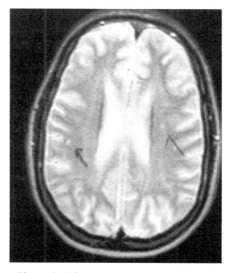

Figure 31-4C.

CASE HISTORY PROBLEM SOLVING PART V—A: IMAGING CORRELATION

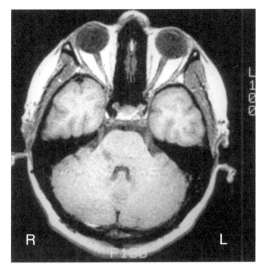

Figure 31-5A.

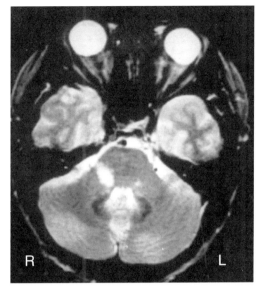

Figure 31-5B.

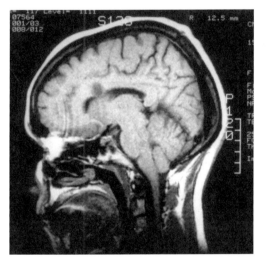

Figure 31-5C.

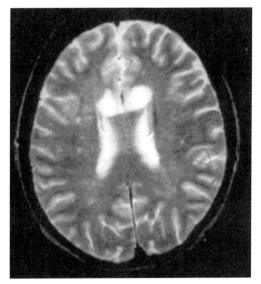

Figure 31-5D.

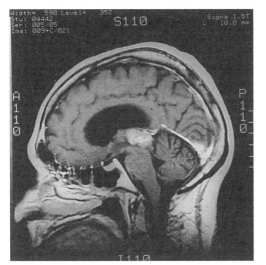

Figure 31-6A.

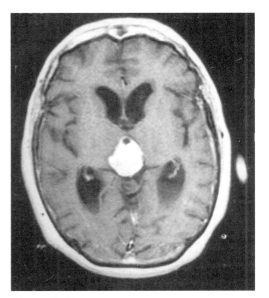

Figure 31-6B.

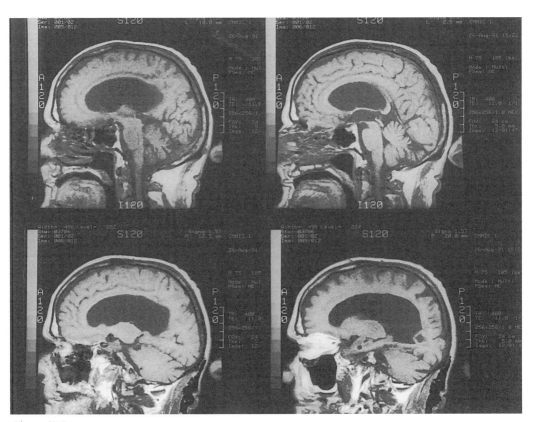

Figure 31-7.

CASE HISTORY PROBLEM SOLVING PART V—A: IMAGING CORRELATION

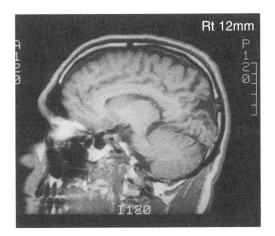

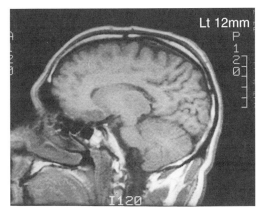

Figure 31-8.

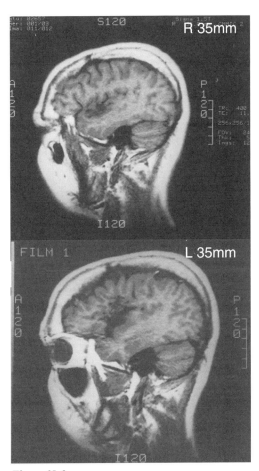

Figure 31-9.

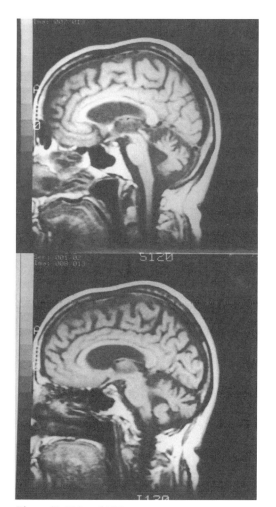

Figure 31-10A and 10B..

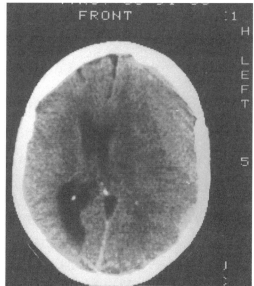

Figure 31-12.

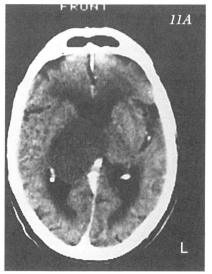

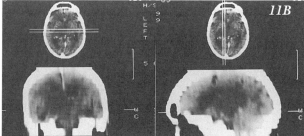

Figure 31-11A and 11B.

CASE HISTORY PROBLEM SOLVING PART V—A: IMAGING CORRELATION

Case 31A-1: This 43-year-old right-handed white male, two months prior to admission had the acute onset of blurring of vision and possible diplopia. Two weeks prior to admission, he had the acute onset of ataxia, vomiting, dizziness, and left face numbness. He improved, then developed severe problems with balance, slurring of speech and difficulty in the use of the left arm, which he labeled as "weakness". The patient was also aware that he had become emotionally labile. Cranial nerve examination demonstrated a paralysis of conjugate lateral gaze to the left. A marked left peripheral facial weakness was present. He had a marked decrease in whisper perception in the left ear. Strength was actually intact. There was a marked dysmetria on the left finger-to-nose and left heel-to-shin tests. The patient was ataxic in walking. Deep tendon reflexes were symmetric, plantar responses were extensor bilaterally. Sensory examination was normal.

QUESTIONS:

1. What is the most likely diagnosis in terms of localization and pathology
2. What study would be most diagnostic or confirmatory of your clinical diagnosis?
3. Select the most appropriate photograph.

Case 31A-2: This 68-year-old right-handed married white female and retired clerical worker was referred for evaluation of "weakness in both lower extremities" of approximately two years duration. She had been falling for at least one year. Although the patient and her husband initially denied any bladder symptoms, her daughter, a nurse, was able to indicate that urinary incontinence had occurred on several occasions during the last year. The patient indicated her memory had been "bad" for 1 to 2 years. Her husband minimized this symptom. Her daughter subsequently was able to relate a more significant impairment in mental functions. The patient's mother, age 88, had senile dementia and problems walking.

NEUROLOGIC EXAMINATION:

1. *Mental status:* The patient was oriented to time, place and person. Delayed recall was 0 / 5 in 5 minutes without assistance: 4 /5 with assistance.
2. Cranial nerves: the following findings were present:
   a. Tremor of head.
   b. Hyperactive jaw jerk.
   c. Positive glabellar sign.
3. *Motor system:* the following findings were present:
   a. Variable generalized weakness that at times disappeared.
   b. Gait: Small steps, at times waddling, unsteady on the turns.
   c. Sitting: Often tended to fall backwards on the examining table.
4. *Reflexes:*
   a. Deep tendon reflexes were everywhere hyperactive.
   b. Plantars responses were equivocally extensor bilaterally.
   c. Grasp reflex was released bilaterally.
   d. Palmomental responses were positive bilaterally.
5. *Sensory system:* Intact.

QUESTIONS:

1. Considering the combination of ataxia of gait, impairment of memory, and urinary incontinence, indicate the most likely syndrome.
2. How do you interpret the bilateral release of grasp, the palmomental reflex, the bilateral extensor plantar responses, and the hyperactive jaw jerk?
3. Provide a differential diagnosis of this syndrome.
4. Which study or studies would you obtain to confirm your diagnosis?
5. Select appropriate illustrations (several may be appropriate).
6. If this patient also had severe impairment of upward gaze and headache, what would be the most likely diagnosis and which would be the most appropriate illustration.

7. If this patient also had a past history of head trauma with resultant coma and recovery 10 years previously, what would be the most likely clinical diagnosis and which illustration would then best correspond?

8. What is the most appropriate therapy?

Case 31A-3: This 47-year-old right-handed married white female farm manager awoke one week before evaluation with numbness and "novocaine-like" sensation involving the entire trigeminal distribution on the right, including the tongue, lip, mucosa, etc. She had no pain and no other neurologic symptoms. Five to six years previously, she had intermittent unsteadiness that was attributed to an "inner ear infection". She also was aware that during the last year, if she sat in a hot tub, she would be totally exhausted.

NEUROLOGIC EXAMINATION:

This was entirely within normal limits except for a selective decrease in touch sensation involving the entire distribution of the right trigeminal nerve, including mucous membranes, gums, upper and lower right half of the tongue and upper and lower lip. Touch sensation was also decreased over the right cornea. In all areas, pain sensation was intact. Graphesthesia over the face was intact.

SUBSEQUENT COURSE:

All numbness disappeared approximately two months after onset. She returned five months later with complaint of one-week duration of "feeling slightly woozy". Neurologic examination was not remarkable. Symptoms were reproduced by rotation.

QUESTIONS:

1. What is the location of the lesion corresponding to the sensory symptoms involving the face?
2. What is the appropriate neurodiagnostic study? What other studies might be obtained
3. Select the appropriate illustration.
4. What is the most likely clinical diagnosis?
5. Which therapy would you recommend?

Case 31A-4: This 70-year-old right-handed white widow and retired telephone company clerk was referred for evaluation of unsteadiness in walking, which had been present for 23 years. The symptom had begun in relationship to a hospitalization for a myelogram. At that time, the patient had been involved in an automobile accident and had developed lumbar radicular pain. The evening after the myelogram, she had developed "coma, convulsions, and a temperature of 110°.

The medical records of that hospitalization were not complete but did indicate she had a prolonged, markedly elevated temperature (> 104°F.) secondary to meningitis.

At that time, the patient also had the onset of episodes of tinnitus, vertigo, olfactory hallucinations of roasting meat, a sense of de'ja vu and the sensation of hearing a symphony. She would be observed to be glassy-eyed and inattentive.

Past history indicated the following:
1. Cervical disc surgery at age 30.
2. Lumbar disc surgery at age 38 and age 60.
3. Hyperthyroidism had been treated with radioactive iodine. She was receiving thyroid replacement.
4. Diabetes mellitus had been treated with insulin for 5 years.

NEUROLOGIC EVALUATION:

1. *Mental status:* Intact.
2. *Cranial nerves:* Intact.
3. *Motor system:*
  a. Strength intact.
  b. Stance: Unsteady on a narrow base and only slightly worse with eye closure.
  c. Gait: Unable to walk a tandem gait, unsteady on the turns.
  d. Finger-to-nose test and heel-to-shin tests were not remarkable.
4. *Reflexes:*
  a. Deep tendon reflexes: Achilles reflexes were absent bilaterally.
    b. Plantar responses were flexor.
5. *Sensory system:*

a. Pain sensation was decreased to the knees bilaterally.

b. Vibration sensation was absent at toes and decreased at ankles.

c. Position sensation was intact.

SUBSEQUENT COURSE:

The neurologic examination did not change over the next two years.

QUESTIONS:

1. As regards the ataxia of stance and gait present since the hospitalization 23 years previously:

a. Indicate the location of lesion.

b. Indicate the most likely pathology.

c. Indicate best study to demonstrate a. & b.

d. Select the appropriate illustration.

2. As regards the episodes characterized by sensory perception of "roast meat odor", "music of a symphony", tinnitus and vertigo, etc.:

a. Indicate diagnosis.

b. Indicate most likely etiology.

c. Select the best test to support diagnosis.

d. Indicate the most appropriate treatment.

3. As regards the absent Achilles reflexes, the decreased vibratory sensation and distal decrease in pain sensation:

a. Indicate diagnosis.

b. Indicate the most likely pathology.

Case 31A-5: This 70-year-old single right-handed white male priest had been unsteady for at least 1.5 years. He had been stumbling over names for the same period but had no other problems in memory. Urinary symptoms were denied.

Past history indicated atrial fibrillation in the past. Hypertension had been under treatment for 10 to 20 years. There was no definite history of cerebrovascular disease.

Family History: A niece had hydrocephalus that was shunted when she was a child. Two of four sisters had cerebral aneurysms.

NEUROLOGIC EXAMINATION:

1. *Mental status:* Oriented for time, place and person. Delayed recall without assistance 0 out of 5 objects in 5 minutes, with assistance 3 out of 5 objects.

2. *Cranial nerves:* Intact, including fundi, except for hyperactive jaw jerk.

3. *Motor system:*

a. Strength intact.

b. Stance: He was relatively steady until he closed his eyes.

c. Gait: He walked on a broad base but was able to do tandem gait. He was unsteady on the turns.

d. Finger-to-nose test demonstrated a slight terminal tremor.

4. *Reflexes:*

a. Deep tendon stretch reflexes were 2 in the upper extremities, 3 at the patella, and 2-3 at Achilles tendon.

b. Plantar responses were extensor bilaterally

c. There was no release of instinctive grasp.

5. *Sensory system:*

a. Position sensation was intact at fingers and toes.

b. Vibration sensation was absent at toes, decreased at ankles, and markedly decreased at fingers compared to wrist, and at wrist compared to elbows.

c. Pain sensation was decreased at toes to mid arch and over fingers.

QUESTIONS

1. As regards ataxia of stance and gait: Localize the lesion. Indicate the pathology.

2. Indicate best test to support the diagnosis.

3. Select the two best illustrations that would correspond to this case.

4. Does this patient also have another neurologic disorder?

SUBSEQUENT COURSE:

The patient had periodic neurologic follow-up evaluations. He was relatively stable over the next 4 years; then he reported a minor increase in the degree of unsteadiness. Two years later, he developed a sudden onset of diplopia in looking to the left and a minor ptosis of the left lid. Examination now demonstrated:

1. Chronic atrial fibrillation.
2. Moderate ptosis of the left lid.
3. Weakness of the left lateral rectus movements.
4. Decreased hearing in the left ear.
5. Deep tendon reflexes increased on the left.
6. Pain sensation decreased over the entire right side, including the face.
7. Patient stood on broad base and walked with small steps on broad base.

His diplopia cleared, but previous neurologic symptoms and signs remained.

QUESTIONS

5. As regards this last sudden episode, localize the pathology. Indicate the nature of the pathology.
6. Indicate the best study to support the clinical diagnosis.
7. Does any illustration correspond to this episode?

Case 31A-6: This 36-year-old right-handed white male lecturer was referred for evaluation of symptoms related to a motor vehicle accident that occurred 2.5 years earlier. He had initially symptoms in the neck and right arm, but these essentially cleared. A more persistent area of symptomatology related to the lumbar area. Occasional sharp pains extended from the back to the left buttock, the left posterior thigh, the posterior calf and the ankle. Back pain improved but left leg pain increased. In addition to the shooting pains in the left leg, he complained of intermittent numbness in the left posterior thigh. He denied weakness, bladder or sexual dysfunction. There were no symptoms in the right leg, except for occasional burning in the anterior tibial area.

NEUROLOGIC EXAMINATION:

Totally intact except for the following:
1. *Reflexes:* slight decrease in the left patellar reflex and a marked depression of both Achilles deep tendon stretch reflexes.
2. *Back:* The patient could forward flex to 60° to the vertical plan, at which point he complained of pain shooting down the left posterior thigh.
3. *Straight-leg-raising:* This was limited to 50 to 60° on the left, with pain in the left posterior thigh. On the right, straight-leg-raising could be carried out to 90°.
3. *Sciatic nerve:* Tenderness was present on the left.

QUESTIONS:

1. What is the clinical diagnosis? Be specific as to the localization and pathology.
2. Assuming that this patient had *not* responded to "conservative" measures, which diagnostic study would you obtain? Be specific.
3. Select the illustration - corresponding to this case.
5. Define "conservative measures".
6. What therapy would be recommended if these measures had failed and pain progressed?

Case 31A-7: This 27-year-old right-handed female college maintenance worker developed progressive numbness in the left arm extending from the elbow to all of the fingers. Initially, this was intermittent, but it soon became constant. She also developed persistent discomfort in the posterior cervical area extending to the left supraclavicular area. There was no history of trauma. Her primary physician obtained initial X-ray studies and, subsequently, referred the patient to a neurosurgeon.

QUESTIONS:

1. What is the differential diagnosis at this time?
2. What initial study would you, as the primary physician, obtain?
3. Select the appropriate illustration.
4. Based on that illustration, what is the most likely diagnosis?

SUBSEQUENT COURSE:

The neurosurgeon found the deep tendon reflexes absent at the left biceps and triceps. He requested a more definitive neuroimaging study. Based on that study, the neurosurgeon performed definitive neurosurgical procedures. The patient had an excellent result, except for residual numbness in the fingers of the left hand and a left Horner's syndrome.

QUESTIONS:

1. Which definitive study would you, as the neurosurgeon, have requested?
2. Select the appropriate illustration.
3. Based on the illustration, what is the most appropriate diagnosis?
4. Which neurosurgical approaches (procedures) were utilized?
5. Why did the patient have residual sensory deficit in the thumb and index finger?
6. Why did the patient have a residual Horner's syndrome?

SUBSEQUENT COURSE:

Six months after her last surgery, she returned to the neurosurgeon, complaining of burning pain in the right scapular area and some intermittent tingling of the fourth and fifth fingers of the right hand.

Examination 6 months later demonstrated:
1. *Cranial nerves:*
   a. Mild ptosis of the left eyelid, a smaller pupil on the left, and decreased sweating over the left side of the face.
   b. Whisper perception was decreased bilaterally.
2. *Motor system:* There was mild weakness in the right triceps muscle.
3. *Reflexes:* The triceps and biceps deep tendon reflexes were decreased on the left, but the radial and finger jerks were increased on the left. Plantar responses were flexor.
4. *Sensory system:* Pain sensation was decreased over the thumb and index finger and to a lesser degree over the middle finger of the left hand.
5. *Neck and supraclavicular areas:* She was tender over the right supraclavicular area, with pain extending into the right arm.

QUESTIONS:

7. What is the diagnosis of this new problem involving the right arm?
8. Which diagnostic procedure would you request?
9. What is the explanation for the smaller left pupil, mild left lid ptosis and decreased sweating over the left side of the face?
10. What condition might concern you regarding the bilateral deficit in hearing? How would you evaluate this problem?

Case 31A-8: This 20-year-old left-handed white male was the product of a prolonged delivery. The mother claimed, "the head was held back." The day after birth, a seizure occurred. At age 10 months, he was noted to have problems in movement of the right arm and leg. He was still unable to stand at 14 months of age. When he was evaluated at the Children's Hospital Medical Center at age 2 years, he was described as having a spastic and dystonic right hemiparesis, with severe delays in expressive speech. At age 4 to 5 years, he began to have frequent seizures, which began with tingling of the right hand and face, and focal movements of the right hand and lips. These would secondarily generalize 3 to 4 times per day.

Family History:

The mother had a long-standing mild right hemiparesis, with focal sensory seizures involving right face and arm. A brother carried a diagnosis of schizophrenia.

QUESTIONS:

1. What is your clinical diagnosis (location of lesion, type of pathology) at this point?
2. What study would have best demonstrated the location and nature of the pathology?
3. How would you manage the seizures?
4. How would you manage the spastic dystonic right hemiparesis and expressive speech delay?

SUBSEQUENT COURSE:

He had undergone, at age 11, a heel-cord-lengthening and toe-tendon-stretching operation.

At school, he had achieved third grade proficiency in reading and writing and tenth grade proficiency in math.

The patient was seen again 10 years later for a second opinion regarding seizure control. Over the intervening years, seizures had occurred at variable frequency. His seizures were primarily focal motor seizures beginning in the right arm. At times, secondary generalization might occur, particularly when flurries of focal seizures occurred. Often a warning occurred, which he could not precisely identify but which had components of a dreamy, strange, or familiar sensation. Over the years, he had been treated with phenytoin, phenobarbital, valproic acid, carbamazepine and primidone. Despite this, seizures were still occurring in flurries 3 to 4 days per week.

NEUROLOGICAL EXAMINATION:

1. *Mental status:* He was oriented for time, place and person. Delayed recall was five out of five in five minutes. Later, neuropsychological testing indicated that visual non-verbal learning was performed at a much higher level than auditory verbal learning. Reading, spelling, and arithmetic were all severely impaired.

2. *Cranial nerves:* There was smallness of the right side of the face.

3. *Motor system:*
  a. There was smallness of the right hand, thumb, arm, and foot.
  b. There was a right-sided weakness of the distal hand muscles, wrist extensor, and triceps.
  c. Patient walked with a hemiplegic posture. The right arm was flexed at elbow and there were dystonic movements of the right hand. The leg was extended at hip, with external rotation and circumduction.
  d. There was marked resistance to passive motion at the right shoulder and ankle ("tight Achilles tendon"). With rotation of the head, significant alterations occurred in the degree of spasticity in the flexors and extensors at the right elbow.

4. *Reflexes:*
  a. Deep tendon reflexes were increased on the right.
  b. Plantar responses were extensor on the right.

5. *Sensory system:* Intact for all primary and cortical modalities.

QUESTIONS:

5. Indicate your diagnostic approach.
6. Today, what would be the best study to indicate the location, extent, and nature of the lesion?
7. What would you expect the electroencephalogram to demonstrate?
8. Indicate your therapeutic approach at this point.
6. Discuss the alteration in the spasticity of flexors and extensors at the elbow on rotation of the neck.

Case 31A-9: This 18-year-old right-handed white male was the product of a prolonged delivery that apparently was complicated by "umbilical cord around the neck". "He did not breathe well at birth and may have aspirated." Three generalized convulsive seizures occurred at birth, and generalized convulsive seizures recurred with vaccination at 9 months of age and then once every 6 to 12 months. No seizures occurred from age 7 to age 15 years. Seizures then recurred in relation to reduction of medication. Warning symptoms preceded most seizures. He would sense a trembling of the extremities, and then vision would black out for 1 to 2 minutes while he was still standing and able to think. Then he would fall down and would be tonic/clonic for 1 to 2 minutes, with grater involvement of the left side. He would be "sleepy and confused afterwards". He was described as having average performance in school but reading had always been slow.

NEUROLOGICAL EXAMINATION:

Intact except for:
1. Slight clumsiness in movements of the left hand.

2. Deep tendon reflex increased on the left.

SUBSEQUENT COURSE:

The patient did well with no seizures for five years. Then he had the onset of "trance-like episodes." According to the patient, he would note slurring of speech and then would be unaware of the environment. According to his father, he would appear to stare for one minute and would not speak for one minute. There would be slight rigidity of both arms, but he would not fall to the ground, and there would not be a generalized convulsive seizure. At the end of the episode, he would be "fully aware of his environment." No automatisms would be noted during or after the episode. With adjustment of medications, the episodes were again controlled. Episodes recurred seven years later and then, after eight more years, more frequently (one to three episodes per week).

Several years later, he had a more prolonged episode, beginning with a "seasick sensation and problem with vision of the left eye" and continuing with prolonged confusion. He was "incoherent for 30 minutes and confused for an additional 30 minutes." Eventually, with additional adjustment of medication, control was again eventually achieved.

QUESTIONS:

1. What is your diagnosis?
2. Where is the pathology located?
3. What is the nature of the pathology?
4. This patient had multiple electroencephalograms. Indicate the possible findings.
5. Which medication would you utilize in management of this patient?
6. Which neuroimaging study would best demonstrate the location and nature of the pathology?
7. Select the illustration, which best correlates with this case.

Index

INDEX

A

Abnormal Development of the Nervous System 4-15
  Malformations resulting from abnormalities in growth and migration
    Anencephaly 4-17
    Agenesis of corpus callosum 4-19,
    Clinical Case 4-1: bilateral subcortical band heterotopias; double cortex 4-17.
    Clinical Case 4-2: Agenesis of the corpus callosum 4-19
      Heterotopias 4-15
      Holoprosencephaly 4-19 4-17
      Lissencephaly 4-18
      Micropolygri 4-18 , 4-14
      Macrogyria 4-18
      Microencephaly 4-19
      Porencephaly 4-19
      Schizencephaly 4-19, 4-15
      Cerebellar Agenesis 4-19
    Malformations Resulting from Chromosomal Trisomy and Translocation 4-20
  Malformations resulting from defective fusion of dorsal structures 4-20
      Spinal Bifida 4-20
      Cranial Bifida 4-20
      Arnold Chiari 4-21

Acetylcholine,
  basal ganglia 19-2, 19-5
Action Potential 5-12, 5-13
  Ionic mechanisms 5-13, 5-12
Action potential cycle 5-18
Active Transport 5-8, 5-21
Agnosia, 24-15
Akinetic (atonic seizures), 29-12
Akinetic mutism, 29-2
Alcoholic cerebellar atrophy, 20-8
Alzheimer's disease, 30-12
  and Down's syndrome, 30-15
  and basal nucleus of Meynert, 30-18
  cholinergic hypothesis ,30-18
  clinical findings and course, 30-17
  evaluation, 30-18
  genetics, 30-15
  hippocampus , 30-14
  prevalence, 30-12
  pathology, 30-13
  treatment, 30-19
Amnesia
  Amnestic-confabulatory syndrome, 30-5
  Traumatic, 30-11
Amygdala 22-6, 22-8
      Olfactory nuclei 22-6
        Corticomedial
        Basolateral
      Limbic nuclei 22-6
        Central
        Basolateral
      Ventral amygdalofugal pathway 22-7, 22-14
Amyotrophic lateral sclerosis, 9-19
Aneurysms, 13-7,26-26,26-27,*26-28*
Anticonvulsant medications
  Choice of medication, 29-22
  Molecular basis, 29-21
  Pharmacologic properties, 29-21
  Side effects, 29-19
  Teratogenic effects, 29-19
Aphasia and related disorders, 24-3,*24-3*
  Alexia without agraphia
    anomic or amnesic or nominal aphasia,24-14
    anterior type aphasia, 24-6
    Broca's area and Broca's aphasia, 24-6
      transcortical motor aphasia, 24-15
    Clinical cases
      Case 24-1, acute anterior (Broca's), 24-7
      Case 24-2, chronic anterior (Broca's), 24-8
      Case 24-3, chronic nonfluent &apraxia, 24-9, *24-9*
      Case 24-4, fluent, Wernicke's, 24-10
      Case 24-5, fluent, disconnection,& Gerstmann's,24-12
      Case 24-6, fluent, disconnection (conduction or repetition type), 24-13
      Case 24-7, dyslexia due to AVM, 24-15
    Fluent aphasias, 24-6
    conduction type fluent aphasia, 24-11
    disconnection type fluent aphasia,24-11
    dominant inferior parietal type fluent aphasia, 24-11
    dysgraphia, 24-12
    dyslexia, 24-12
    Gerstmann's syndrome, 24-12
    pure word deafness, 24-10
    transcortical sensory aphasia,24-15
    Wernicke's area and Wernicke's type of fluent aphasia, 24-9
    global aphasia, 24-14
    isolation of speech areas aphasia, 24-14
    mixed transcortical aphasia (isolation of speech areas),24-14
    selective dyslexia (a type of disconnection),24-15
    visual agnosia, 24-1
Apo E, 30-17
Apoptosis 4-4
Apraxia, 24-17
Arcuate fasciculus and role in language function, 24-5
Arsenic, 8-17, 27-28, and chapter 27 on CD ROM supplement

Index

Arteries
    Vertebral 1-17
    Basilar 1-17
Arterial supply 1-16, 1-19, table 1-13
        Anterior circulation-Internal carotid, Middle cerebral, anterior cerebral 1-16
        Posterior circulation-Vertebral, Basilar, Posterior cerebral 1-16
        Circle of Willis 1-17 1-19
Arteriovenous malformations, 26-31
Ataxia of standing body posture, 20-10
Ataxia telangiectasia, 20-18
Atrophic changes 3-21
Autonomic Nervous System 16-10, 16-12
        Cutaneous & deep vessels, glands and hair. 16-15
        Dual Innervation of Specific Structures 16-14 to 16-15
            The Eye 16-14
            Lacrimal Glands 16-14
            Salivary Glands 16-14
            The Heart 16-14
            The Lungs 16-14
            Abdominal Viscera 16-15
            Pelvis 16-15
    Enteric Nervous System 16-11
    Localization of preganglionic para-and sympathetic neurons 16-11
    Parasympathetic Nervous System and the Cranial Nerves 16-11
    Parasympathetic Nervous System and the Sacral Plexus 16-11
    Postganglionic Neuron 16-10
    Preganglionic Neuron 16-10
    Sympathetic Ganglia
        Cervical Sympathetic Ganglia 16-13
        Lumbar Sympathetic Ganglia 16-14
        Thoracic Sympathetic Ganglia 16-13
Autosomal dominant cerebellar ataxias (ADCA), 20-19
Autosomal recessive cerebellar ataxias, 20-17
Axon and Axon Origin 3-8
Axon hillock 3-8, 3-16
Axonal Guidance 3-24
    Nerve Growth Factors 3-24
Azorean disease, 20-20

B

Babinski Response 7-22, 9-2
Basal ganglia, 19-1
  anatomy, 19-1, *19-1*
  circuits, 19-1,19-3, *19-1*
  Clinical cases
    Case 19-1, Parkinson's disease, 19-12
    Case 19-2, hemichorea, 19-16, 19-18
    Case 19-3, hemiballismus, 19-17
    Case 19-4, Huntington's disease, 19-20
    Case 19-5, Huntington's disease, 19-21 (CD ROM)
    Case 19-6, hepatolenticular degeneration, 19-22 (CD ROM)
  direct pathway,19-1, 19-2
  indirect pathway,19-1, 19-3
  microanatomy of striatum, 19-4,19-5
  symptoms of dysfunction :akinesia , dyskinesia,19-4
  Syndromes of dysfunction, 19-7
    chorea, 19-14
    double athetosis, 19-21
    dystonia, 19-22
    familial paroxysmal dyskinesia, 19-24
    generalized chorea, 19-17
    hemiballism, 19-15,19-16, *19-19*
    hemichorea, 19-15, 19-16,*19-18*
    hepatolenticular degeneration, 19-21
    Huntington's disease, 19-18,*19-19,19-21*
    neuroleptic induced movement disorders,19-23
    Parkinsonian syndrome, 19-7, 19-4
    Parkinson's disease, 19-7, *19-9*
    Parkinsonism plus syndromes, 19-14, 19-15
    secondary Parkinsonism, 19-14
    tics,19-24
    Wilson's disease, 19-21
  transmitters, 19-1, 19-2, *19-1*
  Basal nucleus of Meynert, 30-18
Bacterial endocarditis, 26-30
Basilar artery syndromes,13-13,13-14,13-15,13-16
Beta amyloid, 30-15
Bladder, 9-2,18-3
    Autonomic bladder 16-16
      Sensory paralytic bladder 16-16
      Reflex bladder 16-16
Blood-Brain Barrier 3-24
Brain abscess, 27-22, *27-23,27-24*
Brain death, 29-2
Brain stem, 13-1 Brain stem 11-1 11-1, 11-2 mri
    Brain stem and eye movements 11-30
    Cases effecting eye movement 11-31
    Case 11-1 tumor in pons 11-31
    Case 11-2 pineal tumor 11-33
    Tegmentum 11-1
        Zones 11-2
        Ventricular
        Lateral
        Medial
        Central
        Basilar
    Clinical signs after lesions in tegmentum of brain stem 11-29
    Tectum 11-2
Medulla 11-3

Index

Level:
Spinomedullary junction & motor decussation 11-4, 11-4
Lower medulla at sensory decussation 11-4, 11-5
Inferior olive,11-7, 11-6
Pons 11-10
Level:
Lower pons at Facial colliculus 11-10, 11-7
Upper pons at motor and main sensory of V 11-13, 11-8
Midbrain 11-15
Level:
Inferior colliculus 11-15, 11-9
Superior colliculus and pontine basis 11-18, 11-18
Cerebral peduncles 11-21   11-11
Clinical cases in Brain Stem
Case 13-1, platybasia, 13-7 (CD ROM)
Case 13-2, total infarct PICA, 13- 11
Case 13-3, lateral medullary infarct, 13-11 (CD ROM)
Case 13-4, anterior inferior cerebellar artery territory infarct, 13-14, (CD ROM)
Case 13-5, paramedian pontine infarct, 13-15,13-16
Case 13-6, Weber's syndrome, paramedian midbrain infarct, 13-17
Case 13-7, extensive posterior cerebral artery infarct, 13-17, (CD ROM)
Case 13-8, basilar artery thrombosis, 13-17,13-15 (CD ROM)
Case 13-9, brain stem glioma, 13-19,13-24
Case 13-10, brain stem glioma, 13-19 (CD ROM)
Case 13-11, multiple sclerosis, brain stem-cerebellum predominant, 13-20,13-21
Brain tumors, 27-7 (see ependymomas, glioblastomas, gliomas, lymphomas, meningiomas, medulloblastomas, metastatic tumors ,neoplasms)
Brown Sequard syndrome, 9-2

C

Ca Channels 5-9
CASE HISTORY PROBLEM SOLVING
CHAPTER 10:SPINAL CORD, NERVE   ROOT, PERIPHERAL NERVE AND MUSCLE:
Case 10-1, cervical 7 radiculopathy secondary to acute ruptured disc, 10-2
Case 10-2, T3-4 spinal cord compression extrinsic tumor (meningioma) left lateral producing a partial Brown Sequard syndrome, 10-3
Case 10-3, Guillain Barré syndrome, 10-4
Case 10-4, Cervical spondylosis with C7 cord compression, 10-5
Case 10-5, Combined system disease, B12 deficiency myelopathy and peripheral neuropathy, 10-5
Case 10-6, Glioma of cervical spinal cord, 10-6
Case 10-7, polymyositis, 10-7
Case 10-8, syringomyelia, 10-8

CHAPTER 14: BRAIN STEM AND CRANIAL NERVES

Case 14-1, Amyotrophic lateral sclerosis (classical), 14-2
Case 14-2, Lateral (dorsolateral) medullary syndrome, 14-3
Case 14-3, Pretectal-midbrain syndrome, lacunar, 14-4
Case 14-4, brain stem infarct lower pons primarily territory of anterior inferior cerebellar artery (plus) due to vertebral basilar ischemia, 14-5
Case 14-5, Cerebellar pontine angle tumor, vestibular Schwannoma, 14-6
Case 14-6, Weber's syndrome with prior basilar vertebral TIA's and ischemia, 14-7
Case 14-7, Myasthenia gravis, 14-8
Case 14-8, Syringomyelia and syringobulbia. Coma due to CO_2 retention and low O_2 saturation

CHAPTER 25: CEREBRAL CORTEX: CORTICAL LOCALIZATION

Case 25-1, Glioblastoma rt. occipital extending to right parietal, 25-1
Case 25-2, Glioma right parietal, 25-2
Case 25-3, Prefrontal/premotor metastatic tumor, 25-3
Case 25-4, Embolus inferior division Rt. MCA, Non dominant syndrome,25-4
Case 25-5, Focal seizure Rt motor/sensory cortex secondary to metastatic tumor. Jacksonian march, and Todd's postictal palsy, 25-5
Case 25-6, Right temporal lobe seizures simple and complex partial. Epidural hematoma at age 18 months, 25-6
Case 25-7, Onset of seizures in left premotor cortex(tonic abduction of arm

with head and eye turning). Stroke at age 9 months ? arterial occlusion, 25-7
Case 25-8, Glioma left orbital frontal-anterior temporal. Ictal aggression. 25-8

CHAPTER 28: DISEASES OF THE CEREBRAL HEMISPHERES:
Case 28-1, Posterior communicating aneurysm with subarachnoid hemorrhage, 28-3
Case 28-2, Carotid TIA's involving retina and carotid border zone, 28-4
Case 28-3, MCA aneurysm with SAH. Post op bacterial meningitis and focal seizures, 28-4
Case 28-4, Subacute right sided subdural hematoma, 28-6
Case 28-5, Glioma with seizures originating right Rolandic area, 28-7
Case 28-6, MCA lenticulostriate penetrating branch infarct left internal capsule. Pontine lacune is less likely. 28-8
   Contrast case #1 Carotid TIA's borderzone and retina 28-8
   Contrast case #2 cardiogenic embolus main stem MCA 28-9
Case 28-7, Brain abscess, left temporal plus left peripheral facial paralysis, 28-9
Case 28-8, Emboli left MCA and left arm, 28-10

CHAPTER 31 GENERAL CASE HISTORY PROBLEM SOLVING:

Case 31-1, Parkinson's disease, and dementia plus old polio, 31-1
Case 31-2, Pituitary adenoma with extrasellar compression of optic chiasm, post op diabetes insipidus, 31-2
Case 31-3, Carotid TIA's hemisphere and retinal , 31-3
Case 31-4, Idiopathic epilepsy (primary generalized epilepsy),31-3
Case 31-5, Meningococcal meningitis with infarct right hemisphere and focal seizures, 31-4
Case 31-6, Alzheimer's disease, 31-5
Case 31-7, Hemichorea and hemiballismus,31-6
Case 31-8, Left cerebellar metastatic tumor, 31-7
Case 31-9, Right temporal lobe mass with herniation: brain abscess, 31-8
Case 31-10, Huntington's chorea, 31-9

Case 31-11, Wernicke's encephalopathy, 31-10
Case 31-12, Multiple complications of alcoholism and diabetes: cerebellar degeneration and peripheral neuropathy,31-12

CHAPTER 31 PART I GENERAL CASE HISTORY WITH ILLUSTRATION CORRELATION
Case 31A1, Acute intrinsic lesion of multiple sclerosis at left cerebellar inferior pontine angle, 31-A7
Case 31A2, Hydrocephalus (secondary to a pineal region tumor), 31A-7
Case 31A3, Multiple sclerosis right mid pons, 31A-8
Case 31A4, cerebellar atrophy due to hyperthermia, temporal lobe seizures, diabetic peripheral neuropathy, 31A-8
Case 31A5, hydrocephalus, peripheral neuropathy, inferior pontine tegmental infarct, 31A-9
Case 31A6, Ruptured lumbar disc ?L5-S1 31A-10
Case 31A7 cervical radiculopathy, Schwannoma, 31A-10
Case 31A8 porencephaly left Sylvian, 31A-11
Case 31A9 seizure disorder of focal origin right occipital, 31A-12

KEY TO THE PHOTOGRAPHS FOR CHAPTER 31

31-1 Ruptured disc L5-S1, 31A1
31-2 Enlarged LT cervical neuro foramina at C5-C7, probable Schwannoma 31A1
31-3 Schwannoma LT C6-C7, 31-A2
31-4 Multiple sclerosis , major lesion left pontine tegmentum, 31-A2
31-5 Multiple sclerosis Right mid pontine tegmentum and corpus callosum 31-A3
31-6 Pineal region tumor producing hydrocephalus, 31-A4
31-7 Hydrocephalus, possible aqueductal stenosis, 31-A4
31-8 Atrophy right occipital, 31-A5
31-9 Atrophy left peri sylvian, 31-A5
31-10 Cerebellar atrophy, 31-A6
31-11 Glioma (Glioblastoma) of right thalamus, 31-A6
31-12 Left frontal subdural hematoma, 31-A6

Index

Cell Membrane 5-1, 5-1
  Chemical potential 5-2
  Osmotic strength 5-2
  Osmolarity 5-4
  Osmotic Pressure 5-4
  Diffusion 5-3
Central Nerve Regeneration 3-23
Cerebral Hemispheres 1-9 1-8, 1-9
  Correlation of Structure and Function 17-10, 17-10, 17-11,22-20
    Frontal lobe 1-9 1-10
      Motor Areas , 17-11
        4   17-11
        6   17-11
        8   17-13
        44 & 45   17-13
    Lateral surface 22-21
    Orbital surface 22-21
      Clinical Case 22-4 Right frontal lobe tumor 22-4 , 22-24
    Parasagittal lesions
      Prefrontal 22-21
        Prefrontal lobotomy 22-23 22-19
    Parietal lobe 1-9   1-11, 17-14
    Postcentral areas 3,1, 2 17-14
    Functional localization
    Temporal lobe 1-10 1-12
    Temporal Lobe Overview 17-14
        Cytoarchitecture 17-1
      Study of Functional Localization
        Stimulation 17-8
        Ablation   17-9
          Auditory Associational   areas 42 & 22 17-14, 17-15
          Wernicke's Areas area 22 17-15
          Middle temporal area 21   17-14
          Inferior temporal area 20   17-14
          Posterior temporal area 37 17-14
        Entorhinal areas 27, 28, 35 17-14
          Lateral neocortical areas 22-14
          Auditory and auditory association area 22-15   17-14, 17-15
  Symptoms following stimulation of temporal lobe 22-17
  Symptoms of disease of temporal lobe 22-15
      Effects on hearing 22-19
        Aphasia   22-19
        Visual defects 22-19
        Kluver-Bucy Syndrome 22-19
        Memory 22-21
      Psychiatric Disturbances 22-20, 22-26 (Table 22-6), Supplemental CD
          Aggressive behavior   22-20
        Clinical Cases from the temporal lobe:
          Case 22-1 Simple and complex seizures from left temporal 22-17
          Case 22-2 focal seizures originating in right temporal lobe 22-18
          Case 22-3 Complex partial seizures 22-19 Occipital Lobe 17-16
        Visual receptive (calcarine cortex) area 17 17-16
        Visual associational areas 18, 19 17-16
        Confirmation of Location of pathology 17-10
    Occipital lobe 1-10 1-12
      Anatomy
      Functional localization
    Cingulate gyrus 1-11 1-13
      Anatomy
      Functional localization
      Insular Gyri 1-11 1-14
      Basal nuclei 1-11 1-15
      white matter
        Association fibers 11-12
        Commissural fibers 11-12
        Subcortical fibers 11-12
        Corticofugal 11-12
        Corticopetal 11-12
Cerebral Neuron types 17-2
  Spiny neurons 17-2
    Pyramidal 17-2, 17-2
    Stellate   17-3 , 17-2
  Smooth neurons 17-3
  Fundamental Organization
    Homogenetic 17-4
    Heterogenetic 17-4
      6-layered = molecular, external granular, external pyramidal, inner granular, inner pyramidal, multiform 17-4
Von Economo Categories; Agranular, Homotypical, Midpoint, Polar, Granular 17-6, 17-5, 17-6
Cerebrum Subcortical white matter
    Associational   17-17
    Short subcortical 17-17
      U fibers

6   Index

Long fiber bundles 17-18
Uncinate fasciculus
Superior longitudinal fasciculus
Cingulum
Inferior longitudinal fasciculus
Inferior frontal occipital fasciculus
Choroid plexus 3-30; fig 3-35

Cingulum 22-14
Circadian rhythms 16-10
Cordotomy 7-19
Corticofugal efferent fibers-corticospinal,corticobulbar
Corticopetal afferents –thalamocortical 17-16
Cortical Commissural
   Corpus callosum 17-17, 17-13,17-14
   Anterior comissure
   Hippocampal
Cortical Projectional 17-16
CRANIAL NERVES
   Components 12-2
   Cranial Nerve I, Olfactory 12-2, 12-2
   Cranial Nerve II, Optic   12-4, 12-3
   Cranial nerve III, Oculomotor 12-4, 12-4
   Clinical Case 12-1; 12-6 12-4
   Cranial Nerve IV, Trochlear 12-7 12-4
   Cranial Nerve V, Trigeminal 12-7, 12-
      Trigeminal   15-10
      Ganglia
      Descending/spinal nucleus
      Chief sensory nucleus
      Mesencephalic nucleus
   Sensory nuclei
     Taste solitary nucleus 12-20
     General sensation   12-20
Lesions of Peripheral Branches of Vagus
  Superior laryngeal 12-21
  Recurrent laryngeal 12-21
  Supranuclear lesion 12-21
Cranial Nerve XI, Spinal Accessory 12-21, 12-12
Cranial Nerve XII, Hypoglossal 12-22, 12-13
  Dysfunction of cranial nerves
    Motor 11-28
    Sensory 11-28
  Effects of Extrinsic Lesions on Cranial Nerves 12- Table 12-5 &Cases on CD
  Cavernous Sinus syndrome
    Case 12-6 on CD

Cerebellopontine Angle Syndrome
    Case 12-7 on CD
Vestibular
    Case 12-8 Vestibular Schwanoma
Jugular Foramen Syndrome
    Case 12-9 on CD
Voluntary Control of Cranial Nerves
  Corticobulbar-Voluntary Control of Cranial Nerves V, VII, IX-XII 12-23
  Corticomesencephalic -Voluntary Control of Eye movements via Cranial nerves III,IV and VI 12-23
    Motor nucleus
  Clinical Case 12-2 ;Trigeminal Neuralgia 12-10
  Clinical Case 12-3 Ophthalmic Herpes 12-11
  Cranial Nerve VI, Abducens 12- 11 12-4
  Cranial Nerve VII, Facial 12-11 12-6
    Parasympathetic ganglia 12-13
      Submandibular
      Pterygopalatine
  Clinical Case 12-5: Bell'sPalsy 12-14
  Cranial nerve VIII,
  Vestibuloaccoustic 12-15, 12-7
    Cochlear 12-15, 12-8
      Olivocochlear bundle   12-16
    Vestibular 12-17, 12-7
    Vestibular nuclei 12-9
  Cranial Nerve IX, Glossopharyngeal 12-18 12-10
  Cranial Nerve X, Vagus 12-19, 12-11
    Motor nuclei
      Dorsal motor nucleus of X 12-20
      Ambiguous nucleus of X   12-20
Cranial Nerve Differentiation 4-10
  Innvervation of muscles of somite origin CN III, IV, VI, XII 4-9
  Innervation of muscles originating in pharyngeal arches 4-9
    First arch –trigeminal 4-10
    Second arch – facial 4-10
    Third arch - glossopharyngeal 4-10
    Arches 4-6 vagus 4-10
  Innervation of preganglionic parasympathetic smooth muscles 4-10
    Cranial nerves III, VII, IX, X
  Innervation of Special Sense organs 4-10
    Cranial nerves I, II, VIII

Index

Extrinsic syndromes, 13-2
   Cerebellar pontine angle,13-2
   Developmental abnormalities at the foramen magnum,13-7
   Foramen magnum syndromes, herniations meningiomas, 13-7
   Lateral cerebellum, 13-4
   Midbrain, pretectal, pinealoma, 13-5
   Midbrain, tentorial syndromes, herniations meningiomas,13-5
   Midline cerebellum, 13-3
  Intrinsic syndromes
   Gliomas of brain stem, 13-19
   Multiple sclerosis, 13-19,13-20, 13-21
   Pontine hemorrhage, 13-18
   Vascular syndromes, 13-8
Brain Stem Centers
     Respiration 11-23
     Cardiovascular 11-23
     Coughing 11-25
     Deglutition 11-24
     Vomiting 11-24
     Emetic 11-24
CAG trinucleotide repeats
 in Huntington's disease, 19-19
 in spinocerebellar atrophy (SCA), 20-19
     in x linked recessive bulbospinal neuronopathy, 9-18
Carcinomatosis of the meninges, 27-18,*27-22*
Cataplexy, 29-26
Caudate nucleus, 19-1
Central pattern generators, 18-1
Central pontine myelinolysis, 27-27, (see fig 13-28 on case history CD ROM)
Cerebellar peduncles
    Superior cerebellar peduncle 15-16
    Middle cerebellar peduncle 15-16
    Inferior cerebellar peduncle15-16
    Lesions of cerebellar peduncles 11-27
Cerebellum, 20-1
   anatomy, 20-1
   afferents, 20-3
     cytoarchitecture, 20-2
    Dysfunction 11-26
   efferents, 20-3
   longitudinal divisions, 20-1
   transverse divisions, 20-1,20-2
 Cerebellum Clinical Cases
  Case 20-1, medulloblastoma, floccular nodular syndrome, 20-7
  Case 20-2, midline, metastatic, 20-7
  Case 20-3, alcoholic cerebellar degeneration, 20-9, (CD ROM)
  Case 20-4, anterior lobe syndrome, spinocerebellar degeneration, 20-9

  Case 20-5, lateral hemisphere syndrome, hemangioblastoma, 20-5, (CD ROM)
  Case 20-6, cerebellar infarct, (PICA), 20-16
  Case 20-7, arteriovenous malformation, 20-16, CD ROM)
 effects of disease, 20-5
 functional correlations, 20-4
 major syndromes, 20-6
  alcoholic cerebellar atrophy, 20-8
  anterior lobe, 20-8
  atrophy related to fever and status epilepticus,20-9
  floccular-nodular lobe, 20-6
  hemorrhage, 13-8
  lateral cerebellar hemisphere, 20-10
  midline cerebellar tumors, 20-7
  nuclei, 20-1
  neocerebellar or middle-posterior lobe, 10-10
  paraneoplastic subacute cerebellar degeneration, 20-21
   peduncles, 20-13
  spinocerebellar degenerations, 20-9,20-16
  vascular syndromes,20-14, 20-17
 Topography of representation,20-4
Cerebellar ataxia with myoclonus, 20-17
Cerebral dominance, 24-1
Cerebral embolism, 26-20
Cerebrospinal fluid,
 content, 2-26
 examination, (lumbar puncture),2-26
Cervical spondylosis, 9-4
Chiari ,(Arnold Chiari) malformation, 9-15,9-16
Chorea, 19-14
Claustrum (norepinephirine) 17-9
Climbing fibers, 20-2,20-3
Cocaine, 27-27,26-6, (see chapter 27 supplement on CD ROM)
Colloid cyst of third ventricle, 27-18,*27-19,27-20*
Columnar organization of cerebral cortex, 21-1
Coma, 29-1,29-28
 Acute onset 29-27,29-28
 Care of comatose patient, 29-28
 Prognosis,29-28
 Prolonged disturbance of consciousness, 29-28
Compression fractures, 9-3
Computerized axial tomography scanning, (CT scans), 2-13, 2-16, 2-17,2-19,2-28
Concept of level of lesion at spinal cord level, 2-3
Concept of extrinsic versus intrinsic, 2-8,9-1,9-10,13-1,13-2
Confusion, 29-2
Consciousness,29-1
Corpus callosum
   Role in learning, 30-4

Index

transfer of information, 24-18
Cortical shock, 18-12
Corticobasal degeneration (CBD), 19-14,19-15
Corticorubral spinal system, 18-16
Craniopharyngioma, 27-18,*27-19*
Creutzfeldt-Jakob disease, *30-18,30-20*

D

Decerebrate rigidity 11-27
Deep tendon stretch reflexes, 2-4
Degeneration 3-20
Delayed or disordered motor development, 18-25
Delayed response test and delayed alternation, 18-24
Delirium, 29-2
Dementia, 30-1,30-11
  Alzheimer's disease 30-12
  Creutzfeldt-Jakob, 30-18,30-20
  Frontal-temporal, 30-20
  General paresis, 30-19
  Lewy body, 30-20
Dentate gyrus 22-11
        Cytoarchitecture 22-11, 22-12A
  Pick's, 30-20
Dermatomes – radicular patterns, 2-2, *2-4, 2-5,2-6*
Dermoids, 27-18 Development of Blood vessels in the brain 4-4
Development /Differentiation 4-5, 4-4
       of ventricular system 4-5
       Spinal Cord 4-5
       Mesencephalon 4-7
        Prosencephalon 4-8 4-8, 4-9
       Rhombencephalon 4-7
Development of dominance, 24-2
Development of motor function, 18-9, 18-25
Development of speech, 24-3
Diagnostic studies in neurology, 2-12
  Angiography, (arteriography), 2-25
  Cerebrospinal fluid examination, 2-26
  Computerized axial tomography scanning, (CT scans), 2-13, 2-16, 2-17,2-19,2-28
  Duplex scans, 2-26
  Electroencephalography, (EEG), *2-21*
  Electromyography, (EMG), 2-13
  Evoked potentials, 2-15,2-17,2-19, 2-25
  F response or wave, 2-15
  Functional MRI, 2-21
  H reflexes, 2-15
  Lumbar puncture, 2-26
  Magnetic resonance angiography, (MRA), 2-26,*2-26,2-27*
  Magnetic resonance imaging, (MRI scans), 2-13,*2-16,2-18,2-20*
  Myelography, 2-14
  Nerve conduction velocity, 2-13,*2-14*
  Pneumoencephalography, 2-17,2-19
  Positron emission tomography (PET scan), 2-20,*2-22*
  Radioactive brain scans and flow studies, 2-20
  Single photon emission computed tomography, (SPECT), 2-20,*2-21*
  Transcranial doppler, 2-26
  Ventriculography, 2-18
  X-rays, 2-13,2 –15
Degenerations,27-26, see each clinical chapter and chapter 27 on CD ROM supplement
Devic's syndrome, 9-13
DIENCEPHALON
        Epithalamus 16- 7
            Metathalamus 15-8
        Thalamus 15-3
            Anterior nuclei 15-3
            Anterior tubercle 15-4, 15-2
            Medial nuclei   15-3
                Magnocellular 15-4
                Parvicellular 15-4
            Midline nuclei 15-3, 15-4
            Intralaminar   15-3, 15-5
                Parafascicular 15-5
            Lateral nuclei
              Ventrobasal 15-3
              Pulvinar   15-5
                Medial division
                Lateral division
                Inferior division
Differential diagnosis, 2-1
  Clinical case
    Case 2-1, amyloid angiopathy leg area 2-27
  Location of lesion, overview 2-1
    basal ganglia, 2-7
    brain stem, 2-4
    cerebellum, 2-8
    cerebral cortex, 2-7
    muscle, 2-1
    nerve root, 2-2
    neuromuscular junction. 2-2
    peripheral nerve, 2-2
    spinal cord, 2-2
  Type of pathology, 2-8
Disc disease, 8-21
  Cervical, 8-19,8-22,9-4
  Lumbar, 8-19, 8-20
Diseases of Muscle
  Muscular Dystrophies 6-11
    Duchenne' Muscular Dystrophy 6-11 , 6-14, Case 6-1pp 6-5
    Becker's Muscular Dystrophy   6-11
    Myotonic Myopathies   6-13
    Congential myopathies   6-14
    Metabolic Myopathies   6-14
    Periodic Disorders of Muscles

Index

Familial Periodic Paralysis 6-15, Case 6-2pp 6-15
Acquired Disorders of Muscle 6-16
    Polymyositis 6-16, 6-15, Case 6-3 ; pp 6-17
Disease of the Neuromuscular Junction
  Postsynaptic Disorders
    Myasthenia gravis 6-17, Case 6-4; pp 6-20
      Anti cholinesterases,
      Thymectomy 6-20
      Pharmacological & toxic 6-21
      Reversible agents 6-21
      Neurotoxic agents 6-22
        pesticides organophosphates 6-22
          snake toxins 6-22
    Pre Synaptic Disorders 6-22
      Eaton-Lambert 6-22
      Botulism   6-23
      Antibiotics   6-23
      Spider venom (black widow) 6-23
Disorders of the Autonomic Nervous System
    Eye – Edinger-Westphal nucleus,
    Horner's, Argyll-Robertson pupil 16-15
    Blood Vessels- Raynaud's 16-15
      Heart & GI   16-15
       Bladder 16-15 to 16-16
        Micturition 16-16
        Uninhibited bladder 16-16
Dissociated sensory loss in syringomyelia, 9-14
Domoic acid, 30-10
Dopamine receptors, 19-1, 19-3
Dopaminergic systems, 19-2,19-5
Double athetosis, 19-21
Drop attacks, 29-12
Duplex scans, 2-26
Dysarthria, 24-1,20-13
Dyssynergia cerebellaris myoclonica, 20-17
Dystonia, 19-22
Dystrophic neurites, 30-14

E

Early experience and brain development, 30-4
Effectors 3-15
Eye movements 11-30
    Clinical Cases effecting eye movements 11-31
      Case 11-1 tumor in pons 11-31
      Case 11-2 pineal tumor 11-33
Electroencephalography, (EEG), 2-21
EEG normal, 2-21,*2-22 and CD ROM supplement*
EEG abnormalities, 2-21,2-22,*2-23,2-24*,29-3
  Focal slow waves, 2-21,2-23
  Focal spikes and sharp waves, 29-3,*29-4,29-5*
  Focal voltage suppression, 2-21,*2-23*
  Generalized slow wave activity, 2-22,*2-24*

  Generalized seizure discharges, 29-4,29-7, *29-9,29-10,29-13,29-17,29-18*,
  Generalized burst suppression and isolated cortex , 17-20
  Generalized suppression, see brain death and neocortical death
Electromyography, (EMG), 2-13
Encephalitis, 27-26,*27-26* (see chapter 27 on CD ROM Supplement)
Encephalomalacia, (infarction), 26-3,*26-4*
Endocrine disorders affecting nervous system, 27-27(see chapter 27on CD ROM supplement)
Enteric Nervous System 16-11
Entorhinal region 22-10
Ependymal Cells 3-19, 3-27
Epidermoids, 13-3,27-18
Epilepsy and seizures , focal or partial, 29-4,29-5
  Causes, 29-5
  Multifocal discharges, 29-6
  Pathophysiology, 29-5
  Regional variations in capacity for discharge 29-5
  Secondary epileptogenesis,29-6
  Spread of discharge, 29-6
Epilepsy, management, 29-19
  Use of and side effects of anticonvulsant medications, 29-19, 29-20,29-21 29-22,
Epilepsy, status, 29-19 ,29-22
Epilepsy, primary generalized or idiopathic, 29-7, 29-8
  Classification 29-7
    Childhood absence epilepsy, 29-8
    Generalized tonic clonic, 29-12,29-14
    Juvenile absence epilepsy, 29-9
    Juvenile myoclonic epilepsy(JME),29-9
    Overlap of syndromes, 29-15
    Photosensitivity and genetics, 29-10,29-11
  Molecular biology, 29-13,29-17
  Neuropathology, 29-15
  Pathophysiology, 29-15,29-16,29-18

Epilepsy, secondary or symptomatic generalized, 29-7
Ependymoma, 20-7,*13-4,27-14*
Epidural abscess, 9-7
Evoked potentials, *2-15,2-17,2-19, 2-25*
Excitatory Postsynaptic Potential EPSP 7-8, 7-16, 7-17
Extracellular Space   3-25
Eye
  Structure of Eye 23-1, 23-1
      Outer tunic cornea and sclera
      Middle tunic vascular and pigmented
      Inner neural tunic 22-3
        Photoreceptors 23-3 23-2

Index

Rods – vision in dim light and night vision 23-4
    Cones - color vision 23-5
Lens 22-3
Pupil
    Pupillary muscles 23-2
    Pupillary reflexes 23-2
    Light reflex
    Accommodation
    Fixed pupil

F

Fetal Alcohol Syndrome 4-20
F response or wave, 2-15
Familial paroxysmal dyskinesia, 19-24
Fixation system for eye movement, 18-20
Fornix 22-13, 22-14
    Fimbria 22-14
    Portio alveus 22-14
    Portio tenia 22-14
    Portio corpus 22-14
    Portio columnaris 22-14
Fracture dislocations, 9-3
Friedreich's ataxia, 20-17
Frontal eye fields,18-19, 18-21
Functional MRI, 2-21

G

Gait disorders of the elderly, 18-25
Gamma System 7-12 to 7-14, 7-23, 7-24
Generalized chorea, 19-17
Glands Associated with the Brain ,Pineal and Pituitary 1-14
Glia-Supporting cells of the CNS 3-16
    Astrocytes 3-17, 3-23, 3-24
    Oligodenrocytes 3-17, 3-23, 3-24
    Endothelial cells 3-17
    Ependymal Cells 3-19, 3-27
    Microglia 3-18, 3-26
        Ameboid microglia 3-18
        Resting microglia 3-18
        Activated microglia 3-18
        Reactive microglia 3-18
        Multinucleated cells 3-18, 3-31
    Mononuclear cells 3-18; table 3- 10
    Monocytes 3-18
    Pericytes 3-18
Golgi neuron method 3-1 3-1, 3-2
Golgi Tendon Organs 7-14
Golgi type I neuron 1-2, 3-1
Golgi type II neuron 1-2, 3-2
Gliomas and glioblastoma multiforme, 18-18,27-10,27-11,*18-19,27-12,27-13,27-14*
Globus pallidus, 19-1
Goldman Equation 5-7
Golgi cells of cerebellum. 20-2
Granule cells of cerebellum,20-2
Guidelines
    Localizing disease in brain stem 11-25

H

H reflexes, 2-15
Hallervorden Spatz disease, 19-14
Headaches, 27-8
  Brain tumors, 27-8
  Cluster, 27-8
  Migraine, 27-8
  Muscle tension, 27-8
  Occipital/cervical neuralgia, 27-8
Hemangioblastoma, 20-12, 20-13
Hemorrhage, intra cerebral, 2-28,13-18
Hemiballism, 19-15,19-16,19-19
Hemichorea, 19-15, 19-16,19-18
Hereditary spastic paraplegia, 9-21
Herpes simplex
  Encephalitis, 30-11
  Sensory ganglion, 8-24
Herpes zoster
  Dorsal root ganglion, 8-23
Hippocampus 22-8 22-10
 Cytoarchitecture 22-11
 And dementia
 And memory, 30-7
 Herpes simplex encephalitis, 30-11
 Selective vulnerability, 30-10
 Stimulation, 30-10
 Vascular syndromes, 30-8
 Vulnerability 22-11
 and Memory 22-13
 and Cingulate gyrus 22-13
Histogenesis 4-2
 Hensen's node 4-1
 Primitive streak
 Neuroblasts 4-1,4-2
  Neural plate 4-1
  Neural tube 4-2
  Spinal cord 4-3
  Brain stem 4-5
 Marginal zone 4-2
Hormones & Releasing factors produced in Hypothalamus 16-7; table 16-2
    Adrenocorticotropic Hormone ACTH 16-7
    Corticotropin releasing factor 16-7
    Growth Hormone -Somatotropic factor 16-7
    Lactogenic Hormone (Prolactin) 16-7
    Lactogenic inhibitory hormone 16-7
    Leutinizing releasing factor 16-7
    Melanocyte Stimulating Hormone 16-7
    Melanocyte Inhibitory Hormone 16-7
        MSH releasing factor 16-7
    MSH inhibitory factor 16-7

Index

Thyrotropic releasing factor 16-7
Hormones Produced in Hypothalamus
    Vasopressin (supraoptic) anti ADH 16-6
    Oxytocin (periventricular nucleus) 16-6
Hormones Produced in Adenohypophysis
    adenohypophysis 16-7; 16-11
    Gonadotropins; leutinizing hormone LH 16-7
    Follicle stimulating hormone 16-7
    Growth hormone somatotropin(STH) 16-7
    Thyrotropic Stimulating Hormone 16-7
    Thyrotropic Inhibitory Hormone 16-7
Intermediate/mantle zone 4-3
  Subventricular zone 4-2
  Ventricular zone 4-3
  Germinal Cells 4-3
  Growth Cone guidance 4-3
  Programmed Cell Death 4-4
Huntington's disease, 19-18,*19-19,19-21*,
Hydrocepahlus 4-22, 4-19
Hypothalamic nuclei
    Anterior region: 16-2, 16-3
      Preoptic,
      Supraoptic
      Pariventricular
      Anterior
      suprachiasmatic
    Middle region: 16-2
      Dorsomedial
      Ventromedial
      Lateral
      Arcuate
    Lateral zone
      Lateral nucleus
      Lateral tuberal nucleus
    Posterior region
      Posterior
      Mammillary nuclei
Hypothalamic-Hypophyseal tract 16-5, 16-8
  Supraoptic andparaventriuclar origin 16-3
Hypothalamic-Hypophyseal portal system 16-5, 16-10
  Supraoptic nucleus of hypothalamus, and osmoregulation 16-5
  ADH and vasopressin 16-3
    Suprachisamatic nucleus of hypothalamus and biological clock 16-3
  Tuberal nuclei of the hypothalamus forms tuberoinfundibular pathway

  Tuberoinfundibular pathway and releasing factors 16-5, 16-6
  Ventromedial hypothalamic nuclei and feeding 16-3, 16-9
Hypothalamic Pathways
    Medial forebrain bundle 16-4, 16-5
    Dorsal longitudinal fasciculus 16-4
    Retinohypothalamic fibers to suprachismatic nucleus 16-4
    Fornix to mammillary nuclei 16-4
    Periventricular system 16-4
    Mammillotegmental tract 16-4
Hypophysiotrophic area 16-6
Hypothalamus and body temperature 16-8
Hypothalamus center for heat loss 16-8
Hypothalamus center for heat production and conservation 16-8
  And Water Balance and Neurosecretion 16-9
  And Food intake 16-9
    And Sexual Behavior 16-9
    And Sleep Cycle 16-9
    and Emotions 16-9
    and light levels 16-10
Clinical Case 16-1.Adenoma of pituitary gland with acromegalic features 16-1, 16-1
Hypothalamic dopaminergic system, 19-6

I

Imprinting, 30-1
Inhibitory Postsynaptic Potentials IPSP 7-9, 7-16
Infections of the cerebral hemispheres, 27-18
  Clinical cases
    Case 27-5, 27-22, subdural empyema (CD ROM)
  Focal, 27-19, acute (See Brain abscess, subdural empyema)
  Focal chronic 27-23
  Diffuse ,generalized, 27-23 (See meningitis and encephalitis)
Inhibitory surround (surround inhibition), 29-6
Instinctive behavior, 30-1
INTERNAL CAPSULE 15-5, 15-6
    Anterior limb
    Genu
    Posterior limb
      Retrolenticular 15-7
      Sublenticular portion 15-7
      Thalamolenticular portion 15-7
    Thalamic radiations
      Anterior radiations 15-8
        Inferior radiations 15-8
        Posterior radiations 15-8
        Superior radiations 15-8

Interrelation primary motor, premotor, prefrontal, 18-9

K

Kindling, 30-4

L

Lamination of major pathways 7-20
Language function evaluation, 24-4
Lateral (dorsolateral) medullary syndrome, 13-10
Lateral premotor area (PMA), 18-17
Lead toxicity,8-17, 27-28 and chapter 27 on CD ROM supplement
Learning,30-1
 declarative (explicit), 30-1
 non declarative (implicit), 30-1
Leptomeningeal anastomosis, 26-3
Leukodystrophies, See chapter 27 on CD ROM supplement
Lhermitte's sign , 9-29
Limbic system 22-1, 22-3, 22-4, 22-5
Lipidosis, See chapter 27 on CD ROM supplement
Localization of disease process, 2-1
Locked in syndrome,(pseudo akinetic mutism or pseudo coma), 29-2
Long term potentiation, 30-4
Lumbar puncture, 2-26
Lymphomas, primary central nervous system, 27-15

M

Machado-Joseph disease, 20-20
Magnetic resonance angiography, (MRA), 2-26
Magnetic resonance imaging, (MRI scans), 2-13,2-16,2-18,2-20
Malformations
 -characterized by excessive growth of
 ectodermal and mesodermal tissue
 affecting skin, nervous system and other tissues
  Tuberous sclerosis 4-21
  Neurofibromatosis 4-21
  Cutaneous angiomatosis with associated
Malformations of the CNS 4-21
  Sturge-Weber Syndrome 4-22
Malformations resulting from abnormalities in the ventricular system
  Syringomyelia 4-22
  Syringobulbia 4-22
Mammillothalamic pathway 16-4
Manganese toxicity producing Parkinsonian syndrome, 19-10
Medullary vascular syndromes, 13-10
Medulloblastoma, 13-3, 20-7, *13-4, 20-8,27-10*
Membrane Basis of Integration 7-8
Memory, 30-1
 Clinical cases Effecting Memory
  Case 30-1, Wernicke Korsakoff, 30-5
  Case 30-2, transient global amnesia, 30-9
  Case 30-3, temporal lobe seizures, 30-10 (CD ROM)
  Case 30-4, Alzheimer's disease, 30-17
  Case 30-5, general paresis, 30-19 (CD ROM)
  Case 30-6, probable Lewy body dementia, 30-20 (CD ROM)
  Case 30-7, thalamic dementia –glioblastoma, 30-18
 Neurobiological mechanisms, 30-3
 read out phase, 30-2
 stages of human memory, 30-2
  immediate –short term, 30-2
  long term –labile stage, 30-2
  long term –remote stage, 30-2
  disorders,30-4
  recent memory, 30-4
Meningiomas, 27-15,27-17,*27-16,27-18*
 Cerebellar pontine angle,2-18 13-3
 convexity, *27-16*
 foramen magnum13-6
 intradural, ,compressing spinal cord, 9-8,*9-5*
 parasagittal, 1- ,18-13,*18-14,18-15*
 tentorial , 13-5
Mental retardation, 30-1
Meningitis, bacterial 27-23,,27-25,27-26,*27-24*
Meningitis ,carcinomatous,,27-18, *27-22*
Meningitis ,non bacterial, (see chapter 27 on CD ROM supplement)
Mercury, 27-28 and chapter 27 on CD ROM supplement
MERRF(myoclonus epilepsy with ragged red fibers), 20-18
Mesolimbic dopaminergic system, 19-5
Mesocortical dopaminergic system, 19-5
Metastatic tumors, 27-17,*27-20, 27-21,27-22*
Micturition 16-16
Midbrain vascular syndromes, 13-15,13-16,13-17
Mini-mental status examination (MMSE), 2-12
Mirror focus, 30-4
Mononeuropathies, 8-4, (see under nerve), (for cases see peripheral nerve)
Mossy fiber, 20-2
Motivational drives and learning, 30-3
Motor cortex, 18-9,18-10
 recovery of function after damage, 18-12
Motor development, 18-9
Motor Pathways in Spinal Cord 7-19
  Upper Motor Neuron Syndrome 7-19
   Babinski respone 7-32
  Lower Motor Neuron Syndrome 7-19
Motor system, 18-1
 Clinical cases:
  Case 18-1, focal seizures –parasagittal meningioma. 18-13

Index

Case 18-2, focal seizures-premotor-oligodendroglioma, 18-18
Case 18-3, normal pressure hydrocephalus 18-21, (CD ROM)
MPTP induced Parkinson syndrome, 19-10
Multiple sclerosis
   Spinal cord 9-27, *9-28,9-29*
   Brain stem, 13-19, *13-20,13-21*
   Cerebral hemispheres,27-27,*27-27,27-28* and CD ROM supplement for Chapter 27
Multisystem atrophy (MSA), 19-15
Myelography, 2-14
Myelin sheath 3-9, 3-17, 3-28
Myelination 3-9
Muscle spindle 7-11, 7-20
Myoclonus
  Anatomical basis,29-10
  Etiological and prognostic classification, 29-10

N

Narcolepsy, 29-26
Negative motor response, 18-23
Neglect syndrome, 21-8
Neocortical death, 29-3
Neoplasms of the cerebral hemisphere, 27-7 (see gliomas, glioblastomas, meningiomas, metastatic tumors, oligodendrogliomas)
  Clinical cases
    Case 27-3, grade 2 astrocytoma,27-11 (CD ROM)
    Case 27-4, Glioblastoma multiforme,27-14,*27-12,27-13*
Nernst Equation 5-6
Nerve conduction velocity, 2-13,2-14
Nerves
  brachial plexus, 8-4
  lateral femoral cutaneous nerve of thigh, 8-8
  long thoracic, 8-7
  lumbar plexus, 8-8,8-9,8-12
  median, 8-7
  obturator, 8-9
  peroneal, 8-10,8-14
  radial, 8-8
  sacral plexus, 8-8,8-10,8-13
  sciatic, 8-9,8-14
  tibial, 8-11,8-14
  ulnar, 8-7
Nerve Growth Factors 3-24
Nerve Muscle Junction
  Endplate 6-7, 6-10
    Acetylcholine, Ach 6-8
    Denervation sensitivity 6-9
    Anticholinesterase, neostigmine, endrophonium, eserine 6-9
    Reinnervation 6-10
Nerve root, 8-19
Neural Crest cells 3-20, 4-2

Neurulation 4-1, 4-1
Neurofibrillary tangles, 30-13,*30-14 30-15*
Neurofibromatosis, (von Recklinghausen's disease), 9-10
Neuroleptic induced movement disorders, 19-23
Neurological history in diagnosis, 2-9
Neurological examination, abbreviated, 2-12
Neurological examination, detailed, 2-9
Neuromyelitis optica, 9-13
Neurosyphilis, 9-22, 9-23
Neurofibrillar tangles. 3-8
Neuron
    Dendrites 3-2
    Soma 3-2
    Nucleus 3-7, 3-8, 3-9
    Barr Body 3-9
    Endoplasmic reticulum 3-4
    Rough Endoplasmic reticulum/ Nissl Substance 3-10, 3-11. 3-12
    Smooth endoplasmic reticulum 3-4 3-10
    Lysosomes 3-4, 3-10, 3-11
    Perixosomes 3-5
    Mitochondria 3-5, 3-10, 3-11, 3-12
    Centrosomes 3-6
     Inclusions 3-6 3-11A
    Glycogen 3-6, 3-11B
    Lipid Droplets 3-6, 3-11
    Neurosecretory Granules 3-6, 3-13B
Neuronal Cytoskeleton 3-7, 3-15
    Microtubules 3-7
    Microfilaments 3-7
    Intermnediate filaments 3-7
Neuronal migration 4-14
Neuronal Organelles 3-8
Neuron types
    Unipolar, Bipolar,Multipolar 1-1, 1-1
    Pyramidal 17-2, 17-2
    Smooth neurons 17-3
    Spiny neurons 17-2
    Stellate 17-3 , 17-2
Neurophysiology of Cerebral Cortex 17-21
    Basis of EEG found on CD Rom for chapter 17
    Activity of Isolated Thalamus 17-21
Activity of Isolated Cerebral Cortex 17-21
    Evoked Potentials 17-21, 17-17
      Primary evoked response
        Augmenting response
      Recruiting response
        Arousal 17-22
    Capacity for Focal Discharge 17-24, 17-21
    Capacity for Contralateral spread of discharge 17-25, 17-22

Capacity for bilateral synchronous discharge, experimental epilepsy 17-27, 17-26, 17-27
Neurotransmitters 3-15
Nigral striatal dopaminergic system, 19-2
Nociception and Pain 7-17 to 7-18
    Receptors 7-17
    Modulation of pain transmission 7-18
    Chordotomy to relieve pain 7-18, 7-29
Nondominant hemisphere functions, 24-17,21-8
Normal pressure hydrocephalus, 18-26
Nucleus acumbens 22-5
Nutritional disorders, 27-26,27,27 See chapter 27 on CD ROM Supplement

O

Olfactory system 22-2, 22-2
    Hypothalamus
        Mammillary nuclei 22-5
        Mammillothalamic tract
    Epithalamus 22-5
Olivopontocerebellar atrophy (OPCA), 19-15

P

Pain 15-15 15-11
    From the body 15-15
    From the head 15-11
    Gate 7-14
    Referred pain from the viscera 15-15
Papez Circuit 15-15, 15-42, 2-13
Parallel fibers, 20-2
Paraneoplastic subacute cerebellar degeneration, 20-21
Parietal lobe,21-1
  Clinical cases
    Case 21-1, focal sensory seizure in face secondary to a glioblastoma, 21-3,*21-4,21-5*
    Case 21-2, metastatic tumor to postcentral gyrus with pain syndrome, 21-5
    Case 21-3, metastatic to non dominant parietal lobe with nondominant syndrome ,21-8,*21-9,21-10*
    Post central gyrus, 21-1,21-3,*21-2*
    Parietal lobules, dominant, 21-7
    Parietal lobules, non dominant, 21-8
    Parieto-occipital eye fields, 18-19
Parkinsonian syndrome, 19-7, 19-4
Parkinson's disease, 19-7
  clinical symptoms and signs, 19-8
    Clinical Case 1-1 parasagittal meningioma 1-17
  differential diagnosis, 19-13
  etiology, 19-9
  management, 19-11
  pathology, 19-8, *19-9* .
Parkinsonism plus syndromes, 19-14, 19-15
Parkinsonism-dementia –ALS complex, 19-14

Patch Clamp 5-9
Pathways
    Basic organization of pathways 12-13
        Cells of origin
        Location of tract in the brain
        Site of termination of pathway Function
        Motor control of hand the corticospinal pathway 1-13
    Motor
        Corticobulbar 15-9, effects of disruption 11-25
        Corticonuclear 15-9)
        Corticospinal 15-9 effects of disruption 11-25
        Descending autonomics 15-9
        Dorsal longitudinal fasciculus 15-9
        Medial longitudinal fasciculus (MLF)
        Monoamine containing 15-19
        Rubrospinal 11-9
        Tectospinal 11-15
    Sensory-
        Anterolateral 15-11
        Auditory 15-17
        Lateral lemniscus 15-17
        Cuneocerebellar 15-16
        Spinothalamic 15-15 15-11
            Effects of disruption 11-27
        Solitary tract
        Dorsal spinocerebellar
        Ventral spinocerebellar
        Medial lemniscus 15-12 15-9
            Effects of disruption 11-27
        Pain 15-15 15-11
            From the body 15-15
            From the head
            Referred
        Posterior columns 15-12
            Fasciculus gracilis 15-12
            Fasciculus cuneatus 15-12
        Trigeminal
            Thalamic 15-14
        Vestibular 15-17
        Spinocerebellar 15-12
        proprioception 15-12
        tactile discrimination 15-12
Hemiballismus 15-9
Subthalamus 15-12
Ansa lenticularis 15-12
Subthalamic fasciculus 15-12
Thalamic fasciculus 15-12
Peripheral vs Central Nerve Structure 3-11, 3-19
    Perineurium 3-11
    Epineurium 3-11
    Endoneurium 3-11
    Sheath of Schwann 3-11
    Central Nerve Structure 3-12

Index

Peripheral nerves and nerve roots, chapter 8
  Clinical cases
    Case 8-1, brachial plexopathy, Pancoast tumor, 8-6
    Case 8-2, long thoracic nerve, 8-7, (CD ROM)
    Case 8-3, median nerve, carpal tunnel, 8-7
    Case 8-4, diabetic mononeuropathy, lumbar-sacral plexopathy, 8-12
    Case 8-5, Guillain Barré syndrome, 8-16
    Case 8-6, Charcot Marie Tooth disease, HMSN type I, 8-18, (CD ROM)
    Case 8-7, lumbar radiculopathy, 8-20
    Case 8-8, cervical radiculopathy, 8-22
  Disorders of, 8-1 (see mononeuropathies and polyneuropathies)
  Distribution, compared to radicular segmental, 2-5, 2-6
Peripheral Nerve Regeneration 3-22, 3-34
Perfusion, 26-2
Pes cavus, 9-26
Physiology of Dendrites and cell body
    Decremental conduction 5-22, 5-22
      Calcium channels 5-22
        Transient calcium channels 5-23
        Long lasting calcium channels 5-23
        Ligand gated calcium channels 5-24
        Regulation 5-24, 5-24
Pineal body 16-10
Pinealoma, 13-5, *13-5*
Pituitary tumors, 27-16, *23-10, 2-18*
Plasticity of motor and sensory cortex, 18-12, 30-4
Pneumoencephalography, 2-17, 2-19
Poliomyelitis, 9-17
Polyneuropathies, 8-15, (for cases, see peripheral nerve)
  Acrylamide, 8-17
  Amyloid, 8-18
  Antineoplastic, 8-17
  Arsenic, 8-17
  Charcot Marie Tooth disease, 8-18
  Chronic inflammatory demyelinating, (CIDP)
  Diabetic, 8-17
  Diphtheritic, 8-17
  Familial dysautonomia, 8-18
  Guillain Barré syndrome, 8-15
  Hereditary sensory motor, (HMSN), 8-18
  Hexa carbon industrial solvents, and other agents, 8-17
  HIV, 8-18 (see also chapter 27)
  Isoniazid, 8-17
  Miller Fisher variant or syndrome, 8-15
  Lead, 8-17
  Monoclonal gammopathy, 8-18
  Nutritional deficiencies of B vitamins, 8-17
  Porphyria, 8-17
  Refsum, 8-18
  Remote effect of malignancy, 8-18
  Thalidomide, 8-17
  Trichloroethylene, 8-17
  Tri ortho cresyl phosphate, 8-17
  Uremic, 8-17
Pontine vascular syndromes, 13-13, 13-14, 13-15, 13-16
Positron emission tomography (PET scan), 2-20, 2-22
Post central gyrus and sensory representation, 21-1, *21-2*
Posterior inferior cerebellar artery syndrome, 13-10, 13-12
Posterior Root fibers 7-16
Potassium Channels
    Chemically controlled 5-9, 5-10
    Inactivation 5-16
    Refractory Period 5-20
    Equilibrium potential 5-5
    Resting membrane voltage 5-5
    Control 5-8, 5-4
    Chemically controlled 5-9, 5-10
    Threshold 5-15
Prefrontal cortex, 18-23
Premotor cortex, 20-16
Presenilin, 30-16
Primary motor cortex, 18-9, 18-10
Primary lateral sclerosis, 9-20
Progressive bulbar palsy, 9-20
Progressive muscular atrophy, 9-20
Progressive pseudobulbar palsy, 9-20
Progressive supranuclear palsy (PSP), 19-15
Proprioception 15-16
Purkinje cell, 20-2
Putamen, 19-1
Pyramidal tract, 18-14

R

Radiculopathies, 8-19, (for cases see peripheral nerve)
Radioactive brain scans and flow studies, 2-20
Receptors
    Sensory endings 3-15
    Free nerve endings 3-15
    Encapsulated sensory endings 3-16
    Meissner's corpuscles 3-16
    Pacinian copuscles 3-16
    End Bulb of Krause and Golgi-Mason 3-16
    Muscle and tendon spindles 3-16
Reciprocal Innervation of a Joint 7-7
Reflexes, 18-2
  developmental changes, 18-8, 18-9
  extension or extensor, 9-2, 18-2
  flexion, 18-2, 18-3
  grasp, 18-6

16 Index

instinctive tactile avoiding reaction, 18-7, 18-8
instinctive tactile grasp reaction, 18-6,18-7
instinctive visual grasp and avoiding
Reactions, 18-8
  interlimb, 18-4,18-5
  local sign in flexion reflexes, 18-3
  mass reflex, 9-2, 18-2
  optical righting, 18-5
  placing reaction (visual and tactile), 18-6
  righting, 18-4,18-5,18-6
  stretch, 9-2,
  tonic labyrinthine, 18-3
  tonic neck, 18-2
Reflexive Response to Pain   7-6
Remote effects of malignancy, 20-21, 27-27, (See chapter 27 on CD ROM supplement)
Response of the Nervous System to Injury 3-20
Reticular Formation 11-21
    Non-cerebellar nuclei 11-21
        Raphe
        Central
        Lateral
    Cerebellar nuclei   11-21
        Lateral of medulla
        Paramedian of medulla
        Tegmental of pons
    Cholinergic nuclei 11-23
    Monoamine nuclei Norepenephrine
    Serotinergic 11-21
    Pathways
        Lateral reticulospinal 11-22
        Medial reticulospinal 11-22
        Cenral tegmental   11-22
        Ascending reticular   11-22
Retrograde changes 3-20, 3-30
Rexed's Lamina of the Spinal Cord 7-17 , 7-18
Rhinencephalon 22-1
Romberg sign, 9-23

S

Saccadic eye movements, 18-19
Sacral sparing, 9-3
Saltatory Conduction 5-20
Schwannomas,
  Intradural with spinal cord compression, 9-9,9-9
  Nerve root, 8-23, *2-15,2-16,8-25*
  Neurofibromatosis, 9-10
  Vestibular,13-2, *13-3,27-20*
Secondary Parkinsonism, 19-14
Segmental sensory patterns,2-3,
Segmental motor innervations, 2-3
Segmental reflex innervation, 2-4
Seizures and seizure disorders, 29-1 (refer to epilepsy)
Senile plaques, 30-15, *30-14,30-16*
Senses
    Aristotles five senses – balance, hearing, smell, sight, taste 1-2
    General senses – pain, temperature, touch, pressure 1-2
    Special senses - balance, vision, hearing, taste, smell 1-2
Sensory receptors 1-2
Sensory receptors and effectors in skin 1-3
Sensory modalities, primary compared to cortical/discriminative,21-2,21-3
Septum 22-5
Shy-Drager syndrome, 19-15
Single photon emission computed tomography, (SPECT),2-20,2-21
Skeletal muscles 6-1
    Contraction 6-1
    Molecular Architecture of Contraction 6-5
    Muscle proteins 6-5
    Actin 6-5
    Mysoin 6-5
    Sarcomeres 6-2 6-4
    Motor units 6-1, 6-1
    Filaments 6-2
    Excitation-Contraction Coupling 6-2
    Filament interaction 6-5
    Cross Bridges 6-5
    Force Velocity 6-6
     Active State 6-6
    Reticular structures   6-2
    t-SR coupling
    T system 6-5, 6-5
    Threshold   5-15
    Toxins and local anesthetics 5-19
Sleep, 29-23
  Anatomical basis, 29-23,29-24
  Multiple sleep latency test, (MSLT), 29-29,*29-27*
  Polysomnography, *29-14*
  Sleep stages, 29-23
    Rapid eye movement stage (REM), 29-23,29-24
    Slow wave, 29-23,29-24
  Abnormalities of sleep
    Excessive daytime sleepiness, 29-26
    Insomnia, 29-25
    Narcolepsy, and narcolepsy/cataplexy syndrome, 29-26
    Sleep apnea, 29-25
    Sleep paralysis and hypnagogic hallucinations, 29-27
Smooth pursuit eye movements, 18-20
Sodium Channels 5-7, 5-13
Sodium selectivity 5-14
Special Senses see CD
    Vestiuloacoustic

Index

Visual
Spinal cord, 9-1
  Acute epidural abscess, 9-7
  Amyotrophic lateral sclerosis (ALS), 9-19,9-16
  Anterior horn cell disorders, 9-17
  Anterior poliomyelitis, 9-17
  Anterior spinal artery syndrome, 9-10
  Brown Sequard hemisection syndrome, 9-2,9,3
  Bulbospinal neuronopathy, Kennedy's disease 9-18
  Clinical cases,
    Case 9-1, ruptured cervical disc with compression , 9-4,9-6
    Case 9-2, epidural carcinoma with T4-T5 compression, 9-6
    Case 9-3, Schwannoma with compression, 9-10, (CD ROM)
    Case 9-4, transverse myelitis, 9-13, (CD ROM)
    Case 9-5, syringomyelia, 9-16
    Case 9-6, syringomyelia and syringobulbia, 9-16,9-17, (CD ROM)
    Case 9-7, poliomyelitis in a non immunized adult, 9-18, (CD ROM)
    Case 9-8, classical amyotrophic lateral sclerosis, 9-20
    Case 9-9, tabes dorsalis, 9-22
    Case 9-10, combined system disease, 9-25
    Case 9-11, Friedreich's ataxia, 9-27, (CD ROM)
    Case 9-12, multiple sclerosis with cervical myelopathy onset, 9-29
    Case 9-12, multiple sclerosis, 9-13, (CD ROM)
  Cervical spondylosis, 9-4
  Compression, acute, 9-1,9-6
  Compression, chronic, anterior midline, 9-2
  Compression fractures and fracture dislocations, effects on, 9-3,9-4
  Compression from epidural metastatic carcinoma and lymphoma, 9-5
  Compression from vertebral tuberculosis, 9-7
  Dissociated sensory loss, 9-11
  Friedreich's ataxia, 9-25,9-26
  Hematomyelia, 9-3
  Hereditary spastic paraplegia, 9-21
  HTLV associated progressive myelopathy, 9-13
  Multiple sclerosis, 9-27,9-28,9-29
  Neuromyelitis optica, Devic's syndrome, 9-13
  Primary lateral sclerosis, 9-20
  Progressive muscular atrophy,9-20
  Spinal muscular atrophies,9-18
  Spinal shock, 9-1,18-2,18-3
Satellite Cells 3-19, 3-28
Schwann Cells 3-19, 3-29
Spinal Cord 7-1

  Gross Anatomy 7-1, 7-2, 7-3, 7-5
  Cross Section Anatomy 7-3, 7-6
  Anterior Horn Cells   7-4, 7-8
    Medial nuclear group   7-5
    Lateral nuclear group   7-5
    Preganglionic autonomic nuclei   7-5
    Interneurons 7-14
  Organization of Gray Matter –Rexeds 10 lamina 7-15 to 7-16, 7-28
  Lamina in Spinal Cord gray matter after
Spinal Cord Pathways Table 7-1   7-21
Synaptic Mechanisms 7-10
  Slow Potentials 7-11
Stretch Receptors 7-11
Stria terminalis 16-4 ,22-14
Subacute combined degeneration, 9-23,9-24
Supporting cells of the CNS 3-16
  Oligodenrocytes 3-17, 3-23, 3-24
  Endothelial cells 3-17
  Mononuclear cells 3-18;table 3- 10
  Microglia 3-18, 3-26
    Ameboid microglia 3-18
    Resting microglia   3-18
    Activated microglia 3-18
    Reactive microglia   3-18
    Multinucleated cells 3-18, 3-31
  Astrocytes 3-17, 3-23, 3-24
  Endothelial cells 3-17
  Mononuclear cells 3-18;table 3- 10
  Oligodenrocytes 3-17, 3-23, 3-24
  Pericytes 3-18
Synapse   3-14 ;fig 3-17
Synaptic transmission 3-14
Synaptic Types
    Electrical synapses 3-13
    Chemical synapses 3-13
Synaptic Vesicles 3-14, 3-17, table 3-18
Syringomyelia, 9-14,9-16,9-17
Stretch reflexes, 7-6, 7-10, 7-11, 9-2
Stimulation of cerebral cortex: focal seizures
  frontal eye fields, 18-21, 18-22
  hippocampus or temporal lobe, 30-10,30-10
  occipital eye fields, 18-21,24-16
  post central gyrus, 21-2
  premotor, 18-17, 18-17
  primary motor, 18-10, 18-14, 18-15
  speech areas, 24-5
Striatal –nigral degeneration, 19-15
Subarachnoid hemorrhage, 26-26
Substantia nigra, 19-1,19-2
  compacta, 19-1, 19-2
  reticularis, 19-1, 19-3
Subthalamic nucleus, 19-1, 19-3
Supplementary motor cortex (SMA), 18-17
Suppressor areas for motor activity, 18-23

18 Index

Syncope, 29-1
Syphilis, 9-22
Syringomyelia, 9-14
Syringobulbia, 9-15
T
Tabes dorsalis, 9-22,9-24
  Transverse myelitis, 9-12,9-13
  Tumors, epidural, metastatic, 9-5
  Tumors, intradural, meningiomas and Schwannomas, 9-8
  Tumors, intrinsic, astrocytomas and ependymomas, 9-13
Tactile Discrimination-posterior columns 7-18
Telencephalon 4-11
   Differentiation of primary sulci 4-12, 4-11
           Development of Cerebral cortex 4-12
                   Gray matter differentiation via migration of neuroblasts 4-13
                   Commissures   4-14
           Myelination   4-14
           Migration Disorders of cerebral cortex
                   X-linked 4-16
                   Miller-Dieker 4-16
                   Bilateral nodular periventricular heterotopias 4-16
                   Microdysgensis 4-16
           Neuronal maturation 4-15
Tentorial herniation, 13-5,*13-6*
Teratomas, 27-18
Thalamus
         Thalamic Afferents 15-7, 15-8
                   Ventral anterior 15-3
                   Ventral lateral   15-3
                   Ventral posterior 15-3
                   Medial nucleus 15-4
                   Lateral nucleus 15-4
                           Lateral dorsal 15-4
                           Lateral posterior 15-3
                   Reticular 15-4
                   Metathalmus
                   Medial geniculate nucleus
                           Audition 15-6
                   Lateral geniculate nucleus
                           Vision   15-6
                           Posterior   15-3
                   Dorsal thalamus
                           Specific nuclei 15-3
                           Non-specific nuclei 15-4
Thalamic Input: Noradrenergic, Serotoninergic, Dopaminergic,Cholinergic, GABAergic 15-8
  And dementia, 30-19
  And memory, 30-6
Thalamic Syndrome (of Dejerine) 15-9
Tonsillar herniation, 13-6
Tics,19-24
Transcranial doppler, 2-26

Transient global amnesia syndrome, 30-9
Trauma, brain 27-1,*27-1,27-4*
  Clinical cases
     Case 27-1, epidural hematoma, 27-2
     Case 27-2, subdural hematoma,27-5
  Concussion, 26-1
  Contusions, 26-1
  Epidural hematoma, 27-2
  Lacerations, 27-2
  Late effects ,27-6
  Management of severe head trauma, 27-5
  Prevention,27-6,27-7
  Skull fractures,27-2
  Subdural hematoma, 27-3, *27-4*
Tremors, 20-21
Trimesters
       First 4-1
       Second Trimester 4-1
       Third trimester 4-1
Toxic disorders 27-26,27-28 (see chapter 27 on CD ROM supplement),
V
Vascular disease, 26-1
   Brain stem, 13-8, (see under brain stem)
   Cerebral hemispheres, 26-1
   Anterior cerebral artery (ACA) syndromes, 26-17,*26-17*
   Anterior choroid artery syndrome, 26-18
   Anterior communicating artery (ACOM), 26-18
   Border zone, 26-5,*26-6*
   Cerebral embolism, 26-20,*26-20*
   Clinical cases
   Case 26-1, infarct MCA, ACA territory, following ICA stenosis,26-6,*26-7*
    Case 26-2, carotid TIA's
    Case 26-3, internal carotid occlusion, infarct, MCA,PCA 26-9,*26-10,26-12*
       Case 26-4, total territory MCA infarct, 26-15,*26-16*
       Case 26-5, embolic occlusion calcarine branch PCA, 26-19,*26-20*
       Case 26-6, cortical posterior cerebral artery syndrome, embolic, 26-20 (CD ROM)
       Case 26-7, intracerebral hemorrhage, 26-23 (CD ROM)
       Case 26-8, subarachnoid hemorrhage, PCOM aneurysm, 26-27 (CD ROM)
       Case 26-9, subarachnoid hemorrhage, MCA aneurysm, 26-27 (CD ROM)
       Case 26-10, subarachnoid hemorrhage, ACOM aneurysm, 27-27 (CD ROM)
       Case 26-11, bacterial endocarditis, mycotic aneurysm, ICH, 26-30 (CD ROM)
       Case 26-12, arteriovenous malformation, 26-31 (CD ROM)

Index

　Ischemic occlusive, 26-1
　Internal carotid artery syndromes, 26-5,*26-6,26-7,26-10,26,12*
　Intracerebral hemorrhage (ICH) 26-22, 26-23,*26-22,26-24,26-25*
　Lacunar /penetrating branch syndromes, 26-11,26-13,*26-12,26-13,26-14*
　Middle cerebral artery (MCA) syndromes, 26-11,*26-12,26-13,26-14,26-15,26-16*
　Posterior cerebral artery (PCA) syndromes, 26-18,26-19,*26-19,26-20*
　Subarachnoid hemorrhage, 26-26,26-27,26-30,26-31,*26-28,26,2,26-32*

Vascular supply, 9-10,9-11
Vegetative state, persistent, 29-3
Ventriculography, 2-18
Vertebral artery syndromes, 13-9,*13,10,13-11*,
Vertebral tuberculosis, 9-7
Visual System 23-1
　　Visual pathway 23-6
　　　　Lateral geniculate 23-11
　　　　Optic radiation 23-11
　　Occipital lobe
　　　　Area 17, 18, 19 23-12
　　　　Ocular dominance columns 23-12
　　　　Stimulation of areas 17,18,19 23-13
　　　　Occipital lobe and eye movements 23-15
　　　　Lesion in visual receptive striate area 17 23-16
　　　　Lesions of extrastriate areas 18 and 19 23-16
　　　　Effects of vascular lesions in calcarine cortex 23-20
　　Physiology of visual system 23-13
　　　　Simple cell 23-13
　　　　Complex cell 23-13
　　　　Hypercomplex cell 23-13
　Watershed, 26-5
　　　　Lesions in visual pathway 23-8, 23-7
　　　　Clinical Cases in visual system
　　　　Clinical Cases 23-1 subfrontal meningioma compressing optic nerve 23-9
　　　　Case 23-2 pituitary adenoma compressing optic chiasm 23-10
　　　　Case 23-3 mass lesion in right occipital lobe 23-17
　　　　Case 23-4 seizures of focal origin left occipital 23-18
Vitamin B12, 9-23
Vitamin E deficiency and cerebellar ataxia, 20-18
Von Economo's encephalitis and Parkinson's disease,19-11

W

Wallenberg syndrome, 13-10
Wallerian Degeneration 3-21, 3-32, 3-33
Watershed, 26-5
Weber's syndrome, 13-19
Wernicke's encephalopathy, 30-5
Wernicke-Korsakoff syndrome, 30-5
Wilson's disease, 19-21
Working memory, 18-24

X

X-rays, 2-13,2 -15,